PROTEIN METABOLISM OF THE NERVOUS SYSTEM

PROTEIN METABOLISM OF THE NERVOUS SYSTEM

Edited by Abel Lajtha

New York State Research Institute for
Neurochemistry and Drug Addiction
Ward's Island, New York City

PLENUM PRESS • NEW YORK–LONDON • 1970

Library of Congress Catalog Card Number 74-85373

ISBN 978-1-4684-8111-2 ISBN 978-1-4684-8109-9 (eBook)
DOI 10.1007/978-1-4684-8109-9

Softcover reprint of the hardcover 1st edition 1970
A Division of Plenum Publishing Corporation
227 West 17th Street, New York, N.Y. 10011

United Kingdom edition published by Plenum Press, London
A Division of Plenum Publishing Company, Ltd.
Donington House, 30 Norfolk Street, London W.C.2, England

Contributors

Bernard W. Agranoff, Mental Health Research Institute; Department of Biological Chemistry, University of Michigan, Ann Arbor (page 533).

Stanley H. Appel, Division of Neurology, Duke University Medical Center, Durham, North Carolina (page 621).

L. Austin, Department of Biochemistry, Monash University, Clayton, Victoria, Australia (page 271).

Julio M. Azcurra, Mental Health Research Institute, The University of Michigan Medical Center, Ann Arbor, Michigan (page 517).

Samuel H. Barondes, Departments of Psychiatry and Molecular Biology, Albert Einstein College of Medicine, Bronx, New York (page 545).

Claude F. Baxter, Veterans Administration Hospital, Sepulveda, California; Division of Neurosciences, City of Hope Medical Center, Duarte, California; Department of Physiology, UCLA School of Medicine, Los Angeles, California (page 439).

Ya. V. Belik, Institute of Biochemistry, Academy of Sciences, Ukrainian SSR, Kiev (page 77).

Samuel Bogoch, Foundation for Research on the Nervous System; Boston University School of Medicine, Boston Massachusetts (page 555).

S. C. Bondy, Department of Biological Chemistry, School of Medicine; Brain Research Institute, University of California Center for the Health Sciences, Los Angeles, California (page 3).

J. J. Bray, Department of Biochemistry, Monash University, Clayton, Victoria, Australia (page 271).

Eric G. Brunngraber, Illinois State Psychiatric Institute, Chicago, Illinois (page 383).

Wendy Cannon, Department of Pharmacology, University of British Columbia, Vancouver, B.C., Canada (page 219).

Li-Pen Chao, Department of Neurology, University of California School of Medicine, San Francisco, California (page 643).

Doris H. Clouet, New York State Research Institute for Neurochemistry and Drug Addiction, Ward's Island, New York, New York (page 699).

Carl W. Cotman, Division of Psychobiology, University of California, Irvine, California (page 151).

Hatim F. Daginawala, Department of Biochemistry, Nagpur University, Nagpur, India (page 459).

Ranajit Kumar Datta, Research Division, Beth Israel Medical Center, New York, New York (page 185).

Herman C. B. Denber, Research Division Biochemistry Laboratory, Manhattan State Hospital, Ward's Island, New York, New York (page 685).

Brian D'Monte, New York State Research Institute for Neurochemistry and Drug Addiction, Ward's Island, New York, New York (page 185).

Bernard Droz, Départment de Biologie, Commissariat à l'Energie Atomique, Gif sur Yvette 91, France (page 93).

Elizabeth Roboz Einstein, Department of Neurology, Department of Biochemistry and Biophysics, University of California School of Medicine, San Francisco, California (page 643).

Stanley E. Geel, Department of Physiology–Anatomy, University of California, Berkeley, California (page 335).

P. L. Ipata, Institute of Biological Chemistry, Free University of Chieti, Chieti, Italy (page 409).

Monique Jacob, Centre de Neurochimie du C.N.R.S., Faculté de Médecine, Strasbourg, France (page 129).

Stanley Jacobson, Neurology Service, VA Research Hospital, Chicago, Illinois (page 491).

D. Karcher, Department of Neurochemistry, Born-Bunge Foundation Berchem-Antwerp; Institute Pasteur, Brussels (page 589).

Marian W. Kies, Laboratory of Cerebral Metabolism, National Institute of Mental Health, Bethesda, Maryland (page 659).

Edward Koenig, Department of Physiology; Neurosensory Laboratory, State University of New York, Buffalo, New York (page 259).

Harold Koenig, Neurology Service, VA Research Hospital; Department of Neurology, Northwestern University Medical School, Chicago, Illinois (page 491).

Herbert L. Koenig, Laboratoire de Cytologie, Faculté des Sciences, Paris, France (page 93).

Steven Edward Kornguth, Department of Neurology; Department of Physiological Chemistry, University of Wisconsin Medical Center, Madison, Wisconsin (page 631).

Abel Lajtha, New York State Research Institute for Neurochemistry and Drug Addiction, Ward's Island, New York, New York (pages 39, 185, and 355).

Fritz Lipmann, The Rockefeller University, New York, New York (page 305).

A. Lowenthal, Department of Neurochemistry, Born-Bunge Foundation Berchem-Antwerp; Institute Pasteur, Brussels (page 589).

Chung Y. Lu, Neurology Service, VA Research Hospital; Department of Neurology, Northwestern University Medical School, Chicago, Illinois (page 491).

H. R. Mahler, Chemical Laboratories, Indiana University, Bloomington, Indiana (page 151).

Paul Mandel, Centre de Neurochimie du C.N.R.S., Faculté de Médecine, Strasbourg, France (page 129).

F. A. Manzoli, Institute of Histology and General Embryology, University of Bologna, Bologna, Italy (page 409).

Neville Marks, New York State Research Institute for Neurochemistry and Drug Addiction, Ward's Island, New York, New York (pages 39 and 185).

I. G. Morgan, Department of Biochemistry, Monash University, Clayton, Victoria, Australia (page 271).

M. R. V. Murthy, Department of Biochemistry, Faculty of Medicine, Laval University, Quebec, Canada (page 109).

David Nachmansohn, College of Physicians and Surgeons, Columbia University, New York, New York (page 313).

Rajinder Nayyar, Neurology Service, VA Research Hospital; Department of Neurology, Northwestern University Medical School, Chicago, Illinois (page 491).

S. Ochs, Department of Physiology, Indiana University Medical Center, Indianapolis, Indiana (page 291).

A. V. Palladin, Institute of Biochemistry, Academy of Sciences, Ukrainian SSR, Kiev (page 77).

N. W. Penn, Willowbrook State School, Neuroendocrine Research Unit, Staten Island, New York, (page 355).

Giuseppe Porcellati, Department of Biological Chemistry, University of Pavia, Pavia, Italy (page 601).

Donald A. Rappoport, Division of Molecular Biology, Department of Pediatrics; The Marine Biomedical Institute, The University of Texas Medical Branch, Galveston, Texas (page 459).

M. Reiss, Willowbrook State School, Neuroendocrine Research Unit, Staten Island, New York (page 355).

Derek Richter, Medical Research Council Neuropsychiatric Research Unit, Carshalton and Epsom, England (page 241).

Sidney Roberts, Department of Biological Chemistry, School of Medicine; Brain Research Institute, University of California Center for the Health Sciences, Los Angeles, California (page 3).

Panna Sanghavi, Neurology Service, VA Research Hospital, Chicago, Illinois (page 491).

Otto Z. Sellinger, Mental Health Research Institute, The University of Michigan Medical Center, Ann Arbor, Michigan (page 517).

Eric M. Shooter, Departments of Genetics and Biochemistry; Laboratories for Molecular Medicine, Stanford University School of Medicine, Stanford, California (page 419).

Louis Sokoloff, Laboratory of Cerebral Metabolism, National Institute of Mental Health, Bethesda, Maryland (page 367).

S. Takahashi, Willowbrook State School, Neuroendocrine Research Unit, Staten Island, New York (page 335).

David N. Teller, Research Division Biochemistry Laboratory, Manhattan State Hospital, Ward's Island, New York, New York (page 685).

Sujata Tewari, Neurochemistry Laboratories, Veterans Administration Hospital, Sepulveda, California (page 439).

L. Thiry, Department of Neurochemistry, Born-Bunge Foundation Berchem-Antwerp; Institute Pasteur, Brussels (page 589).

Paola S. Timiras, Department of Physiology–Anatomy, University of California, Berkeley, California (page 335).

Georges Ungar, Baylor University College of Medicine, Houston, Texas (page 571).

Silvio Varon, Department of Biology, University of California at San Diego, La Jolla, California (page 419).

R. Vrba, Department of Pharmacology, University of British Columbia, Vancouver, B.C., Canada (page 219).

Isabel J. Wajda, New York State Research Institute for Neurochemistry and Drug Addiction, Ward's Island, New York, New York (page 671).

I. Wegelin, Institute of Histology and General Embryology, University of Bologna, Bologna, Italy (page 409).

W. Zeman, Department of Neurochemistry, Born-Bunge Foundation Berchem-Antwerp; Institute Pasteur, Brussels (page 589).

Claire E. Zomzely, Department of Biological Chemistry, School of Medicine; Brain Research Institute, University of California Center for the Health Sciences, Los Angeles, California (page 3).

Preface

Few can deny the paramount importance of the neurosciences, undoubtedly one of the most challenging fields in contemporary science. Recent years have witnessed the awakening of interest in brain research by many distinguished investigators from other branches of science, which has made possible the multidisciplinary approach needed for the complex problems of this field. The present book, which deals with one aspect of this research, is the result of the symposium held under the auspices of the New York State Research Institute for Neurochemistry and Drug Addiction in April 1968.

It has become clear that brain proteins are involved in all aspects of mental function and dysfunction, and the present volume documents the latest advances in our knowledge (advances made to a large extent by contributors to this volume). The chapters not only convey some of the enthusiasm and wonderful, cooperative spirit of the many excellent scientists exploring the brain, and their wealth of ideas; they also illustrate the many approaches from which cerebral proteins can be studied in a meaningful manner. In some areas even preliminary evidence is worth discussing: e.g., it is an exciting achievement that we can begin to apply the disciplines of biochemistry to phenomena of learned behavior and information handling.

Not many years ago all of the available information in this field could have been summarized in a brief review article. Now even an entire book may not be enough to cover all aspects of the multifaceted, perplexing problems of brain research. Perhaps this book will serve to provide some background and some ideas for furthering such research. We can look with great expectation toward important future advances in understanding of brain function and perhaps in ability to ameliorate brain dysfunction. It is a pleasure to thank all those who helped to make the symposium such a success, and those who contributed to the publication of this book. A special note of thanks is owed to the contributors, who, many months after the symposium, were willing to revise their papers to make them suitable for a book. The symposium would not have been possible without the continuing aid and support of the Department of Mental Hygiene of New York State, and it is a pleasure to thank all the members of the Institute and also Dr. O. K. Dlamond, Director of Manhattan State Hospital, Dr. H. Denber, Director of Research at the Hospital, and Dr. P. Wexler, Assistant Commissioner for Education of the Department of Mental Hygiene, for their support and cooperation. In addition, I would like to thank the following companies for their generous support: Abbott Laboratories, Geigy Corporation, Hoffman LaRoche, Lederle Laboratories, Eli Lilly and Company, Merck and Company, Inc., Sandoz Pharmaceutical, Smith Kline and French, Wallace Pharmaceutical, and Wyeth Laboratories. It is unfortunate that Dr. H. Waelsch, who was the founding spirit of the Institute, and who contributed so significantly to the subjects discussed, could not participate because of his premature death. He, who was among the first to study cerebral protein metabolism, would have approved of the excellent papers and discussions of the ever-widening horizons in this area.

Studies of protein metabolism have contributed perhaps the most important advance in modern biochemistry. The work described in this volume shows the application of molecular biology to neurochemistry. A summary of present knowledge usually points out a need for further work. It is hoped that this book will serve as some help in this direction.

New York, New York
April 1969

Abel Lajtha

Introduction

To the casual bystander the nature of this book on the protein metabolism of the nervous system may not immediately be apparent. But for those who are working in the field and who are aware of its many-sided implications, the appearance of this first comprehensive treatise on the proteins of nervous tissues is an event of quite unusual significance. It stands out as a milestone on a path of research which, only a few years ago almost totally unexplored, now is rapidly becoming a main highroad of scientific advance.

For most of us the central interest in the subject is its relevance to the working of the brain. That is the impetus behind much of the research and the reason for the special interest it evokes. Proteins and lipoproteins make up more than 90% of the organic matter of the brain, and they are the basis of all the complex functions it performs. Yet, despite their obvious importance, they have hitherto received little attention and we still know less about them than about the proteins of blood, liver, and other organs. Only within the last few years has any serious attempt been made to isolate and learn about the properties of the first few proteins, such as the S-100 protein of Blake Moore, which are truly characteristic of nervous tissues. The study of their distribution, metabolism, and functional significance is only just starting, and this is now a field of tremendous promise.

The reasons for the past neglect are not easy to understand. Later generations looking back on us may find it incomprehensible that with all the scientific resources at our command we should have chosen to devote our energies to so many dubious and unrewarding projects, while ignoring the greatest challenge in the whole biological field—to unravel the basic workings of the brain. One factor which is partly responsible is the previous lack of suitable techniques; and certainly in the protein field our research potential has been revolutionized in recent years by developments such as those in specific carbon labeling and multiple isotope procedures, as also by progress in methods of chromatography, disc electrophoresis, differential centrifugation, and many other essential techniques. Less excusable perhaps has been an overlong adherence by some to a sterile one-dimensional concept of "pure" neurophysiology, that sought to describe all the complex phenomena of living tissues solely in terms of the electrical potentials they produce. Here, then the role of neurochemistry is sufficiently clear—to add a new biochemical dimension to the gleanings of anatomy, physiology, and behavioral science, and so to combine them in developing a new and broader concept of neurobiology as a whole.

The protein metabolism of the brain is closely concerned with processes of maturation and with various aspects of behavior denoted by such terms as instinct, motivation, and memory. But the practical implications of brain research extend far more widely. This is evident in the field of medicine, where mental disorder is coming to be recognized as one of the most pressing problems of the present time. The study of mental illness is probably the most backward branch of medical research, but already we are learning of disorders such as homocystinuria and arginino-succinaturia, where specific proteins of the brain are involved. Here there is a failure to synthesize a necessary protein with enzymic activity. The effect of a protein-deficient diet in stunting the early growth and development of the brain is now well established, and there is a need for further study of the effects of protein malnutrition on the mental development of the human child. Damage to the protein structure of the brain is believed to be involved in behavioral syndromes produced by alcoholic intoxication and drug abuse. The relation of protein metabolism to schizophrenia is still a matter for speculation, but it is known that genetically determined predisposing factors are involved, and these must operate through the synthesis of specific proteins transcribed from the individual genetic endowment of DNA.

The protein metabolism of the nervous system covers so wide a range of different disciplines that a book of this kind could only be produced through the cooperation of a group of scientists representing a number of different aspects of research. The plan of arranging a prior conference at which the different contributions could be discussed has much to commend it, and this has been of value in clarifying some of the basic facts and resolving many minor differences of interpretation and emphasis. The contributors are indebted to Abel Lajtha and his colleagues, not only for the general planning of the conference, but also for the excellent arrangements that were made in the recently established Institute of Neurochemistry and Addiction on Ward's Island and in the beautiful surroundings at Wave Hill. The factors determining the success of a meeting of this kind are hard to define: to find the right people and the right setting is not enough—it means also choosing the right topic for discussion at the right time. In this case the choice of protein metabolism could not have been bettered, and its topical relevance to so many lines of current thinking accounts for the lively interest it inspired.

Scientific discussions have to be impersonal, but scientists are human and therefore conscious, not only of the hopes and aspirations inherent in their work, but also of the many hardships and frustrations that are all too often involved, the underlying "blood, sweat, toil and tears." If, in a story of achievement and advance it is permitted to dwell for a moment on a more somber note, mention should be made of the awareness in the minds of many attending the conference of the still-pervading personality and courageous spirit of an intellectual giant, the much-loved Heinrich Waelsch. To him we are indebted, not only for the creation of the Institute of Neurochemistry, but also, as a pioneer in the field of brain research, for much of the basic thinking and advance recorded in this book. Here in these pages the old indomitable spirit seems to rise again, and if we must bemoan the cruel fate that robbed him through his untimely death of the final satisfaction of seeing his labors come to fruition, at least he would probably have wished for no more fitting an epitaph.

Returning now to the laboratory and to thoughts of proteins and brain research, can we take a forward-looking glance and ask: what of the future? Where do we go from here? Clearly, there are many current lines of work that need to be extended as well as new things that ought to be tried, but the very diversity of possible lines of investigation makes it a difficult choice. Should we concentrate on the isolation and characterization of more individual proteins, or try applying a multidisciplinary approach to the study of changes associated with maturation or with different varieties of functional state? Once again, the most important problem in research is the choice of a problem; and probably for everyone there is a different answer, depending on individual experience, talents, and skills. Here this book may be of service in presenting a number of different approaches that can be viewed and assessed objectively by those who are interested in working in the field. Those of us who are already involved can only wish the newcomers well. The road is wide open and there are many prizes ahead.

Carshalton, England
February 1969

Derek Richter

Contents

I: Metabolism Related to Turnover

I

Metabolism Related to Turnover

Chapter 1

Protein Synthesis in the Nervous System*

Sidney Roberts, Claire E. Zomzely, and S. C. Bondy

Department of Biological Chemistry
School of Medicine and the Brain Research Institute
University of California Center for the Health Sciences
Los Angeles, California

INTRODUCTION

Within the past few years concepts of the nature and regulation of processes involved in the synthesis of proteins in the nervous system have undergone revolutionary alterations. To some extent these changes reflect expanding knowledge of the intimate details of the diverse amino acid-incorporating systems present in most cells. However, the discovery of specialized and, possibly, unique mechanisms in neural elements has been of equal significance for an understanding of these processes. The functional concomitants of the various facets of protein synthesis in the nervous system pose one of the most fascinating problems of modern biological research.

The principal site of protein synthesis in neural tissues is undoubtedly the cytoplasmic ribosome. However, protein synthesis probably also occurs on ribosomes or RNA-containing structures in the nucleus, mitochondria, and other subcellular locations of these tissues. Perhaps to a greater extent than in other eukaryotic cells, functional alterations in the nervous system are acutely dependent upon coordinated variations in the protein-synthesizing activities of different subcellular organelles.

The discussion which follows will be concerned principally with the special properties of cytoplasmic ribosomes of cerebral origin. However, certain salient features of other amino acid-incorporating systems which have been described in the nervous system will be noted. Where possible,

*The original investigations reported in this review were supported by research grants from the Division of Research Grants, National Institutes of Health, US Public Health Service (NB-07869) and the United Cerebral Palsy and Educational Foundation. Mrs. Carole Feingold provided valuable secretarial and bibliographic assistance.

these characteristics will be related to developmental status, regional area, cell type, and intracellular location. Interactions of these various systems with the environment will also be outlined. Finally, the possible contribution of each mechanism to the structure and function of neural tissues will be considered. A comprehensive review of protein synthesis in the nervous system has recently been prepared.[1]

CYTOPLASMIC RIBOSOMES

General Aspects

Cytoplasmic ribosomes of the mature neuron occur concentrated in the perinuclear region, in close proximity to the endoplasmic reticulum of the Nissl substance.[2] However, these ribosomes may exist largely in the free form; i.e., unattached to cellular membranes.[3,4] Ribonucleoprotein particles are also scattered throughout the cytoplasm of the cell body[4] and are present in proximal dendrites,[5] but have not been identified in the distal axon. A relatively high proportion of neuronal ribosomes occurs in the monomeric form or in small aggregates. Larger polyribosomes are also present, particularly in the smaller neurons. In glia, as in other nonneuronal cells,

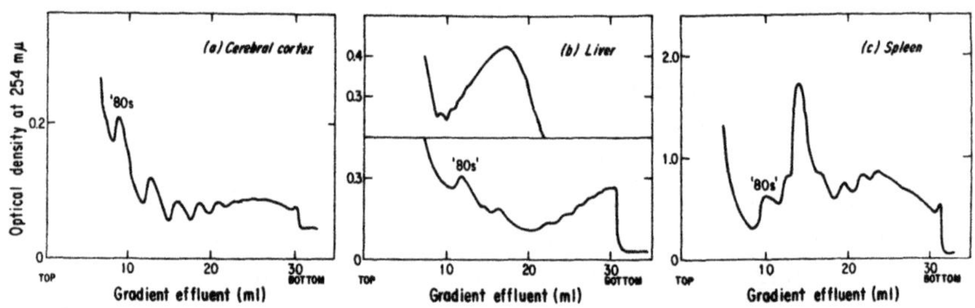

Fig. 1. Sucrose density-gradient profiles of post-mitochondrial supernatant fractions prepared from different rat tissues. The post-mitochondrial supernatants were prepared in 12 mM $MgCl_2$, 100 mM KCl, and 50 mM Tris-HCl buffer (pH 7.6). The hepatic preparation was treated with sodium deoxycholate (final concentration 1.5%). An aliquot of each suspension (2 ml), equivalent to approximately 0.65 mg wet wt of tissue, was layered on a linear sucrose density gradient (35–15%) containing buffer and salts in the concentrations noted above. The gradients were centrifuged in a Spinco SW25 swinging bucket rotor at 25,000 rpm and 0°C. Time of centrifugation was 4 hr, except for the upper of the two curves shown in Fig. 1 for the hepatic preparation, where the time of centrifugation was only 2 hr. Absorbance of the effluent collected from the top of the gradient was monitored continuously at 254 nm. The peak corresponding to the monoribosomal form in each tissue preparation is labeled "80S." (a) cerebral cortex, (b) liver, (c) spleen. (Reprinted with modifications from Zomzely *et al.*[7])

TABLE I

Influence of Magnesium Concentration in Preparative Medium on Sedimentation Properties of Cerebral and Hepatic Polyribosomal Preparations[a]

Mg^{2+} (mM)	Ribosomal species			
	Cerebral cortex		Liver	
	$S_{20,w}$	%	$S_{20,w}$	%
5	76	25	73	1
	113	22	109	2
	142	17	150	3
	168	11	>150	94
	>168	26	—	—
12	74	9	77	1
	108	12	112	1
	137	9	149	1
	162	10	290	97
	173	10	—	—
	>173	52	—	—

[a] Reprinted with modifications from Zomzely *et al.*[8]

ribosomes are more uniformly distributed throughout the cytoplasm and are bound to a greater extent to the endoplasmic reticulum.[6]

The characteristics of neuronal ribosomes of mature neurons *in situ* appear to be reflected in the properties of cell-free preparations derived from cerebral tissue. Thus, a smaller proportion of ribosomal aggregates heavier than the dimer was present in post-mitochondrial supernatant preparations from rat brain compared to similar preparations from other tissues. (Fig. 1).[7] Moreover, relatively high concentrations of Mg^{2+} ions (10–12 mM) were required to produce the maximum yield of heavy polyribosomes from homogenates of rat cerebral cortex (Table I). Presumably, Mg^{2+} ions function to stabilize cerebral messenger RNA-ribosome complexes by decreasing the tendency to dissociation and inhibiting endogenous ribonuclease activity.[9] In these studies, tissues were lightly homogenized in a Tris-buffered medium (pH 7.4–7.6) containing 25 mM sucrose, KCl, and the concentration of Mg^{2+} indicated in Table I. The post-mitochondrial supernatant was treated with sodium deoxycholate (final concentration 1.0–1.3%), then layered over 0.5 M sucrose-Tris buffer which had previously been layered over 1.8–2.0 M sucrose-Tris buffer. Both media contained the same Mg^{2+} concentration as the appropriate preparative

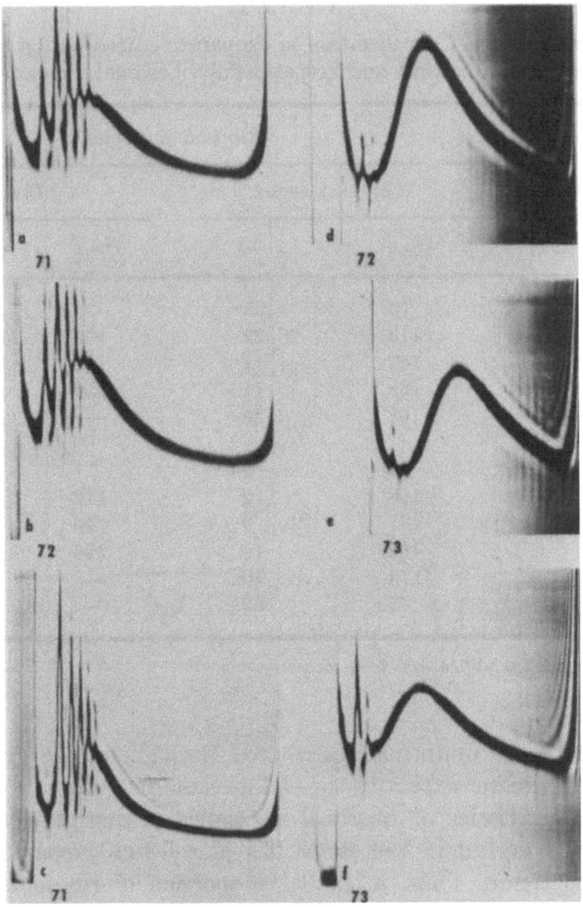

Fig. 2. Influence of ion concentrations in suspending medium on sedimentation patterns of cerebral and hepatic polyribosome preparations. Polyribosomes were isolated by zone centrifugation in sucrose containing 10 mM magnesium acetate, 40 mM NaCl, 100 mM KCl, and 20 mM Tris-HCl buffer (pH 7.6).[12] Sedimentation analyses of the resuspended pellets were carried out as described in Table I, but in the media indicated below. The $S_{20,w}$ values are shown for the monoribosomal components. (a) Cerebral polyribosomes analyzed in 10 mM magnesium acetate, 40 mM NaCl, 100 mM KCl, and 20 mM Tris-HCl buffer (pH 7.6). (b) Cerebral polyribosomes analyzed in 5 mM $MgCl_2$, 25 mM KCl, and 50 mM Tris-HCl buffer (pH 7.4). (c) Cerebral polyribosomes analyzed in 1 mM $MgCl_2$, 25 mM KCl, and 50 mM Tris-HCl buffer (pH 7.4). (d) Hepatic polyribosomes analyzed in 10 mM magnesium ace-

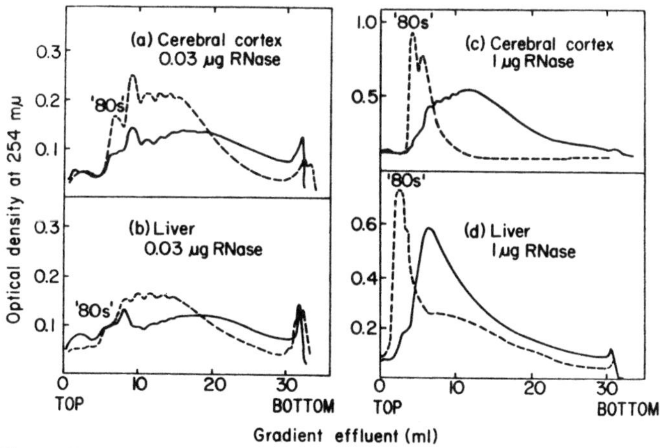

Fig. 3. Influence of pancreatic ribonuclease on cerebral and hepatic polyribosomes. Polyribosomes were suspended in medium containing 10 mM magnesium acetate, 40 mM NaCl, 100 mM KCl, and 20 mM Tris-HCl buffer (pH 7.6). Pancreatic RNase (Sigma Chemical Co.) was added to some of the samples, which were then incubated under various conditions. The sample was subsequently layered on a linear sucrose gradient containing the same buffer and salts as the suspension medium and centrifuged in a Spinco SW25 swinging bucket rotor at 25,000 rpm and 0°C. Absorbance of the effluent collected from the top of the gradient was monitored continuously at 254 nm. (a) Cerebral polyribosomes, centrifuged for 1.5 hr in a 25–5% linear sucrose gradient. Solid curve: untreated; dashed curve: incubated with 0.03 μg RNase/mg RNA for 30 min at 0°C. (b) Hepatic polyribosomes from rats fasted 18 hr, centrifuged for 1.5 hr in a 25–5% linear sucrose gradient. Solid curve: untreated; dashed curve: incubated with 0.03 μg RNase/mg RNA for 30 min at 0°C. (c) Cerebral polyribosomes, centrifuged for 1.5 hr in a 35–15% linear sucrose gradient. Solid curve: untreated; dashed curve: incubated with 1 μg RNase/mg RNA for 30 min at 0°C. (d) Hepatic polyribosomes, centrifuged for 0.5 hr in a 35–15% linear sucrose gradient. Solid curve: untreated; dashed curve: incubated with 1 μg RNase/mg RNA for 30 min at 0°C. (Reprinted from Zomzely *et al.*[7])

---

tate, 40 mM NaCl, 100 mM KCl, and 20 mM Tris-HCl buffer (pH 7.6). (e) Hepatic polyribosomes analyzed in 5 mM $MgCl_2$, 25 mM KCl, and 50 mM Tris-HCl buffer (pH 7.4). (f) Hepatic polyribosomes analyzed in 1 mM $MgCl_2$, 25 mM KCl, and 50 mM Tris-HCl buffer (pH 7.4). (Reprinted from Zomzely *et al.*[8])

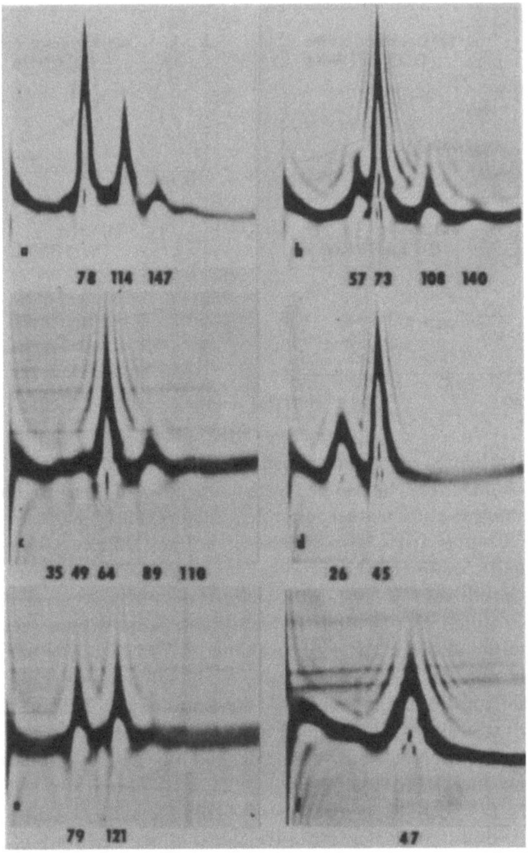

Fig. 4. Sedimentation patterns of cerebral mixed ribosomes obtained by deoxycholate treatment of the microsomal fraction of rat cerebral cortex. Sedimentation analyses were carried out as described in Table I. $S_{20,w}$ values are shown: (a) Analyzed in 4 m*M* $MgCl_2$, 25 m*M* KCl, and 50 m*M* Tris-HCl buffer (pH 7.4). (b) Analyzed in 100 m*M* KCl, 50 m*M* Tris-HCl buffer (pH 7.4). (c) Dialyzed 16 hr against 1 m*M* Tris-HCl (pH 7.6); analyzed in the same medium. (d) Analyzed in 50 m*M* KCl, 1 m*M* Tris-HCl buffer (pH 7.6), containing 3 μmoles EDTA/mg ribosomal protein. (e) Duplicate of sample (d) dialyzed 5 hr at 4°C against 12 m*M* $MgCl_2$, 100 m*M* KCl, 1 m*M* Tris-HCl buffer (pH 7.6) and analyzed in the same medium. (f) Dialyzed 40 hr at 4°C against 100 m*M* KCl, 50 m*M* Tris-HCl buffer (pH 7.4) and analyzed in the same medium. (Reprinted from Zomzely *et al.*[8])

medium. The zonal preparation was centrifuged at 0°C in a Spinco No. 40 rotor at 40,000 rpm for 4 hr. Sedimentation analyses were carried out on the resuspended pellets in the same media (lacking sucrose) in a 12-mm cell in the Spinco model E analytical ultracentrifuge. Concentrations of ribosomal protein varied from 2 to 4 mg/ml. Photographs were obtained with schlieren optics at appropriate intervals after the rotor reached full speed of 29,500 rpm.

Polyribosomes have also been isolated from brain in reasonable yield by the use of endogenous ribonuclease inhibitors[10] or bentonite[11]. These various conditions were not necessary to produce a maximum yield of polyribosomes from liver or spleen, even though ribonuclease activities were higher and endogenous inhibitor concentrations were lower in these tissues

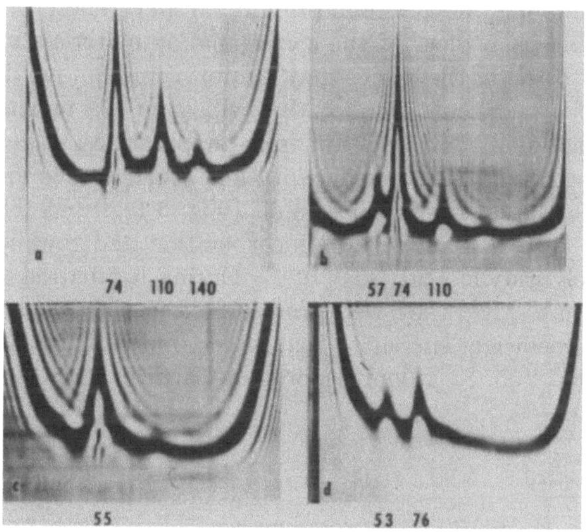

Fig. 5. Sedimentation patterns of ribonucleoprotein particles isolated from rat cerebral cortex. Sedimentation analyses were carried out as described for Table I. The $S_{20,w}$ values are shown. (a) Mixed ribosomes, analyzed in a medium containing 4 m*M* $MgCl_2$, 25 m*M* KCl, and 50 m*M* Tris-HCl buffer (pH 7.4). (b) Mixed ribosomes, analyzed in a medium containing 0.05 m*M* $MgCl_2$, 100 m*M* KCl, and 50 m*M* Tris-HCl buffer (pH 7.4). (c) 55-*S* particles prepared from mixed ribosomes by differential centrifugation under the ionic conditions described in (b) and analyzed in the same medium. (d) 55-*S* preparation after incubation for 30 min at 37°C in a medium containing 12 m*M* $MgCl_2$, 100 m*M* KCl, and 50 m*M* Tris-HCl buffer (pH 7.4). (Reprinted from Zomzely *et al.*[14])

than in cerebral cortex.[7] In line with the foregoing observations was the finding that cerebral polyribosomes exhibited unusual instability in the presence of very low concentrations of Mg^{2+} ions (Fig. 2)[8] or pancreatic ribonuclease (Fig. 3).[7] Polyribosomal lability in cerebral cell-free preparations appears to be a property of the ribosome population of neurons to a greater extent than that of glia (C. E. Zomzely, S. Roberts, and S. Peache, unpublished observations).

The instability of cerebral ribosomal aggregates could be due to sensitivity of the attached messenger RNA or to unusual properties of the ribosomes *per se*. Cerebral monoribosomes exhibited properties which were generally comparable to those of the analogous species derived from other mammalian tissues, including sedimentation behavior and association-dissociation reactions (Fig. 4).[8] Similarly, the physical and chemical characteristics of ribosomal RNA[13] and protein[1] from cerebral cortex appeared indistinguishable from those of the corresponding materials from liver tissue. However, cerebral ribosomes may exhibit some unique properties. In the presence of low concentrations of Mg^{2+}, ribosome-like moieties have been isolated by fractional centrifugation from preparations of cerebral mixed ribosomes which exhibited a sedimentation constant significantly lower than that of the larger ribosomal subunit [Fig. 5 (c,d)].[14] These 50–55 *S* particles were active in amino acid incorporation and revealed a marked response to polyuridylic acid (Fig. 6)[15]. During incubation in a medium containing 12 m*M* Mg^{2+} they were transformed in part to particles with a sedimentation coefficient similar to that of cerebral monoribosomes (i.e., 76 *S*). The cerebral "55 *S*" ribonucleoprotein particles may have consisted

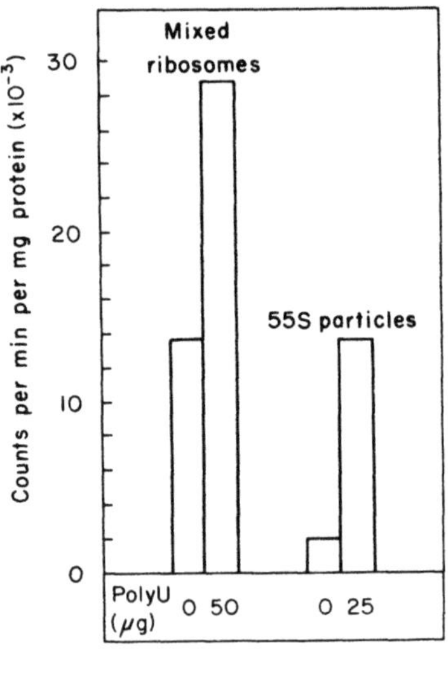

Fig. 6. Amino acid incorporation by ribonucleoprotein particles isolated from rat cerebral cortex. The incorporation system contained 1.5 mg ribonucleoprotein, 2 m*M* NaATP, 0.25 m*M* NaGTP, 20 m*M* creatine phosphate sodium salt, 0.1 mg creatine phosphokinase, 0.05 *M* Tris-HCl buffer (pH 7.4), 0.25 M sucrose, 12 m*M* $MgCl_2$, 100 m*M* KCl, and 1 μCi of uniformly-labeled *L*-[^{14}C]-phenylalanine (2.8 nmoles). Optimal amounts of polyuridylic acid (poly U) were added where indicated. Final volume was 1.4 ml. Incubation was carried out for 60 min at 37°C. The mean is shown for three analyses in each instance. (Reprinted from Roberts and Zomzely.[15])

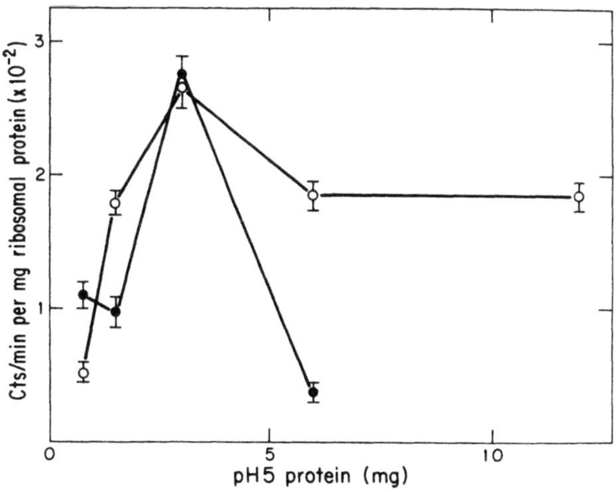

Fig. 7. Effect of varying concentrations of pH 5 enzymes on incorporation of uniformly labeled *L*-[^{14}C]leucine into proteins of rat cerebral microsomes and ribosomes *in vitro*. Conditions of incubation were similar to those described in the legend to Fig. 6, except that 1 μCi of uniformly labeled *L*-[^{14}C]leucine (1.6 nmoles) was present instead of phenylalanine. Incubation time was 30 min. Microsomal protein (open circles); ribosomal protein (darkened circles). Each value represents the mean $\pm$S.E.M. of three determinations. (Reprinted with modifications from Zomzely *et al.*[17])

of a mixture of true "60 *S*" subunits and dimers of "40 *S*" subunits. Particles exhibiting similar sedimentation properties have also been noted in ribosomal preparations from rat liver at low Mg^{2+} concentrations.[16] However, these particles probably represented an alternate form of the hepatic monoribosome.

The protein-synthesizing properties of cytoplasmic ribosomes from brain revealed certain distinctive features which appeared to be related primarily to the unique instability of cerebral messenger RNA-ribosome complexes. Although basic requirements for optimal protein synthesis in ribosomal systems from brain were similar to those noted for other tissues, the cerebral preparations were more sensitive to minor variations in these factors.[15,17] For example, amino acid-incorporating activity of ribosomes derived from rat cerebral cortex exhibited striking alterations in response to changes in concentrations of activating enzymes in the form of the microsomal supernatant or pH 5 enzymes (Fig. 7).[17,18] In part, these findings may be attributed to a possible deficiency of transferases in cerebral ribosomal systems[10] and to destruction of endogenous cerebral messenger

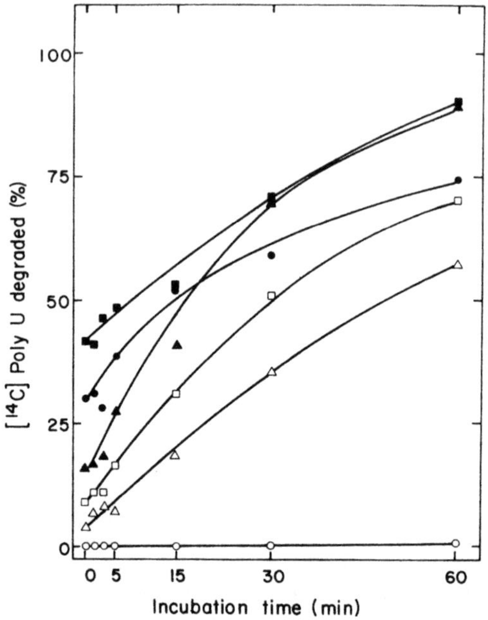

Fig. 8. Ribonuclease activity of cerebral and hepatic polyribosomal systems. Measurements were made by the method of Barondes and Nirenberg,[20] using 50 nmoles of [^{14}C]poly U as polynucleotide phosphorus. The medium in all cases contained 12 mM $MgCl_2$, 100 mM KCl, and 50 mM Tris-HCl (pH 7.6). The protein content was 0.5–0.6 mg for polyribosomes and 1.3 mg for the pH 5 enzyme fractions. Each point represents the average of three closely agreeing determinations. Open circles, cerebral polyribosomes; open triangles, cerebral pH 5 fraction; open squares, complete cerebral system; darkened circles, hepatic polyribosomes; darkened triangles, hepatic pH 5 fraction; darkened squares, complete hepatic system. (Reprinted from Zomzely *et al.*[7])

RNA by ribonucleases which are concentrated in the pH 5 enzyme fraction of brain.[7] However, ribonuclease activities of cerebral extracts and subcellular fractions were low compared to those of the analogous preparations from other tissues (see, for example, Fig. 8).

The optimal concentrations of ATP, amino acids, and ions for maximum incorporation of amino acids into protein of cerebral ribosomal systems were also critical.[15,17] Several investigators, using various conditions, have reported that incorporation of amino acids into protein by cerebral ribosomes was selectively inhibited *or* stimulated by the presence of certain other amino acids.[17,18,21–25] In one study, a mixture of 19 *L*-[^{12}C]amino acids, each at 10^{-3} or 10^{-2} m*M* concentration, stimulated incorporation of *L*-[^{14}C]leucine into protein of cerebral ribosomes.[26] Concentrations of unlabeled amino acids at 10^{-1} m*M* levels or higher were inhibitory. In experiments where amino acid mixtures markedly enhanced incorporation, a high proportion of the stimulatory activity was due to the dicarboxylic acids, γ-aminobutyric acid, and certain other amino acids,[23,25] which occur in brain in high concentrations.[27] However, it must be noted that a wide

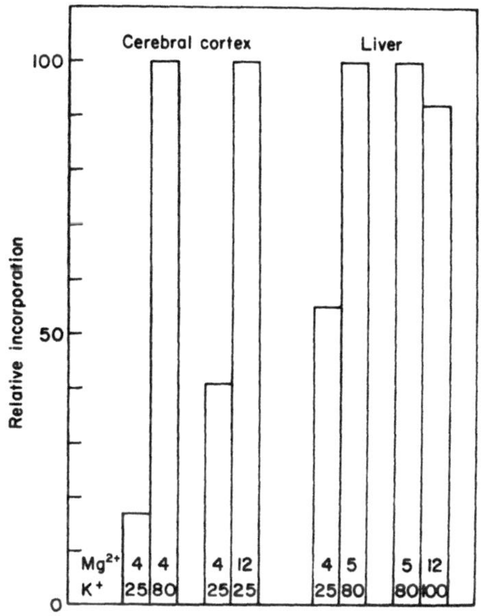

Fig. 9. Effect of varying concentrations of $MgCl_2$ and KCl on incorporation of uniformly labeled *L*-[^{14}C]leucine into particulate proteins of rat cerebral and hepatic ribosomes *in vitro*. Conditions of incubation were similar to those described in the legend to Fig. 6. The ratio of pH 5 protein to ribosomal protein was 2:1 for cerebral ribosomes and 1:1 for hepatic ribosomes. Relative incorporation values are based on counts per min per mg ribosomal protein. The value of 100 was assigned to the higher value of each of the paired analyses. Bar heights represent the mean of 2–4 determinations at each ionic (mM) concentration. (Reprinted from Roberts and Zomzely.[15])

variety of physiological and pharmacological agents have also been shown to alter cerebral protein synthesis *in vivo* and *in vitro*.[22,23,28,29] The physiological significance of these various findings has not been established.

Endogenous amino acid-incorporating activity of cerebral ribosomes was maximal in the presence of 10–12 m*M* Mg^{2+} and 80–100 m*M* K^+ and was

TABLE II

Influence of Incubation on Sedimentation Properties of Cerebral and Hepatic Polyribosomes[a]

Incubation time (min)	Ribosomal species			
	Cerebral cortex		Liver	
	$S_{20,w}$	%	$S_{20,w}$	%
[b]	71	6	72	< 1
	100	10	105	< 1
	123	10	132	1
	152	9	>132	98
	172	7	—	—
	>172	58	—	—
0	46	5	54	< 1
	62	17	79	< 1
	89	9	99	1
	114	9	av 176	98
	153	8	—	—
	>153	52	—	—
15	63	30	48	< 1
	88	9	71	< 1
	110	10	95	1
	130	12	av 151	98
	147	9	—	—
	>147	30	—	—
30	59	31	58	< 1
	89	10	76	< 1
	112	9	101	1
	129	8	av 152	98
	av 148	42	—	—

[a] Conditions of incubation were similar to those described in the legend to Fig. 6. The ratio of pH 5 enzyme protein to ribosomal protein was 2:1. After incubation for the times indicated in Table II the suspensions were subjected to sedimentation analysis as described in the legend to Fig. 2. Reprinted from Zomzely *et al.*[7]

[b] Data for cerebral and hepatic polyribosomes suspended in medium not containing pH 5 enzymes or cofactors.

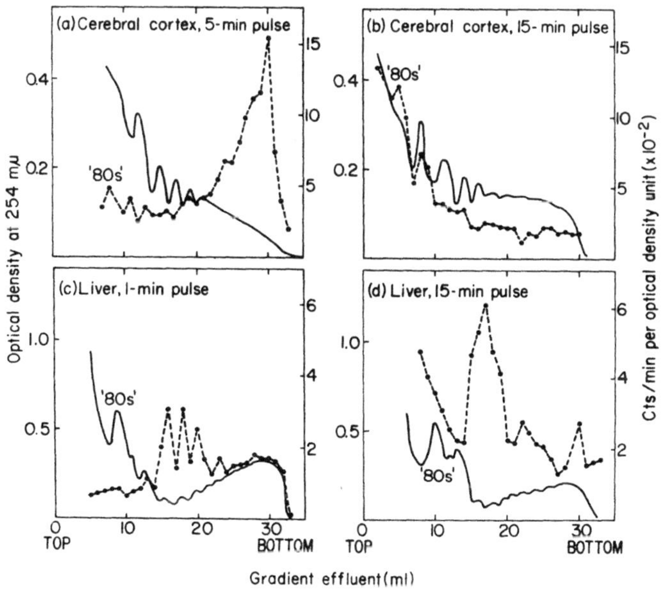

Fig. 10. Protein-synthesizing activities of cerebral and hepatic polyribosomes *in vivo*: relationship to state of aggregation. (a) cerebral cortex, 5 min after injection of [^{14}C]leucine; (b) cerebral cortex, 15 min; (c) liver, 1 min; (d) liver, 15 min. Solid lines, optical density; darkened circles, cts/min per optical density unit ($\times 10^{-2}$). (Reprinted from Zomzely *et al.*[7])

greatly reduced by alterations in these levels (Fig. 9).[15] To some extent NH_4^+, Na^+, and K^+ could be substituted for one another provided that the total concentration of these ions remained approximately constant.[22,25,30] The analogous hepatic preparations were much less sensitive to ionic variations. High concentrations of Mg^{2+} may exert their stimulatory effects on protein synthesis in cerebral ribosomal systems in part by inhibiting ribonuclease activity and thereby maintaining polyribosomal structure.[7]

As a consequence of the instability of cerebral messenger RNA-ribosome complexes, cell-free preparations from this tissue not only contained a relatively small proportion of polyribosomes, but these disaggregated more readily during incubation in an amino acid-incorporating system than analogous preparations from liver (Table II).[7] Comparable data were obtained under *in vivo* conditions of protein synthesis. Cerebral post-mitochondrial supernatants were prepared from rats given an intracisternal injection of uniformly labeled *L*-[^{14}C]leucine (10 μCi) 5 or 15 min before autopsy. Hepatic post-mitochondrial supernatants were prepared from rats given 20 μCi of [^{14}C]leucine *via* the portal vein 1 or 15 min earlier. The animals were generally given these injections under ether anesthesia and killed by decapitation. Post-mitochondrial supernatants were prepared in 12 m*M* $MgCl_2$, 100 m*M* KCl, and 50 m*M* Tris-HCl buffer (pH 7.6) and treated with sodium deoxycholate (final concentration 1–1.5%). An aliquot of each suspension (2 ml), equivalent to approximately 0.65 mg wet wt of tissue, was layered on linear sucrose density gradients (35–15%) containing buffer and salts in the concentrations above. The gradients were centrifuged for 4 hr in a Spinco SW25 swinging bucket rotor at 25,000 rpm and 0°C. The effluents from the top of the gradient were monitored continuously for absorbance at 254 nm and collected as 1-ml samples for determination of radioactivity. Radioactivity incorporated into protein from injected [^{14}C]amino acids was initially found mostly in the larger polyribosomes present in post-mitochondrial supernatants prepared from either cerebral cortex or liver of the adult rat (Fig. 10).[7] However, within 15 min after

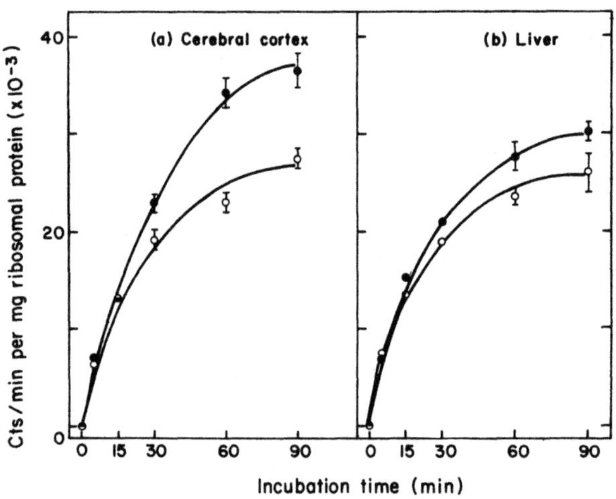

Fig. 11. Kinetics of incorporation of *L*-[^{14}C]phenylalanine by cerebral and hepatic polyribosomes in the presence and absence of poly U. Each system contained 1 mg ribosomal protein, 2 mg pH 5 enzyme protein, 2 m*M* NaATP, 0.25 m*M* NaGTP, 20 m*M* creatine phosphate sodium salt, 0.1 mg creatine phosphokinase, 50 m*M* Tris-HCl buffer (pH 7.6), 12 m*M* $MgCl_2$, 100 m*M* KCl, and 1 μCi (10^6 cts/min) uniformly labeled *L*-[^{14}C]phenylalanine. Final volumes were 1 ml. Each value represents the mean of three closely agreeing analyses. Where appropriate, the S.E.M. is also shown. (a) cerebral system, (b) hepatic system (fasted rats). Open circles: without poly U; darkened circles: with poly U (300 μg/mg ribosomal protein). (Reprinted from Zomzely *et al.*[7])

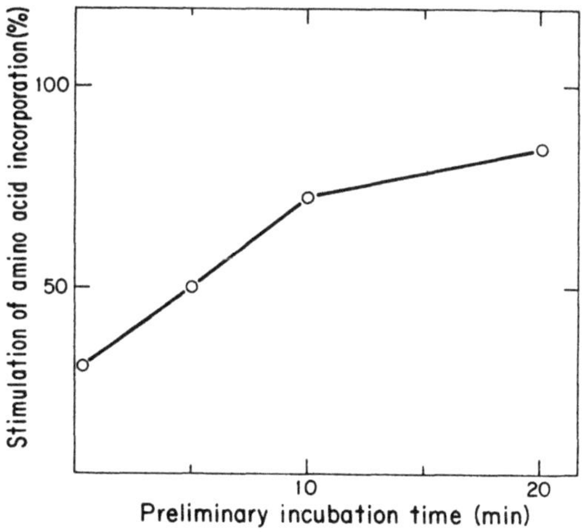

Fig. 12. Effect of time of preliminary incubation on stimulation of amino acid incorporation into protein of brain ribosomes due to brain nuclear RNA. Conditions of incubation were similar to those described for Table V, but the preliminary incubation time varied as shown. (Reprinted from Samli and Roberts.[26])

administration of the precursor, most of the radioactivity in *cerebral* polyribosomes now appeared in regions of the gradient where lighter ribosomes were to be found, whereas little change in distribution occurred in the liver preparations. Moreover, radioactivity in the smaller cerebral aggregates was initially relatively high compared to that found in the corresponding hepatic species. A proportion of these aggregates may normally be attached to native messenger RNA molecules of relatively small size and participate in the formation of proteins serving specialized functions in the brain. The presence of low-molecular-weight RNA species in cerebral nuclear preparations, with high stimulatory activity on amino acid incorporation in homologous ribosomal systems, has been noted.[26] However, it is possible that these RNA species represented partially degraded molecules of natural messenger RNA which still retained some template activity.

Cerebral ribosomes obtained from adult rats were unusually responsive to polyuridylic acid compared to similar preparations from other mammalian tissues.[7,11,15,31] A significant but delayed increase in amino acid incorporation was observed even in cerebral polyribosomal preparations (Fig. 11).[7] This responsiveness was associated with disruption of certain cerebral messenger RNA-ribosome complexes during incubation and the resulting appearance of ribosomal monomers and subunits (Table II) to which added messenger RNA molecules bind most readily.[14,32,33] In a comparable fashion, preincubation of cerebral ribosomes in the presence of an amino acid-incorporating system containing the pH 5 fraction markedly sensitized these preparations to the stimulatory action of polyuridylic acid during a subsequent incubation.[7] As noted earlier, the pH 5 fraction is the major source of ribonuclease activity in cerebral systems. The relatively small stimulation of amino acid incorporation normally observed with preparations of nuclear RNA from adult rodent brains upon addition to an homologous ribosomal system[31,34,35] was also greatly enhanced by prior incubation of the ribosomes with the pH 5 fraction (Fig. 12).[26] Hepatic ribosomal systems appeared to be less responsive to cerebral and hepatic nuclear RNA preparations under similar circumstances.[35]

The roles of genetic and environmental factors in the establishment of the unusual characteristics of cerebral messenger RNA-ribosome systems are unknown. However, brain-barrier mechanisms may be important in this regard. The development of these barriers could impose restrictions upon the biosynthetic capacities of the brain and result in adaptive alterations in the control of protein synthesis at the transcriptional and translational levels.[15,36,37]

Cytological and Regional Differences

Conflicting data exist which bear on the relative protein-synthesizing activities of different cell types and diverse areas of the brain. The average turnover rate of brain proteins in the adult monkey *in vivo* did not vary significantly in cerebral cortex, cerebellum, thalamus-hypothalamus, or medulla-pons.[38] Although proteins of white matter (corpus callosum) seemed to possess a faster turnover rate than those of the other brain regions investigated, the suggestion was advanced that the presence of different compartments of radioactive amino acid precursor in white matter might be responsible for this result. In contrast, cell suspensions from cerebral cortices of young mice (5–11 days old) were more active in protein synthesis than the analogous preparations from cerebellar or basal areas.[39] Moreover, slices of gray matter from adult guinea pig brain were slightly more active than white matter in the uptake of [^{14}C]glycine into protein.[40] Finally, crude microsomal fractions derived from gray matter of adult rat brain were reported to incorporate *DL*-[1-^{14}C]lysine into protein twice as rapidly as similar preparations from white matter.[41] The difference in activity appeared to be located in the amino acid-activating enzymes (see also Takahasi and Abe[42]) as well as in the microsomal fraction. However, the microsomal fraction from white matter was contaminated with myelin. Recent investigations indicate that purified polyribosomes from gray or white matter of rat brain may have similar capacities for protein synthesis (C. E. Zomzely, S. Roberts, and S. Peache, unpublished observations).

The available data suggest that regional differences in the overall rate of protein synthesis in the brain may be dependent to a significant extent upon variations in ribosomal content. Analyses carried out on isolated cells from various areas of the nervous system revealed that the RNA concentration of the neuronal population was several times that of glia.[43] The actual ratio may vary from values exceeding 10:1 in regions of large neurons to only about 3:1 in areas where small neurons predominate (see, for example, Landolt *et al.*[44] Moreover, relative numbers of neurons fluctuate widely even from one area of the cerebral cortex to another. Base composition analyses of neuronal and glial RNA supported the concept that the differences in RNA content of neurons and glia and of diverse regions of the nervous system were related primarily to variations in ribosomal concentrations.[45] This conclusion was further documented by morphological evidence. Thus, although glia contain fewer ribosomes than neurons, glial ribosomes appear to be more highly aggregated *in situ.*[46] Moreover, evidence has been obtained that glial polyribosomes may be at least as active in protein synthesis as the analogous populations of neuronal ribosomes (C. E. Zomzely, S. Roberts, and S. Peache, unpublished observations).

Developmental Changes

Cellular variations in concentration and activity of amino acid-incorporating systems in the brain have particular significance for an analysis of developmental changes in protein synthesis in this organ. Overall protein synthesis in the brain appears to be more active in the immature animal than in the adult. However, the relative contributions of neurons and glia and of different subcellular organelles to this picture have not been delineated. Development of the central nervous system is associated with striking alterations in relative proportions of neurons and glia and in the morphological and functional characteristics of these cells. The glial population, which may be relatively small in the cortex of immature rats, far exceeds that of the neurons in adult animals.[47]

The average half-life of brain proteins in the rat was found to increase with age as a consequence of the accentuated formation of proteins with long turnover rates.[48] In cerebral cortex the major decrease in protein synthesis may occur soon after birth and be succeeded by a lesser and more gradual decline.[49] At varying intervals after parenteral administration of radioactive amino acids, proteins of all brain subcellular fractions were

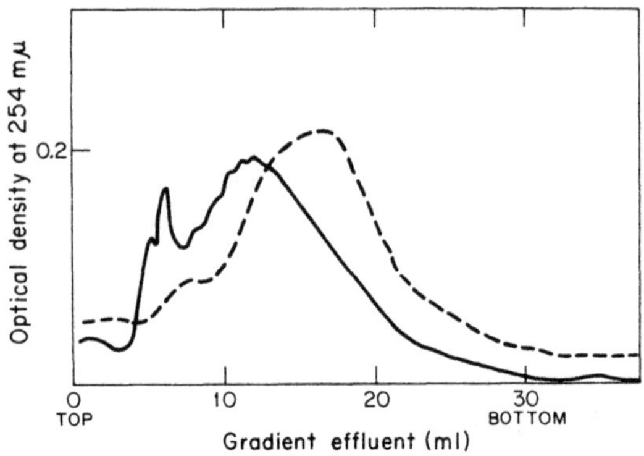

Fig. 13. Sucrose density-gradient profiles of polyribosomes isolated from cerebral cortices of infant and adult rats. Polyribosomes were prepared by the method of Campagnoni and Mahler[25] and were layered on linear sucrose density gradients (35–15%) containing 12 mM $MgCl_2$, 100 mM KCl, and 50 mM Tris-HCl buffer (pH 7.6). The gradients were centrifuged for 1.5 hr in a Spinco SW25 swinging bucket rotor at 25,000 rpm and 0°C. The effluent from the top of the gradient was monitored continuously for absorbance at 254 nm. Solid curve: adult (six-week) cerebral cortex; broken curve: infant (two-day) cerebral cortex. (C. E. Zomzely, S. Roberts, and S. Peache, unpublished observations.)

found to be more highly-labeled in newborn or very young rats than in older animals.[50,51] Except in the case of the nuclear fraction, the decrease in labeling occurred mainly within one week after birth. Interpretation of these data is complicated by incomplete knowledge of possible variations in the precursor pool for protein synthesis, resulting in part from development of the blood-brain barrier.

In vitro experiments generally support the conclusion that the average rate of brain protein synthesis declines progressively with age. Brain cortical slices from 16-day-old rats incorporated [^{14}C]valine into protein at a rate which was only one-sixth the value found in seven-day-old animals.[52] Amino acid-incorporating activities of cerebral ribosomal preparations have also been reported to decrease during early development.[11,30,31,41,53] Evidence has been presented that alterations in both ribosomes and activating enzymes may be responsible for the apparent decline in protein synthesis in the brain during development.[31,41,54] However, these observations may have been related in part to a decrease in the ratio of neurons to glia and to variations in purity of components of the amino acid-incorporating systems at different stages of development. For example, cerebral polyribosomes selectively isolated by zonal centrifugation from rats of various ages appeared to possess comparable activities in protein synthesis (C. E. Zomzely, S. Roberts and S. Peache, unpublished observations).

Nevertheless, alterations undoubtedly occur in neural protein-synthesizing systems during development, including an early loss of the heaviest ribosomal aggregates[11] (C. E. Zomzely, S. Roberts, and S. Peache, unpublished observations). Polyribosomal preparations from cerebral cortex of newborn or infant (two-day) rats contained significantly higher proportions of heavy aggregates than the corresponding preparations from immature (22-day) or adult (six-week) animals (Fig. 13). Developmental changes in the size distribution of ribonucleoprotein particles in mixed ribosomal preparations from brain were accompanied by alterations in amino acid-incorporating activity.[11,31] Variations in the rate of synthesis of certain membrane proteins, including myelin proteins, may be an important aspect of the developmental process in brain.[41,55] However, membrane proteins of nervous tissue are probably manufactured to a considerable extent in mitochondria or other nonmicrosomal sites. These proteins appear to exhibit somewhat different patterns of synthesis during development. Thus, the maximum rate of synthesis of myelin proteins presumably occurs at the time of most active myelination rather than at birth (see below).

Nature of Proteins Synthesized

Evidence has been presented for the synthesis of certain proteins unique to the nervous system by cytoplasmic ribosomes. This may be the case for the "*S*-100" protein, which is highly acidic and possesses a molecular weight of approximately 20,000–25,000.[56,57] Rubin and Stenzel[21] reported that a significant fraction of the proteins labeled after incubation of ribosomes from rabbit cerebral cortex with [^{14}C]leucine was precipitated by antibody to "*S*-100" protein. This protein may have the highest turnover rate of all soluble proteins extracted from rat brain.[58] Hydén and McEwen[59] reported that "*S*-100" protein was primarily concentrated around (but not in) the nuclei of oligodendrocytes and scattered throughout the glial membrane system. It was also found in neuronal nuclei. Moore and his co-workers[57,60] concluded that "*S*-100" protein is distributed in all parts of the nervous system and that its cellular localization could not be deduced from available data. However, clonal glial cells derived from rat glial tumor have recently been reported to produce "*S*-100" protein in tissue culture.[61]

Ribosomal formation of membrane proteins has also been reported. Cerebral nuclear RNA from four week-old mice, added to a homologous

TABLE III

Stimulation of the *Escherichia coli* (S-30 Fraction) Amino Acid-Incorporating System by Cerebral and Hepatic RNA Preparations[a]

Addition	Radioactivity incorporated (cts/min per mg protein)
None	1480
Cerebral nuclear RNA (51 μg)	4880
Cerebral ribosomal RNA (68 μg)	1842
Hepatic nuclear RNA (68 μg)	3240

[a] Reprinted with modifications from Bondy and Roberts.[35]

ribosomal system, enhanced the production of a basic encephalitogenic protein associated with brain myelin.[34] Moreover, Lim and Adams[55] noted that microsomal preparations from immature rat brain incorporated *L*-[^{14}C]valine into membrane proteins. This property was lacking in adult animals.

Regulation

Protein synthesis in all eukaryotic cells is principally directed by messenger RNA synthesized on DNA templates in the nucleus proper. Nucleolar DNA codes for the synthesis of ribosomal RNA.[62] Nuclear DNA is much more heterogeneous than nucleolar DNA and it serves as template for the formation of many more RNA species.

RNA synthesis in brain may be oriented toward the synthesis of messenger RNA to a greater extent than in liver. Thus, nuclear RNA from adult rat brain exhibited greater stimulatory activity in an *E. coli* amino acid incorporating system than the analogous hepatic preparation (Table III).[35] The reaction mixture (0.25 ml) contained 100 m*M* Tris-HCl (pH 7.8), 50 m*M* KCl, 12 m*M* magnesium acetate, 1 m*M* NaATP, 0.015 m*M* NaGTP, 10 m*M* creatine phosphate sodium salt, 6 m*M* mercaptoethanol, 0.12 mg creatine phosphokinase, 0.1 m*M* concentrations of each of 20 *L*-amino acids except phenylalanine, 0.25 μCi of uniformly labeled *L*-[^{14}C]phenylalanine (392 mCi/mmole) and 1.17 mg of *S*-30 protein. Incubation was carried out for 40 min at 37°C.). Moreover, the proportions of RNA pulse-labeled *in vivo* or randomly-labeled by methylation *in vitro* which hybridized to DNA were greater in cerebral nuclear and cytoplasmic preparations from newborn or adult rats than in the corresponding hepatic preparations (see, for example, Table IV).[37] Also consistent with this conclusion was the finding that RNA

TABLE IV

Hybridizable RNA in Rat Brain[a]

Tissue	Hybridizable RNA	
	Nucleus	Cytoplasm
Adult brain	2.1	0.65
Adult liver	1.2	0.48
Newborn brain	2.2	0.88
Newborn liver	1.5	0.40

[a] Values are expressed as percent of total radioactivity hybridizing to homologous DNA. RNA was labeled by exposure to [^{3}H]dimethylsulfate. The labeled RNA (8–12 μg) was incubated for 18 hr with 25–40 μg of denatured rat liver DNA. The average of three separate experiments is shown. Reprinted with modifications from Bondy and Roberts.(37)

polymerase activity was higher in brain than in liver of the adult animal.(63) The relative contribution of neurons and glia to messenger RNA production in the brain is not clear. Thus, although more of the RNA in neurons appears to be of the cytoplasmic type compared to that in glia,(45) and turnover of RNA is probably less active in the neuron,(64–66) the neuron forms a greater quantity of RNA.(67,68) Therefore, the overall rate of synthesis of messenger

TABLE V

Sedimentation Characteristics and Stimulatory Activities of Nuclear RNA Fractions from Rat Brain[a]

Addition	Radioactivity incorporated (cts/min per mg protein)	Stimulation (%)	RNA in fraction (μg)	Stimulatory activity in fraction (% of total)
None	6,300	—	—	—
Nuclear RNA				
>34*S*	7,870	25	95	2
"28*S*"	6,170	0	525	0
"18*S*"	8,430	34	600	22
<12*S*	17,580	180	400	76

[a] Nuclear RNA was fractionated on a linear sucrose density gradient; the RNA in each fraction was precipitated with ethanol and purified. Portions (80 μg) of each fraction were tested for stimulatory activity in the presence of 0.02 μCi of a mixture of 20 uniformly labeled *L*-[^{14}C]amino acids, 0.04 mg brain ribosomal protein, and 0.3 mg pH 5 enzyme. Incubation was carried out for 30 min at 37°C. The ribosomes and pH 5 enzymes had been subjected to prior incubation for 10 min. Reprinted from Samli and Roberts.(26)

RNA may be significantly higher in neurons than in glia. Available information on the RNA polymerase activity of cerebral gray and white matter supports this deduction.[63]

Cerebral nuclei may produce an unusually high proportion of messenger RNA species of relatively small molecular size, including species with sedimentation coefficients of 12 *S* or lower which stimulate amino acid incorporation into ribosomal proteins *in vitro* (Table V).[26, 34, 35] In support of this conclusion was the finding that cerebral polyribosomal preparations contained smaller aggregates than those present in most other mammalian tissues.[8] In part, this pattern reflected the greater sensitivity of cerebral

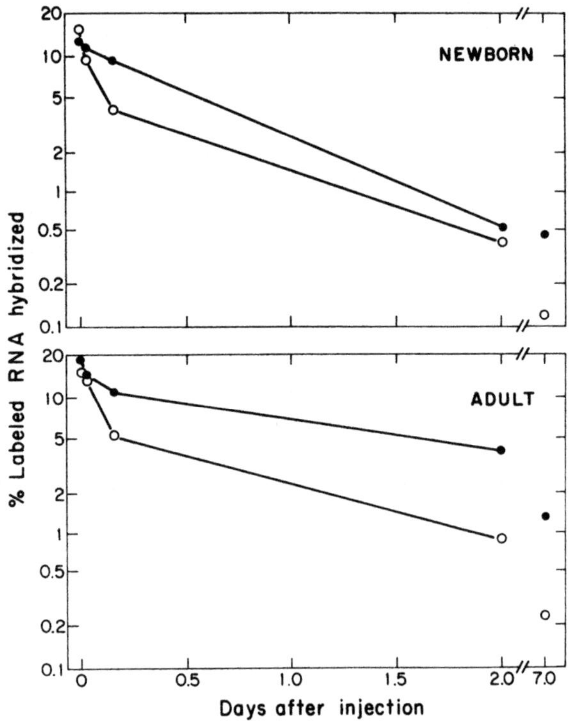

Fig. 14. Turnover of the hybridizable fraction of pulse-labeled nuclear and cytoplasmic RNA from brains of adult and newborn rats. RNA fractions were isolated at varying intervals after intracervical injection of 1 mCi [5-^{3}H]-uridine. Incubation was carried out for 18 hr with 3.8–10 μg RNA and 30–50 μg denatured rat liver DNA. (a) adult rats; (b) newborn rats. Darkened circles: nuclear RNA; open circles: cytoplasmic RNA. (Reprinted from Bondy and Roberts.[37])

messenger RNA-ribosome complexes to disruptive effects of ribonuclease and other influences.[7] However, the formation of small polypeptides which serve special functions in the brain may also be indicated.

Developmental changes have been observed in the properties of messenger RNA preparations isolated from whole rat brain. Cerebral nuclear RNA from newborn rats exhibited greater stimulatory activity on amino acid incorporation in a homologous ribosomal system than similar preparations from adult animals.[31] In addition, the turnover rates of hybridizable RNA in nuclear and cytoplasmic preparations from the newborn rat brain were significantly higher than the turnover rates of the analogous fractions obtained from the adult animal (Fig. 14)[37] This finding may be related to a greater template activity of chromatin in the newborn rat brain (S. C. Bondy and S. Roberts, unpublished observations). As noted earlier, a greater proportion of large ribosomal aggregates with high protein-synthesizing activity *in vivo* was present in newborn and infant rat brain compared to the adult. However, the proportion of hybridizable RNA relative to the total RNA present in rat brain did not appear to be markedly altered during development (Table IV)[37] These findings suggested that short-lived messenger RNA molecules with high template activity were replaced by longer-lived species with lesser activity. This transformation appeared to be reflected in alterations in the spectrum of turnover times exhibited by brain proteins in the developing rat.[48]

MITOCHONDRIA

General Aspects

The ubiquitous occurrence of nucleic acids as integral components of the mitochondrion seems well established.[69] Purified mitochondrial fractions obtained from either cell bodies or synaptosomes of rat cerebral cortex contained significant concentrations of DNA and RNA.[70,71] Evidence has been presented that nucleic acids of mitochondria may be closely associated with the inner membrane. Mitochondrial DNA differs in several important characteristics from the corresponding fractions in other parts of the cell. Thus, mitochondrial DNA occurred as circular filaments about 5 μ in length in cells of all tissues including brain.[72] Mitochondrial DNA obtained from mouse brain renatured more readily after heating than nuclear DNA from the same source and was therefore considered to be less heterogeneous.[73] Hybridization experiments also indicated that DNA from mouse-brain mitochondria possessed limited informational content.[74] A large fraction of the hybridizable RNA molecules in these mitochondria appeared to be formed on nuclear DNA templates. Additional components of mitochondrial protein-synthesizing systems have been isolated from mammalian cells. These components include ribosomes, transfer RNA, amino acid-activating enzymes, RNA polymerase, and DNA polymerase.[75] In common with mitochondrial DNA, ribosomes and ribosomal RNA from mitochondria may have certain distinctive features compared to the analogous fractions elsewhere.[69]

Klee and Sokoloff[76] first presented evidence that brain mitochondria may be active in protein synthesis. Optimal conditions for amino acid incorporation by cerebral mitochondria appear to be markedly dependent upon mitochondrial state. No requirement for exogenous pH 5 enzymes or ATP existed with purified, intact preparations.[22,77] Amino acid incorporation was stimulated by ADP, GTP, and certain substrates of the citric acid cycle (e.g., α-ketoglutarate). Puromycin and chloramphenicol inhibited protein synthesis by cerebral mitochondria, whereas ribonuclease, cycloheximide, and actinomycin D were ineffective in this regard.[22,76–79] However, interpretation of the actions of inhibitors and pharmacological agents on protein synthesis in the intact cerebral mitochondrion is limited by the sensitivity of this organelle to environmental changes and by the relative impermeability of the mitochondrial membrane.

Cytological and Regional Differences

Synaptosomal and cell-body mitochondria from rat cerebral cortex both actively incorporated [^{14}C]leucine into protein *in vitro*.[70,79,80] These findings seemed to implicate synaptosomal mitochondria in local axonal synthesis of certain proteins (see below). Conflicting evidence has been obtained relating to the possibility that synaptosomal mitochondria may normally derive a significant proportion of their proteins from the neuronal cell body by axoplasmic flow.[81–83] Very little information is available concerning the relative protein-synthesizing activities of mitochondria from different parts of the nervous system (see, however, von Hungen *et al.*[83]). Moreover, neurons and glia have not been compared in this respect.

Developmental Changes

The synthesis of mitochondrial protein may be especially important in the brain during maturation. Thus, the increase in mitochondrial protein accounted for two-thirds of the total elevation in brain protein from the first to the twenty-first day of age in the rat.[84] Crude mitochondrial fractions from rat brain exhibited maximal incorporation of *DL*-[1-^{14}C]leucine into total protein at about 10 days after birth.[76] This corresponded to the period of most active myelination and appeared to be associated with rapid synthesis of proteolipid (see also Mokrasch and Manner[85]). The specific activity of the proteolipid fraction declined precipitously thereafter.[76] In contrast, incorporation of isotope into residual protein (insoluble in a chloroform–methanol mixture) by the mitochondrial preparation was maximal at birth and declined less during maturation. Mitochondrial protein synthesis may be more responsive to environmental control in immature brain than in adult brain (see below).

Nature of Proteins Synthesized

Existing evidence suggests that the proteolipids of myelin are primarily of mitochondrial origin. Both crude and purified mitochondrial preparations have been reported to be active in the synthesis of these proteins.[76,78,86] Cerebral proteolipids appear to be formed selectively by oligodendroglia and Schwann cells,[87] to turn over at a significant rate only in the maturing brain,[88] and to differ from similar proteins of nonneural origin.[89]

Certain proteins of the cerebral mitochondrion are probably synthesized within this organelle.[90] Thus, the water-insoluble proteins of brain mitochondria exhibited an average half-life of 26 days. This value was not significantly different from that of total mitochondrial protein. In contrast, the water-soluble proteins of brain mitochondria had an average half-life of only 18 days. It seems likely that a major portion of the latter proteins, including most mitochondrial enzymes, are manufactured elsewhere.[69]

Regulation

Little is known concerning the regulation of mitochondrial protein synthesis in the nervous system. However, an important site of control may be at the translational level. Factors which affect energy transformations in the mitochondrion are probably of special significance. The capacity of brain mitochondria to provide energy and other factors required for protein synthesis was greater in immature animals than in adults.[91] In addition, mitochondria from immature rats responded uniquely to thyroid hormone with the production of factors which functioned in microsomal as well as mitochondrial protein synthesis. These phenomena may reflect an important hormonal control of protein synthesis in the developing brain.

AXONAL STRUCTURES

Axoplasmic flow undoubtedly carries preformed proteins and nucleic acids from their site of origin in the neuronal cell body to the distal portions of the axon.[81,92] However, these substances appear to be formed locally in the axon as well.[83,93–96] Supporting this conclusion were the observations that the axonal portions of sciatic and spinal nerves were active in incorporating radioactive amino acids into protein *in vitro*.[94,96–98] Incorporation was stimulated by nerve damage.[97] Conflicting data were obtained relating to the actions of various antibiotics on axonal protein synthesis.

Synaptosomal structures other than mitochondria appear to contain appreciable quantities of RNA and may be capable of synthesizing certain proteins of significance to synaptic function.[70] Fractionation of synaptosomes after incubation in an amino acid-incorporating system revealed that proteins of the membrane and soluble fractions were most highly labeled.[79] However, synaptic vesicle proteins did not become labeled in these experiments. Moreover, the turnover rate of synaptic vesicle proteins *in vivo* was relatively slow, suggesting that vesicular breakdown and renewal did not accompany synaptic transmission of neural impulses.[83] Incorporation of amino acid by synaptosomal membranes after incubation of synaptosomes was inhibited by cycloheximide (which depresses protein synthesis in cytoplasmic ribosomes), but not by chloramphenicol (which is inhibitory in mitochondrial preparations).[79] Ribonuclease did not inhibit amino acid incorporation by the membrane fractions. Membranous structures of the axon and nerve terminal may contain occult ribosomes which are protected from the action of certain inhibitors by the membranes. Alternatively, membrane RNA may be present in a nonribosomal form and yet be capable of functioning at a translational level in protein synthesis.[99]

Proteins which may be formed locally in the axon include certain membrane proteins and acetylcholinesterase. The latter appears to be present in synaptic vesicles.[79,100,101] Regeneration of acetylcholinesterase activity in cat hypoglossal nerves after irreversible inactivation or nerve transection occurred earliest in the distal axon.[93,102,103] Restoration was partially inhibited by puromycin,[93] but was actually stimulated by actinomycin D.[104] In cultures of embryonic chick brain, the level of acetylcholinesterase was unaffected by addition of chicken-liver histone, puromycin, or polylysine, and was elevated by actinomycin D.[105] In contrast, synthesis of proteins of microsomal origin in this system was stimulated by histone and inhibited by the other additives. These characteristics suggest that acetylcholinesterase may be formed in axonal membranes but do not rule out the possible intervention of synaptosomal mitochondria.

PROTEIN SYNTHESIS AND SPECIALIZED FUNCTION IN THE NERVOUS SYSTEM

The relationship between protein synthesis and specialized function in the nervous system has not been clarified. This fact may be attributed to lack of knowledge concerning those proteins which participate in the unique activities of neural tissue, their cellular and subcellular sites of formation, and the details of regulatory mechanisms which control their synthesis.

Quantitatively, the most important site of protein synthesis in the nervous system is the cytoplasmic ribosome. However, mitochondria in neurons and glia, as well as synaptic and other axonal membranes in neurons, may synthesize proteins which play special roles. The control of protein synthesis in these latter sites appears to be interwoven with processes which regulate protein synthesis by cytoplasmic ribosomes. The suggestion has been advanced that amino acids employed for synthesis of proteins in the distal axon and at the synaptic terminal of the neuron may be derived principally from lysosomal digestion of proteins formed at ribosomal sites in the cell body.[80,95] According to this hypothesis, the maintenance of axonal structure and function would be largely dependent upon ribosomal protein synthesis and axoplasmic flow for formation and delivery of proteins and derived amino acids. Moreover, mitochondria in both neurons and glia may depend upon protein synthesis in cytoplasmic ribosomes for formation of all but structural proteins of the inner membrane[90,106] and possibly proteolipids of the myelin sheath.[76,78,86]

Analysis of regulatory processes in the central nervous system which relate protein synthesis to specialized function may logically focus upon the properties of brain messenger RNA. The messenger RNA population of brain cells appears to exhibit special characteristics which change during development and in response to alterations in the environment. Moreover, a relatively high proportion of RNA synthesis in the brain may be oriented toward formation of messenger RNA.[37] Presumptive evidence has been obtained for a decrease during brain maturation of certain of the larger messenger RNA species present in very young rats. Thus, the proportion of large polyribosomes in cerebral cell-free preparations was less in the adult rat than in the newborn or infant animal[11] (C. E. Zomzely, S. Roberts, and S. Peache unpublished observations). The change in the messenger RNA population during development may involve partial replacement of species exhibiting rapid turnover rates,[37] which are highly active in protein synthesis,[31] with molecules possessing slow turnover rates and lesser template activity.

The adult brain contains a high proportion of unstable messenger RNA-ribosome complexes which may be largely of neuronal origin[7,8] (C. E. Zomzely, S. Roberts, and S. Peache, unpublished observations). The possible intervention of certain of these unstable polyribosomes in adjustments to environmental alterations affecting the nervous system has been suggested.[7,8] Several physiological and pathological phenomena appear to support this hypothesis, including the process of chromatolysis following nerve injury,[68] the disaggregation of brain polyribosomes induced by electroconvulsive treatment in the rabbit,[107] and the reaggregation of brain polyribosomes in rats exposed to ambient light and sound for 15 min after three days in the dark.[108] In addition, numerous reports have dealt with changes in RNA and protein synthesis in central and peripheral nerves resulting from specific and nonspecific stimuli, as well as behavioral experiences. The unique lability of certain messenger RNA-ribosome complexes in the nervous system may provide a mechanism which facilitates the coding, selection and transmission of information during the passage of neural impulses.[7,8,15]

REFERENCES

1. S. Roberts, Protein synthesis, *in* "Handbook of Neurochemistry" (A. Lajtha, ed.), Vol. V, Plenum Press, New York (to be published).
2. S. L. Palay and G. E. Palade, The fine structure of neurons, *J. Biophys. Biochem. Cytol.* **1,** 69–88 (1955).
3. R. Ekholm and H. Hydén, Polysomes from microdissected fresh neurons, *J. Ultrastructure Res.* **13,** 269–280 (1965).
4. C. Sotelo and S. L. Palay, The fine structure of the lateral vestibular nucleus in the rat. I. Neurons and neuroglial cells, *J. Cell Biol.* **36,** 151–179 (1968).
5. D. Bodian, A suggestive relation hip of nerve cell RNA with specific synaptic sites, *Proc. Natl. Acad. Sci. U.S.* **53,** 418–425 (1965).
6. E. Mugnaini and F. Walberg, Ultrastructure of neuroglia, *Ergeb. Anat. u. Entwicklungsgeschichte* (*Abt. III Z. ges. Anat.*) **37,** 194–236 (1964).
7. C. E. Zomzely, S. Roberts, C. P. Gruber, and D. M. Brown, Cerebral protein synthesis. II. Instability of cerebral messenger ribonucleic acid–ribosome complexes, *J. Biol. Chem.* **243,** 5396–5409 (1968).
8. C. E. Zomzely, S. Roberts, D. M. Brown, and C. Provost, Cerebral protein synthesis. I. Physical properties of cerebral ribosomes and polyribosomes, *J. Mol. Biol.* **20,** 455–468 (1966).
9. M. L. Petermann, "The Physical and Chemical Properties of Ribosomes," Elsevier Publishing Company, Amsterdam (1964).
10. Y. Takahashi, K. Mase, and H. Sugano, Preparation of polysomes from rat brain tissue, *Biochim. Biophys. Acta* **119,** 627–629 (1966).
11. M. R. V. Murthy, Protein synthesis in growing rat tissues. II. Polyribosome concentration of brain and liver as a function of age, *Biochim. Biophys. Acta* **119,** 599–613 (1966).
12. A. J. Munro, R. J. Jackson, and A. Korner, Studies on the nature of polysomes, *Biochem. J.* **92,** 289–299 (1964).
13. D. Schneider and S. Roberts, Base compositions of 18*S* and 28*S* RNA fractions from rat cerebral cortex, *J. Neurochem.* **15,** 1469–1471 (1968).

14. C. E. Zomzely, S. Roberts, D. M. Brown, and D. Rapaport, Isolation of 55-*S* ribonucleoprotein particles with amino acid-incorporating activity, *Biochim. Biophys. Acta* **103,** 529–531 (1965).
15. S. Roberts and C. E. Zomzely, Regulation of protein synthesis in the brain, *in:* "Protides of the Biological Fluids" (H. Peeters, ed.), Vol. 13, pp. 91–102, Elsevier Publishing Company, Amsterdam (1966).
16. H. Bielka, H. Welfle, M. Böttger, W. Förster, I. Schneiders, and A. Henske, Strukturveränderungen und dissoziation von leberribosomen in abhängigkeit von der Mg^{++}-konzentration, *European J. Biochem.* **5,** 183–190 (1968).
17. C. E. Zomzely, S. Roberts, and D. Rapaport, Regulation of cerebral metabolism of amino acids—III. Characteristics of amino acid incorporation into protein of microsomal and ribosomal preparations of rat cerebral cortex, *J. Neurochem.* **11,** 567–582 (1964).
18. M. R. V. Murthy and D. A. Rappoport, Biochemistry of the developing rat brain. VI. Preparation and properties of ribosomes, *Biochim. Biophys. Acta* **95,** 132–145 (1965).
19. D. H. Clouet, M. Ratner, and N. Williams, [^{14}C]leucine incorporation into brain ribosomes, *Biochim. Biophys. Acta* **123,** 142–150 (1966).
20. S. H. Barondes and M. W. Nirenberg, Fate of a synthetic polynucleotide directing cell-free protein synthesis. I. Characteristics of degradation, *Science* **138,** 810–817 (1962).
21. A. L. Rubin and K. H. Stenzel, *In vitro* synthesis of brain protein, *Proc. Natl. Acad. Sci. U.S.* **53,** 963–968 (1965).
22. M. K. Campbell, H. R. Mahler, W. J. Moore, and S. Tewari, Protein synthesis systems from rat brain, *Biochemistry* **5,** 1174–1184 (1966).
23. K. H. Stenzel, R. F. Aronson, and A. L. Rubin, *In vitro* synthesis of brain protein. II. Properties of the complete system, *Biochemistry* **5,** 930–936 (1966).
24. Y. Takahashi, K. Mase, and S. Abe, The protein biosynthesis in a ribosomal system of the brain, *J. Biochem. (Tokyo)* **60,** 363–371 (1966).
25. A. T. Campagnoni and H. R. Mahler, Isolation and properties of polyribosomes from cerebral cortex, *Biochemistry* **6,** 956–967 (1967).
26. M. H. Samli and S. Roberts, Properties of brain nuclear RNA fractions which stimulate amino acid incorporation in homologous ribosomal systems, *J. Neurochem.*, in press.
27. S. Roberts, Regulation of cerebral metabolism of amino acids—II. Influence of phenylalanine deficiency on free and protein-bound amino acids in rat cerebral cortex: Relationship to plasma levels, *J. Neurochem.* **10,** 931–940 (1963).
28. M. R. V. Murthy, Protein synthesis in growing rat tissues. I. Effect of various metabolites and inhibitors on phenylalanine incorporation by brain and liver ribosomes, *Biochim. Biophys. Acta* **119,** 586–598 (1966).
29. D. H. Clouet and M. Ratner, The effect of morphine administration on the incorporation of [^{14}C]leucine into protein in cell-free systems from rat brain and liver, *J. Neurochem.* **15,** 17–23 (1968).
30. S. C. Bondy and S. V. Perry, Incorporation of labeled amino acids in the soluble protein fraction of rabbit brain, *J. Neurochem.* **10,** 603–609 (1963).
31. S. Yamagami, R. R. Fritz, and D. A. Rappoport, Biochemistry of the developing rat brain. VII. Changes in the ribosomal system and nuclear RNA's, *Biochim. Biophys. Acta* **129,** 532–547 (1966).
32. A. Gierer, Function of aggregated reticulocyte ribosomes in protein synthesis, *J. Mol. Biol.* **6,** 148–157 (1963).

33. J. O. Bishop, Reticulocyte ribosome fraction with an exceptional capacity for polyphenylalanine synthesis, *Nature* **208,** 361–365 (1965).
34. I. D. Herriman and G. D. Hunter, Cytoplasmic protein synthesis in mouse brain, *J. Neurochem.* **12,** 937–947 (1965).
35. S. C. Bondy and S. Roberts, Messenger ribonucleic acid of cerebral nuclei, *Biochem. J.* **105,** 1111–1118 (1967).
36. S. Roberts and B. S. Morelos, Regulation of cerebral metabolism of amino acids—IV. Influence of amino acid levels on leucine uptake, utilization and incorporation into protein *in vivo, J. Neurochem.* **12,** 373–387 (1965).
37. S. C. Bondy and S. Roberts, Hybridizable ribonucleic acid of rat brain, *Biochem. J.* **109,** 533–541 (1968).
38. S. Furst, A. Lajtha, and H. Waelsch, Amino acid and protein metabolism of the brain—III. Incorporation of lysine into the proteins of various brain areas and their cellular fractions, *J. Neurochem.* **2,** 216–225 (1958).
39. T. C. Johnson and M. W. Luttges, The effects of maturation on *in vitro* protein synthesis by mouse brain cells, *J. Neurochem.* **13,** 545–552 (1966).
40. K. Mase, Y. Takahashi, and K. Ogata, The incorporation of [^{14}C]glycine into the protein of guinea pig brain cortex slices, *J. Neurochem.* **9,** 281–288 (1962).
41. K. Suzuki, S. R. Korey, and R. D. Terry, Studies on protein synthesis in brain microsomal system, *J. Neurochem.* **11,** 403–412 (1964).
42. Y. Takahashi and S. Abe, Distribution of amino acid activating enzymes in rabbit's brain, *Experientia* **19,** 186–187 (1963).
43. H. Hydén, Behavior, neural function, and RNA, *in* "Progress in Nucleic Acid Research and Molecular Biology" (J. N. Davidson and W. E. Cohn, eds.), Vol. 6, pp. 187–218, Academic Press, New York (1967).
44. R. Landolt, H. H. Hess, and C. Thalheimer, Regional distribution of some chemical structural components of the human nervous system—I. DNA, RNA and ganglioside sialic acid, *J. Neurochem.* **13,** 1441–1452 (1966).
45. H. Hydén and E. Egyházi, Glial RNA changes during a learning experiment in rats, *Proc. Natl. Acad. Sci. U.S.* **49,** 618–624 (1963).
46. E. G. Gray, Tissue of the central nervous system, *in* "Electron Microscopic Anatomy" (S. M. Kurtz, ed.), pp. 369–417, Academic Press, New York (1964).
47. K. R. Brizzee, J. Vogt, and X. Kharetchko, Postnatal changes in glial/neuron index with a comparison of methods of cell enumeration in the white rat, *Prog. Brain Res.* **4,** 136–149 (1964).
48. A. Lajtha, S. Furst, A. Gerstein, and H. Waelsch, Amino acid and protein metabolism of the brain—I. Turnover of free and protein bound lysine in brain and other organs, *J. Neurochem.* **1,** 289–300 (1957).
49. R. J. Schain, M. J. Carver, J. H. Copenhaver, and N. R. Underdahl, Protein metabolism in the developing brain: Influence of birth and gestational age, *Science* **156,** 984–986 (1967).
50. A. A. Abdel-Latif and L. G. Abood, *In vivo* incorporation of L-(^{14}C) serine into phospholipids and proteins of the subcellular fractions of developing rat brain, *J. Neurochem.* **13,** 1189–1196 (1966).
51. S. S. Oja, Studies on protein metabolism in developing rat brain, *Ann. Acad. Sci. Fenn.* **131,** 1–81 (1967).
52. F. Orrego and F. Lipmann, Protein synthesis in brain slices. Effects of electrical stimulation and acidic amino acids, *J. Biol. Chem.* **242,** 665–671 (1967).
53. M. R. V. Murthy and D. A. Rappoport, Biochemistry of the developing rat brain. V. Cell-free incorporation of L-[1-^{14}C]leucine into microsomal protein, *Biochim. Biophys. Acta* **95,** 121–131 (1965).

54. D. H. Adams and L. Lim, Amino acid incorporation by preparations from the developing rat brain, *Biochem. J.* **99,** 261–265 (1966).
55. L. Lim and D. H. Adams, Microsomal components in relation to amino acid incorporation by preparations from the developing rat brain, *Biochem. J.* **104,** 229–238 (1967).
56. B. W. Moore, A soluble protein characteristic of the nervous system, *Biochem. Biophys. Res. Commun.* **19,** 739–744 (1965).
57. B. W. Moore and V. J. Perez, Specific acidic proteins of the nervous system, *in*: "Physiological and Biochemical Aspects of Nervous Integration" (F. D. Carlson, ed.), pp. 343–359, Prentice-Hall, Englewood Cliffs, New Jersey (1968).
58. B. S. McEwen and H. Hydén, A study of specific brain proteins on the semimicro scale, *J. Neurochem.* **13,** 823–833 (1966).
59. H. Hydén and B. McEwen, A glial protein specific for the nervous system, *Proc. Natl. Acad. Sci. U.S.* **55,** 354–358 (1966).
60. B. W. Moore, V. J. Perez, and M. Gehring, Assay and regional distribution of a protein characteristic of the nervous system, *J. Neurochem.* **15,** 265–272 (1968).
61. P. Benda, J. Lightbody, G. Sato, L. Levine, and W. Sweet, Differentiated rat glial cell strain in tissue culture, *Science* **161,** 370–371 (1968).
62. R. P. Perry, The nucleolus and the synthesis of ribosomes, *in*: "Progress in Nucleic Acid Research and Molecular Biology" (J. N. Davidson and W. E. Cohn, eds.), Vol. 6, pp. 219–257, Academic Press, New York (1967).
63. S. C. Bondy and H. Waelsch, Nuclear RNA polymerase in brain and liver, *J. Neurochem.* **12,** 751–756 (1965).
64. B. Daneholt and S.-O. Brattgård, A comparison between RNA metabolism of nerve cells and glia in the hypoglossal nucleus of the rabbit, *J. Neurochem.* **13,** 913–921 (1966).
65. S. H. Appel, Turnover of brain messenger RNA, *Nature* **213,** 1253–1254 (1967).
66. P. Volpe and A. Giuditta, Biosynthesis of RNA in neuron- and glia-enriched fractions, *Brain Res.* **6,** 228–240 (1967).
67. H. Hydén and A. Pigon, A cytophysiological study of the functional relationship between oligodendroglial cells and nerve cells of Deiters' nucleus, *J. Neurochem.* **6,** 57–72 (1960).
68. S.-O. Brattgård, J.-E. Edström, and H. Hydén, The chemical changes in regenerating neurons, *J. Neurochem.* **1,** 316–325 (1957).
69. D. B. Roodyn and D. Wilkie, "The Biogenesis of Mitochondria," Methuen and Co., London (1968).
70. L. Austin and I. G. Morgan, Incorporation of ^{14}C-labeled leucine into synaptosomes from rat cerebral cortex *in vitro*, *J. Neurochem.* **14,** 377–387 (1967).
71. R. Balázs and W. A. Cocks, RNA metabolism in subcellular fractions of brain tissue, *J. Neurochem.* **14,** 1035–1055 (1967).
72. J. H. Sinclair, B. J. Stevens, N. Gross, and M. Rabinowitz, The constant size of circular mitochondrial DNA in several organisms and different organs, *Biochim. Biophys. Acta* **145,** 528–531 (1967).
73. H. G. Du Buy, C. F. T. Mattern, and F. L. Riley, Comparison of the DNA's obtained from brain nuclei and mitochondria of mice and from the nuclei and kinetoplasts of *Leishmania enriettii*, *Biochim. Biophys. Acta* **123,** 298–305 (1966).
74. D. G. Humm and J. H. Humm, Hybridization of mitochondrial RNA with mitochondrial and nuclear DNA in agar, *Proc. Natl. Acad. Sci. U.S.* **55,** 114–119 (1966).
75. M. E. Pullman and G. Schatz, Mitochondrial oxidations and energy coupling, *Ann. Rev. Biochem.* **36,** 539–610 (1967).

76. C. B. Klee and L. Sokoloff, Amino acid incorporation into proteolipid of myelin *in vitro*, *Proc. Natl. Acad. Sci. U.S.* **53**, 1014–1021 (1965).
77. H. S. Bachelard, Amino acid incorporation into the protein of mitochondrial preparations from cerebral cortex and spinal cord, *Biochem. J.* **100**, 131–137 (1966).
78. L. C. Mokrasch, Incorporation of [^{14}C]amino acids into the proteolipid of subcellular preparations of rat brain *in vitro*, *J. Neurochem.* **13**, 49–58 (1966).
79. I. G. Morgan and L. Austin, Synaptosomal protein synthesis in a cell-free system, *J. Neurochem.* **15**, 41–51 (1968).
80. M. K. Gordon, K. G. Bench, G. G. Deanin, and M. W. Gordon, Histochemical and biochemical study of synaptic lysosomes, *Nature* **217**, 523–527 (1968).
81. S. H. Barondes, On the site of synthesis of the mitochondrial protein of nerve endings, *J. Neurochem.* **13**, 721–727 (1966).
82. S. H. Barondes, Further studies of the transport of protein to nerve endings, *J. Neurochem.* **15**, 343–350 (1968).
83. K. von Hungen, H. R. Mahler, and W. J. Moore, Turnover of protein and ribonucleic acid in synaptic subcellular fractions from rat brain, *J. Biol. Chem.* **243**, 1415–1423 (1968).
84. F. E. Samson, Jr., W. M. Balfour, and R. J. Jacobs, Mitochondrial changes in developing rat brain, *Am. J. Physiol.* **199**, 693–696 (1960).
85. L. C. Mokrasch and P. Manner, Incorporation of ^{14}C-amino acids and [^{14}C]-palmitate into proteolipids of rat brains *in vitro*, *J. Neurochem.* **10**, 541–547 (1963).
86. A. J. Tolani and L. C. Mokrasch, Incorporation of ^{14}C-amino acids into proteolipid protein of subcellular fractions from rat brain, heart and liver, *Life Sciences* **6**, 1771–1774 (1967).
87. H. Koenig, An autoradiographic study of nucleic acid and protein turnover in the mammalian neuraxis, *J. Biophys. Biochem. Cytol.* **4**, 785–792 (1958).
88. J. Folch-Pi, Proteolipids, neurokeratin, neurosclerin and copper proteins, *in*: "Protides of the Biological Fluids" (H. Peeters, ed.), Vol. 13, pp. 21–34, Elsevier Publishing Company, Amsterdam (1966).
89. F. Wolfgram, The amino acid compositions of some non-neural proteolipid proteins, *Biochim. Biophys. Acta* **147**, 383–385 (1967).
90. D. S. Beattie, R. E. Basford, and S. B. Koritz, The turnover of the protein components of mitochondria from rat liver, kidney, and brain, *J. Biol. Chem.* **242**, 4584–4586 (1967).
91. C. B. Klee and L. Sokoloff, Mitochondrial differences in mature and immature brain. Influence on rate of amino acid incorporation into protein and responses to thyroxine, *J. Neurochem.* **11**, 709–716 (1964).
92. P. Weiss and H. B. Hiscoe, Experiments on the mechanism of nerve growth, *J. Exp. Zool.* **107**, 315–396 (1948).
93. E. Koenig, Synthetic mechanisms in the axon—I. Local axonal synthesis of acetylcholinesterase, *J. Neurochem.* **12**, 343–355 (1965).
94. E. Koenig, Synthetic mechanisms in the axon—IV. *In vitro* incorporation of [^{3}H]precursors into axonal protein and RNA, *J. Neurochem.* **14**, 437–446 (1967).
95. L. Austin, J. J. Bray, and R. J. Young, Transport of proteins and ribonucleic acid along nerve axons, *J. Neurochem.* **13**, 1267–1269 (1966).
96. A. Edström, Amino acid incorporation in isolated Mauthner nerve fibre components, *J. Neurochem.* **13**, 315–321 (1966).
97. Y. Takahashi, M. Nomura, and S. Furusawa, *In vitro* incorporation of [^{14}C]-amino acids into proteins of peripheral nerve during Wallerian degeneration, *J. Neurochem.* **7**, 97–102 (1961).

98. A. Edström, Inhibition of protein synthesis in Mauthner nerve fibre components by actinomycin-D, *J. Neurochem.* **14,** 239–243 (1967).
99. V. Shapot and H. C. Pitot, Isolation and fractionation of ribonucleic acid from the smooth endoplasmic reticulum of rat liver, *Biochim. Biophys. Acta* **119,** 37–45 (1966).
100. V. P. Whittaker, I. A. Michaelson, and R. J. A. Kirkland, The separation of synaptic vesicles from nerve-ending particles ("synaptosomes"), *Biochem. J.* **90,** 293–303 (1964).
101. P. R. Lewis and C. C. D. Shute, The distribution of cholinesterase in cholinergic neurons demonstrated with the electron microscope, *J. Cell Sci.* **1,** 381–390 (1966).
102. D. H. Clouet and H. Waelsch, Amino acid and protein metabolism of the brain—VIII. The recovery of cholinesterase in the nervous system of the frog after inhibition, *J. Neurochem.* **8,** 201–215 (1961).
103. E. Koenig and G. B. Koelle, Mode of regeneration of acetylcholinesterase in cholinergic neurons following irreversible inactivation, *J. Neurochem.* **8,** 169–188 (1961).
104. E. Koenig, Synthetic mechanisms in the axon—III. Stimulation of acetylcholinesterase synthesis by actinomycin-D in the hypoglossal nerve, *J. Neurochem.* **14,** 429–435 (1967).
105. B. C. Goodwin and I. W. Sizer, Histone regulation of lactic dehydrogenase in embryonic chick brain tissue, *Science* **148,** 242–244 (1965).
106. D. S. Beattie, R. E. Basford, and S. B. Koritz, The inner membrane as the site of the *in vitro* incorporation of L-[^{14}C]leucine into mitochondrial protein, *Biochemistry* **6,** 3099–3106 (1967).
107. C. Vesco and A. Giuditta, Disaggregation of brain polysomes induced by electroconvulsive treatment, *J. Neurochem.* **15,** 81–85 (1968).
108. S. H. Appel, W. Davis, and S. Scott, Brain polysomes: Response to environmental stimulation, *Science* **157,** 836–838 (1967).

DISCUSSION

MANDEL. I am of course very interested in your results, but we disagree on several points. First of all, we found a lower amount of hybridizable RNA and DNA in brain than you did. Thus, about 1.2% of rat DNA hybridized with total brain RNA. The amount of cytoplasmic RNA which hybridized with DNA was much lower than this value. The second point is that we have been unable to get a significant stimulation of ribosomal protein synthesis with pure RNA and have found that there is a possibility of producing artifactual stimulation. I feel that this is because the messenger that sticks to ribosomes in the living animal is not pure RNA, but RNA contaminated with something else. Perhaps the only pure RNA that can stick on ribosomes is viral RNA. I don't believe purified RNA from animal sources has this capacity.

ROBERTS. I don't really think there is a great deal of disagreement between us. These are areas that have not been thoroughly investigated as yet. The actual percentage of hybridization achieved depends to a great extent on the precise procedures used during the random labeling of RNA and on the hybridization measurements. I wouldn't be surprised if this percentage varied in different laboratories. For example, the intrinsic instability of the hybrid itself may lead to variations with different methods. In this connection, it has been demonstrated that all of the hybridizable material can be removed by ribonuclease treatment of the DNA/RNA complex. The other point that you make relates to the degree of stimulation of amino acid incorporation possible with nuclear RNA preparations in ribosomal systems of animal origin. Even within our laboratory, variations have occurred with different methods. Dr. Bondy's nuclear RNA preparations consistently produced only a 25–30% increase in amino acid incorporation by cerebral ribosomes. However, the stimulation was markedly increased when activating enzymes were omitted from this system. Presumptive evidence for the conclusion that the stimulation was due to messenger RNA was forthcoming from the finding that the degree of stimulation of incorporation varied with different amino acids. In similar studies in our laboratory, Dr. Samli was able to achieve a striking stimulation of ribosomal protein synthesis with preparations of cerebral nuclear RNA when she employed preincubated cerebral ribosomes. This stimulation persisted in the presence of excesses of transfer RNA or pH 5 enzymes. In these latter studies many of the factors which could have caused artifacts have been eliminated. However, I agree that the messenger RNA content of a preparation cannot be determined without some independent measure of its presence, such as hybridizability. I should also note that there have been other reports of stimulation of the order of magnitude we reported here—the work of Herriman and Hunter, for example.

MURTHY. Were the ribosomes you used obtained by deoxycholate treatment of the microsomes or of the mitochondrial supernatant?

ROBERTS. In some cases cerebral polysomes were prepared with deoxycholate and in some cases without detergent. Preparations obtained either in the presence or absence of deoxycholate revealed the same properties; for example, the same sensitivity to magnesium variations.

MURTHY. In our laboratory we have found that we do not get the same sensitivity of amino acid incorporation to magnesium with microsomes as with ribosomes. In fact, microsomal incorporation seems to be quite resistant to low magnesium concentrations and even to mild concentrations of EDTA.

ROBERTS. Microsomal preparations are contaminated to a varying extent with ions, amino acids, and pH 5 enzymes. For this reason microsomes generally reveal much less sensitivity to everything. For example, various amino acid mixtures that will stimulate or inhibit incorporation into ribosomal protein affect microsomal preparations to a much lower degree. As I recall, most of your preparations were prepared in the presence of high concentrations of magnesium and may have had a greater magnesium contamination, which would reduce the sensitivity to variations in other factors.

MURTHY. Don't you feel that pretreatment with deoxycholate may predispose cerebral ribosomes to breakdown?

ROBERTS. No, not necessarily. Exactly the same gradient profile could be obtained in the presence or absence of deoxycholate.

MURTHY. Regarding the stimulation that you obtain upon addition of nuclear RNA to the incorporation medium, we have done similar work and obtained very variable results, but if we extracted the nuclei first with 1 *M* sodium chloride and then extracted the two fractions obtained with phenol, the RNA fraction from the sodium chloride-soluble material inhibited incorporation, whereas the RNA fraction from the sodium chloride-insoluble material increased incorporation by about 50 to 100%. Perhaps the differences obtained in different laboratories were due to differences in the mixture of nuclear RNA.

MAHLER. You have shown, and very convincingly, that there is a great deal of breakdown of the brain polysomal population in the course of an incorporation experiment and that there is a fair proportion of monomers. Also, in your nuclear RNA there is an indication of low-molecular-weight material with stimulatory activity. You have tried to correlate these findings in very interesting fashion, suggesting that there may be cerebral synthesis of low-molecular-weight polypeptides specified by the smaller messengers which are presumably on the monosomes. Have you in fact ever run an incorporation experiment with isolated monosomes, and do they incorporate?

ZOMZELY. Ribosomes stripped of messenger RNA show hardly any basal incorporation, but a tremendous response to poly U (about 50 fold). In the presence of added transferases this response increases even further.

MAHLER. This incorporation may be an artifact.

ROBERTS. A few years ago Dr. Zomzely demonstrated that a stripped preparation of brain ribosomal subunits incorporated radioactivity and also responded to poly U. The amount of incorporation was relatively low. During the incorporation procedure the subunits in this preparation became monoribosomes. The point to be emphasized is that there was incorporation apparently in the absence of polysomes.

RAPPOPORT. Returning to the question that Dr. Mandel raised about the stimulatory activity of nuclear RNA preparations, we have found great variations in this property during development in the rat. Protein synthesis by cerebral ribosomes begins to drop after 30 days of age and levels off at 50 days. Brain nuclear RNA preparations from the very young animal produced markedly enhanced incorporation. In contrast, adult cerebral RNA was relatively ineffective. Chromatin from young rat brain incubated with RNA polymerase produced an RNA which, on the basis of the base ratio, was rich in messenger RNA. We have not measured its messenger properties, however. With brain nuclear RNA from the young animal we obtained relatively marked stimulation of not just one amino acid, but of groups of amino acids. I don't know why adult RNA shows only a 20–30% stimulation whereas the young RNA shows over 100% stimulation. One more point. Since the polysome population depends upon the amount of messenger released, the population of polysomes must be related to the activity of RNA polymerase. The total number of polysomes in rat brain preparations is much greater in the very young animal, where the polymerase activity is very much higher.

ROBERTS. The influence of development on messenger RNA formation and protein synthesis in the brain is a topic that requires additional investigation. Most of the studies to date have been carried out on whole brain without regard to the changing proportions of neurons, glia, and undifferentiated cells.

Chapter 2

Developmental Changes in Peptide-Bond Hydrolases*

Neville Marks and Abel Lajtha

New York State Research Institute for Neurochemistry and Drug Addiction Ward's Island, New York, N.Y.

INTRODUCTION

The rapid growth of brain during ontogeny affords a valuable experimental tool for studying the factors affecting protein breakdown and turnover. Despite the wealth of information documenting the remarkable changes during development,[1–5] the correlation between chemical events and functional changes is poorly understood. Incorporation experiments with labeled amino acid precursors have established that brain proteins are in a state of dynamic equilibrium. The aspects concerned with the synthetic mechanisms are covered by Murthy[6] and by Roberts *et al.*[7] The purpose of this report is to comment on the areas concerned with protein breakdown, which is a necessary sequel to synthesis to maintain turnover. Other studies concerned with development are an earlier report from this laboratory,[3] the report of Oja,[8] and the contribution of Palladin and Belik.[9]

Like other tissues, brain contains a diversity of proteins with variable half-lives. The average half-life for the majority of proteins is about 15 days, although minor components show a range extending from less than 1 hr to several months.[10] These values must be regarded as estimates, since in a rapidly changing population of heterogeneous proteins precise information is required for the rates of synthesis of specific proteins, together with knowledge of their rates of breakdown. During development the rates of protein turnover are considerably higher than in the adult,[10] and a relatively small change in the differential rates of synthesis compared to breakdown can materially affect the rate of protein deposition and ultimately the matura-

*This study was supported in part by grant NB-03226, US Public Health Service. We also wish to acknowledge the help of Dr. R. K. Datta, Mary O'Mara, and Charles Bellman.

tional process. For bacterial and yeast cells there are several reports showing a lower rate of breakdown in the growing compared to the nongrowing organisms,[11–13] and Rittenberg *et al.*[14] reported that growth in regenerating liver is the result, not of an increased synthesis, but of a decreased breakdown. The role of proteolytic enzymes has received attention in relation to the increased cathepsin activity during early embryogenesis[15,16] and the changed levels as a result of regeneration in different tissues.[17] Many factors can be considered to influence the rate of protein breakdown in animals, such as the levels of hormones,[19,20] pathological states, states of excitation and exhaustion,[21,23] nutritional status,[154] and availability of essential amino acids.[10,22,24] The transport of protein precursors undoubtedly plays an important role in maturation, as illustrated by the changed pool size and the rate of transport for certain amino acids.[10] There appears to be a complex relationship between the levels of circulating amino acids and tissue amino acids, partly as a consequence of interactions in membrane transport, and of variations in intracellular metabolism.[25,26] The increased demands during development may require considerable reutilization of liberated amino acids, and the degree of reutilization may parallel other changes seen in the maturational process.[27] The rates of degradation cannot be computed directly from rates of amino acid incorporation, since the rates of synthesis and degradation are not always in balance; also, additional complicating factors are the local reutilization of newly liberated amino acids and the heterogeneity of the amino acid pool.[8,10] The prelabeling of proteins in long- and short-term experiments has provided an estimate of the potentiality of catabolic systems, as illustrated by the higher rates for release of radioactivity from rat-brain homogenates of young animals fed with labeled amino acids compared to adults.[28,29] In addition to alterations in protein breakdown, there is an increase in brain protein content accompanied by a change in solubility and extractability; enzymes follow characteristic patterns during growth (some increase and some decrease), and there is a change in the relative proportions of proteins with fast and slow turnover.[10] Development, furthermore, is accompanied by changes in the extracellular space and the amino acid pool,[8,30] and by a decreased incorporation *in vitro* of labeled amino acids in brain tissue and peripheral nerve.[31–34,152]

We extended an earlier study[3] to include the measurement of representative peptide hydrolases (E.C. 3.4) and peptidyl-peptide hydrolases (E.C. 3.4.4) during the critical phases of rat-brain maturation. To give some perspective to this study, the findings are preceded by a brief review of the hydrolases selected and a summary of previous knowledge in relation to protein breakdown. The exact mechanism of protein breakdown is unknown; we have suggested that degradation, like synthesis, is an orderly process, in

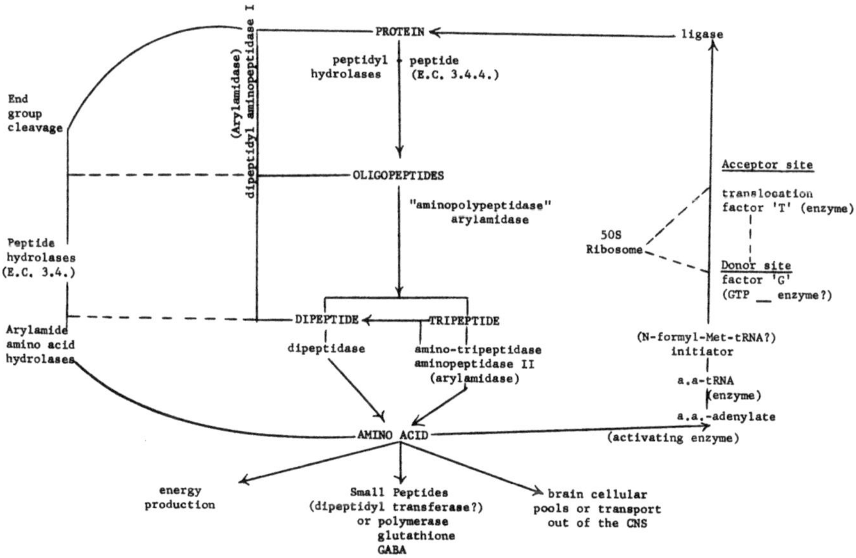

Fig. 1. Postulated scheme for brain peptide turnover. The synthetic pathway is based on the proposed scheme in *E. coli*.[158]

which different batteries of tissue peptide hydrolases are linked together to form a pathway leading to the liberation of smaller peptides or amino acids.[35] At present only simple schemes can be postulated for such pathways (Fig. 1). It must be stressed that with the exception of the endocrine glands, closely linked with brain function, there is no large pool of free peptides other than glutathione in the brain. This may indicate a rapid breakdown or reutilization of the free polypeptides in protein turnover. The existence of other free peptides recently reported in trace quantities may represent intermediates in the degradative pathway or simply artifacts of extraction.

CLASSIFICATION

On the basis of the synthetic substrates introduced by Bergmann[36] the following major groups of peptide hydrolases (E.C. 3.4) can be distinguished: aminopeptidases (specific for substrates with free -NH_2 end-groups, E.C. 3.4.1); carboxypeptidases (free -COOH end-groups, E.C. 3.4.2); and dipeptidases (free -NH and -COOH end-groups, E.C. 3.4.3) (Table I). Despite the many advances in the characterization and purification of such

TABLE I

Classification of Peptide and Peptidyl-Peptide Hydrolases[a]

E.C. listing	Systematic or (trivial name)	Major substrate	References
3.4	*Peptide hydrolases*		
3.4.1.2	*L*-leucyl peptide (leucine aminopeptidase)	Leu-NH_2; Leu-Gly	35, 38, 45
3.4.1.2	aminoacyloligopeptide (aminopeptidase)	R-R[b] (R-R-R-Rn)	35, 52, 53
3.4.1.3	aminoacyldipeptide (aminotripeptidase)	Leu-Gly-Gly R-R-R	45, 47, 90
3.4.2.1	Carboxypeptidase A	Z-Gly-Phe	44, 45
	Carboxypeptidase B	Hippuryl-Arg	35
3.4.3.1	Glycyl-glycine dipeptidase (diglycinase)	Gly-Gly	54, 55
3.4.3.3	Aminoacylhistidine	β-Ala-His	58
3.4.4	*Peptidyl-peptide hydrolases*		
—	Cathepsin B	BANA	92, 40, 91
3.4.4.9	Cathepsin C (dipeptidyl transferase)	Glu-Tyr-NH_2	97, 40, 91 86, 89
3.4.4.23	Cathepsin D (acid proteinase)	Hemoglobin (proteins)	39, 40, 102, 104–107
[c]	Neutral proteinases	Casein proteins	40, 110, 115, 117, 118

[a] Most enzymes included in this table are listed by the I.U.B.[36]

[b] This enzyme has not been isolated from brain, but since it acts on di- and tripeptides, it overlaps the specifity of other aminopeptidases in crude extracts.

[c] Enzymes are not listed, but are reported as present in brain extracts.

enzymes, some caution must be exercised in relating the simple hydrolysis of synthetic substrates to the functional activity *in vivo*. Two examples serve to illustrate the manner in which synthetic substrates can be misleading: (1) the chromogenic substrates used for the histochemical detection of leucine aminopeptidase are also known now to represent a completely new class of peptide hydrolases, known by the trivial name of arylamidase;[35,37,38] (2) the substrates introduced to measure intracellular proteolytic enzymes (cathepsins, see below) failed to detect the most important hemoglobin-degrading enzyme in liver, spleen, muscle, and brain.[39,40] This enzyme has been named cathepsin D, and is inactive with the synthetic substrates for cathepsins A–C.[39] It is therefore clear that the simple hydrolysis of peptides, esters, or amides *in vitro* does not represent the complex situation within the cell; it cannot, for example, take into account (a) the nature of the true physiological substrate, (b) the accessibility of native substrate to the enzyme and the effect of cellular cofactors, and (c) the possibility of zymogen or isoenzyme transformations.

The study of autolysis in homogenates of various tissues,[66] including that of the brain,[28] has revealed two active systems for the degradation of proteins: one group of enzymes is active at acid pH (acid proteinases), and differs in properties from a second group of enzymes that have not been completely identified but are active at neutral pH (neutral proteinase and peptide hydrolases).[40]

No systematic classification is available for endopeptidases (peptidyl-peptide hydrolases, E.C. 3.4.4).[36] Based on the properties of the active center, the three groups commonly recognized are serine (trypsin), sulfhydryl (papain), and acid (pepsin) hydrolases.[41] An alternative and more arbitrary scheme is based on biological properties, and includes extracellular digestive enzymes, intracellular proteolytic enzymes (cathepsins, see below), and circulatory enzymes such as thrombin and plasmin.[41] Cathepsins are considered the most important endopeptidases concerned with intracellular breakdown, and receive the greatest emphasis in present ontogenetic studies. The lack of adequate purification, together with the existence of multiple enzymatic forms (isoenzymes) and ambiquities concerning the specificity, has hindered a rational classification. Indeed, the I.U.B. notes in a postscript that other proteolytic enzymes undoubtedly exist, but must await additional characterization before inclusion in its list.[36]

PEPTIDE HYDROLASES

Aminopeptidases

Comparatively little is known concerning brain aminopeptidases because of the difficulties encountered in the isolation, and lability. Leucine aminopeptidase has attracted considerable interest as a potential tool for protein sequence determination, since all amino acids in peptide linkage are, to varying extents, susceptible to hydrolysis.[42] Even in the case of this well-studied aminopeptidase there is some question of its purity, since many preparations contain endopeptidase contaminants.[43] Amides at one time were considered the most specific substrates, and Brecher[44] has reported the hydrolysis of Leu-, Tyr-, and Phe-NH_2 in crude cerebral mitochondrial, microsomal, and supernatant fractions, with activation in some instances by divalent metal ions Mn^{2+}, Co^{2+}, and Mg^{2+}. Metal activation and the considerable esterase activity associated with this enzyme are features of outstanding interest, but have not received detailed studies in brain. The dual peptidase and esterase function of leucine aminopeptidase is not understood, but is a property shared by many other proteolytic enzymes.

TABLE II

Distribution of Aminopeptidase, Carboxypeptidase, and Arylamidase in Rat-Brain Subcellular Fractions[a]

Substrate	Relative enzyme activities				
	H	*N*	Mt.	Mc.	Supt.
Aminopeptidase					
Leu-Gly	(100)	4	7	6	61
Gly-Gly[b]	18	2	11	5	1
Leu-Gly-Gly	100	3	10	8	55
Leu-Leu-Leu	30	3	5	5	25
Carboxypeptidase					
Z-Leu-Tyr	5	2	2	0	1
Arylamidase					
α-Asp-β-NA	3	0	0	0	1
γ-Glu-β-NA	3	0	1	0	1
Lys-β-NA	83	5	10	6	37
Arg-β-NA	70	3	6	4	35
Leu-β-NA	50	3	11	7	26
Ala-β-NA	77	5	46	12	21
Met-β-NA	63	4	15	6	20
Gly-β-NA	16	1	5	2	16
Phe-β-NA	13	1	3	1	3
Ser-Tyr-β-NA	15	1	1	1	5

[a] Values are relative to activity (100) in the homogenate (*H*) with Leu-Gly as substrate. Assays were done at pH 7.6 with 2 m*M* peptide or 0.5 m*M* arylamide and incubated 30 min at 37°. From Marks *et al.*[94,96].

[b] 1 ml Co^{2+}.

It has been apparent for some time that most tissues, including the brain, contain a variety of peptidyl hydrolases active in the hydrolysis of Leu-Gly and other leucyl peptides.[45,46] Patterson *et al.*[38] attempted to characterize leucine aminopeptidase from liver by the ratio of hydrolysis with Leu-Gly, Leu-Gly-Gly, and Leu-NH_2; different ratios with these three substrates are reported for muscle, lens, renal, and brain tissues,[35] pointing to the considerable heterogeneity of leucyl-peptide hydrolases (Table II). The brain enzyme, like that of other tissues, is intracellular, with a pH optimum of about 7.0,[45] and differs from the less stable extracellular mucosal enzyme with a reported pH optimum of 8.5.[66] Recently Binkley *et al.*[46] purified from renal tissues three hydrolases on the basis of Leu-Gly hydrolysis; the enzyme in the soluble supernatant fractions differed markedly in its spec-

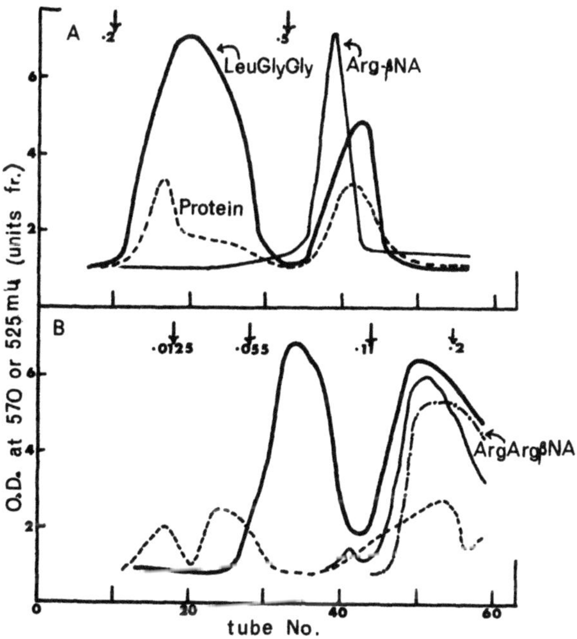

Fig. 2. Distribution pattern of hog-brain soluble peptide hydrolases after passage through (A) DEAE-cellulose and (B) DEAE-Sephadex at pH 7.6, Tris-HCl buffer. Peaks were eluted by stepwise addition of NACl as indicated. The first peak in both cases represents aminopeptidase with Leu-Gly-Gly as substrate (heavy solid curve), the second represents a mixture of monoacyl (light solid curve) and dipeptidyl arylamidases (dot-dash curve) together with aminopeptidase. (From Marks.[96])

ificity from that of two enzymes present in the microsomes (endoplasmic membrane fraction) in splitting Leu-βNA, Leu-NH_2, Leu-Gly-Gly, and a large number of other di- and tripeptides containing leucine.

The only other aminopeptidases that have received attention in the brain are enzymes displaying a marked specificity for tripeptides.[45,47] Our recent studies point to considerable heterogeneity of such enzymes in brain, with the possibility of species differences between the rat and the pig.[45,96] Aminotripeptidases can be differentiated from other aminopeptidases by different chromatographic profiles on elution with salt gradients from DEAE-cellulose or DEAE-Sephadex (Fig. 2) and on the basis of the sensitivity to metal-ion addition or the effect of certain antibiotics. In this manner two distinct groups of aminopeptidases were differentiated from whole rat-brain

TABLE III

Substrate Specificity of Rat-Brain DEAE-Cellulose Fractions in Presence of Aminopeptidase Substrates[a]

	Relative specific activity	
	Arylamidase fractions (Peak II)	Aminotripeptidase fractions (Peak I)
Ala-Gly-Gly	0	100
Leu-Gly-Gly	0	47
Gly-Gly-Leu	0	31.5
Leu-Leu-Leu	0.5	24
Lys-Lys-Lys	0	17.5
Leu-Leu	5	9.5
Isoleu-Leu	0	2.5
Lys-Gly-Gly	0	2.2
Pro-Gly-Gly	0	2.2
α-Glu-Gly	0	1.1
Gly-Try	4	0
Gly-Gly	1.25[b]	0
Poly-Leu	0.1	0
Leu-Gly	1.5	0
Gly-Gly-Gly	0	0
Gly-Gly-Gly-Gly	0	0
D-Leu-Gly-Gly	0	0
D-Leu-Gly	0	0
Poly-Phe	0	0

[a] Incubation, 1 ml containing 0.25 mM, except poly-amino acids, which were 0.05 mM Tris-HCl buffer at pH 7.6. Incubated 20 min at 37° and then determined by the modified ninhydrin method of Matheson and Tattrie.[157] Enzyme used for tripeptidase was 5–10 μg, and for other fractions 25 μg. From Marks *et al.*[94,96]

[b] In the presence of 1 mM Co^{2+}.

extracts.[45] The first enzyme eluted with low-salt gradients was inactive with arylamide substrates, but hydrolyzed a variety of peptides with a marked specificity for tripeptides (Table III). This activity was not altered by dialysis against EDTA or by metal-ion addition and was unaffected by puromycin or its analogs. The second group of rat-brain aminopeptidases was eluted at higher salt gradient, was inactive with added peptides, but rapidly cleaved a series of arylamide substrates, particularly those with Arg-, Lys- and Leu-substituents. For some experiments it was possible to isolate, in small yield, enzyme with a greater specificity for basic- compared to neutral-substituted analogs. Peak-II enzyme activity was reduced or lost in prolonged dialysis against EDTA solutions containing dithiothreitol to

TABLE IV

Substrate Specificity of Purified Hog-Brain Peptide Hydrolases[a]

	Relative enzyme activity	
	DEAE Peak I (Aminopeptidase)	DEAE Peak II (Aryl- and aminopeptidase)
Naphthylamide		
L-Leu-	0	(100)
D-Leu-	0	0
Arg-	0	140
Ala-Ala	0	50
Lys-Lys	0	75
Arg-Arg-	0	60
Tripeptide substrates		
Met-Met-Ala	(100)	(100)
Met-Met-Met	56	97
Ala-Ala-Ala	32	268

[a] Conditions of incubation and testing identical to Tables II and III. Results demonstrate an overlap in specificity with tripeptide substrates, in contrast to purified rat brain extracts in Table III. The relative values are expressed in relation to the values in parentheses. From Marks.[90, 96]

stabilize activity; activity was restored and in some cases increased by metal-ion addition (Co^{2+}, Mn^{2+}, etc). The inhibition by PCMB and activation by mercaptoethanol indicated that brain arylamidases are dependent on a free —SH group. This enzyme was readily distinguished from peak-I activity by the potent inhibition upon addition of puromycin or actinomycin. Both peaks of aminopeptidase were present to a large extent in soluble supernatant fractions (60%), but some activity (10% or more) was associated with subcellular fractions, particularly in the synaptosomal regions. A small but significant level was associated with purified mitochondrial and microsomal membranes.[101]

A considerably more complex situation emerged on the purification of aminopeptidases from the soluble hog-brain post-mitochondrial fractions. The same two peaks were observed on elution from DEAE-cellulose or Sephadex columns (Fig. 2), but the second peak showed a wider specificity than that of the purified rat-brain enzyme. This latter peak, in addition to the hydrolysis of monacyl arylamides, also hydrolyzed dipeptidyl arylamides and tripeptides (Table IV). The hydrolysis of all arylamides and certain tripeptides (except Leu-Gly-Gly) was potently inhibited by puromycin or its analogs (Table V). Storage of the mixed aminopeptidase peak II at

TABLE V

Comparison of Purified Hog-Brain Aminopeptidase with Arylaminopeptidase[a]

Addition (0.05 mM) or other property	Per cent change in activity or other property			
	Aminopeptidase	Aryl- and aminopeptidase		
	Peak I	Peak II		
	Met-Met-Ala	Arg-βNA	Arg-Arg-βNA	Met-Met-Ala
K_m	10^{-4}	10^{-5} M	10^{-5} M	10^{-5} M
pH optimum	7.0	7.9, 9.0	7.0, 9.0	6.5
Puromycin (methoxy Phe)[b]	0	−100%	−100%	−100%
Ki-puromycin	—	10^{-5} M	10^{-5} M	10^{-6} M
Puromycin-Tyr	0	−100%	−100%	−100%
-Leu	0	− 36%	− 91%	—
-β-Ala	0	− 61%	− 91%	—
-methoxy-Tyr-Gly	0	− 72%	− 80%	—
-Tyr-methoxy-Gly	0	− 50%	− 84%	—

[a] Enzyme was prepared from hog brain by the procedure previously described for rat brain (Marks *et al.*[45]). Peak I refers to enzyme eluted from DEAE-cellulose with 0.05 M NaCl, and peak II to the peak eluted with 0.11 M NaCl. The other amino acid-substituted derivatives were kindly donated by Dr. Leon Goldman of Lederle Labs. Inc. From Marks.[96]

[b] Puromycin, 9-(methoxyphenylalanine)-3-deoxy-β-D-ribofuramosyl-6-dimethylamino-9H-purine.

0° or 20° led to loss of the puromycin-sensitive hydrolase activity without affecting the splitting of Leu-Gly-Gly. The fact that pig brain contains a variety of aminopeptidases differentiated by the hydrolysis of arylamides or tripeptides, lability, and sensitivity to puromycin is consistent with the postulate expressed earlier, that brain, like other tissues, contains a battery of intracellular enzymes for the hydrolysis of peptides. Inhibition by the antibiotics known to interfere with the mechanisms of protein synthesis is particularly intriguing in view of their known behavioral changes *in vivo*.[48] McDonald *et al.*[49] concluded that inhibition of pituitary peptidases by puromycin was competitive in nature and related to cationic size—smaller cations, such as the aminonucleoside itself, or amicetin, Tris, Li, K, and Na, were progressively less inhibitory. Puromycin with a methoxyphenylalanine side chain was the most effective inhibitor, without marked alteration in potency on substitution with monoacyl and dipeptide amino acids groups (Table V). Puromycin is considered to act by early termination of peptide elongation during protein synthesis,[50] but the possibility cannot be excluded that inhibition of degradative systems play a role in the alteration of behavior and associated phenomena.

Dipeptide Hydrolases

Particular care must be exercised in the isolation and identification of dipeptidase, since dipeptides also serve as substrates for other peptidases. Generally, dipeptidases are differentiated from each other and from other enzymes on the basis of activation with different metal ions.[35,51] This method of classification is complex and must await purification of the enzymes involved for final verification. The histological findings of Pope[52] with *DL*-Ala-Gly can be interpreted as being due to a variety of peptide hydrolases rather than a specific alanyl-glycine dipeptidase. He reported a higher activity in human frontal isocortex, compared to the rat, with the highest concentration in the intralaminar layers II, IV, and Vb, but reinterpretation of the data on a dry weight basis would indicate high aminopeptidase activity in cytoplasmic extensions of both neurons and glial cells.[53] The only dipeptidase to receive detailed study is glycyl-glycyl dipeptidase (diglycinase) present in glycerol extracts of histological slices and strongly activated by Co^{2+}.[54,55] Additional purification has not been attempted in brain, but diglycine is hydrolyzed by brain particulate and supernatant fractions and in some cases activated by metal ions (Table II). Such an enzyme may be of importance in view of the inhibitory effects reported for glycine in cat spinal cord preparations: glycine occurs in relatively high concentration in cat spinal cord and in invetebrate nerve tissue.[56,57] Relatively little is known about other cerebral dipeptidases except for the report of Abraham *et al.*[58] that ^{14}C-β-alanine is released from ^{14}C-alanylhistidine in brain slices, possibly indicating a specific carnosinase.

Carboxypeptidases

These enzymes occur in relatively low concentrations in brain and have not been the subject of any intensive investigation (Table II). Since highly purified preparations are available from other tissues, carboxypeptidases constitute the best studied class of peptide hydrolase, leading to notable advances in our knowledge of enzyme structure, zymogen formation, and mechanisms of catalysis. Carboxypeptidase A is specific for peptides with neutral amino acids on the C-terminal end, except for proline and hydroxyproline. The chief means of identification of carboxypeptidases are the synthetic substrates with a protected N-terminal group. Crude brain extracts readily hydrolyze the peptide sequences present in Insulin B chain (Z-Phe-Phe, Z-Val-Phe, Z-Gly-Phe) with the highest activities largely particulate bound.[60,61]

There are several reports that brain tissue can hydrolyze acetylated peptides such as S-acetylglutathione, S,N-diacetylcysteine, N-acetylaspartate,[62,63] and N-acetyl-Asp-Glu.[64,65] Enzymes hydrolyzing peptides substituted in the N-terminal position with acyl, N-halogen acyl, or ester groupings probably belong to the general grouping of carboxypeptidases, but others have considered them as a separate category termed "acylases."[66] Recently Goldstein[67] isolated two enzymes from rat cell sap fractions that corresponded with acylase I and II obtained from kidney.[63] Considerable significance can be attached to such enzymes, since brain contains a high concentration of N-acetylaspartate and trace levels of other acyl peptides.

Activation of zymogens has long constituted an effective approach to the elucidation of the mechanisms of enzyme catalysis responsible for some aspects of protein breakdown.[68] For example, in the case of trypsinogen the release of a peptide fragment from the NH_2-terminus results in conformational changes accompanied by enzyme activation. In the case of procarboxypeptidase A the picture is more complicated because of the existence of distinct subunits: one subunit is the actual precursor of the peptide hydrolase; a second gives rise to an endopeptidase; the third remains to be identified.[69] The endopeptidase subunit can be combined with the active exopeptidase to yield a dimeric complex with carboxypeptidase activity. A second feature of particular interest is the differential alteration of bovine carboxypeptidase A to an esterase activity, depending on the treatment. Alteration of activity can occur on replacement of metal prosthetic group Zn^{2+} with Cd^{2+} or Hg,$^{2+}$ acylation, UV, etc.[70] Peptidases can be activated by addition of chemical "modifiers," generally N-substituted hydrolysis products of synthetic peptides or esters, or inhibited by addition of peptides, which compete with the metal-binding sites.[71] All these alterations in peptidase activity pose interesting questions, especially concerning the mechanisms of protein breakdown in the brain.

Carboxypeptidase B is specific for peptides with a C-terminal basic residue and has attracted interest because of its property of inactivating the biologically active peptides with basic end-groups known collectively as the "kinins." Kinin production itself represents an interesting mechanism of activation of serum protein precursor by specific proteolytic enzymes. Krivoy and Kroeger[72] reported that brain contained an enzyme capable of inactivating bradykinin, which was almost identical in its properties to pancreatic carboxypeptidase B. Recently, Iwata *et al.*[125] studied the subcellular distribution of an enzyme degrading bradykinin: the enzyme gave optimal activity in rat-brain fractions at pH 7.6 and increased in nuclear fractions following pentetrazol-induced convulsions. The brain enzyme has not been purified, but enzymes inactivating kinins in other tissues have been the subject of intensive research.[73]

Arylamide Amino Acid Hydrolases

In early studies it was assumed that enzymes hydrolyzing leucyl-β-napthylamide (Leu-βNA) were identical with leucine aminopeptidase, but subsequently it was demonstrated by Patterson *et al.*[37,38] that these two activities represent different classes of hydrolase. These enzymes, tentatively termed "arylamidases," are distributed generally in all plant and animal tissues and are present in high concentration in brain[35,74] and peripheral nerve tissue.[18,57] There is good evidence to suggest that specificity within this group of enzymes is conferred by the amino acid substituent: separate groups hydrolyzing acidic and basic arylamides have been reported, but the evidence for a separate neutral group of arylamidases is less convincing. More recently it has been claimed that enzymes hydrolyzing dipeptidyl arylamides represent yet another class of peptidases, although in some cases (see below) the properties are akin to some classical peptide hydrolases. None of these enzymes are available in a highly purified state from the brain, but as an aid to description, a tentative classification and summary is attempted in Table VI. The nomenclature employed is that adopted by the respective authors.

Hydrolysis of Acidic Arylamides

Glenner *et al.*[75] first reported a specific Ca^{2+}-activated enzyme catalyzing the hydrolysis of α,β-Asp- and γ,α-Glu-β-NA in rat-kidney microsomal fractions. This enzyme was termed aminopeptidase A and would be of considerable interest with respect to brain because of its exceptionally high level of free and protein-bound acidic amino acids.[76] Brain homogenates and, to a lesser extent, the soluble supernatant fractions hydrolyzed γ,α-Glu- and α,β-Asp-β-NA, but the activity was less than 5% compared to the cleavage of neutral and basic surrogates (Table II). This activity was largely tissue bound, with only slight activation by the addition of Ca^{2+} ions. Khairallah and Page[77] recently reported a Ca^{2+}-activated enzyme in plasma specific for N-terminal Asp-Angiotensin II peptide similar in properties to aminopeptidase A. Although this enzyme has been termed angiotensinase, it must be stressed that this and other kinin hormones are polypeptides and are thus susceptible to hydrolysis by a variety of exo- and endopeptidases. The hydrolysis of γ-Glu-β-NA noted for brain extracts in the present study (with liberation of β-naphthylamine) could have occurred by a transpeptidation mechanism: transpeptidases specific for the γ-glutamyl moiety have been reported in brain, but at levels 100-fold lower than in the kidney. This enzyme can be differentiated from glutamyl transferase,

TABLE VI

Tentative Classification and Summary of Enzymes that Split Arylamides[a,c]

Major substrate (arylamide)	Nomenclature	Alternative substrates[b]	Inhibition	Tissue
Acidic				
α-Glu-	Arylamidase A; aminopeptidase; angiotensinase	α-Asp-; β-Asp-; γ-Glut; angiotensin	—	Brain;[35,45,94] serum;[35,77] kidney[75]
Neutral				
Leu-	Arylamidase N	Met-Phe-; Arg-; Ser-Tyr; (paranitroanilides)	Cu^{2+}; Cd^{2+}; pCMB; puromycin; Actinomycin TPCK	Brain;[35,45,74] ant. pituitary;[49,83–88,148,149] post-pituitary;[35,148,149] pineal;[35,138,148] peripheral;[18,57] nerve myelin (spinal cord);[94,147] c.s.f.[35,147]
Basic				
(a) monoacyl				
Arg-	Arylamidase B	Lys-; dipeptidyl arylamides; nitroanilides; tripeptides	As above	Ant. pituitary;[84] Brain;[45,74,93] peripheral;[57] nerve c.s.f.[147] ant. pituitary[49,84,88]

Lys	Lysyl arylamidase; aminopolypeptidase	Arg-; Met-; Leu-; Lys-Val; Ala_4-Ala_5; Lys_5	Di-tripeptides; bradykinin; puromycin; histones; bases; EDTA	
(b) dipeptidyl				
Ser-Try	Dipeptidyl aminopeptidase I; (dipeptidyl transferase) (cathepsin C)	His-Ser-; Ala-Ala-; Gly-Phe-; Ser-Met; Ala_4; Ala_2; Gly-Phe-NH_2; Ser-Tyr-Ser-Met		Kidney;[89,123] ant. pituitary;[86] brain;[45,93,94] spleen[155]
Lys-Ala	Dipeptidyl aminopeptidase II; (carboxytripeptidase) dipeptide esterase	Arg-Arg-; Leu-Ala-; Ala-Ala-; Lys-Ala-OCH_3; tripeptides	Puromycin; Tris; Na^+; K^+; Li^+	Ant. pituitary;[88,85] brain[45,94]
Arg-Arg	Dipeptidyl aminopeptidase III	Ala_4–Ala_2 Ala_5–Ala_2 + Ala_3 Ala_6–3Ala_2 Lys_4–2Lys Val-Leu-Ser-Glu-Gly	Arg-Arg; pCMB; Lys-Lys-βNA	Ant. pituitary;[87] brain[45,94]

[a] The nomenclature is that adopted by the individual authors and is not officially listed by the I.U.B.[36] Since the enzymes are only partially purified from nerve tissue the overlap in specificity may indicate contamination by other peptide hydrolases.

[b] Substrates with a dash represent the arylamide substrate.

[c] Compiled from recent literature and also two recent reviews.[35,51]

which is considered to represent a reversal of the glutamine synthetic pathway with specific binding sites for NH_3, ATP, and glutamine, and from transglutaminase, which is involved in the exchange of the polypeptide γ-glutamyl amide moiety with other amine acceptors.[76,78,79] Transpeptidases may be involved with cysteinylglycine dipeptidase in the turnover of glutathione in brain;[80] such peptide transformations may be important in brain because of the high concentration of glutamyl tripeptides and the presence of trace quantities of glutamyl dipeptides.[81]

Hydrolysis of Basic Arylamides

Based on highly purified enzyme from liver, there is very good evidence for the existence of a "basic arylamidase" or "aminopeptidase B" specific for Arg- and Lys-β-NA.[82] In anterior pituitary Ellis *et al.*[83,84] reported some specificity differences between enzymes hydrolyzing Arg- compared to Lys-β-NA (Table VI), but these differences have not been confirmed for other tissues. As noted previously, we succeeded in isolating from rat-brain extracts an enzyme that hydrolyzes basic arylamides.[45] This enzyme was separated from other aminopeptidases by salt-gradient elution from DEAE-cellulose columns and was differentiated by its sensitivity to metal-ion activation after dialysis against EDTA, and by the potent inhibition in the presence of selected antibiotics. Application of the same separation techniques to hog-brain supernatant fractions led to the isolation of an enzyme peak from DEAE with a considerably wider specificity when tested against monacyl- and dipeptidyl arylamides, and tripeptides (Fig. 2; Table IV). As in the case of rat-brain enzyme, activity for all arylamides and most tripeptides was potently inhibited by puromycin and its analogs (Table V).

The hydrolysis of dipeptidyl arylamides has been the subject of considerable interest since two mechanisms might be involved: (1) N-terminal amino acid release, as in the case of monoacyl dipeptides, or (2) hydrolysis by removal of the dipeptide moiety with release of β-naphthylamine. In the case of partially purified anterior pituitary enzyme, Ellis *et al.*[49,83–88] have suggested the presence of a family of related dipeptidyl arylamidases or aminopeptidases for the hydrolysis of *even*- or *odd*-numbered polypeptides. Like other aminopeptidases, most enzymes recently described are labile, with the consequent limitation in characterization studies; thus the seemingly complex relationships between these enzymes and other peptidases in peptide turnover cannot be fully evaluated at the present time. A summary of such enzymes present in brain, pituitary, and selected tissue is included in Table VI, and the postulated pathways involving such enzymes are illustrated in Fig 1. Based on the hydrolysis of dipeptidyl substrates, the following groups of enzymes have been delineated (Table VI): (I) a sulfhydryl- and chloride-dependent enzyme active at pH 4.0 in the hydrolysis of Ser-Tyr, His-Ser-, Ala-Ala, Gly-Phe, Ser-Met-β-NA,[86] (II) a sulfhydryl- and halide-dependent enzyme active at pH 5.5 in the hydrolysis of Lys-Ala, Lys-Ala-OMe, and certain unsubstituted tripeptides,[49,88] and (III) an enzyme active at pH 9.0 in the hydrolysis of Arg-Arg-β-NA.[87] The similarities of these enzymes to other characterized peptide hydrolases is indicated in the table. The first enzyme, for example, bore many similarities to cathepsin C and has been purified from hog kidney.[84,123] Crude brain extracts also hydrolyzed Ser-Tyr-β-NA, but the activity was 15% compared to the basic monoacyl analogs, which is consistent with earlier observations that cathepsin C activity is low in brain.[40,91] Like cathepsin C, the enzyme is sulfhydryl- and chloride-dependent, but there is some question concerning the hydrolysis of other substrates (Ala-Ala, His-Ser, etc.) that do not fulfill the classical structural requirements formulated by Fruton.[92] McDonald *et al.*[86,155] provided some indication of a functional role for this enzyme in the pituitary, by the reported cleavage of terminal dipeptide fragments Ser-Tyr and Ser-Met from the decapeptide sequence corresponding to the N-terminal end of ACTH. This enzyme was reported to be present exclusively in lysosomes and was distinct in properties from an analagous enzyme in beef liver shown capable of splitting the terminal dipeptide His-Ser from glucagon.[59]

The second class of pituitary dipeptidyl aminopeptidases showed a wider specificity when tested with dipeptides, dipeptide esters, and tripeptide substrates[49,88] (Table VI). The rate of hydrolysis in the partially purified preparations from pituitary of Met-Met-Ala, trialanine, and Lys-Ala-OMe actually exceeded that of the dipeptidyl arylamides, suggesting that this category of enzyme may have properties akin to tripeptidases. Enzymes hydrolyzing Lys-Ala-β-NA are widely distributed in tissues, with the highest concentration in thyroid, followed by the kidney, pituitary, thymus, adrenal, liver, testis, pancreas, and finally brain, which has about 12% of the thyroid level.[88] Purified enzyme from hog brain (DEAE peak II, Fig. 2) hydrolyzed Lys-Ala-, Arg-Arg, Ala-Ala-β-NA (Table IV), but the methods applied to separate the individual activities were unsuccessful. The hydrolysis of the basic dipeptidyl arylamides was sensitive to inhibition by low concentrations of puromycin or its analogs (Table V) and gave a double pH optimum, at 7.0 and 9.0. The puromycin-sensitive hydrolase activities were considerably more labile on storage at 0° and −20° compared to the enzymes hydrolyzing Leu-Gly-Gly. These data are an indication that as in the case of rat-brain enzyme, there is a distinct category of pig-brain hydrolases specific for arylamides or their cogeners. In brain the enzyme was present in the soluble supernatant fractions,[93,94] but it is reported to be largely lysosomal in the pituitary.[49] In connection with the localization, Smith and Farquhar[95] previously showed that lysosomal-like granules in the pituitary actively assimilate and degrade hormone-bearing granules during lactation.

The third enzyme of unusual interest reported in the pituitary[87] catalyzed the specific cleavage of Arg-Arg-β-NA at pH 9.0, and in addition polypeptides with a minimum of four amino acids (Ala_4, Ala_6, poly-*L*-Ala, etc.). Pentapeptides are split by this enzyme into a di- and tripeptide, providing a mechanism for the hydrolysis of *odd*-numbered polypeptides. This enzyme bears greater similarity to the brain enzyme-cleaving Arg-Arg-, and Lys-Lys-β-NA (active at pH 7.0 and 9.0),[96] since it is cytoplasmic in origin, activated by thiol compounds, and inhibited by pCMB. Arylamidase amino acid hydrolases as a group were present in all brain particulates tested, ranging from 60% in the post-microsomal supernatant, to 10% in the crude mitochondrial-synaptosomal fraction, to about 5% in the crude nuclear and microsomal fractions (Table II). In mitochondrial fractions the activity was largely associated with the synaptosomes, but a small activity resided in highly purified preparations of mitochondria. The purification and properties of mitochondrial and microsomal arylamidase are described elsewhere.[101] In summary, there is evidence of acidic and basic arylamidases and possibly dipeptidyl arylamidases with a considerable overlap in specificities. Further purification is required for the full identification of these enzymes in brain and pituitary, but the preliminary data suggest a fascinating array of peptide hydrolases with a range of specificities for the hydrolysis of small and complex polypeptides acting at different intracellular sites. The relationship of these enzymes in the pathways of protein degradation could account for the sequential splitting of many polypeptides and proteins.

INTRACELLULAR PEPTIDYL-PEPTIDE HYDROLASES

Cathepsin C

As noted in the introduction, endopeptidases can be classified according to their biological properties; those enzymes responsible for the breakdown of proteins within the cell are called by the generic name "cathepsins." The most studied enzyme in this group is cathepsin C, which was first purified from beef spleen[92] and appeared to be homospecific with chymotrypsin when tested with dipeptide amide or ester substrates. Extensive studies by Fruton and his group[92,97] showed a more stringent substrate requirement than for the pancreatic enzyme: cathepsin C from beef spleen split substrates of the type $H_2NCH(R)CONHCH(R_2)CO$ . . . X at the point indicated when the amino group R_i and the peptide bond between R_1 and R_2 is unsubstituted, and when R_2 is an aromatic-*L* amino acid. The synthetic substrate of choice is Gly-Phe-NH_2, which is hydrolyzed at pH 5.0 and activated by cysteine. The early preparations also hydrolyzed hemoglobin at pH 5.0, but recently with improved purification methods this enzyme does not readily attack protein substrates and its suggested major role in intracellular protein digestion has been questioned.[98,99] Cathepsin C is exceptional among proteolytic enzymes in its ability to catalyze transamidation reactions at physiological pH; for example, Ala-Phe-NH_2 (*L*-configuration) is converted into the hexapeptide, Ala-Phe-Ala-Phe-Ala-Phe-NH_2, and Gly-Trp is polymerized to form the mixed octapeptide amide.[97,100] Metrione *et al.*[97] have concluded that the major function of cathepsin C is the transfer of dipeptides to suitable acceptors, and they have suggested that this enzyme be renamed dipeptidyl transferase. At one time it was suggested that transamidation reactions of this nature could account for some aspects of protein synthesis,[92,100] but very little evidence has accumulated to support this view. It would seem on kinetic considerations alone that this route does not form a major pathway for peptide production. In our own studies, only trace levels of cathepsin C activity were observed with Gly-Phe-NH_2 at pH 5.0, and with Pro-Tyr-NH_2 in brain extracts, in agreement with other studies.[40] In crude brain homogenates the levels reported are less than 5% of that in the spleen.[91] As noted, hydrolysis of Gly-Phe-β-NA, or Gly-Phe-*p*-nitrophenol,[86,89,155] or Ser-Tyr-β-NA are claimed to bear analogies to cathepsin C activity (Table VI). Brain extracts hydrolyzed Ser-Tyr-β-NA, but the activity was considerably less than other arylamides (Table II), and this activity disappeared during purification procedures.

Cathepsin D

This enzyme was first purified by Press *et al.*[39] and Lapresle and Webb[102] from liver and spleen, and is similar to an older enzyme first described by Anson.[103] With hemoglobin as the substrate, cathepsin D accounted for two-thirds of the proteolytic activity in spleen and some 90% in the brain.[104] Purified preparations of cathepsin D split hemoglobin and albumin at pH 3–4 but were inactive with the synthetic substrates used to define other cathepsins (Fig. 3). The absence of activity with synthetic substrates is probably the reason this enzyme was overlooked earlier, and emphasizes the importance of characterizing endopeptidases with both native and synthetic substrates. Acid proteinase purified from the brain[40,105–107] was similar in most respects to the properties described for purified cathepsin D.[39] It showed no requirement for added metals, or other

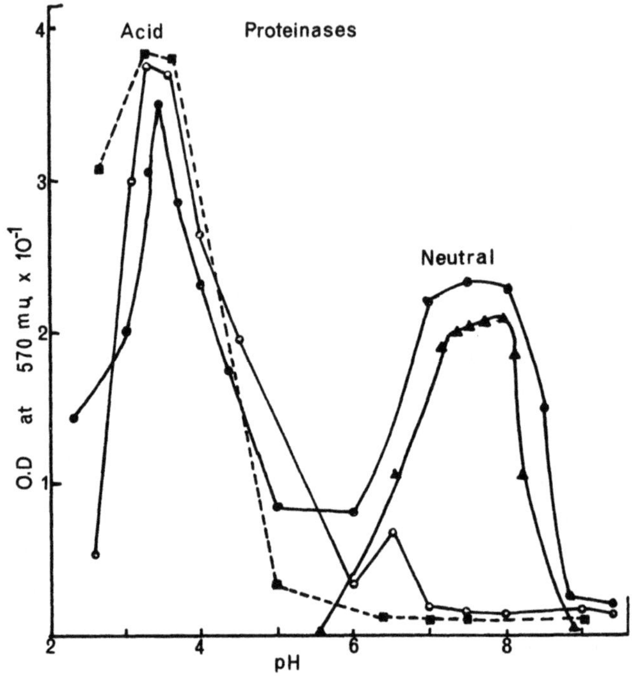

Fig. 3. Comparison of crude and purified rat-brain proteinase pH optima. Endogenous breakdown (closed circles[28]), purified proteinase with hemoglobin (dotted line),[40] or with encephalitogen G basic protein (open circles);[124] purified neutral proteinase (triangles).[40]

cofactors, and appeared to be present in multiple chromatographic peaks.[40] The enzyme did not hydrolyze peptides used for characterizing other acid proteinases or cathepsins and it showed a more restricted bond specificity, similar to that of purified spleen enzyme, when tested with insulin B chain (Fig. 4).[40,60,106]

Neutral Proteinase

Hydrolases active at physiological pH, distinct in properties from cathepsins, are present in plant and animal tissues, and in the case of bacteria have been isolated and purified.[108,119] Similar studies on neutral proteinase from animal tissue have been handicapped by the pronounced lability of these enzymes compared to cathepsins. Ansell and Richter[110] first reported an extremely labile neutral proteinase in brain extracts active for

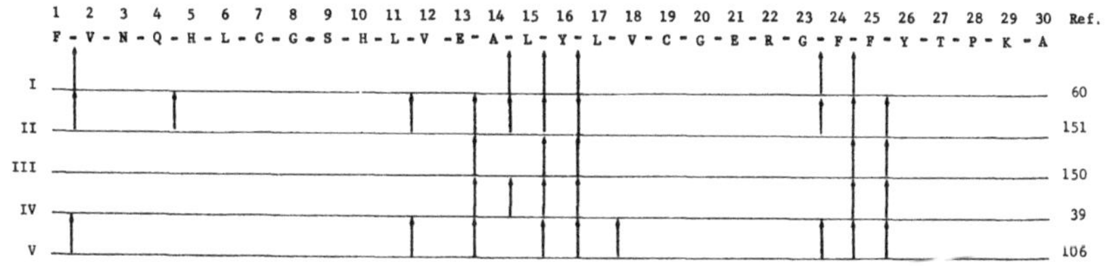

Fig. 4. Specificity of purified brain acid proteinase (I and V) compared to (II) pepsin, (III) purified hog-thyroid proteinase (pH optimum: 4.0), and (IV) spleen cathepsin D. The chain given is the oxidized beef insulin B chain. IUPAC abbreviations are used (the first letter denotes the amino acid, except for Q = Glu-NH_2, E = Glu, Y = Tyr, F = Phe, R = Arg, K = Lys).

about 1½ hr under anaerobic conditions in the absence of added protein substrates. In other tissues labile neutral proteinases have been reported in lens, spleen, kidney, and liver, splitting native substrates such as α_2-crystallin or added proteins such as hemoglobin, casein, or albumin.(40,109,112) No highly purified form of this enzyme is available, but it can exist as a stable contaminant of pituitary growth hormone preparations, resulting in multiple forms of the hormone during preparation and storage.(113) Although neutral proteinases are undoubtedly of great potential physiological importance, their presence in tissues has been questioned on the basis of some assay methods employed. Methods relying on an increase of absorption at 280 mμ in TCA extracts are nonspecific, since they also measure the release of nucleic acids by ribonuclease.(115,116) The methods of choice are specific colorimetric or chromatographic procedures measuring the end-products of protein or peptide degradation. In addition to the earlier evidence that brain homogenates degrade prelabeled protein or show an increase in ninhydrin-positive material at pH 7.6 (Fig. 3), we have isolated and partially purified a labile neutral proteinase from brain capable of degrading casein, hemoglobin, albumin, etc.(40) The enzyme was partially activated by Ca^{2+}, inhibited by pCMB and O-iodosobenzoate, and activated by cysteine, mercaptoethanol, and reduced glutathione (Table VII). This enzyme may therefore require a free sulfhydryl grouping at the active center similar to that of highly purified streptococcal neutral proteinase.(108) Purified brain neutral proteinase showed broad pH optima (max. pH 7.8, Fig. 3) and consisted of more than one component on further purification in DEAE-cellulose, an observation recently confirmed by Riekkinen and Rinne(117). Guroff(118) described an alternative procedure for neutral proteinase purification from rat-brain soluble supernatant fractions active in the degradation of casein, oxidized

TABLE VII

Properties of Cerebral Petidyl-Peptide Hydrolases[a]

	Acid proteinase	Neutral proteinase
pH optimum	3.2	7–8
PCMB 0.5 mM	0	−100%
O-Iodosobenzoate	0	− 60%
GSH[b]	0	25%
β-Mercaptoethanol	0	+ 60%
Ca^{2+} 0.1 mM	0	+ 15%
Co^{2+} 0.1 mM	0	+ 15%
Cd^{2+} 1 mM	0	−100%
Cu^{2+} 1 mM	0	−100%
K_m (Hb)	8×10^{-3} *M*	2×10^{-3} *M*
K_m (casein)	—	3×10^{-2} *M*
Spleen[c]	10	1.5
Kidney[c]	6	3.5
Liver[c]	3	1.3
Muscle[c]	0.3	0.3

[a] Values for results with added materials are expressed as the percentage change compared to the controls. From the data of Marks and Lajtha.[40]

[b] GSH was added in presence of CoA, ATP, and phosphocreatine (see text).

[c] Levels in different rat-tissue homogenates compared to rat brain, which for acid proteinase was 2.2 and for neutral proteinase 0.8 μM/mg protein/hr.

ribonuclease, and insulin B chains (at the Tyr-Leu band), which was characterized by its complete dependence on Ca^{2+} for activity. Studies with highly purified *B. subtilis* neutral proteinase show splitting of peptide bonds in which Leu and Phe contribute the amino group; unlike many other proteolytic enzymes, no esterolytic activity has been observed for this bacterial enzyme with a large number of ester substrates.[119] Neutral proteinase was associated with soluble cytoplasmic fractions such as the microsomal supernatant, followed by the myelin, synaptosomes, and other particulate fractions.[104] There is no evidence that neutral proteinases are associated with lysosomes, although some aminopeptidases active at neutral pH were reported to be present in the pituitary and liver lysosomal fractions.[49,120] The level of neutral proteinase in brain was only ⅔ that of liver, but in spleen and kidney was, respectively, twice and six times that of brain.[104] Stemming from the altered activity in the absence and presence of O_2,[110] and on addition of an ATP-generating system, it has been suggested that splitting of peptide bonds in some tissues required a source of energy.[10,121] This question has not been resolved, although an increased liberation of ninhydrin-positive materials has been reported on addition of coenzyme A, reduced glutathione, and phosphocreatine.[104] These observations could imply formation of activated intermediates; alternatively, glutathione may form complexes with the substrate, as suggested in the case of catheptic hydrolysis of plasma albumin.[122] Umana[115] recently also reported increased release of ninhydrin material on addition of glutathione and coenzyme A and suggested the additional possibility that the materials may repress a neutral proteinase inhibitor. Neutral proteinases are associated with peripheral nerve from a number of species and are also present in lobster ventral cord.[35,57] The significance of their presence in myelinated and unmyelinated fibers has been discussed in relation to the induction of experimental encephalomyelitis and similar disorders.[124]

Mechanisms of Catalysis

Recent rapid advances in determination of both sequence and tertiary structure of endo- and exopeptidases have led to some remarkable discoveries concerning the catalytic mechanism. Earlier studies established a number of common features in the active centers of bovine chymotrypsinogen B, trypsinogen, thrombin, and porcine elastase that led to the expectation that they shared a catalytic mechanism similar to chymotrypsin A. There is now good evidence that chymotrypsin A contains an interior-Asp embedded among hydrophobic residues capable of transferring its charge along the chain of hydrogen bonds *via* histidine to the serine in the active site at the surface of the molecule.[127] Subtilisin BPN, an extracellular protease of *B. amyloliquefaciens*, which has no resemblence to chymotrypsin at the level of primary amino acid sequences, also depends on an active histidine and serine at the active center hydrogen-bonded to an embedded aspartate grouping.[128] A different mechanism may be involved for carboxypeptidase A, a metalloenzyme: the substrate glycyltyrosine binds to the active center with a conformational change, indicating a realignment of a hydrophobic tyrosine at the catalytic site on the surface of the enzyme.[129] Tissue neutral proteinases are unavailable in sufficient purity for comparable studies. Our evidence indicates that a free sulfhydryl group is essential for activity of brain enzyme[40]; for neutral streptoccocal proteinase, Liu and co-workers[108,130] have shown that a single sulfhydryl group associated with an amino acid of unknown structure and a histidine residue is essential for activity.

Development

Many different approaches are available for seeking correlative morphological and biochemical changes during the maturational process.[1,5,111]

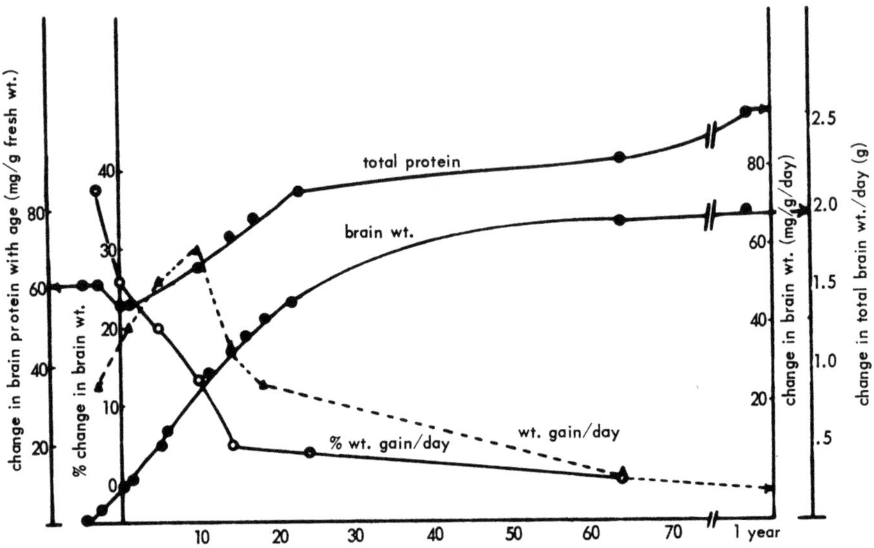

Fig. 5. Evaluation of rat-brain growth by the change in total weight expressed as a percentage of net gain per day compared to the change in protein. The values to the left of the ordinate represent measurements in the foetus at five and three days prior to birth.

Representative of such studies are (1) the changed enzyme patterns characteristic of growth in the brain, (2) the distribution of the synthetic and catabolic systems within the cell or at different brain regions during development, and (3) the neurochemical correlates of behavior. In the preliminary work reported here we studied protein breakdown in rat brain as measured by the levels of representative hydrolases in relation to four well-defined ontogenic periods: (1) the phase of active cell proliferation prior to birth, (2) the period of axonal growth and increased dendritic complexity following birth (0–10 days in the rat), (3) myelination (10–25 days) and (4) the slow deposition of solids such as lipids and proteins in the adult. The rat is particularly suited to such studies since it is born with nearly its full complement of cells in an unmyelinated state. In seeking the critical periods of brain growth in relation to protein turnover and breakdown, we measured the net weight-gain per day and the differential rate of increase evaluated as a percentage in male Cobbs River strain rats. Two sets of data were obtained (Fig. 5) illustrating that the maximum deposition of solids based on weight occurred at the 10th day (coincident with myelination), but that the differential rate was highest at birth, followed by a fall in rate until the 15th day. Such findings are consistent with the many elegant studies conducted in the past that measured ontogenic changes in several different ways, (1) absolute quantitative changes based on total brain weight or some estimate of cell number based on the measurement of DNA, (2) differential growth patterns based on the rate of change of a specific component in relation to the levels of protein or lipids.[1,5] Based on these considerations, the time points selected for the measurement of hydrolases were five and two days prior to birth, in newborns, followed by five-day intervals up to 25 days, at 45 days, and finally in senescent animals of one year or older. Methods for the measurement of individual hydrolases are described in detail elsewhere, and a brief description of the extraction methods is included below. Results are expressed as micromoles of α-amino group (based on a glutamic acid standard) per gram of protein per hour. All measurements were performed on the Cobbs River strain, except in the case of hypophysectomized rats and some very old animals; fetal and young rats were obtained from females with timed pregnancies.

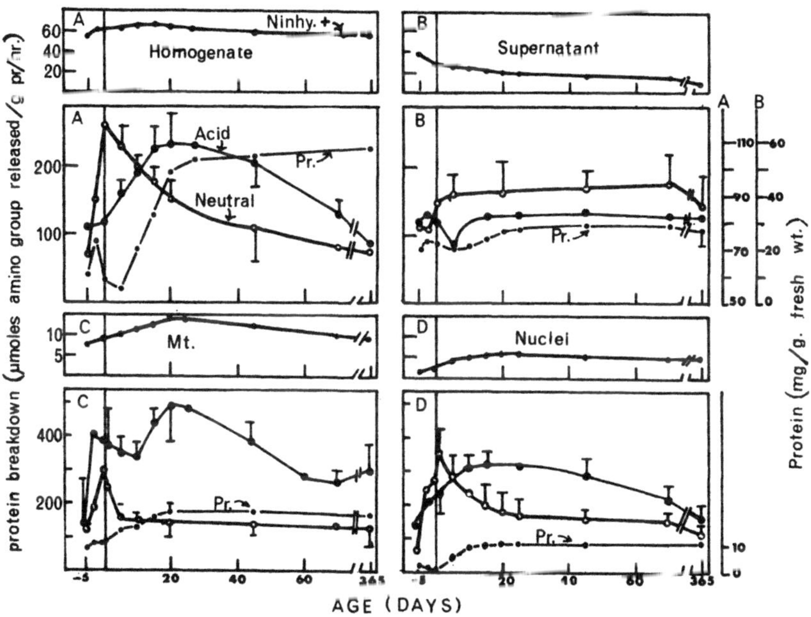

Fig. 6. Alteration in proteinase levels with maturation in Cobbs River rat-brain cellular fractions; A, sucrose homogenate; B, nuclear fraction; C, crude mitochondrial-synaptosomal fraction; D, microsomal-supernatant fraction. The different components analyzed were acid proteinase with hemoglobin as substrate (darkened circles); neutral proteinase with hemoglobin (open circles); protein and the total ninhydrin-positive pool are as shown in the figure. Brain fractions were prepared and analyzed by previously published procedures.[10,104]

Maturation of the nervous system is known to be accompanied by an increased intracellular population of organelles[131] as confirmed by the increased protein levels for brain subcellular fractions (Fig. 6). Lysosomes in particular are known to contain a family of hydrolases (cathepsins, phosphatases, acid DNAase, sulfatases, glucuronidase, etc.), and these enzymes might be expected to accompany in a quantitative fashion the brain growth cycle. There are several reports in the literature that a large number of hydrolases do indeed parallel some aspects of the defined growth periods, but results must await knowledge of the biogenesis of intracellular organelles for a full interpretation. Recently studies in our laboratory showed that other organelles contain a small but significant level of hydrolytic enzymes, sufficient in some cases to permit an autonomous protein turnover.[101,132] It must be stressed that the measurement of an enzyme in a whole brain extract, frequently reported in the literature, can only serve as a guide, since these levels do not reflect regional differences, or the rates of development in specific organelles. Such measurements would not take into account the recent concept of caudal-rostral development in the nervous system: the medulla oblongata and thalamus, for example, are the first to show the most advanced degree of myelination;[133] also, oxygen consumption and succinic dehydrogenase activity are reported to exhibit a caudal-rostral gradient during maturation.[53] A definite caudal-rostral gradient has been reported for proteins, phospholipids, cholineacetylase, and cholinesterase, with levels higher in the lower brain regions compared to the cortex in the three-day old rabbit.[134]

In the study undertaken below, results for whole-brain extracts are compared with the levels in the specified intracellular organelles.

Proteinases and Development

The acid and neutral proteinase activity in whole brain extracts (50 mM tris-HCl buffer pH 7.6 containing 0.2% Triton X-100) increased rapidly in the first three days, reaching maximum levels at 20 days; neutral proteinase fell in senescent rats, but acid proteinase continued to increase (Fig. 7). These and the changes illustrated below represent the breakdown of added denatured hemoglobin prepared by the urea method of Anson[103] (see Marks and Lajtha[40]); the endogenous activity in the absence of added protein substrates followed a parallel pattern, but the results were frequently more variable due to the insensitivity of the testing procedure. Alteration of native (endogenous) proteins during growth is an important aspect that cannot be tackled until methods are available for the characterization and isolation of specific brain proteins. Brain extracts

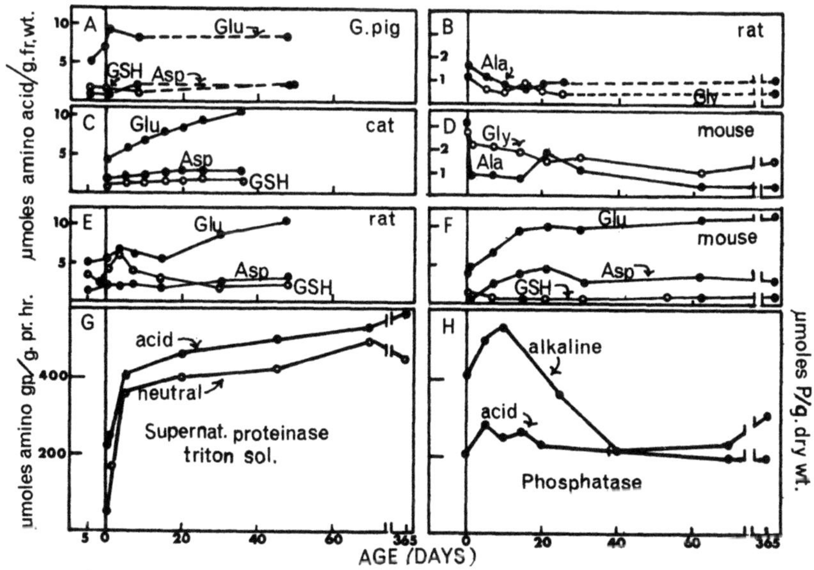

Fig. 7. Alteration in proteinase levels in soluble rat-brain extracts prepared with Tris buffer 50 m*M* pH 7.6 containing Triton X-100 compared to alkaline acid and phosphatase in rat brain mince with age. For comparison with the changed levels of ninhydrin-positive pools with age (Fig. 6), data are included for selected acidic and neutral amino acids and glutathione (GSH) as indicated in the diagram. The phosphatase data is from Cohn and Richter,[146] that for the guinea pig and rat from Oja[24] and Bayer and McMurray,[143] for the cat from Berl,[141] and the mouse from Agrawal *et al.*[142,144]

represent a mixture of heterogeneous proteins, with the possibility of a complex relationship between hydrolases and competing substrates. Different methods of extraction can be expected to yield a different picture, as shown by the hydrolase changes in whole-brain sucrose homogenates compared to Triton-soluble extracts (Fig. 6): acid proteinase reached maximum levels at 10–15 days, but, in contrast to buffer extracts, the neutral proteinase activity fell shortly after birth and continued to fall during later phases of development.

One possible explanation for this finding is that the sucrose homogenate contains protein inhibitors which increase during development; it is well known that plant and animal tissue contain specific proteinase inhibitors, some of which have been characterized and shown to be proteins.[135] Brecher and Quinn[136] have reported the presence in brain extracts of materials that inhibit the tryptic hydrolysis of benzoyl arginylnaphthylamide. In hypophysectomized rats, Lewis *et al.*[137] reported the forma-

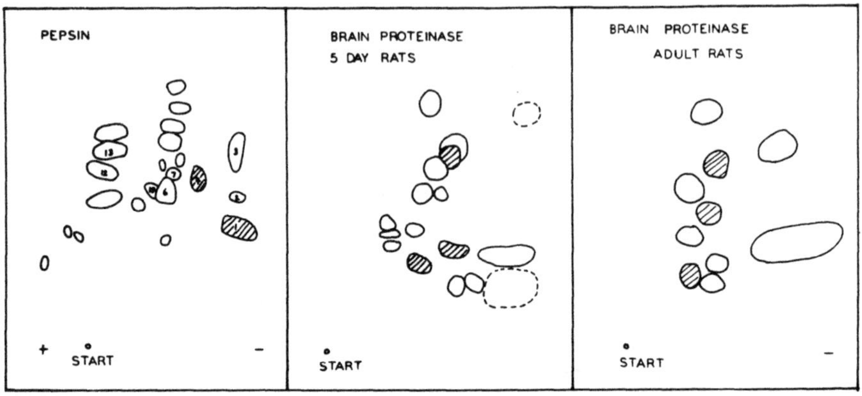

Fig. 8. Peptide maps of oxidized insulin B chain with pepsin or brain proteinase. Separations were made on Whatman No. 3 paper at pH 2.0, 1000 V (formic acid:water 4:96) and submitted to ascending chromatography in butanol:acetic acid:water (50:10:40). Spots were visualized by ninhydrin. Acid proteinase was purified by the methods previously discussed,[40] pepsin was obtained from Worthington Biochem. Corp., New Jersey.

tion of an inhibitor of hypophyseal proteinases, but in our own experiments we were unable to detect any significant changes in soluble proteinases in hypophysectomized animals. It is of interest that hypophysectomy is reported to drastically reduce salivary gland proteases, which appear to be under direct hormonal control and increase during development.[19]

In crude particulate fractions the temporal changes in hydrolase levels are considerably more complex than in hypotonic buffer extracts of brain (Fig. 7). For the mitochondrial-synaptosomal and nuclear fractions acid and neutral proteinase increase *in utero* (0 to 5 days), but at birth the neutral proteinase fell in parallel with that observed in the whole sucrose homogenate (Fig. 6). In the crude mitochondrial fraction acid proteinase fell slightly at birth until the 14th day, then proceeded to rapidly increase coincident with the known onset of myelination. In the microsomal and supernatant fractions acid proteinase decreased *in utero* and at birth but subsequently increased until the sixth day; neutral proteinase followed a slow pattern of increase with age. These different patterns are consistent with an altered localization of proteolytic enzymes, probably associated with the changed protein solubility during development. Only limited information is available on the qualitative changes of different proteins, because of methodological difficulties. A number of possibilities exist in relation to enzymes during development: change in kinetic properties (pH optima, K_m, or V_{max}), in isoenzyme formation, in zymogen formation, and possibly in substrate specificity.[1,2,6,8,139,154] In preliminary studies we obtained some evidence that partially purified enzyme from a five-day-old rat compared to that from adults exhibited a different bond specificity based on peptide maps with insulin B chain, and globin A and B chains derived from hemoglobin (Fig. 8). Data on the ninhydrin-positive pools (TCA-soluble) associated with the different particulate fractions are included in this survey (Fig. 6) because the level of intracellular metabolites (amino acids and small peptides) may parallel and influence protein degradation. In addition, it can be expected from the complex pathways involved in turnover that the end-products of protein degradation act as a control mechanism by limiting the forward reaction rate by end-product inhibition. All the measured ninhydrin pools, except that of the microsomal supernatant fraction, increased following birth, with maximum levels at 10 days for the total sucrose homogenate, at 24 days for the crude mitochondria, and at 30 days for the nuclear fraction; these levels fell in subsequent phases of development (Fig. 6). Although the rat-brain ninhydrin-positive pool contains some 60 components,[140] the preponderance of reacting material can be attributed to acidic amino acids such as glutamate. In the present results the increase in total pool size after birth is consistent with the known increase in glutamate and aspartate (Fig. 7);[153,156] finer analysis of individual amino acids has shown considerable variation in levels with development with some species variation.[24,141–144] During development three groups can be distinguished: (1) acidic components, which increase with age, notably Glu, Asp, GABA, and the tripeptide GSH; (2) neutral amino acids, which decrease (Ala, Gly, Tyr, Tau, etc.); and (3) other amino acids, which do not change significantly, or, alternatively, are present in levels that prevent accurate measurements. As noted, conclusions based on whole-brain extracts do not preclude the resolution of finer temporal changes parallel to development for the pools associated with subcellular or regional fractions, or in relation to caudal-rostral changes.

Peptide Hydrolases

Peptidyl hormones of the pituitary and other endocrine organs play an important role in ontogeny,[1–5,20] but there are no reported studies on the changed levels of peptide hydrolases in those glands or the brain during development. Our studies included a survey of the aminopeptidases present in brain in high concentration utilizing the substrates Leu-Gly-Gly and Arg-and Leu-β-NA. No clear developmental pattern was observed with the aminopeptidase substrate, but arylamidase increased, reaching a maximum at about 15 days in the case of the total homogenate, crude mitochondrial, nuclear, and supernatant fraction (Fig. 9). In the case of Arg-, but not for

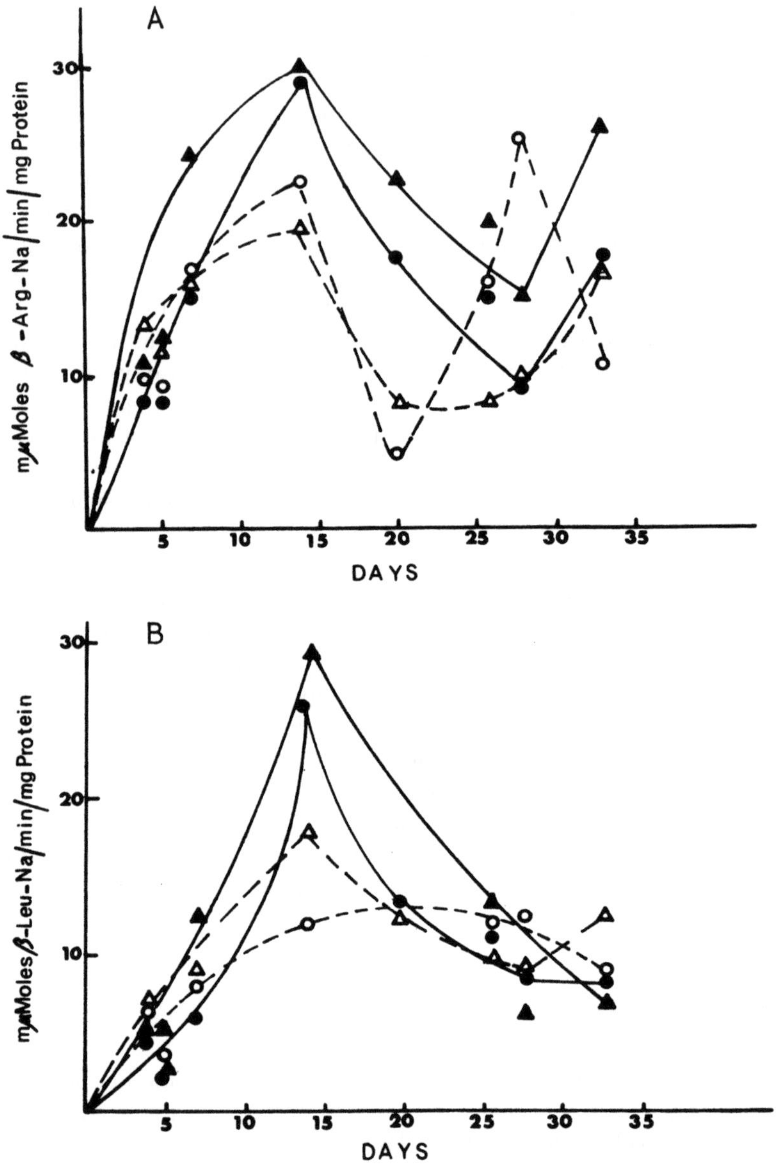

Fig. 9. Changed levels of arylamidase activity in rat-brain cellular fractions (A) with Arg-β-NA as the substrate, (B) Leu-β-NA as substrate. Total homogenate (open circles), nuclear fraction (darkened circles), crude mitochondrial-synaptosomal fraction (open triangles), microsomal-supernatant fraction (darkened triangles).

Leu-, a second peak of activity appeared at 30 days. Enzymes hydrolyzing Leu- and Ala-β-NA are known to be present in myelin,[94,126] and the increased activity noted in the present study corresponds to the onset of myelination. In isolated systems metabolic transformations are possible in the nature of the sequential breakdown of proteins to form physiologically active peptides. Hopsu-Havu *et al.*[145] showed that basic arylamidase acted on the peptides obtained from globulin by tryptic digestion to form bradykinin. It is possible that all the enzymes discussed in this report have important functions, such as the activation or inactivation of enzymes or other physiologically active peptides in the brain.

REFERENCES

1. H. Waelsch, "Biochemistry of the Developing Nervous System" (H. Waelsch, ed.), Academic Press, New York (1955).
2. W. A. Himwich and H. E. Himwich, "The Developing Brain. Progress in Brain Research", Vol. 9 (W. A. Himwich and H. E. Himwich, eds.), Elsevier, Amsterdam (1964).
3. N. Marks and A. Lajtha, *in*: "Variations in Chemical Composition of the Nervous System as Determined by Developmental and Genetic Factors" (G. B. Ansell, ed.), p. 74, Pergamon Press, Oxford (1966).
4. W. A. Himwich, "Developing Brain Newsletter," Nos. 1–33, Galesburg State Research Hospital, Galesburg, Illinois (1964–1969).
5. H. McIlwain, "Biochemistry of the Central Nervous System," Little Brown, Boston (1966).
6. M. Murthy, the present volume, Chapter 5.
7. S. Roberts, C. Zomzely, and S. C. Bondy, the present volume, Chapter 1.
8. S. S. Oja, *Ann. Acad. Sci. Fenn.* **131**, 7–78 (1967).
9. A. V. Palladin and Ya. V. Belik, the present volume, Chapter 3.
10. A. Lajtha, *Int. Rev. Neurobiol.* **6**, 1–98 (1964).
11. J. Mandelstam, *Nature* **179**, 1179–1181 (1957).
12. H. O. Halvorson and S. Spiegelman, *J. Bacteriol.* **64**, 207–221 (1952).
13. M. J. Pine, *J. Bacteriol.* **92**, 746–850 (1966).
14. D. Rittenberg, E. E. Spoul, and D. Shemin, *Fed. Proc.* **7**, 180 (1948).
15. A. A. Neifakh and E. R. Davidov, *Biochemistry* **29**, 235–242 (1964).
16. G. Lundblad and J. Rumnstrom, *Exp. Cell. Res.* **27**, 328–331 (1962).
17. E. Urbani, *in*: "Regeneration and Related Problems" (V. Kiortsis and H. A. L. Thampusch, eds.) pp. 39–55, North-Holland, Amsterdam (1964).
18. C. W. M. Adams and G. G. Glenner, *J. Neurochem.* **9**, 233–239 (1962).
19. L. M. Screebny, *Ann. N.Y. Acad. Sci.* **85**, 182–188 (1960).
20. M. C. Diamond, *Brain Res.* **7**, 407–418 (1968).
21. L. F. Chapman and H. G. Wolff, *A.M.A. Arch. Int. Med.* **103**, 86 (1959).
22. Symposium "Relationship of Nutrition to Central Nervous System Development and Function," *Fed. Proc.* **26**, (1967).
23. G. Ungar, E. Ascheim, S. Psychoyos, and D. V. Romano, *J. Gen. Physiol.* **40**, 635 (1957); **46**, 267–275 (1962).
24. S. S. Oja, *Ann. Acad. Sci. Fenn.* **125**, 7–59 (1966).

25. S. Roberts, *in*: "Brain Barrier Systems" (A. Lajtha and D. Ford, eds.), pp. 235–242, Elsevier, Amsterdam (1968).
26. S. Roberts, *J. Neurochem.* **10,** 931–940 (1963).
27. J. Dobbing, *in*: "Brain Barrier Systems" (A. Lajtha and D. Ford, eds.), pp. 417–427, Elsevier, Amsterdam (1968).
28. A. Lajtha, *in*: "Regional Neurochemistry" (S. S. Kety and J. Elkes, eds.), pp. 25–36, Pergamon Press, Oxford (1961).
29. A. Lajtha and N. Marks, *Dis. Nerv. Sys.* **30** (*Suppl.*), 36–43 (1969).
30. A. Vernadakis and D. M. Woodbury, *Arch. Neurol.* **12,** 284–293 (1965).
31. D. M. F. Greenberg, F. Friedberg, M. P. Schulman, and T. Winnick, Cold Spr. Harb. Symp. Quant. Biol. **13,** 113–117 (1948).
32. S. Gelber, P. L. Campbell, G. E. Deibler, and L. Sokoloff, *J. Neurochem.* **11,** 221–229 (1964).
33. T. C. Johnson and M. W. Luttges, *J. Neurochem.* **13,** 545–552 (1966).
34. C. B. Klee and L. Sokoloff, *J. Neurochem.* **11,** 709–716 (1964).
35. N. Marks, *Int. Rev. Neurobiol.* **11,** 57–90 (1968).
36. M. Bergmann, *Adv. Enzymol.* **2,** 49–68 (1942); "Enzyme Nomenclature," 1964 Recommendations of the International Union of Biochemistry (I.U.B.), Elsevier, Amsterdam (1965).
37. E. K. Patterson, S. H. Hsiao, A. Keppel, and S. Sorot, *J. Biol. Chem.* **238,** 3611–3620 (1963).
38. E. K. Patterson, S. H. Hsiao, A. Keppel, and S. Sorot, *J. Biol. Chem.* **240,** 710–715 (1965).
39. E. M. Press, R. R. Porter, and J. Cebra, *Biochem. J.* **74,** 501–513 (1960).
40. N. Marks and A. Lajtha, *Biochem. J.* **89,** 438–447 (1965).
41. H. Neurath, K. A. Walsh, and W. P. Winter, *Science* **158,** 1638–1644, (1967).
42. E. L. Smith, *in*: "The Enzymes" (P. D. Boyer, H. Lardy, and K. Myrback, eds.), Vol. 4, Pt. A, pp. 1–10, Academic Press, New York (1960).
43. S. R. Himmelhock and E. A. Peterson, *Biochemistry* **7,** 2085–2092 (1968).
44. A. S. Brecher, *J. Neurochem.* **10,** 1–6 (1963).
45. N. Marks, R. K. Datta, and A. Lajtha, *J. Biol. Chem.* **243,** 2882–2889, (1968).
46. F. Binkley, F. Leibach, and F. King, *Biochem. Biophys. Acta.* **128,** 397–405 (1968).
47. A. S. Brecher and R. E. Sobel, *Biochem. J.* **105,** 641–646 (1967).
48. L. B. Flexner, J. B. Flexner, and R. B. Roberts, *Science* **157,** 1377–1383 (1967).
49. J. K. McDonald, J. J. Reilly, B. B. Zeitman, and S. Ellis, *J. Biol. Chem.* **243,** 2038–2037 (1968).
50. D. Nathans and A. Neidle, *Nature* **197,** 1076 (1963).
51. N. Marks *in*: "Handbook of Neurochemistry" (A. Lajtha, ed.), Vol. 4, Plenum Press, New York (1969).
52. A. Pope, *J. Neurochem.* **4,** 31–41 (1957).
53. R. L. Friede, "Topographic Brain Chemistry," Academic Press, New York (1966).
54. L. L. Uzman, M. K. Rumley, and S. Van Den Noort, *J. Neurochem.* **6,** 299–310 (1962).
55. L. L. Uzman, S. Van Den Noort, and M. K. Rumley, *J. Neurochem.* **9,** 241–252 (1962).
56. M. H. Aprison, R. A. Davidoff, and R. Werman, *in*: "Handbook of Neurochemistry," Vol. 3 (A. Lajtha, ed.), Plenum Press, New York (1969).
57. N. Marks, R. K. Datta, and A. Lajtha, *J. Neurochem.* (in press).
58. D. Abraham, J. J. Pisano, and S. Udenfriend, *Arch. Biochem. Biophys.* **104,** 160–165 (1964).

59. S. Kakiuchi and H. H. Tomizawa, *J. Biol. Chem.* **239,** 2160–2164 (1964).
60. N. Marks and A. Lajtha, *Fed. Proc.* **25,** 793 (1966).
61. A. Lajtha and N. Marks, *in*: "Protoides of the Biological Fluids" (H. Peeter, ed.), Vol. 14, pp. 103–114, Elsevier, Amsterdam (1966).
62. H. J. Strecker, P. Mela, and H. Waelsch, *J. Biol. Chem.* **212,** 223–233 (1955).
63. S. M. Birnbaum, L. Levinlow, R. B. Kinsley, and J. P. Greenstein, *J. Biol. Chem.* **194,** 455–470 (1952).
64. E. J. Olsen, H. W. Littleton, and J. V. Auditore, *Arch. Biochem. Biophys.* **119,** 22–28 (1967).
65. K. L. Reichett and E. Kuamme, *J. Neurochem.* **14,** 987–996 (1967).
66. H. Hanson, *in*: "Hoppe-Seyler Thierfelder Handbuch der Physiologisch Analyse," 6C, p. 1, Springer-Verlag, Berlin (1966).
67. F. B. Goldstein, *J. Biol. Chem.* **244** (in press).
68. H. Neurath, *in*: "The Enzymes" (P. D. Boyer, H. Lardy, K. Myrback, eds.), Vol. 4, pp. 11–36, Academic Press, New York (1960).
69. J. R. Brown, R. N. Greenshilds, M. Yamasaki, and H. Neurath, *Biochemistry* **2,** 867–876 (1963).
70. B. L. Valle, *Fed. Proc.* **23,** 8–17 (1964).
71. J. E. Coleman and B. L. Valle, *Biochemistry* **3,** 1874–1879 (1964).
72. W. A. Krivoy and D. Kroeger, *Brit. J. Pharmacol.* **22,** 329–341 (1964).
73. E. G. Erdos and H. Y. T. Yang, *in*: "Hypotensive Peptides" (E. G. Erdos, N. Back, and F. Sicuri, eds.), p. 235, Springer-Verlag, New York (1966).
74. A. S. Brecher and J. B. Suszkiw, *Biochem. J.* **112,** 335–342 (1969).
75. G. G. Glenner, P. J. McMillan, and J. E. Folk, *Nature* **192,** 338–340 (1962); **194,** 867 (1962).
76. H. Waelsch and H. Weil-Malherbe, "Psychiatric der Gegenwart," Vol. IB, pp. 2–75, Springer-Verlag, Berlin (1964).
77. P. A. Khairallah and I. H. Page, *Biochem. Med.* **1,** 1–8 (1967).
78. S. Berl, *Biochemistry* **5,** 916–922 (1966).
79. J. H. Pincus and H. Waelsch, *Arch. Biochem. Biophys.* **126,** 34–52 (1968).
80. P. J. Fodor, A. Miller, A. Neidle, and H. Waelsch, *J. Biol. Chem.* **203,** 991–1002 (1953).
81. Y. Kakimoto, T. Nakajima, A. Kanazawa, M. Takesada, and I. Sano, *Biochim. Biophys. Acta* **93,** 333–338 (1964).
82. V. K. Hopsu, K. K. Mäkinen, and G. G. Glenner, *Arch. Biochem. Biophys.* **114,** 557–566 (1966).
83. S. Ellis, *Biochim. Biophys. Res. Commun.* **12,** 452–456 (1963).
84. S. Ellis and M. Perry, *J. Biol. Chem.* **241,** 3679–3686 (1966).
85. J. K. McDonald, F. H. Leibach, R. E. Grindeland, and S. Ellis, *J. Biol. Chem.* **243,** 4143–4150 (1968).
86. J. K. McDonald, S. Ellis, and T. J. Reilly, *J. Biol. Chem.* **241,** 1494–1501 (1966).
87. S. Ellis and J. M. Nuenke, *J. Biol. Chem.* **242,** 4623–4629 (1968).
88. J. K. McDonald, F. H. Leibach, R. E. Grindeland, and S. Ellis, *J. Biol. Chem.* **243,** 4143–5150 (1968).
89. V. K. Hopsu-Havu and P. Rintola, *Histochem. J.* **1,** 1–17 (1968).
90. N. Marks and A. Lajtha in "*Methods in Enzymology*" (L. Lorand and G. E. Perlmann, eds.), Academic Press, New York (in press).
91. J. M. W. Bouma and M. Gruber, *Biochim. Biophys. Acta.* **89,** 545–547 (1964).
92. J. S. Fruton, *in*: "The Enzymes," (P. D. Boyer, H. Lardy, and K. Myrbäck, eds.), 2nd Ed., Vol. 4, p. 233, Academic Press, New York (1960).

93. A. S. Brecher and S. W. Barefoot, *Arch. Int. Physiol. Bioch.* **75,** 816 (1967).
94. N. Marks, R. K. Datta, and A. Lajtha, *in*: "Macromolecules and the Function of the Neuron" (Z. Lodin, ed.), pp. 220–231, Excerpta Medica, Amsterdam (1968).
95. R. E. Smith and M. G. Farquhar, *J. Cell Biol.* **31,** 319–347 (1966).
96. N. Marks, *J. Biol. Chem.* (submitted for publication).
97. R. M. Metrione, A. G. Neves, and J. S. Fruton, *Biochemistry* **5,** 1597–1604 (1966).
98. R. J. Planta and M. Gruber, *Biochim. Biophys. Acta.* **53,** 443–444 (1961); **89,** 503–510 (1964).
99. S. C. Dhar and S. M. Bose, *Leather Sci.* **11,** 309–320 (1965).
100. H. Wurz, A. Tanaka, and J. S. Fruton, *Biochemistry* **1,** 19–29 (1962).
101. B. D'Monte, N. Marks, R. K. Datta, and A. Lajtha, the present volume, Chapter 8.
102. C. Lapresle and T. Webb, *Biochem. J.* **84,** 455–462 (1962.)
103. M. L. Anson, *J. Gen. Physiol.* **22,** 79 (1938).
104. N. Marks, and A. Lajtha, *Biochem. J.* **89,** 438–447 (1963).
105. A. V. Palladin, M. M. Polyakova, and V. K. Lishko, *J. Neurochem.* **10,** 187–194 (1963).
106. V. K. Lishko, *Ukr. Biokhim. Zh.* **35,** 874–880 (1963); **36,** 565–573 (1964); **37,** 163–168 (1965) (see *Chem. Abstr.* **57,** 5008g; **60,** 12292a; **61,** 14962).
107. N. M. Polyakova, *Sb. Dokl. Erevan.*, pp. 25–38 (1963) (see *Chem. Abstr.* **55,** 12588h; **60,** 12289).
108. T. Y. Liu, W. H. Stein, S. Moore, and S. D. Elliott, *J. Biol. Chem.* **240,** 1143–1149 (1965).
109. R. Van Heyningen and S. G. Whaley, *Biochem. J.* **86,** 92–101 (1963).
110. B. Ansell and D. Richter, *Biochim. Biophys. Acta* **13,** 87–91, 92–97 (1954).
111. D. Richter, *Brit. Med. Bull.* **17,** 118–121 (1961).
112. S. G. Whaley and R. Van Heyningen, *Biochem. J.* **83,** 274–283 (1962).
113. U. J. Lewis, *J. Biol. Chem.* **237,** 3141–3145 (1962); **238,** 3330–3335 (1963).
114. L. A. Tsaryuk, *Ukr. Biokhim. Zh.* **34,** 815–23 (1962); **36,** 334–342 (1964) (*Chem. Abstr.* **61,** 8697).
115. R. Umana, *Anal. Biochem.* **26,** 430–438 (1968).
116. J. Marrink and M. Gruber, *Biochim. Biophys. Acta* **118,** 438 (1966).
117. P. J. Riekkinen and U. K. Rinne, *Brain Res.* **9,** 126–135 (1968).
118. G. Guroff, *J. Biol. Chem.* **239,** 149–155 (1964).
119. J. Feder, *Biochemistry* **6,** 2088–2093 (1967).
120. S. Mahadevan and A. L. Tappel, *J. Biol. Chem.* **242,** 2369–2374 (1967).
121. N. W. Penn, *Biochim. Biophys. Acta* **37,** 55–63 (1960).
122. L. Libensen and M. Jena, *Arch. Biochim. Biophys.* **1964,** 292–296.
123. V. K. Hopsu-Havu, P. Rintola, and G. G. Glenner, *Acta. Chem. Scand.* **22,** 299–308 (1968).
124. E. R. Einstein, J. Csejtey, and N. Marks, *FEB Letters* **1,** 191–195 (1968).
125. H. Iwata, T. Shikimi, and T. Oka, *Biochem. Pharmacol.* **18,** 119–128 (1969).
126. C. S. Beck, C. W. Hasinoff, and M. E. Smith, *J. Neurochem.* **15,** 1297–1301 (1968).
127. D. M. Blow, J. J. Birktoft, and B. S. Hartley, *Nature* **221,** 337–340 (1969).
128. C. S. Wright, R. A. Alden, and J. Kraut, *Nature* **221,** 235–242 (1969).
129. T. A. Steitz, M. L. Ludwig, F. A. Quiocho, and W. N. Lipscomb, *J. Biol. Chem.* **242,** 4662–4668 (1967).
130. T. Y. Liu, *J. Biol. Chem.* **242,** 4029–4032 (1967).
131. F. E. Samson, R. J. Jacobs, *Amer. J. Physiol.* **199,** 693–696 (1960).
132. N. Marks, and B. D'Monte, *Fed. Proc.* **28,** 733 (1969).
133. D. F. DeAlmeida and A. G. E. Pearse, *J. Neurochem.* **3,** 132–138 (1958).

134. R. E. McCaman and M. H. Aprison, *in*: "The Developing Brain," "Progress in Brain Research," Vol. 9 (W. A. Himwich and H. E. Himwich, eds.), pp. 220–231 Elsevier, Amsterdam (1964).
135. R. Vogel, I. Trautschold, and E. Werle, "Natürliche Proteinasen Inhibitoren," Monograph, Biochemie und Klinik, George Thieme, Stuttgart (1966). [Translation: Academic Press (1968).]
136. A. S. Brecher and N. M. Quinn, *Biochem. J.* **102,** 120–121 (1967).
137. U. J. Lewis, J. N. Fisher, and W. P. Vanderlaan, *Acta Chem. Scand.* **17,** (Suppl. 1), 165–169 (1963).
138. L. Thieblot, P. Bastide, S. Blaise, J. Boyer, and G. Dastugue, *Ann. Endocrinol.* **26,** 313–314 (1965).
139. H. P. Von Hahn, *Experientia* **18,** 509–511 (1962).
140. R. K. Shaw and J. D. Heine, *J. Neurochem.* **12,** 151–155 (1965).
141. S. Berl, *in*: "Progress in Brain Research," Vol. 9 (W. A. Himwich and H. E. Himwich, eds.), pp. 178–182, Elsevier, Amsterdam (1964).
142. H. C. Agrawal, J. M. Davis, and W. A. Himwich, *J. Neurochem.* **13,** 607–615 (1968).
143. S. M. Bayer and W. C. McMurray, *J. Neurochem.* **14,** 695–706 (1967).
144. H. C. Agrawal, J. M. Davis, and W. A. Himwich, *J. Neurochem.* **15,** 917–923 (1968).
145. V. K. Hopsu-Havu, K. K. Mäkinen, and G. G. Glenner, *Nature* **212,** 1271 (1966).
146. P. Cohn and D. Richter, *J. Neurochem.* **1,** 166–172 (1956).
147. P. J. Riekkinen and U. K. Rinne, *Brain Res.* **9,** 136–144 (1964).
148. P. Jouan and J. C. Rocaboy, *Compt. Rend. Soc. Biol.* **160,** 859–862 (1966).
149. T. P. J. Vanha-Pertrula and V. K. Hopsu, *Ann. Med. Exp. Fenn.* **43,** 32–39 (1965).
150. L. F. Kress, R. J. Peansky, and H. M. Klitgaard, *Biochim. Biophys. Acta* **113,** 375–389 (1966).
151. F. Sanger, E. O. P. Thomson, *Biochem. J.* **53,** 353 (1953).
152. D. F. Matheson, *J. Neurochem.* **16,** 215–223 (1969).
153. M. K. Gaitonde and D. Richter, *J. Neurochem.* **13,** 1309–1318 (1966).
154. R. T. Shimke, R. Ganshow, D. Doyle, and I. M. Arias, *Fed. Proc.* **27,** 1223–1230 (1968).
155. J. K. MaDonald, B. J. Zeitman, T. J. Reilly, and S. Ellis, *J. Biol. Chem.* **244,** 2693–2709 (1969).
156. W. A. Himwich and H. C. Agrawal, *in*: "Handbook of Neurochemistry." (A. Lajtha, ed.), Vol. 1, pp. 33–52, Plenum Press, New York (1969).
157. A. T. Matheson and B. L. Tattrie, *Can. J. Biochem.* **42,** 95–103 (1964).
158. T. F. Lipmann, *Science* **164,** 1024–1030 (1969).

DISCUSSION

AUSTIN: To what extent are these hydrolases associated with lysosomes of the CNS and peripheral nerve?

MARKS: I do not know of any detailed studies on lysosomes in peripheral nerve. It is possible that they can be detected morphologically in the nerve-cell body and appear in the axons by the mechanism of axonal flow. With regard to the CNS, although lysosomal structures can be identified in electron micrographs and by histochemical tests, they have never been isolated in a satisfactory biochemical manner. Lysosomal-like bodies occur in high concentrations in the crude mitochondrial-synaptosomal fractions, but are not readily separated by sucrose gradients; we are currently examining alternative methods for lysosomal separation as described by D'Monte *et al.* in Chapter 8.

SELLINGER: I agree that there is abundant morphological and histochemical evidence for lysosomes in the brain. A difficult question to answer at this time is whether hydrolases and peptidases occur in the same organelle or bag. We recently reported two populations of lysosomes in cerebral cortex separated by means of sucrose gradients, with different concentrations of marker enzymes, but we did not examine peptidases.

MARKS: Our subcellular fractions show that most brain aminopeptidases and arylamidases occur in the soluble cytoplasmic fractions. About 30% is associated with particulate fractions, chiefly the mixed mitochondrial-synaptosomal fractions and the microsomes. In sucrose gradients peptide hydrolase activities occurred in the regions believed to contain synaptosomes, although it should be stressed that the same sucrose region contains other morphological entities, especially lysosomal-like bodies and mitochondria. Clearly, the establishment of an exclusive location for a particular enzyme, including so-called lysosomal marker enzyme, is greatly dependent on the available preparative methods.

AGRANOFF: Your comment regarding puromycin and memory is most interesting. The best evidence that puromycin acts by the inhibition of protein synthesis is that acetoxycyclohexamide injected intracranially in goldfish is also effective in blocking memory, but at one thousandth of a dose compared to puromycin. Both block protein synthesis. Do these peptidases split puromycin? We know very little about the detoxification of puromycin. There is some information about deamination and demethylation of the sixth amino position. In studies aimed at creating a pulse dose of puromycin, we injected puromycin and a peptidase to block its effect, but we saw no experimental difference. We did not try to see if puromycin blocked peptidase action. Your dose of 10^{-4} M is approximately that used in animal experiments and the level we expect to find in brain following injection of the drug.

MARKS: We have not examined the interesting questions concerning the metabolic transformations of puromycin. Regarding your *in vivo* experiments, it is important to note that puromycin is quite selective and does not inhibit all peptide hydrolases—for example, leucine aminopeptidase. We also tried cycloheximide, but this material was without effect at the same concentration level used for puromycin in blocking the hydrolysis of arylamide substrates. The exact mechanisms of puromycin inhibition are unknown: Ellis and his group[49] reported that it may be related to a cationic effect, since other cations such as even Tris were inhibitory. The aminonucleoside was less inhibitory, indicating that the methoxyphenylalanine side chain may play a role in this mechanism. I think we should take into consideration the possibility that alteration of the degradative pathways in addition to those of synthesis can play a role in memory phenomena, particularly as protein breakdown is essential to turnover.

NEIDLE: Since puromycin is a potential inhibitor of arylamidase activity, it is possible that part of the physiological activity of puromycin is mediated by inhibition of these enzymes. This is consistant with findings that puromycin and acetoxycyclohexamide can inhibit protein breakdown as well as protein synthesis (e.g., tyrosine transaminase*). As Dr. Marks has pointed out, cycloheximide does not inhibit the aryl amidases, nor does acetoxycycloheximide in experiments that I have done. In my view these results indicate that the effects of the antibiotics on protein breakdown may not be directly

*F. G. Kenney, *Science* **156,** 525 (1967).

related to the inhibition of arylamidase activity. It should be kept in mind, however, that puromycin and acetoxycycloheximide may have many sites of action, including effects on protein turnover, and that their biological activity may not be explainable on the basis of protein synthesis inhibition only.

H. KOENIG: I think it is unlikely that neutral proteinases, or peptidases active at neutral pH, are associated with lysosomes. Without exception, all lysosomal enzymes are active at an acid pH and all recent data shows them to be cationic proteins. We have isolated from brain and other tissues acidic lipoproteins that bind basic proteins and may be responsible for the structural latency of many enzymes. The enzymes you described may have an important physiological function, since they are loosely associated with membranes. Lysosomal enzymes are interesting, but nevertheless are active at a nonphysiological pH range.

MARKS: I agree with your comment concerning enzymes active at neutral pH in lysosomes. The fact that some of these enzymes are associated with membranal fractions certainly suggests an important physiological role.

EINSTEIN: We have investigated the change in proteinase levels in rabbit brain with development. In this species acid proteinase reached a peak level at the seventh day, followed by a fall in activity. In contrast, neutral proteinase fell following birth.

UNGAR: Many years ago we reported an enzyme similar in properties to neutral proteinase that is sulfhydryl-dependent and increased by nerve stimulation. This enzyme is inhibited by general and local anesthetics, but is not as stable as the enzyme described by your group and others.

MARKS: Neutral proteinase and arylamidases are very unstable compared to acid proteinase and aminopeptidases hydrolyzing tripeptide substrates. Several groups have reported the purification of a neutral proteinase, but its lability has prevented extension of these studies.

Chapter 3

Breakdown of Proteins: Protamine-Splitting Enzyme

A. V. Palladin and Ya. V. Belik

Institute of Biochemistry
Academy of Sciences
Ukrainian SSR, Kiev

INTRODUCTION

The renewal of tissue proteins in an organ of an adult animal is conditioned by two opposing processes: synthesis and breakdown. While great successes have recently been achieved in studying the mechanisms of biosynthesis in protein–enzyme systems, the enzymatic reactions involved in the breakdown of tissue proteins have been studied insufficiently. Nervous-system tissue is no exception in this respect.

Most studies on the catabolism of nerve-tissue proteins have been devoted to the investigation of enzyme-splitting proteins in the acid range of pH far from the actual physiological conditions of the tissue medium.[1–8] It has long been known that in tissue of the central and peripheral nervous systems there is an enzymatic system with a set, apparently, of exo- and endopeptidases capable of hydrolyzing tissue proteins and their peptides and synthetic derivatives of amino acids or peptides under conditions close to the physiological ones. However, these enzymatic systems still have not been well enough studied. Data on their activity obtained by different methods sometimes do not agree, and in some cases result in false conclusions regarding the presence of proteolytic activity where it is actually absent, or, on the contrary, deny its presence in cases where it is actually present.

Thus, for example, a number of investigators[1,2,15,16] who used gelatin, hemoglobin, and casein as the exogenous substrates failed to detect proteolytic activity in nervous-system tissue in media with pH close to neutral. On the other hand, Adams and Smith,[17] who determined proteolytic activity by an increase of optical density at 280 nm in protein-free filtrates, have found proteinase-splitting hemoglobin in the tissue of pig pituitary at pH 8.3. As was shown by Tsaryuk[18] and somewhat later by Lisowski[19] the increase of density in the protein-free filtrates of brain and pituitary homogenates was, under these conditions, a result of ribonuclease activity rather than proteolysis. Colorimetric determination of an acid-soluble ninhydrin-positive material is not sufficiently specific as an index of proteolytic activity in the investigated tissue. Besides amino acids and peptides, many other compounds involved in the protein-free filtrates of the assays under study also give color with ninhydrin, though less intensive ones.[20] Such functionally significant for nerve tissue and metabolically active amines as histamine and tyramine, pyruvic, α-ketoglutaric, oxalacetic, and ascorbic acids, ammonia and other metabolites are among these compounds. For this reason the results of investigations may be essentially distorted if proteolytic activity is determined in whole-tissue homogenates which contain enzymatic systems leading to the formation of the compounds mentioned, particularly when the assay is carried out under conditions close to the pH optimum of the action of nearly all of these enzymes (pH 6–8).

When dealing with tissue homogenates, the method suggested by Anson[21] for determining proteolytic activity, which is based on measuring tyrosin and tryptophan split from protein, is not free of shortcomings. As shown by Klein *et al.*,[22] tyrosin formed on incubating brain homogenates under aerobic conditions is not completely determinable by this method, probably due to its rapid metabolic transformations. The authors consider that, when determining proteolytic activity by the Anson method, only data obtained under strict anaerobic conditions will be close to the actual intensity of proteolysis in the nerve tissue.

Analysis of the results of various investigations indicates that in measuring proteolytic activity in such complicated enzymatic systems as tissue homogenates and their subcellular fractions it is desirable to have several methods based on different principles, which would permit shortcomings of some of the methods to be compensated by advantages of others.

The above-mentioned considerations prompted us to start developing a sensitive and specific method for determining proteolytic activity in tissue homogenates of the nervous system and its subcellular fractions in media with pH values close to the physiological ones.

MATERIALS AND METHODS

The method is based on the determination of arginine split from a substrate by the effect of tissue proteinases. Used as an exogenous substrate was a commercial preparation of the basic protein protamine in the form of its sulphate. This protein was used to advantage by Veremienko[23] in a micromethod for determining proteolytic activity of trypsin.

To develop conditions for determining proteolytic activity on protamine splitting, adult-rabbit brain was perfused through the carotid arteries with a cooled physiological solution. The brain tissue was homogenized in 0.85% NaCl by a hand-glass homogenizer with Teflon plunger, and 10% homogenate was obtained.

The method of determining the proteolytic activity of the tissue enzyme, the so-called "protamine-splitting enzyme," is as follows: 0.2 ml of 1% protamine solution and 0.3 ml of the studied enzymatic preparation containing 2.5–3.0 mg of tissue protein are added to 0.7 ml of M/7 Veronal–acetate buffer (pH 7.2). The assays are incubated at 37°C for 40 min. Enzymatic reaction is stopped by adding 1.2 ml of 20% trichloroacetic acid (TCA). The TCA is also added to blank tubes prior to the introduction of enzymatic preparation. After incubation the experimental and blank tubes are centrifuged (1000g, 20 min) and arginine content is determined in the protein-free centrifugates by the method of Gilboe and Williams.[24] Experimental assays are subjected to photometry against controls at 508 nm. The amount of arginine split is calculated using a calibrating plot for recrystallized arginine hydrochloride. The activity of the enzyme is expressed in micrograms of arginine split from protamine during 40 min of incubation at 37°C and pH 7.2, per milligram of protein (or per 1 mg of fresh or dry brain tissue). Protein was determined by the method of Lowry *et al.*[25]

RESULTS

The dependence of enzymatic activity of brain homogenate on incubation time, concentration of substrate and enzyme, pH of the medium, and temperature was studied (the effect of each factor was measured while holding the others constant). The stability of the enzymatic system during storage of homogenate and the possible application of this method to determine proteolytic activity in the tissues of other organs were studied as well. In addition, the activity was determined of proteinase in the brain tissue of rabbits differing in age, in various areas of the nervous system, and in various subcellular fractions of the rabbit brain, and as well as in brain tissue of various kinds of animals.

Incubation Time. Figure 1 shows the protamine-splitting activity of the rabbit-brain homogenate as a function of incubation time (average of three experiments, pH 7.5). The protamine hydrolysis has a linear dependence over the whole range investigated, and amounts to 1300–1400 μg (7.5–8.0 μM) of arginine split from 100 mg of protein per hour.

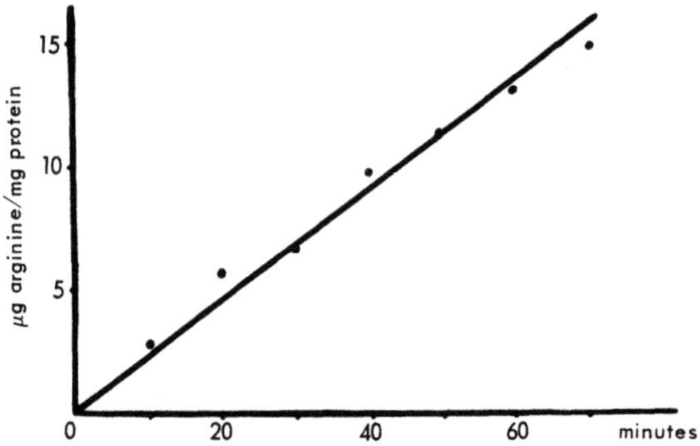

Fig. 1. The dependence of proteolytic activity upon the time of incubation.

Concentration of Substrate. As seen in Fig. 2 (average of three experiments, pH 7.5), at various concentrations of protamine in the samples the linear dependence of the amount of acid-soluble products split, as determined by the reaction of Sakaguchi, remains approximately 0.5–1.0 mg of substrate in the assay. With the increase of protamine concentrations the rate of proteolysis changes negligibly. It should be noted here that the reference optical density of the control assays for the substrate (without enzymatic preparation) against distilled water after incubation varies considerably in different preparations (from 0.03 to 0.45), which testifies to the different amounts of the breakdown products in these preparations. For investigating proteolytic activity only such preparations of protamine were used whose optical density did not exceed 0.05–0.06 (photometry in 1-cm cuvettes).

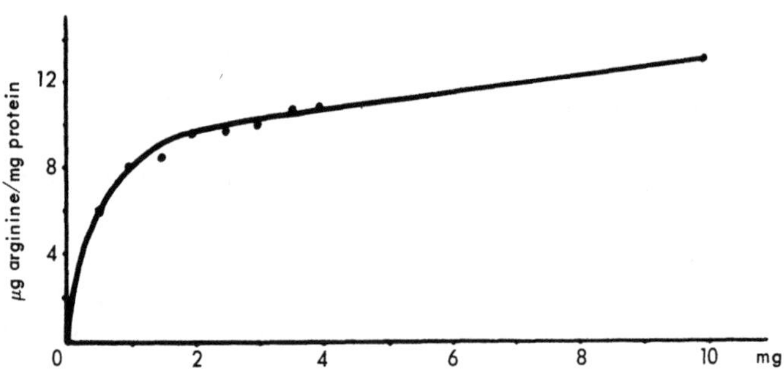

Fig. 2. The dependence of proteolytic activity upon the concentration of substrate.

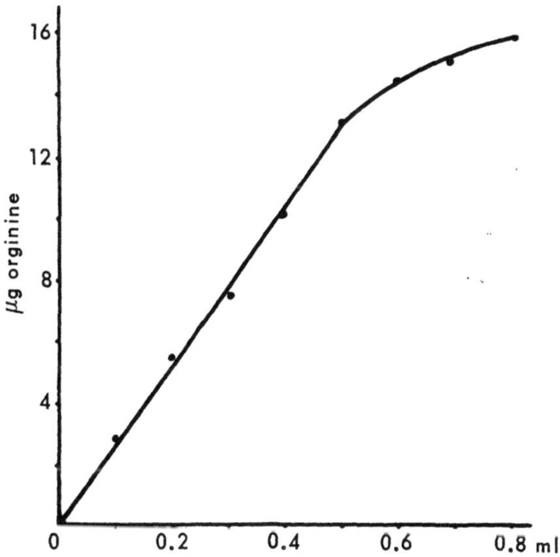

Fig. 3. The dependence of proteolytic activity upon the concentration of enzyme.

Concentration of Enzyme. The investigated assays contained from 0.1 to 0.8 ml of the brain homogenate (0.9–7.2 mg of tissue protein) and were incubated at pH 7.5. A direct dependence of protamine splitting on enzyme concentration is observed up to 0.5 ml of homogenate (4.5 mg of tissue protein) in the investigated assay (Fig. 3, average of three experiments, pH 7.5). At a further increase of enzyme concentration, the increase in the reaction rate reduces. In the control enzyme assays (without adding protamine) containing from 0.9 to 7.2 mg of endogenous protein, after incubating at pH 7.5 and 37°C for 40 min, no acid-soluble products were detected (negative Sakaguchi reaction).

pH of the Medium. Proteolytic activity was determined in assays with various concentrations of hydrogen ions (pH 3.0–9.4). As seen in Fig. 4 (average of four experiments), the optimal proteolysis of protamine is observed in a rather broad range of pH, from 5.3 to 6.9. At pH values below 5.3 and above 6.9 the intensity of protamine splitting essentially slows down, and at pH values 4.6 and 9.4 it amounts to 40 and 50% of the maximal level, respectively; at pH 3.0 enzymatic activity is practically absent. In connection with the fact that at pH 7.5 the rate of arginine splitting considerably decreases in comparison with its maximum level, further proteinase activity was determined at pH 7.2, i.e., in the range closer to the pH optimum of enzyme action.

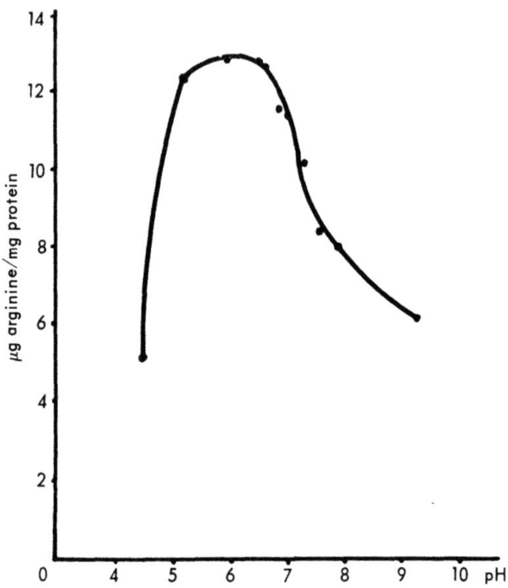

Fig. 4. The dependence of proteolytic activity upon the pH of the medium.

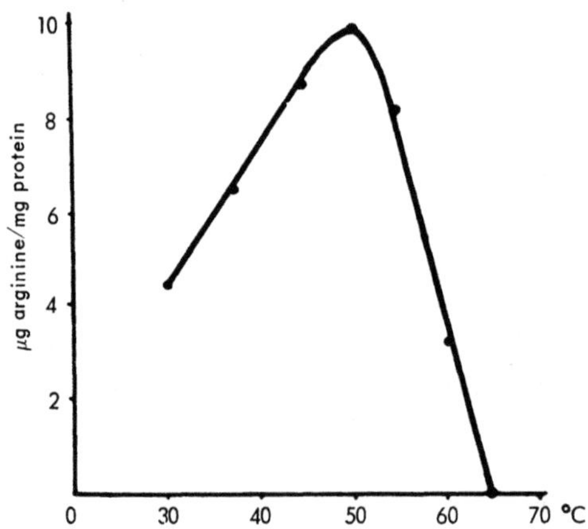

Fig. 5. The dependence of proteolytic activity upon the incubation temperature.

Incubation Temperature. Determination of enzyme activity at pH 7.2 and 20-min incubation at various temperatures (from 30 to 60°C) showed that the intensity of protamine splitting increases with increasing temperature to 50°C, then drops, especially sharply after 55°C, and at 65°C the enzymatic system is inactivated completely (Fig. 5, average of 4–8 determinations).

Stability of Enzymatic System during Storage. The change of proteolytic activity in brain tissue was investigated during the first 3.5 hr after rabbit death (homogenate kept at room temperature) and for long-term storage (homogenate at 4–5°C). The results showed that the activity of the brain homogenates under consideration almost does not vary during the first day of storage; the activity decreases slightly after two days of storage, and by the fifth day it decreases only by 16–18%. It was also shown that the enzyme activity in lyophilized preparations of the brain is only slightly lower than that of freshly prepared homogenates, and remains almost unchanged during long-term storage at of low temperature (4–5°C). In lyophilized preparations of the myelin fraction the enzyme activity is not essentially changed during months of storage at 4–5°C.

The Proteinase Activity in Different Areas of the Nervous System. The activity of the enzymes was investigated in homogenates of the gray and white

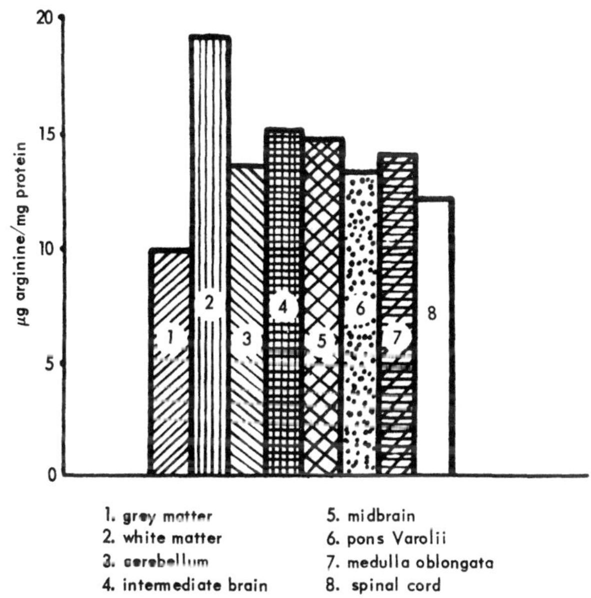

Fig. 6. Proteolytic activity in different areas of the brain.

matter of the cerebral hemispheres, in the cerebellum, pons Varolii, midbrain, diencephalon, medulla oblongata, and spinal cord of adult rabbits. The highest activity of proteinase is found in the white matter of the cerebral hemispheres, where it is twice as high as in the cortex (Fig. 6). The remaining five areas of the brain are close to each other with regard to the level of proteolytic activity and hold a position intermediate position between the gray and white matter of the cerebral hemispheres. The level of proteolytic activity in spinal-cord tissue exceeds only slightly that of the cortex.

The Proteinase Activity in Subcellular Fractions of the Brain. Using differential centrifugation, the following fractions were obtained from 10% homogenates of adult-rabbit brain prepared using a 0.32 M solution of sucrose: the nuclear fraction (980g, 10 min), the total crude mitochondria (11,500g, 20 min), microsome fraction (105,000g, 60 min) and a soluble cytoplasmic fraction (supernatant liquid). By means of centrifugation in a sucrose density gradient according to Whittaker,[26] but with a fraction selection somewhat changed by us (Fig. 7), three main fractions of subcellular structures were obtained from the crude mitochondrial fraction: (A) myelin; (B) nerve endings, synaptic vesicles, and deformed membranes; and (C) mitochondria; with four intermediate zones (I, II, III, and IV). With such a selection the possibility of contamination of the principal subcellular fractions A, B, and C with other structures was decreased.

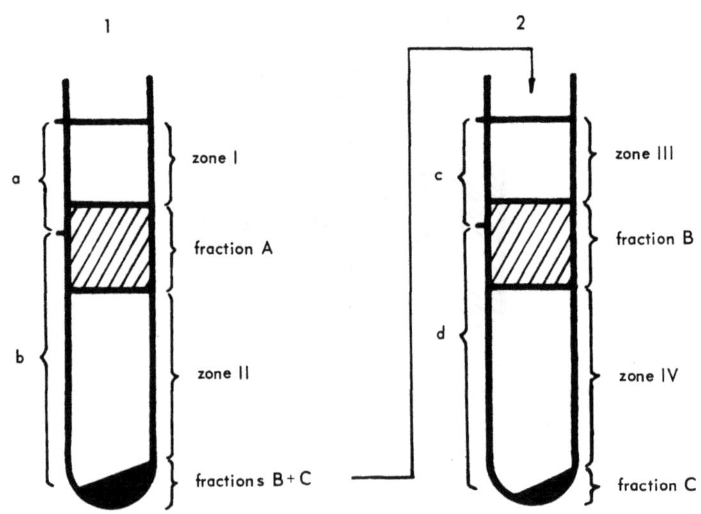

Fig. 7. A scheme of crude-mitochondria fractionation.

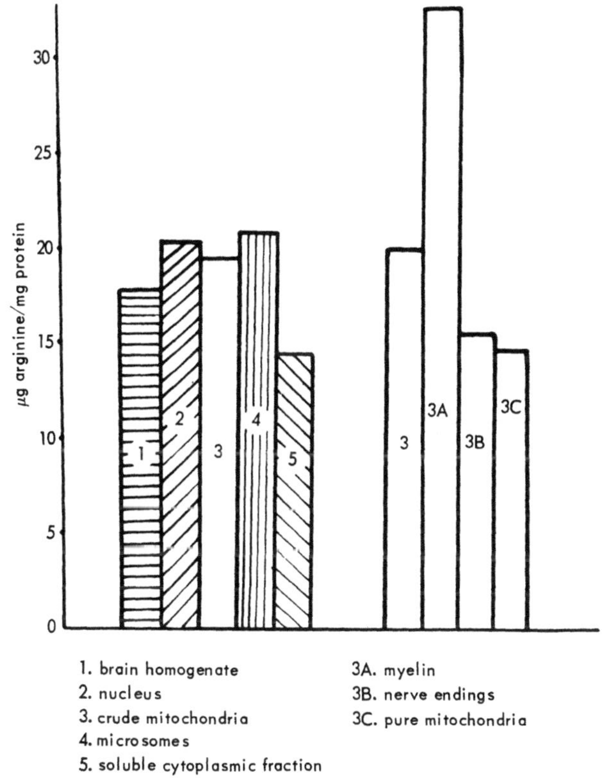

Fig. 8. Proteolytic activity in subcellular fractions of nerve tissue.

As seen in Fig. 8, the fractions of brain subcellular particles isolated by means of differential centrifugation differ very slightly among themselves in the level of proteinase activity (calculated per 1 mg of protein), except for the soluble cytoplasmatic fraction, in which the proteolytic activity per 1 mg of protein is considerably lower than in homogenate, nuclear, mitochondrial, and microsomal fractions ($P < 0.001$).

The fractions isolated by the centrifugation of crude mitochondria in a sucrose density gradient are characterized by unequal proteinase activity. The activity of enzyme in fraction A (myelin) is more than twice as high as that in fractions B (nerve endings) and C (mitochondria), and is 80% higher than the activity of the initial brain homogenate. The myelin fraction contains 33% of the protein, but 55% of the total proteinase activity of the initial crude mitochondrial fraction.

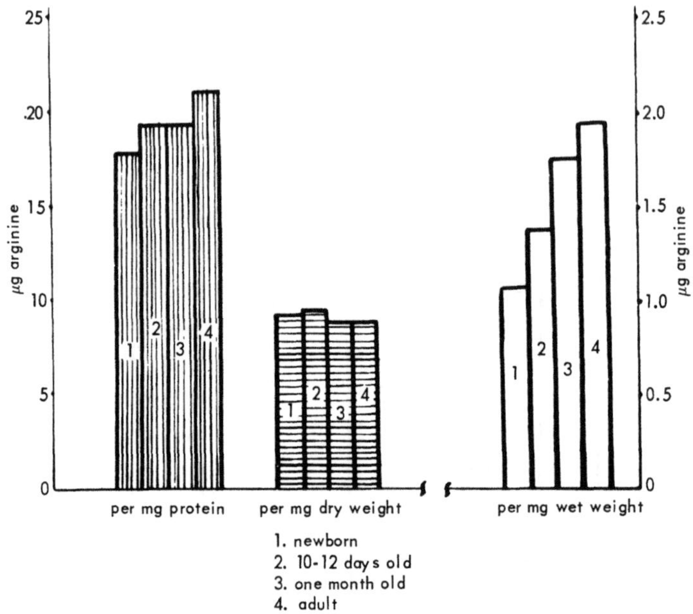

Fig. 9. Proteolytic activity in rabbit brains of different ages.

Changes in the Proteinase Activity with Age. The proteolytic activity of brain homogenates was investigated in rabbits of four age groups: newborn, those beginning to see (11–12th day of post-embryonal development), those one month old, and adult (6–7 months). As seen from Fig. 9, the proteolytic activity of the brain increases with age (calculating per wet weight and tissue protein). The proteinase activity of brain in rabbits beginning to see, those one month old, and adults was (calculating per wet weight) 127, 162, and 180%, respectively, of the level in newborn rabbits (in all the age groups $P < 0.001$). Calculating per tissue protein, the enzymatic activity increases with age less intensively than when calculating per wet weight of tissue (the difference is statistically authentic for all the age groups, except between the rabbits begining to see and those one month old) A different type of age change is detected when calculating the activity per dry weight of brain tissue. In this case the highest activity is found in the brain tissue of rabbits beginning to see, and the lowest in adults. The differences between newborn and adult rabbits ($P < 0.05$), between those beginning to see and those one month old ($P < 0.05$), and between those beginning to see and adults ($P < 0.01$) are statistically authentic. The most intensive increase of proteolytic activity is observed during the first 10–12 days of post-natal development.

Proteinase in Brain Tissue of Other Animals. The activity of enzyme in brain tissue of cat (12.6 ± 0.21 μg of arginine/1 mg of protein, $n = 7$) and *Citellus suslicus* (12.6 ± 0.28, $n = 8$) coincides with that of rabbit brain (12.6 ± 0.44, $n = 14$). In rat brain protamine-splitting activity is somewhat lower (10.4 ± 0.24, $n = 10$), and in brain tissue of pig (5.2 ± 0.12, $n = 6$) and cattle (5.4 ± 0.16, $n = 13$) it is approximately twice as low.

The Activity of the Enzymatic System in Other Organs. With the aim of checking the possibility of using this method for the determination of proteolytic activity in other tissues of an animal organ, three experiments were carried out in which homogenates of kidney, spleen, pancreas, liver, brain, heart, skeletal muscles, and blood were taken for investigation. The highest proteinase activity is detected in the kidneys (62 μg of arginine per 1 mg of protein separated during 40 min of incubation at pH 7.2 and 37°C); the level is considerably lower in spleen (35 μg) and in pancreas (26 μg); and still lower in liver (20 μg), brain (13 μg), and heart (7 μg); in skeletal muscles and blood the proteolytic activity is very negligible (1–2 μg).

DISCUSSION

The results stated above indicate that the suggested method of colorimetric determination of tissue proteolytic activity using the basic protein protamine as a substrate is sufficiently simple to carry out and possesses high sensitivity and accuracy. It gives well-reproducible results when using protamine preparations not containing a great quantity of breakdown products. The sensitivity of the method is practically determined by the minimum quantity of arginine (2 μg) which is found by the method of Gilboe and Williams.[24] The fact that in assays without the addition of protamine proteolytic activity is not determined with separated arginine permits one to conclude that endogenous substrates which under the given incubation conditions could have formed breakdown products giving the positive Sakaguchi reaction are absent in brain tissue. This is probably due both to the comparatively low arginine content in brain proteins and to their resistance to the effect of the tissue enzymatic system. As commercial protamine preparations are nonstandard, the results obtained with various preparations can differ considerably among themselves. To obtain reproducible results it is necessary to use a substrate of the same output series and to store it at low temperature. It is desirable to develop a method for obtaining standard protamine preparations.

As seen from the data obtained, the proteolytic enzymes of brain protamine intensively split over a sufficiently wide range of pH (from 5.3 to 6.9). This may be considered as a confirmation of the concept[10,13,14] of the presence of some enzymatic systems of proteolysis in brain tissue. Besides, the high protamine-splitting activity of the brain over a sufficiently wide range of pH may be a result of the presence of the same type of peptide bonds in this substrate and their spatial accessibility. It is possible that while hydrolyzing such a substrate the same enzyme serves both exo- and endopeptidase. Starting from the fact that protamine consists mainly of arginine residue and does not contain aromatic amino acids,[27] and taking into account the specificity of the known peptide hydrolases,[28] one may consider that the specificity of the effect of the investigated enzymatic system is similar to carboxypeptidase B (3.4.2.2.) which is known to hydrolyze peptides with a C-end residue of arginine or lysine and belongs to the group of exopeptidases. The specificity of the enzymatic system, as well as the theoretical possibility of the participation of some enzymes in splitting protamine, still requires experimental solution.

As noted previously, when investigating proteolytic activity in tissue homogenates and their fractions containing many enzymatic systems, it is desirable to have a series of methods which will cancel the shortcomings of one another. In this respect, the determination of tissue proteolytic activity according to arginine separated from protamine possesses the advantages that this amino acid is predominant in the exogenous substrate (in endogenous proteins its content is comparatively low), its determination is based on a sufficiently specific reaction, and the scale of the metabolic transformation of this amino acid during the incubation period considerably overlaps the intensity of its separation from protamine.

The data on brain proteinase show the stability of this enzymatic system both at the first hours after animal death and during long-term storage of homogenate or lyophilized preparation at low temperature. The enzymatic system under investigation seems not to be identical to the unstable neutral proteinase which was described previously by Ansel and Richter[10] and was then investigated by Klein and co-workers.[22] The difference between these enzymatic systems consists in the fact that the investigated proteinase is active under aerobic conditions, while the unstable proteolytic system mentioned manifests its activity only under conditions of strict anaerobiosis.

To the present our knowledge of the physiological role of intracellular proteinases in nervous-system tissue is still very limited. Proteolytic systems manifesting their activity both in acid and neutral media need to be studied. In keeping with the subject of this report, we shall concern ourselves mainly with the role of neutral proteolytic activity.

An age increase of the activity of protamine-splitting enzyme (Fig. 9) and acid proteinase[7] in brain tissue at first sight does not agree with the data on the decrease with age of self-renewal processes of proteins in this organ.[29]

However, in comparing these results one should take into account that at the early stages of individual development so-called synthesis of growth is predominant; the intensity of this sharply decreases at the early post-natal period; the synthesis of self-renewal during this period is noticably activated.[30] The latter ensures renewal of protein structures which were formed at earlier stages of devolopment; it is this synthesis that is probably realized with the participation of enzymatic systems splitting proteins to free amino acids.

However, some data concerning the activity of brain proteolytic systems are not always in good agreement with limiting the role of proteolysis enzymes to processes of protein renewal, and permit one to assume their possible participation in other more specific functions of nervous tissue. If, for example, proteinase took part only in the renewal of tissue proteins, then in nuclear and microsomal fractions, distinguished by their high intensity of protein metabolism, this enzyme should have been expected to be more active than in other intracellular fractions.

From the same considerations the enzyme activity in the gray matter of cerebral hemispheres should have been higher than in the white matter. However, as seen from the data presented, the highest activity of proteinase is found in the myelin fraction (when investigating subcellular fractions) and in the white matter of cerebral hemispheres (in comparing activities of different brain areas), i.e., just in the structures that in intensity of protein renewal are considerably inferior to the microsome and nuclear fractions[31] and to brain gray matter,[32] respectively. In microsomes and nuclei of the brain the activity of neutral proteinase was found to be even lower than in mitochondria.[13] A higher activity of proteolysis in the white matter of cerebral hemispheres as compared with the gray matter was also established by other methods.[10,13]

The available data enable one to assume a possible connection between the studied proteolytic system and the function of the myelin of nerve fibers. Such an assumption agrees with both the data presented on a lower activity of proteinase in the brain of newborn rabbits and those beginning to see, in which the process of myelination is not yet completed, and the data of Adams and co-workers[33–35] on a high catheptic and aminopeptidase activity of myelin structures of nervous tissue.

It is of interest that Marks and Lajtha,[14] using other methods of determination, found a higher activity of neutral proteinase in the fraction of nerve endings, and on this basis they advanced an assumption of some still unknown role of these enzymes in synaptic function. Elucidation of these possible specific functions of proteolytic systems in nervous system tissue is of great interest and must be an object of further investigations.

The data on a higher activity of proteinase (calculating per dry weight) in brain tissue of animals beginning to see are worth noting (Fig. 9). Acid proteinase calculated per dry weight of rabbit brain tissue is also more active during the period when the rabbit is beginning to see.[7] The intensification of proteolysis in visual centers of the cortex during irritation of the retina by light,[36] activation of an anabolic phase of protein metabolism[29,37,38] and nucleic acids,[37,38] as well as the change in chemical composition of proteins[39] in the brain of animals during the performance of the visual function, were previously shown. All these data, as well as the results presented above, may be considered to testify to the presence of a connection between the biochemistry of nervous tissue and its functional state.

SUMMARY

By means of a simple and sensitive method for determination of proteinase activity in brain tissues and by using protamine as an exogenous substrate it was shown that rabbit brain tissue contains a stable enzymatic system with a pH optimum of action from 5.3 to 6.9 which is active in splitting protamine under aerobic conditions.

The highest rate of protamine splitting at pH 7.2 is detected in the white matter and the lowest rate in the gray matter of the cerebral hemispheres; the cerebellum, diencephalon, medulla oblongata, midbrain and spinal cord, as well as the pons Varolii, hold an intermediate position with respect to the level of proteolytic activity.

The activity of the enzyme in the myelin fraction is considerably higher than in other subcellular fractions of the brain obtained by differential and gradient centrifugation.

Proteolytic activity of brain tissue increases with the age of the animal, and is especially intensive during the first days of post-natal life.

The proteinase activity is essentially lower in the brain tissue of cattle and pig than in the brain of the rabbit, cat, *Citellus suslicus*, and rat.

REFERENCES

1 M. W. Kies and S. Schwimmer, *J. Biol. Chem.* **145,** 685 (1942).
2. G. B. Ansell and D. Richter, *Biochim. Biophys. Acta* **13,** 87 (1954).
3. N. M. Polyakova, Ya. V. Belik, and L. A. Tsaryuk, *Ukr. Biokhim. Zh.* **32,** 623 (1960).
4. N. M. Polyakova and V. K. Lishko, *Ukr. Biokhim. Zh.* **34,** 208 (1962).
5. A. V. Palladin, N. M. Polyakova, and V. K. Lishko, *J. Neurochem.* **10,** 187 (1963).

6. A. Lajtha, *Int. Rev. Neurobiology* **6**, 1 (1964).
7. Ya. V. Belik and V. I. Tyulenev, *Zh. Evolyutsionnoi Biokhim. i Fiziol.* **2**, 333 (1966).
8. Ya. V. Belik and V. I. Tyulenev, *Ukr. Biokhim. Zh.* **39**, 343 (1967).
9. A. K. Upirvitskaya, *in*: "Biochemistry of the Brain," p. 35, Gor'kii (1941).
10. G. B. Ansell and D. Richter, *Biochim. Biophys. Acta* **13**, 92 (1954).
11. G. Porcellati and B. Curti, *J. Neurochem.* **5**, 277 (1960).
12. L. L. Uzman, M. K. Rumley, and S. van den Noort, *J. Neurochem.* **6**, 299 (1961).
13. A. Lajtha, *in*: "Regional Neurochemistry," p. 25, Pergamon Press, Oxford (1961).
14. N. Marks and A. Lajtha, *Biochem. J.* **89**, 438 (1963); **97**, 74 (1965).
15. S. Edlbacher, E. Goldschmidt, and V. Schlappi, *Z. Physiol. Chem.* **227**, 118 (1934).
16. A. Beloff and R. A. Peters, *J. Physiol.* **103**, 461 (1965).
17. E. Adams and E. L. Smith, *J. Biol. Chem.* **191**, 651 (1951).
18. L. A. Tsaryuk, *Ukr. Biokhim. Zh.* **34**, 815 (1962); **36**, 334 (1964).
19. J. Lisowski, *Arch. Immunol. Therap. Exp.* **12**, 645 (1964).
20. E. D. Schilling, P. I. Burschill, and R. A. Clayton, *Analyt. Biochem.* **5**, 1 (1963).
21. M. L. Anson, *J. Gen. Physiol.* **22**, 79 (1938).
22. G. E. Klein, E. G. Kurtzkhaliya, and N. V. Gvaliya, *Soobshch. Akad. Nauk Gruz. SSR* **44**, 331 (1966).
23. K. M. Veremienko, *Ukr. Biokhim. Zh.* **35**, 294 (1963).
24. D. D. Gilboe and J. N. Williams, *Proc. Soc. Exp. Biol. Med.* **91**, 535 (1956).
25. O. H. Lowry, N. J. Rosebrough, A. L. Farr, and R. J. Randall, *J. Biol. Chem.* **193**, 265 (1951).
26. V. P. Whittaker, *Biochem. J.* **72**, 694 (1959).
27. K. Felix, *Adv. Prot. Chem.* **15**, 1 (1960).
28. "Nomenklatura Fermentov" ("Enzyme Nomenclature"), Moscow (1966).
29. A. V. Palladin, Ya. V. Belik, and L. S. Krachko, *Biokhimiya* **22**, 359 (1957).
30. V. N. Nikitin, L. A. Dryuchina, and Z. L. Semenova, *Ukr. Biokhim. Zh.* **21**, 296 (1949).
31. Ya. V. Belik, Ya. T. Terletskaya, and L. S. Smerchinskaya, *in*: "The First All-Union Conference on the Biochemistry of the Nervous System," Tartu (1968).
32. A. V. Palladin and N. Vertaimer, *Dokl. Akad. Nauk SSSR* **102**, 319 (1955).
33. C. W. M. Adams and N. A. Tugan, *J. Neurochem.* **6**, 334 (1961).
34. C. W. M. Adams and O. B. Bayliss, *J. Histochem. Cytochem.* **9**, 473 (1961).
35. C. W. M. Adams and G. G. Glenner, *J. Neurochem.* **9**, 233 (1962).
36. G. Gorodis'ka, *Nauk. Zap. Ukr. Biokhim. Inst.* **1**, 105 (1926).
37. S.-O. Brattgård, *Acta Radiol. Suppl.* **96**, 1 (1959).
38. K. Schreier, *Monatschrift für Kinderheilkunde* **110**, 290 (1962).
39. N. V. Sadikova and G. K. Kudryashova, *in*: "The Nervous System," Vol. 5, p. 16, Izd. Leningr. Univ., Leningrad.

Chapter 4

Localization of Protein Metabolism in Neurons*

Bernard Droz

Département de Biologie
Commissariat à l'Energie Atomique
Gif sur Yvette 91, France

and

Herbert L. Koenig

Laboratoire de Cytologie
Faculté des Sciences
Paris, France

INTRODUCTION

The localization of the cell structures involved in the metabolism of proteins is one of the most delicate problems that neurochemists have to face. The problem is rendered difficult by the structural complexity of nervous tissue and by the heterogeneity of its cell population.

Histologically, the nervous system is composed of a great variety of cells with multiple processes, which are closely associated and intermingled; indeed, neurons of various categories, astrocytes, oligodendrocytes, and microgliocytes, differ from each other in morphology, metabolism, and function.

Furthermore, at the cellular level one part of a neuron differs from another part; as a consequence of its high specialization, the axon does not possess the same synthetic capacity as the perikaryon. Thus, the great variety of cell organelles and their intracellular arrangements reflect the metabolic compartmentalization of the neuronal cytoplasm.

Among the various techniques proposed for tracing protein in neurons from their sites of synthesis through their pathways of transfer, radio-

*The authors wish to thank Mrs. J. Boyenval for her fine technical assistance and Mrs. M. Lucarain for the photography.

autography, and especially quantitative radioautography, can provide fruitful information and supplement biochemical data.

The radioautographic localization of protein is based on the following principles: when labeled amino acids are introduced in a system which is synthesizing proteins, labeled amino acid residues are incorporated into growing polypeptide chains. In the course of tissue processing it has been shown that the free amino acids are washed out, whereas proteins are preserved: about 96% of the radioactivity retained in processed tissues is produced by the newly synthesized proteins.[1-3] Therefore, these newly-formed proteins can be simultaneously detected and localized by the radioautographic reaction that they produce.

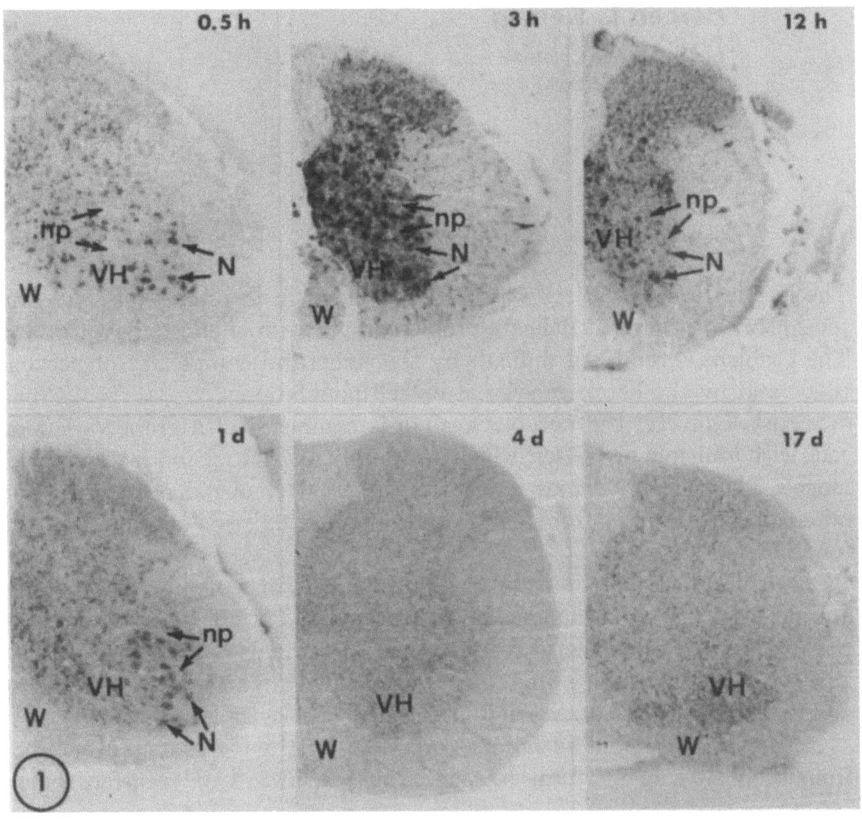

Fig. 1. Unstained radioautographs of the spinal cord of adult mice intraperitoneally injected with a single dose of ^{35}S-methionine and sacrificed at various time intervals thereafter.

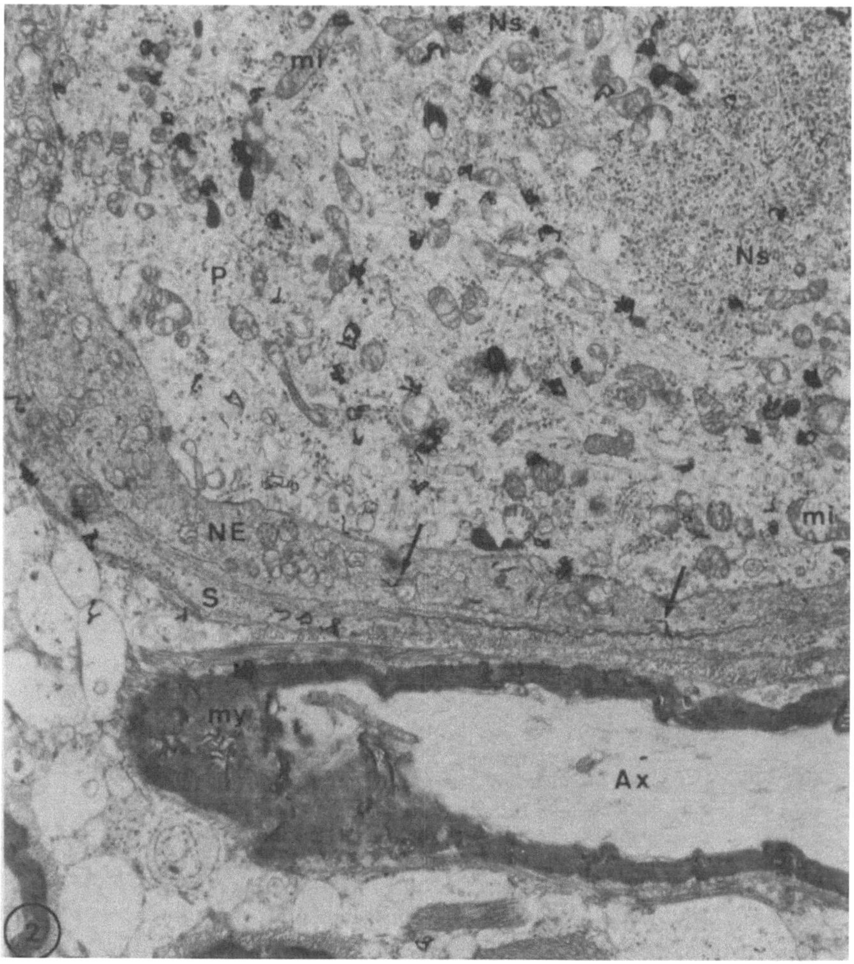

Fig. 2. Electron-microscope radioautograph of the perikaryon of a ganglion cell in the ciliary ganglion of a chicken intravenously injected with ^{3}H-leucine and sacrificed 5 min later. The ciliary ganglion was fixed in formaldehyde, postfixed in osmium tetroxyde, and embedded in epon; thin sections of regular thickness were radioautographed with Ilford L4 emulsion.[9] In the perikaryon (P) most of the silver grains of the radioautographic reaction are related to the clumps of Nissl substance (Ns), to scattered ribosomes of the neuroplasm, and to a few mitochondria (mi). A giant nerve ending (NE) encompasses this perikaryon; two silver grains (arrows) overlay this presynaptic structure. The thin cytoplasm of a satellite cell (S) displays a definite reaction. The lower part of the figure shows that the axoplasm (Ax) and the myelin sheath (my) of an axon are free of reaction.

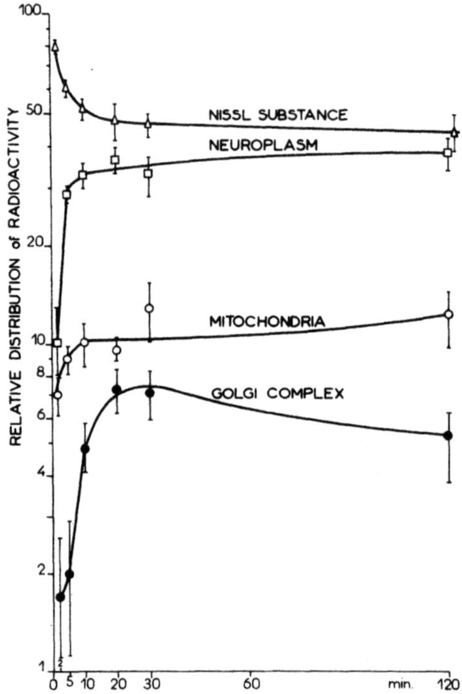

Fig. 3. Distribution of radioactivity in various cell organelles of the perikaryon of spinal ganglion cells in rats intravenously injected with ^{3}H-leucine, as estimated from electron-microscope radioautographs. The Nissl substance appears to be the first and main site of incorporation of the label, which rapidly declines, probably by transfer to other organelles. In the Golgi complex the early uptake of label is exceedingly low. A peak of radioactivity is reached in the Golgi region at 20–30 min, presumably by transfer from the Nissl substance. Both neuroplasm and mitochondria show a moderate incorporation very soon after injection. Then the label increases progressively in these organelles, probably by migration from the Nissl substance.

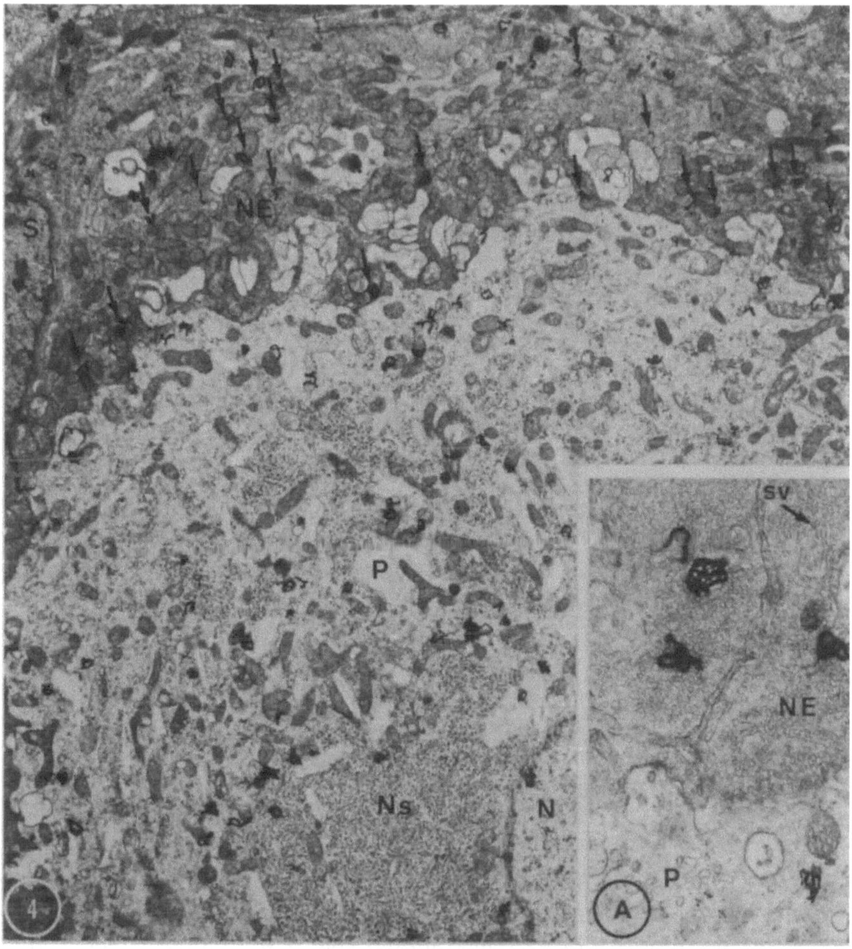

Fig. 4. Electron-microscope radioautograph of a ganglion cell in the ciliary ganglion of a chicken injected five days before with ^{3}H-leucine. Same technique as Fig. 2. In the nerve cell body a few silver grains persist over the nucleus (N) and the perikaryon (P), indicating the presence of "sedentary" proteins. Numerous silver grains (arrows) point to the presence of the label in the giant nerve ending (NE), which encompasses the ganglionic cell. Compare the intensity of the reaction at five days with that of the Fig. 2 at 5 min. The satellite cell (S) is still labeled. Inset A: A higher magnification of the synaptic region, showing that the silver grains mainly overlay the areas rich in synaptic vesicles (sv) present in the nerve ending (NE).

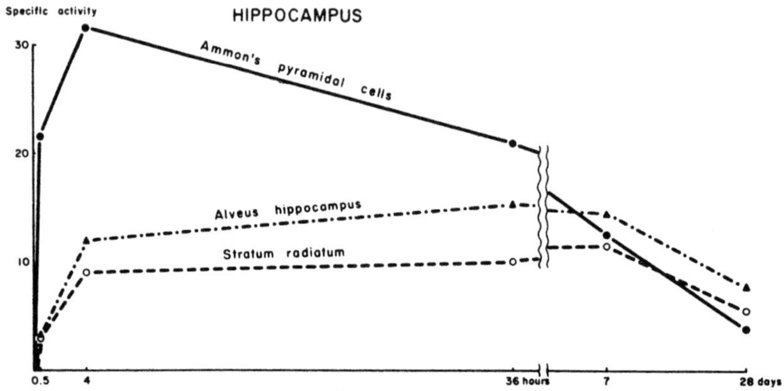

Fig. 5. Time curves of specific activity in the hippocampus of rat as measured by light-microscope radioautography after an intraperitoneal injection of ^{3}H-leucine and by microspectrophotometry. The specific activity of protein (conc. labeled/conc. total) was determined in various areas of the hippocampal section and expressed as the ratio: (number of silver grains per 100 μ^2)/(absorption of Millon chromophore). A first peak appears at 4 hr in the nerve cell body of the pyramidal cells of the cornu Ammonis. Later, after 36 hr, a second maximum is observed in the alveus (mainly composed of the axons of the pyramidal cells) and in the stratum radiatum (containing the dendrites of the pyramidal cells). Such curves are in agreement with the conditions required for a transfer of protein from the nerve cell bodies into regions made up with nerve cell processes. The analysis of decay curves of labeled proteins in the pyramidal nerve cell bodies indicates the presence of at least two main classes of proteins turning over in 0.9 and 17.3 days, respectively.

THE NERVE CELL BODY

The turnover of protein in individual neurons may be visualized using light-microscope radioautographs (Fig. 1) obtained at different time intervals after the injection of a labeled amino acid.[4–6] The nerve cell bodies (N), especially in the ventral horn, display an early and strong reaction (0.5 hr), which reaches its maximum of intensity around 3 hr. Then the radioactivity declines progressively with time until the nerve cell bodies can no longer be distinguished from the surrounding structures. In the neuropil (np) the radioactivity is first weak (0.5 hr), then increases between 3 and 12 hr after the injection. Finally, the radioactivity of the neuropil declines slightly in the following days, and seems to persist longer in the ventral horn (17 days). In the white matter (W) the discrete reaction seen at early time intervals is due to glial cells and blood vessels. Later (4 and 17 days) the intensity of the reaction is enhanced by the presence of radioactive axons.

Analysis of the rate of disappearance of radioactivity from the peri-

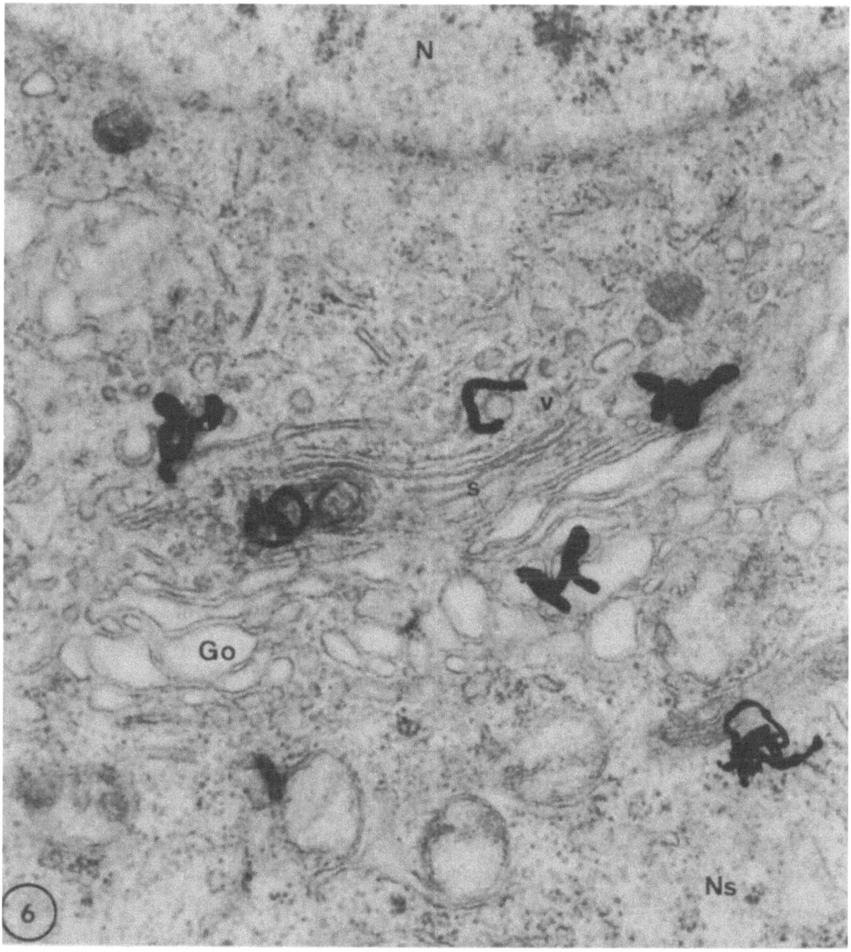

Fig. 6. Electron-microscope radioautograph of the Golgi complex of a spinal ganglion cell in rat 30 min after incubation in a medium containing ^{3}H-glucosamine. Same technique as Fig. 2. At the lower right corner silver grains lie at the periphery of the Nissl substance. Most of the label accumulates in the flattened sacs and vesicles of the Golgi region (Go). It is presumed that one fraction of the newly formed protein is coupled with sugars in the course of their transfer from the Nissl substance to the Golgi complex and therefore gives rise to glycoproteins. (From A. Rambourg and B. Droz, unpublished.)

karyon of various neurons points to the existence of two main classes of proteins[7]: the "sedentary" proteins, which spend about two weeks in the perikaryon, and the "migratory" proteins, which leave the nerve cell body in one day and enter the axon. Similar results obtained from isolated nerve cells previously labeled with ^{35}S-methionine confirm the validity of the turnover values calculated from quantitative radioautographs.[8]

Furthermore, newly formed proteins may be traced at the subcellular level by means of electron-microscope radioautography after the administration of an intravenous dose of tritium-labeled amino acid. The data collected by high-resolution radioautography indicate that the synthesis of protein is prominent in the Nissl substance of nerve cells (Figs. 2 and 3). The mitochondria, nucleus (chromatin and nucleolus), and ribosomes scattered in the neuroplasm of the perikaryon and axon hillock also elaborate protein, but to a lesser extent. The analysis of the kinetics of labeled porteins (Fig. 3) suggests that proteins synthesized in the Nissl substance follow various pathways[9]: a first fraction may persist in the Nissl substance as "sedentary" protein (Fig. 4); a second fraction is probably exchanged with nucleus and mitochondria; a third fraction, presumably the greatest one, would share in the elaboration of neuroplasmic components such as neurofilaments and neurotubules; a fourth fraction is transferred to the Golgi complex, where they accumulate (Fig. 5). From the Golgi region protein gives rise to lysosomes and vesicles of various types.

When labeled sugars are administered instead of amino acids, radioautographs point to the smooth extremity of the endoplasmic reticulum of the Nissl substance and to the flattened sacs of the Golgi complex as the respective sites of incorporation of ^{3}H-glucosamine[10] and ^{3}H-galactose.[11] Then labeled sugars and labeled proteins behave and migrate in a similar manner after accumulation in the Golgi zone (Fig. 6). Since the presence of glycoprotein has been cytochemically demonstrated in the Golgi complex, lysosomes, and "coated" vesicles of the nerve cell bodies,[12] it can be assumed that proteins are modified in some way during their detour through the Golgi zone and may give rise to glycoproteins.

The newly formed "migratory" proteins therefore appear to be transfered into various cell organelles with which they move into the axon. In consequence, the nerve-cell processes are continuously supplied with new proteins and organelles by the nerve cell body.

NERVE CELL PROCESSES

Dendrites

The regions of the brain which consist mainly of dendritic arborizations exhibit only a slight capacity to synthezize protein (Fig. 5), except for the initial portion of large dendrites containing clusters of Nissl substance and therefore corresponding to expansions of the nerve cell body.[6,7] After a certain time elapses, labeled proteins invade dendritic arborizations, and then disappear slowly (Fig. 5).

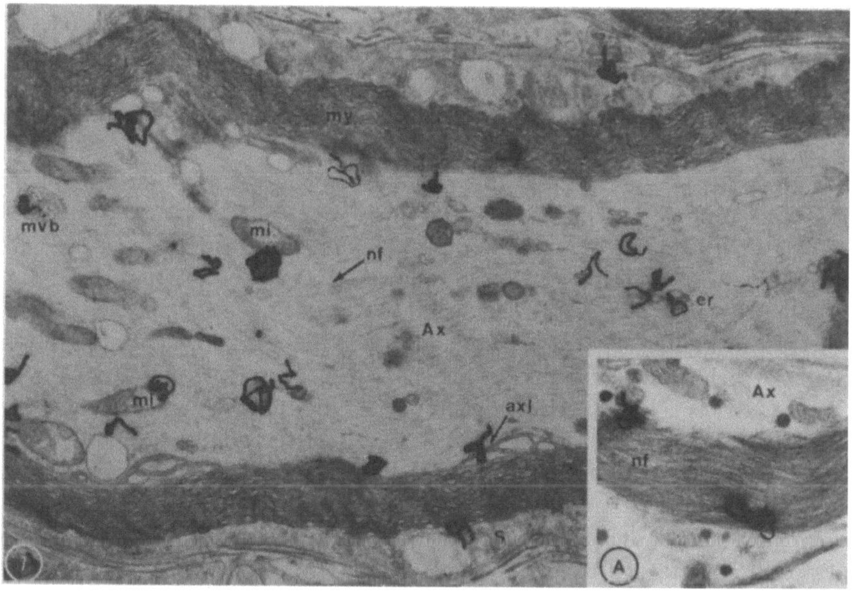

Fig. 7. Electron-microscope radioautograph of the axon of a ganglionic cell in the ciliary nerve of a chicken injected 24 hr before with ^{3}H-leucine. Same technique as Fig. 2. Axons which were unlabeled a few minutes after the injection (see Fig. 2) are highly radioactive within a hew hours. The presence of the label is detected in the axoplasm, which contains numerous neurofilaments (nf) and a few neurotubules. Silver grains are seen in close relation with multivesicular bodies (mvb), mitochondria (mi), elements of the endoplasmic reticulum (er), and the axolemmal region (axl). Inset A: Electron-microscope radioautograph of the axon of a leech (*Hirudo medicinalis*) injected 13 days before with ^{3}H-leucine. Contrary to the axons of vertebrates, in which the neurofilaments are loosely distributed throughout the axoplasm, the axons of the leech contain neurofilaments which are bunched together in a unique bundle (nf). Most of the axonal radioactivity is found to be associated with these packed neurofilaments as compared to the rest of the axoplasm.

Axons

In contrast to the nerve cell body, the axon proper is apparently devoid of ribosomes except in the initial portion.[13] Radioautographic results obtained from myelinated or unmyelinated axons indicate that, on the one hand, the incorporation of labeled amino acid into axonal protein[14,15] is either absent or at least too negligible (Fig. 2) to account for the replacement of the axoplasmic components,[10] and on the other hand, the delivery of Schwann cell protein to the axon[16] would be, if it occurs at all, a very restricted process.[18]

The third possibility is an axoplasmic transport of proteins and cell organelles moving from the nerve cell body toward the nerve endings.[19] Radioautographic data collected from *individual* axons show that labeled proteins rapidly leave the nerve cell body and invade the initial portion of the axon, whereas the distal regions are not yet radioactive;[7,14,20–24] with time the labeled protein moves progressively toward more distal regions at a rate of 0.6–0.9 and 2.0–2.5 mm/day, respectively, in adult and young growing rats.[18] A more rapid protein flux has been detected (100 mm/day) after local injection of labeled amino acids.[22,23,25,26] When the radioactivity reaches these distal regions the label declines in the proximal regions of the axons.[16,18,20] In the axon most of the label is distributed in the axoplasm rich in neurofilaments (Fig. 7 inset A) and neurotubules, whereas a minor part is related to the axolemmal region, multivesicular bodies, the smooth endoplasmic reticulum, and mitochondria (Fig. 7). While Ochs *et al.*[23] state that labeled protein engulfed in the axon should be free to move in the fluid part of the axoplasm, electron microscopy indicates unequivocally that labeled proteins are also closely associated with the structural components of the axon.

The degradation of axonal proteins is probably ensured all along the axons and at the nerve endings by the various proteolytic enzymes studied by Lajtha and Marks.[27] Nevertheless, with time a certain amount of "migratory" protein reaches the nerve endings (Fig. 4).

TABLE I

Concentration of the Radioactivity in Preganglionic Nerve Endings and Perikaryons of the Ciliary Ganglion of Chickens Injected with ^{3}H-Leucine and Sacrificed at Various Time Intervals[a]

Group	Time after injection	Number of counted silver grains	Nerve endings (*NE*)	Perikaryons (*P*)	Ratio *NE*/*P*
I	5 min	4472	12.1	65.1	0.18
	20 min	6081	20.4	132.9	0.15
	60 min	6308	22.2	96.0	0.23
II	1 day	6385	69.1	91.9	0.75
	2 days	3771	77.4	59.2	1.30
	5 days	3992	70.2	53.1	1.32

[a] The radioactivity concentration is expressed as the number of silver grains per 100 μ^2 after subtracting the background.

Nerve Endings

Barondes[28–30] has shown that nerve-ending fractions of mouse brain injected with labeled leucine were first poorly labeled; then, after several days the specific activity of protein was tremendously enhanced, presumably by the arrival of labeled migratory protein. The possibility that nerve endings synthesize a slight amount of protein has recently been suggested,[30–33] but in brain fractionation it is difficult to prevent the nerve endings from containing contaminant capable of synthesizing protein.[30] Here again radioautography helps to establish whether or not the label is confined to nerve endings. After an intravenous injection of ^{3}H-leucine to chicken several types of nerve endings were examined at successive time intervals.[34–35] A small but significant fraction of labeled leucine was found in nerve endings as early as 5 min after the injection (Table I and Fig. 2). Only a minor part of the label (about 10%) was related to mitochondria, whereas most of the radioactivity appeared in regions containing synaptic vesicles. However, the nature of the linkage by which the labeled amino acids are retained (i.e., by means of a peptide bond or artifactual retention) remains to be elucidated. Nevertheless, even if a true synthesis of protein takes place in axonal endings, the net output would be exceedingly small and therefore could not account for the replacement of the whole presynaptic protein. Indeed, the late accumulation of labeled protein in nerve endings (Fig. 4; Table I), at a time at which the free labeled precursor has disappeared from the tissue, indicates that the renewal of protein in nerve endings is mainly ensured, if not exclusively, by a transport of migratory protein along the axon.

CONCLUSION

The intraneuronal pathways of protein metabolism, as derived from radiautography, are tentatively summarized in Fig. 8.

The synthesis of neuronal protein is located in the nerve cell body and at the base of dendrites. Most of the newly formed proteins originate from the ribosomes of the Nissl substance. Then a fraction of the newly synthesized proteins is transfered to the Golgi complex and seems to give rise to protein and glycoprotein components of lysosomes and "coated" vesicles. Another fraction probably bypasses the Golgi region and might participate in the elaboration of neurofilaments and neurotubules. Nucleus and mitochondria, which are able to synthesize protein, would also receive protein components by transfer from ribosomes. A local synthesis of protein in axonal processes and nerve endings remains questionable. However, pro-

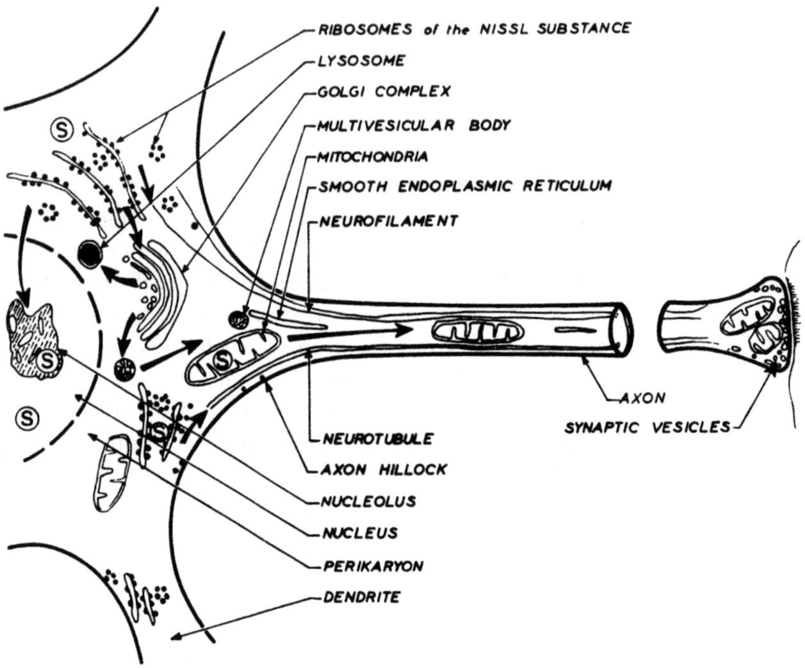

Fig. 8. Diagrammatic representation of the sites of synthesis and of the pathways of transfer of protein in the neuron. The main sites of protein synthesis (S) are the ribosomes of the perikaryon. Mitochondria also seem able to edify protein. In the nucleus both chromatin and nucleolus incorporate early amino acids into proteins. Protein elaborated in the ribosomes may follow various pathways[9]: a first fraction would migrate to the nucleus and would account for the late wave of labeled protein appearing in this structure; a second fraction would share in the elaboration of mitochondrial components, such as enzymes. Mitochondria may be engulfed into the axon and transported toward nerve endings; a third fraction, probably the greatest one, would be responsible for the edification of the subunits making up neurofilaments and neurotubules; it is speculated that most of the newly produced protein replaces and repairs neurofilaments and neurotubules worn out by proteolytic enzymes;[20] a fourth fraction is transferred to the Golgi complex. In the course of this migration sugars (glucosamine, galactose, etc.) may be added to a protein moiety and would give rise to glycoprotein. From the Golgi complex protein would move in the neuron associated to lysosomes, multivesicular bodies, elements of the smooth-surfaced endoplasmic reticulum, and "coated vesicles." Some of these organelles enter the axon and move along its length. Thus the axon is submitted to an incessant traffic of various types of proteins which move to nerve endings.

tein definitely turns over in the axons and nerve endings, where used axoplasmic proteins are continuously balanced by new "migratory" proteins. These "migratory" proteins are elaborated in the nerve cell body and transported along the neuronal extensions in close association with structual components of the axon. This incessant convection of new protein from the cell body toward the nerve endings would ensure both the integrity and the adaptability of neural circuits.

REFERENCES

1. B. Droz and H. Warshawsky, *J. Histochem. Cytochem.* **11,** 426–435 (1963).
2. T. Peters, Jr. and C. A. Ashley, *J. Cell Biol.* **33,** 53–60 (1967).
3. M. Bergeron and B. Droz, *J. Microscopie* **7,** 51–62 (1968).
4. C. P. Leblond, N. B. Everett, and B. Simmons, *Am. J. Anat.* **101,** 225–271 (1957).
5. D. Richter, M. K. Gaitonde, and P. Cohn, *in*: "Structure and Function of the Cortex" (D. B. Tower and J. P. Schade, eds.), pp. 340–347, Elsevier Publishing Company, Amsterdam (1960).
6. B. Schultze, W. Oehlert, and W. Maurer, *Beitr. Pathol. Anat.* **120,** 58–84 (1959).
7. B. Droz and C. P. Leblond, *J. Compar. Neurol.* **121,** 325–346 (1963).
8. B. Jakoubek, E. Gutmann, J. Fischer, and A. Babicky, *J. Neurochem.* **15,** 633–642 (1968).
9. B. Droz, *J. Microscopie* **6,** 201–228 (1967).
10. A. Rambourg and B. Droz, *J. Microscopie* **8,** 79a (1969).
11. B. Droz, *J. Microscopie* **6,** 419–424 (1967).
12. A. Rambourg, *Compt. Rend.* **268,** 1426–1428 (1967).
13. S. L. Palay, C. Sotelo, A. Peters, and P. M. Orkand, *J. Cell Biol.* **38,** 193–201 (1968).
14. A. Edström, *J. Neurochem.* **13,** 315–321 (1966).
15. E. Koenig, *J. Neurochem.* **14,** 437–446 (1967).
16. B. Droz, *Intern. Rev. Cytology* **25,** 363–390 (1969).
17. M. Singer and M. M. Salpeter, *J. Morphol.* **120,** 281–316 (1966).
18. B. Droz, *in*: "The Use of Radioautography in Investigating Protein Synthesis" (C. P. Leblond and K. B. Warren, eds.), Vol. 4, Academic Press, New York (1965).
19. P. Weiss, *Neurosciences Res. Progr. Bull.* **5**(4), 371–400 (1967).
20. B. Droz and C. P. Leblond, *Compt. Rend. Assoc. Anat.* **121,** 106–114 (1964).
21. T. Utakoji and T. C. Hsu, *J. Exp. Zool.* **158,** 181–202 (1965).
22. R. J. Lasek, *Neurosci. Res. Progr. Bull.* **5**(4), 314–317 (1967).
23. S. Ochs, J. Johnson, and M. H. Ng, *J. Neurochem.* **14,** 317–331 (1967).
24. J. Francoeur and J. Olszewski, *Neurology* **18,** 178–184 (1968).
25. R. J. Lasek, *Exp. Neurol.* **21,** 41–51 (1968).
26. B. Grafstein, *Science* **157,** 196–198 (1967).
27. A. Lajtha and N. Marks, *in*: "Protides of the Biological Fluids" (H. Peeters, ed.), Vol. 13, pp. 103–114, Elsevier Publishing Company, Amsterdam (1966).
28. S. H. Barondes, *Science* **146,** 779–781 (1964).
29. S. H. Barondes, *J. Neurochem.* **13,** 721–727 (1966).
30. S. H. Barondes, *J. Neurochem.* **15,** 343–350 (1968).
31. L. Austin and I. G. Morgan, *J. Neurochem.* **14,** 377–387 (1967).
32. I. G. Morgan and L. Austin, *J. Neurochem.* **15,** 41–51 (1968).

33. L. A. Autilio, S. H. Appel, and P. Pettis, *Fed. Proc.* **27**, 399 (1968).
34. H. L. Koenig and B. Droz, *in*: "Proc. IVth European Regional Conf. on Electron Microscopy," pp. 525–526, Tipografia polyglotta Vaticana, Rome (1968).
35. R. W. Young and B. Droz, *J. Cell Biol.* **39**, 169–184 (1968).

DISCUSSION

AUSTIN: In more recent experiments we found a rapid movement of protein, and also of free amino acids. Your demonstration of a rapid labeling in the nerve ending in Fig. 4 could still be due to local synthesis from labeled amino acid that has passed down as free amino acid and been incorporated into protein at the nerve ending.

DROZ: You ask why it is necessary for the labeled amino acids to travel along the axon when the labeled amino acids may be picked up by nerve endings from the extracellular compartment. We have incubated the isolated ciliary ganglion in a medium containing labeled amino acids; the nerve endings, which are severed from the nerve cell body, are still able to retain labeled amino acids.

BOGOCH: About the labeling with carbohydrates, Rambourg showed some pictures from a previous publication of synaptic junctions which appear to contain glycoprotein material. What is the time course of that labeling, and the earliest time that one sees a label of the sugar?

DROZ: Rambourg devised a cytochemical reaction showing a glycoprotein carbohydrate moiety which is located in the synaptic cleft. We have performed an experiment by incubating the ciliary ganglion in ^{3}H-glucosamine and in ^{3}H-galactose. A slight incorporation takes place in nerve endings, but incorporation is mainly in nerve cell bodies.

FORD: Is the rate of labeling of the different components, as seen by electromicroscopy, the same in different types of nerve cells, or does it vary? In some regenerating cells, which we examined by light microscopy with autoradiography, we noted that the distribution of label was different in the facial nucleus as compared to the hypoglossal nucleus, which perhaps implied a difference in the rate at which label would, say, be showing up in endoplasmic reticulum and then migrating to the Golgi complex and so on. I was wondering if you have observed in normal cells a difference in the rate at which your radioactive material is appearing in endoplasmic reticulum and migrating toward the axon.

DROZ: No, we have examined only two types of cells: the spinal ganglion cells and the ciliary ganglion cells in chicken. For young growing animals and adults there is no difference in the time sequence of the events; it's only a question of intensity.

BARONDES: As for the autoradiography shown here, it certainly seems very clear that most of the protein comes from upstairs; this technique has the virtue of being able to resolve things very clearly, and there's no contamination problem, which is of course the great problem with the subcellular fractionation procedures. The question about glucosamine is quite different, however; if one measures labeled glucosamine incorporation into nerve-ending protein, one finds that there is tremendous incorporation of labeled glucosamine into a soluble protein of nerve endings and into a particulate protein of nerve endings locally; that is, the kinetics of incorporation of glucosamine into protein at the nerve endings is completely different from that of leucine. Leucine is largely put in upstairs and shipped down; then these polypeptides can have glucosamine added to them locally at the nerve ending.

KOENIG: About the Golgi apparatus: Lysosomal enzymes have now been purified and the amino acid composition determined, such as in the case of DNAse, which is a glycoprotein. In recent studies on purified renal lysosomes we have found that apparently all of the acid hydrolases are glycoproteins: purified enzyme fractions have abundant hexosamine and hexose and also contain very basic proteins. I think this sits very nicely with the idea that the incorporation of hexosamine and hexose resides in the Golgi apparatus. It's generally thought that so-called primary or pure lysosomes are formed by the Golgi apparatus. This interpretation is consistent with the new knowledge regarding the only glycoprotein enzymes that have thus far been described. There may be others. Now, one other point regarding the Golgi: we found anatomical continuity between the rough ER on the one hand and the Golgi on the other. As is well known, thiaminpyrophosphatase or nucleoside phosphatase activity are excellent morphological markers for the Golgi. In fluorocitrate-poisoned nerve cells one of the very striking cytochemical alterations at the EM level is the accumulation of nucleoside diphosphatase activity in the cisternal of the rough ER, indicating that these two membrane-bound systems (rough ER and Golgi) are intercommunicated.

AGRANOFF: This is a suggestion for experiments in which an isotope is administered and then some period of time allowed to pass before killing, and it might help resolve some issues which still may exist in terms of reutilization *vs. de novo* synthesis. In our laboratory Dr. Lim recently completed some experiments in goldfish in which we used the pulse-chase-type of experiment. After a small amount of labeled leucine was given for a period of time, the animal was given several thousand times the dose of unlabeled leucine, which essentially stopped further incorporation and also increased the excretion of metabolites of leucine. Since one can also use antibiotic antimetabolites to begin the chase, one might be able to apply this sort of approach if or when one could keep up the chase for several days without any toxicity.

DROZ: When living animals injected with "hot" amino acids are given a chase, how can we be sure that the large amount of "cold" amino acid injected into the animal readily penetrates the cell? I think this is not always the case in the brain of living animals. However, this chase may be carried out by incubating isolated ciliary ganglia: a chase with cold amino acid at a concentration one thousand times higher than that of the labeled one may be combined with antimetabolites such as puromycin. I would also like to answer Dr. Koenig's question dealing with the Golgi complex; the use of a cytochemical technique (PTA at low PH) which was devised by Rambourg[12] in my laboratory visualizes the distribution of glycoprotein in the individual sacs of the Golgi complex; the inner ones contain the highest concentration of carbohydrate. By means of this technique, both vesicles and lysosomes are detected, indicating that these organelles contain a high proportion of carbohydrate.

OCHS: I think it's thoroughly clear now that there are two different rates of axonal transport. In your radioautography in all probability you're measuring what we call the low rate, and your rates would fit very well with our low rates and those of many others. It's also true that in the high rate we do get free leucine coming down, and we have evidence that it is inside the axon. I think part of the answer to whether the protein that you see comes from this component is that in a mouse this is relatively much lower.

DROZ: There is another possibility; you use a technique which does not extract the small polypeptide molecules, and perhaps these polypeptides migrate faster. In our case I can retain only high-molecular-weight protein.

OCHS: There is a polypeptide fraction which comes down.

BOGOCH: I just want to draw attention to the early labeling of the carbohydrate units: I believe that's 10 to 20 min in the Golgi apparatus. Is it correct that carbohydrate added came within a 10–20 min period?

DROZ: Yes. In the case of ^{3}H-galactose or ^{3}H-glucosamine we examined the tissue 5 min after the incubation or the injection. Galactose probably combined with a protein fraction that had been made 5 or 10 min before.

Chapter 5

Membrane-Bound and Free Ribosomes in the Developing Rat Brain*

M. R. V. Murthy

Department of Biochemistry
Faculty of Medicine
Laval University
Quebec, Canada

INTRODUCTION

The most striking morphological feature of the neuron is the tremendous accumulation within its cytoplasm of ribonucleoprotein particles. Some of these are found attached to a well-developed endoplasmic reticulum and some are present free in the surrounding matrix.[1] In this respect the neurons resemble certain protein-secreting glandular cells such as those of pancreatic acini and the salivary glands.[2] That a continuous and rapid synthesis of protein occurs in the neuron is evident from its large size and from rapid changes in the RNA and protein metabolism during the regeneration of axons.[3]

Although the ribosomes and the endoplasmic reticular membrane are almost universally present in cells, their form, quantity, and distribution are found to vary during cell differentiation and growth. As a rule, the endoplasmic reticulum is not a prominent feature in the relatively undifferentiated cells.[4] On the other hand, highly differentiated cells fulfilling specialized functions always have a highly developed endoplasmic reticulum.[1,5,6] In tissues which synthesize protein for secretion, such as liver and pancreas, the majority of ribosomes are seen to be attached to the membrane,[4] whereas reticulocytes which do not export protein lack endoplasmic reticulum and carry out protein synthesis primarily with "free"

*The author is grateful to the Medical Research Council of Canada for a research grant for carrying out these investigations. Thanks are due to Mlle. N. Chiasson, Mlle. Helene Allard, and Mme. M. Lamarche for their skilled technical assistance.

agregates of ribosomes.[7] By altering the intracellular or environmental conditions of the tissue, it is possible to affect the rate of protein synthesis, and this is accompanied by changes in the relative proportions of "free" monomeric ribosomes, "free" polysomes, and the membrane-bound ribosomes.[8–10]

It also appears that different kinds of proteins are synthesized by ribosomes depending on their attachment to *m*-RNA and to membrane. Thus, ribosomes attached to membrane are reported to be responsible for the synthesis of protein for secretion, while the ribosomes free in the cytoplasm synthesize protein for intracellular purposes.[11] Specialized proteins such as antibodies are also found to be synthesized by membrane-bound ribosomes.[12,13] It has been shown recently[8] that in liver the most stable variety of *m*-RNA is associated with ribosomes that are attached to the membrane. It is also these same ribosomes that appear to direct the synthesis of albumin.

In the normal embryonic growth and differentiation of the neuronal system of chick embryos a three-to-fourfold increase in the quantity of the endoplasmic reticulum has been noted in the progression from embryonic stage 4 to stage 36.[4] Electron-microscope studies have shown that during the maturation of the motoneurons of the chick spinal cord there is a progressive increase in the membrane structure and also in the number of membrane-bound ribosomes.[15] Changes in the quantity and distribution of "free" and membrane-bound ribosomes have also been noted during post-natal growth. There is a high concentration of free ribosomes in the liver of foetal and post-natal rats, but five days after birth the proportion of free ribosomes falls rapidly to the adult level, which is only one-fourth of the original concentration.[16]

In addition to their attachment to the membrane, ribosomes active in protein synthesis are also associated with other macromolecular structures, such as the nascent polypeptide,[17] sRNA,[18] and *m*-RNA[19] at specific sites on their surface. The mechanism of interaction between the membrane and ribosomes and the actual roles played by the two direct participants as well as the other associated molecules in this reaction are of great interest in view of the functional uniqueness attributed to the membrane-bound ribosomes.

In our earlier studies it was found that the concentration of both the microsomes and polysomes decreased in rat brain during post-natal development. Thus, the adult brain contained only one-fourth as much microsomal protein[20] and one-third as many polysomes as the newborn.[21] In these studies polysomes were prepared by deoxycholate treatment of the post-mitochondrial supernatant and therefore no distinction was made between those that were "free" in the cytoplasm and those that were bound to the membrane. Our recent experiments have been concerned with the separation of membrane-bound and free ribosomes in brains of rats of different age groups and examination of any structural and functional differences that may exist in these two types. Some of this work will be reported in the following pages.

RESULTS AND DISCUSSION

Brains from two-week-old, male, Sprague-Dawley rats were homogenized in a solution containing 0.25 *M* sucrose, 4 m*M* $MgCl_2$, 25 m*M* KCl and 35 m*M* Tris, pH 7.8 (solution A) (Fig. 1). Nuclei, debris, and mitochondria were removed by centrifugation at 10,000g for 15 min. Microsome fractions were obtained from the post-mitochondrial supernatant by centrifuging successively at 30,000 rpm in the 30 head of the Spinco preparative ultracentrifuge for various periods of time. The highest sedimentation of RNA and protein occurred during the first few minutes (Fig. 1). As the centrifugation was continued, there was a further sedimentation of RNA and protein, but the actual quantities sedimented per unit time declined. Since the sedimentation rate was not the same for protein and RNA, the microsomal fractions obtained at different intervals had different RNA:protein composition, reflecting the relative proportions of membrane and ribonucleoprotein components present in each fraction. These fractions were further examined as follows (Table I): Microsome fractions M_1, M_2, M_3, and M_4 were prepared as in Fig. 1. They were each suspended in solution A, to which 10 μg/ml of rat-liver ribonuclease inhibitor[22] was added. The suspensions were treated with deoxycholate at a final concentration of 1% of the detergent. The solutions were clarified by a preliminary centrifugation at 35,000g for 15 min. The supernatants from this step were again centrifuged at 105,000g for 90 min to obtain the polysomes P_1, P_2, P_3, and P_4 from the corresponding microsomal fractions. The polysomes were each suspended in a buffer containing 4 m*M* $MgCl_2$, 25 m*M* KCl, and 35 m*M* Tris buffer (pH 7.8) and treated with 0.2μg/ml of RNase at 4°C for 30 min. The suspension was centrifuged at 105,000g for 20 min and the pellets containing the monomers were washed twice in the same buffer, to which 10 μg/ml of rat-liver ribonuclease inhibitor was added. Amino acid incorporation was determined using a reaction mixture which contained 10 μM of Tris buffer (pH 7.8), 2.5 μM of phosphoenol pyruvate, 7.5 μg of pyruvate kinase, 0.25 μM of ATP, 0.062 μM of GTP, 10 μM of KCl, 2.5 μM of $MgCl_2$, a mixture of amino acids to give 0.025 μM of each amino acid, 1.2×10^5 dpm of ^{14}C-phenylalanine (U.L., 1.6 mC/μM), 2.5 μg of RNase-inhibitor, ribosomes or microsomes equivalent to 0.35 mg RNA, pH-5 enzyme from two-week-old rat brains equivalent to 0.5 mg protein, and water

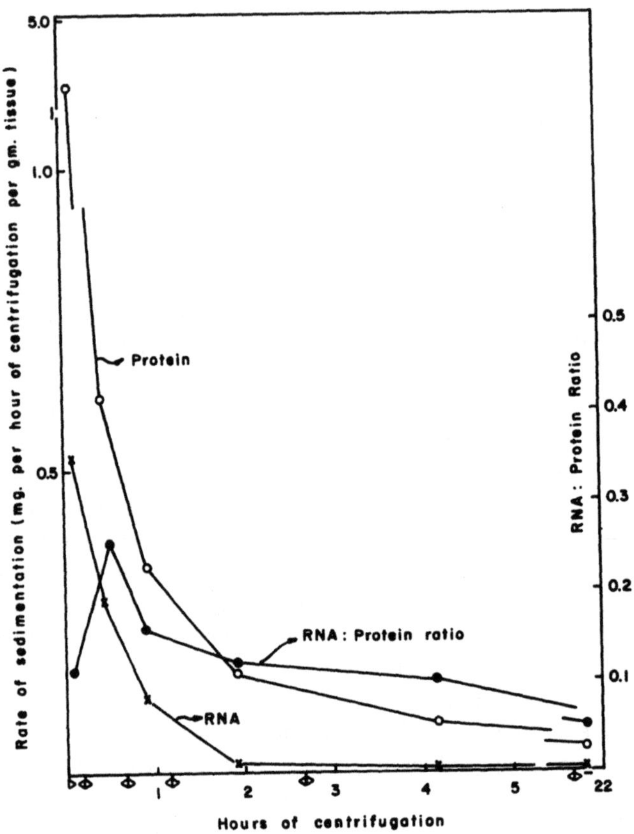

Fig. 1. Rates of sedimentation of RNA and protein from the post-mitochondrial supernatant of rat brain during ultracentrifugation. Six fractions were obtained, termed M_1, M_2, M_3, M_4, M_5, M_6, corresponding, respectively, to sedimentations during the time intervals 0–10 min, 40–70 min, 70 min–2.7 hr, 2.7–5.7 hr, and 5.7–21.7 hr.

to a final volume of 0.25 ml. The reaction mixture was incubated at 37° for 20 min. At the end of incubation the reaction was stopped by the addition of 5% TCA. After washing the precipitate with hot 5% TCA (90°C for 15 min), ethanol, and ether, the radioactivity in the total residue was determined using a liquid scintillation counter. It was found that: (1) when these fractions were tested for amino acid incorporation in the presence or in the absence of poly U the activities were different for each fraction; (2) when these fractions were treated with deoxycholate in the presence of

TABLE I

Phenylalanine Incorporation by Various Microsome and Ribosome Preparations

Fraction	RNA:protein ratio	Phenylalanine incorporation (DPM)		
		Minus poly U	Plus poly U	Per cent increase of incorporation
M_1	0.11	743	930	25
M_2	0.27	2410	3976	65
M_3	0.14	1335	2536	90
M_4	0.12	594	684	15
P_1	0.55	510	740	45
P_2	0.51	1211	2155	78
P_3	0.56	855	1881	120
P_4	0.55	183	192	5
R_1	0.53	158	365	131
R_2	0.55	140	361	158
R_3	0.51	172	382	122
R_4	0.48	172	285	65

RNase inhibitor the ribosomes so obtained had the same RNA:protein ratios, but they differed in their ability to incorporate phenylalanine depending on the presence or the absence of poly U; their activites were lower than those exhibited by the corresponding microsomes from which they were isolated; (3) when the microsomes were treated successively with deoxycholate and RNase, the monomeric ribosomes so produced were very similar, in that they had the same RNA:protein ratios, all were comparatively inactive in amino acid incorporation, and all regained the capacity to incorporate phenylalanine upon addition of poly U.

The above results indicate that the individual ribosomal monomers of rat brain do not differ greatly among themselves in regard to their composition or amino acid incorporation; however, they acquire functional differences by virtue of their interaction with *m*-RNA and the membrane. In order to examine this question further, sucrose-density-gradient centrifugation was used to separate rat brain microsomes into three different classes based on their attachment to *m*-RNA and to the membrane.

The post-mitochondrial supernatant prepared as in Fig. 1 was centrifuged at 105,000g for 90 min to obtain the total microsomes. Fractionation was performed in a linear sucrose density gradient (27 ml; 5–20% w/v) layered over a 5-ml cushion of 50% sucrose.[23] The microsomes as well as the sucrose gradient were in a solution containing 0.15 mM $MgCl_2$, 25 mM KCl, and 35 mM Tris-HCl pH 7.8 (solution B). From the profile of the

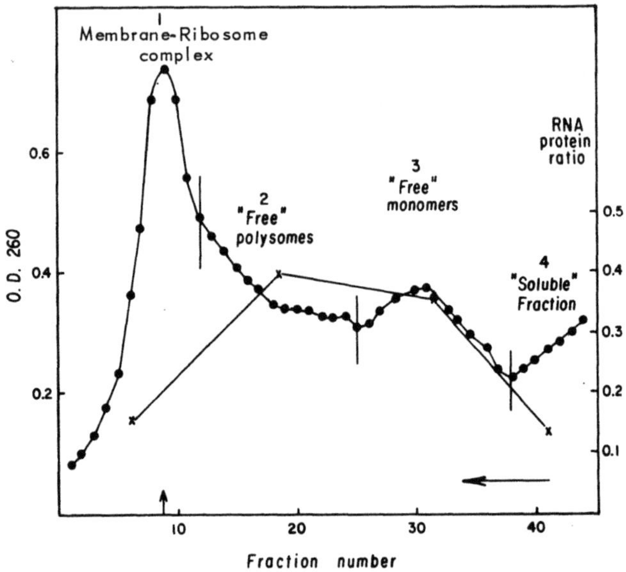

Fig. 2. Fractionation of rat-brain microsomes on sucrose density gradient. Centrifugation: 51,500g for 90 min. The small arrow indicates the position of the interphase separating the gradient and 50% sucrose. The large arrow indicates the direction of centrifugation. Fractions were collected by piercing the bottoms of the tubes. Circles, Sedimentation profile; crosses, RNA:protein ratio.

sucrose density gradient four different regions could be identified, numbered 1–4 in Fig. 2. The RNA: protein ratios of the contents of these regions indicate that the bulk of the material in regions 1 and 4 consisted of protein, whereas regions 2 and 3 contained ribonucleoprotein particles. The four groups of fractions were then treated with deoxycholate in the presence of RNase inhibitor and the ribosomes so obtained were tested for phenylalanine incorporation activity with and without added poly U. The results are shown in Table II. Ribosomes from regions 1 and 2 appeared to be predominantly polysomes, since they showed a high activity for phenylalanine incorporation in the absence of poly U and showed no significant enhancement in phenylalanine incorporation after addition of poly U. Ribosomes in region 3 appeared to be mostly in monomeric state, since incorporation in the absence of poly U was very little, and addition of poly U resulted in a threefold increase in incorporation. Region 4 contained few intact ribosomes, since phenylalanine incorporation was not significant either in the presence or absence of poly U.

TABLE II

Incorporation of Phenylalanine by Ribosomes Prepared from Sucrose Density Gradient Fractions of Rat-Brain Microsomes[a]

Fractions	Phenylalanine incorporation (DPM)		Per cent increase in incorporation
	Minus poly U	Plus poly U	
Region 1	362	405	12
Region 2	308	340	10
Region 3	92	275	200
Region 4	41	52	—

[a] Preparation of ribosomes and determination of phenylalanine incorporation were carried out as described in discussion of Table I.

On the basis of RNA: protein ratio, phenylalanine incorporation, and the position occupied in the sucrose density gradient, the constituents of the four regions were tentatively classed as follows: region 1, consisting possibly of a mixture of polysomes attached to membrane fragments of the endoplasmic reticulum and "free" polysomes (referred to hereafter as the membrane–ribosome complex), region 2, consisting mainly of "free" polysomes, and region 3, consisting mainly of single monomeric ribosomes. The light material in region 4 was probably composed of soluble proteins and ribonucleic acids and fragments of ribonucleoprotein particles degraded from ribosomes.

In order to determine what kind of structural or functional differences exist between the two types of polysomes in regions 1 and 2, the microsomes were tested for their response toward various degradative procedures such as incubation and treatments with RNase and EDTA. These procedures affect the microsomal structure in different ways and lead to the release of ribosomes and ribosomal components. Thus, incubation under conditions of amino acid incorporation is reported to release monomers from polysomes in synchrony with amino acid incorporation, after completing their movement relative to *m*-RNA.[24] Addition of RNase (pancreatic ribonuclease) at low concentrations has been used to degrade *m*-RNA selectively from polysomes giving rise to monomeric ribosomes.[25] EDTA (ethylenediaminetetraacetic acid) disrupts the ribosomes into subunits, presumably by chelating with magnesium ions, which stabilizes the attachment between the subunits.[26] Changes in the size distribution of ribosomes due to the action of these reagents can be followed by determining the optical density profile in the sucrose density gradient. The attachment or release of newly synthesized protein as a result of these treatments can be followed by measuring the distribution of radioactivity in different regions of the gradient after subjecting amino acid-labelled microsomes to sucrose density gradient centrifugation.

Incubation of brain microsomes under conditions of amino acid incorporation resulted in a slight decrease of OD_{260} from the polysomal region 1 and a simultaneous increase of OD_{260} in the "free" monomers. Microsomes prepared as in Fig. 2 were added to a reaction medium for incorporation of ^{14} C-phenylalanine (Table I). One-fourth of this mixture was kept at 0°C, while the rest was incubated at 37°C for 20 min. The labeled microsomes were recovered by centrifugation and were divided into three portions. One portion was immediately layered over the sucrose density gradient for fractionation. The second portion was suspended in solution B (Fig. 2) to which RNase was added at a concentration of 0.5 μg/ml and the incubation was continued at room temperature for an additional 3 min. The third portion was taken up in a solution containing 25 mM KCl, 10 mM Tris-HCl (pH 7.8), and 10 mM sodium EDTA and maintained at 4°C for 10 min. Sucrose density gradient fractionation of these microsomes were

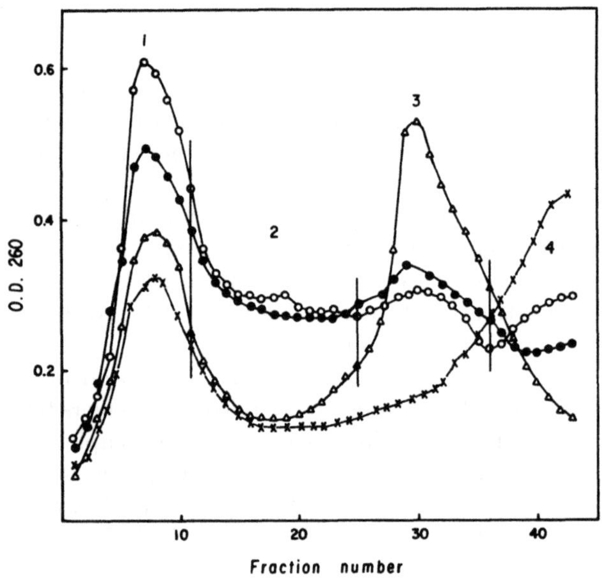

Fig. 3. Release of ultraviolet-absorbing material from rat-brain microsomes on incubation under conditions of amino acid incorporation and on treatment with RNase or EDTA. Open circles: control, unincubated; darkened circles: incubated only; triangles: incubated and then treated with RNase; crosses: incubated and then treated with EDTA.

performed as in Fig. 2. The results (Fig. 3) can be interpreted in terms of the release of monomers from polysomes in synchrony with amino acid incorporation.[24] A random fragmentation of *m*-RNA by the action of RNase in the medium can be discounted, since the incubation medium contained RNase inhibitor. However, it is not possible to determine from these data whether the monomers originated from the membrane-bound polysomes or from the "free" polysomes or both. The observation that the concentration of free polysomes in region 2 was unaltered after incubation does not necessarily mean that these polysomes were not degraded, since it is probable that polysomes which were degraded were compensated for by the release of intact polysomes from region 1. These intact polysomes may have been originally bound to the membrane, or they may just represent the heavy polysomal contaminants which sedimented along with the membrane fraction.

When preincubated brain microsomes were treated with a low concentration of RNase, the sedimentation profile shows that approximately 35–40% of ultraviolet absorbing material was released from the membrane-ribosome complex (Fig. 3). Almost all of the "free" polysomes were degraded. Only the peak corresponding to the "free" monomers increased, indicating that the action of RNase on the membrane–ribosome complex as well as on "free" polysomes consisted in degrading the *m*-RNA and releasing the ribosomes in the monomeric form. RNase treatment also had the effect of releasing newly synthesized protein from the membrane–ribosome complex and the "free" polysomes, as measured by a fall in radioactivity in these two regions (Fig. 4). The protein which was thus released was still attached to the ribosomes, since it accumulated in the region of monomers. The membrane–ribosome complex was less susceptible to the action of RNase than the "free" polysomes as measured by the migration profiles of optical density and radioactivity. A large part of OD_{260} and almost all of the original radioactivity remained in the membrane-ribosome complex even after RNase treatment, indicating that a majority of ribosomes containing the newly formed polypeptide remained attached to the membrane. This can be interpreted in two ways: (1) RNase destroyed the *m*-RNA just as effectively from the polysomes bound to the membrane as from the "free" polysomes, and the monomeric ribosomes so produced were still held to the membrane surface in spite of the loss of *m*-RNA; (2) all the *m*-RNA associated with the membrane-bound ribosomes was not degraded by RNase, being protected against this enzyme because of its association with the membrane–ribosome complex. The OD_{260} material remaining with the membrane contained undegraded bound polysomes. The second interpretation is favored in view of the observation to be discussed subsequently that *in vitro* polysomes interact with the membrane much

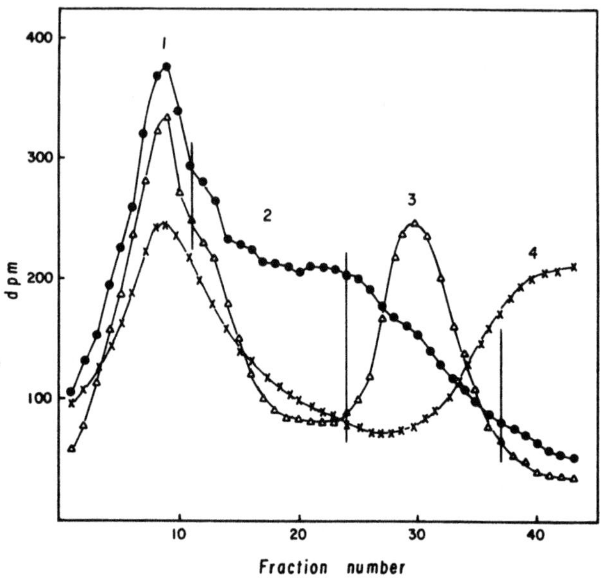

Fig. 4. Release of radioactivity from rat-brain microsomes labeled with ^{14}C-phenylalanine, on treatment with RNase or EDTA. The procedures were as described for Fig. 3. For determination of radioactivity 1 mg of bovine serum albumin was added to each fraction as a carrier and the total protein was precipitated with TCA at a final concentration of 5% of the precipitant. Radioactivity in the total precipitate was then assayed as for Table I. Darkened circles: incubated only; triangles: incubated and then treated with RNase; crosses: incubated and then treated with EDTA.

more strongly than the monomers. The resistance of membrane-bound ribosomes to RNase under conditions where the free ribosomes become degraded has also been reported for rat-liver ribosomes by Blobel and Potter.[27] These authors, however, suggest that the binding of the ribosomes to the membrane may mask the site of attack of RNase in the ribosome or may cause a conformational change in the ribosome structure making it resistant to RNase attack.

When rat-brain microsomes labeled with ^{14}C-phenylalanine were treated with EDTA the following observations were made from the sucrose density gradient (Fig. 3): (1) there was a considerable reduction in the OD_{260} of the membrane–ribosome complex, suggesting a release of ribosomes; (2) there was almost a complete disappearance of "free" polysomes and monomers, and (3) there was an accumulation of OD_{260} material in the soluble fraction. Since the only region of the gradient gaining in optical density was the "soluble" fraction, it is assumed that ribosomes both bound and free were dissociated into subunits. These subunits would remain in the "soluble" fraction under conditions of centrifugation employed here. However, a considerable amount of OD_{260} material remained in the membrane–ribosome complex. Studies using pig-liver microsomes have shown that at low concentrations of EDTA such as that used in this study, only the smaller subunits of the ribosomes are detached, and the larger subunits remain attached to the microsomal membrane.[28] The loss of OD_{260} from the membrane–ribosome complex after EDTA treatment may be assumed to be at least partly due to the dissociation of the smaller subnits from the bound polysomes. A part of this loss must also be due to the release of both subunits of "free" polysomes trapped in the membrane–ribosome complex. This is indicated by the distribution of radioactivity in the sucrose density gradient after EDTA treatment (Fig. 4). There was a loss of radioactivity from both the membrane–ribosome complex and from the "free" polysomes, with a simultaneous increase in the "soluble" fraction. Since the newly synthesized polypeptide chain is presumed to be attached to the larger subunit,[17] this suggests that the loss of radioactivity from the membrane–ribosome complex is to be attributed to the dissociation and release of contaminating "free" polysomes. However, a considerable proportion of the labeled polypeptide was still held by the membrane–ribosome complex, as evidenced by the presence of large radioactivity in this fraction. The fact that the larger subunits were also retained at the same time by the membrane would indicate that these subunits, the newly synthesized polypeptide, and the membrane may all be mutually involved in the observed resistance of the membrane–ribosome complex to EDTA. Similar conclusions have been reached by Sabatini *et al.*[28] on the basis of their experiments with pig-liver microsomes.

In order to check the possibility of whether ribosomes exhibited differences in susceptibility to EDTA at the site of attachment of the subunits, in addition to differences in susceptibility at the junction of the membrane and the large subunit, the following experiment was carried out. Membrane-bound ribosomes were released from the membrane–ribosome complex by means of deoxycholate treatment. These ribosomes as well as the "free" polysomes were treated with RNase to obtain the monomers and then with EDTA to dissociate the monomers to the subunits. The products of EDTA treatment were separated using sucrose density gradient centrifugation. The results are presented in Fig. 5. It is seen that monomeric ribosomes from either of the two sources were completely dissociated into the subunits, and the relative proportions of the two subunits were also the same.

In the above studies optical density measurements at 260 mμ have

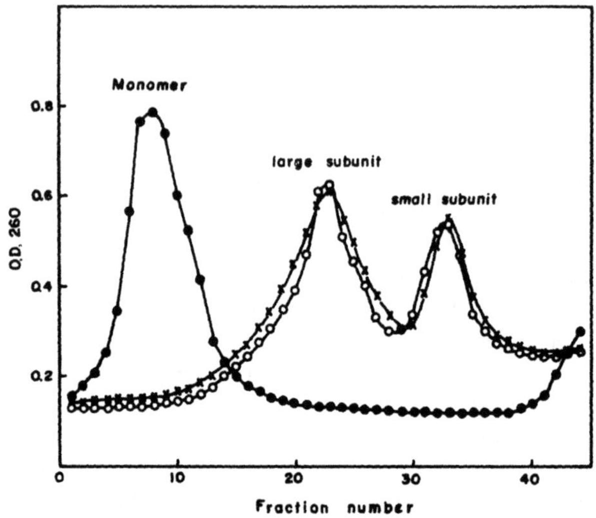

Fig. 5. Formation of ribosomal subunits from monomeric ribosomes. The ribosomes and their subunits were separated from each other by centrifugation in a sucrose density gradient (5–20% w/v) in 50 m*M* KCl plus 1 m*M* Tris-HCl, pH 7.6;[28] 2 ml of EDTA-treated ribosomes were layered over 30 ml of the sucrose gradient. Centrifugation: 63,500g for 5 hr. Darkened circles: monomers from "free" polysomes; open circles: ribosomal subunits from the polysomes of the membrane–ribosome complex; crosses: ribosomal subunits from "free" polysomes.

been used to follow the migration of ultraviolet-absorbing substances from one region of the sucrose gradient to another. Thus, by comparing the extent of loss in OD_{260} from the membrane–ribosome complex with the gain in OD_{260} in other regions of the density gradient, it is possible roughly to identify whether the material released by a particular treatment consisted of polyribosomes, monomeric ribosomes, or soluble proteins and nucleic acids. But the identification of ultraviolet-absorbing material retained in the complex is not easy, since the optical density may be due to either nonribosomal protein or nonribosomal RNA reported to be present in the membrane[29] or to the proportion of membrane-bound ribosomes which are resistant to the action of the reagents employed, or to a combination of these factors. The following experiment was conducted in order to determine directly the content of polyribosomes in the membrane–ribosome complex after treatment with RNase and EDTA. The fraction of microsomes corresponding to the membrane–ribosome complex was isolated from fresh microsomes and from microsomes subjected to the above treatments. The ability of these fractions, in amounts of each equivalent to 0.35 mg of RNA, to incorporate ^{14}C-phenylalanine was examined. The results are presented in Table III. A considerable proportion of amino acid-incorporating activity, amounting to approximately 60% of the original, remained bound to the membrane–ribosome complex after treatment with low concentrations of RNase or EDTA. Assuming that the amino acid incorporation was due to active polysomes, it can be inferred that the membrane–ribsome complex contained a large number of polysomes which were not released by the action of the above reagents and that at least a part of OD_{260} in this fraction was due to these bound polysomes.

The concentration of polysomes in the membrane fraction as revealed by the above studies suggested that *m*-RNA may, in some manner, participate in the attachment of ribosomes to the membrane. In order to examine this possibility further, the interaction between the microsomal membrane and the ribosomes *in vitro* was measured in the presence and in the absence of artificial polynucleotide messengers using a procedure described by Süss *et al.*[30] Membranes free of ribosomes were obtained by washing the microsomes with a buffer containing 10 m*M* sodium pyrophosphate. For the preparation of radioactive ribosomes, polysomes were incubated with ^{14}C–phenylalanine under conditions of amino acid incorporation and the labeled polysomes were treated with a low concentration of RNase to liberate the monomers. Membranes and ribosomes each containing 1 mg of protein were suspended in 4 ml of a solution containing 0.2 *M* sucrose, 10 m*M* Tris-HCl (pH 7.4), and 5 m*M* $MgCl_2$. The mixture was kept at 0°C for 30 min and then it was layered over 1 ml of 2.0 *M* sucrose containing 10 m*M* Tris and 5 m*M* $MgCl_2$. The tubes were centrifuged at 125,000g for 16 hr in a swinging bucket rotor. The supernant containing the membranes was diluted with an equal volume of water and spun for 60 min at 105,000g. Radioactivity in the resulting pellet was determined using a liquid scintillation counter.

TABLE III

Incorporation of Phenylalanine by the Membrane–Ribosome Complex of Rat Brain after Treatment with RNase or EDTA

Treatment	Phenylalanine incorporation (DPM)
Nono	012
EDTA	360
RNase	350

Table IV shows that the ability of the ribosomes to bind to the microsomal membrane *in vitro* was doubled by the addition of poly U. Poly A also enhanced the binding, but to a smaller degree. This suggests that attachment to *m*-RNA may be a favorable prerequisite for binding to the membrane.

Assuming the validity of the published data, that the ribosome is attached to the membrane by means of its larger subunit[28] and that *m*-RNA is in association with the smaller subunit[19] which is further away from the membrane, the role of *m*-RNA in the attachment of membrane and ribosome may consist in stabilizing this attachment by an orderly arrangement of the ribosomes and by orienting the ribosomes in such a manner that the correct surface is presented for interaction with the membrane.

Figure 6 shows the results of an experiment in which the microsomes from newborn and adult rat brains were fractionated on a sucrose density gradient and the distribution of OD_{260} material in the gradient compared between the two tissues. It is seen that the newborn rat brain contains a larger proportion of the membrane–ribosome complex and a smaller proportion of monomeric ribosomes as compared to the adult. Similar results were obtained even when the fractions were extracted with hot 5% perchloric acid and the optical density of the perchloric acid extract determined at 260 mμ. Even though the contribution of protein to the optical density profile was eliminated in this manner, the values obtained for the membrane–ribosome complex in this procedure constitutes a combination of undetermined proportions of membrane-bound polysomal RNA, free polysomal RNA sedimented along with the membrane as a contaminant, and membrane RNA. Blobel and Potter[27] have described a method for quantitative separation of membrane-bound and free ribosomes from rat liver. This procedure was used with the following modifications in the preparation of the post-nuclear supernatant, to isolate quantitatively the two kinds of ribosomes from newborn and adult rat brains. Brains were

TABLE IV

Interaction *in Vitro* between Ribosomes and the Membrane Fraction

Additions	Radioactivity bound to the membrane (DPM)
None	115
poly U (50 μg)	238
poly A (50 μg)	170

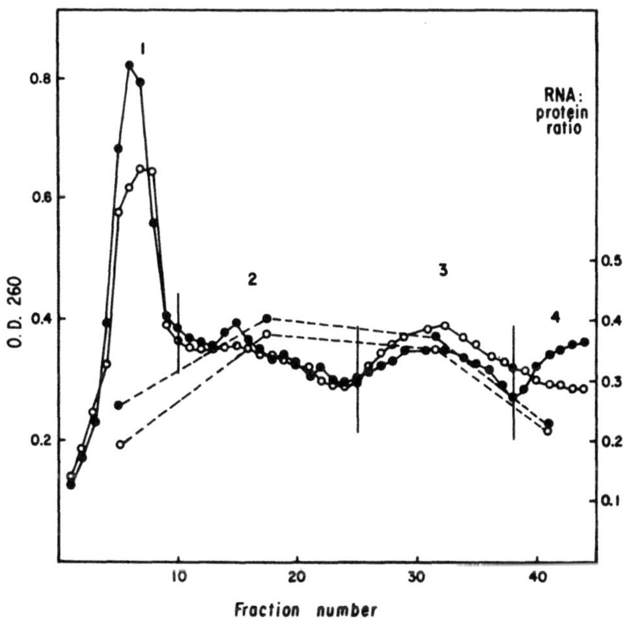

Fig. 6. Sucrose density gradient profiles of newborn and adult brain microsomes. Procedures for the preparation and fractionation of microsomes were as for Fig. 2. The solid and broken lines represent, respectively, the optical density at 260 mμ and the RNA:protein ratios of the fractions. Darkened circles: newborn; open circles: adult (three-month-old rats).

homogenized in 5 vol of 0.25 *M* sucrose solution containing 25 m*M* Tris (pH 7.8), 35 m*M* KCl, and 4 m*M* $MgCl_2$ (0.25 *M* sucrose–TKM). The homogenate was centrifuged at 5000g for 10 min and the supernatant (supernatant 1) containing most of the cytoplasm was carefully removed. The sediment containing the nuclei was resuspended by homogenization in 2.3 *M* sucrose–TKM of the same volume as the original homogenate and was centrifuged at 78,500g for 1 hr in Spinco No. 30 rotor. The supernatant from this centrifugation was added to supernatant 1 from the first centrifugation and the two were mixed by homogenization in a glass Teflon homogenizer. The resulting extract corresponds to the post-nuclear supernatant of Blobel and Potter. The amount of RNA in the "bound" ribosomes was calculated as the difference between the amounts of RNA in the "total" cytoplasmic ribosomes and "free" ribosomes. The results are presented in Table V. The free and membrane-bound ribosomes from either of the two sources had the same composition in regard to RNA and protein. However,

TABLE V

Proportion of Membrane-Bound Ribosome to Free Ribosomes in the New-Born and Adult Rat Brain

	RNA:protein ratio		
	Free ribosomes	Bound ribosomes	Ratio of bound:free ribosomes (based on RNA)
New-born	0.45	0.47	0.42
Adult	0.46	0.50	0.30

the ratio of bound to free ribosomes was about 40% higher in the newborn as compared to the adult.

The results of the experiments so far described can be summarized as follows. Individual ribosomal monomers of rat brain do not differ greatly among themselves in regard to their composition or amino acid incorporation; however, they acquire structural and probably functional differences by virtue of their interaction with *m*-RNA and also the microsomal membrane. Some of the ribosomes in the rat brain occur in attachment with the membrane, and these ribosomes are largely in the polysomal form. Polysomal structure may even be a favorable factor in the binding of the ribosomes to the membrane. The ribosomes associated with the membrane are more resistant than the "free" polysomes to the action of degrading agents which affect their structure. This resistance may be a result of a close interrelationship between several structural elements with which the ribosome is associated, the membrane, the *m*-RNA, and the newly synthesized polypeptide. The newborn rat brain differs from the adult tissue in having a higher proportion of membrane-bound ribosomes and a lower proportion of monomeric ribosomes.

REFERENCES

1. G. E. Palade, *J. Biophys. Biochem. Cytol.* **1,** 59–68 (1955).
2. S. L. Palay and G. E. Palade, *J. Biophys. Biochem. Cytol.* **1,** 69–88 (1955).
3. H. Hydèn, *in*: "The Cell," Vol. 4, Part 1, pp. 215–323 (J. Brachet and A. E. Mirsky, eds.), Academic Press, New York (1961).
4. K. R. Potter, *in*: "The Cell," Vol. 2, pp. 621–675 (J. Brachet and A. E. Mirsky, eds.), Academic Press, New York (1961).

5. J. P. Dadoun, quoted by Y. Moulé, *in*: "Cellular Membranes in Development," p. 124, (M. Locke, ed.), Academic Press, New York (1964).
6. F. S. Sjostrand, *in*: "The Exocrine Pancreas, Normal and Abnormal Functions," p. 1 (A. V. S. de Reuck and M. P. Cameron, eds.), Little, Brown Co., Boston (1962).
7. E. C. Henshaw, T. B. Bojarski, and H. H. Hiatt, *J. Mol. Biol.* **7,** 122–129 (1963).
8. S. H. Wilson and M. B. Hoagland, *Biochem. J.* **103,** 556–566 (1967).
9. T. E. Webb, G. Blobel, V. R. Potter, and H. P. Morris, *Cancer Res.* **25,** 1219–1224 (1965).
10. P. Cammarano, G. Giudice, and B. Lukes, *Biochem. Biophys. Res. Commun.* **19,** 487–493 (1965).
11. M. S. C. Birbeck and E. H. Mercer, *Nature* **189,** 558–560 (1961).
12. M. Kern, E. Helmreich, and H. N. Eisen, *Proc. Natl. Acad. Sci. U.S.* **45,** 862–867 (1959).
13. E. L. Kuff, M. Potter, K. R. McIntyre, and N. E. Roberts, *Fed. Proc.* **21,** 154 (1962).
14. R. Bellairs, *J. Embryol. Exp. Morphol.* **7,** 94–115 (1959).
15. J. Eschner and P. Glees, *Experientia* **19,** 301–303 (1963).
16. I. T. Oliver, W. E. C. Blumer, and L. J. Witham, *Comp. Biochem. Physiol.* **10,** 33–38 (1963).
17. Y. Tashiro and P. Siekevitz, *J. Mol. Biol.* **11,** 166–173 (1965).
18. M. Cannon, P. Krug, and W. Gilbert, *J. Mol. Biol.* **7,** 360–378 (1963).
19. M. Takanami and T. Okamoto, *J. Mol. Biol.* **7,** 323–333 (1963).
20. M. R. V. Murthy and D. A. Rappoport, *Biochem. Biophys. Acta* **95,** 121–131 (1965).
21. M. R. V. Murthy and D. A. Rappoport, *Biochem. Biophys. Acta* **119,** 599–613 (1966).
22. K. Shortman, *Biochem. Biophys. Acta* **51,** 37–49 (1961).
23. P. N. Campbell, G. Serck-Hanssen, and E. Lowe, *Biochem. J.* **97,** 422–431 (1965).
24. H. Noll, T. Staehelin, and F. O. Wettstein, *Nature* **198,** 632–638 (1963).
25. M. Jacob, J. Samec, J. Stevenin, J. P. Garel, and P. Mandel, *J. Neurochem.* **14,** 169–178 (1967).
26. Y. Tashiro and P. Siekevitz, *J. Mol. Biol.* **11,** 149–165 (1965).
27. G. Blobel and V. R. Potter, *J. Mol. Biol.* **26,** 279–292 (1967).
28. D. D. Sabatini, Y. Tashiro, and G. E. Palade, *J. Mol. Biol.* **19,** 503–524 (1966).
29. C. Bergeron-Bouvet and Y. Moulé, *Biochem. Biophys. Acta* **123,** 617–619 (1966).
30. R. Süss, G. Blobel, and H. C. Pitot, *Biochem. Biophys. Res. Commun.* **23,** 299–304 (1966).

DISCUSSION

ZOMZELY: We have carried out similar experiments with nowborn, 21-day-old, and 42-day-old rats. We prepared polysomes from the cortices of these animals in medium containing high Mg^{2+} (10 m*M*) and in the presence and absence of detergent. We too found a higher proportion of heavier polysomes in the newborn cortices, which decreased somewhat by 21 days and was even less in the adults (42 days). However, the amino acid-incorporating activities of the 21-day and 42-day rat polysomes were the same. In fact, the activity of the newborn polysomes, with the *highest* proportion of heavy components, was also the same as that of the polysomes from the adult rat, with the *lowest* proportion of heavy components. Moreover, polysomes from the various age groups showed similar amino acid-incorporating activities regardless of the age of the rats supplying the pH 5 fraction. I wonder whether the discrepancy between our results may be related in part to the fact that we were using cortex rather than whole brain. Thus, polysomes from white matter of adult rats have a lower proportion of heavy polysomes than those from gray matter. This factor is not involved in preparations from newborn rats, where little white matter is present.

MURTHY: If you isolate the polysomes from the post-mitochondrial supernatant of the newborn or the adult brain by deoxycholate treatment, the amino acid-incorporation activity would be expected to be the same in both preparations unless the polysomes in one of the preparations are grossly contaminated by monomers. As shown in Table I, when we isolate the total ribosomes from different fractions of brain microsomes, the amino acid incorporation is different in these ribosomes. Incorporation in the presence and absence of poly U, however, shows that the differences in activity are attributable to the relative proportions of the polysomes and monomers in the preparations. What is being suggested in this communication is that even if the specific activity *in vitro* of the polysomes isolated from the newborn and adult brain may be the same, it is the binding of the polysomes to the membrane that in part makes the difference between the *in vivo* activity of the newborn and the adult. As regards possible differences in the activities of pH 5 enzymes, our results[20] reported two years ago did not indicate such differences. But at that time we were not dealing with ribosomes themselves, but with microsomes. In the microsomes it is difficult to demonstrate even an absolute requirement for pH 5 enzyme unless the microsomes are thoroughly washed beforehand.

ZOMZELY: Our studies have dealt only with purified polysomes, and here, in spite of the appreciable difference in polysomal components in newborn and adult rat brain, incorporating activity is similar *in vitro*. By contrast, you have reported greater activity with newborn polysomes than with adult.

ROBERTS: I think one must make a very clear distinction among the types of cell populations which are present in different preparations during development. Dr. Murthy's polysomal preparations have always been made with whole rat brain, while Dr. Zomzely is speaking primarily of studies with cerebral cortical tissue. In the newborn rat the cerebral cortex consists almost entirely of neurons or undifferentiated cells. As development proceeds the proportion of glia tends to exceed that of neurons. There may be no fundamental disagreement, then, with any of the results which have been presented, because Dr. Zomzely also finds a depression in the proportions of heavy polyribosomes in cerebral cortex, although not to the extent that Dr. Murthy indicates in the whole brain. The activating-enzyme picture may also be related to the development of the white matter and the preponderance of white matter in the adult. Suzuki, Korey, and Terry* reported that activating enzymes as well as microsomes of gray matter had the same protein-synthesizing activity in newborn and adult animals. However, the activity of the activating enzymes was altered very greatly in the white matter during development. It is obviously very difficult to compare protein synthesis in brains of animals of different ages without taking into account the different population of cells.

MANDEL: I think data dealing with so-called membrane-bound and free ribosomes must be evaluated very carefully. Thus, attachment of ribosomes *in vitro* to membranes may represent nonspecific binding. We found that we could separate ribosomes from their apparent attachment to membranes simply by centrifugation. I think it is very difficult to obtain from a homogenate that which preexisted in the cell.

* K. Suzuki, S. R. Korey, and R. D. Terry, Studies on protein synthesis in brain microsomal system, *J. Neurochem.* **11**, 403–412 (1964).

MURTHY: I agree. The distinction between bound and free ribosomes in my experiments was made not only using ordinary differential centrifugation, but also using the discontinuous gradient described by Blobel and Potter. They found that after centrifugation of the gradient for about 8 hr no more free ribosomes sedimented to the bottom of the tube. The material remaining suspended was presumably held in the membrane by stronger bonds than those ribosomes which sedimented.

MANDEL: Sedimentation depends on the density of the sample as well as on the molecular weight. If you have lipid membranes bound to the ribosomes, it would be more difficult to sediment them.

FORD: I'd like to emphasize the remark which Dr. Roberts made. Folch-Pi has noted that when you compare brains of animals which are two or three days old with those of the adult you are looking at two different animals and two different tissues. We recently measured the volume of the gray matter occupied by nerve cells in the ventral horn of young rats. We found that at seven days of age the nerve cells (which at this age are very small) occupied approximately 20% of the total volume. Between 25 and 35 days of age, when volume changes have essentially ceased in this area, they occupied only 8% of the total volume. So there is a very large difference in the surrounding matrix of neuropil. In ribosomal preparations from gray matter or whole brain the mixture changes drastically with different stages in the life of the animal.

MURTHY: This is a consideration that should be taken into account in interpreting studies made with any tissue with a nonhomogeneous population of cells.

SELLINGER: I wonder whether you would care to speculate about the proposition that messenger RNA may help in binding ribosomes to membranes, in view of the recently published observation of Blobel and Potter† that messenger RNA is not involved in binding the ribosome to endoplasmic reticulum.

MURTHY: I imagine you are referring to their recent paper in the *Journal of Molecular Biology*. In our experiments, poly U was added to a mixture of membranes and monosomes and produced increased binding as determined by centrifugation. Assuming that ribosomes are bound readily to polynucleotides to form polysomes, then one can say that polysomes are in some manner bound more readily to the membrane than are monosomes. However, this might just be a physical artifact, because the polysomes are much larger and may be trapped more easily in the membrane fraction than are the monosomes. Therefore, unless we investigate very carefully the mechanisms of binding from various points of view, we cannot define what kind of binding is involved in any given case.

† G. Blobel and V. R. Potter, *J. Mol. Biol.* **26**, 293 (1967).

Chapter 6

Regulation of Transcription in Nervous Cells

Paul Mandel and Monique Jacob

Centre de Neurochimie du C.N.R.S.
Faculté de Médecine
Strasbourg, France

INTRODUCTION

The evolution from bacteria to the association of specific cells in tissues and in organisms leads to a specialization of cell types. Supplementary control mechanisms appear in order to regulate the relation between cells and organs in a symbiotic way.

Since the number of functions has considerably increased, and since every function is served by several biochemical reactions, the number of genes has to increase in the same proportion, so that the final quantity of DNA per cell and the molecular weight of DNA in mammalian cells are considerably higher than in bacteria. The higher cellular activity at the genetic level needs a central control, which mainly lies in the nucleus; and the existence of the nucleus creates special problems in RNA synthesis:

1. How are the precursors for RNA synthesis supplied, since there is a requirement for the four nucleoside triphosphates?
2. How is the nuclear RNA biosynthesis organized for the specific regulations in differentiated mammalian cells?
3. What is the mechanism of RNA transport from the nucleus to the cytoplasm, where protein synthesis occurs?

We will analyze these problems and report some related observations.

THE SUPPLY OF PRECURSORS FOR RNA SYNTHESIS

The supply of precursors for RNA synthesis (Fig. 1) in bacteria is not subject to limitations like those in eukariotic cells, since the bacterial cells

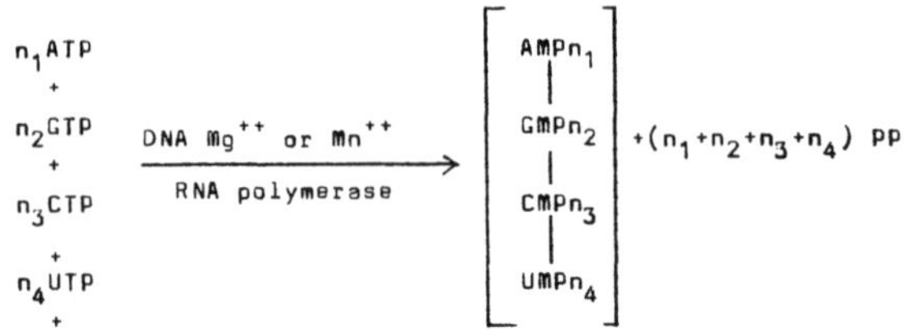

Fig. 1. Chemical reactions of RNA synthesis.

have no internal barriers for free cytoplasmic nucleotides. In mammalian cells the nuclear pool of free nucleotides is very low, and the question arises as to how the precursors are supplied for the rapid RNA synthesis that follows the signals arriving at the nucleus. There are three possibilities: (1) the nucleus starts to synthesize the four nucleoside triphosphates (NTP) *de novo* according to its requirements as soon as the signal arrives; (2) the free nucleotides are transferred from the cytoplasm to the nucleus; (3) there are some storage molecules in the nucleus which are able to liberate nucleotides for RNA synthesis. We will discuss these three possibilities.

Intranuclear Biosynthesis of Nucleoside Triphosphates when the Signal for RNA Biosynthesis Arrives

This mechanism seems unlikely. The biosynthesis of the four NTP's needs at least 13 reactions for every purine nucleotide and eight or nine for every pyrimidine nucleotide. Only little is known about NTP synthesis in nuclei. The enzymatic equipment, essentially glycolytic, does not present optimal conditions for the production of high quantities of ATP, nor for the transphosphorylation to form GTP, CTP, and UTP. Moreover, a considerable disproportion exists between the nucleotide content of the nucleus and the speed of their utilization in RNA synthesis. Thus, an intranuclear biosynthesis does not seem compatible with the velocity of RNA synthesis.

Transfer of Free Nucleotides from the Cytoplasm to the Nucleus

Until recently the general opinion was that only nucleosides pass across the nuclear membrane.[1–3] Simler *et al.*[4] were able to demonstrate that at least AMP could be transferred to the nucleus (Table I). Incubating brain and liver nuclei in the presence of double-labeled α-^{32}P-ATP and ^{14}C-adenine or ^{3}H-adenine, we found that the relationship of labeling in

TABLE I

Synthesis of NAD by Nuclei Incubated *in Vitro* in the Presence of Nicotinamide Mononucleotide and Double-Labeled ATP Adenine[^{14}C, α-^{32}P]

Exp.	Addition	Ratio of radioactivity $^{14}C/^{32}P$	
		ATP	NAD
Rat brain nuclei			
1	None	1.19	1.15
2	None	1.19	1.20
3	PO_4^{3-}	1.19	1.21
4	PO_4^{3-}	1.19	1.20
5	PO_4^{3-}	1.00	1.04
6	PO_4^{3-}	1.00	1.17
7	PO_4^{3-}	1.00	1.025
8	Adenosine	1.61	1.51
Liver nuclei			
1	PO_4^{3-}	1.20	1.10
2	PO_4^{3-}	1.19	1.38
3	PO_4^{3-}	1.19	1.23
4	PO_4^{3-}	1.20	1.26
5	PO_4^{3-}	1.50	1.49
6	Adenosine	1.19	1.17

NAD, which was synthesized only in the nucleus, was similar to that in the extranuclear incubation medium. The addition of nonradioactive phosphate or adenosine did not change the ratio of labeling. This implies a transfer of the nucleotide as a whole, rather than a dephosphorylation of the nucleotide followed by a transfer of the nucleoside and a rephosphorylation.

Since extranuclear free nucleotides are available for intranuclear RNA synthesis, it would be interesting to know how the cytoplasmic pool of free nucleotides is regulated. The problem is not an easy one, the free nucleotides being involved in many metabolic pathways (Fig. 2). A control mechanism between these different pathways is necessary in order to provide the nucleotides at any moment. The problem arises of the regulation of the enzymatic activities involved in the different processes for the production of nucleotides, of substrates for RNA synthesis, and of coenzymes. It seems difficult to accept the idea that feedback inhibition acts in a similar way in animal cells and in bacteria, because of the compartmentation in the former and to the existence of different loci for synthesis, storage, and utilization.

The analysis of the distribution of free nucleotides in different tissues regularly reveals a very low CTP content in resting cells.[5–10] However,

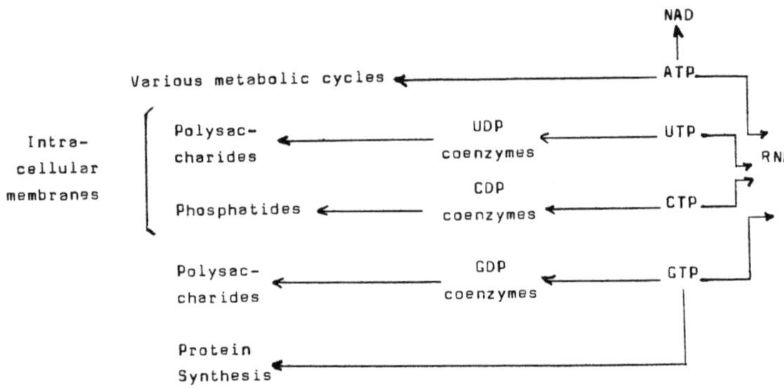

Fig. 2. Summary of the intervention of the nucleoside di- and triphosphates in different metabolic cycles.

in growing or tumor cells which undergo an active RNA synthesis the amount of nucleoside triphosphates, and mainly of CTP, is much higher. In the brain of a young rat CTP quantities are relatively important whereas in an adult brain they are very low (Table II).[5,7–9] Klethi and Mandel[10] observed an important difference in the eye lens, between the CTP amount in the epithelial zone containing proliferating cells and in the whole lens. Hence it appears that an adaptation of the NTP amount to the increasing level of RNA synthesis exists and that the low content of CTP can be a limiting factor for this synthesis. As a matter of fact, the biosynthesis of CTP is the lowest among the four NTP, as it appears from a study on the incorporation of ^{32}P in the α-phosphate of the NTP in a resting rat liver (Table III).[11] Thus, CTP may well be the limiting factor, and

TABLE II

Nucleoside Triphosphate Distribution in Different Rat Tissues (μM/100 g wet wt.)

Organ	Age	ATP	GTP	UTP	CTP
Liver	Adult	277	49.83	42.57	26.07
Lens	Adult	132	25.4	18.6	Traces
Brain	one day old	188	36.8	53.5	13.0
Brain	eight days old	165	30.3	30.4	6.4
Brain	20 days old	210	36.3	28.8	2.4
Brain	30 days old	216	34.4	29.1	1.3
Brain	Adult	240	27.9	20.6	Traces

TABLE III

Specific Activities of α-Phosphates of Free Liver Nucleotides after Intraperitoneal Injection in Rat of 300 $\mu Ci^{32}P$/100 g wt. (cpm/μg α-P)

	2 hr	10 hr	16 hr	24 hr	48 hr	96 hr
CMP	3,041	1,729	3,562	525	660	435
AMP	29,310	11,075	8,407	7,050	4,925	2,530
GMP	7,257	4,986	6,303	4,600	3,640	3,225
UMP	8,850	6,393	7,007	6,400	4,860	4,620

the rate-limiting enzyme could be the UTP aminases, since the level of UTP is relatively high.[5]

The Possibility of an NTP Supply in Nuclei from Storage Molecules

It was demonstrated by several authors that the DNA-dependent RNA polymerase contains polyadenylate-synthesizing activity.[12–19] Edmonds and Abrams[20,21] have shown the presence of a polyadenylate-synthesizing enzyme in calf thymus nuclei, and Burdon[22] has found a polycytidylate-synthesizing enzyme in ascitic tumor cells. The presence of polyadenylate-synthetase activity was also demonstrated in mammal cell

TABLE IV

Poly A Synthesis from ATP by a Soluble Fraction of Brain Nuclei

	$\mu\mu$moles AMP incorporated per mg protein
Complete incubation medium (50 μg polyadenylate)	333.0
−poly A	23.9
−poly A + DNA (60 μg)	26.1
−poly A + ribosomal RNA (100 μg)	193.2
+actinomycin (10 μg)	315.0
+pyrophosphate (10 mM)	0.0
+ethylenediaminetetraacetate (10^{-3} M)	0.0
+inorganic phosphate (10 mM)	299.0
+ammonium sulfate (10 p.100 saturation)	14.7
+preincubation with cold ATP and enzyme extract for 2 min followed by addition of [α-^{32}P]ATP and of $(NH_4)_2SO_4$	153.0
+CTP, GTP, UTP (0.25 μM each)	82.7

TABLE V

Distribution of Radioactivity (per cent) in the Four 2′3′-Nucleoside Monophosphates after Alkaline Hydrolysis of the Soluble Polyadenylate Synthetase Product

Exp.	GMP	AMP	CMP	UMP
1	1.0	94.0	4.5	0.5
2	1.2	96.6	1.6	0.6

cytoplasm by Klemperer[23] and Giron and Huppert[24]; moreover, polyadenosine diphosphate ribose synthetase in liver and kidney nuclei was first described in our laboratory[25] and confirmed by others.[26,27]

We found polyadenylate, polycytidylate, and polyuridylate synthetases in brain nuclei.[28–31]

Evidence for polyadenylate synthesis by a nuclear soluble extract from rat brain is presented in Table IV. We note that this synthesis does not involve the presence of DNA, is strongly stimulated by the presence of a polyadenylate primer, is not influenced by actinomycin, and is, in part, inhibited by the other NTP's. The polyadenylate structure of the polymer was confirmed by alkali and phosphodiesterase hydrolysis; the former demonstrates that 90% of the nearest-neighbor sequences are ApA (Table V), the latter that AMP-α-^{32}P is progressively liberated. The problem arises whether polyadenylate synthetase and RNA polymerase are two different enzymes or correspond to two activities of the same enzyme. The problem is under investigation. The procedure and the results were similar for the demonstration of poly C and poly U synthesis.[29,30]

In order to be used as a nucleotide storage form, the homopolymers should be degraded in the nuclei. In fact, incubation of polyadenylate ^{32}P with a nuclear brain extract liberates ATP, ADP, and AMP, respectively, 8.8, 25.5, and 65.7%. This distribution suggests the possibility of a rapid production of NTP from the homopolymer. The reaction probably consists of two steps, the first liberating ADP, the second involving an adenylate kinase activity present in the nuclei furnishing ATP and AMP.

The characterization of the enzymes hydrolyzing polyadenylate is under investigation. The evaluation of the homopolymer quantities present in the cerebral cell nuclei is not easy, because of the rapid degradation of the polymers by a nuclear enzyme. However, Edmonds and Abrams[32] demonstrated the presence of polyadenylate in thymus nuclei, and polyadenosine diphosphate ribose (ADPR) was found in liver nuclei in our laboratory.[33] We found some evidence in favor of the presence of poly A in liver nuclei,[34] and Hadjivassiliou and Brawerman found polyadenylate in the cytoplasm.[35]

TABLE VI

RNA Polymerase Activity of Brain Nuclei[a]

Reaction mixture	Incorporation GMP α-^{32}P $\mu\mu$M/mg protein
Particulate enzyme	
Complete	110
Omit UTP + CTP	12.0
+ DNASE	13.5
+ Actinomycin (50 μg)	10.0
+ Pyrophosphate (10 m*M*)	11.0
+ Inorganic phosphate (10 m*M*)	112.0
+ RNASE (5 μg)	18.0
+ Ammonium sulfate (0.4 *M*)	420
Soluble enzyme	
Complete	49.1
Omit UTP α-CTP	3.0
Omit DNA	1.0
+ Pyrophosphate (10 m*M*)	2.5
+ Inorganic phosphate	48.5
+ Ammonium sulfate 0.4 *M* at 0°	2.0
+ Ammonium sulfate 0.4 *M* after 3° preincubation enzyme + DNA + NTP	29

[a] The complete reaction mixture (0.25 ml) is 160 m*M* Tris buffer pH 7.2, 4 m*M* $MnCl^2$ or 30 m*M* $MgCl^2$, 8m*M* 2-mercaptoethanol, 1 m*M* CTP, ATP, UTP, and GTP, one labeled in α with ^{32}P, and approximately 0.2 mg protein from the enzyme extract. In the case of soluble enzyme 60 μg of DNA were added. Incubation for 10 min at 37°.

RIBONUCLEIC ACID SYNTHESIS *IN VITRO*

Particulate RNA Polymerase

A chromatin-linked enzymatic activity which catalyzes the synthesis of RNA similar to the "aggregate RNA polymerase" discovered by Weiss[35a] was described in brain, simultaneously by Barondes[36], Bondy and Waelsch,[37] and in our laboratory.[38]

This enzymatic activity is characterized by (1) a dependence on the presence of the four different NTP's and of divalent cations, Mg or Mn; (2) its inhibition by deoxyribonuclease, by pyrophosphatase, and by actinomycin; and the sensitivity of the purified product to ribonuclease (Table VI). Although the preparations from the cortex were in some cases consistently more active than those from white matter[37], a statistically significant difference could not be established.[36] It seems likely that the RNA poly-

TABLE VII

RNA Polymerase Activity of the Aggregate RNA Polymerase of Rat Liver and Brain Nuclei[a]

	nM ^{32}P incorporated/mg DNA	
	17°C	17°C
	Without $(NH_4)_2SO_4$	With $(NH_4)_2SO_4$
Brain	1.2	7.0
Liver	0.75	2.5

[a] For incubation conditions see Table 6. Incubation, 20 min.

merase activity is lower in glial cells than in neurons.[36] However, glial cells do contain quite a high amount of RNA polymerase activity. This finding is consistent with the autoradiographic demonstration of incorporation of RNA precursors into RNA of oligodendroglia.[39]

When brain and liver RNA polymerase activities are measured the results are somewhat variable; the enzymatic activity is sometimes higher, sometimes lower, in brain than in liver. On the other hand, the base distribution of the synthesized RNA was similar to that of ribosomal RNA[40] as it was observed in liver.[41,42] From these two observations a question arises whether the differences observed as well as the unexpected base distribution were related to a ribonuclease activity present in the incubation medium, which degrades the more sensitive single-stranded DNA-like RNA. In fact, it was found that the liver[42] and the brain aggregate RNA polymerase do contain ribonuclease.[43] However, at 17°C these nucleolytic activities are very low[42,43]; in these conditions, in liver about 30% of the synthesized RNA is DNA-like, unlike that found at 37°C.[42] The product synthesized in brain is under investigation.

Since a higher apparent polymerase activity at 37°C could be due to a lower ribonuclease activity, we reinvestigated the RNA polymerase activity at 17°C, where the nucleolytic activities are very low (Table VII). It appeared that the RNA polymerase activity was about 70% higher in brain than in liver at 17°C. Moreover, the difference between the liver and brain polymerase activity cannot be related to a difference in the ribonuclease content, since the degradation by this enzyme at 17°C was about 5% in brain and 10% in liver.[43]

Goldberg[44] observed that an increase of ionic strength with ammonium sulphate or KCl stimulates RNA synthesis catalyzed by the HeLa or

L cell nuclear aggregate RNA polymerase.[44] Several explanations were proposed for this phenomenon: (1) appearance of a distinct latent RNA polymerase activity;[45] (2) dissociation of histones from DNA, thus increasing the template activity of the chromatin[42,46]; (3) inhibition of nucleolytic enzyme.[47,48]

The investigations performed in our laboratory did not show any evidence to support the hypothesis that a latent RNA polymerase of a distinct species is involved in the increase of RNA biosynthesis at a high ionic strength.[42] In addition, it was found by Chambon *et al.*[42] that the template efficiency of rat-liver chromatin increases to values equal to that of deproteinized-liver DNA when aggregate RNA polymerase activity was measured at a high ionic strength. Similar results were found by Marushige and Bonner[49] with deoxyribonucleoproteins. It was also observed in our laboratory that the values obtained in the presence of ammonium sulphate are proportional to the amount of RNA polymerase molecules present in the chromatin.[42,50] We compared the aggregate RNA polymerase activity in rat brain and rat liver at a high ionic strength at 17°C and 37°C.[43] These activities were, respectively, 2.8 and 2.2 times higher in brain than in liver. One can speculate on the fact that at 17°C, when the ribonuclease activity is very low, the measure of the aggregate RNA polymerase activity represents the actual template activity of the chromatin, whereas at a higher ionic strength the whole potentiality of the RNA polymerase present in chromatin is determined. Thus, the actual as well as the potential RNA polymerase activity in brain is higher than in liver. We should, however, keep in mind that by itself the high biosynthetic activity as measured *in vitro* does not allow us to draw definite conclusions concerning their functional significance. However, these data are in agreement with the *in vivo* observations showing a high rate of RNA synthesis in brain.

Soluble RNA Polymerase

Mammalian RNA polymerase has been obtained in a soluble form by Furth and Ho[51] and by Ishihama[52] from tumors, by Ballard and Williams-Ashman from rat testis,[53] and in our laboratory from nuclei of several organs and rat liver.[54] We could also obtain from brain nuclei an RNA polymerase containing an extract without DNA contamination (Table VI).[55] This RNA polymerase is retained in the nuclei if brain tissue is homogenized in a hypertonic sucrose solution (2.2 *M*) and can be released from the nuclei without any DNA extraction. The polymerase activity is completely dependent on the addition of DNA and catalyzes an actinomycin D-sensitive incorporation of α-^{32}P-labeled NTP in polynucleotides only in the presence of all four NTP's and of DNA. After KOH hydrolysis one observes that all four nucleotides are labeled. Phosphodiesterase degradation shows a stepwise liberation of the radioactive material. Thus, it can be concluded that the obseıved phenomenon is not an end attachment but a polyribonucleotide synthesis. All the properties of the soluble RNA polymerase are similar to those of the aggregate RNA polymerase.[42] Thus, it seems likely that RNA polymerase exists in two forms in the nuclei, one free and soluble, the other attached to chromatin. The accuracy of the copy of DNA could be shown for the soluble RNA polymerase by base composition analyses to be similar to that of the DNA template when using different DNA's.[55] This finding indicates that this RNA polymerase may insure the genetic transcription.

There are, of course, still several problems which need more investigation concerning the mechanism of RNA polymerase activity. For instance, we do not know the nature of the signals which start or stop the transcription, nor do we know how completed synthesized polyribonucleotides are released from the template and from the enzyme in animal cells. The mechanism is probably different from that which was demonstrated for bacteria.[56,57] Finally, our knowledge about the control mechanisms of RNA polymerase activity in different functional conditions is mainly limited to speculation.

RIBONUCLEIC ACID SYNTHESIS *IN VIVO*

Determination of the Fractions of the Genome Coding for RNA in Brain

In order to determine the fraction of the genome active in the cells, one can use the sequence complementarity between the RNA synthetized *in vivo* and the part of the DNA that has served as a primer for this synthesis during the transcription process. Molecular hybridization proved to be a useful tool in these investigations. However, in order to detect the RNA fixed to the DNA hybridization, the use of radioactive RNA is required, and its specific activity must be known in order to calculate the corresponding quantity. This implies that RNA of definite species must be used rather than a mixture of RNA's with different labeling rates. We were mainly interested in ribosomal RNA and in dRNA from total cells or from microsomes in rat brain. Ribosomal RNA could be prepared free of messenger activity by extracting it from polysomes previously treated by ribonuclease in such conditions that the messenger RNA strand linking the ribosomes would be the only one to be degraded.[58,59] It was not possible, with the methods available, to obtain dRNA free of ribosomal RNA. However, its

TABLE VIII

Type of RNA	Per cent of DNA complementary to the RNA at saturation	
	Average	Extreme values
Ribosomal	0.15	0.14–0.16
dRNA		
Total	1.24	1.15–1.31
Microsomal	0.78	0.68–0.90
Nuclear-restricted	0.46	Calculated
Giant-size	1.27	1.09–1.40

specific activity could be calculated by assuming that it was equal to that of the γ-P of the nucleotides when the isotopic equilibrium of the nucleotide pool was reached after the injection of ^{32}P-phosphates. Knowing the specific activity and the hybridization characteristics of ribosomal RNA, its quantity could be substracted and that of hybridized dRNA could be calculated. A direct determination of the specific activity of the hybridized RNA showed that our basic assumption was true.[(59)]

Table VIII shows the saturation values of DNA with ribosomal RNA, dRNA from total cells, and from microsomes. Ribosomal RNA reaches its saturation level at low RNA/DNA input (10–20 μg/ml of RNA in the incubation medium); 0.15% of the DNA sites are then occupied.[(59)]

A high RNA/DNA input is required to reach a plateau of saturation with total and microsomal dRNA[(59–61)] (2 mg/ml of RNA in the medium). Then 1.2 and 0.8% of the DNA sites are occupied by total dRNA and microsomal dRNA, respectively. This implies that 0.4% of the DNA codes for nuclear-restricted RNA's.

From these numerical values we can try to evaluate how many cistrons per diploid cell are active in the synthesis of RNA *in vivo*. The average molecular weight of ribosomal RNA can be considered as 10^6, that of dRNA is more uncertain, but lies probably between 10^5 and 10^6. Knowing that a diploid cell contains 6.7×10^{-12}g of DNA, one calculates that ribosomal RNA corresponds to 6000, total dRNA to 50,000–500,000, and microsomal dRNA to 30,000–300,000 cistrons. Nuclear-restricted dRNA's would thus be coded for by 20,000–200,000 cistrons. These figures seem very high if we assume that each cistron corresponding to microsomal dRNA codes for one species of proteins. In fact, a redundancy of cistrons in animal cells is highly probable;[(62)] this is, for instance, illustrated in the case of ribosomal RNA, where 6000 cistrons are coding for two species of RNA. It must also be stressed that the indicated values are probably minimal, since those dRNA which are not synthesized at a high rate or which correspond only to a very small number of cistrons would not be detected with our technique, which is mainly concerned with the most abundant RNA species.

The Transport of dRNA from the Nucleus to the Cytoplasm

It has been shown previously[63] that in brain as well as in other animal cells dRNA's with a high sedimentation coefficient (>28 S) are found in the nucleus, whereas smaller dRNA's are present in polysomes where protein synthesis takes place.[58] The cytoplasmic dRNA's have about the size that would be expected for templates for proteins of an average size, so that it is unlikely that they are the results of an unphysiological degradation. Since all known dRNA's are synthesized with DNA as a template, they are inside the nucleus; and since few small-size RNA can be detected in the nucleus, one could wonder whether there is any relationship between nuclear giant-size dRNA and cytoplasmic messenger RNA. Hybridization techniques could again help to answer this question. We had shown previously (see above) that only 65% of the dRNA species can be found in the cytoplasm, 35% being nuclear-restricted. At this stage there were three possible hypotheses concerning the nature of giant-size dRNA: (1) it could represent a direct precursor of cytoplasmic dRNA; nuclear-restricted dRNA would then be of a different size; (2) it could represent the nuclear-restricted dRNA; the precursors of cytoplasmic dRNA would then be of a different size and; (3) it could contain the polynucleotidic sequences of both cytoplasmic dRNA and nuclear-restricted dRNA.

Giant-size dRNA from rat brain was prepared from total RNA fractionated by sucrose density gradients and hybridized with DNA. The experimental conditions and the determination of the quantity of dRNA hybridized with DNA were the same as described above. The saturation curves (Table VIII) showed that giant-size dRNA was complementary to the same amount of DNA as dRNA extracted from total cells. This means that the same polynucleotidic sequences were present and that the third hypothesis was the right one. Thus, giant-size dRNA is a precursor of cytoplasmic dRNA and also contains the nuclear-restricted sequences. Moreover, experiments with small-size dRNA's have shown that they do not contain significant amounts of nuclear-restricted-RNA.[60,61] It is, however, not yet known whether giant-size dRNA's contain two different types of molecules, some being the precursors of the cytoplasmic RNA, the other corresponding to nuclear-restricted-RNA, or whether some parts of one

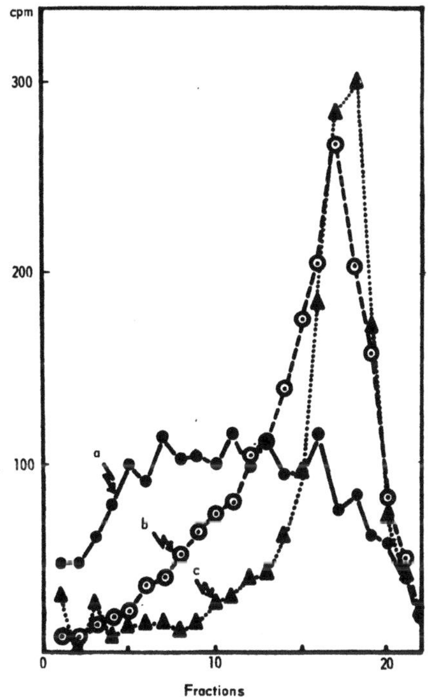

Fig. 3. Action of RNAase on the size of subnuclear particles. A brain nuclear extract was prepared 1 hr after the injection of ^{3}H-uridine to rats. The extract was divided into three fractions which were incubated for 10 min at 3°C. (a) Without RNAase (control); (b) with RNAase 0.09 μg/ml; (c) with RNAase 0.23 μg/ml. After incubation the fractions were immediately submitted to analysis in a 15–30% linear sucrose gradient in a SW-50 rotor. Centrifugation, 1 hr at 49,000 rpm; 40 S at tube 17. Under these experimental conditions the heavier material is found at the bottom of the tube.

molecule are transferred to the cytoplasm, whereas others would remain inside the nucleus.

Either way, we must assume the existence of a mechanism of specific cleavage of the giant molecule (or polycistronic RNA) and of a mechanism of selection of the fragments for their transfer to the cytoplasm.

Seeking the mechanism of specific cleavage, we were led to two possible hypotheses: (1) a ribonuclease with specific requirements for the nucleotidic sequences between the cistrons would perform the task; (2) some parts of the nucleotidic sequences would be protected, for instance, by proteins, against the action of any nuclease. The latter possibility would imply that giant-size dRNA must exist in a particulate form in the nucleus. We have tried to isolate such particles and to study their properties, especially their sensitivity to ribonuclease action.[64,65]

A purified brain nuclear fraction was prepared. The nuclei were lysed at pH 8 in the presence of DOC 0.2%, chromatin was sedimented, and the supernatant extract was examined in sucrose density gradients for the presence of particles (Fig. 3). Rapidly labeled RNA was associated with structures sedimenting between 80 and 300 S; giant-size RNA was present in the heavier structures (Fig. 4). The size of the RNA and of the structures

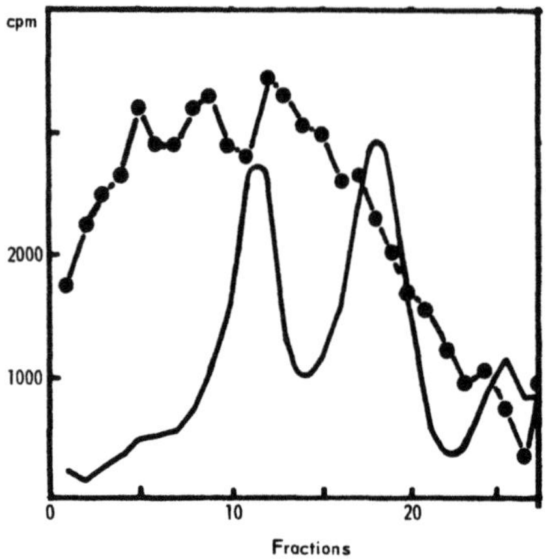

Fig. 4. Sucrose density gradient of RNA from subnuclear particles. Subnuclear-particle RNA was prepared from rat brain in the presence of the cytoplasmic inhibitor of RNAse 1 hr after the injection of tritiated uridine. Analysis was carried out in a linear sucrose gradient 5–20% in a SW-25 rotor of a Spinco centrifuge. Centrifugation, 14 hr at 24,000 rpm. Solid curve: optical density at 260 nm, mainly representing that of carrier nonlabeled RNA; circles: ^{3}H radioactivity.

from which it was extracted were related. We observed that there was an absolute requirement for ribonuclease cytoplasmic inhibitor in the extraction medium if giant-size dRNA was to be obtained.

If the extract was treated with ribonuclease in very mild conditions and at low ionic strength, the size of the particles decreased to 30–50 (Fig. 3) and that of their RNA to 5–9 *S* (Fig. 5). Greater amounts of ribonuclease led to a complete destruction of the particles without the appearance of

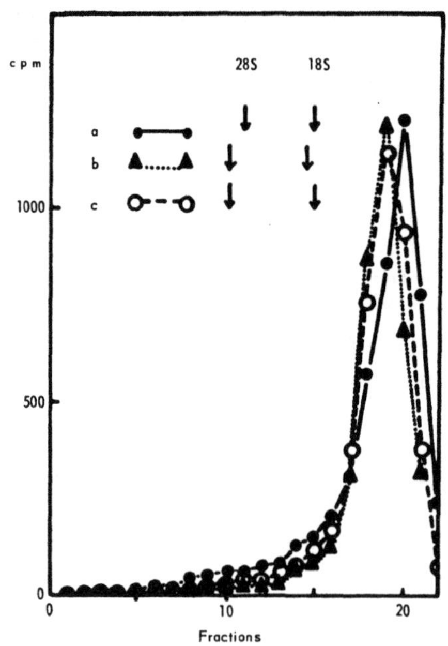

Fig. 5. Action of RNAase on the size of RNA of subnuclear particles. Subnuclear particles were submitted to the action of RNAase under the conditions described in Fig. 3(c). RNA was extracted and analyzed in a linear sucrose gradient 5–20% in a SW-50 rotor. Centrifugation, $2\frac{1}{2}$ hr at 49,000 rpm. Curves (a) and (b): RNAase action carried out in buffer of higher ionic strength; curve (c): same conditions as in Fig. 3(c).

intermediary structures of lower molecular weight. At higher ionic strength, ribonuclease provoked the transformation of the rapidly sedimenting RNA into 5–9-*S* RNA, whereas the size of the particles was much less affected. This probably means that the 30–50-*S* units are linked by bonds other than RNA.

The results are in favor of a polysomal-like structure of the type described by Samarina *et al.*[66] for these particles: 30–50-*S* subunits would be linked by the dRNA chain; a fraction of this chain would be associated with the subunits and probably protected by proteins against the action of nuclease; the nucleotidic sequence between the subunits would be the most susceptible to the enzymatic degradation.

Thus, we succeeded in obtaining particles carrying giant-size dRNA, and some sequences of this dRNA show a differential susceptibility to the attack by ribonuclease. It is difficult to obtain an action of ribonuclease limited *in vitro* as it would be *in vivo*, but it is of interest to note that the 30–50-*S* subunits carry a 5–9-*S* RNA only slightly smaller than the average polysomal messenger RNA. Although the existence of such a mechanism of

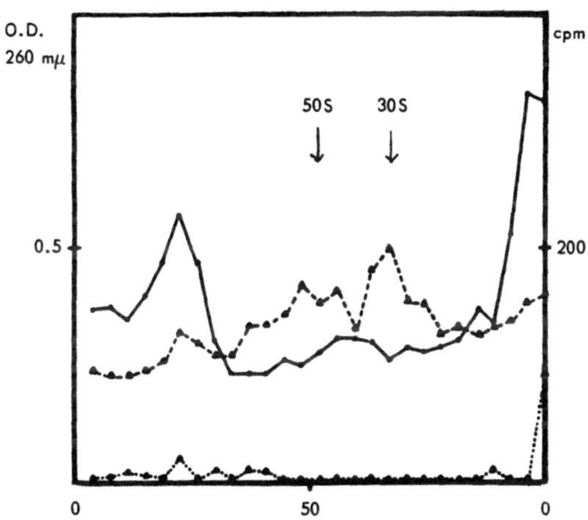

Fig. 6. Incorporation of ^{3}H-orotic acid into 20–60-*S* particles from brain microsomes. A microsomal fraction treated with DOC was analyzed in a 15–30% linear sucrose gradient. Centrifugation, 15 hr at 23,000 rpm in a SW-25 rotor. Labeling time was 30 min with ^{3}H-orotic acid. Solid curve: optical density at 260 nm. Dashed curve, triangles: acid-insoluble ^{3}H radioactivity. Dotted curve, circles: RNAase resistance of acid-insoluble ^{3}H radioactivity.

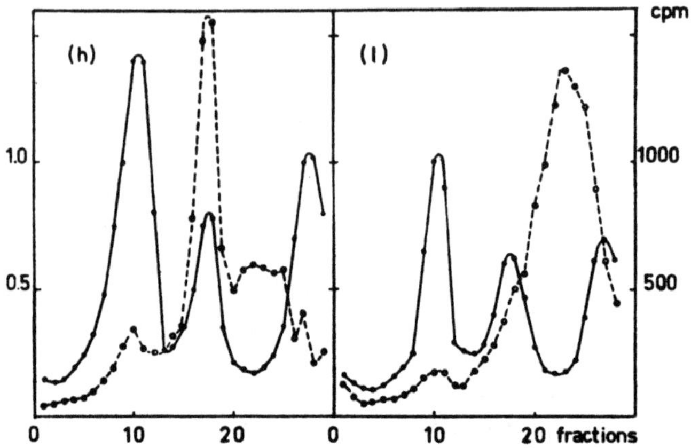

Fig. 7. Sucrose-density-gradient analysis of RNA from microsomal 20–60-S particles. Both 20–40-S (l) and 40–60-S (h) particles were collected, and RNA was extracted and analyzed in a 5–20% linear sucrose gradient. Centrifugation as in Fig. 4. Solid curves: optical density at 260 nm (mainly that of unlabeled carrier RNA). Dashed curves: ^{32}P-acid-insoluble radioactivity.

cleavage has not been demonstrated yet *in vivo*, it could account for the transformation of nuclear polycistronic RNA into cytoplasmic messenger RNA. The 30–50-S subunits, or group of subunits if a supplementary mechanism of regulation limits the *in vivo* process, could thus represent a form of transport of *m*-RNA from the nucleus to the cytoplasm. In this case they should be found in the cytoplasm either free, or bound to cellular sites where protein synthesis occurs.

We first examined a microsomal fraction of rat brain for the existence of ribonucleoprotein particles smaller than 80-S ribosomes.[67,68] As expected, there were no such particles in a free form in this fraction; however, when the microsomes were treated with DOC at the concentration required to release polysomes, we found rapidly labeled RNA associated with structures sedimenting between 20 and 60 S (Fig. 6). At least up to 2 hr after injection of ^{3}H-uridine the specific activity of these particles was higher than that of polysomes. We studied separately the 20–40-S and the 40–60-S structures. The 20–40-S particles contained polydisperse RNA (Fig. 7), with a base composition close to that of DNA and with sedimentation characteristics similar to that of polysomes (maximum at 8–15-S). The buoyant density of the particles measured in CsCl was $\rho = 1.36$. The 40–60-S particles contained a polydisperse RNA also showing a maximum of sedimentation at 8–15-S and 18-S ribosomal RNA (Fig. 7). Examination of the buoyant density showed that two main species were present at $\rho = 1.36$ and $\rho = 1.55$. The results are in favor of the existence of two populations of particles, one at 20–60-S and carrying messenger RNA ($\rho = 1.36$), the other carrying 18-S ribosomal RNA ($\rho = 1.55$) with characteristics close to those of the 45-S precursor of ribosomes described in other cells.[69–73]

We also studied the post-microsomal supernant of brain. We found rapidly labeled polydisperse RNA associated with particles sedimenting at 20–40-S with a buoyant density of $\rho = 1.40$.

The microsomal and post-microsomal particles just described have sedimentation characteristics close to those of the structures obtained by a mild ribonuclease treatment from larger subnuclear particles. Their RNA is also of the messenger type. These preliminary results favor a relationship. But, in the present state of our research, they do not mean that they are identical and that the cytoplasmic particles originate from the nuclear ones. It is clear that the identity of their protein moiety must first be demonstrated.

CELL DIFFERENTIATION AND ONTOGENESIS

Cell differentiation and ontogenesis are closely related to the transcription of the genetic code, that is, to RNA synthesis. Thus, the control of RNA biosynthesis during these periods is one of the main approaches for their understanding. In this connection, I should like to report two observations made in our laboratory.

TABLE IX

Incorporation of ^{32}P in the Free Nucleotides and RNA of the Rabbit Retina during Ontogenesis

Specific activity (cpm/μg)	1 day (Control)	8 days (Control)	21 days	
			Control	Light-deprived since birth
of α ^{32}P				
AMP	10,803 ± 215	6930 ± 170	6280 ± 182	11,460 ± 250
CMP	2020 ± 20	1030 ± 35	1335 ± 55	2390 ± 30
GMP	1256 ± 16	1560 ± 35	1180 ± 40	1840 ± 30
UMP	1132 ± 14	3620 ± 112	1205 ± 55	960 ± 20
of P-RNA	1542 ± 30	1923 ± 17	2450 ± 47	2625 ± 35
Ratio P-RNA/α-P of free nucleotides	0.487	0.680	1.141	0.742

During ontogenesis visual stimuli are necessary for an induction of a normal biosynthesis. A retardation in the evolution of this biosynthesis was observed when the animals were brought up in total darkness.[74] The rate of RNA synthesis was consistently higher in the retina of young animals maintained in light compared to those maintained in darkness (Table IX).[75]

The second observation concerns the early differentiation of chicken embryo brain. Recently it was found in our laboratory[76] that in neuroblasts and spongioblasts of chicken embryo brain, RNA synthesis is mainly limited to the nucleus until about 15 days of embryonic life. Only later, in parallel to the differentiation of nucleolus, does a rapid increase of RNA appear in the cytoplasm. These two examples show two kinds of control, one endogenous, the other environmental, which should be kept in mind during investigations on differentiation and ontogenesis.

In summary, during the last few years great progress has been realized in our knowledge of the mechanism of transcription (RNA synthesis) and of translation (protein synthesis) in bacterial cells. These basic phenomena are similar in anuclear and eukariotic cells, but there are many particularities which result from the morphological and physiological specialization of animal cells in general and of nervous cells in particular. Thus, several problems need more investigation, among them those briefly discussed in this paper: (1) the regulation mechanism of the precursor pool for RNA and protein synthesis; (2) the selection mechanism of specific messengers and their transport to the cell cytoplasm; and (3) the differentiation mechanism, which is closely linked to transcription and translation.

In addition, an answer has to be found to the question of the significance of the high rate of RNA and protein synthesis in nervous cells, which is similar to that of tissues having a high protein secretory activity.

It seems that two hypotheses may be considered: (1) the intensive protein synthesis is linked to the existence of the cell processes—dendrites and axons—for which the cell body produces the structural proteins and peptides; and (2) all neurons are peculiar secretory cells and the secretory proteins follow the axoplasmic flow in order to ensure the turnover of specific structural and functional proteins at the nerve endings.

REFERENCES

1. B. S. MacEwen, V. G. Allfrey, and A. E. Mirsky, *J. Biol. Chem.* **238,** 758 (1963).
2. R. Penniall, Sh. M. Liu, and J. P. Saunders, *Biochim. Biophys. Acta* **76,** 170 (1966).
3. E. Karjalainen, *Acta Chem. Scand.* **20,** 586 (1966).
4. S. Simler, D. Popovic, and P. Mandel, *Bull. Soc. Chim. Biol.* **49,** 1509 (1967).

5. P. Mandel, *Bull. Soc. Chim. Biol.* **49,** 1491 (1967).
6. P. Mandel, M. Wintzerith, N. Klein-Pete, and L. Mandel, *Nature* **198,** 1000 (1963).
7. P. Mandel, *in*: "Progress in Nucleic Acid Research and Molecular Biology," (J. N. Davidson and W. E. Cohn, eds.), Vol. 3, p. 299, Academic Press, New York (1964).
8. S. Edel and G. Poirel, *Bull. Soc. Chim. Biol.* **48,** 935 (1966).
9. S. Edel and P. Mandel, *J. Physiol.* **57,** 606 (1965).
10. J. Klethi and P. Mandel, *Bull. Soc. Chim. Biol.* **50,** 709 (1968).
11. M. E. Ittel, M. Wintzerith, and P. Mandel, *FEBS Letters* **2,** 141 (1969).
12. J. T. August, P. J. Ortiz, and J. Hurwitz, *J. Biol. Chem.* **237,** 3786 (1962).
13. M. Chamberlin and P. Berg, *Proc. Natl. Acad. Sci. U.S.* **48,** 81 (1962).
14. I. H. Goldberg, M. Rabinowitz, and E. Reich, *Proc. Natl. Acad. Sci. U.S.* **48,** 2094 (1962).
15. P. R. Venkataraman and H. R. Mahler, *J. Biol. Chem.* **238,** 1058 (1963).
16. C. F. Fox and S. B. Weiss, *J. Biol. Chem.* **239,** 175 (1964).
17. A. Stevens, *J. Biol. Chem.* **239,** 204 (1964).
18. C. W. Chung, H. R. Mahler and M. Enrione, *J. Biol. Chem.* **235,** 1448 (1960).
19. D. B. Straus and E. Goldwasser, *J. Biol. Chem.* **236,** 849 (1961).
20. M. Edmonds and R. Abrams, *J. Biol. Chem.* **235,** 1142 1960).
21. M. Edmonds and R. Abrams, *J. Biol. Chem.* **237,** 2636 (1962).
22. R. H. Burdon, *Biochem. Biophys. Res. Commun.* **13,** 37 (1963).
23. H. G. Klemperer, *Biochim. Biophys. Acta* **72,** 416 (1963).
24. M. L. Giron and J. Huppert, *Compt. Rend.* **261,** 1451 (1965).
25. P. Chambon, J. D. Weill, J. Doly, M. T. Stroesser, and P. Mandel, *Biochem. Biophys. Res. Commun.* **25,** 638 (1966).
26. Y. Nishizuka, K. Ueda, K. Nakazawa, and O. Hayaishi, *J. Biol. Chem.* **242,** 3164 (1967).
27. S. Fujimura, S. Hasegawa, Y. Shimizu, and T. Sugimura, *Biochim. Biophys. Acta* **145,** 247 (1967).
28. P. Mandel, N. Pete, A. R. Dravid, and M. Ramuz, Abstr. 3rd Meet. Fed. Europ. Soc. Biochem., Varsovie (1966) G-112.
29. P. Mandel, A. R. Dravid, and N. Pete, *J. Neurochem.* **14,** 301 (1967).
30. M. Jacob and P. Mandel, *in*: "Protides of the Biological Fluids," (H. Peeters, ed.), p. 63, Elsevier Publishing Co., Amsterdam (1965).
31. P. Mandel, N. Pete, and A. R. Dravid, *J. Physiol.* **59,** 261 (1967).
32. M. Edmonds and R. Abrams, *J. Biol. Chem.* **238,** 1186 (1963).
33. J. Doly and P. Mandel, *Compt. Rend.* **264** (Série D), 2687 (1967).
34. P. Mandel and S. Simler, unpublished data.
35. A. Hadjivassiliou and G. Brawerman, *J. Mol. Biol.* **20,** 1 (1966).
35a. S. B. Weiss, *Proc. Natl. Acad. Sci. U.S.* **46,** 1020 (1960).
36. S. H. Barondes, *J. Neurochem.* **11,** 663 (1964).
37. S. C. Bondy and H. Waelsch, *Life Sciences* **3,** 633 (1964).
38. N. Pete, M. Wintzerith, L. Mandel, and P. Mandel, *Compt. Rend.* **258,** 5283 (1964).
39. H. Koenig, *J. Biophys. Cytol.* **4,** 785 (1958).
40. A. R. Dravid and P. Mandel, unpublished data.
41. C. C. Widnell and J. R. Tata, *Biochim. Biophys. Acta* **123,** 478 (1966).
42. P. Chambon, M. Ramuz, J. Doly, and P. Mandel, *Biochim. Biophys. Acta* **157,** 504 (1968).
43. D. Munoz and P. Mandel, *Compt. Rend. Soc. Biol.* **162,** 2283 (1968).
44. I. H. Goldberg, *Biochim. Biophys. Acta* **51,** 201 (1961).
45. C. C. Widnell and J. R. Tata, *Biochim. Biophys. Acta* **87,** 531 (1964).

46. P. Chambon, M. Ramuz, and J. Doly, *Biochem. Biophys. Res. Commun.* **21,** 156 (1965).
47. D. D. Cunningham and D. F. Steiner, *Fed. Proc.* **25,** 788 (1966).
48. D. F. Steiner and J. King, *Biochim. Biophys. Acta* **119,** 510 (1966).
49. K. Marushige and J. Bonner, *J. Mol. Biol.* **15,** 160 (1966).
50. P. Chambon, H. Karon, M. Ramuz, and P. Mandel, *Biochim. Biophys. Acta* **157,** 520 (1968).
51. J. J. Furth and P. Ho. *J. Biol. Chem.* **240,** 2602 (1965).
52. A. Ishihama, *Biochim. Biophys. Acta* **145,** 272 (1967).
53. P. L. Ballard and H. G. Williams-Ashman, *J. Biol. Chem.*, **241,** 1602 (1966).
54. M. Ramuz, J. Doly, P. Mandel, and P. Chambon, *Biochem. Biophys. Res. Commun.* **19,** 114 (1965).
55. P. Mandel, A. R. Dravid, and N. Pete, unpublished data.
56. M. Revel, M. Herzberg, A. Becarevic, and F. Gros, *J. Mol. Biol.* **33,** 231 (1968).
57. S. Ochoa, *Naturwiss.* **55** (11), 505 (1968).
58. M. Jacob, J. Samec, J. Stevenin, J. P. Garel, and P. Mandel, *J. Neurochem.* **14,** 169 (1967).
59. J. Stevenin, J. Samec, M. Jacob, and P. Mandel, *J. Mol. Biol.* **33,** 777 (1968).
60. M. Jacob, J. Stevenin, and P. Mandel, *Compt. Rend.* **266** (*Série D*), 1675 (1968).
61. J. Stevenin, P. Mandel, and M. Jacob, *Proc. Natl. Acad. Sci. U.S.* **62,** 490 (1969).
62. R. J. Britten and D. E. Kohne, *Science* **161,** 529 (1968).
63. M. Jacob, J. Stevenin, R. Jund, C. Judes, and P. Mandel, *J. Neurochem.* **13,** 619 (1966).
64. J. Stevenin and M. Jacob, *Abstr. Bull. Soc. Chim. Biol.* (1968), in press.
65. J. Stevenin and M. Jacob, *Europ. J. Biochem.*, submitted for publication.
66. O. P. Samarina, E. M. Lukanidin, and G. P. Georgiev, *Biochim. Biophys. Acta* **142,** 561 (1967).
67. J. Samec, P. Mandel, and M. Jacob, *J. Neurochem.* **14,** 887 (1967).
68. J. Samec, M. Jacob, and P. Mandel, *Biochim. Biophys. Acta* **161,** 377 (1968).
69. R. P. Perry and D. E. Kelley, *J. Mol. Biol.* **16,** 255 (1966).
70. M. Girard, H. Latham, S. Penman, and J. E. Darnell, *J. Mol. Biol.* **11,** 187 (1965).
71. E. C. Henshaw, M. Revel, and H. H. Hiatt, *J. Mol. Biol.* **14,** 241 (1965).
72. E. H. McConkey and J. W. Hopkins, *J. Mol. Biol.* **14,** 257 (1965).
73. H. Ristow and K. Köhler, *Biochim. Biophys. Acta* **142,** 65 (1967).
74. P. Mandel and S. Goswamy, *Orient. Arch. Ophthal.* **4,** 137 (1966).
75. S. Goswamy, P. Mandel, and P. Karli, *in*: "Biochemistry of the Eye," (U. Dardenne and J. Nordmann, eds.), p. 514, S. Karger, Basel (1966).
76. M. Sensenbrenner and P. Mandel, *Z. Zellforsch.* **82,** 65 (1967).

DISCUSSION

FORD: There is a point here on which an anatomist can comment in relation to the distribution of adenine in the brain following an intravenous injection. Presumably, adenine administered in such a fashion must pass through the membranes which are classically referred to as the blood-brain barrier. It has been demonstrated in radioautographic studies that the labeled material within the first ½ hr after injection is primarily in the neuropil.* If one then examines autoradiographs made of brain tissue

*Donald H. Ford, Changes in brain accumulation of amino acids and adenine associated with changes in the physiologic state, *in*: "Progress in Brain Research," Vol. 29 (A. Lajtha and D. H. Ford, eds.), pp. 411–413 (1967).

from animals killed 1 hr after adenine injection, the label is observed to have accumulated in the neurons. By 2 or 4 hr, or at much later periods, the distribution of most of the radioactivity in the brain appears to be in the neurons. The shift of radioactivity from neuropil to neuron more or less fits in with the idea of glial transport from the capillaries to the neurons. One can then dissect out motor neurons of the cord and compare the amount of activity per unit weight with that of liver, and it turns out that the large motor horn cells accumulate more activity than do liver cells per equivalent weight.† This is not perhaps surprising in view of the very high metabolic activity of neurons. The ^{3}H activity in the brain following intravenous injection with ^{3}H-adenine seems to be largely in the phosphorylated derivatives which may be formed from adenine. One may carry this concept of neuronal localization of neucleic acid bases a step further with uracil, which also demonstrates a high accumulation in nerve cells. Again a significant percentage of the labeled material in brain is present in the phosphorylated derivatives of uracil.*

†Donald H. Ford and R. Rhines, H^{3} Accumulation in spinal cord neurons following intravenous injection of adenine-H^{3} as determined by liquid scintillation counting procedures on dissected neurons, *Acta Neurol. Scand.* **43,** 427–439 (1967).

*Donald H. Ford and B. Blatt, ^{3}H-Uracil accumulation in motor neurons compared with that occurring in blocks of nerve tissue, liver, muscle and plasma in the rat, *J. Neurological Sci.* **6,** 9–17 (1968).

Chapter 7

Insoluble Proteins of the Synaptic Plasma Membrane*

H. R. Mahler

Chemical Laboratories
Indiana University
Bloomington, Indiana

and

Carl W. Cotman

Division of Psychobiology
University of California
Irvine, California

INTRODUCTION AND DEFINITIONS

The presynaptic plasma membrane of a CNS neuron is that part of its limiting membrane that confines the terminal knob or bouton of its axonal process.[1,2] Within the space defined by this structure are contained a variety of components believed to be intimately connected with its function. Thus, in chemical synapses† we observe synaptic mitochondria, as

*Research supported by grant No. GB-4855 from the National Science Foundation. The personal support by a Research Career Award (No. PHS GM 05060) to H. R. Mahler, and of a predoctoral fellowship (No. 1-FI-GM30) to C. W. Cotman from the Institute of General Medical Sciences of the N.I.H., U.S.P.H.S. is gratefully acknowledged. The investigations at Indiana University described here were performed by a research group jointly directed by Professor Walter J. Moore and one of us (H.R.M.). We wish to express our appreciation to Professor Moore for many fruitful discussions, to Dr. Kern von Hungen, Dr. W. McBride, and Mr. Fredric White for permitting us to cite some of their unpublished observations, and to Mrs. Sue Hertweck for her very capable electron micrography. This is Indiana University Chemical Laboratories Publication No. 1693.

†Whittaker[1] has recently provided a concise description of the various types of synapses and the morphology of the associated organelles.

well as numerous synaptic vesicles of various types, all bathed by the more soluble components of the appropriate portion of the cytosol. Notable by their absence are ribosomes, free or attached.[3] Separating the presynaptic from the post-synaptic (dendritic or somatic) membrane is a cleft some

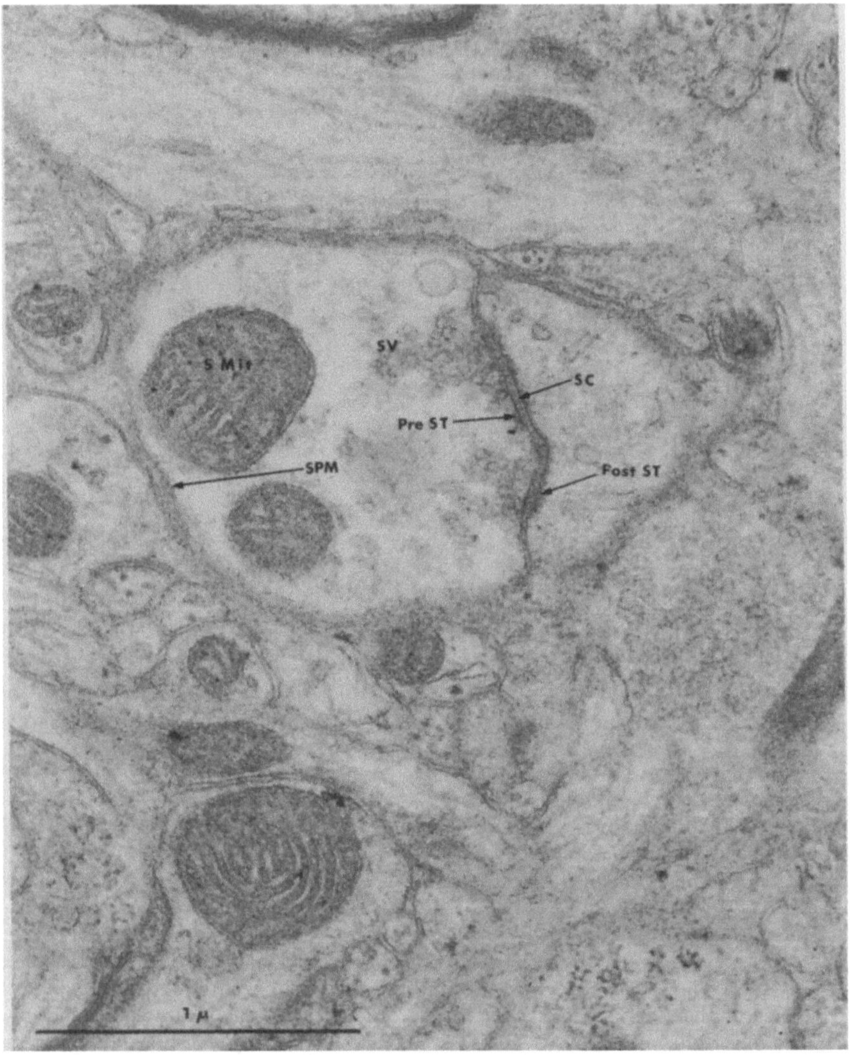

Fig. 1. A representative chemical synapse in the middle visual cortex of the rat. Typical structures seen are: SPM, synaptic plasma membrane; SV, synaptic vesicles; S Mit, synaptic mitochondrion; Pre ST, presynaptic thickening; SC, synaptic cleft; Post ST, post-synaptic thickening. Embedded in Epon; stained with uranyl acetate/lead citrate.

200 Å across and occupied, at least in part, by what appear to be thin connecting fibers or processes. The ultrastructure of such a synapse can be visualized from the electron micrograph of Fig. 1 and is presented in schematic form in Fig. 2.

The synapse is the structure across which neurons must communicate, during the original construction of the link, as well as in its subsequent use for transmission of the electrical impulse. In the chemical synapses of the CNS this is believed, largely by analogy and inductive reasoning, to be mediated by the quantized discharge of a chemical transmitter substance, originally stored within the vesicle, across the cleft, and its reception and eventual destruction on the post-synaptic side.[4,5] Coincident with, or probably preceding, these events there occurs a rapid increase in the permeability and hence a rapid depolarization of the presynaptic membrane.[6]

Considering its crucial position within the scheme of CNS function, remarkably little is known concerning the chemistry of the synaptic membrane*. In general, plasma membranes are composed of complex arrays of lipoproteins further modified at times by acidic carbohydrates.[2,7,8,9] Some of the unanswered problems at the most elementary level, that of the component proteins of this particular member of this class, may be formulated as follows: How many, and what kind of proteins make up the presynaptic membrane? To what extent are these proteins unique to this structure; specifically, how much overlap is there between them and the proteins of other parts of the neuronal membrane, such as the axon, and the post-synaptic portion (dendrite or soma)? How different are neuronal *plasma* membranes from the membrane components of other neuronal organelles: the endoplasmic reticulum, the mitochondria, the synaptic vesicles? How different are they from the plasma membranes of other cells: of the Schwann cells (i.e., myelin), of other glial cells, or, for that matter, of the plasma membrane of the red cell, etc.? What is their mode of biosynthesis and renewal? Do all protein components act synchronously, as a unit, or is there indication of individual insertion and removal? What is the locale of their biosynthesis and assembly; which part of these processes occur proximal to the membrane, within the confines of the bouton; which in the far-distant reaches of the cell body?

As a first, essential step toward a partial resolution of these questions, we have isolated, most recently and most successfully by means of zonal ultracentrifugation, the synaptic plasma membrane (SPM) of rat cerebral cortex.[10–12] In this endeavor we have been aided by the fact, well authen-

*Since the oral presentation of this paper the current state of knowledge concerning the neuronal membrane has been admirably reviewed by Lehninger,[2] thus obviating the necessity of lengthy citations here.

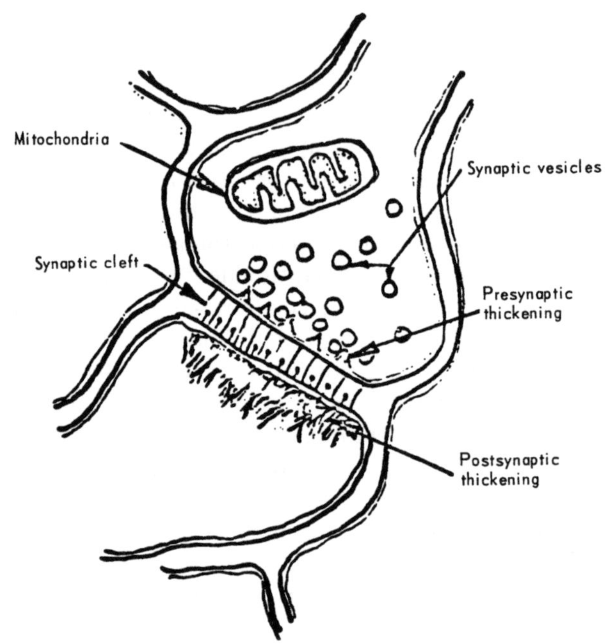

Fig. 2. A diagrammatic representation of synaptic structures.

ticated in the literature,[1,13–15] that in the process of homogenization and fractional centrifugation of CNS tissue, at the shear stresses and the media ordinarily employed, the synaptic knobs, especially of myelinated neurons, are sheared off from their axons, and subsequently vesicularize to form the so-called nerve-end(ing) particles or synaptosomes (Fig. 3). These entities, after lysis by osmotic shock, then provide the source of the membrane preparation of interest.[15–17]

Very frequently many of the connecting and post-synaptic entities remain attached to, and in fact consitute one of the more characteristic morphological features of, the final membrane preparation. For this reason we have preferred the term just mentioned rather than the designations "synaptic membrane," "nerve-ending or synaptosome membrane" or "synaptosome ghost," all frequently used in the literature for analogous preparations.

Here we report on the preliminary characterization of the SPM by

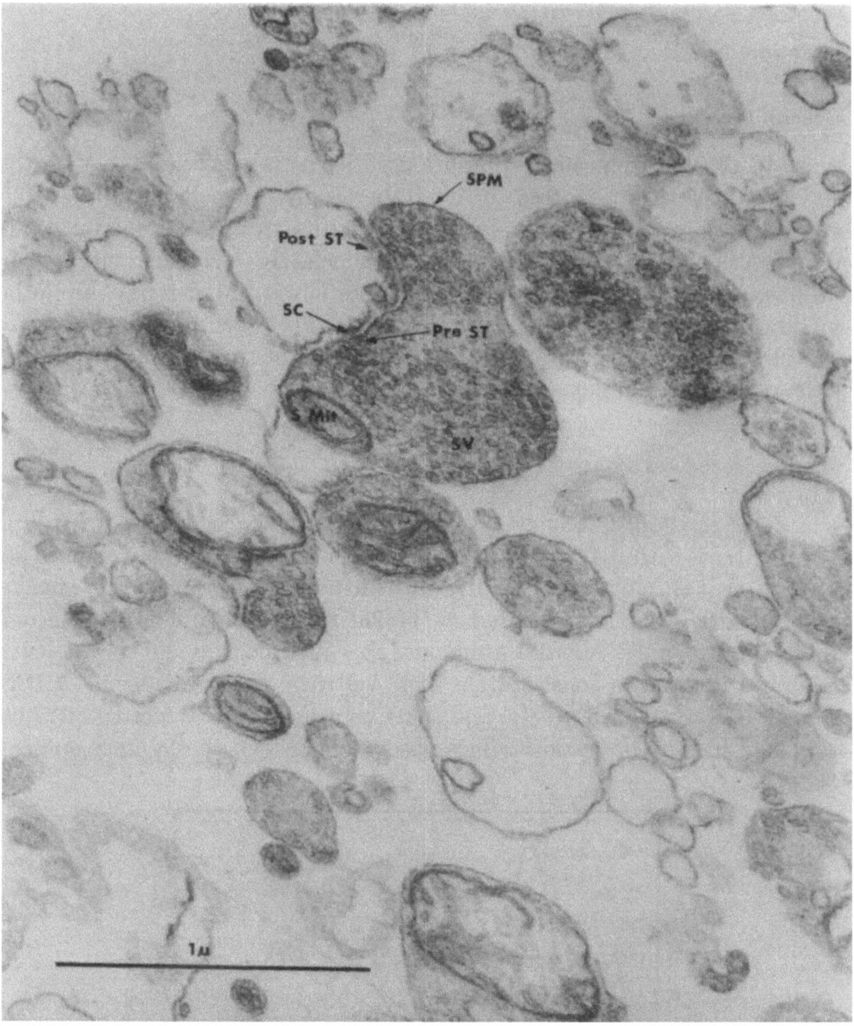

Fig. 3. Pellet from a nerve-end particle (synaptosome) fraction.[15,28] Many pinched-off boutons are seen. The representative one in the upper half shows all the features seen in the synapse *in situ*. See Fig. 1 for the explanation of the symbols used.

various morphological, biophysical, and biochemical criteria, including a definition of its insoluble proteins by means of amino acid composition and electrophoretic mobility, with special emphasis on a group of "structural" proteins. We also describe some data bearing on its biosynthesis and turnover.

ISOLATION AND METHODOLOGY

Isolation

General Principles

Most of the isolation techniques used by us take advantage of the observation originally reported by De Robertis and his collaborators[16,17,17a] that SPM can be isolated by gradient centrifugation of a crude mitochondrial fraction after prior lysis by osmotic shock. However, the optimal conditions for osmotic lysis were found to be different in our study, and other modifications in the isolation procedure have also been introduced. These, as well as the detailed rationale and description of the three types of preparation actually employed, are described by Cotman.[10]

Specific Procedures

Zonal Separation. The procedure yielding the most satisfactory preparation, both with regard to purity and quality, as well as convenience and yield, is that employing zonal ultracentrifugation of a crude, water-shocked mitochondrial fraction (isolated at 14,000 rpm × 20 min in a Spinco 30 rotor) in a 15–50% sucrose gradient. This method was developed in the laboratory of and in collaboration with Anderson at Oak Ridge.[11] A BXV rotor was employed routinely, and centrifugation was of the order of 27,000 rpm × 2 hr. Of the various zones observed (Fig. 4), the lightest one con-

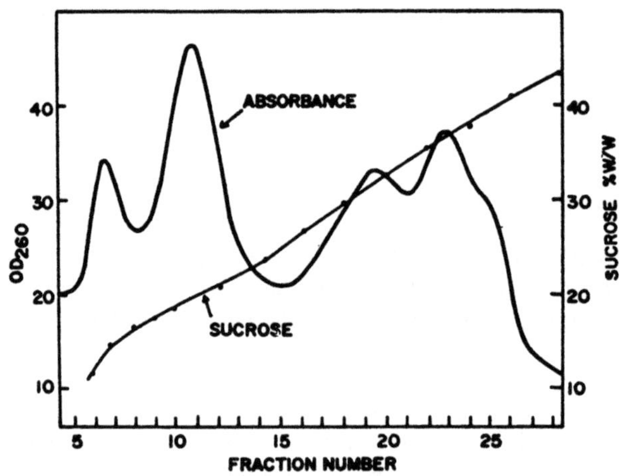

Fig. 4. Fractionation by zonal centrifugation. (BXV rotor, 2.5 hr at 25,000 rpm). Light-scattering profiles are shown.

tains myelin, the one centering on 23% sucrose (w/w) consists mainly of myelin, partially unshocked synaptosomes, and certain membrane fragments; the one located at 38–41% sucrose contains a remarkably homogeneous population of mitochondria (Fig. 5), while SPM are contained within the central zone that bands at 29–34% sucrose. Our best preparations have been obtained from a very narrow zone encompassing the span

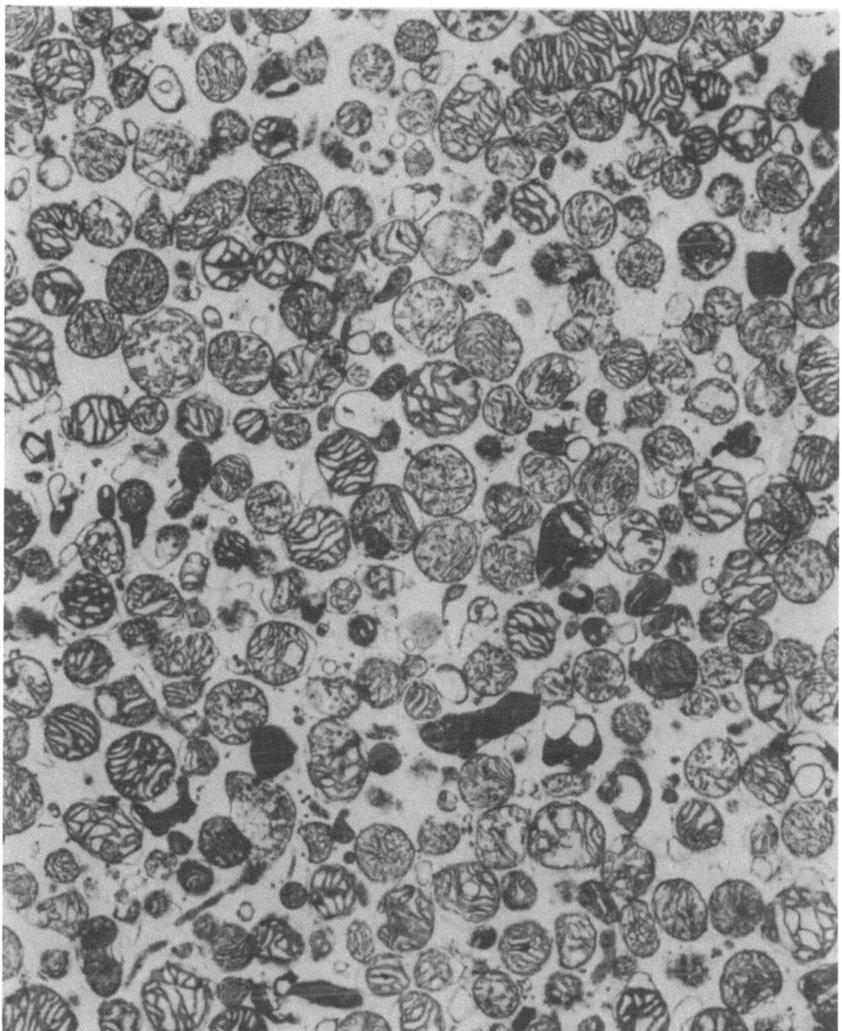

Fig. 5. The mitochondrial fraction (~38% sucrose) from a zonal preparation. Embedding and staining as in Fig. 1.

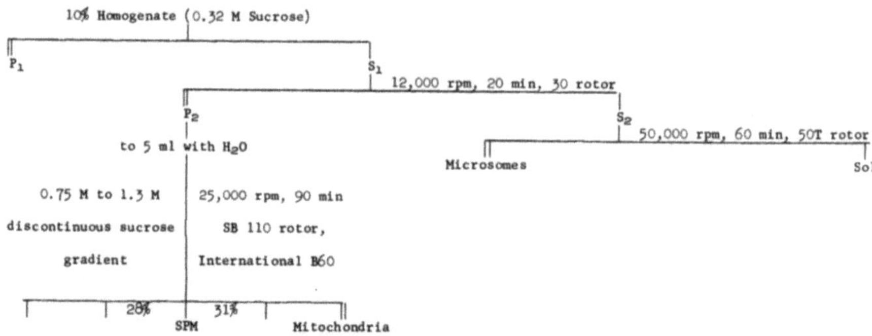

Fig. 6. A short-cut scheme for the isolation of purified SPM fractions by the use of sucrose gradients; DG P_3' = SPM.

between 29.0 and 31.4% sucrose, corresponding to a density difference of 0.013 g/ml. The yield in the SPM zone is of the order of 30 gm protein from each 10 gm wet weight of brain used.

Separation on Sucrose Gradients. The methodology developed for zonal separations formed the basis of a short-cut method particularly suitable for small-scale isolates, for instance, for tracer studies. This procedure is outlined in Fig. 6. In earlier work we also used a modification of the method of Whittaker *et al.*[15]

Separation on Ficoll Gradients. We also found that for certain purposes a fractionation technique, employing purification of synaptosomes prior to

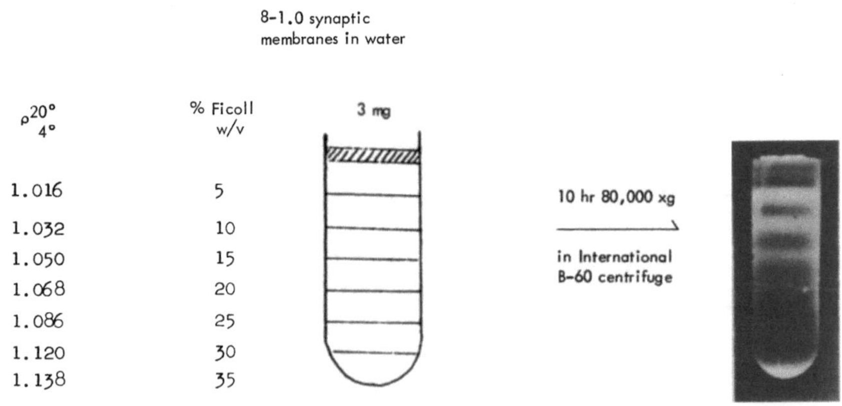

Fig. 7. Purification of SPM by means of isopycnic sedimentation in a Ficoll gradient (pH 6.7, 2 mM Tris HCl; 5 ml/layer). SPM bands in the region with a density of 1.03–1.05 g/ml.

lysis, followed by separation of the various membranous elements on a discontinuous gradient of Ficoll, afforded a satisfactory preparation. Although this technique is a great deal more laborious than our routine zonal preparation, the final SPM fraction banding at 10–20% Ficoll (Fig. 7) is of comparable quality. Yields, however, are not as satisfactory (~5 mg protein from 10 gm brain).

Methodology

Enzyme Assays

Na^+-K^+ activated, ouabain-sensitive ATPase (Na-K ATP phosphohydrolase, E.C. 3.6.1.4) was assayed by determining the difference between the amount of ATP hydrolyzed in the presence and absence of ouabain. The assay mixture contained 1 ml of buffer, which consisted of 50 m*M* Tris, 100 m*M* NaCl, 30 m*M* KCl, 3 m*M* $MgCl_2$, and 0.2 m*M* ouabain at a pH of 7.3 and 30–100 μg of protein in a volume of 0.05–0.10 ml. One set of samples contained no ouabain. The mixtures were preincubated at 37°C for 5 min, after which the reaction was started by the addition of 0.05 ml of 60 m*M* disodium ATP. The reaction was allowed to proceed for 10 min before it was stopped by the addition of 5 ml of a solution of ammonium molybdate (0.2%), sulfuric acid (0.2 *M*), and water. The color was developed by adding 0.2 ml of 1-amino-2-naphthol-4-sulfonic acid.[18] Shortly after addition of the reagent the solutions were cleared of turbidity by centrifuging for 10 min in a clinical centrifuge or in the International PK centrifuge at 3000 rpm. Optical density was read at 600 nm. The color was not stable at room temperature, but was stable for a few hours at +4°C. The reaction was linear over the protein concentrations used, and a comparison of enzyme inactivated by boiling with the reagent blank showed that this method of removing protein was satisfactory. The ATP reagent and the boiled enzyme blanks were of nearly identical magnitude in the samples studied.

NADH oxidase was assayed by following the oxidation of NADH at 350 nm, the isosbestic point of cytochrome c. The assay mixture contained 1 ml 0.05 *M* sodium phosphate buffer, 0.02 ml NADH (0.10 mg/ml), 0.02 ml cytochrome c (10 mg/ml) and enzyme. The reaction was started by the addition of NADH. The blank did not contain NADH.

Antimycin-insensitive NADH oxidase was assayed by the same procedure in the presence of 1 μg/ml of antimycin A. The antimycin was made up to a concentration of 100μg/ml in 95% ethanol and stored as a stock solution at −10°C in the dark.

NADPH oxidase was measured as for NADH oxidase, except NADPH served as electron donor.

NADH cytochrome c reductase (NADH:cytochrome c oxidoreductase, E.C. 1.6.2.1) was determined as described by Dallner *et al.*[19]

NADPH cytochrome c reductase (NADPH:cytochrome c oxidoreductase E.C. 1.6.2.3) was assayed by the procedure of Dallner *et al.*[19]

Cytochrome oxidase (cytochrome c:O_2 oxidoreductase, E.C. 1.9.3.1) was assayed as described by Duncan and Mackler.[20]

Acetylcholinesterase (acetylcholine acetyl-hydrolase, E.C. 3.1.1.7) was assayed as described by Ellman *et al.*[21]

β-N-acetylglucoaminidase (β-2-acetylamino-2-deoxy-D-glucoside acetylaminodeoxyglucohydrolase, E.C. 3.2.1.30) was determined by automatic assay on a Technicon analyzer as described by Corbett.[22] The enzyme was preincubated for 30 min at room temperature in 0.1% Triton X-100 in buffer before assay.

Acid phosphatase (orthophosphoric monoester phosphohydrolase, E.C. 3.1.3.2) was measured with β-glycerolphosphate as substrate (β-glycerolphosphatase). The assay mixture contained 1 ml of 0.05 *M* Na acetate buffer (pH 5.0), 5 m*M* β-glycerolphosphate, 0.02% digitonin and enzyme (0.05–0.3 mg). Incubations were carried out following the conditions of Sellinger *et al.*[23] for 14 hr at 37°C. Boiled enzyme blanks were identical with reagent blanks.

Electron Microscopy

Embedded preparations were observed with a Siemens Elmiscop Ia microscope operated at an accelerating voltage of 80 kV or an RCA model 3A electron microscope operated in the same manner. Fixation was carried out by a number of methods. The most effective way to preserve nerve-end particles for embedding in Araldite was with Luft potassium permanganate fixative as described by Hosie.[24] One-half milliliter of particle suspension at +4°C was rapidly mixed with at least 10 volumes of the fixative and allowed to remain at that temperature for 10 min, after which the fixed particles were pelleted at 6000 rpm for 5 min in a Sorvall centrifuge. Dehydration was carried out successively in 30% ethanol for 5 min, 50% for 5 min, followed by 70% ethanol, in which it was stored for embedding in Araldite. Synaptic membranes and synaptic vesicles were prepared for embedding in Epon[25] by fixing a small pellet with Caulfield's buffer,[26] i.e., Palade's veronal acetate buffered osmium containing sucrose, or in 2% osmium tetroxide in 0.05 *M* $NaPO_4$ buffer, pH 7.2.

Separation, Analysis, and Electrophoresis of Membrane Proteins

The techniques used have been described in previous publications from this laboratory.[11,27]

Tracer Methodology

Most of the techniques employed have already been described.[28,29]

PROPERTIES

Properties of the SPM

Morphology (Figs. 8 and 9)

The general morphology of representative entities contained in the SPM fraction is that expected of a plasma membrane: linear or vesicular structures composed of unit membranes, with the anticipated spacing—two electron dense lines each 20 Å in diameter and separated by a clear space approximately 35 Å across. In addition, closer examination, especially at higher magnification, reveals the presence in a significant number of bodies of *all* the features (except for the removal of mitochondria) characterizing the synaptic junction *in vivo* (compare Fig. 1 and Fig. 9). We thus have available characteristic morphophological, as well as enzymatic, markers (see below) that set apart the membrane we wish to study, and in this regard we are, perhaps, somewhat more fortunate than other investigators interested in plasma membranes (see, for example Parsons,[7] Wallach,[8] and Warren *et al.*[9]). Furthermore, most, if not all, af the membranes visible in this fraction appear to be "thick" (90–100 Å diameter) rather than "thin" (50–60 Å diameter) membranes. Since both the membranes of the endoplasmic reticulum and of the mitochondria are reported to be of the latter type,[7] the morphological details provide some evidence for the purity of the preparation.

Isopycnic Density

The buoyant density of the SPM, as determined by its isopycnic banding in a variety of media, is summarized in Table I and compared to similar studies on other membranes. A bouyant density of 1.125–1.140 in sucrose appears to be somewhat characteristic of SPM, shared only by the plasma membrane of the Ehrlich ascites carcinoma cell, and by the outer membrane of liver mitochondria. Particularly significant is the observation that electroplax plasma membrane appears to band at a different density. Current views suggest that isopycnic banding densities are principally a reflection of the lipid composition of the membranes involved and that a density in the range observed corresponds to a content of approximately 50% phospholipid by weight.[6] This prediction is borne out by actual analysis in the studies of Lapetina *et al.*[37]

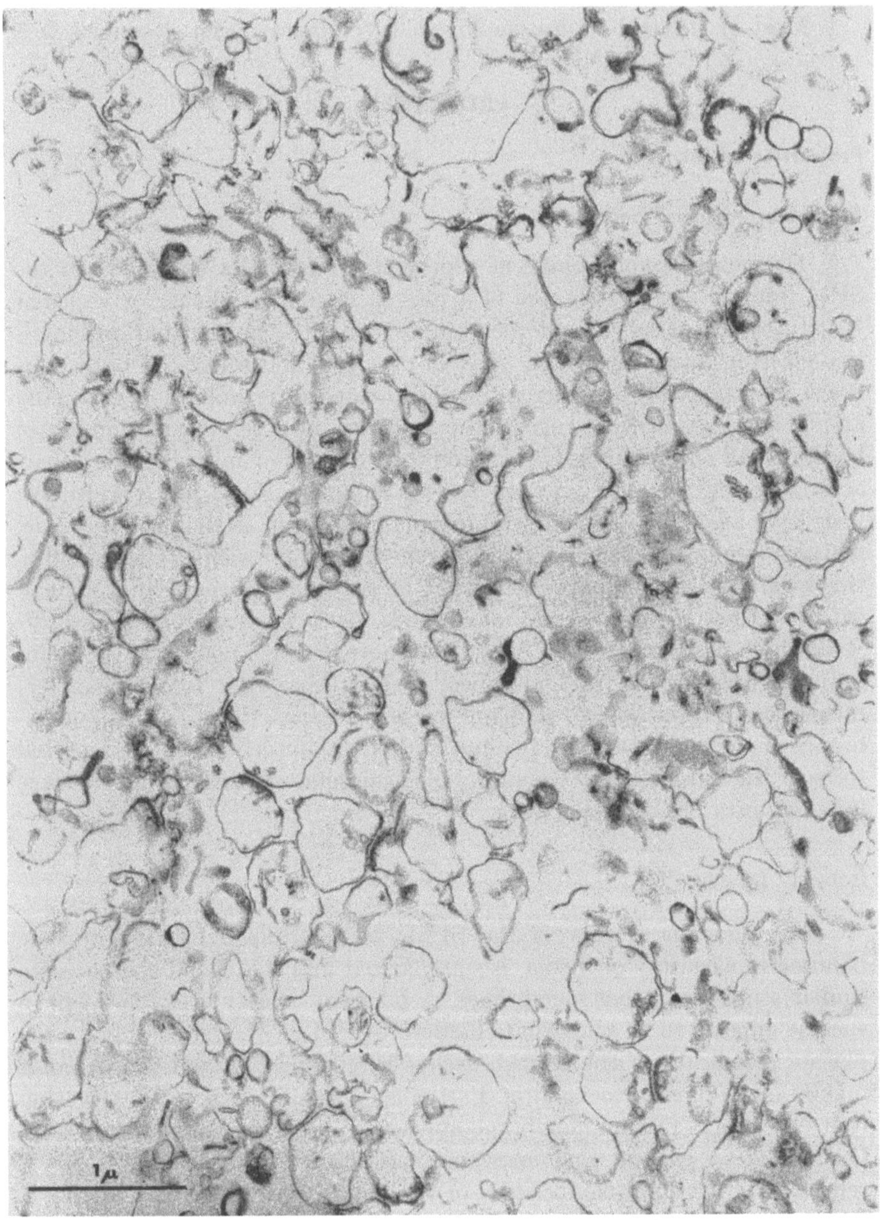

Fig. 8. SPM fraction from a zonal gradient (29.1–31.7% sucrose).

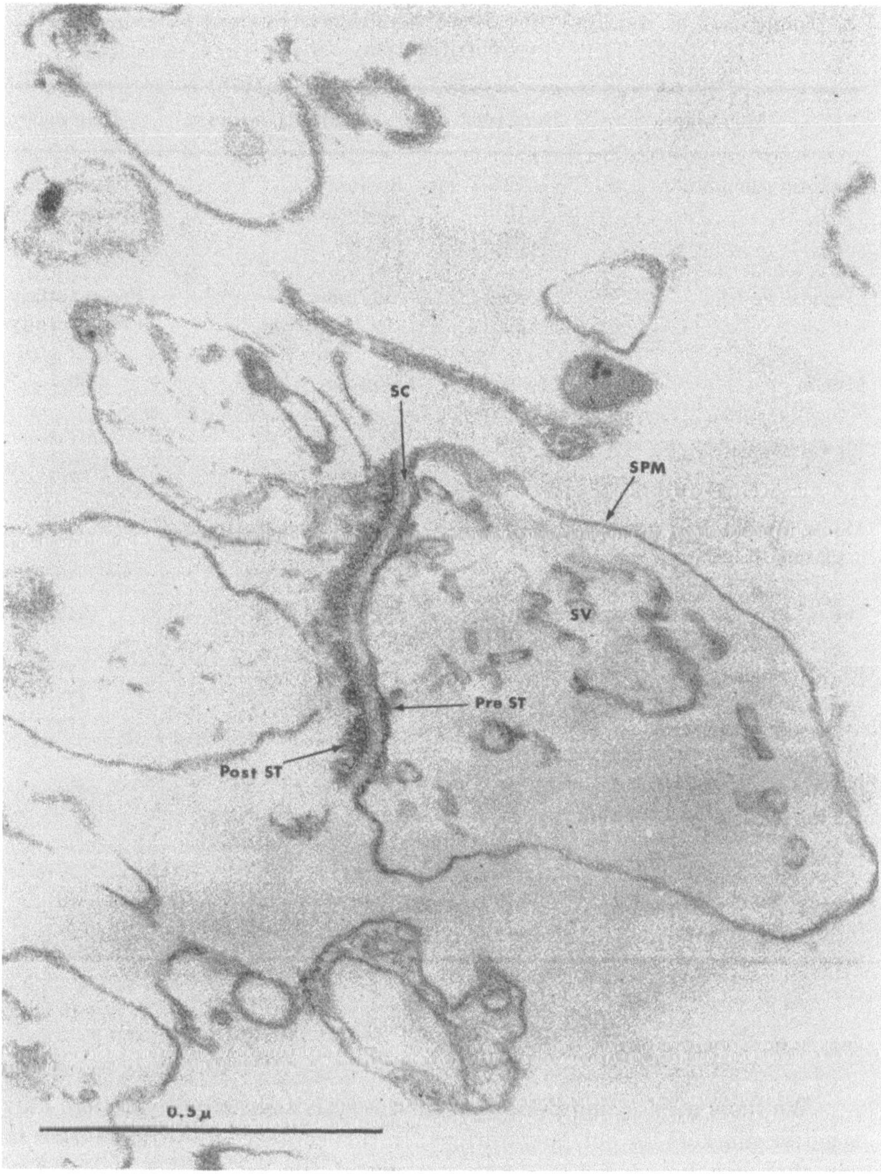

Fig. 9. Appearance of the SPM fraction shown in Fig. 8, at higher magnification. Many structures similar to those characteristic of the synapse *in vivo* are shown. Compare with Figs. 1 and 3; see legend to Fig. 1 for an explanation of symbols used. Sample preparation described in the text.

TABLE I
Comparison of Densities of Various Membranes Obtained by Isopycnic Centrifugation

Membrane	Density at 4°C	Gradient medium	Reference
Synaptic membrane	1.125–1.140	Sucrose	Present study
	1.15	CsCl-sucrose	Present study
	1.04	Ficoll	Present study
Synaptic vesicles	1.08	Sucrose	Present study
	1.15	CsCl-Sucrose	Present study
Myelin	1.10	Sucrose	(30)
Liver plasma membrane	1.18	Sucrose	(31)
	1.22	Sucrose	(32)
Mouse fibroblast (L cell) plasma membrane	1.21	Sucrose, 5 m*M* $MgCl_2$	(33)
HeLa cell plasma membrane	1.19	Sucrose	(34)
Ehrlich ascites	1.13	Sucrose	(7)
carcinoma cell	1.15	NaBr	(7)
plasma membrane	1.04	Ficoll	(7)
Eel *Electrophorus* electroplax plasma membrane	1.20	Sucrose	(35)
Liver mitochondria:			
outer membrane	1.12–1.14	Sucrose-20 m*M* $NaPO_4$	(36)
inner membrane	1.21	Sucrose-20 m*M* $NaPO_4$	(36)

Enzymatic Composition

We have used a number of enzymatic activities as both positive and negative markers for our preparations. In this context a positive marker is one that can be considered characteristic of the subcellular fraction or organelle of interest, a negative marker one that is to be regarded as absent in this fraction but present in and characteristic of another—potentially contaminating—fraction or organelle (see, e.g., De Duve[38]). When compared to the original homogenate the total activity of a positive marker in

TABLE II

Comparison of Enzyme-Specific Activities in Various Fractions

Fraction	NaK ATPase	NADH oxidase, antimycin-insensitive	Cytochrome oxidase
Homogenate, total	1.0	1.0	1.0
SPM[a]	7.3 ± 1	0.4 ± 0.2	0.5 ± 0.2
Mitochondria	—	2.4 ± 0.5	3.5 ± 1.0
Microsomes	—	5.5 ± 1.5	—

[a] Data were compiled from three experiments. SPM values are for the fraction at about 30% sucrose. The SPM value for NADH oxidase, antimycin-insensitive, was corrected for contamination by mitochondrial antimycin-insensitive NADH oxidase based on the ratio cytochrome oxidase/antimycin-insensitive NADH oxidase in the two fractions.

a purified fraction would be expected to remain essentially constant* and its specific activity to increase by the reciprocal of the fractional content,† while both the total and specific activity of negative markers should tend to zero with increasing purification. We have chosen Na^+-K^+ activated, ouabain-sensitive ATPase as a positive marker[16,17,24] for SPM,‡ and the following negative markers for evaluation of the extent of contamination by other membranes: Antimycin-insensitive NADH oxidase has been reported to occur in liver in the outer membrane of mitochondria;[7,36]§ in liver and brain in various components of the microsomal fraction (smooth and rough endoplasmic reticulum);[39,40] and in liver also in the limiting plasma membrane.[31,41] It is therefore an excellent negative marker for the most likely contaminants. Cytochrome oxidase is considered a marker exclusively confined to the inner membrane of mitochondria[7,36] and is the preferred indicator for mitochondria and mitochondrial fragments. Finally, β-N-acetylglucoaminidase and β-glycerolphosphatase (acid phosphatase)

*This is true *only* for the idealized case in which the enzymatic activity under consideration occurs exclusively in the fraction of interest *and* the recovery of that fraction is quantitative.

†Defined as (amount of protein in relevant fraction)/(total protein in homogenate).

‡This activity would also be expected to be present in various other membrane systems, such as the endoplasmic reticulum and various plasma membranes other than the SPM. However, these other membranes might also be expected to contain antimycin-insensitive NADH oxidase, and therefore appropriate corrections can be applied by means of the use of this negative marker.

§While in liver, antimycin-insensitive NADH cytochrome c reductase appears to have a distribution identical to that of the corresponding NADH oxidase, this does not appear to be true in brain. On analysis by zonal ultracentrifugation we find the former enzyme to be associated *exclusively* with mitochondria, presumably in their outer membranes, while the latter exhibits the expected multimodal distribution.

are representative of a large group of enzymes localized in the lysosmes of liver[42,43] and, probably, of brain.[44,45]

Preliminary data on SPM prepared by any of the procedures reported in the literature indicated an enhancement of specific activity of the Na^{+}-K^{+}-activated ATPase between five and sevenfold, with a concomitant decrease of specific activity of cytochrome oxidase to approximately 10–15% of the starting material. Content of antimycin-insensitive NADH oxidase, however, indicated intolerably high levels of contamination of SPM by other (probably reticular) membranes (specific activity anywhere from equal to twice that of the whole homogenate). It was principally these

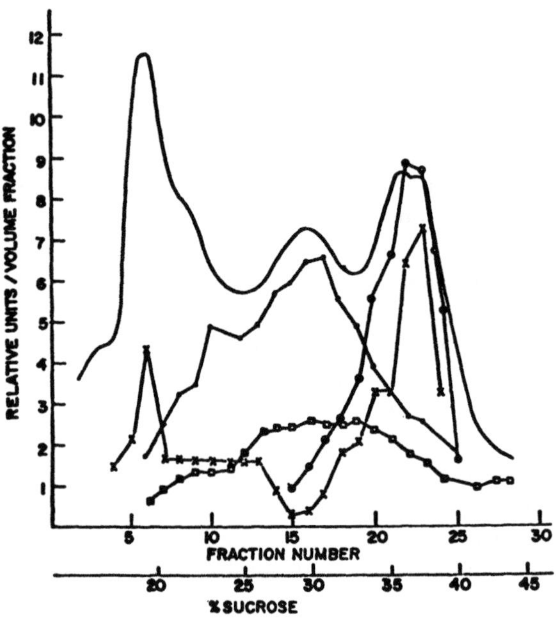

Fig. 10. Enzyme distributions in zonal gradients (BXV rotor, 2.5 hr at 27,000 rpm). The nature of the various isolated zones is described in the text. SPM band around 30% sucrose. Smooth curve: light-scattering OD profile; dots: NaK ATPase (average of two experiments); crosses: NADH oxidase, antimycin-insensitive (average of two experiments); squares: glucoaminidase (average of three experiments; glycerolphosphatase showed identical total activity profiles); circles: cytochrome oxidase (average of two experiments). Recovery of enzyme activity from P_2 total was: NAK ATPase, 97%; NADH oxidase, antimycin-insensitive, 80%; glucoaminidase, 90%; cytochrome oxidase, 50%.

findings that prompted us to search for a more satisfactory purification procedure. The data of Table II indicate that we have been successful in this quest, and that our zonal preparation appears to be of adequate purity by these criteria ($\leq 10\%$ contamination by either mitochondrial or microsomal membranes). Similar, but much less extensively documented, results have also been obtained with Ficoll gradients. One of the interesting results of these studies has been the recognition that part of both N-acetylglucoamidinase and glycerolphosphatase activity appears to be closely associated with the SPM preparation. They band in identical positions in zonal sucrose and other gradients, and hence are localized in entities of identical buoyant density (Fig. 10). ATPase and N-acetylglucoamidinase activities appear to be affected about equally by a variety of treatments, and no additional enrichment or purification with regard to *either* activity can be obtained on rate sedimentation. Therefore, particles with which they are associated possess virtually identical sedimentation coefficients as well. We do not know whether this behavior is indicative of some peculiarity of neuronal, or possibly synaptosomal, lysosomes, or of the SPM. Also interesting is the virtual absence of a NADH oxidase (or even of any enzyme system capable of oxidizing NADH or NADPH) in the SPM. This finding tends to rule out Skou's postulate[46] of an obligatory link between electron transport and ATPase-mediated alkali-cation transport.

Electrophoretic Pattern of Constituent Proteins

One of the more powerful techniques available for the characterization of a membrane in terms of its constituent proteins is provided by their electrophoretic mobility on polyacrylamide gels. The pattern observed not only provides a ready check on the reproducibility and purity of any given preparation relative to others in the same class, but also provides some insight into the complexity and individuality of one *type* of membrane compared to others derived from the same, or any other, kind of cell. By this means we had already established that the patterns of myelin, synaptic vesicles, various microsomal membranes, and a cruder SPM fraction all were reproducible and distinct from one another, and so provided a useful fingerprint for these various membranes.[27] When we extended the method to the purified SPM preparations just described, we observed the characteristic pattern shown in Fig. 11. It consists of five major bands (1–5) approximately equal in intensity, and an equal number of minor bands. It is free of artifacts except, perhaps, for a diffuse background throughout the region of interest, and it is reproducible: at least five different zonal preparations, as well as preparations made by means of the other two procedures, all generated similar patterns.

Figures 12 and 13 display the electrophoretic mobilities of other membrane preparations determined by the same technique. Of particular interest are comparisons between synaptic vesicles and SPM (Fig. 12) and between three different plasma membranes (Fig. 13): SPM, myelin (Schwann cell plasma membrane), and erythrocyte plasma membrane. All three are distinct, even with respect to their major (presumably structural) components. The myelin pattern is considerably less complex, while that of the

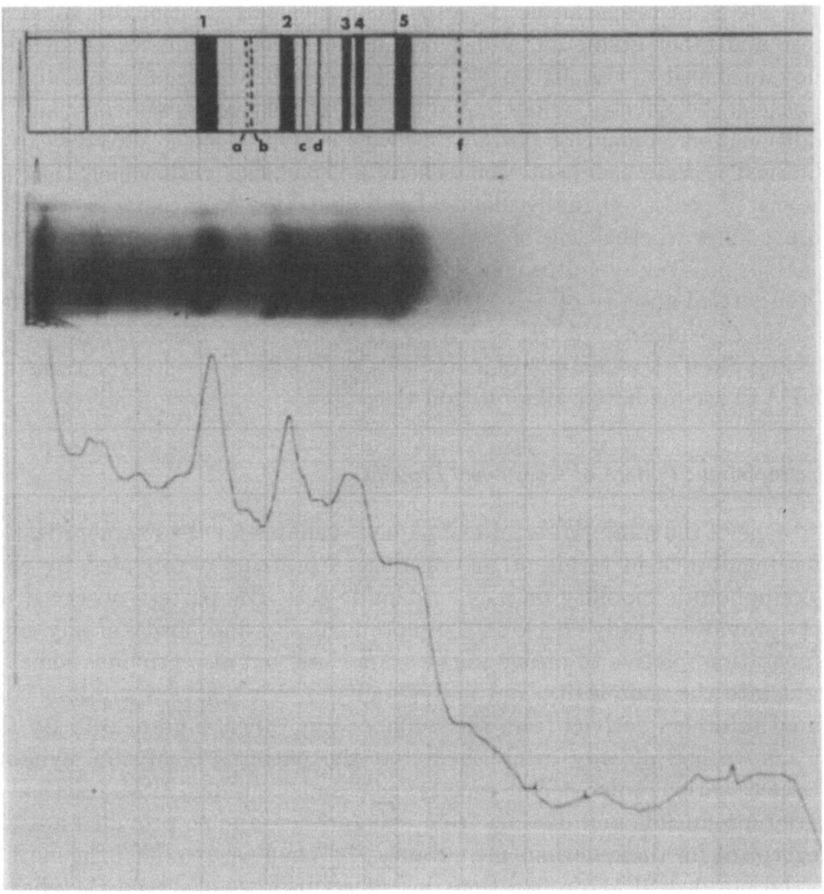

Fig. 11. Electropherogram of the insoluble proteins of SPM in phenol–acetic acid–urea. Migration is toward the cathode from left to right; 90 min at constant current (5.5 mA/tube). Proteins are stained with Amido Black. Three representations are shown: A photograph of the actual gel is in the center, a diagrammatic sketch is at the top, and a scan recording by means of a Joyce–Loebl microdensitometer is at the bottom.

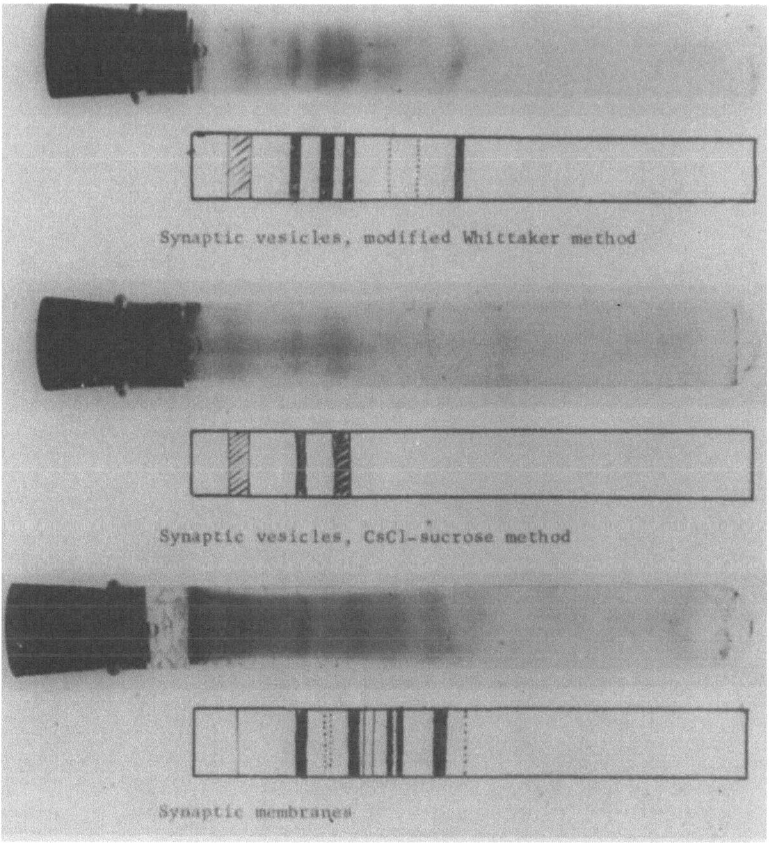

Fig. 12. Comparison of the electropherograms of the SPM with those given by synaptic vesicle preparations. Conditions as described in the legend to Fig. 11.

red cell membrane is apparently of an order of complexity similar to that of SPM. This latter finding is of a significance, since this preparation was obtained from a single and homogeneous cell type.* Inferentially then, the multiplicity of bands in the SPM may also be an intrinsic property, rather than being due to a mixture of membranes in the preparation originating from different cell types. Rats from two strains (Sprague-Dawley and Wistar) yielded SPM preparations indistinguishable in their electrophoretic patterns. On the other hand, we observe at least one major differ-

*This was a highly purified preparation from a single Rh-negative donor, kindly provided us by Dr. S. Bakerman.

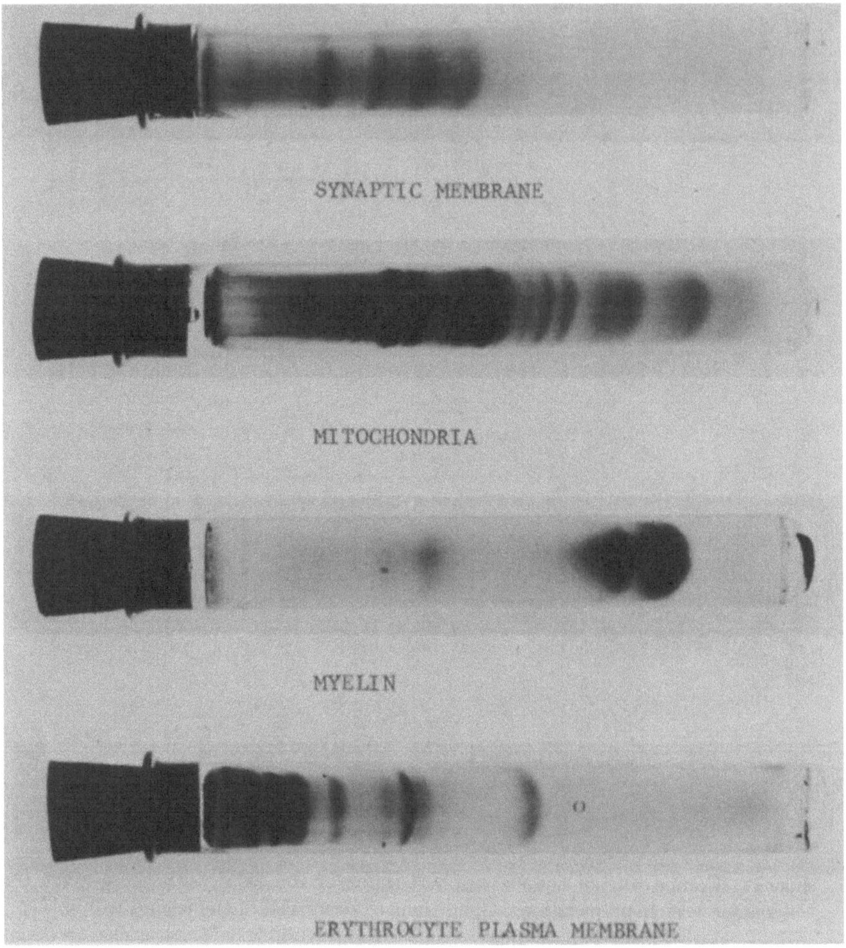

Fig. 13. Comparison of the electropherograms given by the insoluble proteins of various membrane fractions.

ence when comparing the mobilities of proteins from SPM of rat and guinea pig (Fig. 14).

Amino Acid Composition

Additional clues concerning variable and invariant features in membrane structure can be derived from an examination of the amino acid composition of their insoluble proteins. For instance, any postulate of uni-

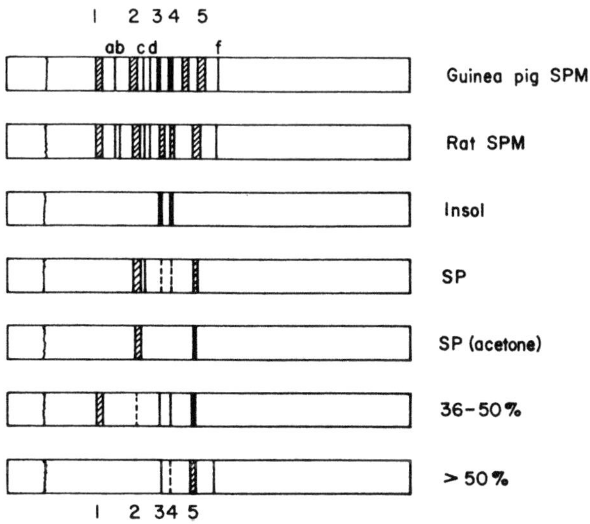

Fig. 14. Schematic comparison of electrophoretic patterns of, from top down, the insoluble proteins of SPM for guinea pig cerebral cortex; SPM from rat cerebral cortex; the insoluble fraction remaining after the latter has been treated with a mixture of cholate, deoxycholate, and sodium dodecyl sulfate (15 mg protein and 30 mg, 15 mg, and 22.5 mg, respectively); the structural protein fraction precipitated from the mixture solubilized with detergents at an ammonium sulfate concentration of 18% of saturation; the structural protein fraction washed by extraction with acetone; the fraction precipitated from the mixture solubilized with detergents between ammonium sulfate concentrations 36 and 50% of saturation; and the fraction remaining in solution at an ammonium sulfate concentration greater than 50% of saturation.

formity of membrane structure would demand identity, or at least close homology, in amino acid composition of the major, structural proteins of different membranes, particularly when derived from the same organism. This model can clearly be ruled out on the basis of the data summarized in Table III. The compositions of SPM and myelin are strikingly different, not only with regard to important structure determinants such as Pro and (Gly + Ala), but even with respect to their total content of highly polar and non-polar residues. Clearly, these two plasma membranes of the same tissue bear very little resemblance to one another—and similar differences obtain between SPM and other intracellular membrane systems such as the mitochondria and the endoplasmic reticulum. On the other hand, SPM and

TABLE III

Amino Acid Composition of Membranes

(all in no. of residues per 1000 residues)

Amino acid	SPM, rat brain[a]	Mito-chondria, rat brain[a]	Rough reticulum, rat liver[b]	Myelin, bovine[c]	Eryth-rocyte, human[d]	Plasma mem-brane, HeLa[e]	Smooth mem-brane, HeLa[e]
Lysine	58.5	63.0	82.6	58.8	50.0	104.8	78.7
Arginine	45.9	45.8	46.8	41.2	49.3	49.6	49.8
Histidine	21.4	21.2	—	23.2	25.5	38.2	22.2
Aspartic	101.7	88.5	112.2	65.2	82.9	73.2	101.4
Glutamic	120.3	106.0	124.0	73.2	125.8	114.4	119.6
Serine	70.8	57.5	47.2	100.8	62.7	70.8	80.0
Threonine	53.9	53.5	56.5	63.9	58.0	52.4	60.0
Cystine (½)	17.1	11.6	1.5	37.6	14.8	—	—
Glycine	70.1	77.5	88.7	96.8	67.1	92.9	83.9
Alanine	75.1	84.5	91.0	90.5	82.3	84.0	84.9
Methionine	30.4	21.6	—	11.6	24.3	15.5	18.9
Isoleucine	49.4	51.5	55.4	41.8	47.7	33.7	36.6
Leucine	93.1	95.0	111.6	88.7	123.2	78.0	76.4
Valine	61.0	58.0	84.3	59.0	57.7	60.4	59.3
Phenylalanine	46.2	95.3	28.5	52.9	46.6	25.4	34.4
Tyrosine	32.1	33.6	—	33.1	26.0	22.8	16.7
Tryptophan	1.5	—	3.0	11.8	4.5	—	—
Proline	51.2	60.5	66.5	39.6	51.1	83.7	76.9
Total basic	126	130	83	123	125	183	150
Total acidic	222	194	236	138	208	188	221
Total nonpolar	311	305	282	296	326	236	242
Ser + Thr	124	111	104	164	111	123	140
Gly + Ala	145	162	180	187	149	177	169

[a] Present study.
[b] V. C. Manganiello and A. H. Phillips, *J. Biol. Chem.* **240**: 3951 (1965).
[c] F. H. Hulcher, *Arch. Biochem. Biophys.* **100**: 237 (1963).
[d] S. Bakerman and C. Wasemiller, *Biochemistry* **6**: 1100 (1967).
[e] M. B. Bosmann and E. H. Eylar, *Arch. Biochem. Biophys.* **128**: 51 (1968).

erthrocyte plasma membrane, although distinct entities with characteristic individual features even in their protein electropherograms, do appear to have in common certain invariants. They contain virtually identical numbers not only of basic residues, but of proline and of (Gly + Ala) as well. They greatly resemble one another in their content of total polar and total nonpolar residues, and in approaching a ratio close to unity between these two classes. Interesting also is the remarkable constancy of the sum (nonpolar + Ser + Thr).

Resolution

Solubilization and Fractionation

The rationale behind these investigations was really twofold: (1) to determine whether the problem of the complexity and individuality of membranes with regard to their constituent proteins was susceptible to further simplification and analysis by the separation, purification, and characterization of one or more of the major components, and (2) to subject to a critical test the then current hypothesis that all membranes contained as one of their major structural determinants a monodisperse entity, the so-called structural protein.[47] These structural proteins of different membranes were each believed to be composed of a single polypeptide chain of molecular weight about 25,000, closely resembling one another in amino acid composition.[47–52] In an even more restricted version of the hypothesis as applied to a single-celled eukaryote, i.e., *Neurospora crassa*, this similarity was to be replaced by a complete identity of appropriate protein components isolated from mitochondria, other membranes, and even the soluble portion of the cytoplasm.[51,52]

In order to accomplish these aims, we applied, with suitable modifications, the methodology first developed by Green and his collaborators for mitochondria[48–50] to purified preparations of SPM. A fraction conforming to the operational definition of a "structural protein" could indeed be isolated by ammonium sulfate fractionation (0–18% of saturation) of an SPM solubilized by means of detergents (2 mg deoxycholate, 1 mg cholate, and 1.5 mg sodium dodecyl sulfate per milliliter). It is distinct from other fractions salted out at higher ammonium sulfate concentrations (Fig. 14) and represents a significant, reproducible, and interesting protein component of the SPM.

Properties of the Structural Protein

The entity so isolated accounts for some 20–30% of the total insoluble proteins of the SPM, and, as shown in Figs. 14 and 15, is composed of two bands previously identified as major components (mainly band 2 plus some of band 5) of the unmodified membrane. It is thus not an artifact of the treatment with some rather potent detergents, in the sense of being, for instance, a denatured version of some other (catalytic?) proteins of the SPM. Band 2 appears unique to this protein fraction and absent from other

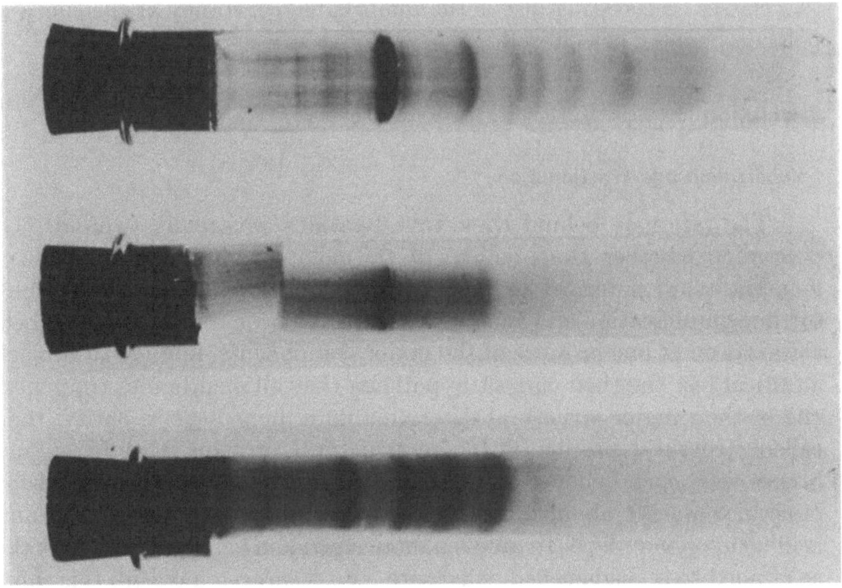

Fig. 15. Comparison of the electropherogram of structural proteins from mitochondria (top), SPM (center), with that of the whole SPM (bottom).

such fractions (Fig. 14). When compared to a structural protein fraction isolated from the mitochondria of the same preparation it is this band also that appears to be present in both organelles (Fig. 15). In addition, the degree of complexity of the mitochondrial structural protein appears greater, but neither of the two patterns can be regarded as representative of a homogeneous polypeptide.

The amino acid compositions of these two structural protein fractions are compared with each other on the one hand, and with the global insoluble proteins of the SPM on the other (Table IV). Although statistically significant differences are apparent between the structural protein and the totality of the insoluble proteins of the membranes, these are neither very profound nor any more pervasive than are the differences between the two structural protein fractions themselves. As already mentioned, mitochondrial structural protein appears more heterogeneous than does its counterpart in the SPM, and it is therefore not certain whether the differences observed are referrable to intrinsic differences in compositon of the polypeptides responsible for the overlap between mitochondria and SPM (Fig. 15), or are due to contributions by additional components in the former. In any event, we can conclude (1) that structural protein fractions from

TABLE IV

Amino Acid Composition of "Structural Proteins"

Amino acid	SPM	"Structural protein" of SPM[a]	"Structural protein" of brain mitochondria
Lysine	58.5	61.5	63.6
Arginine	45.9	(51.5)	49.3
Histidine	21.4	22.4	22.4
Aspartic	101.7	[94.5]	86.6
Glutamic	120.3	[113.2]	102.0
Serine	70.8	[61.9]	51.2
Threonine	53.9	50.0	48.6
Cystine ($\frac{1}{2}$)	17.1	7.2	8.7
Glycine	70.1	[72.5]	83.0
Alanine	75.1	[76.6]	85.0
Methionine	30.4	[36.7]	46.7
Isoleucine	49.4	54.5	56.3
Leucine	93.1	(99.6)	100.5
Valine	61.0	62.5	66.5
Phenylalanine	46.2	48.1	45.4
Tyrosine	32.1	35.0	35.6
Tryptophan	1.5	—	—
Proline	51.2	[55.6]	48.7
Total basic	126	135	136
Total acidic	222	208	189
Total apolar	311	312	318
Ser + Thr	124	112	100
Gly + Ala	145	149	168

[a] Statistically significant differences (90% confidence level) between SPM and structural protein from the same source are shown by parentheses, those between structural protein of SPM and the homologous fraction from rat-brain mitochondria are indicated by brackets.

different membrane systems appear to be distinct; (2) that they are heterogeneous with regard to their constituent polypeptides and their amino acid composition; and (3) that the latter does not appear to differ qualitatively from that of other major insoluble proteins contributing to the structure and physical stability of these membranes. Somewhat similar conclusions have recently been reached quite independently by Green and his collaborators.[53,54]

Biosynthesis

Turnover Studies

Von Hungen *et al.*[28] have investigated protein turnover in various subcellular and subsynaptosomal fractions of rat-brain cortex. In these experiments mature animals (100–120 days old) were used, and radioactive leucine was injected as shown in Fig. 16. The time course of label decay in various subsynaptosomal fractions is shown in Fig. 17. Several interesting features have emerged from studies such as these. One is the remarkable similarity in turnover among the various insoluble components of the synapse (including the SPM), each showing contributions by a rapid component or components with a half-life of the order of one week, and of a slow component, accounting for some 20% of the total protein, with a half-life of 21 days. The second is that the soluble proteins of the synapse give clear indication of having been synthesized elsewhere (i.e., in the perikaryon) and transported to the nerve ending, presumably by classical, "slow" axoplasmic flow.[54–58] Conversely, the insoluble parts of the organelle give no indication of a lag, and thus their proteins must have been synthesized either in the immediate vicinity of, or been transported much more efficiently to, the synaptic region (see also below). Third, these data show that synaptic vesicles are not characterized by a turnover rate that is either unusually high or at least asynchronous with that of other membranous components of the synapse. These findings, together with the electrophoretic data summarized earlier, permit us to rule out a model for synaptic

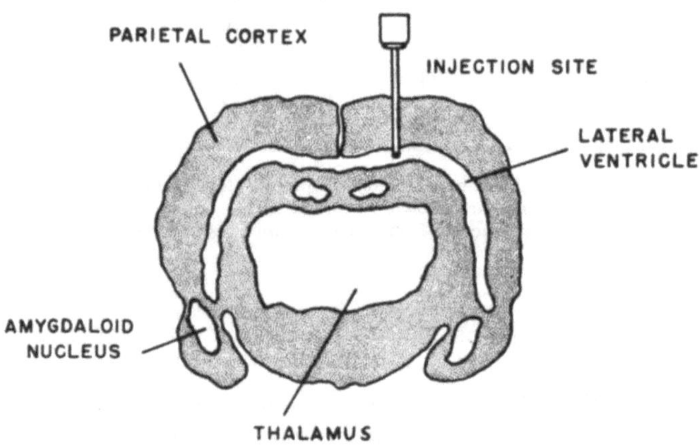

Fig. 16. Mode of administration of label for turnover experiments.[28] (With permission of the *Journal of Biological Chemistry*.)

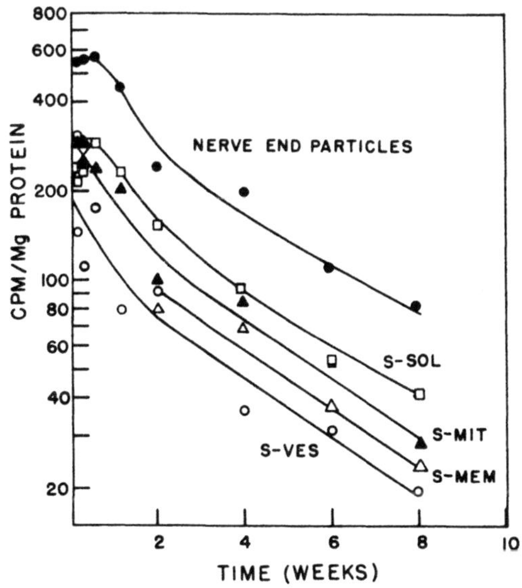

Fig. 17. Turnover of labeled leucine in the proteins of the synaptic complex.[28] Sol: soluble; mit: mitochondrial; mem: SPM; ves: vesicles. (With permission of the *Journal of Biological Chemistry*.)

function that postulates the discharge of transmitter from synaptic vesicles coincident with their fusion with the SPM (and requires, for mass balance, either subsequent reformation of vesicles from the SPM by budding, or destruction of some part of it, presumably by controlled proteolysis).

Amino Acid Incorporation in Vivo

In order to gain further insight into the dynamics of protein synthesis as it applies to the SPM, we have turned our attention to a study of the incorporation of labeled leucine shortly after administration by the same route as described earlier. In agreement with the turnover experiments, protein in all membrane fractions, i.e., the microsomes, mitochondria, and SPM, all became maximally labeled over a relatively short time ($\leq$15 hr) and thereafter decayed with characteristic kinetics. In order to evaluate significant differences in labeling kinetics, these are expressed as relative specific activities defined as (specific activity of fraction of interest) : (specific activity of total homogenate) (Figs. 18 and 19). From the data of

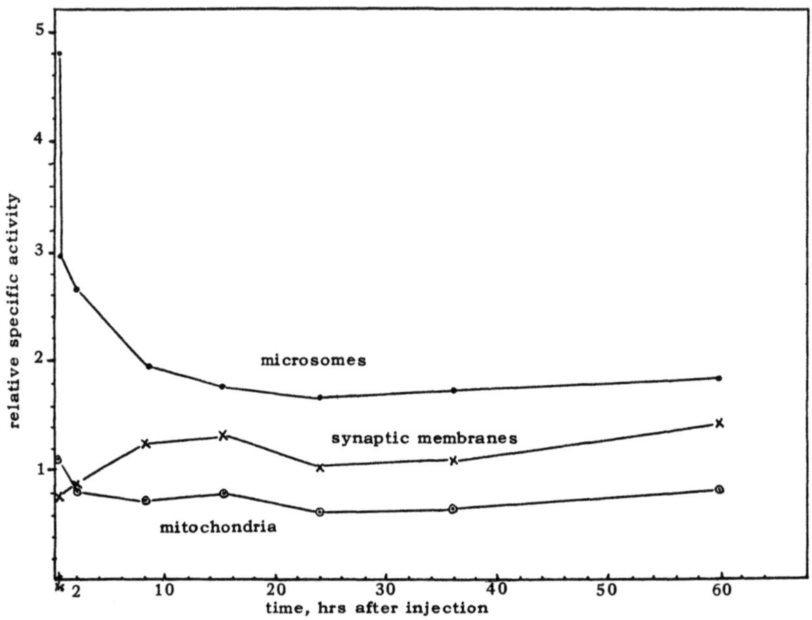

Fig. 18. Incorporation of 4,5-^{3}H-leucine into insoluble proteins of membranes. Each animal was injected in the right ventricle with 7.5 μCi of leucine (1 Ci/mMole) in 0.15 μl of 0.9% NaCl. Each time-point corresponds to three animals. SPM were isolated from shocked, crude mitochondrial fractions (17,000g $\times$ 20 min) on sucrose gradients as the band between 28 and 31% sucrose (see Fig. 6). Microsomes were isolated by centrifugation at 104,000g $\times$ 60 min. All fractions were extracted with water and 0.9% KCl. Data are presented as relative specific activities.

Fig. 19 it is evident that both microsomes and mitochondria, known sites of protein synthesis, do qualify kinetically as possible sources of newly synthesized protein, while the SPM appears to qualify, at least in part, as a potential sink. In other words there is a relatively slow flow of label into that fraction. The data are, however, not precise enough—and their resolution, because of the very low activity levels attainable, is not aided greatly by going to even shorter times—to rule out unequivocally *any* contribution of the membrane to its own biosynthesis or to distinguish between different modes of accumulation for protein synthesized elsewhere.* Likely candidates

*Barondes' most recent data[58] on *in vivo* transport of proteins containing newly incorporated leucine to nerve endings in mice showed both a rapid and a slow component in the labeling patterns of relatively crude SPM fractions. In contrast, the experiments of Ray *et al.*[66] suggest that for hepatocyte plasma cell membrane the site of protein synthesis is in the perikarya, distal from the membrane.

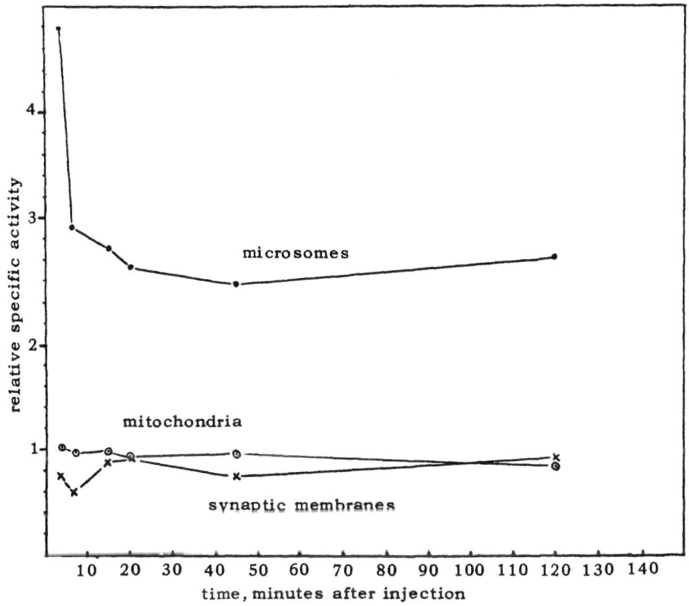

Fig. 19. Incorporation of ^{3}H-leucine into insoluble proteins after a short exposure. Experiments were performed as described in the legend to Fig. 18. For the first three time-points homogenates were fractionated in the presence of 0.025 *M* *D,L*-leucine as a nonradioactive chase.

for the latter process would appear to be rapid axoplasmic flow,[58–62] "reverse" transfer from the perikarya of post-synaptic neurons (or, more trivially, incorporation only into the post-synaptic component of the SPM), transfer from surrounding glia, and intersynaptic transfer from synaptic mitochondria to the SPM.[63–65]

Amino Acid Incorporation in Vitro

The time course of amino acid incorporation at very short intervals after administration of label therefore becomes crucial to an understanding of the mode of biosynthesis of the SPM. Since, as just indicated, the amount of information obtainable *in vivo* is severely hamstrung by experimental limitations, we are currently attempting to extend our investigations to the use of more highly resolved *in vitro* preparations. These have the added advantage of greater flexibility in the employment of several tried and true techniques of tracer methodology, such as pulse-chase experiments and the use of selective inhibitors. We are currently employing two such preparations, cortex slices as described by Orrego and Lipmann,[67] and intact synaptosomes, similar to those used by Autilio *et al.*,[68] and by Morgan and Austin[69] (also see Austin *et al.*[70]). Preliminary exploration with the first technique suggests that there may be transfer of label into the SPM during a chase period with unlabeled amino acid subsequent to a 15–min pulse labeling period. Conversely, we tentatively infer from very cursory studies by the second technique that, at least with our mature animals, the contributions to *in vitro* incorporation made by extramitochondrial, and especially by extramicrosomal, elements in the synaptosome preparation is probably not quantitatively significant.

Since, at least to a very large part, synaptic proteins, including the insoluble components of the SPM, appear to be of somatic origin, it may not be amiss to summarize here some of the key features, particularly those with unique and perhaps regulatory significance, that we have previously established for protein synthesis by neuronal polyribosomes.[29] In general, most of the properties of this system are reminiscent of those established for analogous preparations from other sources. The system is highly active and apparently capable not only of completing but even initiating the synthesis of polypeptide chains for which it had been preprogrammed *in vivo.*[70] Its optimal Mg^{+2} concentration is 10mM; when the total divalent-metal concentration is held constant at that value while some of the Mg^{+2} is substituted by Ca^{+2}, there is at first a stimulation [25% at a $(Ca^{+2})/(Mg^{+2})$ ratio of $\frac{1}{4}$], followed by a pronounced inhibition that reaches 90% when all the Mg^{+2} has been replaced by Ca^{+2}. At a ratio of $3Ca^{+2}/2Mg^{+2}$ the system is poised at the control value and can swing over either toward stimulation or inhibition, depending on whether the ratio is increased or decreased. Similar poising is observed when the concentration ratio of monovalent cations are altered. Optimal activity here is observed at a $[Na^+]$ plus $[K^+]$ of 140 mM. The system is poised at approximately 110 mM $[K^+]$ plus 30 mM $[Na^+]$. Increasing or decreasing the $[K^+]$ contribution by 20 mM stimulates activity by some 25%, while a further decrease in $[K^+]$ to 20 mM produces 50% inhibition.

CONCLUSIONS

Synaptic plasma membranes from rat-brain cortex can be isolated in a highly purified form. They appear as unique in their morphology as they are in a number of biophysical and biochemical properties. They are composed of some five major, presumably structural, polypeptides, that appear to be susceptible to individual isolation and characterization. A structural protein fraction composed of two of these components has been isolated. On the basis of a comparison of electrophoretic mobilities and amino acid composition with those of analogous fractions from brain mitochondria, as well as of other insoluble proteins, we conclude that such structural proteins, as isolated, are neither homogeneous nor unique. Synaptic vesicles do not turn over by a process involving fusion with and reformation from the synaptic plasma membrane. The proteins of the latter do appear, however, to be in a dynamic state (i.e., capable of turnover) with respect to their amino acid precursors, even in adult animals, with a half-life for this process of approximately 21 days. Some preliminary evidence suggests that at least a major part of the synthesis responsible takes place at sites distal from the membrane itself.

REFERENCES

1. V. P. Whittaker, *Proc. Natl. Acad. Sci. U.S.* **60,** 1081 (1968).
2. A. L. Lehninger, *Proc. Natl. Acad. Sci. U.S.* **60,** 1069 (1968).
3. B. L. Palay and G. E. Palade, *J. Biophys. Biochem. Cytol.* **1,** 69 (1955).
4. P. Fatt and B. Katz, *J. Physiol.* **117,** 109 (1952).
5. B. Katz, and R. Miledi, *J. Physiol.* **168,** 380 (1963).
6. K. Kusano, D. R. Livengood, and R. Werman, *J. Gen. Physiol.* **50,** 2579 (1967).
7. D. F. Parsons, *in*: "Proc. 7th Canadian Cancer Res. Conf." p. 193, Pergamon Press, Oxford (1966).
8. D. F. H. Wallach, *in*: "Specificity of Cell Surfaces" (B. D. Davis and L. Warren, eds.), pp. 129–163, Prentice-Hall, Englewood Cliffs, New Jersey (1967).
9. L. Warren, M. C. Glick, and M. K. Nass, *in*: "Specificity of Cell Surfaces" (B. D. Davis, and L. Warren, eds.), pp. 109–127, Prentice-Hall, Englewood Cliffs, New Jersey (1967).
10. C. W. Cotman, Structure and function of the synaptic complex from rat brain, PhD Dissertation, Indiana University (1967).
11. C. Cotman, H. R. Mahler, and N. G. Anderson, *Biochim. Biophys. Acta* **163,** 272 (1968).
12. C. W. Cotman, H. R. Mahler, and T. E. Hugli, *Arch. Biochem. Biophys.* **126,** 821 (1968).
13. E. C. Gray and V. P. Whittaker, *J. Anat. (London)* **96,** 79 (1962).
14. E. De Robertis, A. Pellegrino de Iraldi, G. Rodriguez de Lores Arnaiz, and L. Salganicoff, *J. Neurochem.* **9,** 23 (1962).
15. V. P. Whittaker, I. A. Michaelson, and R. J. A. Kirkland, *Biochem. J.* **90,** 293 (1964).
16. E. De Robertis, G. Rodriguez de Lores Arnaiz, L. Salganikoff, A. Pellegrino de Iraldi, and L. M. Zieher, *J. Neurochem.* **10,** 225 (1963).
17. G. Rodriguez de Lores Arnaiz, M. Alberic, and E. De Robertis, *J. Neurochem.* **14,** 215 (1967).

17a. E. G. Lapetina, E. F. Soto, and E. De Robertis, *Biochim. Biophys. Acta* **135,** 33 (1967).

18. O. Lindberg and L. Ernster, *in*: "Methods of Biochemical Analysis" (D. Glick, ed.), Vol. 3, Interscience Publishers, New York (1958).
19. G. Dallner, P. Siekevitz, and G. E. Palade, *J. Cell Biol.* **30,** 97 (1966).
20. H. M. Duncan and B. Mackler, *J. Biol. Chem.* **241,** 1694 (1966).

21. G. L. Ellman, K. D. Courtney, V. Andres, and R. M. Featherstone, *Biochem. Pharmacol.* **7,** 88 (1961).
22. J. R. Corbett, private communication.
23. O. Z. Sellinger, R. L. Rucker, and F. de Balbian Verster, *J. Neurochem.* **11,** 271 (1964).
24. R. J. A. Hosie, *Biochem. J.* **96,** 404 (1965).
25. J. H. Luft, *J. Biophys. Biochem. Cytol.* **9,** 409 (1961).
26. D. P. Pease, "Histological Techniques for Electron Microscopy," 2nd Ed., p. 37, Academic Press, New York (1967).
27. C. W. Cotman and H. R. Mahler, *Arch. Biochem. Biophys.* **120,** 384 (1967).
28. K. von Hungen, H. R. Mahler, and W. J. Moore, *J. Biol. Chem.* **243,** 1415 (1968).
29. A. T. Campagnoni and H. R. Mahler, *Biochemistry* **6,** 956 (1967).
30. L. A. Autilio, W. T. Norton, and R. D. Terry, *J. Neurochem.* **11,** 17 (1964).
31. P. Emmelot, C. J. Bos, E. L. Benedett, and P. H. Rumke, *Biochim. Biophys. Acta* **90,** 126 (1964).
32. D. M. Nevelle, *J. Biophys. Biochem. Cytol.* **8,** 413 (1960).
33. L. Warren, M. C. Glick, and M. K. Nass, *J. Cell Physiol.* **68,** 269 (1966).
34. B. Attardi and G. Attardi, *Proc. Natl. Acad. Sci. U.S.* **58,** 105 (1967).
35. A. Karlin, *J. Cell Biol.* **25,** 159 (1965).
36. D. F. Parsons, G. R. Williams, and B. Chance, *Ann. N.Y. Acad. Sci.* **137,** 643 (1966).
37. E. G. Lapetina, E. F. Soto, and E. de Robertis, *J. Neurochem.* **15,** 437 (1968).
38. C. DeDuve, *in*: "Enzyme Cytology" (D. B. Roodyn, ed.), pp. 1–26, Academic Press, New York (1967).
39. A. Giuditta and H. J. Strecker, *J. Neurochem.* **5,** 50 (1959); *Biochim. Biophys. Acta* **67,** 317 (1963).
40 A. Inouye and Y. Shinagama, *J. Neurochem.* **12,** 803 (1965).
41. I. M. Vassiletz, E. F. Derkatchev, and S. A. Neifakh, *Exp. Cell Res.* **46,** 419 (1967).
42. W. Straus, *in*: "Enzyme Cytology" (D. B. Roodyn, ed.), pp. 244–245, Academic Press, New York (1967).
43. P. L. Sarvant, S. Shibko, U. S. Kamta, and A. L. Tappel, *Biochim. Biophys. Acta* **85,** 82 (1964).
44. P. B. Gaban, *in*: *Int. Rev. Cytol.* (G. H. Bourne, and J. F. Danielli, eds.), Vol. 21 (1967).
45. M. K. Gordon, K. G. Bench, G. G. Deanin, and M. W. Gordon, *Nature* **217,** 523 (1968).
46. J. C. Skou, *Physiol. Rev.* **45,** 596 (1965).
47. S. H. Richardson, H. O. Hultin, and D. E. Green, *Proc. Natl. Acad. Sci. U.S.* **50,** 821 (1963).
48. R. S. Criddle, R. M. Bock, D. E. Green, and H. D. Tisdale, *Biochemistry* **1,** 827 (1962).
49. S. H. Richardson, H. O. Hultin, and S. Fleischer, *Arch. Biochem. Biophys.* **105,** 254 (1964).
50. R. S. Criddle, D. L. Edwards, and T. G. Petersen, *Biochemistry* **5,** 578 (1966).
51. D. O. Woodward and K. D. Munkres, *in*: "Organizational Biosynthesis" (H. J. Vogel, J. O. Lampen, and V. Bryson, eds.), p. 489, Academic Press, New York (1967).
52. D. O. Woodward, *Fed. Proc.* **27,** 1167 (1968).
53. D. E. Green, N. F. Haard, G. Lenaz, and H. I. Silman, *Proc. Natl. Acad. Sci. U.S.* **60,** 277 (1968).
54. G. Lenaz, N. F. Haard, A. Lauwers, D. W. Allmann, and D. E. Green, *Arch. Biochem. Biophys.* **126,** 746 (1968).

55. B. Droz and C. P. Leblond, *J. Comp. Neurol.* **121,** 325 (1963).
56. B. Droz., *J. Physiol. (Paris)* **58,** 511 (1966).
57. S. H. Barondes, *Science 146*, 779 (1964); *J. Neurochem.* **13,** 721 (1966).
58. S. H. Barondes, *J. Neurochem.* **15,** 343 (1968).
59. B. Graftstein, *Science* **157,** 196 (1967).
60. B. S. McEwen and B. Graftstein, *J. Cell Biol.* **38,** 494 (1968).
61. J. A. Bray and L. Austin, *J. Neurochem.* **15,** 731 (1968).
62. S. Ochs, this volume, Chapter 13.
63. M. L. Campbell, H. R. Mahler, W. J. Moore, and S. Tewari, *Biochemistry* **5,** 926 (1966).
64. H. S. Bachelard, *Biochem. J.* **100,** 131 (1966).
65. M. W. Gordon and G. G. Deanen, *J. Biol. Chem.* **243,** 4222 (1968).
66. T. K. Ray, I. Liebermann, and A. L. Lansing, *Biochem. Biophys. Res. Commun.* **31,** 54 (1968).
67. F. Orrego and F. Lipmann, *J. Biol. Chem.* **242,** 665 (1967).
68. L. A. Autilio, S. H. Appel, P. Pettis, and P. L. Gambetti, *Biochemistry* **7,** 2615 (1968).
69. I. G. Morgan and L. Austin, *J. Neurochem.* **15,** 41 (1968).
70. L. Austin, I. G. Morgan, and J. J. Bray, this volume, Chapter 12.
71. H. R. Mahler and B. J. Brown, *Arch. Biochem. Biophys.* **125,** 387 (1968).

DISCUSSION

FOLCH-PI: Did you ever prove that what you call the insoluble and structural proteins are definite entities, i.e., that given the same starting material, you end up with something with the same yield and composition?

MAHLER: Yes, the preparations are quite reproducible. The SPM isolated by zonal centrifugation accounts for a reproducible fraction of the total homogenate protein; about 40% of it is recovered in the insoluble protein component, and of the latter about 20% finally is found in the structural protein fraction. We have analyzed seven different preparations with respect to the parameters reported, particularly with regard to their electropherograms and amino acid composition, and reproducibility has been very satisfactory.

TELLER: You mention that you can't account for the rate of synthesis based on the idea of breakdown of synaptic vesicles. The vesicles don't have to be totally destroyed in order to release their contents and the whole vesicle would not have to be resynthesized in order to be refilled.

MAHLER: The only model we feel we can definitely exclude is one that postulates virtually complete destruction, or merging with the SPM, of the vesicles. We cannot, of course say anything against a sequence of opening, refilling, and closing of the structures.

AUSTIN: Do you think your methods are sensitive enough to pick up proteins in membranes developing at a regenerating nerve cell?

MAHLER: No, I shouldn't think so, not with the ones reported here. However, Dr. Ruth Gurd in our laboratory is developing a micromodification of the same technique, which will use $\leq 10\%$ of the material, which might be adequate for this purpose.

BOGOCH: I am wondering whether the peak of incorporation for synaptic membrane proteins has occurred prior to the first few minutes or much later, of the order of days. For microsomes you get a peak earlier than the first point shown; for your membranes it must have been either before 10 min or between 10 min and two weeks.

MAHLER: What we have tried to do in this presentation is to compare incorporation rates into the SPM with those observed in other fractions capable of autonomous protein synthesis, to detect similarities and differences. We believe we see a difference: microsomes and mitochondria approach the steady state with a negative slope, the SPM with a positive one. To us this means that at least some of the SPM protein is made elsewhere. We have no idea for synaptic vesicles. Synaptic membranes reach a steady state—with a positive slope—after about 15 hr, just like all other fractions.

E. KOENIG: Have you analyzed these synaptic membranes for RNA?

MAHLER: Dr. Burnel in our laboratory has performed such analyses. They contain, on the average, 7 μg of RNase-resistant RNA per mg of protein.

KOENIG: This is the same type of observation that was made on axonal RNA as well. Have you also tried mild acid hydrolysis, because I find that this RNA is rather resistant to this treatment as well.
MAHLER: No we have not.

OCHS: I'd just like to make a comment with regard to the points raised by Dr. Bogoch. A student of mine is working on the cortex to redefine this rapid labeling, and I believe that we do find a fairly rapid labeling, within 10 min, and in the synaptosome fraction at that. Now here, as you rightly point out, we have a problem of fitting this in with our data on rates of axoplasmic flow.

MAHLER: We have really shown two completely different types of experiments. In the turnover experiments of von Hungen's one concentrates on those proteins that turn over or are synthesized relatively slowly, while in the incorporation experiments we are primarily dealing with ones that are synthesized rapidly.

BARONDES: If one looks at the intermediate points one finds that some proteins belonging to all these cell fractions are transported with a delay. In fact, although this effect is most prominent with the soluble proteins of nerve endings, for which it is virtually complete, all proteins, even those sedimenting with mitochondria, with the membranes, or with the vesicles, show it to some extent. We do find a great deal of local incorporation also, in mitochondria, and perhaps into other fractions, which is real.

MAHLER: Our data agree with this so far as many of the fractions, particularly the soluble ones, are concerned. It is for this reason that we extract all soluble proteins and concentrate our efforts on the insoluble, tightly membrane-bound ones. Furthermore, particularly in the case of mitochondria, we know that a great deal even of the insoluble protein, of the order of 80% or better, is specified and synthesized elsewhere.

BARONDES: And if you do the kinetics, between, say 10 min and five days, which is the period when a great deal of this transport occurs, then you would see a progressive increase in the specific activity of the mitochondria as well.

MAHLER: What we hope to do next is to move from "protein" to individual electrophoretic bands and establish their individual kinetics.

FOLCH-PI: Before we go too far in interpreting these results we must remember that Dr. Mahler gauges his best preparation to be 80% pure. The other 20% may well obscure any of the fine points we are trying to seek.

Chapter 8

Protein Turnover in Membranous Fractions*

Brian D'Monte, Neville Marks, Ranajit Kumar Datta,† and Abel Lajtha

New York State Research Institute for Neurochemistry and Drug Addiction
Ward's Island, New York, New York
and
The College of Physicians and Surgeons
Columbia University
New York, New York

INTRODUCTION

Protein metabolism in the brain proceeds at a rate which is comparable to that in other organs, such as the liver. Although this dynamic state for cerebral protein was known some time ago, together with information on protein synthesis, very little is known about the mechanisms of protein breakdown involved in turnover. Previous studies have established that the brain contains a family of proteolytic enzymes, such as acid and neutral proteinases,[1–4] and peptide hydrolases, especially the subgroup amino-peptidases.[5,6] In other tissues intracellular proteolytic enzymes (cathepsins) are known to be present in lysosomes, but their precise localization in brain remains to be established.[7] It has been reported that a significant proportion of brain acid proteinase and other hydrolytic enzymes are present in lysosomal structures, but these previous studies failed to establish an unequivocal correlation between the hydrolytic activity and the morphological structures involved.

The physiological role of brain enzymes hydrolyzing peptide bonds is not fully known. The work of Eichholz[8] points to a role for peptidases

*We are indebted to Dr. George D. Pappas and Dr. Martin Beson of the Albert Einstein College of Medicine, New York City, for the preparation and interpretation of the electron micrographs. We thank Mr. Charles Bellman and Miss Mary O'Mara for their technical assistance. This work was supported in part by U.S.P.H.S. Grant No. NB-03226.

†Present address: Beth Israel Medical Center, Research Division, New York, New York.

in the transport of di- and tripeptides across the mucosal lining of the intestine. The microvilli of brush borders are known to contain enzymes hydrolyzing substrates similar to those split by the neutral arylamidases in the brain.[9] A report of arylamidase-containing morphological units on the surface of hepatoma plasma membranes[10] supports the suggestion that these peptidases may be associated with brain membranes: such enzymes may be involved in peptide transport across the cellular membrane or the membranes of the subcellular particulates. It is important to point out that proteinases may also be involved in the conversion of proteins from the inactive to the active state (activation of zymogens)[11]. Alteration of proteins may be a factor in disease: acid proteinase can degrade isolated myelin proteins to form encephalitogens.[12]

In previous studies on the subcellular distribution of peptidases many of the cerebral enzymes were shown to be present in the crude mitochondrial fraction.[3,4] Hydrolytic activity in such fractions cannot be meaningfully attributed to mitochondria themselves, since this fraction is known to consist of a heterogeneous population of myelin, synaptosomes, nerve endings, and mitochondria.[13] Recently it has been shown that brain mitochondria can synthesize protein,[14,15] and in the case of liver, Work[16] has reported that mitochondria contain sufficient DNA to code for at least 50 different proteins. In addition to the synthesis of structural or contractile proteins,[17–21] mitochondria may have the capacity to synthesize proteins with the catalytic functions involved in oxidative phosphorylation[22] or in protein synthesis and breakdown.[11]

The subfractionation of mitochondria into membrane components has recently been a focus of interest and a number of methods have been reported for the separation of inner and outer membranes and soluble matrix components.[23–25] Subfraction E showed the greatest degree of homogeneity of mitochondrial subfractions from sucrose gradient fractions, and hence was chosen as the starting material for an initial study of protein turnover in terms of three parameters: (1) hydrolase content (proteinase and aminopeptidase), (2) incorporation of amino acids into membrane proteins, and (3) protein composition of submitochondrial fragments (by means of polyacrylamide gel electrophoresis).

Microsomes are the site for a large number of biosynthetic mechanisms, including both anabolic and catabolic processes. Liver microsomes, for example, are known to be the site for drug hydroxylation, steroid synthesis, lipid peroxidation, and various phosphatases;[26] in brain they contain acid and neutral proteinases, arylamidases, and aminopeptidases.[6] The role of microsomal hydrolases in peptide and protein turnover is not yet fully understood, but recent work with *E. coli* has indicated that hydrolases are involved in the release of substituted amino acids or oligopeptide–nucleic acid complexes such as di-Phe-tRNA, N-acetyl-aminoacids, and N-substituted oligopeptidyl-tRNA.[27,28] It is believed that the enzymes involved are similar in properties to some peptidases and acylases; acylases are considered to be akin in properties to carboxy-peptidases, since they hydrolyze N-substituted peptides, especially those with an N-acetyl grouping.[11] The possibility also exists that proteolytic enzymes affect the different protein fractions reported necessary for the binding of aminoacyl-tRNA or RNA to ribosomal fractions,[29,30] and in particular for the binding of N-formyl methionine-tRNA required for chain initiation. There are no previous detailed studies of peptidases in mammalian microsomal and ribosomal fractions except for the reports of Binkley[31,32] that rat and kidney ribosomes, in contrast to those of hepatic tissue or *E. coli*, contain dipeptidases hydrolyzing cysteinyl-glycine and leucyl-glycine. Thus, alteration of the accessory protein factors required for protein synthesis may prove to be a sensitive control mechanism for protein turnover.

Liver microsomal preparations are known to consist of different membrane structures, some of which are smooth, and others studded with ribosomes.[33] Palade and his group[34,35] have shown that the intracellular secretion and transport of chymotrypsinogen involved both the ribosomes and the membranous components of the endoplasmic reticulum. Different mechanisms have been postulated for both the ribosomal attachment to cytomembranes and the intracellular transport of newly synthesized proteins. One suggested mechanism involves an interconversion of smooth and rough membranes in combination with the binding of specific metal ions such as Mg.[36] More recently the concept has been advanced that binding may also involve accessory protein factors;[37–39] for the reasons already stated metabolic transformation of such factors by hydrolytic enzymes may be considered to play a role in protein transport and turnover.

SUBFRACTIONATION METHODS

Mitochondria were previously prepared by centrifugation on discontinuous sucrose gradients consisting of manually prepared layers of sucrose.[13] This method has led to notable advances in the study of subcellular particulates. Results with this method were variable because different time periods were adopted by different workers for equilibration of the sucrose layers before centrifugation and because the interfaces formed between such layers are liable to produce artifacts of sedimentation. For some purposes the continuous sucrose gradient prepared mechanically is the method of choice; the sucrose gradient range recently employed varies

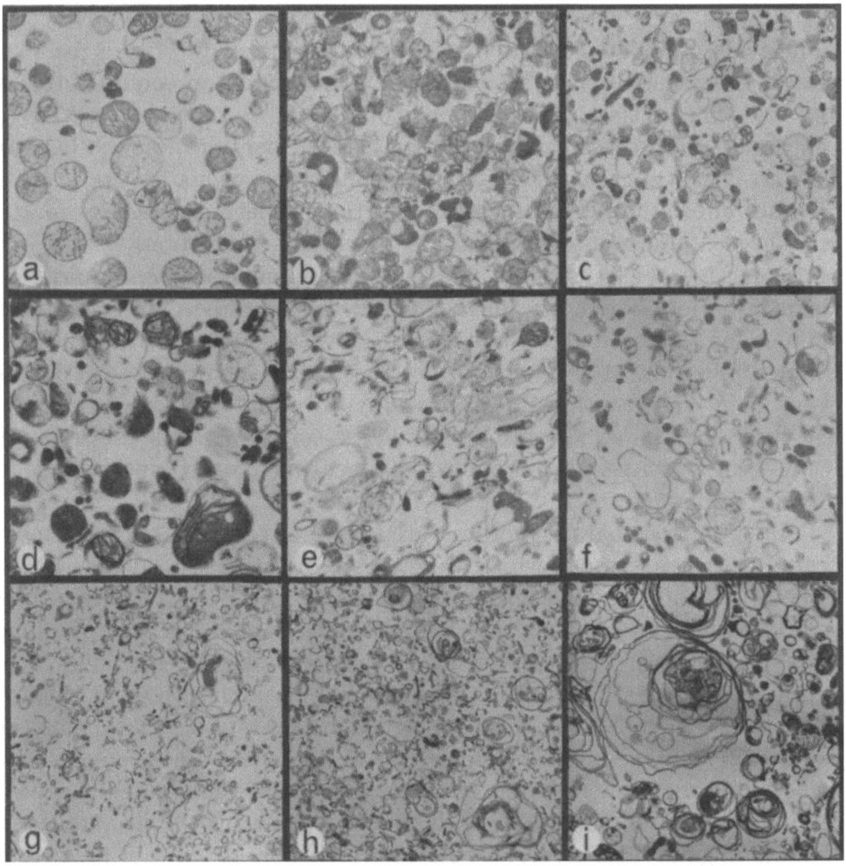

Fig. 1. Electron micrographs of mitochondrial subfractions separated by continuous sucrose-density-gradient (0.8–1.4 *M*) centrifugation; fixed in osmium tetroxide, double-stained with lead nitrate and uranyl acetate. Subfraction E is shown in Fig. 2. Sections were prepared on an LKB ultratome and examined under a Philips EM-200 electron microscope. (a)–(c), (g)–(i): ×4920. (d)–(f): ×9408.

from 0.8 to 1.4 or 1.6 *M*.[40,41] This method resulted in three zones rather than five, with larger yields of purified mitochondria (subfraction E) available for the fractionation into membranes.

Morphology of Gradient Subfractions

Mitochondria were shown to be heterogeneous, and this was confirmed by our finding that mitochondria are distributed in all regions of the sucrose gradient, mainly in fractions E (Fig. 2), and in (a–c) of Fig. 1.

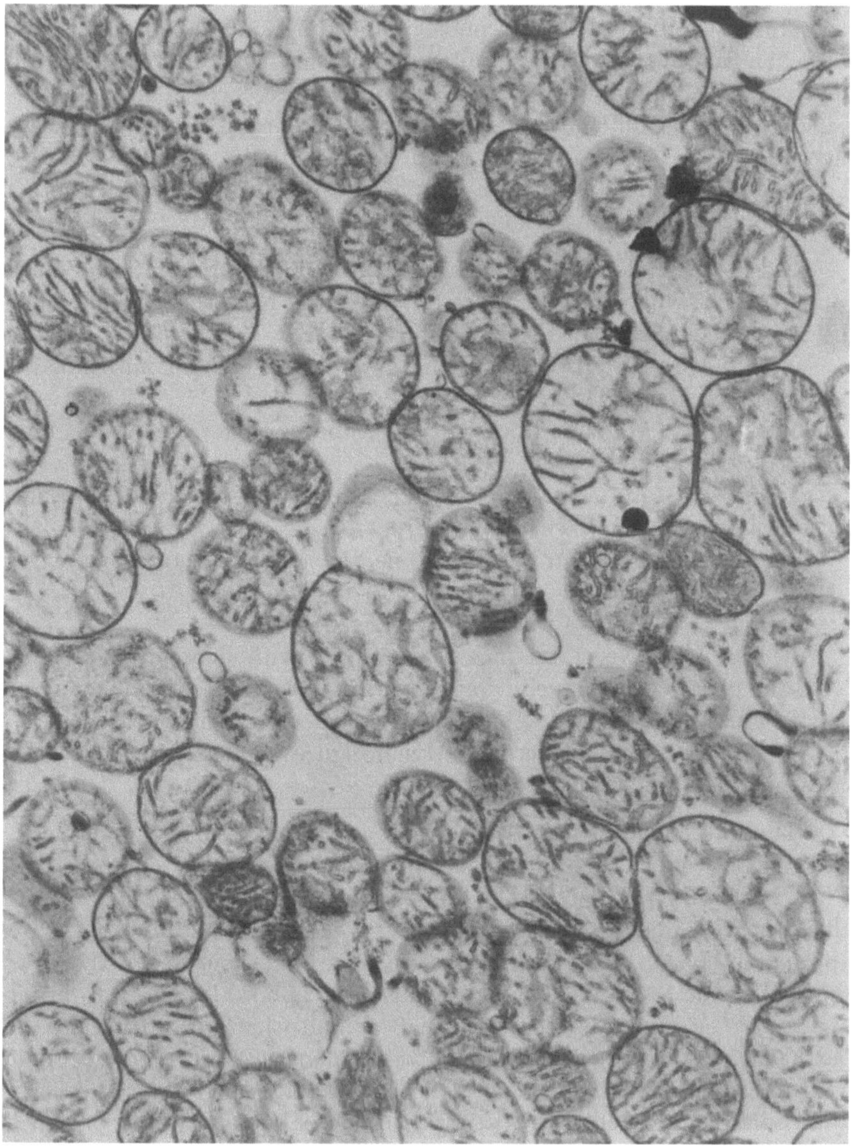

Fig. 2. Electron micrograph of subfraction E. Homogeneous mitochondria with intact cristae and double-membrane structure. Note the absence of lysosomal-like bodies. (×19,500.)

This may indicate the presence of different populations of mitochondria in brain that differ in both morphology and function. The mechanisms of mitochondrial biosynthesis are unknown; but mitochondria are known to increase in the developing brain, and therefore differences in mitochondrial size may reflect different phases in organelle breakdown and replacement.[42]

In addition to subfraction E (Fig. 2), the following different fractions of the continuous sucrose gradient (Fig. 8) were obtained:

1. Fig. 1(a). (1.4 *M* sucrose). Intact mitochondria with a few dense, round bodies similar in appearance to lysosomes. Note that the mitochondria appear smaller than those in subfraction E.
2. Fig. 1(b). Compact mitochondria with a few larger mitochondrial "ghosts" and lysosome-like particles.
3. Fig. 1(c). Mostly mitochondrial outlines with very few whole mitochondria; dense round bodies.
4. Fig. 1(d). An increased proportion of lysosomes with synaptosomal structures with entrapped mitochondria.
5. Fig. 1(e). Vesicular structures, nerve endings, synaptosomes, lysosomes.
6. Fig. 1(f). Vesicles with dense, round bodies.
7. Fig. 1(g). Nerve endings and vesicles with dense round bodies. Note the presence of a myelin fragment with an entrapped mitochondrion.
8. Fig. 1(h). Myelin fragments and vesicular structures.
9. Fig. 1(i). (0.8 *M* sucrose). Myelin figures with entrapped particulates predominate the field.

The mitochondria in fraction E were without visible contamination and represented the most homogeneous fraction in the gradient (Fig. 2). They retained the double-membrane structure characteristic of mitochondria with intact cristae. Both inner and outer membranes measured approximately 80–90 Å and they were separated from each other by a space of about 130 Å. The homogeneous mitochondrial fraction E did not appear to be contaminated by other nonmitochondrial structures—in particular, lysosomes, which are known as densely stained round bodies with a single membranous lining.[7] The virtual absence of lysosomes in the mitochondria reported in this study is of special interest, since it can be concluded that the measureable hydrolytic activity was intrinsic to the mitochondrion. Evidence for brain lysosomes is based largely on histological studies,[13,43] and there have been no biochemical methods reported to yield homogeneous preparations capable of satisfying morphological criteria. It was concluded on the basis of enzymatic studies that crude mitochondrial-synaptosomal fractions contain lysosomes,[3,43–47] but methods for the separation of lysosomes from mitochondria are currently unavailable. A major problem arises from the closeness of the lysosomal density to that of small mitochondria and synaptosomes in sucrose gradients as illustrated in the electron micrographs (Fig. 1). It is unlikely that sucrose gradients with a wider range extending to 1.6 *M* could afford a better resolution: Neidle *et al.*,[41] in a recent study using sucrose gradients of 0.8–1.6 *M*, showed considerable heterogeneity of mitochondrial enzymes in the middle of the gradient, which is also the gradient section reported recently by Sellinger and Hiatt[48] to contain the largest concentrations of lysosomes. Another approach, clearly, is required for the preparation of brain lysosomes, and currently we are attempting to separate these organelles by the prior injection of materials known to alter their density. For example, Triton WR-1339 was successfully employed to alter liver lysosomal density so as to facilitate their separation from mitochondria.[49] Other methods utilizing histochemical dyes as an aid to separation have been documented.[47,50,51]

MITOCHONDRIAL MEMBRANES

Preparation and Morphology

By analogy to the successful methods for splitting liver mitochondria into their constituent membranes, three methods were applied to brain mitochondria: (1) the large-amplitude swelling method with albumin added to the medium;[23] (2) treatment with phospholipase A;[52,53] and (3) treatment with digitonin.[54] As noted, the continuous sucrose-gradient method led to preparations of mitochondria judged to be morphologically free of contamination and therefore suitable for fragmentation studies. The yields of membrane, although low, were sufficient to conduct preliminary studies on their morphology, protein composition, and enzymology. In a prior study, Tipton[24] reported on enzyme distribution in mitochondrial membranes, but these results cannot be fully evaluated or compared to the current study, since he employed crude mitochondrial-synaptosomal preparations as the starting material.

Rat-brain mitochondria exposed to hypotonic medium (20 m*M* phosphate buffer, pH 7.2, containing 0.02% albumin) gave rise to mitochondrial ghosts free of matrix material, although in some cases intact mitochondria were visible in the sample fields (Fig. 3). Since mitochondria from brain are reported to be less susceptible to swelling compared to those of liver,[55] it was expected that the yield of membranes would be increased with a longer period of contact with the swelling medium. For most experiments the mitochondria were swollen in the hypotonic phosphate medium

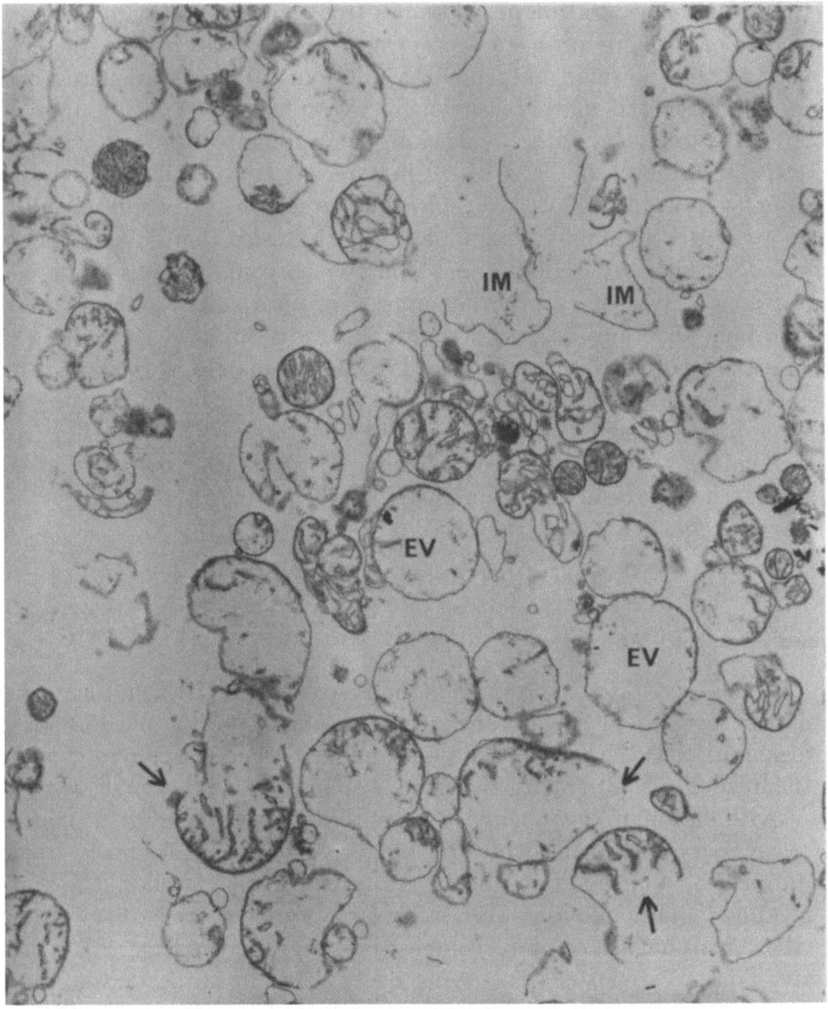

Fig. 3. Electron micrograph of mitochondrial membrane fraction A prepared by the large-amplitude swelling method. A few mitochondria have not been affected by the hypotonic buffer and remain intact; the majority appear in the form of ghosts or empty vesicles (EV). The arrows indicate mitochondria in the actual process of splitting; in some instances the fully detached inner membranes (IM) are visible. (×5160.)

for 20 hr; the results showed that there was a significant increase in the number of mitochondria with partial splitting of their outer membranes. Parsons *et al.*[23] reported that for liver only 15% of the mitochondria showed evidence of a stripped-off outer membrane. This method is only

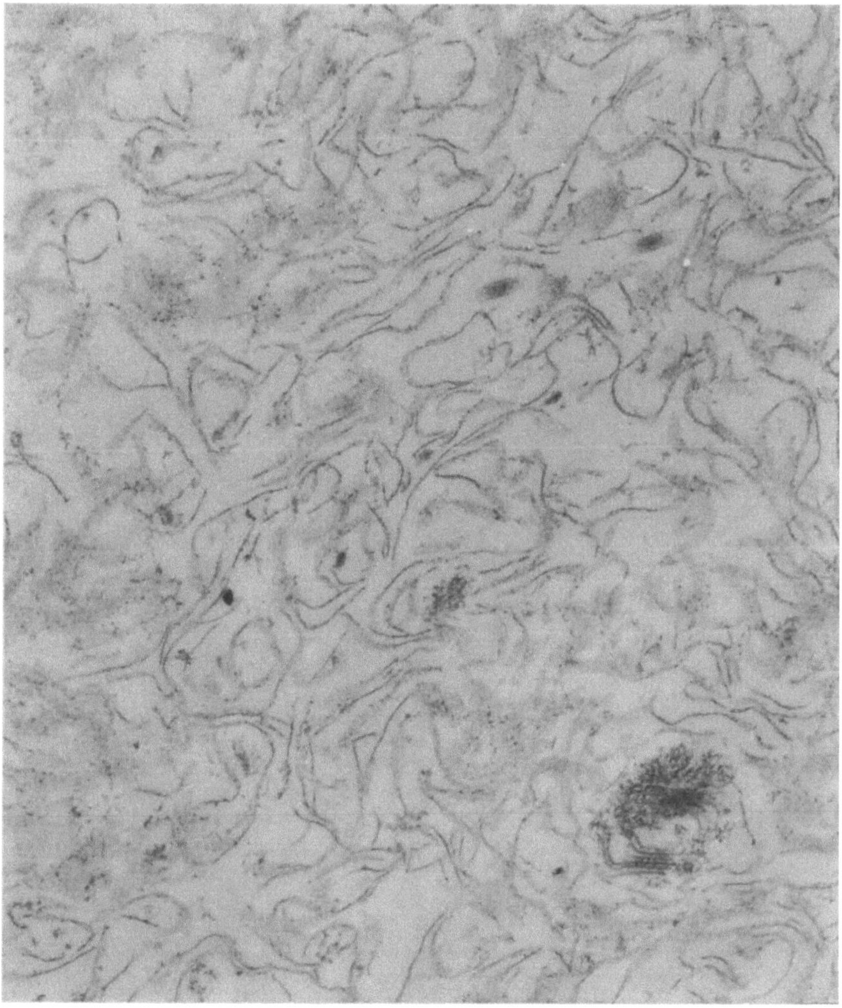

Fig. 4. Electron micrograph of inner membrane prepared by the phospholipase method. Sample field shown consists entirely of inner membranes arranged in the form of sheets without vesicle formation. (×29,600.)

satisfactory if additional centrifugation procedures are employed for the separation of the outer membrane from other mitochondrial fragments. In our early experiments with rat brain the yields did not permit these additional steps. This method is of great potential, since it does not involve the addition of detergents known to interfere with a number of enzymes.

Studies are in progress to adapt the full procedure to brain mitochondria with improved procedures utilizing hog brain and high-capacity rotors (Spinco SW27 and zonal heads).

Mitochondria fractionated by the phospholipase method[52] yielded better preparations of inner and outer membranes. The inner membranes

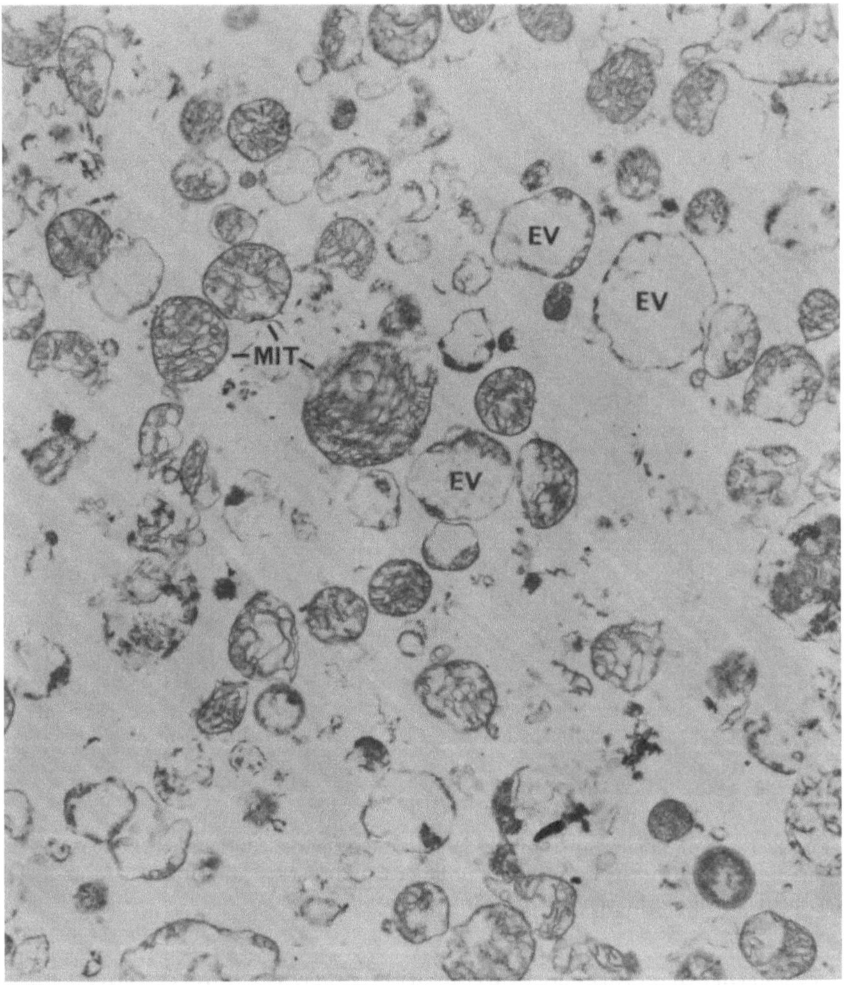

Fig. 5. Electron micrograph of mitochondrial membrane fraction 1. Mitochondria were prepared by digitonin treatment according to the procedure of Schnaitman *et al.*[54] (1.03 mg/10 mg protein) and centrifuged at 9500g. EV represent empty mitochondrial "ghosts" devoid of matrix material, while the unfragmented mitochondrial matrix has a swollen and vesicular appearance (MIT). (×11,760.)

were homogeneous (Fig. 4) and were arranged in the form of sheets without any vesicular formation. The thickness of these membranes was in accord with that shown in intact mitochondria (Fig. 2). Mitochondrial outer membrane appears to be exceptionally fragile, and all methods led to fragmented appearances as judged by electron microscopy. These results are in agreement with the experience of other groups in the preparation of outer membrane in the brain and, to a lesser extent, from the liver (H. R. Mahler, personal communication).

Digitonin treatment of brain mitochondria by a procedure similar to that used by Schnaitman *et al.*[54] in liver (1.03 mg digitonin per 10 mg mitochondrial protein) led to the formation of three fractions; (1) inner membrane and matrix material sedimented at 9500g for 10 min (Fig. 5); (2) fragmented mitochondrial membranes sedimented at 40,000g for 10 min (Fig. 6); and (3) outer membrane sedimented at 144,000g for 1 hr.

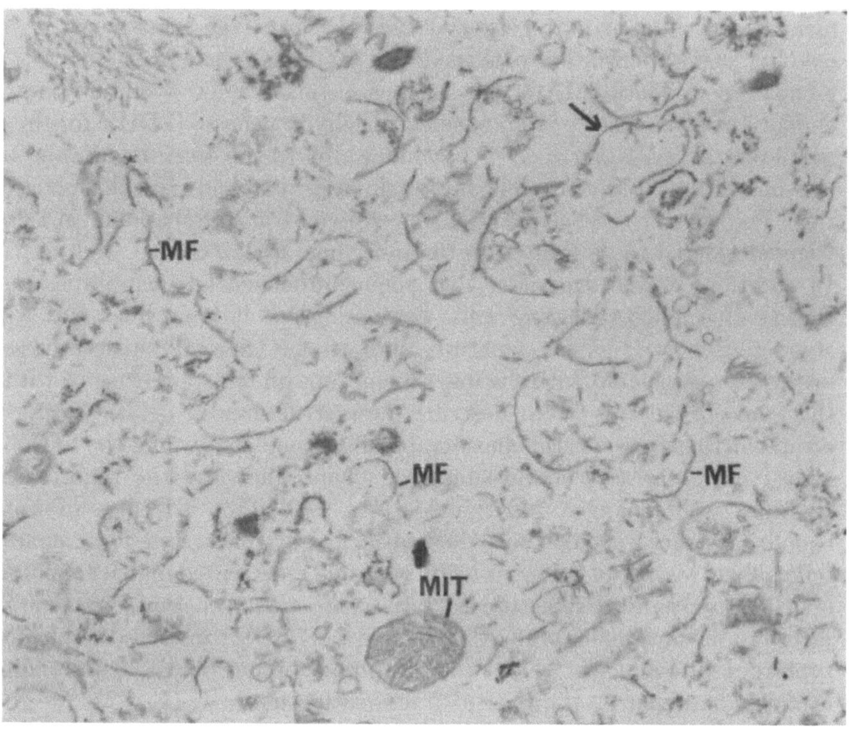

Fig. 6. Electron micrograph of mitochondrial membrane fraction 2 prepared by digitonin treatment. The field shows isolated double-membraned fragments (MF) similar to that seen in a typical intact mitochondrion (MIT). The arrow depicts outer membrane partially stripped off the inner membrane. (×12,740.)

The first fraction contained about 50% of unbroken mitochondria, suggesting that either a higher concentration of digitonin is required for brain, or that additional procedures such as ultrasonic vibration be employed.[56] Fragmented mitochondrial membranes are clearly visible in fraction 2 (Fig. 6) with partial or complete separation of inner and outer membranes; in a few cases single membranes have reformed to form vesicle structures. The outer membrane material obtained by the final centrifugation at 144,000g led to preparations with a highly fragmented appearance similar to those seen with the other procedures. As with the Parsons procedure, this method is shown capable of partially separating inner and outer membranes.

Marker Enzymes

An abundance of data now exists concerning the nature and location of membrane-bound enzymes, especially in relation to the components of the oxidative phosphorylation pathway. For example, it has been shown that monoamine oxidase (MAO) can serve as a marker for liver outer mitochondrial membrane[54] and succinate dehydrogenase (SDH) for inner membrane.[52] Allmann *et al.*[57] have allocated to the outer membrane the enzymes of the citric acid cycle, fatty acid oxidation, and transphosphorylation. The enzymes associated with the electron-transfer chain and ATPase were considered to be located in the inner membrane, cristae, and matrix. In addition, it was suggested that some mitochondrial membranes may consist of smaller repeating units containing specific enzymes, and with characteristic anatomical structures such as the tripartite structures and base pieces associated with the inner membrane and cristae. It is not within the scope of this report to describe these structures in detail, except to comment that there is still considerable discussion concerning the morphological characteristics of mitochondrial membranes and the reported alterations with changed functional states.[58–62] Two extremes of mitochondrial structure have been described: (1) a contracted form characterized by a highly folded inner membrane and a decreased matrical volume, and (2) an expanded structure with the inner membrane compartment adjacent to the outer membrane. The latter structure is considered to be more typical of mitochondria *in vivo* and was similar in morphology to the mitochondria obtained by the present fractionation methods (Fig. 2).[61,63,64]

At one time it was considered that biochemical determinants, such as membrane-specific enzymes, would act as useful adjuncts in specifying mitochondrial membranes. The early claims concerning a number of enzymes have been challenged and the question of enzyme location has not been resolved. To cite a few examples: the results of Schnaitman *et al.*[54] for MAO location are different from those of Green *et al.*,[25] but it has been suggested that the differences can be accounted for by the assay procedures. The recent results of Beattie[56] for membranes prepared by a combination of digitonin treatment with sonication suggest that the fatty acid-oxidizing enzymes are located on the inner membrane rather than the outer membrane. Finally, Green *et al.*[25] have claimed that the rotenone-insensitive NADH-dependent cytochrome c reductase is largely present in the microsomal fraction and is not a good marker for the outer membrane. It is therefore clear that the status of specific "marker" enzymes is unsatisfactory in its present form.

Since studies of "marker enzymes" have not been pursued in the brain, we took the opportunity to investigate this aspect in membranes prepared by the first (hypotonic swelling) and second (phospholipase) procedures described earlier. These two procedures were compared, since it was felt that the addition of detergent may adversely affect some enzyme activities. In the hypotonic-medium method MAO (assayed by the procedure of Cerletti *et al.*[65]) and SDH (by the method of McCaman *et al.*[66]) were present at higher levels in the heavier rat-brain membranes of fraction A that would be expected to contain mainly inner-membrane material (Table I). In the phospholipase procedure no differences were observed in the distribution of MAO. Based on these preliminary data, it would seem that MAO and SDH may not be the best enzymes to specify brain mitochondrial membranes. This aspect is currently being reinvestigated using other

TABLE I

Distribution of Mitochondrial Membrane Marker Enzymes

Fraction[a]	Total units of enzyme activity[b]	
	Monoamine oxidase (MAO)	Succinate dehydrogenase (SDH)
Subfraction E	1.58	3.56
Inner membrane	0.31	0
Outer membrane	0.11	0
Fraction A	2.44	3.99
Fraction B	0.99	1.51

[a] Inner and outer fractions were prepared by the phospholipase method; fractions A and B refer to the heavier and lighter fractions obtained by large-amplitude swelling.

[b] MAO and SDH were assayed as outlined in the text. For MAO a unit refers to number of nanomoles tyramine oxidized per minute; for SDH, nanomoles O_2 per minute.

procedures[67] to split mitochondria that may be less deleterious to the specific marker enzymes.

HYDROLASES

Sucrose Gradient Distribution

The heterogeneity of organelles present in the crude mitochondrial fraction was confirmed by a study of the distribution of various hydrolases in a continuous sucrose gradient (Fig. 7). Acid deoxyribonuclease, acid phosphatase, and N-acetyl-β-D-glucosaminidase show a trimodal pattern of distribution, whereas arylsulfatase has a single peak of activity. Previous studies[43,48] demonstrated a bimodal distribution of most lysosomal enzymes. The ubiquitous presence of these hydrolases within the gradient accentuates the difficulties involved in the isolation of a pure fraction of lysosomes. It is probable that there is more than one lysosomal population.[7] The question of whether any organelle apart from lysosomes can contribute to mechanisms of protein breakdown can only be answered following the isolation of a pure mitochondrial fraction. The continuous sucrose

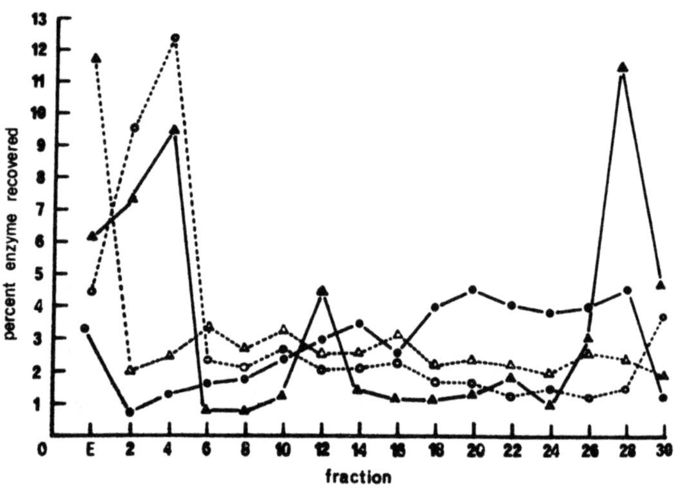

Fig. 7. Distribution of hydrolases in mitochondrial subfractions. Crude mitochondria were layered on a continuous sucrose gradient (0.8–1.4 *M*) and centrifuged at 24,000 rpm for 2 hr in a SW 25.1 rotor. Thirty fractions were collected by tube puncture and assayed for the following enzymes after six cycles of freezing–thawing: closed triangles, acid deoxyribonuclease, closed circles, acid phosphatase, open triangles, N-acetyl-β-*D*-glucosaminidase, open circles, arylsulfatase. Data based on the results of Marks and D'Monte.[105]

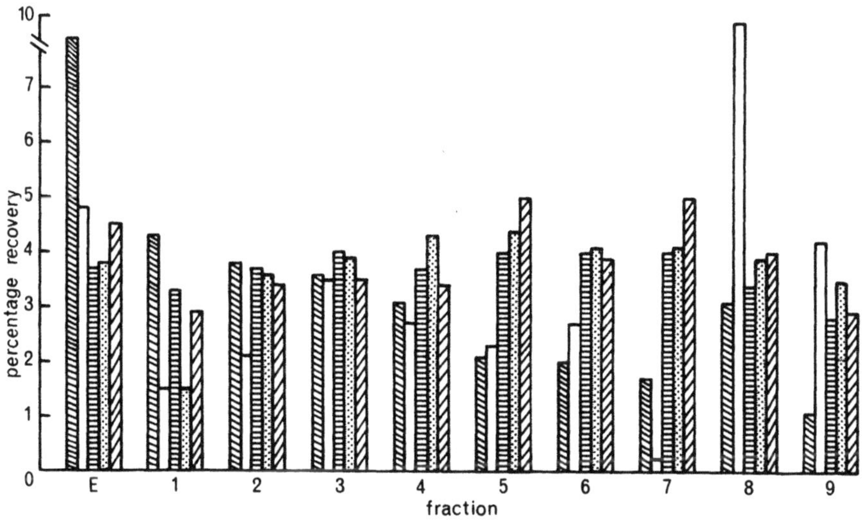

Fig. 8. Distribution of endo- and exopeptidases in 10 subfractions of a crude mitochondrial fraction (containing 270 mg protein) layered on a continuous 0.8–1.4 *M* sucrose gradient and centrifuged at 24,000 rpm for 2 hr in a Spinco SW 25.1 rotor. Fractions were collected by tube puncture, and enzyme assays carried out as described earlier.[3,6] Enzyme activities expressed as total units (nanomoles per minute) enzyme recovered per subfraction as a percentage of that present in the crude mitochondrial fraction. ▧ acid proteinase (Hb, pH 3.8); □ neutral proteinase (casein, pH 7.6); ▤ Arg-β-NA (pH 7.6); ▩ Leu-β-NA (pH 7.6); ▨ Leu-Gly-Gly (pH 7.6). Data from the results of Marks and D'Monte.[105]

gradients were also divided into ten fractions, which were assayed for endo- and exopeptidases by methods described previously (Fig. 8).[3,4,6] Again a distribution of activities similar to that for the other hydrolases was noted. Activities for all enzymes were in the same range throughout the middle region of the gradient (fractions 1–7); there was a sharp peak for acid proteinase in fraction E and another for neutral proteinase in myelin. Aminopeptidases exhibiting degradation of dipeptidyl arylamides have been located in lysosomal fractions isolated from the pituitary.[68]

Membrane Distribution

The higher levels of proteolytic enzymes in the fractions prepared by large-amplitude swelling presumably relate to the deleterious effect of phospholipase A discussed above (Table II). It is of interest that mitochondria themselves are reported to contain an endogenous phospholipase A; the effect

TABLE II

Distribution of Endo- and Exopeptidases in Mitochondrial Membranes

Fraction[a]	Total protein (mg)	Units of enzyme activity[b]				
		Acid prot.	Neutral prot.	Arg-β-NA	Leu-β-NA	Leu-Gly-Gly
Homogenate	1407	3265	1183	15038	12572	9792
Crude mitochondria	240	1355	147	1982	1711	1252
Subfraction E	10.3	186	11	62	41	58
Inner membrane	1.5	0.9	[c]	1.0	0.9	1.8
Outer membrane	1.4	2.4	[c]	0.7	0.7	8.3
Fraction A	2.3	0.3	1.4	3	1.4	2
Fraction B	1.0	8.0	0	0.3	1.0	1

[a] Inner and outer membrane fractions were prepared by the phospholipase method; fractions A and B refer to the heavier and lighter fractions obtained by large-amplitude swelling.

[b] Units of enzyme activity are nanomoles amino acid or β-naphthylamine released per minute. Enzymes were assayed according to the methods described previously.[2,6] Protein was measured using the Folin–Ciocalteau reagent.[103]

[c] Not measureable due to the high endogenous neutral proteinase activity of the snake venom used as the source of phospholipase A.

of phospholipase A both on membranes and enzymes may indicate a role for this enzyme in oxidative phosphorylation.[69–72] In particular, phospholipase A is known to inhibit the NADH-CoQ region of the respiratory chain by removal of essential lipids and the formation of inhibitory phosphatides.[73]

Caution must be exercised in the discussion of results with low enzyme recovery, since cross-contamination of membrane fragments, or contamination by other sources (microsomal) cannot be ruled out. As noted earlier, our mitochondrial fractions, as well as the derived membranes, were checked for morphological purity and then subjected to additional purification procedures. In spite of these steps, the highly purified preparations still exhibited a low but significant proteolytic activity when tested with protein or peptide substrates. Recently Fitzpatrick and Pennington[74] reported that purified liver mitochondria separated from lysosomes by prior injection of Triton WR-1339 and gradient centrifugation still exhibited proteolytic activity when incubated with myoglobin as the substrate. They showed that treatment with Triton X-100 led to a fall in the pH 3.8 activity without affecting the mitochondrial proteolytic activity measured at neutral pH. Alberti and Bartley[74a] also observed proteinases in highly purified liver mitochondria active at pH 7.4 and 5.8, which appeared in the high-speed supernatant of mitochondria after fragmentation with Triton X-100.

In our study on brain mitochondria the proteinase levels were considered to be highly significant in relation to the known rates of protein turnover shown by the following calculations. If it is assumed that 1 mg of mitochondrial protein contains 10^4 nM of glutamic acid equivalents, the values of 18 and 1.1 nM per min per mg protein for acid and neutral proteinase, respectively, in purified mitochondria would indicate half-lives of 6 hr (acid proteinase) and five days (neutral proteinase); similar estimates for the membrane components indicate half-lives of four (outer) and eight days (inner). It should be emphasized that no statements can be made regarding the preferential location of neutral proteinase, because of the inherent difficulties created by heat-resistant proteinase activity at pH 7.6 in the samples of venom employed.

In other tissues it is interesting that Binkley[31] recently reported membrane-bound aminopeptidases in renal tubular brush border particulates that hydrolyzed both Leu-Gly and Leu-β-NA, an enzyme that is distinct in its properties from those hydrolyzing the same substrates in microsomes. Earlier studies showed that the microvilli of brush borders contain enzymes hydrolyzing Leu-β-NA, Arg-β-NA, Leu-Gly, and Leu-Gly-Gly.[8] Globular repeating units containing arylamidase activity have also been reported on the plasma membrane of liver and hepatoma cells.[9] These studies, along with the present results, put forward a case that hydrolytic enzymes are associated with some membranes which, together with the other components of the synthetic and degradative pathways, permit an autonomous protein turnover within the organelle. The possibility also cannot be excluded that these enzymes play a role in transport mechanisms in these particulates.

AMINO ACID INCORPORATION

In Sucrose Gradient Fractions

A heterogeneous distribution of label was observed in sucrose gradient fractions following *L*-^{14}C-lysine administration in the living animal (Fig. 9). Intracisternal injection was chosen for most studies to increase the level of label within 30 min. Previous studies have shown a peak of incorporation into brain proteins[75] and into liver mitochondria[76] within this time period. The two major peaks of activity shown occur in the mitochondrial region of the gradient. These results suggest a higher rate of turnover from proteolipid than hitherto supposed; incorporation *in vivo* is greater than that *in vitro*. The close proximity of particulates active in the synthesis of protein may be responsible for the high incorporation noted in this fraction of the gradient. It is unlikely that the exceptionally high rates of incor-

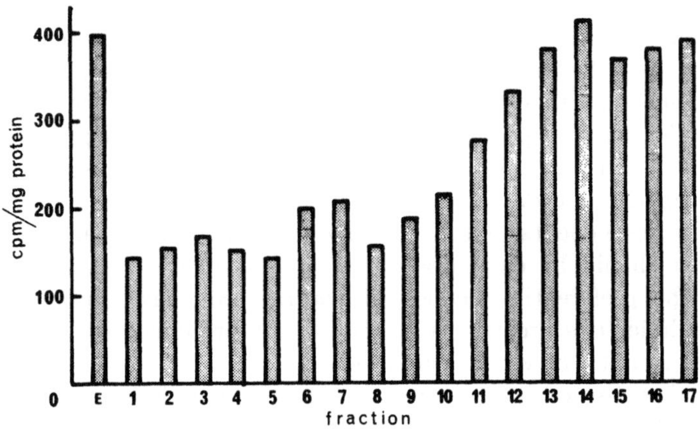

Fig. 9. Amino acid incorporation into mitochondrial subfractions. A total of 1.0 μCi of *L*-^{14}C-lysine was injected intracisternally into each rat, which was sacrificed after 30 min. Crude mitochondria were fractionated on a continuous sucrose gradient as described before (see Fig. 7). Results based on the data of Marks and D'Monte.[105]

poration observed in this study are due to trapped particulates, since these occur at low levels compared to other areas of the gradient. Klee and Sokoloff[77] reported high incorporation into myelin obtained from young rats (three days of age) with *D,L*-^{14}C-leucine *in vitro,* which decreased with age. The rats selected for the present study were 5–6 weeks old, an age at which myelinization is completed. It is known that in ox-heart mitochondria the major incorporation of *D,L*-1-^{14}C-leucine injected intravenously into rats (incorporation times of 6–90 min) occurs in a lipoprotein material, suggesting a possible role in the turnover of the membranous components.[78]

Incorporation Into Mitochondrial Membranes

In our study in rat-brain mitochondrial membranes 70% of the label (*L*-^{14}C-lysine) was incorporated into the membrane fractions, with a 35% higher incorporation into the inner membrane (Table III). A similar differential distribution was noted by Neupert *et al.*[79] and Beattie *et al.*[80] for liver mitochondrial membranes after injection of *L*-^{14}C-leucine *in vitro* and separation of the membranes by hypotonic buffer treatment. These results raise the possibility that inner- and outer-membrane proteins are synthesized by two different systems.

TABLE III

Incorporation of L-^{14}C-Lysine into Mitochondrial Membranes Prepared by the Phospholipase Method

Fraction	Specific radioactivity[a]	
	In vivo	*In vitro*
Subfraction E	474	7712
Inner membrane	549	8444
Outer membrane	407	3289
Soluble matrix	140	116

[a] In the *in vivo* experiments each animal was given an intracisternal injection of 1.0 μCi ^{14}C-lysine which was allowed to incorporate for 40 min. The *in vitro* labeling experiments consisted of incubating the mitochondrial suspension in the medium used by Campbell *et al.*[14] with 2.5 μCi ^{14}C-lysine. Incorporation was stopped after 40 min by addition of cold lysine and the fractionation of membranes carried out according to the method described previously.[53] Specific radioactivity is expressed as counts per minute per milligram protein extracted with lipid solvents as described earlier.[11]

The ability of the mitochondrion to synthesize all endogenous structural and soluble protein components has been questioned.[81,106] Bronsert and Neupert[82] reported that locust-muscle mitochondria during moulting appear to synthesize most of their structural protein, and Beattie *et al.*[83] recently showed that soluble brain mitochondrial proteins (half-life of 18 days) may be extramitochondrial in origin. The question of penetration of proteins or partially completed peptide chains into mitochondria is particcularly intriguing and hitherto has not received explanation.[84] It has been suggested that some synaptosomal proteins are supplied by mitochondria,[85] but it is unlikely that the mitochondrion could form the major source for other cellular constituents. As noted earlier, the half-lives of the proteins for the inner and outer membranes were eight and four days, respectively, based on the calculations derived from their rates of degradation. Interestingly, Brunner and Neupert[86] reported half-lives of 3–11 days for liver mitochondrial membranes, with the outer membrane turning over three times faster than the inner membrane. There is, therefore, very good evidence for the heterogeneity of turnover of different mitochondrial components, with the possibility of different biogenetic origins. Evidence for the heterogeneity of turnover also exists for the incorporation of phospholipid precursors, with a higher incorporation of choline and serine into the outer membrane compared to the inner membrane.[87,88] The relationship between phospholipid and protein turnover in membranes represents an important aspect, since these compounds are intimately linked together in membrane structure, and probably in function.

PROTEIN COMPOSITION

If organelle membranes arise by different biogenetic mechanisms, then they might be expected to differ in protein and amino acid composition. Qualitative differences were observed in the electrophoretic patterns of inner- and outer-membrane proteins solubilized by phenol–acetic acid–urea solvents on disc-gel electrophoresis. The question of protein solubility has been studied in plant-leaf proteins,[89] myelin proteolipids, and particulate fractions.[22,90–93] Membrane proteins tend to be hydrophobic and to resist solubilization by normal solvents. The phenolized solvent proved to be an excellent one for dissolving membrane protein, and yielded the patterns shown in Fig. 10. Of the 19 bands obtained for E, similar in number to that reported elsewhere,[93] the outer membrane was more complex, with 13 bands, compared to 10 for the inner membrane. The fine structure of proteins on disc gel cannot be easily determined, due to the multiple character of the bands. Each protein band requires further separation on another gel column for finer analysis. Some of these proteins may be complements of the electron-transfer chain, as has been shown in the case of the liver mitochondrial membranes.[22]

The amino acid composition of the membranes was similar to that of the starting material, fraction E (Table IV); each had high contents of as-

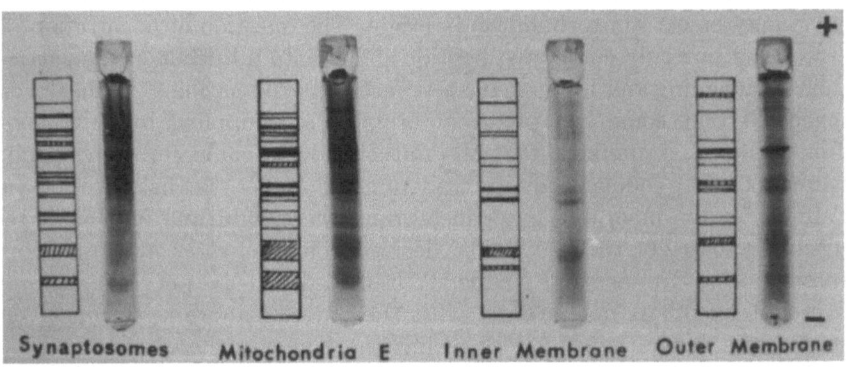

Fig. 10. Disc-gel electrophoresis patterns of intact mitochondria (purified fraction E) compared to a sucrose gradient fraction from the center of the gradient [synaptosomal region, see Fig. 1(e)] and to the derived inner and outer membranes prepared by phospholipase treatment. Samples containing 40 μg protein were solubilized in phenol–acetic acid–urea (2:1:1) and loaded onto gels equilibrated in the same solvent. Electrophoresis was carried out for 2.5–3 hr at 3 mA per tube at pH 2.7. The schematic diagrams are drawn to scale: different intensities of bands are indicated by varying shades. Results based on previous data.[105]

TABLE IV

Amino Acid Contents of Various Brain Proteins
(Residues per 100 residues)

Amino acid	Subfraction E[a]	Inner membrane[a]	Outer membrane[a]	Soluble matrix[a]	Chicken brain mito.[b]	Rat brain mito.[b]	Rat brain mito.[c]
Asp	9.3	9.7	9.2	12.0	8.9	8.8	8.9
Thr	4.3	4.9	5.5	7.0	5.2	5.5	5.4
Ser	6.6	6.5	6.3	7.0	6.8	7.4	5.8
Glu	10.1	10.0	7.8	7.8	10.8	10.7	10.6
Pro	7.0	5.3	4.7	0.2	5.3	4.6	6.1
Gly	9.4	9.6	9.2	9.1	9.0	8.5	7.8
Ala	8.3	9.2	8.9	7.9	9.7	9.2	8.5
Val	6.5	6.3	6.8	6.6	6.7	6.4	5.8
Cys (½)	1.3	1.4	3.0	1.4	—	—	1.2
Met	0.3	1.4	0.3	0	2.7	2.3	2.2
Ile	5.6	6.9	6.4	6.5	4.9	4.9	5.2
Leu	8.6	8.5	8.8	6.8	9.3	9.3	9.5
Tyr	2.3	3.0	2.7	2.5	3.1	2.9	3.4
Phe	4.0	4.4	4.5	9.5	4.6	4.4	4.5
NH_3	—	—	—	2.0	10.0	10.6	—
Lys	7.5	8.1	10.0	7.8	5.9	6.5	6.3
His	1.8	2.3	1.5	1.9	2.4	2.7	2.1
Arg	5.2	5.1	4.3	4.3	4.6	6.0	4.6

[a] Present study.
[b] Burkhard and Klaassen.[104]
[c] Cotman *et al.*[93]

partic acid, glycine, and glutamic acid. This is in close agreement with previous results.[93]

MICROSOMAL MEMBRANES

Preparation

A survey of fragmentation methods showed that deoxycholate (DOC) at 0.5% concentration afforded the best yield of endoplasmic reticulum membranes. In this respect DOC was superior to other methods of fragmentation employing Triton X-100 or GR-7 and digitonin.[94,95] DOC has previously been used for the extraction of enzymes from other tissue microsomal preparations: for example, pancreatic amylase,[35] liver catalase,[96] and, recently, dipeptidases from renal microsomes hydrolyzing Leu-β-NA, Cys-Gly, and Leu-Gly.[31,32] The ribosomal and endoplasmic membrane fractions prepared by DOC treatment (Fig. 11) were employed for the morphological, chemical, and enzymological studies described below.

Morphology

Microsomes prepared for electron microscopy by the standard procedure (see Fig. 1) consisted largely of smooth membranes, but rough mem-

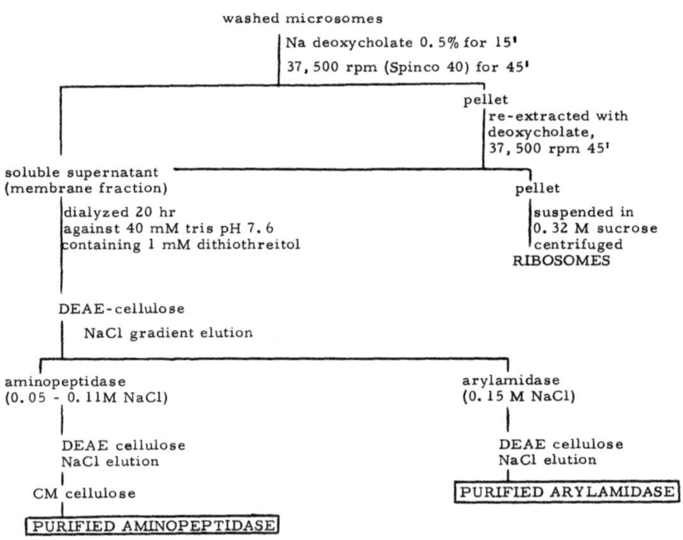

Fig. 11. Method of purification of microsomal membrane peptide hydrolases.

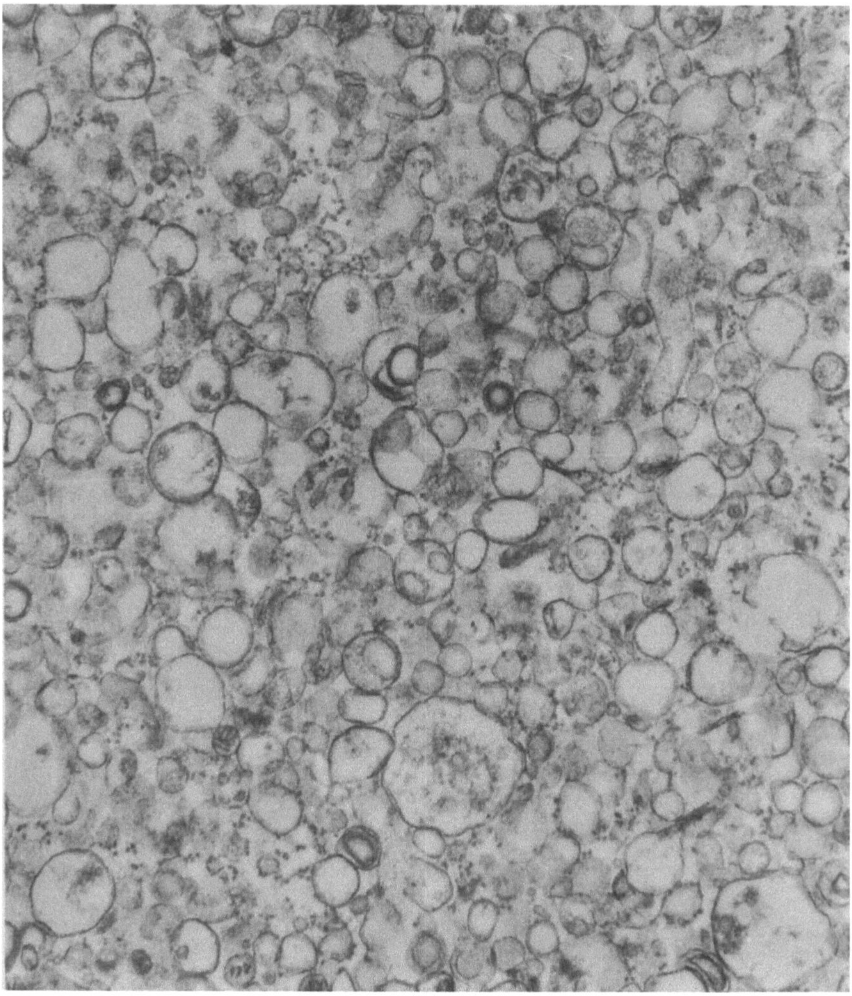

Fig. 12. Electron micrograph of rat-brain microsomal pellet (37,500 rpm Spinco 40 for 45 min) embedded in Epon and stained with lead nitrate and uranyl acetate. (×20,325.)

branes with ribosomes attached to one surface of the membrane were visible in sample fields (Fig. 12). Most membranes appeared to be vesicular in shape, and in some cases they enclosed smaller vesicles in the manner previously described by Ernster *et al.*[97] for liver microsomes. The ribosomal fractions consisted of the typical rosette formation characteristic of this

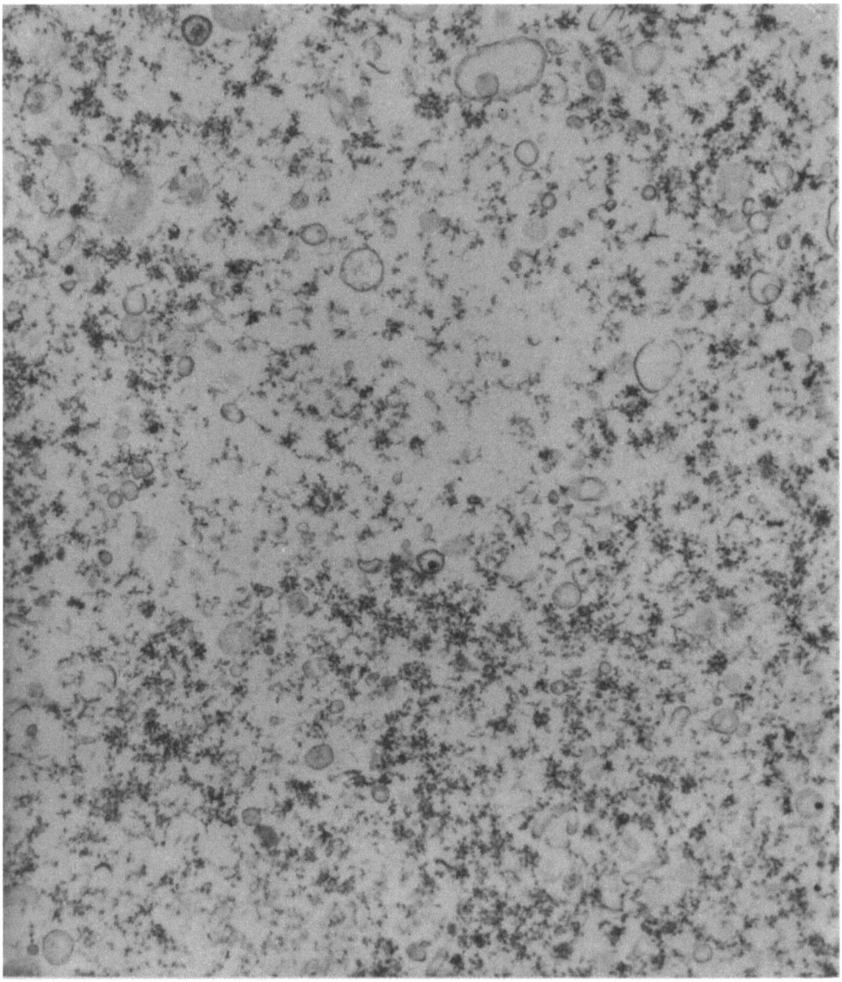

Fig. 13. Electron micrograph of rat-brain ribosomes embedded in Epon and double stained with lead nitrate and uranyl acetate. (×39,195.)

organelle, together with a slight contamination by smooth circular membranes (Fig. 13). The soluble supernatant containing the microsomal membranes (Fig. 11) formed a precipitate after dialysis to remove the detergent; this precipitate consisted entirely of partially solubilized membranous components devoid of ribosomal contamination. This fraction was used chiefly for the measurement and purification of brain hydrolases described in the next section.

Hydrolases

Distribution

The major portion of the microsomal peptide hydrolases (some 5–10% of the total homogenate activity[98]) resided in the DOC-soluble or endoplasmic membrane fraction, with only trace activities associated with ribosomes (Table V). In agreement with earlier studies, only low levels of neutral and acid proteinase were observed with hemoglobin as the substrate, and no activity with the carboxypeptidase substrates could be detected. The highest peptide hydrolase activities for the intact microsomes and derived membrane fractions occurred with the arylamides containing Leu- and Arg- and with the peptide substrates, tri-leucine, Leu-Gly, and Leu-Gly-Gly.

TABLE V

Distribution of Enzyme Activities in Microsomal Fractions[a]

	Microsomes, specific activity[b]	Per cent recovered: Microsomal membranes	Per cent recovered: Ribosomes
Arylamidases			
Leu-β-NA	4.8	65	0
Met-β-NA	3.7	36	1
Arg-β-NA	6.3	97	3
Peptide hydrolases			
Leu-Gly	3.6	11	1
Leu-Gly-Gly	3.5	65	3
Leu-Leu-Leu	6.0	75	2
Z-Leu-Tyr	0	0	0
Z-Glu-Tyr	0	0	0
Peptidyl-peptide hydrolases			
casein (pH 7.6)	1.2	0	0
Hb (pH 7.6)	<0.5	0	0
Hb (pH 3.8)	<0.5	0	0
poly-Leu	0	0	0
Z-Phe-Phe-NH_2	0	0	0

[a] The incubation mixtures contained 50 μg enzyme protein with 0.5 mM substrate in 1.0 ml buffer; the assays were performed as described earlier.[2,6]

[b] Specific activity is expressed as nanomoles amino acid or β-naphthylamine released per minute per milligram of protein.

Purification and Properties

The methods chosen for purification were based on those previously described for whole brain extracts.[6] Washed microsomes were extracted twice with DOC, and after dialysis to remove detergent the membrane fraction was submitted to salt-gradient elution from DEAE-cellulose columns (Figs. 11, 14). The first peak eluted by low salt concentration hydrolyzed a series of peptides with marked specificity for tripeptides; it was further purified by passage through CM-cellulose. This peak was inactive with arylamide substrates, in contrast to the DEAE peak II eluted at a higher salt concentration (Fig. 14). The different specificities are shown in Table VI: the aminopeptidase showed the highest activity with Ala-Gly-Gly and Leu-Gly, the arylamidase with Ala- and Arg-β-NA as substrates. Purified aminopeptidase was inactive with Pro-Gly-Gly, Arg-Gly-Gly, and tri-glycine, and the arylamidase was inactive with tryptic substrates. The two enzyme groups were readily distinguished by metal sensitivity and the effect of puromycin. Arylamidases rapidly lost activity on dialysis

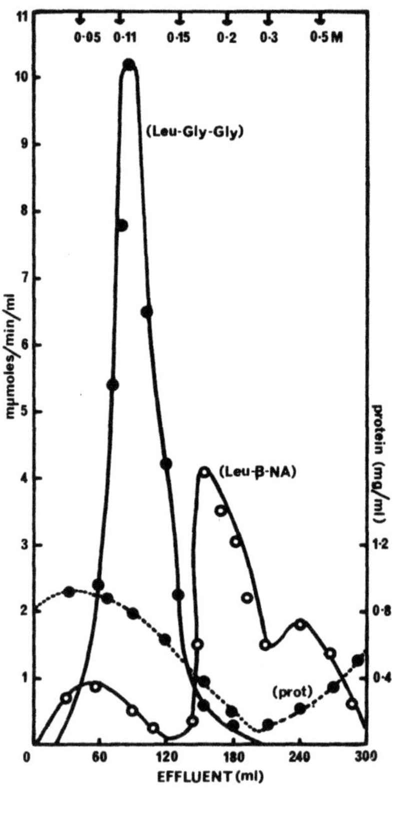

Fig. 14. Separation of aminopeptidase (Leu-Gly-Gly) from arylamidase (Leu-β-NA) on a DEAE-cellulose eluted with NaCl gradient: aminopeptidase eluted by 0.11 *M* NaCl, arylamidase with 0.15 *M* NaCl. Yield of enzyme varied from 55 to 75%. (Results in Figs. 12–14 are based on the unpublished data of Marks, Datta, and Lajtha.)

TABLE VI

Substrate Specificities of Purified Microsomal Membrane Peptide Hydrolases[a]

Substrate	Aminopeptidase[b]	Arylamidase[c]
Leu-Gly-Gly	100	0
Gly-Gly-Leu	50	0
Ala-Gly-Gly	275	0
Leu-Leu-Leu	33	0
Leu-Gly	225	0
Poly-Leu	0	0
Z-Leu-Tyr	0	0
Z-Phe-Phe-amide	0	0
Leu-β-NA	0	100
Ala-β-NA	0	220
Arg-β-NA	0	210
Benzoyl-*DL*-Arg-β-NA	0	0

[a] The assay mixture contained 50 μg of enzyme protein with 0.5 m*M* substrate in a volume of 1 ml of 40 m*M* Tris-HCl buffer, pH 7.6.
[b] Further purified by fractionation on CM-cellulose, then dialyzed against EDTA.
[c] Further purified by another DEAE-cellulose fractionation.

against EDTA, but activity was restored and, in some cases, activated by the addition of metal ions (Table VII); the sensitivity to different concentrations of metal ions for neutral and basic arylamides may indicate that different enzymes are involved, as shown in earlier studies.(6,99,100) In

TABLE VII

Effect of Metal Ions on Purified Microsomal Arylamidase[a]

	Leu-β-NA[b]		Arg-β-NA[b]	
Addition	10^{-3} *M*	10^{-4} *M*	10^{-3} *M*	10^{-4} *M*
Undialyzed	100	100	100	100
Dialyzed	15	15	32	32
Co^{++}	185	32	133	219
Mn^{++}	141	113	95	81
Mg^{++}	—	21	—	69
Ca^{++}	—	35	—	67

[a] Purified enzymes were dialyzed overnight against one liter 40 m*M* Tris-HCl buffer (pH 7.0) containing 1 m*M* EDTA and 0.1 m*M* dithiothreitol. The assays were performed as described earlier.(6)
[b] Percent response of arylamidase activity for metal-ion concentrations of 10^{-3} *M* and 10^{-4} *M*.

TABLE VIII

Properties of Purified Microsomal Arylamidase and Aminopeptidase

Property or addition[a]	Per cent change with added materials or other properties[b]	
	Arylamidase (Leu-β-NA)	Aminopeptidase (Leu-Gly-Gly)
KCN	+ 74	+18
Dithiothreitol	+ 67	+ 5
p-CMB	− 60	−42
Thiosulfate	+ 65	+35
Ascorbic acid	+ 68	+10
o-Iodosobenzoate	− 72	−72
Puromycin	−100	0
K_i (puromycin)	10^{-5} *M*	—
K_m	7×10^{-4} *M*	6×10^{-4} *M*
pH optimum	6.5	7.0

[a] K_i and K_m were calculated by the method of reciprocal plots.
[b] In experiments with added materials the conditions of assay were the same as in Table V. Percentages calculated with undialyzed preparation activity as 100%.

contrast to aminopeptidases, arylamidases were potently inhibited by puromycin ($K_i \times 10^{-5}$) (Table VIII). The effect of other added agents indicated that the endoplasmic aminopeptidases may contain enzymes with a sensitive -SH group, similar to that shown for the arylamidases.

CONCLUSIONS

The present report formed part of a general study from this laboratory on the enzymatic mechanisms involved in brain-protein breakdown and synthesis. For this purpose we studied the levels of representative peptide or peptidyl-peptide hydrolases together with observations on the rate of synthesis in purified subcellular fractions or in their derived membranous components. Protein turnover requires both degradative and synthetic mechanisms, and these were shown to be intrinsic to highly purified mitochondrial fractions.

Mitochondria prepared by the continuous sucrose gradient were judged to be free from lysosomal contamination on the basis of electron microscopy. Considerable heterogeneity for mitochondria and lysosomes was demonstrated by morphological and enzymatic criteria in other areas of the sucrose gradient employed. Studies on the role of degradative enzymes in protein turnover were extended to membrane fractions prepared from the purified mitochondrial fraction E and with washed microsomes. A survey of the fractionation conditions, such as large-amplitude swelling, and digitonin, phospholipase A, deoxycholate, and similar detergents, revealed that the best preparation of mitochondrial membranes was obtained with phospholipase, and of endoplasmic membranes with deoxycholate. All methods for the preparation of brain mitochondrial outer membrane led to the formation of a highly fragmented membrane material; the swelling and digitonin procedures showed promise in splitting the double membrane and may prove a suitable first step for outer-membrane preparation accompanied by alternative methods for separation. In all preparations the marker enzymes studied (MAO and SDH) were found to be unsuitable to specify mitochondrial inner and outer membrane in the manner indicated in studies with liver mitochondrial membranes. The use of marker enzymes as an adjunct to morphological findings in brain particulates may be dependent on the methods for membrane preparation and will require further exploration.

Studies on differential localization showed an association of the aminopeptidase and, to a lesser extent, acid proteinase with the outer mitochondrial fractions, but statements concerning other peptide hydrolases cannot be made, due to the poor enzyme recoveries in the membrane fractions. In the case of microsomes the enzymes were clearly differentiated, with most of the activity associated with the endoplasmic membranes and only trace activities with the ribosomal fractions. The calculated rates of protein breakdown in mitochondria match those of protein synthesis, thus reinforcing the view that this organelle is autonomous with respect to protein turnover. Also, the calculated rates for the membrane components themselves showed considerable heterogeneity in turnover, supporting the concept that they arise by different biogenetic mechanisms. This data was supported by different spectra of the protein solubilized with phenol solvents and separated by disc-gel electrophoresis. Mitochondrial subfractions separated on sucrose gradients also displayed considerable heterogeneity after administration of *L*-^{14}C-lysine *in vivo*, with the highest incorporation rate into fraction E and myelin proteins.

Many questions remain unanswered concerning the role of particulate and membrane-bound proteolytic enzymes. In addition to protein degradation and turnover, such enzymes may be involved in the activation of hormone or enzyme precursors or in the etiology of some diseases. Their association with endoplasmic membranes may be consistent with a role in the attachment and release of peptidyl–nucleic acid complexes as already cited in the case of bacteria. A membranous site for peptidases may facilitate the transport of newly synthesized peptides or proteins in smooth and rough cytomembranes as described by Palade and his group.[34,35,101,102] The absence of marked acid and neutral proteinase activity in microsomes and endoplasmic membranes indicates an absence of an autonomous protein turnover, in contrast to the mitochondrion. Microsomes are considered to be the major site for protein synthesis, with export of the protein to other cellular sites. Thus, the association of peptide hydrolases with endoplasmic membranes may facilitate the transport of synthesized materials and contribute to the turnover of protein in the nervous system.

REFERENCES

1. G. B. Ansell and D. Richter, *Biochim. Biophys. Acta* **13,** 87–91 (1954).
2. G. B. Ansell and D. Richter, *Biochim. Biophys. Acta* **13,** 92–97 (1954).
3. N. Marks and A. Lajtha, *Biochem. J.* **89,** 438–447 (1963).
4. N. Marks and A. Lajtha, *Biochem. J.* **97,** 74–83 (1965).
5. M. W. Kies and S. Schwimmer, *J. Biol. Chem.* **145,** 685–691 (1942).
6. N. Marks, R. K. Datta, and A. Lajtha, *J. Biol. Chem.* **243,** 2882–2889 (1968).
7. C. de Duve and R. Wattiaux, *Ann. Rev. Physiol.* **28,** 435–492 (1966).
8. A. Eichholz, *Biochim. Biophys. Acta* **135,** 475–482 (1967).
9. A. Eichholz and R. K. Crane, *Biochim. Biophys. Acta* **135,** 959–965 (1967).
10. P. Emmelot, A. Visser, and E. L. Benedetti, *Biochim. Biophys. Acta* **150,** 364–375 (1968).
11. N. Marks, *Intern. Rev. Neurobiol.* **11,** 57–97 (1968).
12. E. R. Einstein, J. Csejtey, W. Davis, N. Marks, and A. Lajtha, *Intern. Arch. Allergy Appl. Immunol.* (1969), in press.
13. E. DeRobertis, A. P. de Iraldi, G. R. De Lores Arnaiz, and L. Salganicoff, *J. Neurochem.* **9,** 23–35 (1962).
14. M. K. Campbell, H. R. Mahler, W. J. Moore, and S. Tewari, *Biochemistry* **5,** 1174–1184 (1966).
15. M. W. Gordon and G. G. Deanin, *J. Biol. Chem.* **243,** 4222–4226 (1968).
16. T. S. Work, *Bull. All-India Inst. Med. Sci.* **1,** 72–90 (1967).
17. R. S. Criddle, R. M. Bock, D. E. Green, and H. Tisdale, *Biochemistry* **1,** 827–842 (1962).
18. R. S. Criddle, D. L. Edwards, and T. G. Peterson, *Biochemistry* **5,** 578–582 (1966).
19. S. Fleischer, W. L. Zahler, and H. Ozawa, *Biochem. Biophys. Res. Commun.* **32,** 1031–1038 (1968).
20. G. Lenaz, N. F. Haard, A. Lauwers, D. W. Allmann, and D. E. Green, *Arch. Biochem. Biophys.* **126,** 746–752 (1968).
21. S. Puszkin, S. Berl, E. Puszkin, and D. D. Clarke, *Science* **161,** 170–171 (1968).
22. K. Takayama, D. H. MacLennan, A. Tzagoloff, and C. D. Stoner, *Arch. Biochem. Biophys.* **114,** 223–230 (1966).
23. D. F. Parsons, G. R. Williams, and B. Chance, *Ann. N.Y. Acad. Sci.* **137,** 643–666 (1966).
24. K. F. Tipton, *Biochim. Biophys. Acta* **135,** 910–920 (1967).
25. D. E. Green, D. W. Allmann, R. A. Harris, and W. C. Tan, *Biochem. Biophys. Res. Commun.* **31,** 368–378 (1968).
26. G. Dallner and L. Ernster, *J. Histochem. Cytochem.* **16,** 611–632 (1968).

27. N. Cuzin, N. Kretchmer, R. E. Greenberg, R. Hurwitz, and F. Chapeville, *Proc. Natl. Acad. Sci. U.S.* **58,** 2079–2083 (1967).
28. H. Kossel and U. L. Rajbhandary, *J. Mol. Biol.* **35,** 539–560 (1968).
29. J. C. Brown and P. Doty, *Biochem. Biophys. Res. Commun.* **30,** 284–291 (1968).
30. J. Lucas-Lenand and A. L. Haenni, *Proc. Natl. Acad. Sci. U.S.* **59,** 554–559 (1968).
31. F. Binkley, *J. Biol. Chem.* **236,** 1075–1082 (1961).
32. F. Binkley, F. Leibach, and N. King, *Arch. Biochem. Biophys.* **128,** 397–405 (1968).
33. K. R. Porter and R. D. Machado, *J. Biophys. Biochem. Cytol.* **7,** 167–180 (1960).
34. P. Siekevitz and G. E. Palade, *J. Biophys. Biochem. Cytol.* **7,** 619–644 (1960).
35. C. M. Redman, P. Siekevitz, and G. E. Palade, *J. Biol. Chem.* **241,** 1150–1158 (1966).
36. C. M. Redman and D. D. Sabatini, *Proc. Natl. Acad. Sci. U.S.* **56,** 608–615 (1966).
37. G. Manner and B. S. Gould, *Nature* **205,** 670–671 (1965).
38. W. Chefurka and Y. Hayashi, *Biochem. Biophys. Res. Commun.* **24,** 633–638 (1966).
39. S. Comorosan and P. Lugojan, *Europ. J. Biochem.* **6,** 460–466 (1968).
40. F. Fonnum, *Biochem. J.* **106,** 401–412 (1968).
41. A. Neidle, C. J. Van den Berg, and A. Grynbaum, *J. Neurochem.* **16,** 225–231 (1969).
42. F. E. Samson, Jr. and R. J. Jacobs, *Amer. J. Physiol.* **199,** 693–696 (1960).
43. H. Koenig, D. Gaines, T. McDonald, R. Gray, and J. Scott, *J. Neurochem.* **11,** 729–743 (1964).
44. H. Beaufay, A. Berleur, and A. Doyen, *Biochem. J.* **66,** 32P (1957).
45. O. Z. Sellinger, D. L. Rucker, and F. De Balbian Verster, *J. Neurochem.* **11,** 271–280 (1964).
46. J. Mordoh, *J. Neurochem.* **12,** 505–514 (1965).
47. H. Koenig, *in*: "Handbook of Neurochemistry," (A. Lajtha, ed.), Vol. II, Plenum Press, New York (1969).
48. O. Z. Sellinger and R. A. Hiatt, *Brain Res.* **7,** 191–200 (1968).
49. R. Wattiaux, M. Wibo, and P. Baudhin, *in*: "Lysosomes, Ciba Foundation Symposium" (A. V. S. de Reuck and M. P. Cameron, eds.), pp. 176–200, Little, Brown and Co., Boston (1963).
50. J. T. Dingle and A. J. Barrett, *Biochem. J.* **105,** 19P–20P (1967).
51. J. T. Dingle and A. J. Barrett, *Biochem. Soc. Agenda Papers*, July 1968.
52. E. Bachmann, D. W. Allmann, and D. E. Green, *Arch. Biochem. Biophys.* **115,** 153–164 (1966).
53. D. W. Allmann and E. Bachmann, *in*: "Methods in Enzymology" (S. P. Colowick and N. O. Kaplan, eds.), Vol. 10, pp. 438–443, Academic Press, New York (1967).
54. C. Schnaitman, V. G. Erwin, and J. W. Greenawalt, *J. Cell Biol.* **32,** 719–735 (1967).
55. D. F. Tapley and C. Cooper, *Nature* **178,** 1110–1120 (1956).
56. D. S. Beattie, *Biochem. Biophys. Res. Commun.* **30,** 57–62 (1968).
57. D. W. Allmann, E. Bachmann, N. Orme-Johnson, W. C. Tan, and D. E. Green, *Arch. Biochem. Biophys.* **125,** 981–1012 (1968).
58. I. Boime, E. E. Smith, and F. E. Hunter, Jr., *Arch. Biochem. Biophys.* **128,** 704–715 (1968).
59. R. A. Harris, J. T. Penniston, J. Asai, and D. E. Green, *Proc. Natl. Acad. Sci. U.S.* **59,** 830–837 (1968).
60. C. R. Hackenbrock, *J. Cell Biol.* **30,** 269–297 (1966).

61. C. R. Hackenbrock, *J. Cell Biol.* **37,** 345–369 (1968).
62. C. R. Hackenbrock, *Proc. Natl. Acad. Sci. U.S.* **61,** 596–605 (1968).
63. J. T. Penniston, R. A. Harris, J. Asai, and D. E. Green, *Proc. Natl. Acad. Sci. U.S.* **59,** 624–631 (1968).
64. J. M. Wrigglesworth and L. Packer, *Arch. Biochem. Biophys.* **128,** 790–801 (1968).
65. P. Cerletti, R. Strom, M. G. Giordano, F. Balestreno, and M. A. Giovenco, *Biochem. Biophys. Res. Commun.* **14,** 408–412 (1964).
66. R. E. McCaman, M. W. McCaman, J. M. Hunt, and M. S. Smith, *J. Neurochem.* **12,** 15–23 (1965).
67. K. H. Byington, J. M. Smoly, A. V. Morey, and D. E. Green, *Arch. Biochem. Biophys.* **128,** 762–773 (1968).
68. J. K. McDonald, F. H. Leibach, R. E. Grindeland, and S. Ellis, *J. Biol. Chem.* **243,** 4143–4150 (1968).
69. I. Honjo, K. Ozawa, O. Kitamura, A. Sakai, and T. Ohsawa, *J. Biochem.* (*Tokyo*) **63,** 311–320 (1968).
70. J. I. Salach, P. Turini, J. Hauber, R. Seng, H. Tisdale, and T. P. Singer, *Biochem. Biophys. Res. Commun.* **33,** 936–941 (1968).
71. P. M. Vignais and J. Nachbaur, *Biochem. Biophys. Res. Commun.* **33,** 307–314 (1968).
72. J. Nachbaur and P. M. Vignais, *Biochem. Biophys. Res. Commun.* **33,** 315–320 (1968).
73. S. Fleischer, A. Casu, and B. Fleischer, *Fed. Proc.* **23,** 486 (1964).
74. K. Fitzpatrick and R. J. Pennington, *Biochem. J.* **111,** 29P–30P (1969).
74a. K. G. M. M. Alberti and W. Bartley, *Biochem. J.* **111,** 763–776 (1969).
75. A. Lajtha, *Intern. Rev. Neurobiol.* **6,** 1–98 (1964).
76. D. S. Beattie, R. E. Basford, and S. B. Koritz, *Biochemistry* **5,** 926–930 (1966).
77. C. B. Klee and L. Sokoloff, *Proc. Natl. Acad. Sci. U.S.* **53,** 1014–1021 (1965).
78. D. E. S. Truman, *Exp. Cell Res.* **31,** 313–320 (1963).
79. W. Neupert, D. Brdiczka, and Th. Bücher, *Biochem. Biophys. Res. Commun.* **27,** 488–493 (1967).
80. D. S. Beattie, R. E. Basford, and S. B. Koritz, *Biochemistry* **6,** 3099–3106 (1967).
81. T. S. Work, J. L. Coote, and M. Ashwell, *Fed. Proc.* **27,** 1174–1179 (1968).
82. U. Bronsert and W. Neupert, *in*: "Regulation of Metabolic Processes in Mitochondria" (J. M. Tager, S. Papa, E. Quagliariello, and E. C. Slater, eds.), pp 426–437, Elsevier Publishing Company, Amsterdam (1966).
83. D. S. Beattie, R. E. Basford, and S. B. Koritz, *J. Biol. Chem.* **242,** 4584–4586 (1967).
84. D. Haldar, K. Freeman, and T. S. Work, *Nature* **211,** 9–12 (1966).
85. L. A. Autilio, S. H. Appel, P. Pettis, and P. L. Gambetti, *Biochemistry* **7,** 2615–2622 (1968).
86. G. Brunner and W. Neupert, *FEBS Letters* **1,** 153–155 (1968).
87. W. Kaiser and F. L. Bygrave, *Europ. J. Biochem.* **4,** 582–585 (1968).
88. F. L. Bygrave and Th. Bücher, *Europ. J. Biochem.* **6,** 256–263 (1968).
89. T. S. Work, *J. Mol. Biol.* **10,** 544–545 (1964).
90. R. L. M. Synge, *Biochem. J.* **65,** 266–273 (1957).
91. M. Bagdasarian, N. A. Matheson, R. L. M. Synge, and M. A. Youngson, *Biochem. J.* **91,** 91–105 (1964).
92. C. W. Cotman and H. R. Mahler, *Arch. Biochem. Biophys.* **120,** 384–396 (1967).
93. C. W. Cotman, H. R. Mahler, and T. E. Hugli, *Arch. Biochem. Biophys.* **126,** 821–837 (1968).
94. E. Reid, *in*: "Enzyme Cytology" (D. B. Roodyn, ed.), pp. 321–406, Academic Press, New York (1967).

95. D. Elson, *in*: "Enzyme Cytology" (D. B. Roodyn, ed.), pp. 407–473, Academic Press, New York (1967).
96. T. Higashi and T. Peters, Jr., *J. Biol. Chem.* **238,** 3945–3951 (1963).
97. L. Ernster, P. Siekevitz, and G. E. Palade, *J. Cell Biol.* **15,** 541–562 (1962).
98. N. Marks, R. K. Datta, and A. Lajtha, *in*: "Macromolecules and the Function of the Neuron" (Z. Lodin and S. P. R. Rose, eds.), pp. 220–231, Excerpta Medica Foundation, Amsterdam (1968).
99. V. K. Hopsu-Havu, K. K. Mäkinen, and G. G. Glenner, *Arch. Biochem. Biophys.* **114,** 557–566 (1966).
100. S. Ellis and M. Perry, *J. Biol. Chem.* **241,** 3679–3686 (1966).
101. J. D. Jamieson and G. E. Palade, *J. Cell Biol.* **34,** 597–615 (1967).
102. J. D. Jamieson and G. E. Palade, *J. Cell Biol.* **39,** 580–588 (1968).
103. O. H. Lowry, N. J. Rosebrough, A. L. Farr, and R. J. Randall, *J. Biol. Chem.* **193,** 265–275 (1951).
104. R. K. Burkhard and D. H. Klaassen, *Comp. Biochem. Physiol.* **18,** 231–235 (1966).
105. N. Marks and B. D'Monte, *Fed. Proc.* **28,** 733 (1969).
106. M. M. K. Nass, *Science* **165,** 25–35 (1969).

DISCUSSION

MAHLER: Studies in my laboratory and by Dr. Salganicoff at the Johnson Foundation confirm the fragility of the mitochondrial outer membrane. It is possible to detach the outer membrane, but it does not separate; one can see quite clearly by negative staining that this membrane readily shreds, it never becomes free. We also agree that the question of membrane-marker enzymes is in a thoroughly unsatisfactory state, especially in the case of brain. Our studies show a biphasic distribution of MAO in both continuous gradients and in the zonal centrifuge even in respect to intact mitochondria: there is as much activity in the endoplasmic reticulum as there is in mitochondria. This enzyme was thought to be an authentic outer-membrane marker, and results are difficult to interpret for fractions that may contain endoplasmic-reticulum contamination. Inner membrane may have good markers, and it will be interesting to see if peptidases are active at this site.

D'MONTE: There still seems to be a lot of discussion concerning both inner- and outer-membrane markers. For example, what about isobutyric or isocitric dehydrogenase?

MAHLER: That again depends on which papers you read. According to Chance and others*, those enzymes are present in the matrix, and according to Green[52] they may be in membranes or spaces between membranes. Perhaps α-keto dehydrogenases may be more satisfactory.

D'MONTE: I agree that the issue of membrane marker enzymes has to be studied in much greater detail, and, in all probability, is different in the case of brain compared to that in liver. It is also possible that mere differences in assay procedure cannot account for the different locations of MAO in liver and brain mitochondrial membranes, as has been suggested by Green[25]. I feel that additional enzymes need to be investigated to characterize brain mitochondrial membranes, and that these would then substantiate morphological evidence.

* C. Schnaitman and J. W. Greenawalt, *J. Cell. Biol.* **38,** 158–175 (1968).

Chapter 9

Assimilation of Glucose in Rat Brain and Metabolic Activities of Various Groups of Brain Proteins*

R. Vrba and Wendy Cannon

Department of Pharmacology
University of British Columbia
Vancouver, B.C., Canada

INTRODUCTION

We often use the expression "structural proteins of brain." This tacitly implies that these proteins are nonfunctional and are distinguishable from other, or "functional," proteins of brain. The experiments reported here were undertaken with the aim of securing data which might help to clarity the following points: (1) In what way is the pattern of metabolic interaction between glucose and brain proteins similar to or different from that in other organs? (2) Can the division of brain proteins into functional and structural proteins be shown experimentally? (3) What is the range of molecular weights of the cytoplasmic brain proteins, and how is this range similar to or different from proteins of liver, heart, and blood? (4) Is there any relationship between the molecular weight of these proteins and the rate and extent of assimilation of glucose carbon into these proteins *in vivo*?

Our approach is based on the following reasoning: Labeled glucose carbon *in vivo* is very rapidly converted into some α-amino acids in the brain and then incorporated into brain protein.[1–3] Because the possibility exists that brain proteins are related to brain function,[4–6] and the latter is uniquely dependent upon glucose supply,[7] it seemed worthwhile to study the glucose–protein relationship in the brain. The permeability of glucose into the brain is unquestioned.[8] The labeling of brain proteins *in situ* using

*We are grateful to Mrs. A. P. Reynolds for participating in some of these experiments. Our thanks are due to Dr. J. G. Foulks and Dr. M. C. Sutter for discussion, criticism, and help in the preparation of the manuscript. This work has been supported by the Medical Research Council of Canada.

^{14}C-labeled α-amino acids derived *in situ* from ^{14}C-glucose offers the advantage of avoiding the "blood-brain barrier." The latter may present a serious obstacle if labeled amino acids instead of labeled glucose are used as protein precursors.(8)

EXPERIMENTAL

Conversion of Glucose Carbon into Proteins in the Brain in Comparison with Other Organs

[U-^{14}C]glucose was injected subcutaneously into growing rats (60 g each), the organs excised 30, 60, or 120 min later, and the proteins isolated and purified as described previously.(3) It may be seen in Fig. 1 that the velocity of incorporation of ^{14}C into brain protein is comparable to that of the liver. The low content of ^{14}C in the blood proteins indicates that the labeled brain proteins were formed in the brain *in situ*, and not transported to the brain *via* the circulation. This conclusion is further supported by the observation that after subcutaneous injection of [U-^{14}C]glucose in the rat the specific radioactivity of the free amino acid fraction was many times greater in the brain than in the liver, blood, kidney, heart, spleen, lung, or

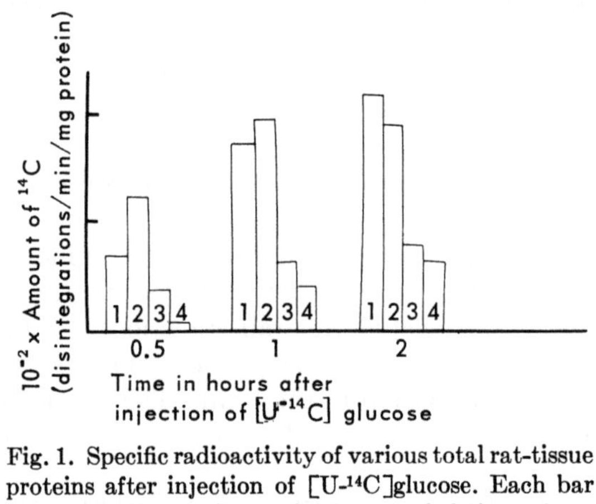

Fig. 1. Specific radioactivity of various total rat-tissue proteins after injection of [U-^{14}C]glucose. Each bar represents the result obtained on pooled tissue from four animals. Each animal received 25 μCi of [U-^{14}C]-glucose (196 mCi/mM, Radiochemical Centre, Amersham, Bucks, batch 38, CFB35) dissolved in 0.5 ml of 0.9% NaCl in presence of 1 mg of carrier glucose. (1) brain, (2) liver, (3) heart, (4) blood.

skeletal muscle.[1,2] It is of special interest to see that the labeling of the proteins of a tissue as metabolically active as the heart is low in comparison to brain or liver.

In other experiments[2] it has been shown that between 2 and 12 hr after injection of [U-^{14}C]glucose the brain proteins had higher specific activity than liver, kidney, heart, spleen, muscle, lung, or blood proteins. The proportion of ^{14}C combined in glutamic and aspartic acids was higher in brain proteins (62–69%) than in the proteins of any other organ examined.

It is not valid to conclude from these data that these processes are more active in the brain than in all other organs. Steele[9] has demonstrated that the retention of ^{14}C from labeled sucrose may be more intensive in bone marrow, colon, rectum, lymph node, salivary gland, stomach, or thymus than in brain. In view of these observations the apparently small absolute amount[3] (fraction of 1%) of injected [^{14}C]glucose carbon assimilated by brain protein acquires a different significance. This assimilation of carbon from administered glucose proceeds to some degree in all or almost all parts of the body. Among the various pathways of disposition of administered glucose, the rate of total assimilation is sufficiently rapid to account for a significant fraction. About 10% of the administered [U-^{14}C]glucose carbon is converted into proteins and lipids in the whole body of the mouse within 1 hr.[10]

Comparison of Various Groups of Brain Proteins, Obtained by Differential Centrifugation

Five hours after injection of labeled glucose, rat-brain homogenates were separated by differential centrifugation into four main fractions: (1) Myelinated structures; (2) nuclei; (3) mitochondria, microsomes, and other cytoplasmic particles; and (4) soluble proteins. The proteins in these main fractions were separated further by extraction with 0.5% deoxycholate solution and fractional precipitation with ammonium sulphate. Measurements were then made of the specific activity of the amino acids isolated from the hydrolysate of the proteins in the various protein fractions which were obtained. The results are shown in Table 1. The degree of labeling of the different fractions was clearly uneven; it was less, for example, in the myelinated structures than in the soluble protein fractions. However, no unlabeled protein could be isolated from the brain 5 hr after injection of glucose, even when various brain proteins were separated using a combination of three methods: fractional centrifugation, solubility in desoxycholate, and precipitation with ammonium sulfate at different concentrations.

TABLE I

Distribution of ^{14}C in Rat-Brain Protein Fractions at 5 hr after Injection of [U-^{14}C]glucose[a]

Protein fraction	Specific radioactivity (counts/min/mg α-amino-N)	
	Expt. 1	Expt. 2
Whole brain protein	268	227
Myelinated structures	126	159
Nuclei		
Residue insol. in 0.5% DOC	—	106
Proteins extracted by 0.5% DOC	286	244
Proteins pptd. by $(NH_4)_2SO_4$, 0–25% sat.[b]	212	241
Proteins pptd. by $(NH_4)_2SO_4$, 50–75% sat.[b]	116	140
Mitochondrial and microsomes		
Residue insol. in 0.5% DOC	116	123
Proteins extracted by 0.5% DOC	162	175
Proteins pptd. by $(NH_4)_2SO_4$, 0–25% sat.[b]	168	151
Proteins pptd. by $(NH_4)_2SO_4$, 25–50% sat.[b]	190	218
Proteins pptd. by $(NH_4)_2SO_4$, 50–75% sat.[b]	97	119
Soluble proteins	—	272
Proteins pptd. by $(NH_4)_2SO_4$, 0–25% sat.	254	298
Proteins pptd. by $(NH_4)_2SO_4$, 25–50% sat.	282	299
Proteins pptd. by $(NH_4)_2SO_4$, 50–75% sat.	294	279

[a] In each experiment four rats were decapitated 5 hr after injecting 10 μCi of [U-^{14}C]glucose with 4 mg carrier glucose in 0.2 ml water. Further experimental details have been described elsewhere.(2) (Reproduced with the permission of *Journal of Neurochemistry*.)

[b] These precipitates were obtained from the desoxycholate (DOC) extracts.

Comparison of Various Groups of Brain Proteins Obtained by Sucrose Density Centrifugation

By applying sucrose density centrifugation, similar results were achieved.(3) Rats were injected with [U-^{14}C]glucose and the content of ^{14}C in proteins of the cerebral P_1 ("nuclear"), P_2 ("mitochondrial"), P_3 ("microsomal"), and high-speed supernatant fractions were measured 7, 22, and 93 hr after injection of labeled glucose. About 40–50% of the ^{14}C assimilated by brain proteins was found in the P_2 ("mitochondrial") fraction. The crude brain mitochondrial fractions were subfractionated on continuous sucrose gradients (0.32–1.80 *M* sucrose) and the ^{14}C content of the proteins of about 20 subfractions was measured. The specific activity of proteins varied considerably in the various separate protein fractions. However, there was a significant labeling of *all* the fractions examined (Fig. 2).

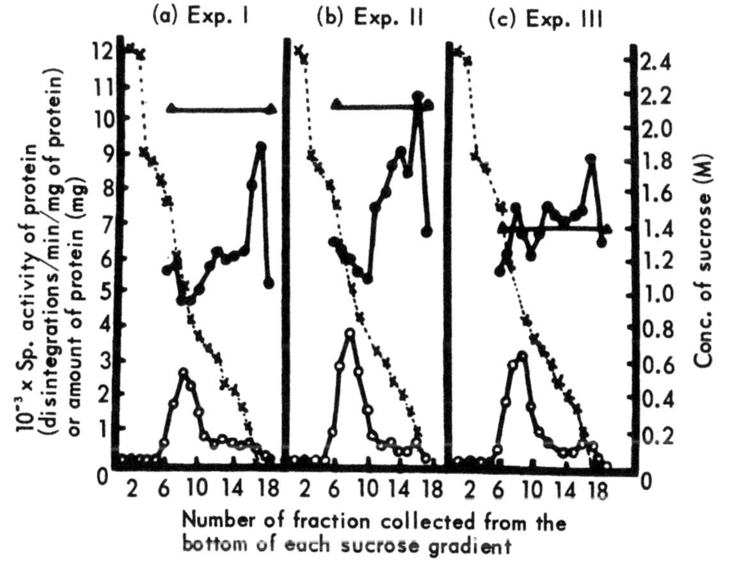

Fig. 2. Subfractionation of the crude rat-brain "mitochondrial" P_2 fractions on continuous sucrose gradients. Darkened circles, specific activity of protein; open circles, amount of protein in each subfraction; crosses, concentration of sucrose along the gradient; darkened triangles, specific activity of the protein of the high-speed brain supernatant of the same experiment. (a) Expt. I, 7 hr after injection of [U-^{14}C]glucose; (b) Expt. II, 22 hr after injection of [U-^{14}C]glucose; (c) Expt. III, 93 hr after injection of [U-^{14}C]glucose. Each point represents the analysis obtained from pooled samples of four brains. Each animal received a subcutaneous injection of [U-^{14}C]glucose over the left flank (81.5×10^6 disintegrations/min; dissolved in 0.7 ml of 0.9% NaClin the presence of 2.17 mg of carrier glucose). (Reproduced with the permission of *The Biochemical Journal.*)

Comparison of Various Groups of Brain and Liver Proteins Obtained by Chemical Fractionation

The proteins of brain and liver were fractionated as outlined in Fig. 3. It should be noted that all protein fractions obtained in this way were finally deproteinized and then subjected to an extensive purification procedure, which consisted of four extractions with large volumes of cold $HClO_4$ (6% w/v), followed by extraction with two different mixtures of organic solvents, two extractions with hot trichloroacetic acid (5% w/v, 90°, 15 min), and repeated extraction with organic solvents; finally, the proteins thus purified were dissolved in 0.5*N* NaOH and their specific activity measured.[3] The results of these experiments are presented in Figs. 4 and 5.

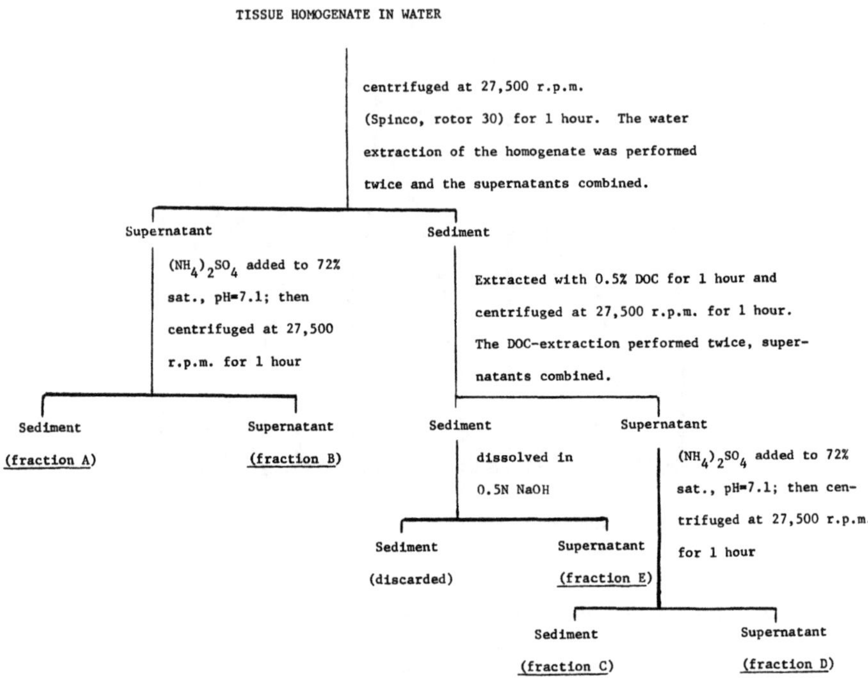

Fig. 3. Summary of fractionation scheme for brain and liver homogenates.

It may be seen that the pattern of labeling of various protein groups is different in brain than in liver (Fig. 4). It has been shown previously[3] that different conclusions may be reached if results of experiments using *in vivo* labeling of proteins are calculated in terms of relative rather than absolute specific activities. The different pattern of labeling of brain proteins is even more apparent if the results are expressed in terms of relative specific activity (Fig. 5) (RSA = 100 × specific activity of proteins of a particular fraction/specific activity of total proteins of the organ). In Figs. 4 and 5 it is evident that in liver the fraction C (DOC-soluble proteins) showed a greater assimilation of glucose carbon per unit weight of protein than a similar protein fraction of brain, whereas the opposite relationship is observed in fraction A (water-soluble proteins). The latter is equally or more labeled in brain than in liver. However, it should be noted that with this method no evidence could be obtained to demonstrate the presence of a significant amount of "metabolically inert protein" in the brain. All isolated groups of brain proteins were labeled and the differences were only quantitative.

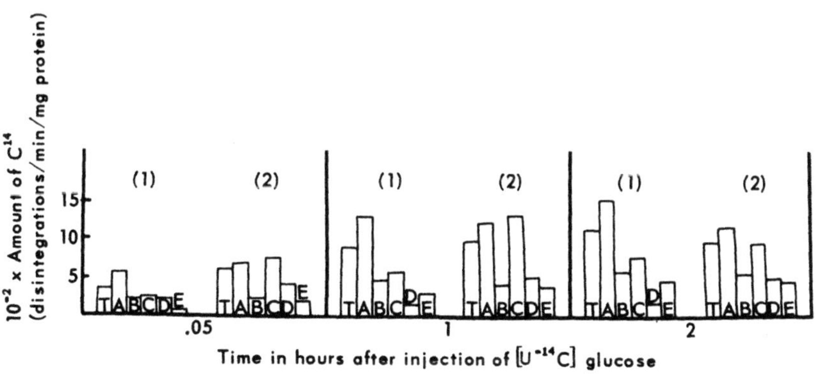

Fig. 4. Specific radioactivity of different groups of rat brain and liver proteins after injection of [U-^{14}C]glucose. (1) brain, (2) liver; capital letters denote various separated protein fractions isolated as described in Fig. 3. T: Specific activity of total brain proteins before fractionation. All other experimental conditions are as described in Fig. 1.

Comparison of Various Groups of Brain Proteins Obtained by Gel Filtration

We then subjected the soluble proteins of the brain high-speed supernatant to further study using gel filtration. The high-speed supernatant proteins represent about 22% of the total proteins in rat brain.[3] A column of Sephadex G-200 (5 cm wide, 1750 ml volume) was equilibrated with

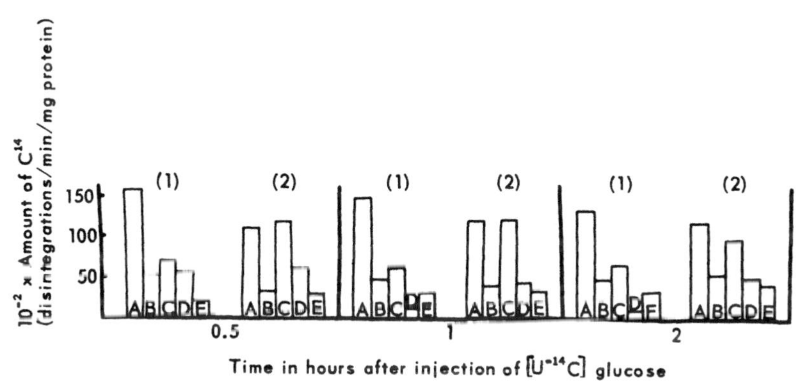

Fig. 5. Relative specific radioactivity of different groups of rat brain and liver proteins after injection of [U-^{14}C]glucose. Specific radioactivity (dpm/min/mg protein) of total tissue proteins = 100. (1) brain, (2) liver; capital letters denote various separated protein fractions isolated as described in Fig. 3. All other experimental conditions are as described in Fig. 1.

0.9% NaCl and calibrated with a set of eight nonenzymic-protein molecular-weight markers. Cube roots of the distribution coefficient were calculated and plotted against the square roots of the molecular weights.[11] A satisfactory linear relationship was observed (Fig. 6) and the same column was subsequently used for separation of rat brain and liver high-speed supernatant proteins.

Six growing rats were injected subcutaneously with [U-^{14}C]glucose and, 22 hr later the animals were decapitated, and the brains, livers, hearts, and whole blood were collected and homogenized in 0.9% NaCl in a Potter–Elvehjem glass and Teflon pestle homogenizer. A portion of the clear high-speed supernatants (Spinco, Model L, Head 40, 29,000 rpm/2 hr) was subjected to gel-filtration on the previously mentioned column (Fig. 6). Blue dextran and glucose were added to the supernatant to mark the void and elution volumes, and 400 fractions (5 ml each) were collected. The method was first checked on a brain high-speed supernatant (Fig. 7). The

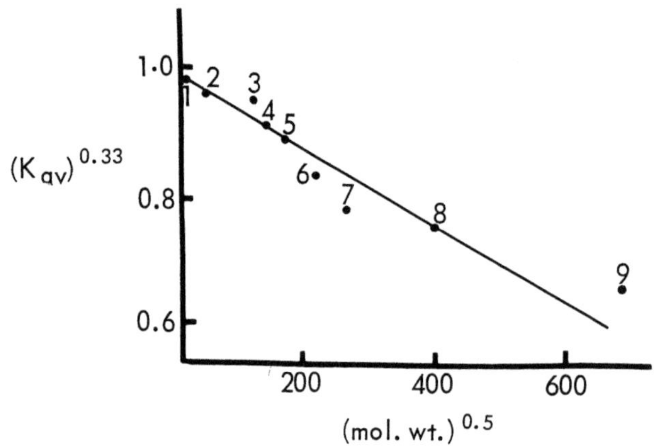

Fig. 6. Calibration of a Sephadex (G-200) column (K50/100) with nonenzymatic molecular-weight markers. Bed volume: 1750 ml. Volume of sample: 20 ml. Elution solvent: 0.9% (w/v) NaCl. The reference substances (their molecular weights and amounts* in parentheses) were: (1) glucose (180; 58 mg); (2) bacitracin (1,450; 16 mg); (3) cytochrome c (horse heart) (12,400; 21 mg); (4) myglobin cryst. (sperm whale), salt-free (17,800; 22 mg); (5) chymotrypsinogen A (beef pancreas) (6 × cryst.), salt-free (25,000; 23 mg); (6) ovalbuman (2 × cryst.) (45,000; 25 mg); (7) albumin (bovine) cryst. (67,000; 20 mg); (8) gamma globulin (human) (160,000; 22 mg); (9) apo-ferritin (horse), amorphous, salt-free (480,000; 22 mg).

* Amounts in mg. contained in 50 ml of a stock solution, suitable portions of which were used for calibration of the column.

recovery of radioactivity expressed as the sum of ^{14}C in the 400 fractions represented 99% of the radioactivity applied to the column. It may be seen in Fig. 7 that the eluate contained a continuous range of proteins with molecular weights ranging from less than 10^3 up to more than 8×10^5. It is obvious from Fig. 7 that ^{14}C was present along the whole elution spectrum.

Results of gel filtration of four different tissues from the same experiment (22 hr after injection of [U-^{14}C]glucose) are presented in Fig. 8. It may be seen that the radioactivity in the high-speed supernatant from all tissues was distributed along the whole range of elution. It also may be seen in Fig. 8 that a large proportion of the total protein of brain and liver high-speed supernatants was recovered in the void volume (eluted together with the "blue dextran" void volume marker). The soluble proteins of brain and liver evidently contain a much higher proportion of large molecules (mol. wt. $\gtrsim 8 \times 10^5$) than the blood. The heart occupies an intermediate position in this respect.

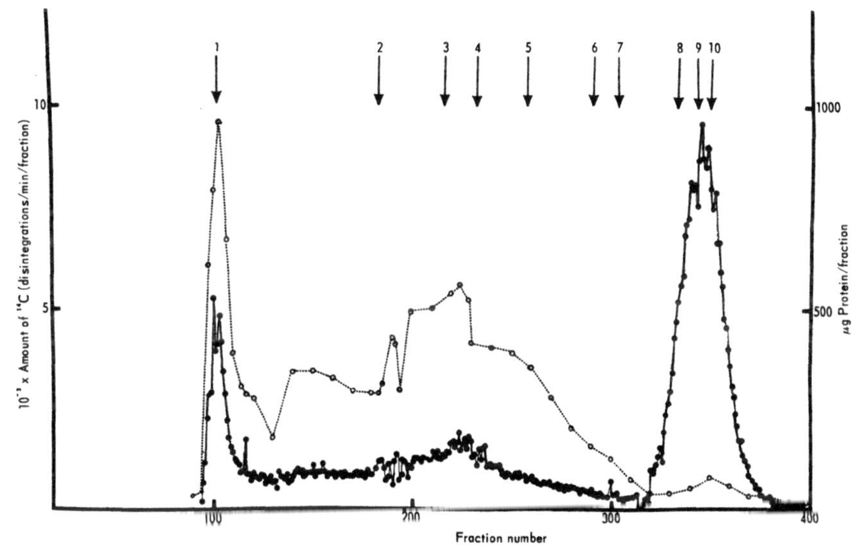

Fig. 7. Gel-filtration chromatography on a Sephadex (G-200) column of brain high-speed supernatant, obtained 22 hr (Expt. A) after injection of [U-^{14}C]glucose. The sample of brain high-speed supernatant applied to the column contained 4.76×10^5 dpm. Recovery of radioactivity (sum of all 400 fractions collected) was 4.71×10^5 (99%). Arrows at the top of the graph indicate the approximate molecular weights based on calibration of the column with markers as indicated in Fig. 6. (1) "blue-dextran;" (2) apo-ferritin; (3) gamma globulins; (4) albumin; (5) ovalbuman; (6) chymotrypsinogen; (7) myglobin; (8) cytochrome; (9) bacitracin; (10) glucose; dashed line: protein; darkened circles: radioactivity.

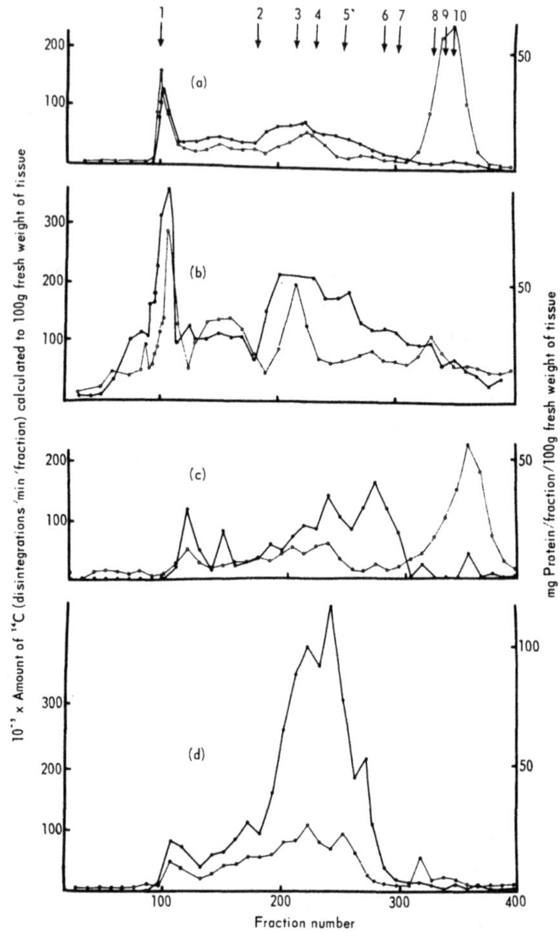

Fig. 8. Gel-filtration chromatography on a Sephadex (G-200) column of brain, liver, heart, and blood high-speed supernatants obtained 22 hr after injection of [U-^{14}C]glucose. Darkened circles: protein; dashed line: radioactivity. 20 ml portions of the high-speed supernatants were applied in each case. Arrows at the top of the graph indicate the approximate molecular weights as indicated in Figures 6 and 7. (a) brain; (b) liver; (c) heart; (d) blood. In each case a sample of the high-speed supernatant, obtained from six pooled organs, was applied. Injected dose of [U-^{14}C]glucose: 1.369×10^8 dpm/animal (60 g).

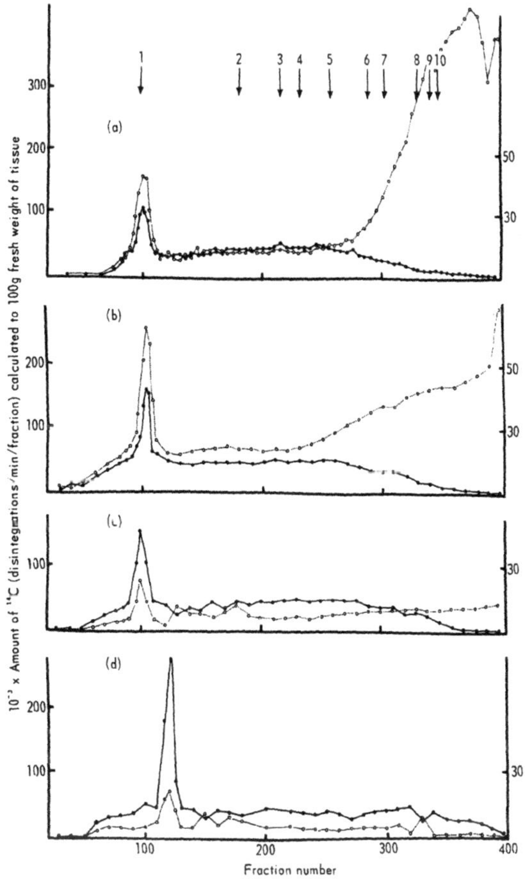

Fig. 9. Gel-filtration chromatography on a Sephadex (G-200) column of brain high-speed supernatants obtained at various time intervals after injection of [U-^{14}C]glucose. Darkened circles: protein; dashed line: radioactivity. In each case 20-ml portions of brain high-speed supernatants were applied. Arrows at the top of the graph indicate the approximate molecular weights as indicated in Fig. 7. Time of decapitation of animals after injection of [U-^{14}C]glucose and amount of protein in the sample applied to the column were as follows: (a) 4 hr (113 mg); (b) 8 hr (101 mg); (c) 12 hr (106 mg); (d) 92 hr (111 mg). In each case a sample of brain high-speed supernatant obtained from six pooled brains was applied to the column. Injected dose of [U-^{14}C]glucose: 1.252 $\times$ 10^8 dpm per animal (60 g).

In Fig. 9 it may be seen that the distribution of ^{14}C along the elution curve does not change very much in the interval of 4–92 hr after the injection of [U-^{14}C]glucose except for a continuous decrease of ^{14}C in the low-molecular part of the eluate. Here, of course, the nonprotein ^{14}C-labeled components of the brain extract are also contained. The incorporation of glucose carbon into the protein is a very rapid process; however, the loss of ^{14}C from the labeled protein is a very slow process, which requires several weeks[(9)] for completion. A similar observation has been made previously by Steele, who followed the fate of carbon from uniformly labeled sucrose in mice *in vivo*. The significance of this phenomenon cannot be fully explained at present, and may indicate the existence of a protein–protein interrelationship. This assumption is also supported by the results presented in Figs. 10 and 11, as described in more detail below.

We must be aware of the fact that under the conditions of gel chromatography of brain high-speed supernatants it is not possible to state if the ^{14}C has been incorporated into protein or into other large-molecular components of the high-speed supernatant, for instance, glycogen, lipid moiety of the lipoproteins, etc. Therefore, suitable numbers of eluated fractions were pooled to provide four groups of proteins with an approximate range of molecular weights as indicated in Fig. 10. The protein fractions obtained in this way were first deproteinized and then subjected to extensive purification procedures as described previously.[(3)] The purified proteins were dissolved in 0.5 N NaOH and their specific activity measured.

The results recorded in Fig. 10 show that the four groups of purified proteins within the investigated ranges of molecular weights were almost uniformly labeled as soon as 4 hr after injection of ^{14}C-glucose. However, it appears that the proteins with a smaller molecular weight were labeled first, and were also the first to show a decrease of ^{14}C. Therefore, an experiment was performed in which the animals were killed as soon as 15 min after injection of labeled glucose and the distribution of labeled carbon among proteins of various molecular weights determined using gel chromatography. The results of this experiment are shown in Fig. 11. It may be seen that 15 min after injection of labeled glucose the large-molecular proteins of the brain high-speed supernatant were absolutely free of any radioactivity; however, some ^{14}C was incorporated into the proteins of low molecular weight. The ^{14}C in the eluate was not only in the nonprotein labeled compounds contained in this part of the eluate, because the isolation and purification of the total protein of the brain high-speed supernatant yields a radioactive preparation. In this particular case the specific activity of the purified protein was found to be 930 disintegrations/min/mg and the amount of protein in the pooled brain high-speed supernatant (Lowry-positive substances[(12)]) was 66.6 mg. Calculation shows that 15

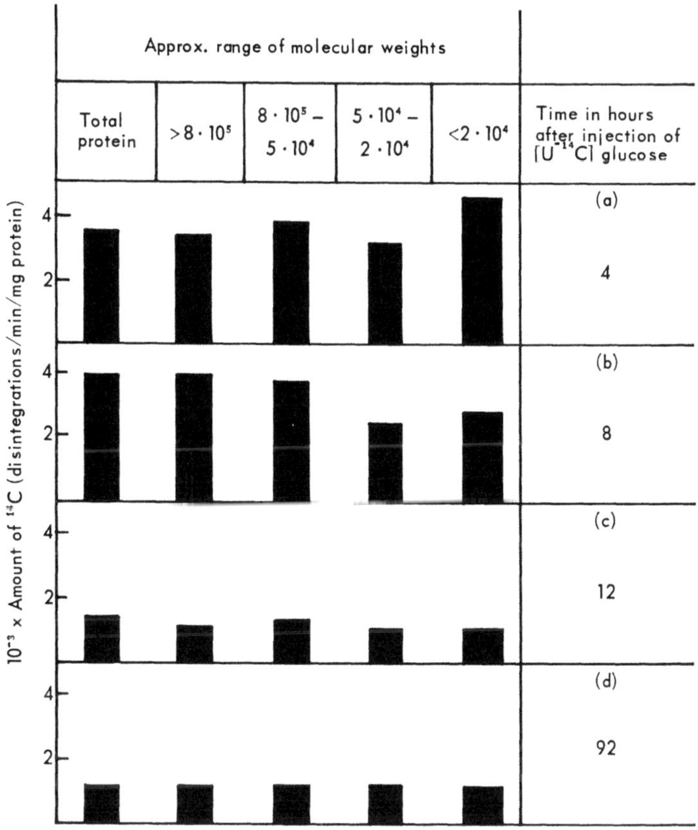

Fig. 10. Specific activities of various groups of proteins of the brain high-speed supernatants at various time intervals after injection of [U-^{14}C]glucose. Fractions of the supernatant eluted from the Sephadex (G-200) column were pooled into four groups according to the approximate molecular weights of the dissolved proteins. The latter were subjected to a purification procedure as described elsewhere,[3] and their specific activities compared to the specific activity of a sample of the total purified protein from each of the original supernatants. Each bar represents specific activities of purified proteins. All other experimental conditions are as described in Fig. 9.

min after subcutaneous injection of [U-^{14}C]glucose approximately 0.005% ^{14}C of the injected dose (1.275×10^9 dpm per four rats, 60 g each) could be recovered from the purified proteins contained in the brain high-speed supernatant. In previous experiments it was shown that (at most) approximately 0.08% of the injected glucose carbon can be incorporated into total brain proteins of the rat.[3] The brain high-speed supernatant contains

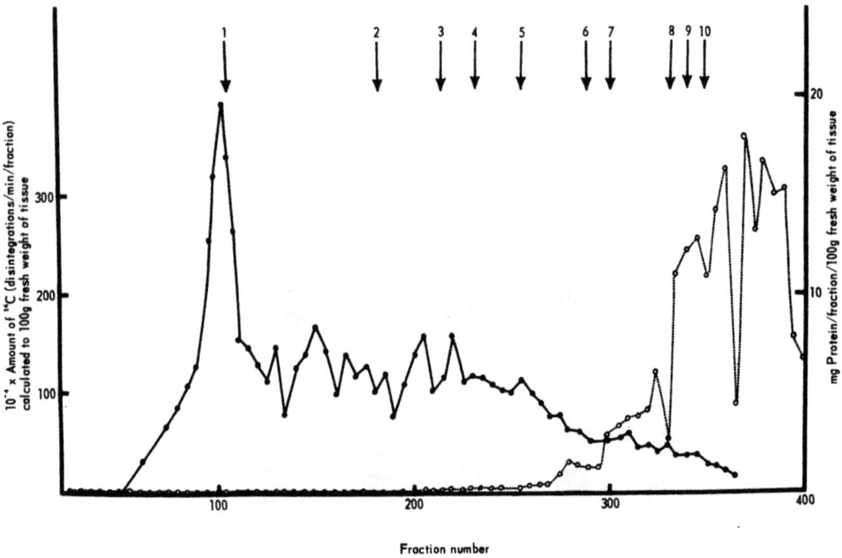

Fig. 11. Gel-filtration chromatography on a Sephadex (G-200) column of brain high-speed supernatant obtained 15 min after injection of [U-^{14}C]glucose. Darkened circles: protein; dashed line: radioactivity. A 20-ml portion of the brain high-speed supernatant containing 43 mg of protein was applied to the column. Arrows at the top of the graph indicate the approximate molecular weights as indicated in Fig. 7. A sample of a high-speed supernatant obtained from four pooled brains was applied to the column. Injected dose of [U-^{14}C]glucose: 3.185×10^8 dpm/animal (60 g).

about 20% of the total brain protein,[3] and the maximum which may be expected to be found in the soluble proteins 7 hr after injection of [U-^{14}C]-glucose is therefore 0.016% of ^{14}C of the injected dose.[3] It is of interest that about 0.005% of the glucose carbon was found in the low-molecular-weight ($<2 \times 10^4$) purified proteins of the brain high-speed supernatant as soon as 15 min after injection of [U-^{14}C]glucose. In this connection it should be noted that in previous experiments an incorporation of ^{14}C into brain proteins has been observed as soon as 5 min after intravenous injection of [U-^{14}C]glucose.[1] In these latter experiments the purified proteins were hydrolyzed in 6 N HCl, and the protein-bound amino acids isolated on ion exchangers. Elliott and Yoshino[42] recently observed that after intravenous administration of [U-^{14}C]glucose into rats radioactivity appeared in amino acids in brain protein between 5 and 15 min after the injection.

DISCUSSION

Physiological Significance of the Appearance of ^{14}C from Glucose in Tissue Proteins

During the past 15 years evidence has accumulated that glucose can be converted into amino acids in the brain. Labeled α-amino acids were isolated from brains of neonatal and young mice[13,14] and from the brains of adult rats after injections of [1-^{14}C]glucose and [6-^{14}C]glucose.[15] Rapid formation of labeled amino acids was found *in vitro* after incubation of brain cortex slices with ^{14}C-labeled glucose[16,17] and after introduction of [U-^{14}C]glucose into "artificial blood" used for perfusion of the brain of a cat with isolated circulation.[18,19] Further, it has been shown[1] that 2 min after intravenous injection of glucose into the circulation about 40% of the glucose carbon retained in the brain is in the form of α-amino acids, mainly glutamic and aspartic acids. The conversion of brain glucose into α-amino acids reaches equilibrium in less than 5 min, and within this interval more than 50% of glucose-derived carbon present in the brain can be accounted for as α-amino acid carbon. The presence of ^{14}C in brain proteins could be demonstrated within 5 min after injection of [U-^{14}C]glucose into the brain of grown rats,[1] and the participation of brain proteins in glucose metabolism in brain has been demonstrated.[2,3] These observations have been confirmed and extended with the use of modified and improved methods by several other authors.[20,24] The distribution, uptake, and the further fate of glucose in various organs may be influenced by mild hypoclycaemia,[25] various drugs,[26,28] hepatectomy,[22] and other pharmacological and physiological factors.[28,29] These and further investigations[10] indicate that the conversion of ^{14}C from labeled glucose into protein is a physiologically significant process. Others have attributed the rapid incorporation of ^{14}C from glucose into amino acids in brain to isotope exchange[30] or to the "crossing over" phenomenon[31] rather than to metabolic incorporation. These opinions also are based in part on the fact that the blood supply of the brain is relatively larger per unit of weight than that of muscle and liver, and that added glucose may enter metabolic processes more rapidly in brain than in other tissues because of the low glucose content of brain (low dilution with endogenous glucose).

The endogenous glucose content of brain indeed represents only a negligible fraction of that of liver.[32,33] However, the specific activity of the total brain proteins after injection of [U-^{14}C]glucose is similar to that of the liver proteins, and significantly higher than that of the heart proteins (Fig. 1). Furthermore, it would be difficult to explain (in terms of glucose content of the organ or of its blood supply) the different pattern of labeling of various groups of proteins in brain and liver (Figs. 4 and 5).

It is also relevant that the vascularization of the heart per unit of weight is considerably higher than that of the brain.[34] The approximate rate of blood flow per 100 g organ per minute is: kidney, 560 ml; liver ($\frac{1}{3}$ arterial, $\frac{2}{3}$ portal), 150 ml; heart (coronary circulation), 100 ml; brain, 55 ml; resting muscle, 3 ml.[35] These figures indicate that there is no correlation between the rate of assimilation of glucose carbon by proteins and the rate of blood flow of the organs studied. Consequently, it appears that the process of uptake of ^{14}C from [U-^{14}C]glucose into protein may be better understood in terms of a dynamic state of equilibrium of various organ proteins[36] and not only in terms of endogenous glucose content, circulation, or rate of blood flow. Lajtha and Toth,[43] who used ^{14}C-labeled *L*-lysine in their experiments, demonstrated the renewal of all cerebral proteins within an experimental period of 150 days. Our experiments confirm their conclusions on the instability of brain proteins and also demonstrate that the metabolic activity of all of the investigated groups of brain proteins may be observed within a few hours after injection of ^{14}C-labeled glucose.

Difference between the Fate of Glucose Carbon Atoms *in Vitro* and *in Vivo*

The physiological significance of the process of incorporation of glucose carbon into amino acids and proteins is even more striking if we compare this process under *in vivo* and *in vitro* conditions. Chain *et al.*[37] incubated samples from various regions of rabbit brain in the presence of 10 μCi of [U-^{14}C]glucose. After 1 hr of incubation they found that whereas 75–90% of the radioactivity was recovered from the acid-soluble fraction, primarily in lactic acid, only about 10% was recovered in the acid-soluble α-amino acids, mainly glutamic acid. The acid-insoluble residue in all samples showed only negligible traces of radioactivity.

Different results were obtained when similar experiments were performed *in vivo*. Within 15 min after intravenous injection of 15 μCi of [U-^{14}C]-glucose into intact rats (100–150 g) more than 60% of the ^{14}C recovered from the brain is in the form of acid-soluble α-amino acids, and both protein and lipids of the brain contain considerable radioactivity within 1 hr after injection of labeled glucose and a measurable amount within 5 min after injection.[1,42] From these studies it may be concluded that in the presence of an intact circulation and normal humoral influence, glucose metabolism *in vivo* consists of an anabolic as well as a catabolic process, the former being characterized by rapid and extensive assimilation of glucose carbon into lipids and proteins, preceded by an extensive incorporation of glucose carbon into some acid-soluble amino acids. In contrast, in *in vitro* preparations the anabolic processes are grossly impaired, much more of the glucose carbon appearing in lactic acid.[37] These observations should influence our views on the possible mechanisms of drug action and hormonal regulation of glucose metabolism. The anabolic and catabolic pathways *in vivo* should not be considered in isolation from each other, since drugs, or hormones, or a particular concentration of intermediates of metabolism, may act to disturb the equilibrium between catabolic and anabolic processes.

The *in vivo* mechanisms responsible for the more active process of assimilation of glucose carbon by proteins in brain than in liver, kidney, blood, heart, spleen, or muscle have not yet been identified. As discussed above, there is no correlation between the rate of this process in various tissues and the endogenous pool of glucose in the organs, the degree of vascularization, or the rate of blood flow. However, the fact that the glutamate/glucose ratio in the brain is considerably higher than in other organs (Table II) may be relevant to this question. Only recently has it been fully grasped that the concentration of small molecules in the cell can control its enzymatic activities.[38] Possibly, it is not the concentration of glucose alone, but a *constellation* of concentrations of several glucose metabolites which may participate in the regulation of glucose metabolism. The latter may be subject to an autoregulatory mechanism, i.e., the products of transformation of the very molecule which is itself the object of the regulatory process may serve as a feedback signal.[39] It is of importance to note that under *in vitro* conditions for studying brain metabolism as performed in most published experiments the concentrations of glucose, glutamate, and other metabolites in the medium were grossly different from *in vivo* conditions, which may be relevant to the different fate of glucose carbon *in vivo*.

TABLE II

Glutamate/Glucose Ratio in Rat Brain and Other Tissues

Tissue	Glucose[32] (mg/100 g)	Glutamic acid[40] (mg/100 g)	Glutamate/glucose ratio	Relative glutamate/glucose ratio (brain = 100)
Brain	8	152	19.0	100.0
Liver	82	49	0.6	3.2
Leg muscle	22	18	0.8	4.2
Kidney	51	96	1.9	10.0
Blood[41]	66	3	0.05	0.3

CONCLUSIONS

1. *In vivo* the rate of incorporation of glucose carbon into proteins of the brain is comparable to the rate of this process in the liver and is much higher than in the heart or in the blood. These results cannot be explained as an artifact caused by the small endogenous glucose pool of the brain or differences in vascularization or in the rate of blood flow of brain as compared to liver and heart.

2. Soluble (0.9% NaCl) rat brain and liver proteins (Lowry-positive substances) display a continuous spectrum of molecular weights, ranging from small molecules (mol. wt. $< 2 \times 10^4$) to large molecules (mol. wt. $> 8 \times 10^5$). The relative amount of the latter is conspicuously small in blood in comparison to brain and liver. The labeling of all soluble proteins of brain and liver, which could be separated from one another by gel filtration (Sephadex G-200), was nearly uniform within 4 hr after subcutaneous injection of [U-^{14}C]glucose. Fifteen minutes after the injection of [U-^{14}C]-glucose the smaller protein molecules (mol. wt. $< 2 \times 10^4$) were significantly radioactive, whereas no ^{14}C could be detected in the larger (mol. wt. $> 2 \times 10^4$) protein molecules. This indicates that glucose carbon is first incorporated into small protein molecules in the brain.

3. It is concluded that: (a) low-molecular brain proteins which are soluble in 0.9% NaCl (mol. wt. $< 2 \times 10^4$) and water-soluble brain proteins which are precipitated with 70% saturated $(NH_4)_2SO_4$ at pH 7.1 assimilate glucose carbon more rapidly than any of the other fractions investigated; (b) separation of various groups of brain proteins using fractional centrifugation, solubility in desoxycholate, precipitation with ammonium sulphate, sucrose-density-gradient centrifugation, and gel filtration of high-speed supernatants provided no evidence to indicate the presence of any significant amount of "metabolically inert protein" in the brain; (c) the dynamic state of body constituents *in vivo* includes the continuous conversion of glucose into protein, and this process is at least as active in the brain as in liver and more active in brain or liver than in heart.

REFERENCES

1. R. Vrba, *Nature* **195,** 663–665 (1962).
2. R. Vrba, M. K. Gaitonde, and D. Richter, *J. Neurochem.* **9,** 465–475 (1962).
3. R. Vrba, *Biochem. J.* **105,** 927–936 (1967).
4. R. Vrba, *Usp. Sovr. Biol. SSSR* **41,** 321–352 (1956).
5. D. Richter, *Brit. Med. Bull.* **21,** 76–79 (1965).
6. A. V. Haroutunian, *in*: "Problems of Brain Biochemistry" (H. C. Buniatian, ed.), Vol. 2, pp. 152–167, Acad. Sci. Armen. SSR (1966).

7. H. E. Himwich, "Brain Metabolism and Cerebral Disorders," pp. 24–25, The Williams and Wilkins Co., Baltimore (1951).
8. J. Dobbing, *Physiol. Rev.* **41,** 130–188 (1961).
9. R. Steele, *J. Biol. Chem.* **209,** 91–103 (1954).
10. R. Vrba, *Biochem. J.* **99,** 367–380 (1966).
11. J. Porath, *Pure Appl. Chem.* **6,** 233–244 (1963).
12. O. H. Lowry, N. J. Rosebrough, N. J. Farr, and B. J. Randall, *J. Biol. Chem.* **193,** 265–275 (1951).
13. J. R. Winzler, K. Moldave, M. E. Rafelson, Jr., and H. E. Pearson, *J. Biol. Chem.* **199,** 485–492 (1952).
14. R. B. Roberts, J. B. Flexner, and L. B. Flexner, *J. Neurochem.* **4,** 78–90 (1959).
15. H. Busch, E. Fujiwara, and L. M. Keer, *Cancer Res.* **20,** 50–57 (1960).
16. A. Beloff-Chain, R. Catanzaro, E. B. Chain, I. Massi, and F. Pocchiari, *Proc. Roy. Soc. B* **144,** 22–28 (1955).
17. M. H. Kini and J. H. Quastel, *Nature* **184,** 252–256 (1959).
18. A. Geiger, Y. Kawakita, and S. S. Barkulis, *J. Neurochem.* **5,** 323–338 (1960).
19. A. Geiger, N. Horvath, and Y. Kawakita, *J. Neurochem.* **5,** 311–322 (1960).
20. M. K. Gaitonde, S. A. Marchi, and D. Richter, *Proc. Roy. Soc. B* **160,** 124–136 (1964).
21. M. K. Gaitonde, D. R. Dahl, and K. A. C. Elliott, *Biochem. J.* **94,** 345–352 (1965).
22. E. V. Flock, G. M. Tyce, and C. A. Owen, Jr., *J. Neurochem.* **13,** 1389–1406 (1966).
23. R. M. O'Neal and R. E. Koeppe, *J. Neurochem.* **13,** 835–847 (1966).
24. P. Lehr and J. Gayet, *J. Neurochem.* **14,** 927–936 (1967).
25. R. Vrba, H. S. Bachelard, and J. Krawczynski, *Nature* **197,** 869–870 (1963).
26. H. S. Bachelard, M. K. Gaitonde, and R. Vrba, *Biochem. Pharmacol.* **15,** 1039–1043 (1966).
27. H. S. Bachelard and J. R. Lindsay, *Biochem. Pharmacol.* **15,** 1053–1058 (1966).
28. E. V. Flock, G. M. Tyce, and C. A. Owen, Jr., *Fed. Proc.* **24,** 423 (1965).
29. M. K. Gaitonde and D. Richter, *J. Neurochem.* **13,** 1309–1318 (1966).
30. R. J. Haslam and H. A. Krebs, *Biochem. J.* **88,** 566–578 (1963).
31. H. A. Krebs, R. Hems, M. J. Weidemann, and R. N. Speake, *Biochem. J.* **101,** 242–249 (1966).
32. K. F. Gey, *Biochem. J.* **64,** 145–150 (1956).
33. J. Krawczynski, H. S. Bachelard, and R. Vrba, *Acta Physiol. Polonica* **15,** 367–372 (1964).
34. S. Cobb and J. Talbott, *Trans. Assoc. Amer. Physicians* **42,** 255–262 (1927).
35. C. L. Evans, "Principles of Human Physiology," p. 635, J. & A. Churchill, London (1952).
36. R. Schoenheimer, S. Ratner, and D. Rittenberg, *J. Biol. Chem.* **130,** 703–732 (1939).
37. E. B. Chain, F. Larsson, and F. Pocchiari, *Proc. Roy. Soc. B* **152,** 283–289 (1960).
38. J. Monod, J.-P. Changeux, and F. Jacob, *J. Mol. Biol.* **6,** 306–329 (1963).
39. F. M. Matschinsky and J. E. Ellerman, *J. Biol. Chem.* **243,** 2730–2736 (1968).
40. H. Waelsch, *Adv. Prot. Chem.* **6,** 299–341 (1951).
41. C. Long, Ed., "Biochemists Handbook," p. 853, E. & F. N. Spon, London (1961).
42. K. A. C. Elliott and Y. Yoshino, *Proc. Canad. Fed. Biol. Soc.* **1968,** 86.
43. A. Lajtha and J. Toth, *Biochem. Biophys. Res. Commun.* **23,** 294–298 (1966).

II

Metabolism Related to Function

Chapter 10

Protein Metabolism and Functional Activity

Derek Richter

Medical Research Council Neuropsychiatric Research Unit
Carshalton and Epsom, England

INTRODUCTION

Studies of brain metabolism *in vivo* have shown that, besides having a high rate of consumption of oxygen and glucose, the brain synthesizes protein at a relatively high rate. During the period of early foetal development protein synthesis is more active in the brain than in most other organs, and the net synthesis of new protein results in a rapid increase in the relative size of the brain. There is a partial falling off in the rate as growth and differentiation come to an end, but it has been shown that active protein synthesis still persists after growth has finished and it continues at a considerable rate throughout the whole of adult life.

Early attempts to measure the rate of protein synthesis *in vivo* led to figures of the order of 14 days for the mean half-life of the proteins of the rat brain,[22,37] but the turnover of individual proteins varies widely, and experiments of short duration indicated a mean half-life of as little as three or four days. There is no direct evidence of the rate of protein synthesis *in vivo* in the human brain, but a conservative estimate based on a mean half-life of 14 days would indicate an overall synthesis of at least 4 g of protein a day. This is the amount of protein contained in 50 g of fresh brain tissue and it represents a rate of protein synthesis of a similar order to that in secreting glandular tissue.

What is the reason for the protein synthesis in the brain? The rate is considerably greater than in muscle or in most other organs and it appears likely that it is related in some way to the functional activity of the neurons. The brain does not secrete protein into the blood, and the protein synthesized must therefore be utilized and broken down locally in the brain; but it is not evident how this is related to cerebral function. The brain is concerned primarily with the analysis of incoming sensory data and the coordination of motor response, using mechanisms which depend on the repetitive discharge of nerve cells, but it is not immediately clear how protein metabolism is involved.

In the following account some of the experimental work relating to the amino acid and protein metabolism of the brain is considered in the light of work carried out in the author's laboratory in recent years.

EFFECTS OF STIMULATION

What is the effect of functional activity on the protein metabolism of the brain? There is some evidence from long-term experiments that stimulation can influence the normal rate of growth and functional development of nervous tissues. Brattgård[10] found that the RNA and protein content of retinal ganglion cells were greater in rabbits exposed to light than in animals reared for 10 weeks in darkness. In a study of rats subjected to light-deprivation for 21 days Maletta and Timiras[40] found that normal animals exposed to light had a higher content of choline acetyltransferase and total protein in parts of the visual pathway of the brain. It has also been reported that the protein content of the cerebral cortex is slightly greater in rats reared under conditions of active sensory stimulation than in animals reared in isolation.[7] Such long-term effects that may be related to the growth or maturation of the brain must be distinguished from more rapid changes that would imply a closer connection between protein metabolism and cerebral function.

Electrical stimulation produces rapid changes in a number of labile metabolites, including the brain lactic acid, phosphate esters, acetylcholine, and ammonia,[51] but it was found that intermittent electrical stimulation of the brain over a period of 3 hr produced little effect on the rate of incorporation of labeled amino acids into brain proteins.[16,23] The interpretation of isotope-incorporation experiments of this kind is difficult, since stimulation of the brain may cause changes in cell permeability, amino acid pools, and specific activity of the tissue pool of labeled precursor; it also affects the circulation of the brain. Another approach was therefore tried by Minard and Richter,[42] who observed the effect of stimulation on the loss of prelabeled protein from rat brain. A total of ten electroshock seizures (two seizures a day) induced in rats, beginning three days after an injection of [U-^{14}C]glucose, had surprisingly little effect on the turnover of brain proteins. It appeared that the ^{14}C-activity remaining from 3–8 days after injection of [^{14}C]glucose was mainly in proteins having a relatively slow turnover and perhaps less easily mobilized than proteins having a rapid turnover. In these experiments it was found that there was a relatively rapid decrease in the specific activity of the protein-bound glutamic acid during the first few days after the injection of [^{14}C]glucose. This gave evidence of the presence of a more rapidly turning-over protein with a high content of glutamic acid. It was suggested that this could be related to the *S*-100 acidic brain protein described by Moore.[43]

Several investigators have reported an effect of physiological stimulation in increasing the activity of the succinate dehydrogenase–cytochrome oxidase system in nerve cells.[27,30] The mechanism involved is not yet clear, but there is suggestive evidence that this may be an adaptive synthesis of new protein.[11] In a study of the effects of stimulation by light on isolated layers from the rat retina Epstein and O'Connor[18] found a rapid increase in succinate oxidase activity in the rods and cones demonstrable within 30 min after exposure to light.

Singh and Talwar[56] have recently been able to define conditions under which physiological stimulation with flickering light increased the rate of incorporation of labeled lysine into proteins of the occipital cortex of the monkey. They found that the effect depended very much on the experimental conditions, since it was not observed with continuous light, which activates some, but inhibits other, cortical neurons.[33] Under optimal conditions stimulation for 45 min produced an increase of 100% in the relative specific activity of the proteins of the occipital cortex. In order to determine what proteins are synthesized in response to sensory stimulation, they separated the soluble proteins by disc electrophoresis on acrylamide gel.[57] The increased radioactivity after stimulation was found to be mainly in a fast-moving band, corresponding to an acidic protein with properties similar to those of the *S*-100 protein. The effect of light stimulation in increasing the content in the occipital cortex of an acidic protein of this type was confirmed by an immunological technique using a complement fixation titration.

What is the physiological function of the acidic protein which is synthesized specifically in nervous tissues? Hydén and McEwen[29] found by a fluorescent antibody technique that the *S*-100 protein could be demonstrated partly in the nuclei of the large neurons, but mainly scattered throughout the glial membrane system. Since it was absent from the perikaryon of the neurons, they concluded that it was mainly a glial protein. However, in their method of stripping the nerve cell bodies of glial cells by free-hand dissection in 0.25 *M* sucrose many dendritic nerve endings are included with the glial cells. The nerve endings are the main site of localization of the catecholamines and other basic transmitter substances which are known to occur in storage granules in association with acidic proteins. Analysis of isolated storage granules from the adrenal medulla has shown that they contain 200 mg of soluble protein per millimole of catecholamine;

the protein is liberated together with the catecholamines on stimulation.[6] Since stimulation increases the release of transmitter substances, it may be expected to cause an increased release and resynthesis of the associated acidic protein.

The soluble protein contained in the catecholamine-containing granules has a high content (25.5%) of glutamic acid.[9] It has not been shown that the protein of the catecholamine-storage granules is identical with the *S*-100 protein, and this does not appear to be supported by recent studies of the distribution of the *S*-100 protein in nervous tissues,[44] but it is likely that more than one acidic protein is concerned in the storage and release of the different transmitter substances. The high content of *S*-100 protein in the molecular and granular layers of the rabbit cerebellum suggests that this protein may be associated rather with a cholinergic transmitter, since acetylcholine is known to be present in these structures, and it has been reported that stimulation causes a release of acetylcholine from the surface of the cerebellum.[48]

AMINO ACID METABOLISM DURING MATURATION

The brain oxidizes glucose, as in other organs, through the tricarboxylic acid cycle, but a striking characteristic of brain metabolism is the high rate of flux of glucose carbon into the free amino acid and protein pools. In perfused cat brain maintained at a functional level corresponding to a normal fast-wave EEG about 35% of the blood glucose is oxidized directly to CO_2 and 25% is converted into the amino acids linked with the tricarboxylic acid cycle.[46] In agreement with this, 22 min after subcutaneous injection of uniformly labeled [^{14}C]glucose more than 70% of the total ^{14}C in the cat brain was found in the amino acids, whereas only 8–16% had entered the amino acids in liver, kidney, lung, muscle, and spleen.[21] The rapid labeling of amino acids in the brain is not due simply to the activity of the aminotransferase in catalyzing isotopic exchange between α-oxoglutarate and glutamate, since this enzyme is also active in liver and other organs.[4] The factors concerned include the compartmentation of metabolites in the tissue and the relative activities of the different enzymes involved in the glycolysis and oxidation of glucose.[20] Glucose is metabolized relatively rapidly in the brain, and the enzymes of the brain mitochondria are geared in such a way that the oxidation of pyruvate leads to an accumulation of α-oxoglutarate, which is promptly converted into glutamate.[35] As a result of this, the glucose carbon is transferred to the glutamate pool, which is relatively large in the brain. The free glutamate exchanges actively with the brain proteins.[52,59]

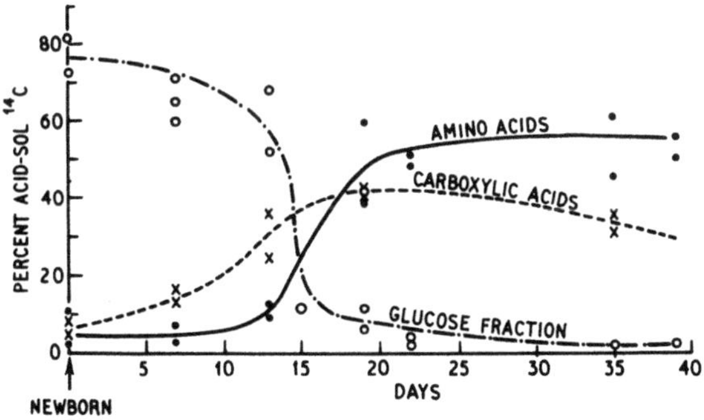

Fig. 1. Distribution of total ^{14}C in glucose, amino acid, and carboxylic acid fractions of the developing rat brain at 20 min after injection of [U-^{14}C]glucose. Values represent percentage of the total ^{14}C in the perchloric acid extract at different ages (Gaitonde and Richter[23]).

If these metabolic characteristics of the brain are of functional significance, it might be expected that this would be revealed by the metabolic changes occurring at the time during development when functional activity first sets in. Flexner[19] defined a so-called "critical period" from 10 to 15 days after birth in the rat, during which the neurons of the cerebral cortex show a sudden outgrowth of their dendrites, and large numbers of new synaptic terminals are formed. At this time the neuronal network of the cerebral cortex comes into functional operation and the electrical activity assumes the pattern seen in the adult brain.

Studies of the utilization of labeled glucose during the growth and development of the brain have shown that in the foetal and immature brain the metabolism of glucose is similar to that in other organs. The rapid flux of glucose carbon into the amino acids which is characteristic of the adult brain is absent up to 12 days after birth, but it develops sharply at the critical period when the cerebral cortex becomes functionally mature (Fig. 1).[23] At this time there is a sudden increase in the synthesis of new protein[12,49] and an increase in activity of enzymes such as the choline acetyltransferase and the monoamine oxidase which are concerned in mechanisms of synaptic transmission.[50] There is also an increase in the overall energy metabolism of the brain.

It is significant that the changes occurring at the critical period include a large increase in activity of a number of enzymes involved in the metabolism of glutamic acid. These include the glutamate decarboxylase, which forms γ-aminobutyric acid (GABA), and which is found uniquely in nervous tissues; they also include the aminobutyrate transaminase and the succinate semialdehyde dehydrogenase, which effect the metabolism of glucose by the GABA pathway.[49,55] At the same time, while the concentration of neutral amino acids remains unchanged, there is a threefold net increase in the content of free dicarboxylic amino acids in the brain.[23] It is evident that the amino acids present in highest concentration in the adult brain are those most readily utilized as a source of energy, since by transamination they yield the ketoacids α-oxoglutarate and oxaloacetate, which can be oxidized through the tricarboxylic acid cycle. Glutamate can also be utilized in the tricarboxylic acid cycle by way of the GABA pathway. These observations point to a special role for glutamate in relation to the functional activity of the brain.

Another characteristic of brain metabolism is the compartmentation of the free amino acid pool.[8,20] Compartmentation is shown, for example, by the observation that after injection of [^{14}C]glutamate the specific activity of glutamine is higher than that of the glutamate in the tissue. This indicates that the formation of glutamine occurs in a small pool containing highly labeled precursor glutamate. It has recently been shown that the metabolic compartmentation of glutamate develops in the brain during the critical period in parallel with other changes in amino acid metabolism.[47] Since the glutamate decarboxylase which catalyzes the formation of GABA is localized particularly in the synaptic nerve endings, it would appear that the amino acid metabolism of the brain is closely connected with the functional activity of these structures.[3,54]

THE ROLE OF GABA IN THE BRAIN

What part does GABA play in the energy metabolism of the brain? The quantitative determination of the percentage of glucose which is oxidized by the GABA pathway is not easily determined by the usual labeling methods, owing to the compartmentation of glutamate in the brain. An attempt was therefore made to get around this difficulty by making use of the fact that when GABA is metabolized the C_1 carbon atom is quickly eliminated as CO_2. The glutamate formed *via* succinic semialdehyde and α-oxoglutarate from [1-^{14}C]GABA contains ^{14}C only in the carboxyl group, which is lost when it is decarboxylated to form GABA; there is therefore no significant reentry of ^{14}C from labeled glutamate into the GABA pool. Control experiments showed that the uptake of labeled GABA into brain cortex slices suspended in physiological saline is relatively rapid (Fig. 2). The turnover rate of GABA could therefore be calculated by following the time-course of the change in specific radioactivity of GABA in cortex slices prelabeled with [1-^{14}C]GABA and oxidizing unlabeled glucose.

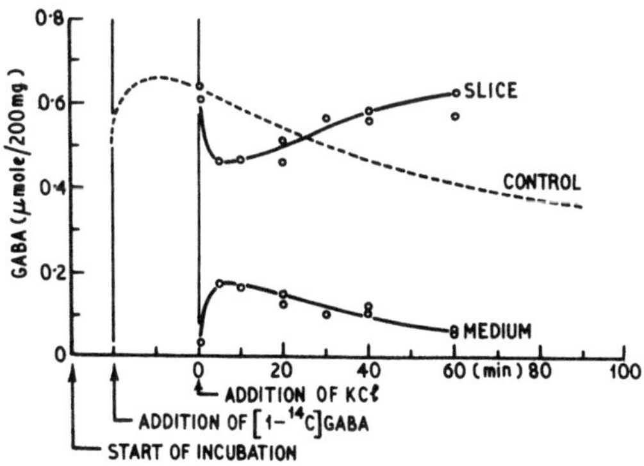

Fig. 2. The effect of stimulation by K^+ ions on the GABA content of guinea pig cortex slices. [1-^{14}C]-GABA (0.135 μM) was rapidly taken up by slices (200 mg) incubated in glucose-saline in O_2 at 37.5°. Addition of KCl (final concentration 66 m*M*) produced a rapid release of GABA from the slices into the medium (Machiyama, *et al.*[39])

Calculation with an analog computer showed that under these conditions the proportion of glucose oxidized through the GABA pathway was approximately 8% of the total oxidized through the tricarboxylic acid cycle.[38] This estimate agrees with one made by an independent method which depended on blocking the GABA pathway with hydroxylamine.[26]

Although 8% is not a high proportion of the overall glucose consumption, it must be remembered that the total glucose consumption of the brain is relatively high. It is also relevant that the glutamate decarboxylase and GABA are not uniformly distributed. Both appear to be localized to the region of certain synaptic terminals.[45,54] It is therefore clear that the metabolism of glucose through the GABA pathway must be considerably higher than 8% at these small but functionally important sites in the brain.

The view that GABA acts as an inhibitory transmitter in the brain has been widely discussed.[36] There is physiological evidence that it imitates qualitatively the action of the cortical inhibitory transmitter, and it has been shown that there is an active uptake mechanism by which GABA is removed from the extracellular spaces and accumulated in the cells.[17,38] However, the only reported experiments claiming to show a release of GABA from the surface of the brain as a result of stimulation could not be confirmed.[31,32]

It has been shown that stimulation by K^+ ions or electrical pulses causes the liberation of transmitter substances (acetylcholine and catecholamines) from brain slices and also from intact nervous tissues. A study was therefore made of the effect of stimulation on the distribution of [1-^{14}C]GABA in the guinea pig cerebral cortex, using slices suspended in physiological saline.[39] It was found that stimulation by the addition of K^+ ions produced a rapid release of GABA into the suspending medium (Fig. 2). The effect appeared to be specific, in that under conditions in which 30% of the GABA in the slices were released the amounts of other amino acids in the slices changed by less than 5%. It has recently been shown that a release of GABA from cortex slices is also produced by electrical stimulation.[35,58] It would thus appear that the main requirements for the acceptance of GABA as a transmitter substance have now been fulfilled.

THE SITES OF PROTEIN SYNTHESIS

In trying to relate protein metabolism to the functional activity of the brain, we need to know where in the brain the protein synthesis occurs. Studies by radioautographic methods in which the proteins were labeled by administering [^{35}S]methionine showed that the incorporation of labeled amino acids is greater in the gray matter than in the white, and most active in regions rich in nerve cell bodies.[14,52] Especially high activity was seen in the hypothalamic nuclei in the regions containing neurosecretory

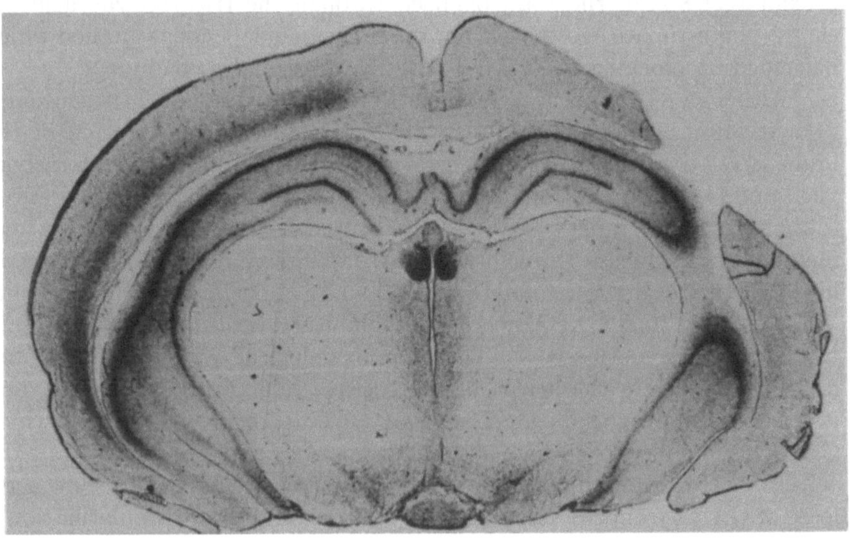

Fig. 3. Radioautogram showing the incorporation of ^{35}S-labeled amino acids into proteins at 180 min after subarachnoid injection of 25 μCi [^{35}S]methionine in a young adult rat.[52]

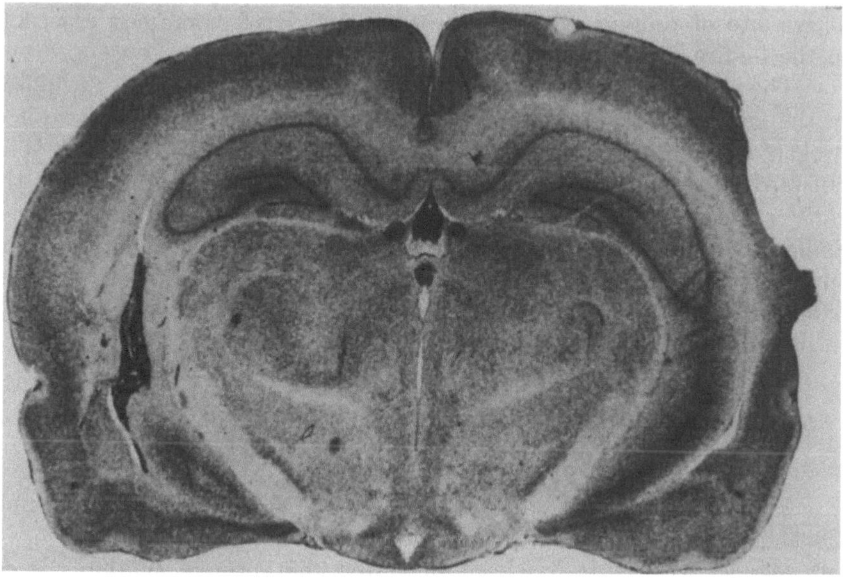

Fig. 4. Radioautogram showing the incorporation of ^{35}S-labeled amino acids into proteins at 24 hr after injection of [^{35}S]methionine in a young adult rat.

cells. In the young growing brain the incorporation is relatively greater and it extends diffusely over all areas; in the adult brain the pattern of incorporation is more restricted and corresponds closely to the distribution of nerve cell layers. In short-term experiments in which the labeled amino acid was administered by subarachnoid injection to rats which were killed after 180 min for isotopic exchange there was a preferential labeling of the proteins of short half-life, which were localized particularly in the nerve cell bodies in regions such as the cerebellum and Ammon's horn (Fig. 3). In long-term experiments, in which the animal was killed after 24 hr, the regions containing labeled protein were seen to have spread into the adjoining regions containing axon material and the dendritic layers (Fig. 4).

Further information about the site of protein synthesis was obtained by studying the incorporation of labeled amino acids into subcellular fractions isolated by differential centrifugation of brain homogenates. Relatively low rates of incorporation were found in the cell nuclei, mitochondria, and myelin. The highest rate of incorporation was found in a liponucleoprotein contained in the microsomal fraction;(13) this corresponded closely in properties to the Nissl material seen in sections stained with basic stains. These observations therefore agreed in indicating as the most active site of protein synthesis the ribosomal RNA contained especially in the bodies of the nerve cells.

The radioautographic evidence suggested that the protein synthesized in the cell body passes out toward the periphery of the cell, where it is broken down into free amino acids by neutral proteinase.[1] The idea of an outward flow of axoplasm toward the periphery was supported by the early experiments of Weiss and Hiscoe,[60] which showed an apparent damming of the flow in artificially constricted nerve fibers; it was again supported by experiments showing an outward flow of isotopically labeled axoplasm,[25,41] and it has been confirmed by many subsequent experiments in other laboratories.

THE RELATION OF PROTEIN SYNTHESIS TO FUNCTION

Four main hypotheses have been put forward to account for the functional significance of the protein metabolism in the brain:

Memory Function. It has been suggested that protein synthesis is directly concerned in the processes of learning or memory storage. The early hypothesis that learning involves the synthesis of specific protein molecules in which each item is encoded at a molecular level has found little experimental support.[57] The view that the formation of memory traces depends on the direct encoding of information at a molecular level is based mainly on analogy with the storage of genetic information in the DNA of the chromosomes; but the argument that the "phylogenetic memory" of the chromosomes is similar to memory based on learning and experience is of questionable validity.[51] While it is true that fairly complex behavioral patterns can be inherited and therefore can be encoded in the DNA, the genetic mechanisms appear to act by determining the growth and structural organization of the neuronal network. Rapid processes, such as those mediated by transmitter substances, need to be effected by small molecules, since the sorting of large numbers of specific macromolecules would take appreciable time. The slow phylogenetic memory of the DNA offers no parallel to the split-second rapidity of recall of a visual image or of a motor skill.

The hypothesis that protein synthesis may be concerned in the consolidation or storage of memory at a synaptic level is another matter. There is now a considerable literature on the behavioral effects produced by inhibitors of protein synthesis, and the evidence generally appears to support the view that protein synthesis is required for memory storage. There is reason to believe that protein synthesis may be involved in local modifications occurring at the synapses in the consolidation of memory traces, but it appears unlikely that this can account for more than a small part of the overall protein metabolism of the brain. It is significant that much of the normal protein synthesis occurs in the bodies of the large motor neurons, which are not the most likely sites for processes immediately concerned with learning and memory.

Membrane Maintenance. The brain differs from other organs in the exceptionally large extent of the lipoprotein membrane extending through the endoplasmic reticulum and in the fine processes of axons, dendrites, and glial cells. The passage of a nervous impulse involves a structural change affecting the ionic permeability of the plasma membrane. Protein synthesis might therefore be attributed to the special requirements of the brain for the repair and maintenance of membranes of cells, some of which are discharging many times a second.

In considering the role of protein synthesis in the maintenance of membrane sturctures, it is relevant that the neuron has by far the highest surface/volume ratio of any cell. However, the proteins of the cell membrane are characterized by their permanence rather than by a high turnover rate; and it is hard to relate the synthesis of membrane proteins to the special features of glutamate metabolism which characterize the amino acid and protein metabolism of the brain.

Synaptic Transmission. Phenylalanine and tryptophan are utilized by the brain as precursors of the catecholamines and 5-HT. Recent work has further suggested that GABA and glycine may have a physiological role as inhibitory transmitters or as modulators of synaptic function.[15,61] The protein metabolism could therefore be concerned in providing the amino acids and other metabolites required for the maintenance of synaptic function in the brain.

While there is evidence that protein synthesis can take place in many different structures, including axons, nerve endings, and glial cells, the most striking characteristic is the high activity of protein synthesis localized in the cell bodies of the motor neurons and in certain neurosecretory cells. The general picture is that there is a regular net synthesis in the nerve cell body of proteins, some of which are known to have a relatively high content of glutamate and aspartate derived from glucose carbon. The proteins flowing out from the cell body into the axoplasm include certain enzymes, such as acetylcholinesterase and cholineacetyltransferase, which are concerned in synaptic transmission (Hebb and Silver,[28]). In the neurosecretory cells a part of the axoplasmic protein is used for the formation of polypeptide hormones, which also serve a transmitter function. Some of the axoplasmic protein in other types of neurons appears to be used for synthesis of transmitter substances such as acetylcholine, noradrenaline, and dopamine, which partly enter the storage granules before their final release at the nerve endings. It is therefore clear that protein synthesis is involved to some extent in the mechanisms of transmission.

Energy Metabolism. The brain has a high energy requirement, but little in the way of carbohydrate reserves. There is evidence that amino acids and protein may be utilized as a source of energy if the supply of glucose fails,[24] and it has been suggested that the brain may be specially adapted for the utilization of amino acids and proteins as an energy reserve.

It has been shown that a considerable part of the axoplasmic material consists of acidic proteins which are also found in association with the basic transmitter substances in the storage granules and which may be expected to give on proteolysis a high yield of glutamic and aspartic acids. The acidic proteins are apparently broken down at the cell periphery, where the utilizable amino acids are oxidized and the others are used to synthesize transmitter substances or returned to the common amino acid pool. The provision at the synaptic terminals of a supply of readily oxidizable substrate by the breakdown of acidic protein might be envisaged as being of significance in relation to the energy metabolism of these functionally important sites. Certainly the acidic protein might serve as an energy reserve that could be utilized if the supply of glucose or other oxidizable substrates should fail; but to utilize protein merely as an oxidizable substrate is so uneconomical that it is unlikely to be a main normal mechanism of energy supply. If, however, glutamate or GABA are specifically concerned in synaptic transmission, the provision at the synaptic terminals of a supply of glutamate could be of functional significance. The acidic proteins could then serve a dual role as a source of GABA and also as carriers of basic transmitter molecules.

Protein synthesis is a normal function of the primitive neurosecretory cells which are found in the nervous system of many invertebrates, and which occur to a small extent even in the mammalian brain. Secretory activity appears to persist, not only in these neurosecretory cells, but also in the highly specialized neurons of the mammalian brain, which can be regarded as modified secretory cells. The primitive neurosecretory cells secrete a polypeptide or protein, while the cortical neurons secrete transmitter substances which are all nitrogenous compounds derived from protein. The adaptation of a neurosecretory cell to produce a transmitter substance of lower molecular weight would be more economical from a metabolic viewpoint. In the higher animals speed of transmission is also an important factor; with a small molecule such as acetylcholine a higher speed of transmission at the synaptic nerve ending could be achieved.

In the mammalian brain, protein synthesis clearly has a number of different functions, and none of the hypotheses that have been discussed can be regarded as incorrect. However, on present evidence it would appear that the need to elaborate the special enzymes, proteins, and other products required for synaptic transmission is the primary reason for the active protein metabolism of the brain.

REFERENCES

1. G. B. Ansell and D. Richter, *Biochim. Biophys. Acta* **13,** 92 (1954).
2. R. Balázs, *J. Neurochem.* **12,** 63–76 (1965).
3. R. Balázs, D. Dahl, and J. R. Harwood, *J. Neurochem.* **13,** 897 (1966).
4. R. Balázs and R. J. Haslam, *Biochem. J.* **94,** 131–142 (1965).
5. R. Balázs, K. Magyar, and D. Richter, *in*: "Comparative Neurochemistry," (D. Richter, ed.), pp. 225–247, Pergamon, Oxford (1964).
6. P. Banks and K. Helle, *Biochem. J.* **97,** 40c–41c (1965).
7. E. L. Bennett, M. C. Diamond, D. Krech, and M. R. Rosenzweig, *Science* **146,** 610 (1964).
8. S. Berl, A. Lajtha, and H. Waelsch, *J. Neurochem.* **7,** 186 (1961).
9. H. Blaschko, A. D. Smith and H. Winkler, *Nannyn-Schmiedebergs Arch. Exp. Path. Pharmak.* **253,** 23 (1966).
10. S. O. Brattgård, *Acta Radiol.* (*Suppl.*) **96,** 1 (1952).
11. I. U. S. Chentsov, V. L. Boroviagin, and B. I. Brodskii, *Biophysics* **6,** 61 (1961).
12. D. H. Clouet and M. K. Gaitonde, *J. Neurochem.* **1,** 126–133 (1956).
13. D. H. Clouet and D. Richter, *J. Neurochem.* **3,** 219 (1959).
14. P. Cohn, M. K. Gaitonde, and D. Richter, *J. Physiol.* **126,** 7P (1954).
15. D. R. Curtis, L. Hösli, and G. A. R. Johnston, *Exp. Brain Res.* **6,** 1–18 (1968).
16. W. Dingman, M. B. Sporn, and R. K. Davies, *J. Neurochem.* **4,** 154 (1959).
17. K. A. C. Elliott and N. M. Van Gelder, *J. Neurochem.* **3,** 28 (1958).
18. M. H. Epstein and J. S. O'Connor, *J. Neurochem.* **13,** 907–911 (1966).
19. L. B. Flexner, *in*: "Biochemistry of the Developing Nervous System" (H. Waelsch, ed.), p. 281, Academic Press, New York (1955).
20. M. K. Gaitonde, *Biochem. J.* **95,** 803 (1965).
21. M. K. Gaitonde, S. A. Marchi, and D. Richter, *Proc. Roy. Soc. B* **160,** 124 (1964).
22. M. K. Gaitonde and D. Richter, *Proc. Roy. Soc. B* **145,** 83 (1956).
23. M. K. Gaitonde and D. Richter, *J. Neurochem.* **13,** 1309–1318 (1966).
24. A. Geiger, N. Horvath, and Y. Kawakita, *J. Neurochem.* **5,** 311 (1960).
25. F. Grande and D. Richter, *J. Physiol. Lond.* **111,** 57P (1950).
26. B. Haber, *Canad. J. Biochem.* **43,** 865 (1965).
27. A. Hamburger, B. O. H. Rosengren, and B. Tengroth, *Acta Opthal.* (*Kbh.*) **42,** 951 ed.), (1964).
28. C. Hebb and A. Silver, in: "Protides of theBiological Fluids," p. 179 (H. Peeters, Elsevier, Amsterdam.
29. H. Hydén and B. McEwen, *Proc. Natl. Acad. Sci. U.S.* **55,** 354 (1966).
30. H. Hydén and A. Pigon, *J. Neurochem.* **6,** 57 (1960).
31. L. L. Iversen and M. J. Neal, *J. Neurochem.* **15,** 1141–1149 (1968).
32. H. M. Jasper, R. T. Khan, and K. A. C. Elliott, *Science* **147,** 1448 (1965).
33. R. Jung, *Exp. Cell. Res.* (*Suppl.*) **5,** 202 (1958).
34. R. I. Katz, T. N. Chase, and I. J. Kopin, *J. Neurochem.* **16,** 961–967 (1969).
35. I. J. Kopin and R. J. Baldessarini, personal communication.

36. K. Krnjević and S. Schwartz, *Nature* **211,** 1372–1374 (1966).
37. A. Lajtha, S. Furst, A. Gerstein, and H. Waelsch, *J. Neurochem.* **1,** 289 (1957).
38. Y. Machiyama, R. Balázs, and T. Julian, *Biochem. J.* **96,** 68P (1965).
39. Y. Machiyama, R. Balázs, and D. Richter, *J. Neurochem.* **14,** 591–593 (1967).
40. G. J. Maletta and P. S. Timiras, *J. Neurochem.* **15,** 787–794 (1968).
41. N. Miani, *J. Neurochem.* **10,** 859 (1963).
42. F. N. Minard and D. Richter, *J. Neurochem.* **15,** 1463–1468 (1968).
43. B. W. Moore, *Biochem. Biophys. Res. Commun.* **19,** 739 (1965).
44. B. W. Moore, V. J. Perez, and M. Gehring, *J. Neurochem.* **15,** 265–272 (1968).
45. M. J. Neal and L. L. Iversen, *J. Neurochem.* **16,** 1245–1252 (1969).
46. S. Otsuki, S. Watanabe, K. Ninomiya, T. Hoaki, and N. Okumura, *J. Neurochem.* **15,** 859–866 (1968).
47. A. J. Patel and R. Balázs, *Biochem. J.*, **111,** 17P (1969).
48. J. W. Phillis and G. C. Chong, *Nature* **207,** 1253 (1965).
49. F. N. Pitts and C. Quick, *J. Neurochem.* **14,** 561–570 (1967).
50. D. Richter, *Brit. Med. Bull.* **17,** 118–121 (1961).
51. D. Richter, "Aspects of learning and memory," pp. 77–99, Heinemann, London (1966).
52. D. Richter, M. K. Gaitonde, and P. Cohn, *in*: "Structure and Function of the Cerebral Cortex," p. 340 (D. B. Tower and J. P. Schadé, eds.), Elsevier, Amsterdam (1960).
53. R. B. Roberts, J. B. Flexner, and L. B. Flexner, *J. Neurochem.* **4,** 78–90 (1959).
54. L. Salganicoff and E. De Robertis, *J. Neurochem.* **12,** 287–309 (1965).
55. K. L. Sims, J. Witztum, C. Quick, and F. N. Pitts, *J. Neurochem.* **15,** 667–672 (1968).
56. U. B. Singh and G. P. Talwar, *J. Neurochem.* **14,** 675–680 (1967).
57. U. B. Singh and G. P. Talwar, *J. Neurochem.* **16,** 951–959 (1969).
58. V. Srinivasan, M. J. Neal, and J. F. Mitchell, *J. Neurochem.* **16,** 1235–1244 (1969).
59. R. Vrba, M. K. Gaitonde, and D. Richter, *J. Neurochem.* **9,** 465–475 (1962).
60. P. Weiss and H. B. Hiscoe, *J. Exp. Zool.* **107,** 315 (1948).
61. R. Werman, R. A. Davidoff, and M. H. Aprison, *Nature* **214,** 681–683 (1967).

DISCUSSION

SHENKEIN: I would like to know how you interpret your radioautographic results as being due to protein synthesis. How do you know that protein synthesis can be correlated with many granules being changed?

RICHTER: That depended on the basic experiment of cutting the brain in half and taking one half for a biochemical experiment in which we separated the proteins, hydrolyzed them, and showed the radioactivity to be in the amino acids combined in the protein, while using the other half of the brain to get the radioautogram. You can show the radioactivity is actually in the [^{35}S]methionine you isolate from the protein.

SHENKEIN: Can you show that this radioactivity is in peptide linkage? How would you distinguish whether it is free [^{35}S]methionine giving you a pattern on the radioautogram, or this being in protein?

RICHTER: The free methionine was washed out by solvents in the processing of the tissue and the radioautogram was obtained from the insoluble fixed protein which remained on the slide.

DROZ: I would also like to answer this question. In the case of injection of tritiated leucine into animals, when you analyze by comparing the histological procedure with the biochemical procedure you can determine stepwise the retention of the labeled amino acid in various components. We found that after injection of leucine about 96% of the radioactivity in histologically processed tissue was due to newly synthesized proteins. About 2–3% seems to be eliminated with the nucleic acids by hot TCA, and only about 1% would be due to adsorption of free amino acid. I would like also to make a few comments concerning the radioautographic reaction. It is very difficult to compare different brain areas because we have to consider not only the intensity of the radioautographic reaction, but also the density of the cell population, which differs from one area to another. But if you compare at the cellular level the activity in sensory neurons with that in motor neurons, it is absolutely confirmed that motor neurons incorporate more and synthesize more proteins than sensory neurons; and also incorporation is greater in the case of young animals than adult ones.

FORD: I would also like to comment on the question about the identification of what the labeled material is combined in after injection. At one time we did some intravenous injections of rats with sulfur-labeled methionine and then did extractions of the brain and found that within 5 min of injection we were able to detect labeled glutathione. Perhaps 5% of the activity in our extract was in glutathione. I have compared the uptake of labeled lysine in ventral horn motor neurons of the rat lumbar spinal cord with the uptake in spinal gray matter of equivalent regions. In the spinal cord gray matter younger rats do as we would expect: they take up more labeled lysine than do the older rats. Then if one examines individual neurons, this difference is much much greater. In the very young animals the neurons are taking up considerably more activity than neurons in older animals. Comparing these figures with other tissues, one finds that in one-, two-, and three-week-old rats the uptake per unit weight of ventral-horn motor cell is greater than in an equivalent weight of liver or any other tissue one might use. At four and six weeks the uptake is comparable to that of liver and is also comparable to that of the adult. Nerve cells of the two- and three-week-old period are fairly close to the period shown in Dr. Richter's report. Enzymes are changing from one sort to another, and this represents a kind of intermediate stage in what is going on in these nerve cells. This also is a point where the volumes of the nerve cells have about reached a plateau, whereas the weights of nerve cells continue to increase up to the levels in 300- or 400-g rats, so there is no plateau at that point.

COHEN: You show that at first most of the label was in glucose and after that a large part was in amino acids, but has there been any change in total glucose utilization?

RICHTER: The actual rate of utilization of glucose goes up all the time during the critical period.

FOLCH: This beautiful representation at the dynamic level of the morphological changes that were studied by other means really deserves great commendation. Now, it is always difficult to express observations, and I wonder whether one could be unwittingly misleading about that? For instance, when you measure protein metabolism during development you are not measuring a steady state. The protein synthesized has at least two different functions; one is to create new cytoplasm and the other one is functional. I think logically one would assume that the protein synthesis related to function is actually smaller during development than in the adult, because the brain is not wholly functional. Do you have any means of distinguishing during that rapidly changing nonsteady-state period of growth, by double-labeling, for example, between what is accumulation of cytoplasm (*de novo* morphogenesis) and what is maintenance? It would be nice to know whether the maintenance of nervous activities always follows the same pattern, along the same metabolic pathways, or whether there is a change in quality with the development of function. That is one question. The second question is about the very interesting novel results shown with the slices. What happens if the GABA is put in competition, for instance, with glutamic acid? if, before adding the GABA to the slices, you flood it with cold glutamic acid?

RICHTER: To deal with the last point first, we did not do that experiment because glutamate itself stimulates brain tissue. If you flooded the slice with glutamate, you might get a release of GABA, and since we would not know how the glutamate was distributed between the different pools, the results would be hard to interpret.

FOLCH: Would that 8% hold if the slice was suspended in a normal amino acid mixture? By normal, I mean the ones usually in the extracellular fluid or in brain tissue, not extracellular.

RICHTER: These slices were suspended in physiological glucose-saline. Because of the transport mechanisms for accumulating amino acids, the slices remove them from the medium and the normal extracellular concentrations are extremely small. They were not in amino acid mixtures. We were studying the metabolism of the endogenous amino acids in the slice.

BERL: It may be of interest to point out a few things and possibly answer some of these questions. It is difficult to calculate the rate of flow or flux from one amino acid to another because we know that the amino acids in the tissue are present in more than one pool. There may be a number of pools of amino acids. If you follow the labeling of amino acids, this will depend of course upon the precursor. That is, if you look at the flow of counts, say, from glutamic acid to glutamine, the result will depend on whether you use glucose, or acetate, or aspartate as your precursor. The labeling will be completely different, indicating that there are a number of different pools of these amino acids and we really don't know anything about the size. In fact, we think that there may even be more than one citric acid cycle in the brain; and perhaps one is involved more in energy metabolism and another in the synthesis of materials that may be used in protein synthesis or in the synthesis of transmitter substances, while a larger pool may be involved in the stabilization of the energy requirements of the tissue.

RICHTER: We were following Garfinkle's sorting out of the pools. He made the first estimate of the pool sizes and our data gave general confirmation of that.

LAJTHA: First I would like to answer Dr. Folch's question, because we have done some of those experiments. The amino acid pool of brain slices as incubated in an amino acid-free but glucose-containing medium is remarkably close to the living brain; it doesn't change, and the endogenous levels seem to be fairly constant. There is a fairly slight proteolysis, and some of the amino acids increase slightly. This state doesn't change even if one amino acid is added, for example, glutamate, to the medium, and the total amino nitrogen tripled or even quadrupled. The rest of the pool still remains very close to the physiological, as if individual controls would be maintained. Now, for my question: is the sudden change in conversion of glucose during development accompanied by the appearance of an enzyme? In particular, does transaminase suddenly appear at that point? How do you explain this sudden change?

RICHTER: I don't think that the transaminase is the operative one, but a whole number of enzymes relating to glucose metabolism change at that time; in particular, hexokinase activity increases, and there is a more rapid utilization of glucose itself, so that I think it is at the earlier stages of glucose metabolism that the effective changes mainly occur. The transaminase activity in the brain is only comparable to that in liver or other organs, and I don't believe that that is the point at which the change takes place.

SHOOTER: During development there are tremendous differences in the pattern of the membrane proteins before this critical 10-day period and after. The proteins become slower in mobility, which suggests that after the critical period, which is between 10 and 25 days, the membranes have larger subunits compared with the earlier stage.

Chapter 11

The Axon as a Heuristic Model for Studying Membrane Protein-Synthesizing Machinery*

Edward Koenig

Department of Physiology
and
Neurosensory Laboratory
State University of New York
Buffalo, New York

It is perhaps no overstatement to say that little is known about the synthesizing machinery concerned with excitable-membrane protein synthesis. With the exception of the mitochondrion, there is a natural inclination to regard all protein synthesis in a cell, including that engaged with membrane renewal (i.e., membrane protein turnover) as being within the exclusive domain of the cytoplasmic, ribosomal machinery. However, one must not lose sight of the fact that the ribosomal system, as we presently conceive of it, synthesizes proteins for "export" (presumably at random) to other regions of the cell interior in order to satisfy local metabolic needs. Such a system, which I will refer to as "the centralized machinery," would seem poorly adapted to fulfill the requirements for neuronal-membrane protein turnover. This may be true for plasma membranes of other specialized cell types as well. The thesis is based upon the fact that the neuronal plasma membrane is functionally differentiated on a regional basis. That is, its functional properties differ in contiguous regions, notwithstanding physical continuity of the membrane structure.

There are reasons for believing that functional differentiation implies local differences in protein makeup of the membrane. An obvious example is the case of the protein controlling Na^+ ion conductance during the rising phase of the action potential, a voltage and time-dependent process[1] which has been referred to as Na^+ activation. According to Hodgkin *et al.*,[2] this process has a Q_{10} of about 3, which is a basis for believing that a specific

* The studies reported here were supported by Public Health Service Research Grant No. NB 04036 from the National Institute of Neurological Diseases and Blindness.

protein controls the selective change in Na^+ permeability. Moreover, recent studies of positive and negative heat associated with the action potential indicate corresponding entropy changes occurring in the membrane during excitation.[3] The Na^+ activation mechanism is localized in regions of the membrane capable of all-or-none action potentials. No doubt other distinctions can be made between delimited areas of the membrane; however, they need not be considered, since they would only reinforce the basic argument.

The fundamental question posed by such considerations is: Are there special mechanisms that direct the spatially discrete incorporation of specific functional proteins synthesized by a centralized machinery, or does the membrane have an intrinsic protein-synthesizing machinery of its own? The terminal region of a cutaneous sensory axon is a good example where graded, receptor potentials are generated in the specialized ending, while action potentials are produced in the preterminal portion. Both regions are remote from a centralized machinery in a physical sense because the axoplasm lacks ribosomes.[4] However, this circumstance makes the axon an obvious experimental object to study the question whether membrane protein turnover is dependent on a centralized machinery. The demonstration of protein synthesis in the axon which can be shown to include axolemmal proteins would be evidence of a *local* membrane protein-synthesizing machinery.

We might review some of the early suggestive evidence that the axon is capable of autochthonous protein synthesis. Some years ago Clouet and Waelsch[5] and Koenig and Koelle[6] independently studied the rate of restoration of acetylcholinesterase (AChE) along peripheral nerves, following irreversible inactivation by organophosphorous inhibitors. The lack of an evident proximo-distal gradient in the rate of restoration of enzyme along the nerves favored the concept of a local synthesis.

Later studies[7] with transected nerves supported the conclusion that AChE is synthesized in the axon. When a peripheral nerve was divided, as in the case of the cat hypoglossal, a striking fivefold increase in AChE levels was produced in the distalmost 3 mm of the central stump only 24 hr after axotomy. No significant change in enzyme levels was evident in the more proximal portions of the nerve in continuity with the cells of origin. If both hypoglossal nerves were transected and one nerve treated locally with puromycin, the AChE level on the treated side was reduced to about half of that of the untreated side 24 hr later. In addition, when AChE activities were compared in a segment outside of the stump region of the decentralized nerve one and four days following diisopropylfluorophosphate (DFP) there was a significant restoration of enzyme at four days *in situ*, notwithstanding decentralization. The restoration was inhibited by 5-fluororotic acid, which also suppressed the restoration of AChE in the contralateral, intact nerve. Thus, it was clear from these experiments that AChE was an example of at least one membrane-bound enzyme which was synthesized locally in the axon.

A logical extension of this work was to study the direct incorporation of labeled amino acids into axonal protein *in vitro*. This undertaking had to await the development of: (1) a technique to isolate myelin-free axons; and (2) a suitable quantitative method for determining nanogram amounts of cellular protein and corresponding radioactivity with high efficiency. The technique for isolating denuded axons entails free-hand exteriorization of ½-cm lengths with a pair of watchmaker's tweezers.[8] Successful isolation depends upon the use of subdural portions of spinal or cranial nerve roots and an appropriate medium.[9] The protein method[10] makes use of concentrated phosphoric acid at elevated temperature to solubilize a cell or axonal clump completely after extraction of lipid and RNA. Tritium radioactivity is assayed after combustion and reduction of organically bound tritium.

I would like to summarize briefly some of the findings with respect to uptake of tritium-labeled leucine into axons of rabbit spinal accessory nerve *in vitro*. Excised nerves (XI cranial nerve root of rabbit) were incubated in a medium containing a full complement of amino acids and salts and ^{3}H-leucine. Axons were isolated after incubation for analysis. A singular characteristic of the axonal system *in vitro* was its stability, as indicated by the relatively undiminished uptake for at least 10 hr, and continued incorporation even beyond 24 hr.[11] Exhaustive trypsin digestion at 37°C left a resistant residue still containing about 30% of the radioactivity incorporated. This is evidence for incorporation into the interior of protein molecules. Puromycin inhibited uptake to the extent of about 70%[12] and represented added proof that incorporation involved peptide-bond formation. Chloramphenicol had no measurable effect, which indicated that axoplasmic mitochondria did not contribute significantly to incorporation, since mitochondrial amino acid uptake is inhibited by chloramphenicol.[13,14]

Actinomycin D also has no effect on ^{3}H-leucine uptake.[11,12] Such negative findings are taken as evidence that protein synthesis in the axon is supported by stable *m*-RNA. This was not found to be the case in the Mauthner axon, however, because Edström[15] found about 50% inhibition of amino acid uptake already after 1 hr incubation. In this regard the disparity between the specialized Mauthner axon of the fish and that of a mammalian motor axon may be related to the very significant difference in RNA content of the two axon types (see below). Nonetheless, there is evidence for a DNA-dependent mechanism operating in the mammalian axon. The evidence is based in part on the finding that there is an incorporation of RNA precursors which is actinomycin D-sensitive (see below). In addition, actinomycin D induces an increased AChE synthesis in a locally

TABLE I

Incorporation of ^{3}H-Amino Acids[a] into Axoplasmic Protein and Axolemmal Fabric and Nonfabric Protein

	Per cent of total protein	Specific activity (cpm/μg)	Number or samples
Fabric protein	23 ± 6%	74 ± 32	5
Axoplasmic and nonfabric protein	77 ± 6%	60 ± 17	5

[a] Reconstituted ^{3}H-protein hydrolysate containing 16 amino acids.

treated portion of peripheral nerve. We shall return to the latter curious effect later, but first let us consider some recent attempts to fractionate the axonal proteins into an axoplasmic portion, and a fabric and nonfabric axolemmal fractions.(9)

A reagent which I have found to be extremely effective in solubilizing the axonal protein residue completely is 90% phenol/glacial acetic acid/ethylenediamine (20:1:1, v/v/v; pH 7.4). If the axons are isolated in a medium containing a strong oxidant (e.g., H_5IO_6), only about 77% of the total axonal protein is solubilized (Table I). The remaining insoluble portion is referred to as fabric protein, which constitutes part of the axolemma.(9) A 1-cm axon segment with an average diameter of 6 μ has about 16–17 ng protein. Assuming a protein concentration of 3% (see Hydén(16)), then 50–55% of the axonal protein mass should be axoplasmic. Therefore, in addition to axoplasmic proteins, about half of the membrane proteins are also extracted by the phenol reagent following treatment with an oxidizing agent. Experiments using carbon particles which adsorb onto the external surface of the axolemma during isolation of myelin-free axons suggest that the external portion of the membrane is spared because the carbon particles remain strongly associated with the membrane residue (i.e., fabric protein).(9) The extractable portion (i.e., nonfabric protein) may represent the internal membrane protein coat. Although an extraction of just half of the membrane proteins may be fortuitous and does not prove the existence of two distinct protein coats as proposed by the classical Davson–Danielli model, the differences in solubility of half the membrane proteins after oxidation is consistent with differences in chemical reactivity of internal and external protein coats of a trilaminar unit membrane (see Robertson(17)).

In experiments using a reconstituted, tritiated protein hydrolyzate of 16 amino acids, incorporation can be demonstrated into the axoplasmic

TABLE II

Efficacy of Releasing Radioactivity from Axons by RNAse and DNAse Digestion Following Incorporation of ^{3}H-Orotic Acid and ^{3}H-Adenine

Exhaustive RNAse digestion	Exhaustive DNAse digestion	Axonal residue	Number of samples
45.5 ± 9.5%	19.3 ± 8.2%	34.0 ± 11.3%	6

and nonfabric portion (i.e., phenol reagent-soluble), as well as into the fabric protein fraction (Table I).[9] This experiment demonstrates unambiguously the incorporation of amino acids into the membrane structure. It is likely that the nonfabric axolemmal fraction is most highly labeled, if the inference is valid that most axoplasmic proteins are derived from the perikaryon (see Droz and Koenig[26]).

Let us now consider axonal RNA. With the use of Edström's microanalytical procedures,[18,19] it was possible to show that ribonuclease (RNAse) digests of isolated axons from the cat spinal accessory nerve contained the characteristic four bases.[8] Earlier, Edström *et al.*[20] had shown that the giant axon of the fish Mauthner neuron contained RNA. The Mauthner axon had 50–100 times more RNA per unit length, or roughly 10–20 times greater concentration, than did the cat or rabbit axons.[8,21] However, the parameter of concentration is misleading because there is a good likelihood that the RNA in the mammalian axon is associated principally with the membrane.

Tritiated orotic acid and tritiated adenine are taken up into an acid-insoluble product in the axon which can be digested in part by RNAse.[11] That proportion (i.e., 10–15%) which is released during the first 1.5 hr of digestion at 37°C is characterized as highly RNAse-sensitive. Most of the radioactivity remains associated with the axonal residue, however, and it is that proportion that is actinomycin D–sensitive (63% inhibition). An exhaustive digestion with RNAse (18–24 hr at 37°C) releases an additional 35% of the incorporated radioactivity. Dilute acid (e.g., 1 *N* HCl), like RNAse digestion, also releases the radioactivity at 37° at two different rates: a rapid rate of release during the first 2 hr, followed by a decreasing rate of release which reaches a limit when about a total of 45–50% of the radioactivity has been rendered soluble after 24 hr.[12] Approximately another 19% of the radioactivity can be released by exhaustive deoxyribonuclease (DNAse) digestion (Table II). The totally obdurate fraction of 34% radioactivity remaining after DNAse and RNAse has now been found to be covalently linked to the nonfabric axolemmal or axoplasmic frac-

TABLE III

A Comparison of *in Vitro* Incorporation of ^{3}H-Orotic Acid and ^{3}H-Adenine between Axons and Purified Whole-Nerve Mitochondria

Exhaustive digestion with RNAse and DNAse (0.4 mg/ml)	Axons (A) (cpm/μg protein)	Number of samples	Nerve mitochondria (M) (cpm/μg protein)	Number of samples	Ratio A/M
−	304 ± 111	(5)	160 ± 78	5	1.9
+	118 ± 32	(5)	29 ± 13	6	4.0
Per cent lost by digestion	62	—	82	—	—

tion.[9] Its acid stability and obduracy to RNAse suggest it to be bound pyrimidine. One is reminded in this context of the brain perfusion experiments of Geiger and Yamasaki,[22] who observed that viability of the perfused preparation depended on the presence of free uridine and cytidine in the perfusion medium. The significance of the protein-bound mononucleotide is obscure at the present.

Since mitochondria are present in the axoplasm, they likely represent axoplasmic organelles where labeled RNA would be found. To ascertain whether the labeled axonal RNA is mitochondrial or extramitochondrial, the following experiment was carried out. A number of cranial nerves were incubated with ^{3}H-orotic acid and ^{3}H-adenine. One nerve was removed at the end of incubation (4 hr) for a source of axons; the remaining nerves were homogenized and fractionated; the crude mitochondrial fraction was purified by discontinuous sucrose-gradient centrifugation, according to a modification of DeRobertis' techniques.[14] The results are shown in Table III. The analyses were carried out simultaneously on axonal clumps and purified whole-nerve mitochondrial clumps placed side by side. The specific radioactivity was based on protein, because what was being determined was the localization of radioactivity (i.e., mitochondrial *vs.* extramitochondrial radioactivity). Since the mitochondria can only constitute a small fraction of the axonal mass, the results in Table III indicate that most of the radioactive RNA product was extramitochondrial in its distribution; this is notwithstanding the residual RNAse-, DNAse-resistent radioactivity, which is now known to be protein bound nucleotide(s) (see above). In any case, if actinomycin D can not penetrate into mitochondria unless they are first caused to swell, as certain experiments indicate,[23] then the inhibition by actinomycin D would suggest that there is an additional source of labeled axonal RNA—namely, from the Schwann cell.

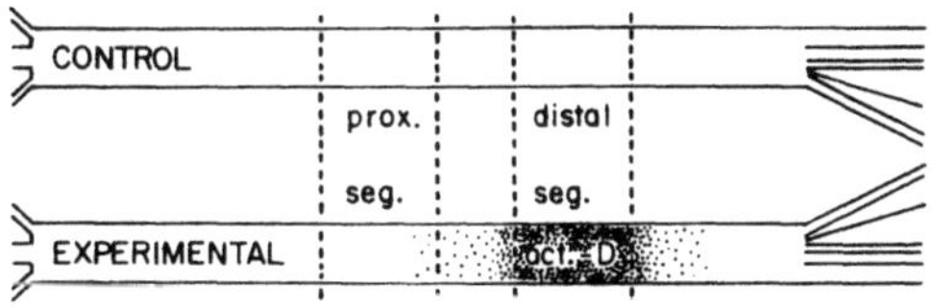

	prox. seg.	distal seg.
(cont.-exp.)/cont.	1.03±.09	1.04±.11
exp./cont.	1.21 ±.23	1.41± .12
exp.(puro.)/cont.	1.08±.33	0.90± .12
exp.(FOA)/cont.	0.99±.17	1.15 ±.12

Fig. 1. A graphic representation of the experiments involving (a) no application; (b) topical application of actinomycin D (act.-D); (c) topical application of actinomycin D and puromycin (puro.); and (d) topical application of actinomycin D and 5-fluororotic acid (FOA). In each instance the results are expressed as the ratio of AChE of experimental to control (see Koenig[24]).

The lack of an apparent effect by actinomycin D on leucine incorporation does not argue in favor of *m*-RNA synthesis locally. This implies that the newly synthesized RNA in the axon may be involved in a regulatory role. Some experiments showing "inductive" effects of actinomycin D on AChE synthesis in nerve also suggest this possibility. These experiments consisted of treating a circumscribed portion of the cat hypoglossal nerve (i.e., distal segment) *in situ* with actinomycin D unilaterally (Fig. 1).[24] AChE was assayed two days later in the distal segments and proximal segments of the treated and control nerves and compared. The actinomycin-treated portion showed 40% higher AChE activity than its corresponding control segment. The segment outside the treated region showed an insignificant increase over its corresponding control. The actinomycin D-induced stimulation was blocked by puromycin or 5-fluororotic acid, both of which had been shown to be capable of blocking local synthesis of AChE.

The action of actinomycin D on AChE synthesis is reminiscent of "derepression" of certain adaptive liver enzymes such as tryptophan pyrolase and tyrosine transaminase.[25] The tentative interpretation of these observations is as follows: AChE is apparently specified by a stable *m*-RNA. The translation of AChE-*m*-RNA, however, would seem to be controlled "locally" by a labile, DNA-dependent product (i.e., RNA), interacting in any of several conceivable selective ways at the translational stage of protein synthesis.

Although ribosomes are not present in the axoplasm, it seems reasonable that there should be a "ribosomal equivalent," carrying out certain functions of the ribosome in the membrane. Ribosome composition and form would not be expected to be identical with that of its cytoplasmic counterpart, since one of its possible functions in the cytoplasm might be to provide a "local" anaqueous environment for the translation process. Hence, proteins serving this function in the ribosomal particle would be superfluous in the membrane, where bulk water is excluded.

I should like to return briefly to the earlier theoretical considerations regarding regional differentiation of the membrane. I would like to make the following proposal. The local or regional differentiation of the neuronal membrane may be brought about by an interaction of the membrane machinery with a local extramembrane DNA-dependent product, such that translation of certain permanent encoded information (i.e., *m*-RNA) becomes selective. Selective suppression or repression at the translation stage (i.e., a negative control) can bring about a corresponding selective omission of specific functional proteins. This is based on the hypothesis that the information encoded in the membrane, specifying its functional proteins, is everywhere redundant. This would mean that the differentiated state at any given membrane region is labile or reversible because the process requires constant renewal of a negative control. Thus, it should be possible to bring about a transduction of functional properties of a given membrane region under appropriate experimental conditions, transforming it, for example, from a "spike" membrane to one that is "graded."

NOTE ADDED IN PROOF

Subsequent analyses of spinal accessory axons have shown that protein mass per unit length is of the order of 8.5 ng/cm, a value half of that reported in the text. The diameter of an average axon from the same nerve has been found to be ca. 5 μ. Assuming a protein concentration of 3%, axoplasmic protein would amount to 6 ng, or about 70% of the total protein mass. Hence, the conclusion that half the membrane protein must have been extracted by the phenol reagent is invalid; however, extraction of some membrane protein is likely. Finally, the phenol reagent must contain a reducing agent (e.g., dithiothreitol) in order to completely solubilize the axonal residue. An oxidizing agent is not necessary to obtain a phenol reagent-resistant residue of ca. 23%; i.e., fabric protein is normally stabilized by intermolecular covalent bonds.

REFERENCES

1. A. L. Hodgkin and A. F. Huxley, *J. Physiol.* **117,** 500–544 (1952).
2. A. L. Hodgkin, A. F. Huxley, and B. Katz, *J. Physiol.* **116,** 424–448 (1952).

3. J. V. Howarth, R. D. Keynes, and J. M. Ritchie, *J. Physiol.* **194,** 745–793 (1968).
4. S. L. Palay and G. Palade, *J. Biophys. Biochem. Cytol.* **1,** 69–88 (1955).
5. D. Clouet and H. Waelsch, *J. Neurochem.* **8,** 201–215 (1961).
6. E. Koenig and G. B. Koelle, *J. Neurochem.* **8,** 169–188 (1961).
7. E. Koenig, *J. Neurochem.* **12,** 343–355 (1965).
8. E. Koenig, *J. Neurochem.* **12,** 357–361 (1965).
9. E. Koenig, Some observations on the stabilization of the neuronal plasmalemma. In preparation.
10. E. Koenig, *J. Cell Biol.* **38,** 562–573 (1968).
11. E. Koenig, *J. Neurochem.* **14,** 437–446 (1967).
12. E. Koenig, *in*: "Macromolecules and the Function of the Neuron" (Z. Lodin and S. P. R. Rose, eds.), pp. 121–127, Excerpta Medica, Amsterdam (1968).
13. A. M. Kroon, *Biochim. Biophys. Acta* **108,** 275–284 (1965).
14. H. S. Bachelard, *Biochem. J.* **100,** 131–137 (1966).
15. A. Edström, *J. Neurochem.* **14,** 239–243 (1967).
16. H. Hydén, *in*: "The Cell" (J. Brachet and A. E. Mirsky, eds.), Vol. IV, pp. 216–323, Academic Press, New York (1960).
17. J. D. Robertson, *in*: "Molecular Organization and Biological Function" (J. M. Allen, ed.), pp. 65–106, Harper and Row, New York (1967).
18. J.-E. Edström, *J. Neurochem.* **3,** 100–106 (1958).
19. J.-E. Edström, *J. Biophys. Biochem. Cytol.* **8,** 39–46 (1960).
20. J.-E. Edström, D. Eichner, and A. Edström, *Biochim. Biophys. Acta* **61,** 178–184 (1962).
21. A. Edström, *J. Neurochem.* **11,** 309–314 (1964).
22. A. Geiger and S. Yamasaki, *J. Neurochem.* **1,** 93–100 (1956).
23. D. Neubert and H. Helge, *Biochem. Biophys. Res. Comm.* **18,** 600–605 (1965).
24. E. Koenig, *J. Neurochem.* **14,** 429–435 (1967).
25. L. D. Garren, R. Howell, G. M. Tomkins, and R. M. Crocco, *Proc. Natl. Acad. Sci. U.S.* **52,** 1121–1129 (1964).
26. B. Droz and H. L. Koening, this volume, Chapter 4.

DISCUSSION

KORNGUTH: I was wondering what an electron micrograph of this preparation would look like.

KOENIG: Let me say that at least under the phase-contrast microscope there is absolutely no evidence of any myelin; are you implying that there may be contamination with myelin?

KORNGUTH: Not necessarily myelin, but ribosomes from the glial cells surrounding the axon.

KOENIG: These axons belong to a cranial nerve root and have Schwann cells. They are myelinated, and there are regions where the Schwann cell cytoplasm comes in fairly close proximity to the axon. These are the nodes of Ranvier and Schmidt-Lanpermann incisures, but I don't think electron micrographs show much in the way of ribosomes there.

RAPPOPORT: Are you implying that there is any DNA in the peripheral or the axonic end? Because when you speak of translation locally, you mean that this would be a polymerase forming RNA. I know from developmental studies that have been done in lower organisms that you can have an RNA protein complex which is messenger

RNA with tremendous stability that does not express itself till days afterwards. So you may have nuclear formation of RNA, which may have long stability, reach a peripheral part of the cell, and act at some later time.

KOENIG: Perhaps you can propose a basis for the RNA incorporation which is actinomycin D-sensitive.

RAPPOPORT: I could not.

KOENIG: Well, this is the only evidence really for a local DNA template, that is, in the vicinity of the axon. It could be in the statellite cell, a very likely possibility, but we can't rule out the possibility of the axon as well.

SHENKEIN: Do you know the RNA polymerase activity in homogenates?

KOENIG: I have not tried this. The amounts of protein we are dealing with in these clumps are on the order of 0.1 μg, and furthermore, I don't know what the effect of the medium would be on the enzymic activity.

SHENKEIN: What is known about half-life of protein?

KOENIG: I don't know; I know on the basis of the restoration of enzyme after irreversible inactivation with DFP that approximately 7% of the enzyme along the hypoglossal nerve is restored per day, but this is on a relative basis only. In this particular nerve, which is not characteristic of all nerves, acetylcholinesterase is fully restored by 10 days or so.

ROBERTS: In cultures of cerebral cortex in which acetylcholinesterase were measured at intervals during culture there was sizable production of acetylcholinesterase compared to other enzymes in the system. Actinomycin had no effect, or stimulated production, whereas it inhibited the production of dehydrogenase and incorporation into a residual protein.

KOENIG: This is very similar to the observation of a local increase of acetylcholinesterase in the nerve. Were inhibitors of protein synthesis used to see whether it was *de novo* synthesis?

ROBERTS: I don't believe this was done; primary interest was in the effects of histones on the production of enzymes and residual protein, and again the acetylcholinesterase activity and amount were refractory to histones, whereas these other proteins I mentioned were stimulated by histone in proper concentrations, so the control of acetylcholinesterase synthesis was anomolous with respect to the control of other proteins in these tissue cultures. If you could assume that the effect of actinomycin D was purely artifactual, would this help your interpretation?

KOENIG: That is very conceivable, except that it is then hard to explain the actions of the antibiotics or the antimetabolites on the actinomycin D-induced increase. Their ability to inhibit would suggest that the increase in acetylcholinesterase is due to a synthesis of protein.

AGRANOFF: In our laboratory in the study of the acrylamide gel patterns of cholinesterase one band comes back much faster than any of the others. It has a half-life probably of something like 4 hr.

AUSTIN: The effect of DFP is not necessarily on acetylcholinesterase; something of the order of 95% of the DFP injected into an animal is taken up by proteins other than cholinesterase. This band that Dr. Agranoff found may not necessarily be the cholinesterase enzyme.

SHNEOUR: There is very interesting work on a pathological condition, Hansen's disease. The most interesting thing that happens at the beginning of this disease is the loss of sensitivity at the terminal and the peripheral nerves. The exact mechanism of this is unknown, but a beautiful work recently published suggested that either cholinesterase or protein synthesis or both may be involved. This disease does not affect the central nervous system, it appears to effect certain localized, nerves in circumscribed patches which are reversible for a while and become irreversible with increasing severity.

MANDEL: I think that Koenig's data can agree with a stable messenger. The effect of actinomycin is a destruction of several membranes; your effect might be explained by a side effect of puromycin, not a direct effect on protein synthesis.

KOENIG: Well, the only evidence that it is on protein synthesis is the puromycin inhibition and the fluororotic acid inhibition of the increase. Do you mean by destroying the membrane you are essentially exposing more enzyme, so that you assay a greater amount? But it is hard to explain the effect by puromycin and fluororotic acid which blocks the increase.

ROBERTS: What is the subcellular site of origin of acetylcholinesterase?

KOENIG: I think that it is synthesized in the axonal membrane; there are obviously also other sites of synthesis. There is an interesting dichotomy that can be seen as far as acetylcholinesterase is concerned in axotomized neurons. You see a loss of the enzyme in the cytoplasm that goes along *pari passu* with the chromotolysis that develops; the enzyme, based on histochemical studies, appears to remain absent in the cytoplasm for a period that could be equated with the outgrowth of the axon. When the axon again reaches the end organ it innervates, there is a restoration of the enzyme in the cytoplasm, which is a rather curious phenomenon. Cytoplasmic enzyme is presumably associated with the endoplasmic reticulum.

MURTHY: In our laboratory we have isolated two types of ribonucleic acids from rat-brain nuclei. They are both rapidly labeled, but they differ in the following respects: RNA_1 is easily extracted by phenol, and when added to ribosomes increases their amino acid-incorporation activity. RNA_2 appears to be firmly bound to DNA and can be released from it only after heat denaturation. This RNA inhibits amino acid incorporation by ribosomes and almost stoichiometrically reverses the stimulation produced by RNA_1. This raises the interesting speculation of whether the DNA-bound RNA may, under certain circumstances, migrate into the cytoplasm and act as one of the controlling factors at the level of translation.

Chapter 12

The Biosynthesis of Proteins within Axons and Synaptosomes*

L. Austin, I. G. Morgan, and J. J. Bray

Department of Biochemistry, Monash University
Clayton, Victoria, Australia

INTRODUCTION

The ability of the axon to synthesize protein has been demonstrated indirectly in a number of ways[1–3] and more directly *in vitro* in Mauthner axons of the goldfish,[4] isolated synaptosomes,[5–7] and giant axons of the squid.[8] Protein synthesis in axons has aroused considerable interest because no demonstrable ribosomes have been found in axons[9] and because, although much of the axonal protein is supplied from the cell body by axoplasmic flow, some axonal proteins, perhaps membranous proteins, can be synthesized *in situ.*

The presynaptic bulb is a specialized area of the axon which, like the main axonal trunk, lacks ribosomes, and the synaptosome, an organelle which includes this presynaptic bulb, can also incorporate amino acids into protein.[5–7] Much attention has been focused on the interrelationship between protein synthesis in the brain and the biochemical understanding of memory and learning.[10–12] Since memory and learning are generally supposed to result from some chemical or physical modification of the synapse, it is not surprising that the synthetic capacity of the synaptic region should receive attention.

The lack of visible ribosomes in both the axon and its ending does not rule out the possibility of protein synthesis in such tissue, since RNA with base ratios consistent with ribosomal RNA has been found in Mauthner axons of fish[13] and in cranial nerve roots in the cat.[14] Furthermore, RNA is present in synaptosomes—in particular, in the nonmitochondrial membranes derived from these synaptic organelles.[5] The synaptosome forms

* We are grateful to Miss L. Edgelow for skilled technical assistance.

when the axon end-foot breaks away during homogenization of brain or spinal cord, and contains the elements visible within the synapse—namely vesicles, mitochondria and pre- and post-synaptic membranes. It is a useful preparation for studying various functions of the axon *in vitro*. It has an advantage over the axon trunk in that it is simpler to isolate and is free from closely associated cells such as Schwann cells and the related myelin sheath. Indeed, with the advent of Ficoll, synaptosomal preparations may be isolated with a greater degree of purity than was possible with the sucrose-gradient technique first advocated by De Robertis *et al.*[15] and Gray and Whittaker.[16] Barondes[17] isolated synaptosomes from the brains of mice which had had ^{14}C-leucine injected intracerebrally in an attempt to show that axoplasmic flow occurs in the central as well as the peripheral nervous system. This he did by studying the rate of labeling of proteins in synaptosomes isolated from these brains. He showed that at least some synaptosomal protein may be derived from the ribosomal system of the cell body, but there was no indication that all of the symptosomal protein is formed in this way.

SYNAPTOSOMAL PROTEIN SYNTHESIS *IN VITRO*

The protein-synthetic capacity of the synaptosome may also be studied *in vitro*. Cerebral cortex from immature rats provides a source of active synaptosomes. This tissue has been studied by us in two ways. In some cases the cortex was chopped into small blocks with a Mickle tissue slicer set at 0.05 mm. The tissue had the properties of slices and many cells were maintained intact. The synaptosomes were isolated from the chopped tissue by homogenization followed by differential and density-gradient centrifugation. The chopped brain actively incorporated leucine into protein from a Krebs phosphate solution in a linear fashion for about 1 hr.[5] All fractions, with the probable exception of myelin, were labeled, and the synaptosomal fraction showed a significant rate of amino acid uptake which was linear for 1 hr.[5] The synaptosomal fraction may be broken down to its components by osmotic shock.[18] Membranes, soluble protein, and mitochondria derived from osmotically shocked synaptosomes were all labeled, with the mitochondria showing the most rapid rate of labeling. The amount of isotope incorporation into vesicles was so low that it can probably be attributed to contamination with soluble proteins. The results strongly suggested that the synaptosome can synthesize protein, and since not only mitochondria, but also synaptosomal membranes and soluble protein were labeled, it seemed probable that a protein-synthetic machinery exists either in the membrane or the cytoplasm of the nerve ending. This

TABLE I

Incorporation of *L*-Leucine into Brain Fractions

Fraction[a]	Specific activity (cpm/mg protein)
Myelin	46
Synaptosomes	296
Mitochondria	901

[a] Each of these fractions was isolated by the method of De Robertis *et al.*[15] They were incubated for 30 min at 37° with [U-^{14}C]-*L*-leucine and the reaction stopped by addition of an equal volume of 10% TCA. Protein was counted in a Packard Scintillation Spectrometer after extraction.[5] Protein was estimated by the method of Lowry *et al.*[47]

mechanism would have to be extramitochondrial, since mitochondria synthesize only their own structural protein and rely on cytoplasmic ribosomal sources for soluble enzymes within the mitochondrion.[39,40]

ISOLATED SYNAPTOSOMES

It has recently been shown that protein may be transported through axons by a rapid (80–120 mm/day or 50–80 μ/min) as well as by the normal slow rate (1–2 mm/day) of axoplasmic flow.[19–24] This means that labeled protein found in nerve endings isolated after the incubation with labeled amino acid could have been formed elsewhere. It was therefore necessary to establish whether or not isolated synaptosomes freed from physical connection with axon or cell body can synthesize protein. Consequently, a crude mitochondrial fraction obtained from cerebral cortex homogenate was incubated with leucine in a medium containing (m*M*) mannitol 300, EDTA 0.2, K_2HPO_4 10, $MgCl_2$ 10, and Tris 10 (pH 7.4). When the synaptosomes were isolated after incubation they were found to be labeled at a rate of about 1 μμM/hr/mg protein, whereas the mitochondria incorporated leucine at three times this rate (Table I).[6] This mitochondrial rate is considerably less than that found by Campbell *et al.*[25] and Bachelard[26] (about 70 μμM/mg protein/hr) and is probably due to inhibitory substances in the preparation as described by Campbell *et al.*[25] When synaptosomes were isolated from the crude mitochondrial fraction and subsequently incubated with labeled leucine they still incorporated the label at the rate of 4 μμM/hr/mg protein, and the incorporation was linear over a period of about 30 min. On disruption by osmotic shock the synaptosomal components were separated on a sucrose gradient[15] and proteins

TABLE II

Incorporation of *L*-Leucine into Protein of Synaptosomal Fractions

Fraction[a]	Specific activity (cpm/mg protein)
Whole synaptosomes	330
Synaptosomal soluble protein	274
Synaptosomal membranes	260
Synaptosomal mitochondria	170

[a] Whole synaptosomes were isolated by the method of Kurokawa *et al.*[48] and incubated with [U-^{14}C]-*L*-leucine for 30 min at 37°. The reaction was stopped by placing all tubes in ice and adding an equal volume of 10 m*M* [^{12}C]-*L*-leucine. After centrifugation the synaptosomes were resuspended in ice-cold water and left for 30 min in order to disrupt them by osmotic shock. The synaptosomal components were then fractionated on a sucrose gradient.[5]

from each component were counted. The soluble proteins and membrane-bound proteins were the most actively labeled (Table II). Vesicles were not detectably labeled at all.

DIFFERENTIAL EFFECTS OF INHIBITORS OF PROTEIN SYNTHESIS

Many antibiotics inhibit protein synthesis in mammalian and bacterial cells.[27] These have differential actions which vary with the organelle or cell being studied. Thus, puromycin inhibits protein synthesis in most systems including mitochondria,[28] while chloramphenicol and certain other antibiotics (lincomycin, tetracycline, and erythromycin) inhibit bacterial and mitochondrial systems, but not ribosomal systems from higher organisms.[29] They have been shown to inhibit mitochondrial systems isolated from brain.[26] Cycloheximide, on the other hand, inhibits ribosomal systems from animal cells, but not mitochondrial or bacterial systems.[30–32] Consequently, these inhibitors may be used to differentiate the biosynthesis of proteins in the synaptosome and, in particular, to determine whether the synthesis is of a mitochondrial or ribosomal type. When synaptosomes were incubated together with these inhibitors the incorporation of leucine was markedly inhibited by cycloheximide and puromycin, but to only a relatively small extent by chloramphenicol, tetracycline, erythromycin, and lincomycin. (Table III). When the synaptosomes were subfractionated after osmotic shock the leucine incorporation into the subfraction showed a very clear-cut difference between the mitochondrial preparation on the one hand and the membrane and soluble cytoplasmic components on the other. Cycloheximide inhibited incorporation into the

TABLE III

The Inhibition of Labeled *L*-Leucine Incorporation into Synaptosomal Proteins by Antibiotics

Antibiotic	Conc. (μM)	Activity in synaptosomal fractions as percentage of control			
		Whole synaptosomes	Soluble	Membrane	Mitochondria
Puromycin	50	24	35	31	23
Cycloheximide	100	29	13	27	92
Chloramphenicol	300	74	98	91	5
Tetracycline	300	72	96	89	3
Erythromycin	300	73	104	95	7
Lincomycin	300	77	101	97	0

latter components substantially, whereas its effect on mitochondria was minimal. Chloramphenicol, tetracycline, erythromycin, and lincomycin had the opposite effect.[6] These results show that the synaptosome has the capacity to synthesize protein and that only part of this capacity is due to synaptosomal mitochondria. Indeed, the low degree of inhibition by chloramphenicol may put an upper limit on the mitochondrial contribution of 25%.

EFFECT OF IONS

Early results showed that the rate of incorporation with synaptosomes was of the order of 3–4 $\mu\mu$M leucine/mg protein/hr when an isotonic medium of sucrose or mannitol was used for the incubation. Deviation from isotonicity, either toward a hyper- or a hypotonic medium, led to a loss of activity (Table IV). Exposure to hypotonic conditions resulted in an irreversible loss of synthetic capacity, whereas synaptosomes, when returned from hypertonic to isotonic conditions, retained their ability to incorporate leucine. The influence of Na^+ and K^+ ions on the rate of incorporation was very marked. Tables V–VII show the effect of substitution of sucrose by NaCl or KCl or both. The KCl doubled the rate of incorporation at 30 mM, but when KCl completely replaced sucrose in the incubation medium the rate fell to less than half (Table V). By contrast, NaCl had no effect when added alone and could substitute for all of the sucrose without affecting incorporation (Table VI). If NaCl replaced sucrose in the presence of 25 mM KCl the rate was increased by about

TABLE IV

The Effect of Hypo- and Hypertonic Conditions on the Incorporation of *L*-Leucine into Synaptosomal Proteins[a]

Sucrose (m*M*)	Specific activity (cpm/mg protein)	
	Incubation A	Incubation B
0	8	6
50	10	12
100	8	9
150	6	40
200	414	424
250	528	528
300	404	512
350	329	520
400	241	503
450	125	495
500	44	478

[a] Synaptosomes were incubated at 37° for 10 min in concentrations of sucrose shown. In incubation A they were incubated in sucrose at the respective concentration, whereas in incubation B they were resuspended in 250 m*M* sucrose prior to the commencement of incubation.

TABLE V

The Effect of Potassium Ions on the Rate of Protein Synthesis by Synaptosomes[a]

Sucrose (m*M*)	250	200	150	100	50	0
KCl (m*M*)	0	30	60	90	120	150
cpm/mg protein	534	1013	948	409	189	160

[a] Synaptosomes were incubated as described for Table I, except that sucrose was replaced by KCl as shown.

TABLE VI

The Effect of Sodium Ions on Protein Synthesis by Synaptosomes[a]

Sucrose (m*M*)	250	210	165	120	80	40	0
NaCl (m*M*)	0	25	50	75	100	125	150
cpm/mg protein (No KCl)	551	578	573	543	531	550	529
cpm/mg protein (+25 m*M* KCl)	1032	1226	1481	1786	2535	2560	2470

[a] Sucrose was replaced in the incubation medium with NaCl. Incubations were carried out either in the presence or absence of 25 m*M* KCl as shown.

TABLE VII

The Effect of Sodium and Potassium Ions when Combined in the Incubation Medium[a]

NaCl (m*M*)	0	30	60	90	120	150
KCl (m*M*)	150	120	90	60	30	0
cpm/mg protein	171	724	1481	2026	2424	531

[a] Incubations were carried out as for Table I except that sucrose was replaced by mixtures of NaCl and KCl as shown.

fivefold. The optimum NaCl concentration is 100 m*M* or greater. Ammonium ions could not substitute for sodium ions. There was also a small stimulation by Mg^{++} and this could be replaced by Ca^{++}. When neither of these ions was added the incorporation rate was still high, but addition of EDTA led to a marked inhibition, suggesting a requirement that was nearly satisfied by traces of Ca^{++} or Mg^{++} in the reagents.

These ionic effects suggest that protein synthesis within the synaptosome is dependent on a transport mechanism for metabolites, linked to the sodium pump. Addition of ouabain to the medium reduced the incorporation by more than 50%, strengthening the possibility of Na^+, K^+-linked adenosine triphosphatase involvement in the membrane transport of metabolites. Autilio *et al.*[(7)] have also found that amino acid is incorporated into soluble and membrane-bound protein of the synaptosome and that the system is activated in the presence of Na^+ and K^+.

ENERGY REQUIREMENTS

ATP and α-oxoglutarate both inhibit leucine incorporation from a sucrose medium into synaptosomal protein, but neither have an effect when used in a medium containing optimum levels of Na^+ and K^+. There was no stimulation on addition of an ATP generating-system containing P-*enol*-pyruvate and pyruvate kinase (Table VIII). The system was resistant to ribonuclease. Osmotically shocked synaptosomes lost all synthetic capacity, but addition of an ATP-generating system restored this activity to about 40% of the control. That this was not due to mitochondria contained in the synaptosomes is shown by the susceptibility of this system to ribonuclease attack.

The experiments with osmotically shocked synaptosomes agree with other evidence (Table IX) that maintenance of synthesis in intact synaptosomes is dependent on an energy source produced by mitochondria within

TABLE VIII

The Effect of an ATP-Generating System and Ribonuclease on Intact and Disrupted Synaptosomes

Addition to medium	Specific activity (cpm/mg protein)	
	Intact synaptosomes[a]	Disrupted synaptosomes[b]
None	2506	12
ATP (5 mM); P-*enol*-pyruvate (10 mM); Pyruvate kinase (0.4 mg/ml)	2493	932
RNase (100 μg/ml)	2556	0
ATP (5 mM); P-*enol*-pyruvate (10 mM); Pyruvate kinase (0.4 mg/ml); RNase (100 μg/ml)	2501	35

[a] Synaptosomes were incubated in a medium containing (mM) NaCl 125, KCl 25, $MgCl_2$ 15, Tris 10, and synaptosomal protein 1–2 mg (pH 7.4).

[b] Synaptosomes were disrupted by suspension in a medium containing (mM) $MgCl_2$ 15 and Tris 10 (pH 7.4) at 4°. Before incubation each component was adjusted to the concentrations shown.

the synaptosome. A number of substances known to block oxidative phosphorylation or electron transport in mitochondria markedly inhibit leucine incorporation. Inhibitors of glycolysis as well as reagents which react with sulfhydryl groups also adversely affect protein synthesis. ATP would be required for both the maintenance of the sodium pump in the synaptosomal membrane and for the activation of amino acids by amino acyl s-RNA synthetases.[33] The ouabain sensitivity of the system suggests that membrane transport is important, but the stimulation of amino acid incorporation in lysed synaptosomes by an ATP-generating system suggests a more direct requirement for ATP. Furthermore, soluble components of the synaptosomes catalyze amino acid-stimulated ATP-pyrophosphate exchange, indicating the presence of amino acid-activating enzymes. The synaptosomal soluble fraction can also replace the whole-brain supernatant fraction in a protein-synthesizing brain ribosomal system.[34]

TABLE IX

The Effect of Various Metabolic Inhibitors on Synaptosomal Protein Synthesis[a]

Inhibitor	Activity remaining as percentage of control
Oligomycin (2 m*M*)	21
Sodium azide (2 m*M*)	23
Sodium amytal (2 m*M*)	51
Antimycin A (0.2 m*M*)	26
2,4-dinitrophenol (0.2 m*M*)	26
Sodium cyanide (5 m*M*)	15
Sodium arsenate (5 m*M*)	31
2-deoxyglucose (20 m*M*)	31
Sodium fluoride (5 m*M*)	7
Sodium iodoacetate (10 m*M*)	4
N-ethyl maleimide (0.2 m*M*)	0
p-chloromercuribenzoate (0.2 m*M*)	3
Ouabain (0.5 m*M*)	48

[a] Synaptosomes were incubated with ^{14}C-labeled *L*-leucine in the complete medium described in Table VIII.

CONTAMINATION

It could be argued that the amino acid incorporation is not into synaptosomes at all, but is merely the result of amino acid uptake by some contaminant such as bacteria, mitochondria, or ribosomes. Despite the unsubstantiated claims of Gordon and Deanin,[35] this incorporation cannot be due to bacterial contamination. All incubations were carried out in the presence of penicillin and streptomycin to minimize bacterial effects. Bacteria were isolated from the incubation mixture and grown on a medium containing ^{14}C-leucine. These labeled bacteria were added to an homogenate of brain and the homogenate fractionated in the usual way. Most of the isotope was found in the nuclear fraction and the remainder with the mitochondrial fraction, and thus experiments conducted with chopped cortex could not be invalidated by bacterial contamination. Autilio *et al.*,[7] using a similar approach, showed that bacteria isolated from the incubation mixture incorporated leucine linearly for at least 5 hr, whereas the rate of incorporation into synaptosomal protein fell off rapidly after 20–30 min. These bacteria were very sensitive to chloramphenicol inhibition, whereas the synaptosomes are not. The synaptosomes are strongly inhibited by cycloheximide, an antibiotic which has little effect on bacteria. The very pronounced ionic effects and the ouabain inhibition also argue in favor of a system derived from mammalian cells rather than bacteria.

Mitochondria from brain synthesize protein,[25,26] but, despite the inclusion of mitochondria within the nerve ending, all of the protein synthesis of this structure cannot be due to mitochondria. The sensitivity of mitochondrial protein synthesis to chloramphenicol has been demonstrated for mitochondria from a wide range of tissues including brain,[26,29,36–39] and yet synaptosomal protein synthesis is inhibited by, at most, 25%, a figure which suggests the contribution made to the system by both synaptosomal and contaminating mitochondria. Cycloheximide is very much more efficient as an inhibitor of protein synthesis in ribosomal systems from nonbacterial sources than of protein synthesis in mitochondrial systems.[30–32] Leucine incorporation into isolated synaptosomes is inhibited by more than 70% by cycloheximide, and this inhibition is higher in nonmitochondrial components of the synaptosome. The mitochondria seem to synthesize only some inner-membrane protein and they import all of their soluble protein requirements from the cytoplasm.[40] In agreement with this, mitochondria derived from synaptosomes show less than 2% of the incorporated isotope in a soluble fraction isolated after sonication.[7] By contrast, the soluble protein isolated from synaptosomes after incubation with leucine has about 20% of the total incorporated label, a surprisingly high figure if it is protein newly formed by mitochondria.

Contamination by free ribosomes released from other structures is not a serious problem in this system. Brain ribosomal protein synthesis is inhibited almost completely by incubation with ribonuclease,[25] but the intact synaptosomal system is unaffected. Furthermore, when labeled microsomes were added to a brain homogenate they did not contaminate the synaptosomal fraction which was subsequently isolated from the homogenate.[5] Ribosomes are dependent on an ATP-generating system, yet synaptosomes incorporate isotope in the absence of such a generating system. Ribosomes would be expected to release labeled protein into the incubation medium, but this does not occur. The rate of incorporation of label into myelin proteins and vesicles is so slow that the level of protein synthesis being considered here could not be due even to gross contamination by either of these two fractions.

SIGNIFICANCE OF SYNAPTOSOMAL PROTEIN SYNTHESIS

The location of the synthetic process in the synaptosome, aside from that portion carried out by the mitochondria, is not known. It would appear probable that the process occurs in membranes, and in partial support of this is the finding that 80% of the incorporated leucine is in particulate structures. If the site of synthesis is the membrane, it could be occurring on either the pre- or post-synaptic membrane, as both are attached to the isolated synaptosome.[15,16,41] However, the mechanism seems to be enclosed within the synaptosome, since newly formed soluble protein remains within the synaptosomal unit, and therefore it would seem to be more likely that it is associated with the presynaptic region of the synapse.

The association of most of the synthetic activity with membranes implies that protein synthesis in the synapse may play some role in synaptic function, and it could be important in a better understanding of the mechanism of action of neurohormones and drugs which alter synaptic functions. It could also be relevant to future studies on the biochemical mechanisms involved in learning and memory.

SYNAPTOSOMAL RNA

If the synapse is capable of protein synthesis, RNA must be available. The synapse does contain RNA, other than that found associated with mitochondria, and this RNA is largely confined to the membrane fraction.[5] Although the type of RNA is not known, comparison with the RNA extracted from axons by Edström *et al.*[13] and Koenig,[14] in both of whose work RNA of ribosomal type was found, implies that the synaptosomal RNA is also ribosomal in nature, since the synapse is a modified portion of the axon. In addition to the presence of RNA, another essential requirement for protein synthesis, i.e., the presence of amino acid-activating systems, has also been demonstrated in the synaptosome.[34]

The origin of synaptosomal RNA is presumably the nucleus, but the nucleus of which cell, the presynaptic neuron, the post-synaptic neuron, or supporting glia?

The transport of proteins through axons by axoplasmic flow has been demonstrated by a number of workers (see, for example, Droz and Leblond,[42] Taylor and Weiss,[43] and Ochs *et al.*[44]). This protein is conducted through the axoplasm by more than one mechanism, since, although the bulk of the protein moves slowly at about 1–2 mm per day, some has been shown to travel at 100–1000 times this rate.[19–24] A similar process might provide RNA, formed in the neuron nucleus, to the axon and axon terminal. In order to test this hypothesis, a mixture of tritiated leucine and ^{14}C-labeled orotic acid was injected into the anterior horn of the chicken spinal cord and labeled protein and RNA was extracted from the sciatic nerve to determine the rate of movement of these components along the nerve.[23,45,46]

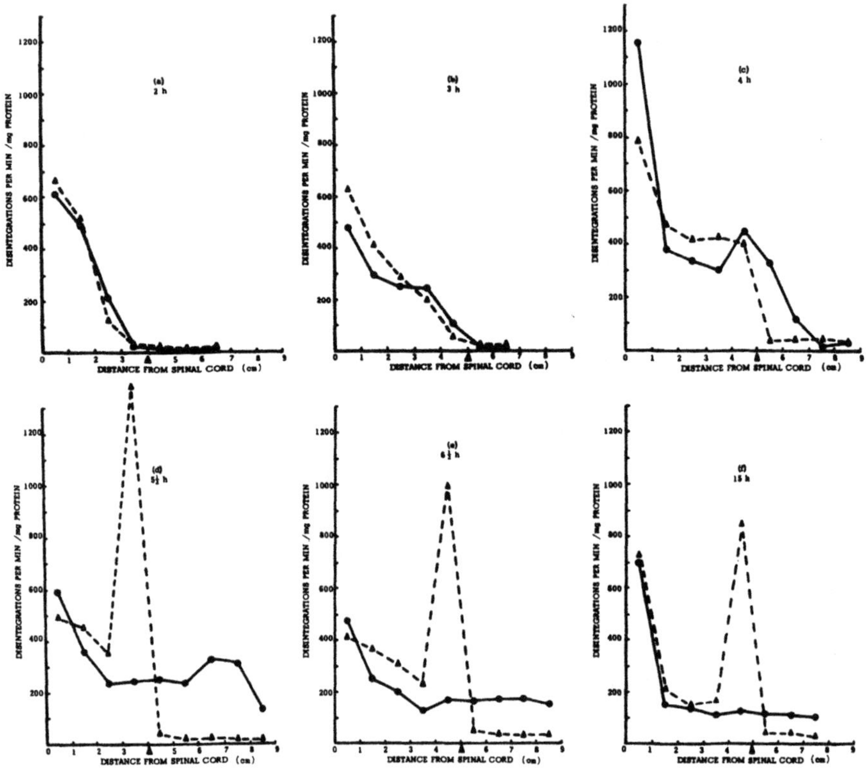

Fig. 1. Fast axoplasmic transport of ^{14}C-labeled proteins in chicken sciatic nerve. Three microliters of a solution containing 3 μCi of [U-^{14}C]-*L*-leucine (311 mCi/mM) was injected into both sides of the lumbar spinal cord of 15-week-old chickens. Typical results of the distribution of labeled proteins up to 15 hrs after injection are shown. The right sciatic nerve (broken line) was ligated in each case at the position marked by the triangle on the abscissa. The left sciatic nerve (solid line) was intact.

Protein moved in the nerve at two rates (Figs. 1 and 2). When the right sciatic nerve was ligated, labeled protein rapidly accumulated at the constriction, but if not so blocked, a peak of activity passed along the nerve at a rate of approximately 300 mm/day. A second, larger peak of protein followed at a much slower rate of 2–3 mm per day. Radioautographs demonstrated that both the fast and slow protein components were inside the axons (Fig. 3). The following radioautographs were taken (Fig. 3):

A. Sciatic nerve control. Scale, 10 μ.
B. Labeled protein in sciatic nerve section, 2 cm from the spinal cord,

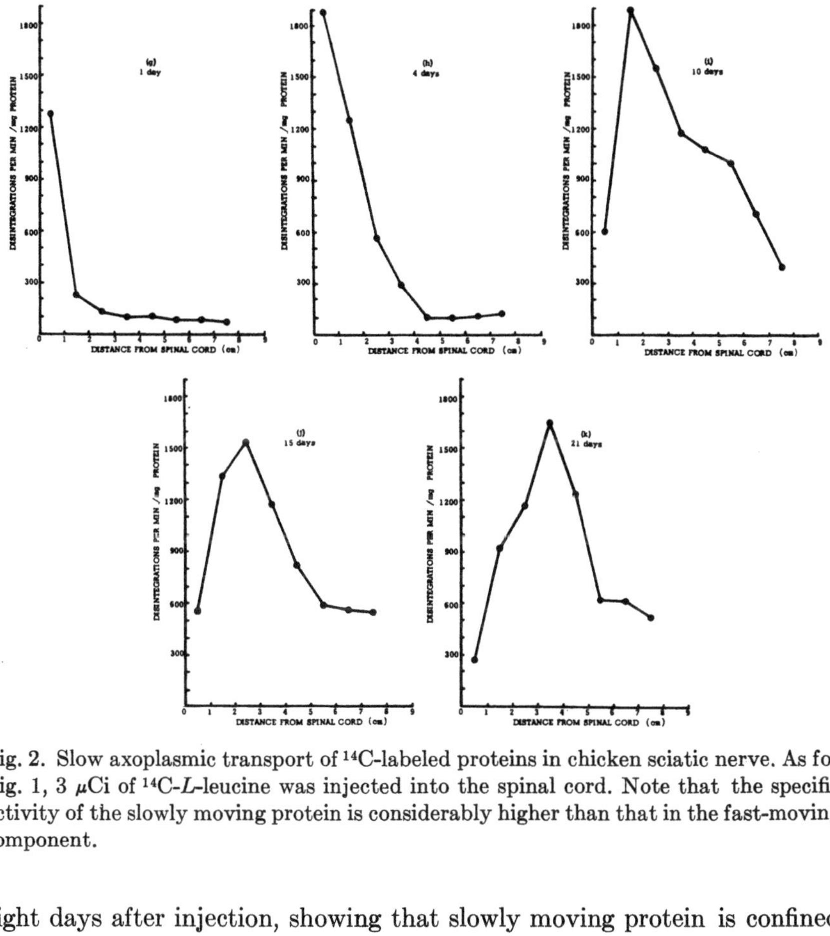

Fig. 2. Slow axoplasmic transport of ^{14}C-labeled proteins in chicken sciatic nerve. As for Fig. 1, 3 μCi of ^{14}C-*L*-leucine was injected into the spinal cord. Note that the specific activity of the slowly moving protein is considerably higher than that in the fast-moving component.

eight days after injection, showing that slowly moving protein is confined within the axons. Exposure time, 10 weeks. Scale, 10 μ.

C. Fast-flowing labeled protein in sciatic nerve, 4 cm from the spinal cord, 5 hr after injection of labeled leucine. This section was taken just proximal to a ligation on the nerve. Exposure time, eight weeks. Scale, 10 μ.

D. Fast-flowing protein in unligated nerve, 4.5 cm from the spinal cord. Both C and D show that the rapidly moving protein is also confined within the axons. Exposure time, 12 weeks. Scale, 10 μ.

E. Labeled RNA in spinal-cord motor neurons after injection of tritiated orotic acid. This section was taken 11 days after injection of the isotope and exposed for 14 days. Scale, 25 μ.

F. Labeled RNA in sciatic nerve, 2 cm from the spinal cord, taken 13 days after injection of orotic acid. The silver grains are both outside and inside the axons. Exposure time, 12 weeks. Scale, 10 μ.

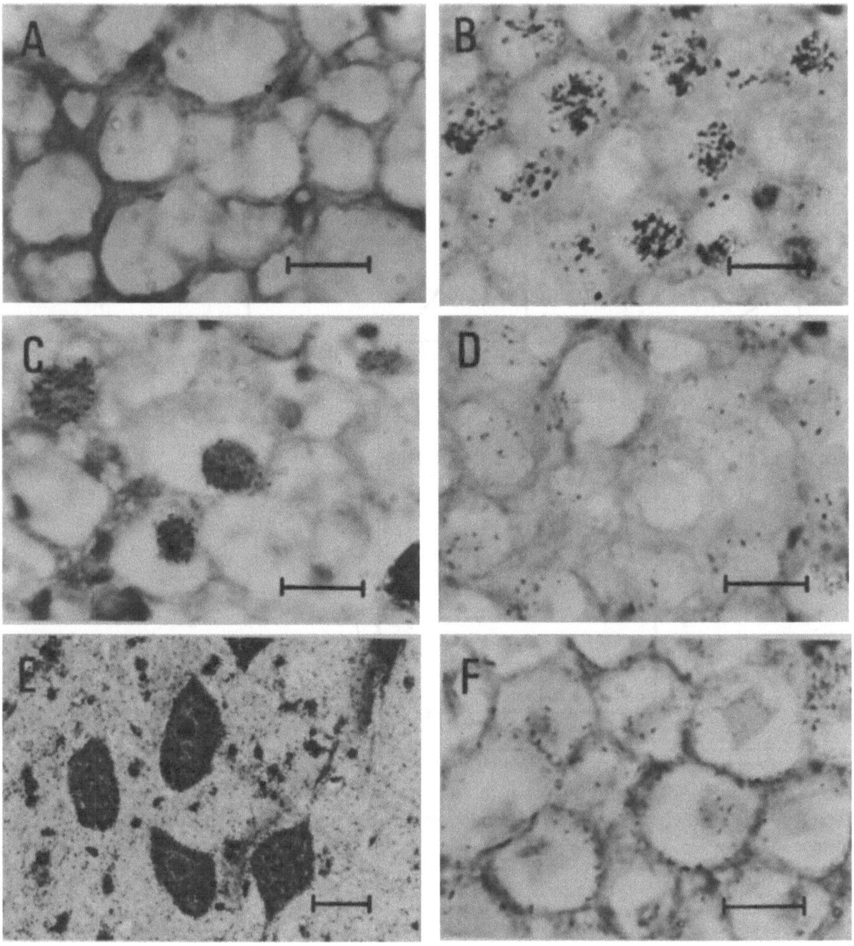

Fig. 3. Radioautographs of chicken spinal cord and sciatic nerve sections after injection of 100–150 μCi of *L*-[4,5-T]-leucine (23 Ci/mM) or 150 μCi of [5-T]-orotic acid (7 Ci/mM) into the spinal cord.

RNA moved along the nerve in a completely different fashion (Fig. 4).[23] A slow spread of RNA appeared along the nerve, but the pattern of movement was more diffuse than that seen with protein, as no peaking was found and no fast movement could be detected. Furthermore, radioautographs showed only a relatively small proportion of the newly formed RNA was within the axons.[46] There was a spread of acid-soluble metabolites of orotic acid along the nerve, and included in these were AMP and UMP.

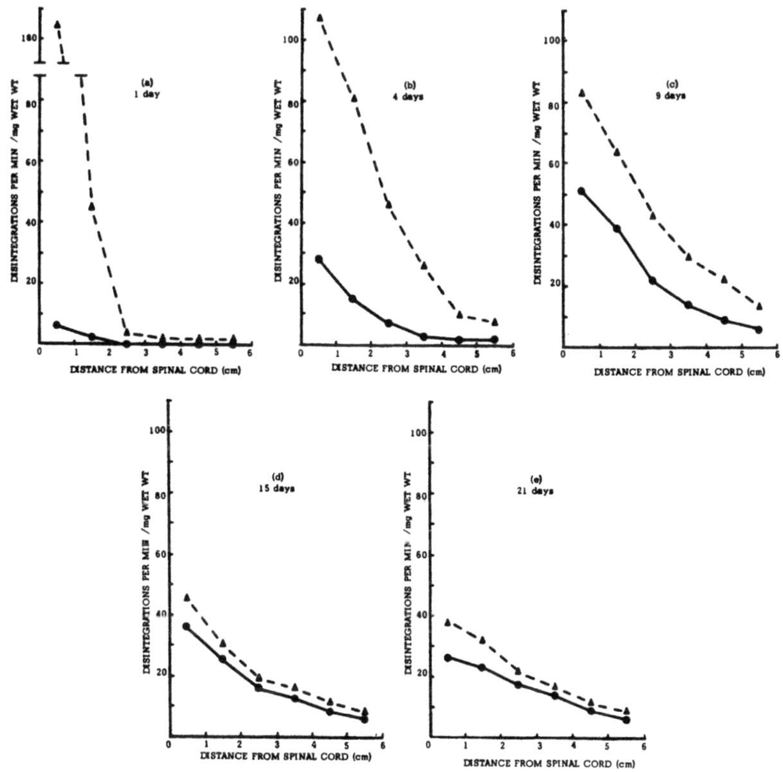

Fig. 4. The distribution of labeled RNA and labeled orotic acid metabolites along the sciatic nerve of the chicken for periods up to 21 days after injection of 5 μCi of [6-^{14}C]-orotic acid (30 mCi/mM) into the left side of the lumbar spinal cord. The label appearing in cold perchloric acid extracts of segments of sciatic nerve is shown by the broken line, and the RNA of the same segments by the solid line.

Thus, the newly synthesized RNA could have been formed from these precursors in the Schwann cells of the sciatic nerve. The RNA appeared to be both ribosomal and transfer-type RNA, since it separated into three peaks on a sucrose gradient with sedimentation values of 28*S*, 18*S* and 4*S*.[23]

The origin of axonal RNA then is unknown. It could be transported from the cell body by a mechanism which is different from that found for proteins and organelles, or it could be formed by Schwann cells or glia along the axon. It does seem, however, that the nerve receives its RNA or RNA precursors in some way by transport along the nerve rather than by direct uptake from the blood, since a gradient of isotope exists along the nerve and the slope of this gradient decreases with time, suggesting some type of passage of material.

The rate of incorporation of amino acid into synaptosomal proteins is low when compared with the rate found for brain ribosomal systems and mitochondria.[25,26] Furthermore, the rate for isolated synaptosomes in a high Na^+, low K^+ medium was 5–10 times that in chopped tissue in a Krebs phosphate medium. As pointed out by Campbell *et al.*[25] for mitochondria, the *in vitro* rate does not necessarily bear any relationship to the *in vivo* rate, and synaptosomal protein synthesis may play a greater role in the intact brain than *in vitro* experiments would indicate. In any case, the relative amount of protein synthesized at the synapse when compared to the rest of the brain is of less importance than the fact that a mechanism for such synthesis does exist, with the implications that this entails for a better understanding of synaptic function.

REFERENCES

1. D. Clouet and H. Waelsch, *J. Neurochem.* **8,** 201–215 (1961).
2. E. Koenig and G. H. Koelle, *J. Neurochem.* **8,** 169–188 (1961).
3. E. Koenig, *J. Neurochem.* **12,** 343–355 (1965).
4. A. Edström, *J. Neurochem.* **14,** 239–243 (1967).
5. L. Austin and I. G. Morgan, *J. Neurochem.* **14,** 377–387 (1967).
6. I. G. Morgan and L. Austin, *J. Neurochem.* **15,** 41–51 (1968).
7. L. A. Autilio, S. H. Appel, P. Pettis, and P.-L. Gambetti, *Biochemistry* **7,** 2615–2622 (1968).
8. A. Giuditta, W. D. Dettbarn, and M. Brzin, *Proc. Natl. Acad. Sci. U.S.* **59,** 1284–1287 (1968).
9. S. L. Palay and G. E. Palade, *J. Biochim. biophys. Cytol.* **1,** 69–88 (1955).
10. L. B. Flexner, J. B. Flexner, R. B. Roberts, and G. de la Haba, *Proc. Natl. Acad. Sci. U.S.* **52,** 1165–1169 (1964).
11. B. W. Agranoff, R. E. Davis, and J. J. Brink, *Proc. Natl. Acad. Sci. U.S.* **54,** 788–793 (1965).
12. S. H. Barondes and H. D. Cohen, *Science* **151,** 594–595 (1966).
13. J.-E. Edström, D. Eichner, and A. Edström, *Biochim. Biophys. Acta* **61,** 178–184 (1962).
14. E. Koenig, *J. Neurochem.* **12,** 357–361 (1965).
15. E. De Robertis, A. Pellegrino de Iraldi, G. Rodriguez de lores Arnaiz, and L. Salganicoff, *J. Neurochem.* **9,** 23–35 (1962).
16. E. G. Gray and V. P. Whittaker, *J. Anat.* (*Lond.*) **96,** 79–88 (1962).
17. S. H. Barondes, *Science* **146,** 779–781 (1964).
18. V. P. Whittaker, I. A. Michaelson, and R. J. A. Kirkland, *Biochem. J.* **90,** 293–303 (1964).
19. B. Grafstein, *Science* **157,** 196–198 (1967).
20. A. M. Kidwai and S. Ochs, *Physiologist* **10,** 220 (1967).
21. R. Lasek, *Brain Res.* **7,** 360–377 (1968).
22. B. S. McEwen and B. Grafstein, *J. Cell Biol.* **38,** 494–508 (1968).

23. J. J. Bray and L. Austin, *J. Neurochem.* **15**, 731–740 (1968).
24. B. G. Livett, L. B. Geffen, and L. Austin, *Nature* **217**, 278–279 (1968).
25. M. K. Campbell, H. R. Mahler, W. J. Moore, and S. Tewari, *Biochemistry* **5**, 1174–1184 (1966).
26. H. S. Bachelard, *Biochem. J.* **100**, 131–137 (1966).
27. E. F. Gale, *Pharmacol. Rev.* **15**, 481–530 (1963).
28. L. W. Wheeldon and A. L. Lehninger, *Biochemistry* **5**, 3533–3545 (1966).
29. G. D. Clark-Walker and A. W. Linnane, *J. Cell Biol.* **34**, 1–14 (1967).
30. H. L. Ennis and M. Lubin, *Science* **146**, 1474–1476 (1964).
31. M. R. Siegel and H. D. Sisler, *Biochim. Biophys. Acta* **103**, 558–567 (1965).
32. D. S. Beattie, R. D. Basford, and S. B. Koritz, *Biochemistry* **6**, 3099–3106 (1967).
33. M. B. Hoagland, *Biochim. Biophys. Acta* **16**, 288–289 (1955).
34. I. G. Morgan, unpublished results.
35. M. W. Gordon and G. C. Deanin, *J. Biol. Chem.* **243**, 4222–4226 (1968).
36. J. Mager, *Biochim. Biophys. Acta* **38**, 150–152 (1960).
37. A. M. Kroon, *Biochim. Biophys. Acta* **72**, 391–402 (1963).
38. D. B. Roodyn, *Biochem. J.* **97**, 782–793 (1965).
39. G. D. Clark-Walker and A. W. Linnane, *Biochem. Biophys. Res. Commun.* **25**, 8–13 (1966).
40. W. Neupert, D. Brdiczka, and W. Sebald, *in*: "Biochemical Aspects of the Biogenesis of Mitochondria" (E. C. Slater, J. M. Fager, A. Papa, and E. Quagliariello, eds.), pp. 395–408, Adriatica Editorice, Bari (1968).
41. D. Bodian, *Proc. Natl. Acad. Sci. U. S.* **53**, 418–425 (1965).
42. B. Droz and C. P. Leblond, *J. Comp. Neurol.* **121**, 325–337 (1963).
43. A. C. Taylor and P. Weiss, *Proc. Natl. Acad. Sci. U.S.* **54**, 1521–1527 (1965).
44. S. Ochs, J. Johnson, and M-H. Ng, *J. Neurochem.* **14**, 317–331 (1967).
45. L. Austin, J. J. Bray, and R. J. Young, *J. Neurochem.* **13**, 1267–1269 (1966).
46. J. A. Peterson, J. J. Bray, and L. Austin, *J. Neurochem.* **15**, 741–745 (1968).
47. O. H. Lowry, N. J. Rosebrough, A. L. Farr, and R. J. Randall, *J. Biol. Chem.* **193**, 265–275 (1951).
48. M. Kurokawa, T. Sakamoto, and M. Kato, *Biochem. J.* **97**, 833–844 (1965).

DISCUSSION

LIM: You have done combination experiments in which you mix labeled microsomes with unlabeled nerve endings and fractionate them. Have you made any calculations to demonstrate that in your procedure even the highest possible contamination by the microsomes in terms of radioactivity can account for all the activity detectable in the nerve endings?

AUSTIN: We have looked for contamination by a number of things. It is an obvious possibility that all of this may be due to contamination by bacteria, by ribosomes, by mitochondria, or by something else. As far as the ribosomes are concerned, the experiment you mentioned was one in which we incubated whole brain, separated off the labeled ribosomes, and added them to a homogenate of unlabeled brain; then we fractionated and found the labeled ribosomes almost entirely sedimented with the unlabeled ribosomes. The contamination of synaptosomes by ribosomes was negligible.

VRBA: You mentioned the movement of protein, and I wonder what sort of protein is, in fact, moving. If it is soluble protein, then I wonder if you would not agree that we might use the word diffusion instead of movement?

AUSTIN: It is difficult to fractionate sciatic nerve. We've homogenized the sciatic nerve, managing to break a number of glass homogenizers in the process, and then fractionated. Now we get various fractions, which we could call mitochondria, etc., but I don't think this is very meaningful. We find a considerable part of the labeled protein is associated with particulate matter. I think it's something of the order of 70% soluble protein, but there is about 30% which is associated with particulate proteins, and this may of course be in particles like mitochondria moving in the nerve.

ROBERTS: In connection with your analyses of the sedimentation properties of RNA, was this RNA derived from the entire nerve or from some single component such as the axon?

AUSTIN: From the entire nerve, and so this could be, for all we know, entirely derived from Schwann cells.

ROBERTS: It could be ribosomal RNA.

AUSTIN: Yes. I wish we knew how to extract the axon itself. What we're trying to do at the moment is to determine the RNA in squid axon, in axoplasm, rather, where we can at least exclude the nerve sheath, and I think we might have a chance of success here. In mammalian axons such as in the sciatic nerve I think the possibility of avoiding contamination with Schwann cells of cytoplasm is very low.

ROBERTS: Didn't Miani and Guiditta extract axons for RNA and find the sedimentation properties similar to those of ribosomal RNA, and this was just from the axon?

AUSTIN: These experiments also involved whole nerve, and the same difficulty would arise with Schwann-cell contamination.

MANDEL: You showed the radioactivity of RNA nine days after the injection of orotic acid.

AUSTIN: This was the time taken to get the maximum amount of labeling into the nerve.

MANDEL: If there is stable messenger, it could be that after a pulse of orotic acid you can stop messenger RNA synthesis and then see if after a chase experiment you will get a stable messenger in nerve endings.

AUSTIN: This is true. The results seem to suggest that there was some sort of movement of RNA down there, but most of this is in the Schwann cell. This could be a transfer from the Schwann cell to the axon or *vice versa*. In fact, we could even interpret this to mean that the axon is there simply to provide the Schwann cell with nutrients, although I don't think anyone would agree with me there.

SELLINGER: You measured RNA in micrograms per milligram in the membranes *vs.* the microsomes, those, of course, had the highest concentration. I wonder if you have separated the smooth from the rough membranes, and if so, have you tested smooth-membrane component for RNA, in view of the fact that membrane RNA has been shown in liver in smooth membranes and its function there has not been determined except to say that it may be responsible for processes which indeed make proteins which remain in the membrane? Would you say that regardless of whether or not the membrane RNA is really synaptosomal membrane RNA, the fact that one finds it there imparts to it this function, that is, to make proteins which are to reside as structural components of the membranes rather than to be exported?

AUSTIN: I am not trying to imply that all of the protein in the synaptosome is made through synthesis within the synaptosome. I would agree with those who support axoplasmic flow of proteins that the majority of protein in the synaptosome probably derives from the cell body; but I believe that within the axon and in the membrane of the axon, and I don't intend to try to define which membrane at this stage, but somewhere in the membrane of the axon there is a capacity to synthesize at least certain proteins. If we can isolate these proteins, I think we'll be a lot further on in being able to discuss this.

MAHLER: You mentioned in passing a recent observation from your laboratory, that the activity was strongly stimulated by sodium and potassium. I wonder whether you'd elaborate a little on this with regard to concentrations.

AUSTIN: The concentrations which gave the best effect, and these are very preliminary results, were 25–75 *mM* potassium and 100 m*M* sodium. The stimulation by potassium was twofold. Sodium alone had no effect, but in the presence of potassium it produces about fivefold stimulation.

Chapter 13

Fast Axoplasmic Transport of Proteins and Polypeptides in Mammalian Nerve Fibers*

S. Ochs

Department of Physiology
Indiana University Medical Center
Indianapolis, Indiana

INTRODUCTION

Using isotopic labeling techniques, it has been verified that in nerve there is a transport system constantly moving materials from the soma out into the axons and likely also into the post-synaptic cell.[1–3] The modern view of neuron is that it is a cell with a high level of protein synthesis,[4] and the preponderance of evidence discussed in the above reviews indicates that most of the synthesis takes place in the soma, where a high level of RNA is present in the ribosomes of the Nissl bodies. Some studies, however, indicate that synthesis is possible in other parts of the cell (see below and Austin and Morgan[5]).

Until recently most of the available isotopic evidence has indicated that nerve materials move at a rate of several millimeters/day, a value originally arrived at by Weiss and Hiscoe[6] from morphological evidence of damming proximal to constrictions of nerve trunks. The increase in the volume of the fibers in the dammed portions above the constriction indicated to Weiss and Hiscoe that all the axoplasmic contents are moving down inside the fibers as a viscous column. Recently, however, this concept has given way to a realization that very much faster rates of axoplasmic flow of several hundred millimeters/day are present in the nerve, as shown, for example, by Dählstrom and Häggendal,[7] who studied the accumulation of catecholamine particles in a dammed region, and by

* Supported by National Science Foundation Grant GB 7234X. Thanks are due to my colleagues, Dr. A. M. Kidwai, Dr. M. I. Sabri, Dr. J. Johnson and N. Ranish, who shared in these investigations, and Mrs. H. Yang for her technical assistance.

Kerkut *et al.*,[8] who found a fast transport of labeled glutamate in frog and snail nerve. Our recent studies of axoplasmic flow in cat mammalian fibers using ^{3}H-leucine as the precursor have also indicated that a fast transport system is present in addition to a slower phase of axoplasmic flow.[9–11] Our recent evidence for fast transport will be presented, as will studies which indicate which labeled components may be related to the fast and to the slow axoplasmic transport mechanisms.

FAST AND SLOW RATES OF AXOPLASMIC TRANSPORT IN VENTRAL-ROOT MOTOR FIBERS

In some earlier studies of axoplasmic flow[1] the labeled precursor was injected systemically. However, systemically injected precursor is taken up by the Schwann cells, and this serves to confound the subsequent analysis of the downflow of labeled materials inside the axons. For this reason it is preferable to inject precursor near the cell bodies. The distribution of activity subsequently found in the ventral roots at various times after injection of ^{32}P-orthophosphate was found to be characteristic of flow, i.e., a proximo-distal gradient of activity with, as more time elapses, an increase in the proportion of activity present in the distal portions of the root.[12] The pattern of outflow is that of an exponentially declining function, a finding also shown by Miani.[13] An estimation of the rate of flow arrived at in our work by taking the proximo-distal movement of the half-activity point of the outflow as a function of time gave 4.5 mm/day as the rate of flow. However, Miani,[13] using the furthest extent of activity found present in nerves at one day, gave 30–70 mm/day as the rate of movement in hypoglossal and vagus nerves, respectively. It is likely that these two apparently different estimations both relate to the same transport process, as can be seen from our data by taking as the measure the furthest extent of activity present in the root one day after injection, where a rate of 30 mm/day may be obtained (cf. Fig. 1 of Ochs *et al.*[9]).

With ^{3}H-leucine as precursor a generally similar pattern was found for the spatio-distal distribution of activity, except, however, that in the first day or so after injection a break was seen in the distribution of activity present in the roots (Fig. 1). The slope of the distal component indicated that it has a much faster rate of flow, one of at least hundreds of millimeters/day.[9] Lasek[14] had also on similar grounds estimated a faster rate for sensory fibers.

In order to more clearly determine the course of the downflow of activity, it was necessary to shorten the interval between cord injection and root removal. When the distribution was examined in ventral roots taken as early as a few hours after injection a break is seen in the distri-

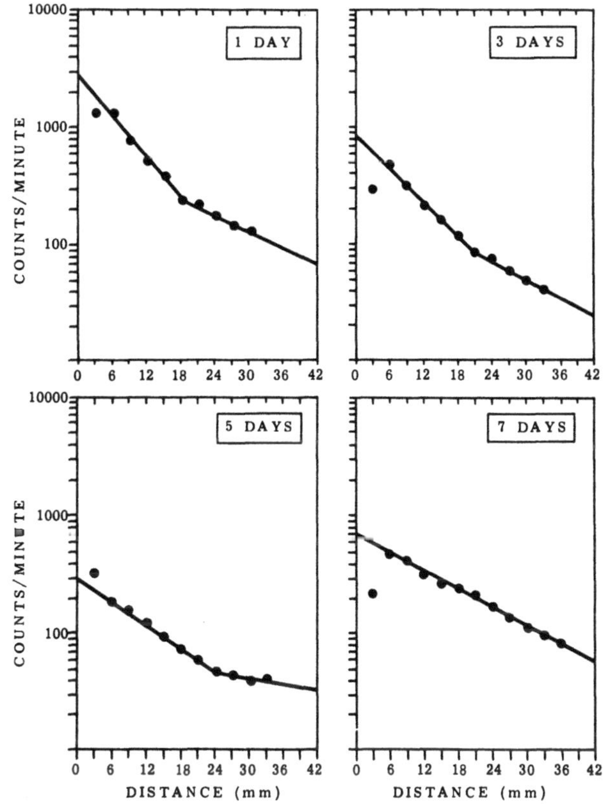

Fig. 1. The distribution of activity found present in 3-mm segments of the ventral root is shown at distances (in mm) from the cord on the abscissa. Activity is plotted on the logarithmic ordinate scale, the points representing the means of four experiments, with the lines drawn through the points by eye. The times given in the panels represent interval between injection of ^{3}H-leucine and root removal. (From Ochs and Johnson.[11])

bution, with evidently two components of axoplasmic flow (Fig. 2). At 1 hr the furthest extent of activity found for the more distal component gave a fast flow rate estimated at 930 mm/day.[10,11] The slow phase of axoplasmic flow given by the component with a steeper slope close to the cord had a rate of several millimeters/day, as found by taking the shift of its half-activity point as the measure.

It was important to determine if the activity found in the roots is in fact inside the axons rather than between the fibers as an endoneurial

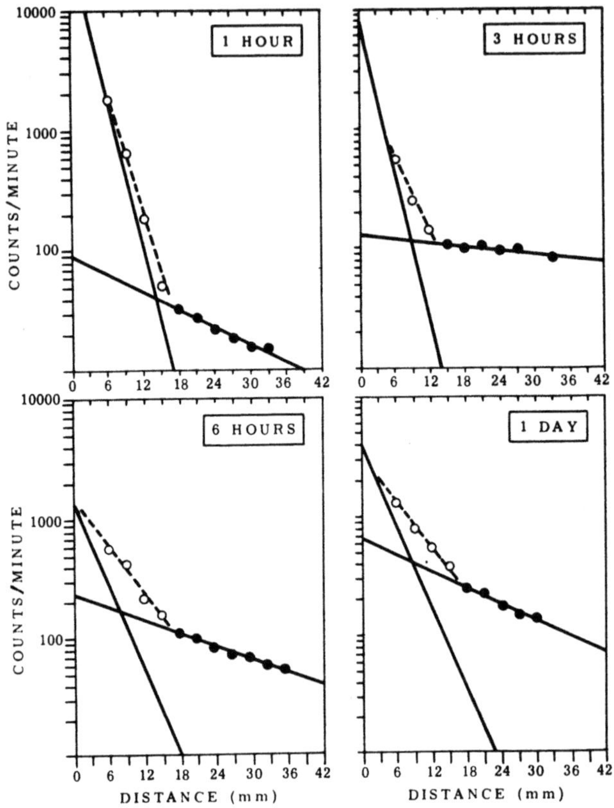

Fig. 2. The curves given are as in Fig. 1, with activity found in ventral roots taken at earlier times after ^{3}H-leucine injection into the cord. The lines fitted to the two components of the curves were fitted by least-mean-square analysis, open circles representing the slow and darkened circles the fast component. (From Ochs and Johnson.[11])

flow,[15] a possibility suggested by Taylor and Weiss.[16] To show an intra-axonal transport, the fibers were interrupted by a method which causes the axons to become sealed off and in effect to be converted into closed cylinders, without a general disruption of the surrounding nerve tissues. The technique used was to freeze a small segment of the root at low temperatures by means of an aluminum bar cooled in dry ice for a brief period of time.[11,12] This initiates a series of changes in the nerve fibers so that within several hours they bead and then close off to become blind-ended tubes. The subsequent block of transport following such a freeze-block was evident by an increase in the activity proximal to the frozen region, with a decrease in activity below the dammed region. Direct evidence for an intra-axonic transport was shown by means of radioautography using the technique originally introduced by Droz and Leblond.(17) An intra-axonic location of labeled material was shown for slow axoplasmic flow,(2) and a similar localization was also found for the fast axoplasmic flow material.(11)

FAST AXOPLASMIC TRANSPORT IN FIBERS OF THE SCIATIC NERVE

Further study of fast axoplasmic flow in the ventral roots is complicated because of their short lengths. Activity is carried down into the sciatic nerve, but the several contributions to the sciatic nerve from the several cord segments which enter the sciatic plexus at different places along its length can cause some alterations in the outflow pattern. These factors may be obviated by an examination of the distribution of activity present in the sensory fibers of the sciatic nerve after a precursor injected into a dorsal root ganglion, e.g., the lumbar 7th (L7). Lasek(14) had examined distributions in the sciatic nerve at times varying from 14 hr to 60 days after injection of the ganglion. However, in order to observe the true form and the full extent of the activity carried by fast axoplasmic transport, it is necessary to examine the nerves at earlier times after ganglionic injection. In our studies small volumes of ^{3}H-leucine having a high concentration (5 mCi/ml) were injected into the L7 ganglia, and then the dorsal roots and sciatic nerves removed 2–8 hr afterwards.(18) The spatial distribution of the activity found at these times clearly indicates the presence of a fast transport system (Fig. 3). The pairs of curves shown are the activities present in the two sciatic nerves taken from the animal at each of the times indicated after injection of their L7 ganglia. In each case a high level of activity remains in the ganglia, as expected. An early indication of axoplasmic flow in nerves 2 hr after injection is seen by the slope of activity spreading outward from the ganglion into the dorsal root and the nerve. At 3 hr and thereafter a much more satisfactory indication of fast transport is shown by the crest of activity present in the sciatic nerve. Beyond the ganglion the activity falls to a plateau, and then more distally it rises to a crest of activity, with a sharp fall beyond it, consistent with a wave of active material being moved down the fibers. The crest appears at increasingly greater distances along the nerve as more time elapses between the injection of the L7 ganglion and the removal of the nerve for measurement. The rate of flow determined by the distal displacement of the crest was found to be 410 mm/day in the adult cat.

It was important to show that this fast moving crest of activity is in fact intra-axonic, and the technique of freeze-block was used for this pur-

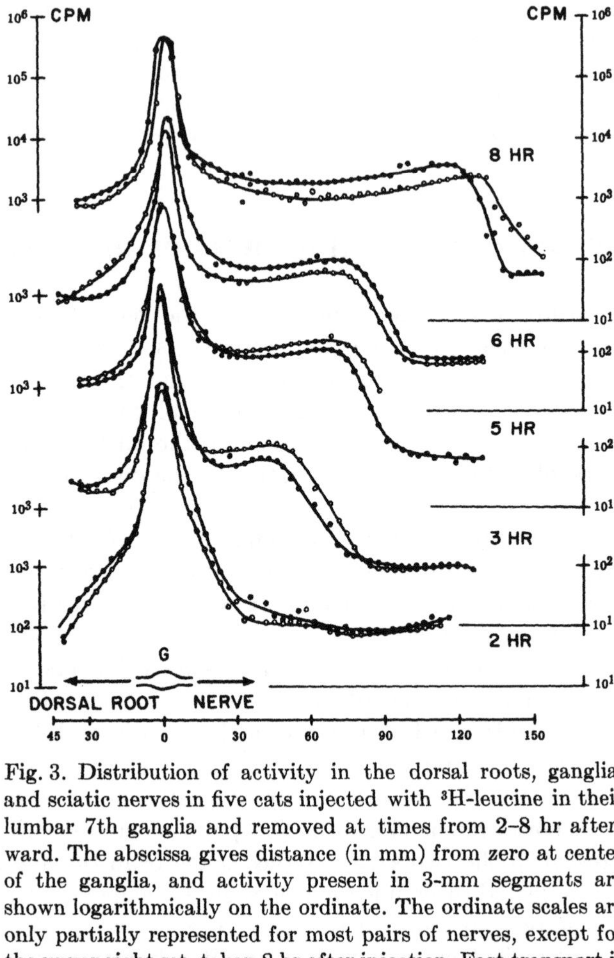

Fig. 3. Distribution of activity in the dorsal roots, ganglia, and sciatic nerves in five cats injected with ^{3}H-leucine in their lumbar 7th ganglia and removed at times from 2–8 hr afterward. The abscissa gives distance (in mm) from zero at center of the ganglia, and activity present in 3-mm segments are shown logarithmically on the ordinate. The ordinate scales are only partially represented for most pairs of nerves, except for the upper right set, taken 8 hr after injection. Fast transport is shown in the nerves from 3 hr onward by the crest of activity present distally and its more distal displacement with time after injection. The rate calculated from the displacement in the adult cat is 410 mm/day. (From Ochs *et al.*[18])

pose. As shown in Fig. 4, damming was found proximal to the zone of freezing, with a low level of activity present beyond the freeze zone. This is to be compared to the pattern of activity found present in the nerve on the opposite control side, where the crest appears at the expected position after a similar ganglionic injection and removal at the same time, at 7 hr.

The rate of 410 mm/day found for the fast transport in sensory fibers[18] is approximately half that previously estimated for fast flow in the

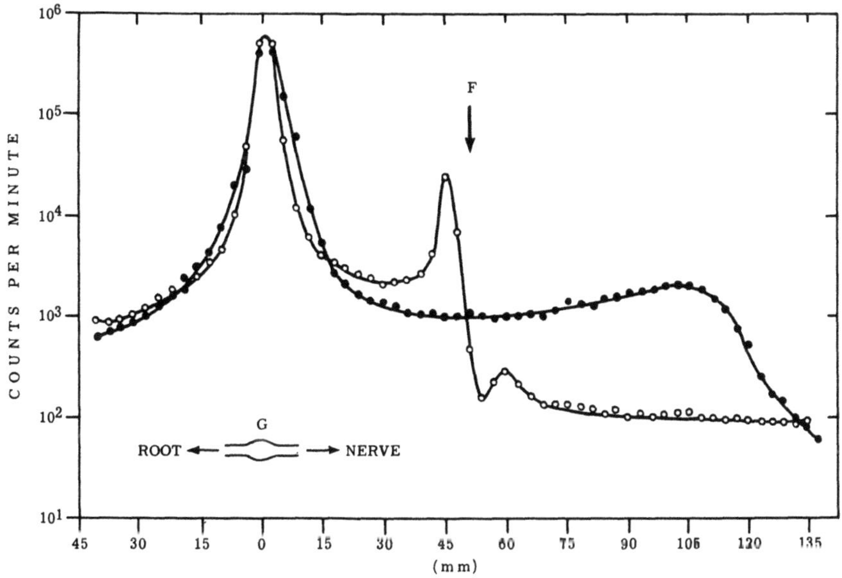

Fig. 4. Effect of freeze-block to prevent fast transport after freezing nerves 16 hr beforehand with a metal bar cooled in dry-ice. Such freezing causes the fibers to become "closed-off." After injection of the ganglia on the control side (darkened circles) a normal pattern for 7 hr is seen. On the frozen side (open circles) activity is present down to the frozen region with an increased level proximal to the dammed region and a fall of activity distal to the frozen zone. The low level present beyond the frozen zone represents leakage of precursor into the general circulation. Abscissa and ordinate as in Fig. 3. (From Ochs and Ranish.[26])

motor fibers of the ventral roots.[10,11] In the case of the ventral roots the measure of rate used was the furthest extent of the spread of activity in 1 hr, and the possibility exists that diffusion could have accounted for some increased extension of activity. Alternatively, there may be a difference in the velocity of motor and sensory nerve fibers or differences in velocity in nerve fibers close to and far from the cell bodies. In some recent experiments a crest of activity was present in the sciatic nerves moving down with the same velocity after injection of ^{3}H-leucine into the ventral-horn region of the L7 segment[26]. This indicates that of these hypotheses, either a contribution by diffusion or a difference in rate in parts of the fibers close to the cord as compared to more distal parts of the motor fibers are likely to be the best possible explanations for the apparently higher rate present in the ventral root.

The difference in the fast and slow axoplasmic flow mechanisms is also shown by the lack of a fast flow component seen after ^{32}P-orthophosphate injection as compared with ^{3}H-leucine injection. When one ganglion was

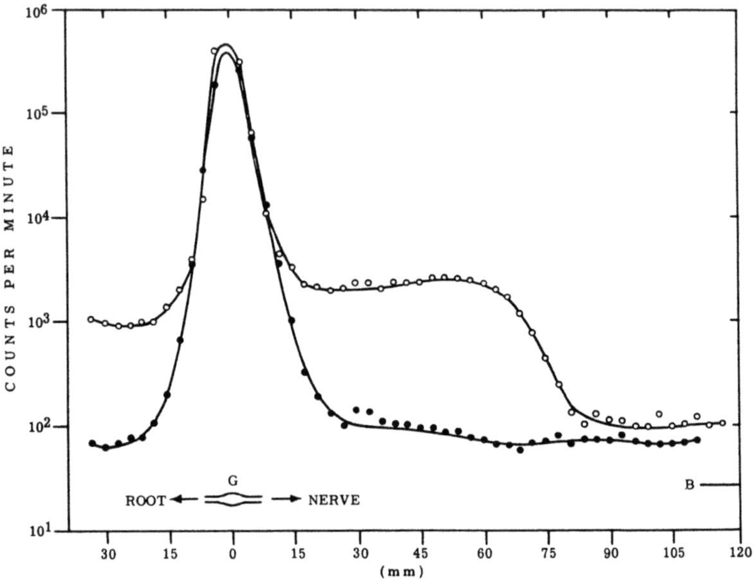

Fig. 5. The downflow of labeled materials after injections of ^{3}H-leucine (open circles) and ^{32}P-orthophosphate (darkened circles) into L7 ganglia of each side of an animal. The pattern of fast transport seen 5 hr after ^{3}H-leucine injection has the usual pattern. On the ^{32}P-orthophosphate-injected side a lack of movement of labeled materials with the fast transport system is found. In other experiments, when several days were allowed, a movement of ^{32}P-labeled material expected of a slow transport is seen. (From Ochs and Ranish.[26])

injected with the precursor ^{32}P-orthophosphate and the other with ^{3}H-leucine different rates of incorporation and downflow between the two are clearly evident (Fig. 5). Leucine is incorporated and moves down as part of the fast flow, while the ^{32}P activity remains close to the ganglion with only a suggestion of a flow. It should be noted that the relative amounts of activity present in the ganglia are similar for both isotope species. Only days after ^{32}P injection is a slow phase of downflow seen with a proximo-distal gradient similar to that previously reported in studies on the motor fibers.[12] This difference in downflow patterns further exemplifies the fact that two different transport mechanisms are present in nerve, one slow and another giving rise to a fast transport of materials.

TRANSPORT OF LABELED COMPONENTS BY FAST AND SLOW AXOPLASMIC MECHANISMS

The incorporation of ^{3}H-leucine into protein and polypeptides is suggested by the results obtained with radioautography (cf. Droz and Leblond[17]) because the unincorporated precursor would likely be washed out in the preparation of the slides. More direct evidence, however, was obtained by our studies of ventral roots which at different times after cord injection were homogenized and subjected to subcellular centrifugation.[9] Activity was found in the "nuclear" (N), "mitochondrial" (M), "particulate" (P), and high-speed supernatant or soluble (S) fractions. The soluble fraction had the highest relative level of activity, and the presence of activity in the soluble protein was shown after TCA precipitation, with most of the activity found in the precipitate. Upon hydrolysis of the TCA precipitate, ^{3}H-leucine was recovered and identified using paper chromatography. Somewhat similar results were later reported by Bray and Austin[19] and McEwen and Grafstein.[20]

In our more recent studies of the soluble protein component Sephadex gel filtration was used to isolate and separate protein and smaller polypep-

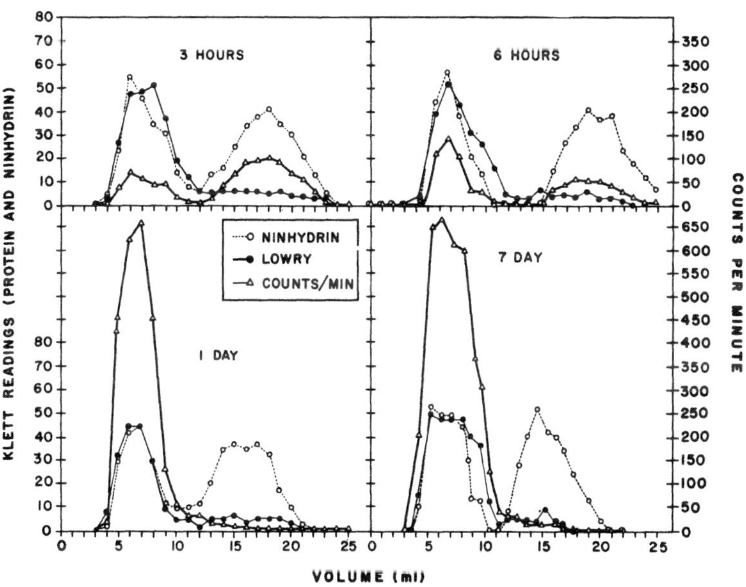

Fig. 6. Tube fractions of eluates of Sephadex G-100 columns of the materials in the supernatant of the ventral-root homogenates prepared by differential centrifugation. The times given in the panels indicate the time of ventral-root removal after cord injection. The first peak after the void volume represents the higher-molecular-weight proteins, the second peak polypeptides and free leucine. These are indicated by Lowry and ninhydrin reactions. Activity given in counts per minute shows a small amount present in the first peak at early times, with more in the second peak. A much greater amount of labeled protein is present in the first peak at one day and later. (From Kidwai and Ochs.[22])

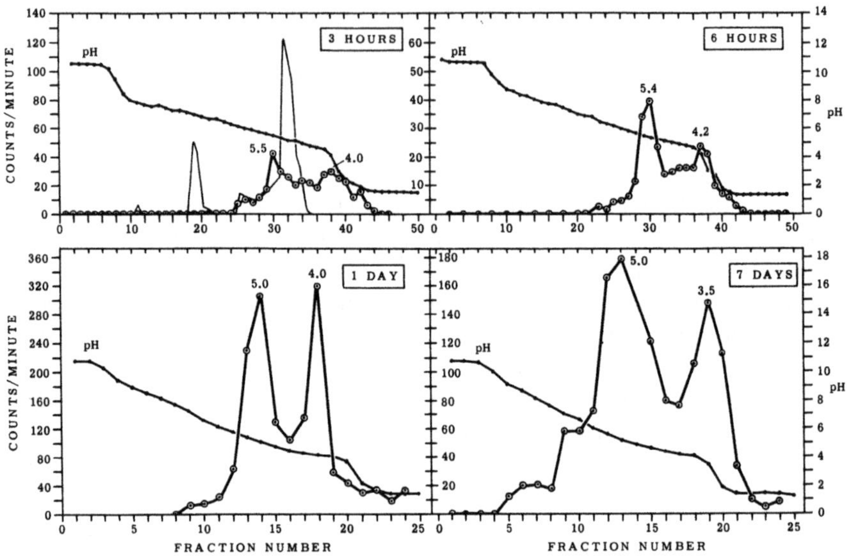

Fig. 7. Isoelectric focusing of the soluble protein isolated from the first peak of the Sephadex eluates (Fig. 6) in roots taken at the times indicated after cord injection. The curves show the pH (dots) and the activity (dotted circles) in counts per minute. In the upper left panel is a characteristic curve (thin line) for unlabeled protein, with its isolectric point of approximately pH 4.5. Labeled protein with an isoelectric point at approximately pH 4–5 increases after 6 hr. (From Kidwai and Ochs.[22])

tide fractions from it. The effluent from Sephadex G-100 columns showed the presence of higher-molecular-weight soluble proteins in the first peak followed by a second peak containing lower-molecular-weight polypeptides and free leucine.[10,21,22]

A point of interest was to determine if changes in the relative amount of labeling of the two peaks of the effluent of Sephadex G-100 gel columns could be related to fast and slow phases of axoplasmic transport. As can be seen in Fig. 6, somewhat more activity was present in the second peak at early times, but at one day and thereafter the bulk of the activity was found present in the higher-molecular-weight proteins contained in the first peak. The later appearance of most of the labeled material in the soluble protein component indicates that the higher-weight proteins are preferentially moved down the fibers by the mechanism responsible for the slow axoplasmic flow.

The soluble protein was further studied using Sephadex G-200 columns. As shown by suitable markers, it was found to be composed of several components, one having a molecular weight of approximately 450,000, the other 68,000. Further characterization of these proteins is in progress. Isoelectric focusing of the protein shows that most of the protein present has an isoelectric point at approximately pH 4.5 and also that the labeled material present ranges from pH 4.5–5.5 and that it appears in greater amount at later times, as expected from its transport by the slow axoplasmic transport mechanism (Fig. 7).

The fast axoplasmic transport mechanism carries down a small amount of labeled soluble protein and a relatively higher component of lower-molecular-weight labeled polypeptide and free leucine. The separation of free leucine and polypeptide from the effluent of the second peak of the Sephadex G-100 columns was accomplished with paper and thin-layer electrophoresis. The free leucine accounts for 50–75% of the activity in the second peak obtained from roots taken 3–6 hr after injection, but soon afterward it rapidly decreases in amount.(22)

The free leucine and other amino acids carried down by the fast axoplasmic flow plus amino acids added to the pool by proteolysis(4) could likely enter into a local synthesis in the axon(23,24) or in the terminals.(5) A local synthesis in the axons, however, cannot account for the relatively large amount of labeled soluble protein carried by the slow phase of downflow. This was shown in experiments, still in progress, where, after L7 ganglion injection with ^{3}H-leucine, following a period of early flow into the fibers a further flow into the nerve fibers was prevented by a ligature made just distal to the L7 ganglion. The free leucine which had gained entrance to the axon did not then become incorporated into soluble protein, as shown by comparing the composition of the materials labeled in the ligated nerves with that present in control nerves of the opposite side, where slow flow had carried down labeled protein.

The particulate (P) fraction has a relatively higher level of labeling at early times, indicating a transport by the fast transport mechanism.(18,22) Possibly, some part of the P fraction may consist of vesicular components or precursor of the vesicles related to synaptic transmission at the terminals. In this context, the accumulation of vesicular-like structures above a dammed region of the nerve as recently described by de Iraldi and De Robertis(25) is of interest. The study of a comparison of the labeling of the P fraction of the sensory fibers after L7 ganglion injection with that of motor fibers now in progress may throw some light on this possibility.

REFERENCES

1. L. Lubińska, *in*: "Mechanisms of neural regeneration" (M. Singer and J. P. Schadé, eds.), *Prog. Brain Res.* **13**, 1–71, Elsevier, Amsterdam (1964).
2. S. Ochs, *in*: "Macromolecules and Behavior" (J. Gaito, ed.), Appleton-Century-Crofts, New York (1966).

3. S. Barondes (ed.), Axoplasmic transport, *Neurosci. Res. Prog. Bull.* **5** (1967).
4. A. Lajtha, *Int. Rev. Neurobiol.* **6,** 1–98 (1964).
5. L. Austin and J. G. Morgan, *J. Neurochem.* **14,** 377–387 (1967).
6. P. Weiss and H. B. Hiscoe, *J. Exp. Zool.* **107,** 315–395 (1948).
7. A. Dählstrom and J. Häggendal, *Acta Physiol. Scand.* **67,** 278–288 (1966).
8. G. A. Kerkut, A. Shapira, and R. J. Walker, *Comp. Biochem. Physiol.* **23,** 729–748 (1967).
9. S. Ochs, J. Johnson, and M.-H. Ng, *J. Neurochem.* **14,** 317–331 (1967).
10. S. Ochs, J. Johnson, and A. M. Kidwai, *Fed. Proc.* **27,** 235 (1968).
11. S. Ochs and J. Johnson, *J. Neurochem.* **16,** 845–853 (1969).
12. S. Ochs, D. E. Dalrymple, and G. Richards, *Exp. Neurol.* **5,** 349–363 (1962).
13. N. Miani, *J. Neurochem.* **10,** 859–874 (1963).
14. R. Lasek, *Brain Res.* **7,** 360–377 (1968).
15. P. Weiss, H. Wang, A. C. Taylor, and M. V. Edds, Jr., *Am. J. Physiol.* **143,** 521–540 (1945).
16. A. C. Taylor and P. Weiss, *Proc. Natl. Acad. Sci. U.S.* **54,** 1521–1527 (1965).
17. B. Droz and C. P. Leblond, *J. Comp. Neurol.* **121,** 325–346 (1963).
18. S. Ochs, M. I. Sabri, and J. Johnson, *Science* **163,** 686–687 (1969).
19. J. J. Bray and L. Austin, *J. Neurochem.* **15,** 731–740 (1968).
20. B. S. McEwen and B. Grafstein, *J. Cell Biol.* **38,** 494–508 (1968).
21. S. Ochs, *in*: Axoplasmic transport (S. Barondes, ed.), *Neurosci. Res. Prog. Bull.* **5** (1967).
22. A. M. Kidwai and S. Ochs, *J. Neurochem.* **16,** 1105–1112 (1969).
23. A. Edström, *J. Neurochem.* **14,** 437–446 (1966).
24. E. Koenig, *J. Neurochem.* **14,** 437–446 (1967).
25. A. P. de Iraldi and E. De Robertis, *Z. Zellforsch.* **87,** 330–334 (1968).
26. S. Ochs and N. Ranish, *J. Neurobiol.*, in press.

DISCUSSION

McEWEN: We independently came across a system similar to the one that Dr. Ochs described, in which two transport rates were found in the optic nerve of the fish after injection of precursor into one eye. The optic nerve crosses completely and the contralateral nerve can be used as a control for background incorporation of label. A slow component having a rate of 0.5 mm/day was found, with activity reaching the tectum 20 to 40 days after injection. In addition, a fast component was present having a rate in excess of 40 mm/day at 20° C (this corresponds to rates of 200 mm/day at 37° C) with activity in the optic tectum found 6–30 hr after injection. The slow component comprises 10–20% of the label we can recover in the eye, whereas the fast component is about one-tenth this. Forty per cent of the slow component consists of soluble protein, the remainder being predominantly particulate, whereas the fast component is 85% insoluble protein. We have concluded that the fast component may consist of a particulate element, such as a transmitter storage granule, while the slow component probably represents many constituents of the axoplasm, including elements of the neurotubules and neurofilaments.

AUSTIN: Would you care to comment on these different rates of flow, Dr. Ochs? We also found a fast component, and also, like you, find free leucine moving along the splanchnic nerve after injection into the coeliac ganglion. A fast component is present, but it is slower than yours, of the order of about 120 mm/day. This is the same rate we find for the movement of norepinephrine granules after the injection of labeled noradrenalin. Also, in the sciatic nerve in the system I described (Chapter 12) we find a fast component which travels as a peak at a rate of the order of 100–150 mm/day.

OCHS: I believe the evidence we obtained from the displacement of the crest of labeled material found in mammalian sensory nerves shows that there is a fast transport of 410 mm/day. This is also true of motor fibers. The apparently higher values obtained by taking the furthest extent of activity in the ventral roots 1 hr after cord injection probably represents an additional effect due to diffusion. There is a slower axoplasmic flow of several millimeters/day (or 30–70 mm/day, depending on how this estimate is made). How these two rates compare with those found in other species—avian, fish, and invertebrate—will have to be determined in comparable ways. The fast transport system is of great interest in this regard.

VRBA: I have no doubt about the reality of the change in position in label along the nerve consistent with an axonal flow, although other factors—diffusion and circulation and some exchange properties—should be considered. What is your definition of axonal flow?

OCHS: I believe the term axoplasmic flow is a general one which could be reserved for a movement of part or all of the axoplasm. Fast axoplasmic flow as we described it should perhaps be more accurately referred to as the fast axoplasmic transport system. It appears most likely that the neurofilaments and/or neurotubules are somehow involved in this fast transport. I wonder if Dr. McEwen considers that the neurotubules are the labeled particulates found related in his faster flow component?

McEWEN: Not exclusively.

OCHS: One possibility is that these linearly organized elements, the neurofilaments and/or neurotubules, are actually being transported down the fiber at these fast rates. We can exclude that hypothesis by using the penomenon of "beading." After a maintained stretch to produce beading, nerves containing transported material were freeze-substituted and sectioned and coated for radioautography. The labeled activity was found present in the expanded part of the beads rather than at the constrictions where the filaments and neurotubules are most concentrated. This shows that labeled materials are free for the most part to move with the fluid part of the axoplasm. Particles, some protein, polypeptide, and free leucine are carried by the fast transport system, protein mainly by the slow.

RICHTER: The flow of materials is clearly of very great physiological importance. I wonder if you got around to testing the effect of such factors as electrical stimulation, narcosis, and so on. Another point, in experiments made in tissue culture some time ago evidence of movement of mitochondria, for example, was found along the axon. I wonder if you can relate that type of movement to either your slow or fast type of transport?

OCHS: So far the effect of stimulation, deafferentation, or the chromatolysis produced by distal nerve section made some time beforehand have all been disappointing in having remarkably little effect on either slow or fast transport rates or amount transported. Until the individual fractions are selectively analyzed we cannot definitely relate a fast transport to specific components such as mitochondria or the small particulates, which are also of special interest because they could be related, at least in part, to transmitters carried down in vesicles or vesicle precursors. In this regard, Johnson working in my laboratory found evidence for a fast downflow of acetylcholinesterase, and this enzyme shows a selective decrease following distal nerve transection.

E. KOENIG: A few years ago I used labeled DFP injected intrathecally and observed labeling coming down the nerve. Have you seen this?

OCHS: We haven't used labeled DFP. It is important to determine which labeled substances are carried by the fast transport and which by the slower axoplasmic flow mechanisms. ^{3}H-cycloleucine, which does not enter normally into protein synthesis, does not move down as part of either the fast or slow axoplasmic flows. This and the evidence that ^{32}P-orthophosphate does not move down as part of fast flow, but does move with slow flow, suggests some control mechanism in the cell for materials transported by the two different transport mechanisms.

Chapter 14

Effect of Electrical and Chemical Stimulation on Protein Synthesis in Brain Slices

Fritz Lipmann

The Rockefeller University, New York, N. Y.

It is only natural that with our long-standing interest in protein synthesis[1] the recent lively discussion of the role of protein turnover in nerve function made it attractive to have another look at protein synthesis in the brain. Many years ago I worked for a while with pigeon brain slices in order to study the metabolic function of thiamine.[2] These experiments left me with the impression that, metabolically speaking, the brain slice gives a good reflection of *in vivo* events. It gave me special confidence that after cutting and exposure to low temperature the slices seemed slightly shocked because, during the first hour, an aerobic glycolysis appeared which, however, disappeared again on further incubation. In other words, the slice normalized during incubation at 37°; it could be used for up to 3 hr without showing any signs of decay.

McIlwain had worked out a method for electrical stimulation of brain slices.[3] Dr. Orrego from the University of Chile was with us on a fellowship and was interested in nerve function. We decided then to study with this preparation the effect of stimulation on protein and RNA synthesis in rat brain slices. Since these studies have been published extensively,[4,5] I do not need to go into the details.

McIlwain's work[6] had shown that electrical stimulation of the slice caused a reversible increase in respiration and the appearance of aerobic glycolysis. These responses were rather quick and could be tested with relatively short stimulation. When we turned to protein synthesis we found that we had to use longer exposure and reasonably high voltage, 1–1.5 V, over periods of 15–30 min. As shown by Mokrasch and Manner[7] in Dr. Folch's laboratory, protein synthesis in brain slices is high after birth and continuously decreases until it reaches a plateau after about 90 days. As shown in Table I, this impressive decrease was confirmed, and Orrego observed a similar decrease in RNA synthesis (Table II).[5] It is important,

TABLE I

^{14}C-Valine Incorporation as a Function of Rat Age

Age (days)	Body weight (g)	No. of experiments	Amino acid incorporation (cpm/mg protein)
Fetus	3.75	3	27,950 ± 4,950
2	10	4	31,240 ± 4,770
4	13	3	34,420 ± 2,360
7	20	8	25,560 ± 13,100
10	30.5	3	8,923 ± 3,610
16	34	2	3,418 ± 1,250
25	72.5	3	3,570 ± 3,110
28	87.5	2	5,755 ± 318
90	177	3	462 ± 242
600	532	4	300 ± 113

therefore, to use animals of a uniform age group for such experiments. We used rats that were 3–4 months old and weighed about 150 g. The effect of 15 min of stimulation on protein synthesis at various voltages is summarized in Table III. The inhibition became nearly maximal at 1.5 V and amounted there to ~30%.

A very similar inhibition of protein syntheses was accidentally observed using chemical stimulation with acidic amino acids: it was then systematically studied. Curtis and Watkins[8] had determined the potency of a large number of acidic amino acids for excitation of neurons by injection into ganglia. We mostly used the strongly active *DL*-homocysteate, nearest to their most potent compound N-methyl-*D*-aspartate, which was not available to us. As shown in Table IV, the amino acid incorporation was

TABLE II

RNA Synthesis as a Function of Rat Age

Age (days)	No. of experiments	Body weight (g)	RNA (cpm/μM DNA-P)
3	4	10.5	26,625 ± 9860
10	4	19.6	26,025 ± 3140
21	3	52	6,735 ± 359
Adult	4	350	5,516 ± 1330

TABLE III

Protein Synthesis as a Function of Voltage Stimulation

Voltage (V)	No. of experiments	Average change (%)	Variation range (%)
0	4	0	+10 to −10
0.5	6	− 5.4	+ 9.5 to −25
1	2	− 9.5	− 4.4 to −14
1.5	7	−27.5	−12.7 to −38.5
2	2	−26.0	−19.5 to −33
3	2	−33.0	−27.0 to −36.6

strongly inhibited on addition of this series. Inhibition was roughly proportional to the *in vivo* activity found in the experiments of Curtis and Watkins.[8]

In searching for an explanation for this inhibition, McIlwain's[6] observation of the appearance of aerobic glycolysis was remembered. This indicated that stimulation might, in some manner, uncouple aerobic phosphorylation. We did confirm the stimulation of aerobic glycolysis by electrical stimulation, but did not find a similar effect with chemical stimulation using *DL*-homocysteate. However, when the ATP content of electrically and chemically stimulated slices was assayed it appeared that there was

TABLE IV

^{14}C-Valine Incorporation in Presence of Acidic Amino Acids

Amino acid	No. of experiments	Protein (cpm/mg/hr)	Inhibition (%)	Relative excitatory potency *in vivo*
N-methyl-*D*-aspartate	—	—	—	100
Control	10	3131		
DL-homocysteate	10	1658	46.7	30–55
Control	4	4650		
L-glutamate	4	2875	33.6	8–20
Control	5	3408		
D-aspartate	5	2140	34.9	4–10
Control	4	4410		
DL-α-aminoadipate	4	3682	16.6	0–6

TABLE V

Influence of Electrical Stimulation and *DL*-Homocysteate on ATP Content in Brain Slices

Method of stimulation	ATP content (μM/g wet wt)	Difference (%)
Controls	0.17 ± 0.11	—
1.5 V	0.12 ± 0.09	−29
Controls	0.27 ± 0.12	—
DL-homocysteate (10^{-3} *M*)	0.13 ± 0.06	−52

a considerable decrease in the ATP concentration of both. Table V shows ATP concentration to be lowered by about 30%, as was protein synthesis: on addition of *DL*-homocysteate an even greater decrease in ATP was found.

In the course of these studies we became more and more aware of a parallel between our present results and earlier reported effects[(9)] of replacing sodium chloride by potassium chloride in the suspension medium for brain slices. There the brain slice exhibits a similar disturbance of energy metabolism, with strongly increased aerobic glycolysis and an increased respiration parallel with uncoupling of phosphorylation and respiration. This potassium effect is specific for brain. Liver or kidney slices, for example, are rather favorably affected by using potassium chloride instead of sodium chloride solution as a medium. We checked for a possible effect of electrical stimulation and acidic amino acids on kidney slices, but found them without effect on protein synthesis (Table VI).

TABLE VI

Influence of Electrical Stimulation and *DL*-Homocysteate on ^{14}C-Valine Incorporation in Kidney Cortex Slices

Manner of stimulation	Protein (cpm/mg/10 min)
Controls	665.8 ± 194
1.5 V	661.1 ± 204
Controls	817 ± 161
DL-homocysteate	919 ± 228

In view of this similarity between electrical and chemical stimulation and potassium ion in their effects on brain slices, we were very interested in a set of experiments that had been undertaken at Rockefeller University by Bennett and Edelman,[10] which they have permitted me to mention here. They tested the effect of spreading cortical depression[11,12] by layering a concentrated potassium chloride solution on the rat brain. Making a bilateral perforation, they used one side of the brain as control, applying sodium chloride there. The results showed that concomitant with a depression of nerve action, there was an inhibition of leucine incorporation on the potassium chloride-injected brain side, which averaged 14–16% in six experiments. Ruscak,[13] in his study on spreading depression, had reported earlier on a metabolic disturbance indicating uncoupling.

I should now like to summarize what we think of the inhibition of protein synthesis with chemical and electrical stimulation, and with exposing the tissue to relatively high potassium concentrations. In both cases it appears there is an uncoupling of the normal energy-yielding metabolism, which leads to a lowering of ATP availability. The effects are specific for brain tissue and are not found in other tissues. The common denominator of all is that effects on the membrane potential are coupled to a disturbance in energy metabolism. This linking of an alteration of the membrane potential in the brain cell with energy turnover appears to be rather singular and specific for the brain.

To put it somewhat differently: a regulation of energy supply obviously is strongly linked in the brain cell to potential changes on the cell membrane. In the experiments we have presented here using brain slices the effect is predominantly negative and in part irreversible. However, the *in vivo* effects of excitatory acidic amino acids and of potassium are reversible. One might therefore conclude that normally the change of potential of the brain cell membrane with excitation also causes a metabolic disturbance. I would just like to alert attention to the point that on the whole this kind of result implicates some linkage, presumably regulatory, between membrane potential and energy coupling that is specific for the brain.

REFERENCES

1. F. Lipmann, *Science* **164,** 1024–1031 (1969); F. Lipmann, *in:* "Metabolism of the Nervous System" (D. Richter, ed.), pp. 329–340, Pergamon Press, London (1957).
2. F. Lipmann, *Scand. Arch. Physiol.* **76,** 255–272 (1937).
3. H. McIlwain and R. Rodnight, "Practical Neurochemistry," Little, Brown and Company, Boston (1959).
4. F. Orrego and F. Lipmann, *J. Biol. Chem.* **242,** 665–671 (1967).
5. F. Orrego, *J. Neurochem.* **14,** 851–858 (1967).

6. H. McIlwain, *in*: "Neurochemistry," (K. A. C. Elliott, I. H. Page, and J. H. Quastel, eds.), p. 212, Charles C. Thomas, Springfield, Ill. (1962).
7. L. C. Mokrasch and P. Manner, *J. Neurochem.* **10,** 541–547 (1963).
8. D. R. Curtis and J. C. Watkins, *J. Physiol.* **166,** 1–14 (1963).
9. P. J. Heald, "Phosphorus Metabolism of the Brain," Pergamon, New York (1960).
10. G. S. Bennett and G. M. Edelman, *Science,* **163,** 393–395 (1969).
11. W. H. Marshall, *Physiol. Rev.* **39,** 239–279 (1959).
12. S. Ochs. *Int. Rev. Neurobiol.* **4,** 1 (1962).
13. M. Ruscak, *J. Neurochem.* **7,** 305–307 (1961).

DISCUSSION

OCHS: I think that spreading depression may not be a good phenomenon to compare with the other work, because the potassium is rather incidental to the spreading depression process the way we conceive of it now. Whatever the mechanism of spreading depression, once it has started in one part of the cortex, it becomes a contiguous, all-or-none process. While potassium may be a starting substance, the depression may be propagated by glutamic acid or perhaps by some other unknown substance. I think we will have to reserve an opinion as to whether potassium is involved in the actual propagation of spreading depression. Another difficulty of spreading depression is a very marked effect of the vasculature, which may show constriction, so that when the model experiment is done, injecting the precursor systemically and then looking at the protein metabolism, we may get a drop due to a barrier change or vascular change which is a concomitant of spreading depression.

LIPMANN: What I can say in this respect is that the metabolic response of brain slices to raising, even moderately, the outside potassium concentration is very similar to that of electrical and chemical stimulation—namely, uncoupling is indicated, since aerobic glycolysis appears parallel with a decrease of protein synthesis. On the other hand, Bennett and Edelman in their *in vivo* experiments mentioned finding a lowering of protein synthesis on the side of the brain exposed to epidural application of 10% potassium chloride. This seems to justify a comparison of *in vivo* and *in vitro* potassium effect with that of electrical and chemical stimulation.

ROBERTS: We have observed that ribosomal preparations obtained from brain tissue are very sensitive to variations in concentration of potassium as well as magnesium, and this applies to their activity in amino acid-incorporating systems, so that this is another site at which the potassium may have an effect.

RAPPOPORT: I think that the idea that the brain is a unique electrical organ is critically important. We exposed the brain in the decapitated catfish, split the corpus callosum so that we had two olfactory systems *in situ,* and used natural olfactory stimulation. By using an olfactory substance, you can cause an increase in RNA synthesis (our data on protein are incomplete). This increase is not enormous, but using one-half of the olfactory system with its brain as a control, the other half being stimulated by an olfactory substance, you definitely get an increase in and *de novo* synthesis of RNA. I think this method bypasses the lack of knowledge as to the media and the impulse and all the other problems that ensue in trying to make a model with a tissue slice. This model, which is more natural, definitively establishes that the electrical impulse, as it naturally evolves from a sensor, does increase the synthesis of this component.

RICHTER: We carried out some experiments some time ago on the effect of electrical stimulation on protein synthesis in the rat brain *in vivo* and were a little disappointed to find that electrical stimulation decreased the synthesis rather than increasing it, which we expected. I wonder if you feel the same explanation holds for these *in vivo* experiments as in your *in vitro* experiments, that in fact this is due to a decrease in the ATP pool.

LIPMANN: Well, I would think so. We can say that we similarly were disappointed to observe decrease rather than increase. Still, it might depend on the extent of stimulation.

MANDEL: Some time ago we looked at the nucleotide triphosphate pool in slices and it appeared that the normal ratio of ATP or ADP in brain is about 15 and in the slices is about 0.3, so that it might be easy to understand that there are troubles in protein synthesis in these conditions of decreased nucleotide triphosphate synthesis.

BOGOCH: In the search for a system in which one could look at increases and decreases of activity we have studied pigeons when they were working and training, and we observed changes in the incorporation of 1-C^{14}-glucose. When the bird is busy working he stops incorporating the sugar into the glycoproteins. This is open to many different interpretations, but I just wonder whether you think that stimulation and work are a different phase from rest and reconstitution, perhaps glycoprotein accumulating during rest rather than during activity. Then it may not be so surprising if electrical stimulation decreases protein synthesis as compared to the resting state.

FOLCH-PI: I think the time sequence of those events is of the essence. It stands to reason that if there is a direct dependence in a synthetic process on the maintenance of the potential due to the metabolic conditions in the brain, an immediately available pool of whatever substrate is necessary must be present for the transmission of the disturbance. If stimulation was continued over a long period of time, apparently protein was depleted, although there may be a steady state which is sufficient for the maintenance of nerve function.

LIPMANN: The shortest time we could use was 15 min and this is probably quite long. Heald used short electrical pulses of 3 min and found an increase in the phosphoprotein turnover. What that means is really not quite clear, but it may be a positive effect. Using glucose as a precursor of amino acids Geiger *et al.* perfused brain with radioactive glucose to produce radioactive amino acids, and thought that electrical stimulation of the brain produced an increase in amino acid incorporation. We checked on that with our slices and found, somewhat to our surprise, that electrical stimulation considerably increased the specific activity of the amino acid pool in the brain cell, due, we believe, to an increase in respiration. Taking the increase in specific activity into account, electrical stimulation with radioactive glucose as the source of amino acid markers turns out to cause an inhibition rather than a stimulation of protein synthesis.[4]

LAJTHA: Isn't there a possibility that the glutamate or the potassium effects are direct and not mediated by ATP levels? Did you find any conditions of stimulation by manipulating glutamate levels in brain slices with ATP levels unchanged? What about other amino acid effects, such as that of lysine, which I would expect to have opposite effects to glutamate?

LIPMANN: Glutamine and basic amino acids had no effect. We did not try to increase ATP levels; it might be difficult to do so.

ROBERTS: It is obvious that the interesting effects of stimulation occur at several levels of cellular organization. One thing happens at the cell surface, perhaps in connection with the electrical and potential changes which are produced, but this may be further complicated by effects within the cell, not only on energy metabolism, but on the activity of the cellular synthesizing systems.

Chapter 15

Proteins in Bioelectricity

David Nachmansohn

College of Physicians and Surgeons
Columbia University, New York, N. Y.

EXCITABLE MEMBRANES

At the turn of the century it was widely accepted that in a fluid system, such as the living cell, ions must be the carriers of electric currents. Since the uneven distribution of Na^+ and K^+ ions between the interior of cells and their outer environment was already known, Overton[63] proposed, on the basis of simple experiments, that during electrical activity Na^+ ions move into the cell interior and an equivalent amount of K^+ ions flow to the outside. It is common knowledge that Overton's assumption was borne out when, after World War II, radioactive ions became available and made it possible to measure ion movements during rest and during activity. But the fundamental question for understanding these processes is that of the control of these ion movements. It was postulated at the turn of the century that the cell membranes surrounding nerve and muscle fibers must have a special ability to change their permeability to ions during activity, although at that time cell membranes were largely a hypothetical postulate. No mechanism was considered for explaining the changes of permeability to ions.

During the last decade, however, a considerable amount of information has accumulated about cell membranes, due to electron microscopy in combination with biochemical and biophysical methods. Although less than 100 Å thick, cell membranes may be the site of many complex enzyme reactions. This is best illustrated by the well-studied mitochondrial membranes (see, e.g., Green and Perdue,[23] Green and Silman,[24] and Racker[65]). The concept of a "unit membrane," which was based on the Danielli–Davson model, assumed that membranes are formed by a bimolecular leaflet of phospholipid surrounded by a protein layer on the inside and on the outside.[66,67] This notion proved to be an oversimplification and has been vigorously criticized by many biochemists and biologists (see, for instance, Elbers[21] and Korn[40]). The molecular organization of membranes is far from being established; they are apparently a mosaic of functional repeating units formed by lipid–protein complexes.[64,79] The lipid seems to be located on the outside of the protein (see, e.g., Benson,[5] Loewenstein,[44] Kennedy[38]).

Let us now return to excitable membranes, i.e., the plasma membranes surrounding nerve and muscle fibers, assumed to control the movements of ions during electrical activity by their ability to change rapidly and reversibly their permeability to ions. The idea of a purely physical process controlling the ion movements during electrical activity, as postulated in the ionic theory, [30,31] is difficult to reconcile with the large amounts of heat produced and absorbed during electrial activity.[1] According to the most recent measurements, it seems that the rising phase of the action current is associated with a marked heat production, the falling phase with a marked heat absorption.[32] The most likely explanation of the results of the heat measurements is the assumption of chemical reactions underlying the permeability cycle in the excitable membrane (Hill;[29] Nachmansohn,[54] pp. 269–270). In view of the central role of enzymes and proteins in cell mechanisms, it appears reasonable to assume that they have an essential function in the elementary processes of bioelectrogenesis.

The well-known theory of neurohumoral transmission associated acetylcholine (AcCh) with nerve activity. It was assumed that, whereas electric currents propagate nerve impulses along nerve and muscle fibers, at the junctions (synapses) chemical compounds, AcCh or others, are released from the nerve ending and, crossing the junctional gap, transmit the impulse to the second cell, nerve or muscle, thereby acting as chemical mediators. This idea was unacceptable to many prominent neurobiologists. Their main objection was based on the evidence that electrical properties of the membranes at junctions and in axons are essentially similar, in spite of some differences. This made it difficult, in their view, to interpret the observations reported in terms of the theory of neurohumoral transmission, i.e., to assume two basically different mechanisms of propagation, a purely physical one in the axonal membranes, and a chemical one between the junctional membranes (see, for instance, Erlanger[22]).

The theory of neurohumoral transmission proposed in the 1930's was based on observations with classical methods of physiology and pharmacology. These methods are essential for many aspects of biology and medicine. They are, however, inadequate for the analysis of the molecular mechanisms responsible for the permeability changes controlling ion movements, events which take place in microseconds in a membrane of less than 100 Å thickness. However, the rapid advances of biochemistry, and especially of protein and enzyme chemistry, the development of highly refined techniques and methods, such as the use of isotopes, electron microscopy, a great variety of optical methods, X-ray analysis, etc., have provided powerful tools for studying cellular mechanisms on cellular, subcellular, and molecular levels. It appeared necessary to apply the new methods for exploring the role of AcCh and to find a more satisfactory answer to the mechanism of nerve-impulse propagation. An analysis of the proteins and enzymes specifically associated with the function of AcCh and their relations to the molecular processes which enable the excitable membranes to change their permeability appeared to be a promising line of approach and a prerequisite for the understanding of such a basic phenomenon of biology as bioelectricity.

However, membranes form an extremely small fraction of the cell mass. Even with the refined methods and techniques available at present, studies of proteins in membranes offer many difficulties. It is therefore fortunate that a material is available which is particularly favorable for studying the proteins and enzymes associated with bioelectricity: the electric organs of electric fish. These organs are the most powerful bioelectric generators created by nature. The discharge of *Electrophorus*, the electric eel, found in the Amazon River, amounts to 600 V. The electrical parameters of a single cell are similar to those in nerve and muscle fibers. It is only their arrangement in series, as in a Voltaic pile, which is responsible for the power of the discharge. In the electric eel there are 5000 to 6000 cells, electroplax, arranged in series. But the most important feature of electric organs is their high specialization for their main function, that is, for bioelectrogenesis. Investigating the occurrence, distribution, and concentration of AcCh-esterase, the enzyme which is relatively specific for hydrolyzing AcCh, in a variety of excitable tissues, nerve, muscle, brain, ganglia, etc., in widely differing species, the writer came across, in 1937, the electric organs of *Torpedo marmorata*. Determinations of the concentrations of AcCh-esterase in the electric tissue of this species showed that 1 kg of tissue (fresh weight) hydrolyzes 3–4 kg of AcCh per hour,[46] although the tissue is formed by 3% protein and 92% water. Similar values were obtained in 1938 with the electric tissue of *Electrophorus*. It was therefore immediately obvious that this material was uniquely faborable for studying the enzymes and proteins associated with AcCh function and their relationship to bioelectrogenesis. In the 30 years since then the material has proven to be of crucial importance for the advances accomplished, and is still at present one of the most useful materials for obtaining one of the specific proteins, AcCh-esterase, in solution or for analyzing this and other proteins in the intact membrane and for exploring their function in bioelectricity. The earlier studies have been summarized in a monograph[50]; the later developments have been described in several review articles.[51,57] In the present chapter some recent advances in the study of the proteins will be presented, with the main emphasis on the progress obtained with AcCh-esterase and -receptor.

PROTEINS IN EXCITABLE MEMBRANES

Role of AcCh in Bioelectricity

The results of the biochemical approach do not support the idea of AcCh being a transmitter of impulses between two cells. The data accumulated over a period of 30 years have provided strong evidence for the assumption that the action of AcCh is an *intra*cellular, or, rather, an intramembranous, process. AcCh acts as the trigger initiating the reactions changing the permeability of excitable membranes to ions; it controls the ion movements during electrical activity. The main features on which this view is based may be briefly summarized:

1. AcCh and the enzymes hydrolyzing and forming it—AcCh-esterase and choline O-acetyltransferase (cholineacetylase)—have been shown to be present in the various types of conduction fibers of nerve and muscle throughout the animal kingdom: in motor and sensory, "cholinergic" and "adrenergic," peripheral and central fibers, in invertebrates and in vertebrates, etc.
2. AcCh-esterase, an enzyme relatively specific for AcCh and distinctly different from other esterases, is localized in the excitable membranes of axons and muscle fibers as well as in those in junctions, i.e., in the membranes of nerve terminal and in the postsynaptic membranes.
3. AcCh-esterase hydrolyzes AcCh in a few microseconds, a prerequisite for attributing to the ester the proposed role as a trigger in the generation of bioelectric currents.
4. The extraordinarily high concentration of the enzyme in the electric organs of electric fish mentioned before appears significant in view of the high specialization of the tissue.
5. Specific and potent inhibitors of AcCh-esterase have been shown to block electrical activity in nerve fibers, physostigmine reversibly, diisopropylphosphofluoridate (DFP) and other organophosphates irreversibly. Thus, electrical activity requires the activity of the enzyme present in the excitable membrane, supporting the assumption of its essential role in the permeability changes.
6. In axons exposed to organophosphates under conditions leading to an irreversible block of electrical activity a powerful specific reactivator of

the phosphorylated enzyme, pyridine-2-aldoxime methiodide, may, under appropriate conditions, restore electrical activity.

7. The presence of a protein distinctly different from AcCh-esterase, the AcCh-receptor, has been demonstrated in pre- and post-synaptic and in axonal membranes. It has been shown that AcCh and its congeners produce effects on the electrical parameters of axons similar to those previously observed on the junctional membranes, provided that the structural barriers for lipid-insoluble quaternary nitrogen derivatives surrounding axonal membranes are inadequate to prevent the compounds from reaching the receptor, or that they are reduced by chemical treatment.

8. Differences between the molecular groups of the active site of the AcCh-receptor and those of the enzyme have been shown using the monocellular electroplax preparation of *Electrophorus*.

9. Lipid-soluble compounds that are closely related in structure to AcCh, such as certain local anesthetics, are inhibitors of the AcCh-receptor. They act as antimetabolites and block electrical activity in all excitable membranes, thus demonstrating the essential role of the AcCh-receptor protein in bioelectrogenesis and supplementing the evidence for the essential role of the enzyme.

On the basis of the data accumulated, the following picture of the role of AcCh seems best to fit the available information. AcCh is released within the membrane by excitation; it acts as a signal recognized within the membrane by a specific AcCh-receptor protein. This reaction induces a conformational change. It is possible and even probable that by this change Ca^{++} ions bound to carboxyl groups of the protein are released. The Ca^{++} ions have been known for a long time to be involved in the excitability of nerve and muscle fibers.[7] The Ca^{++} ions have several remarkable features and properties by which they are distinguished from other divalent cations, as has been discussed by Katchalsky.[37] The Ca^{++} ions released may induce further conformational changes of phospholipids and other polyelectrolytes. The end result of this sequence of chemical reactions is the change of permeability to ions permitting the movements of 20,000 to 40,000 ions or more across the membrane per molecule of AcCh released. These reactions thus act as typical amplifiers of the signal given by AcCh. AcCh-esterase has the function to permit the return of the receptor to its original conformation by rapid hydrolysis of AcCh. The barrier for the ion movements is thereby reestablished. In the meantime, however, the ion movements, the carriers of small electric currents, have stimulated adjacent points. AcCh is thereby released there and the same process is repeated. By the stimulation of successive points of the membrane, the nerve impulse is propagated along the axon. AcCh in its bound (storage) form and

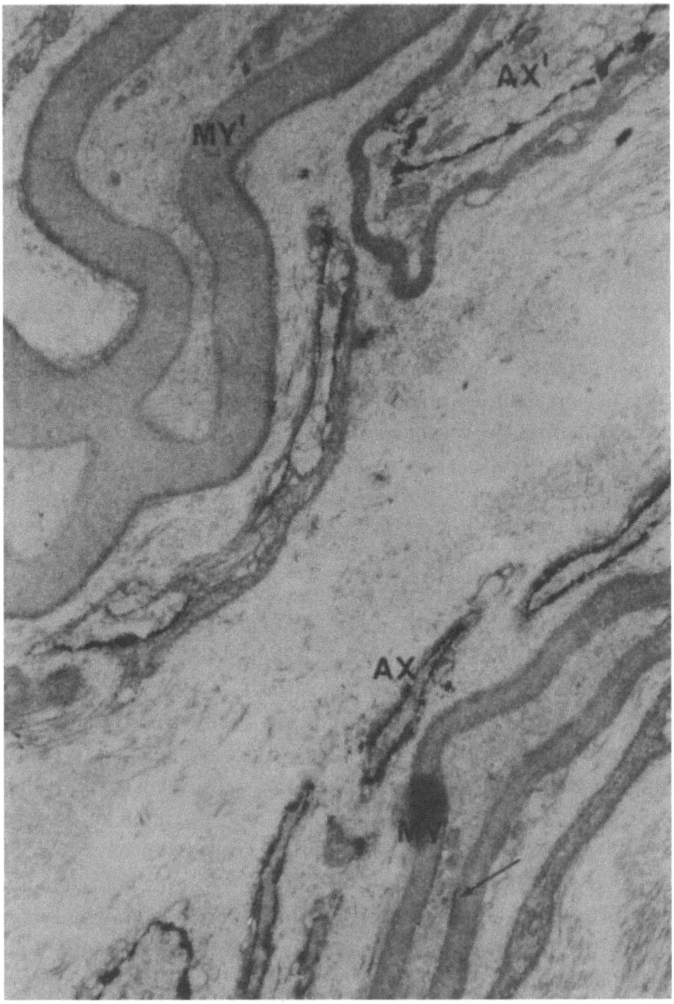

Fig. 1. An electron micrograph of sympathetic nerve fibers of frog. The fibers were incubated with a reaction mixture for demonstrating acetylcholinesterase activity. Acetylthiocholine was used as substrate. The hydrolysis product, thiocholine, reacts with Cu to form Cu-thiocholine, which is converted into CuS. The dense end-product is present at the axolemmal surface of unmyelinated axons (AX). It is present in a myelinated axon (AX^1), in which the membrane is detached from the myelin sheath. (×7759.) For details see Brzin.[8]

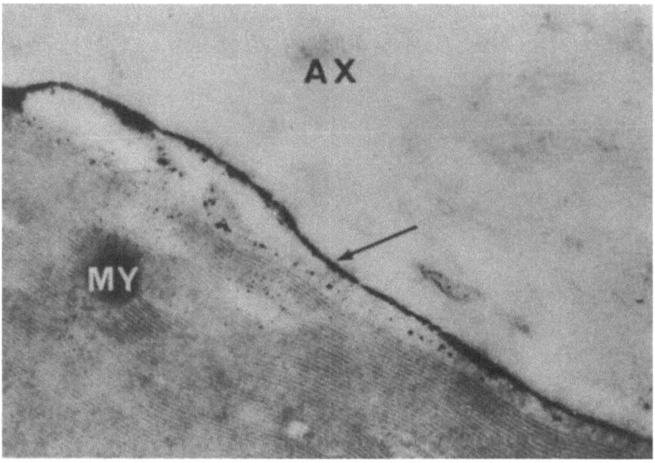

Fig. 2. An electron micrograph of a large myelinated (My) ventral-root axon (AX). Prior to the incubation for acetylcholinesterase activity the fiber was treated with Triton 100-X. Dense end-product is present on the axolemmal membrane (arrow). (×34,676.) Brzin.[8]

the two proteins reacting directly with the specific signal released, AcCh-receptor and AcCh-esterase, are presumably linked together structurally as well as functionally, and form a protein assembly in the excitable membrane in a way comparable to many other complex enzyme systems, for instance, to the electron-transfer system in mitochondrial membranes and to the enzymes involved in fatty acid synthesis. The existence of such protein assemblies is today well established in biochemical thinking and does not need further elaboration. The structural organization of the system may account for the efficiency, the precision, and the speed of the events following the release of AcCh.

AcCh-Esterase

Localization

For more than twenty years it was postulated, on the basis of biochemical data, that the AcCh-esterase is localized in the excitable membranes. This evidence was of necessity indirect. In the last seven years direct evidence for the postulated localization has been obtained in many laboratories, by electron microsocopy combined with histochemical techniques (see, e.g., Barrnett,[2] Torack and Barrnett,[80] Brzin *et al.*,[9] Schlaepfer and Torack,[74] Novikoff *et al.*[62]). In the membranes of unmyelinated

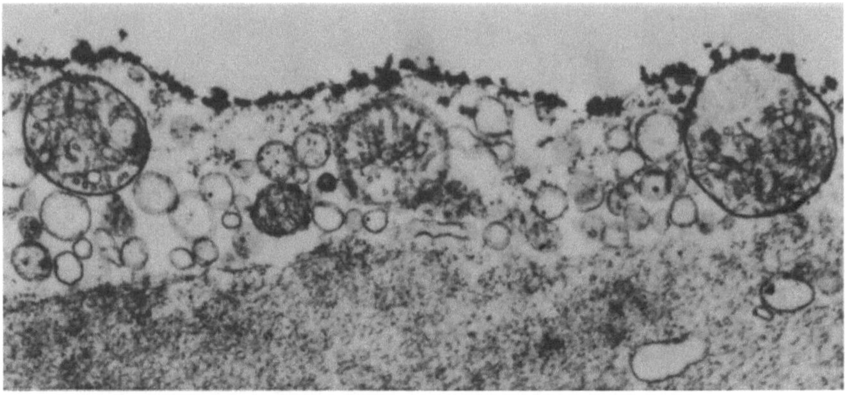

Fig. 3. Normal cholinesterase activity in a plaice muscle. Oblique section of a cell kept at 0°C in isotonic sucrose solution showing black precipitates along the sarcolemma as a result of cholinesterase activity. Substrate: acetylthiocholine for 45 min at 0°C. Small amounts of precipitate can be discerned between the inner and outer membrane of one mitochondrion to the right. (×21,000.) (From Lundin and Hellstroem.[45])

fibers the presence of the enzyme can be readily demonstrated; in those of myelinated ones the presence of the enzyme appeared to be irregular or absent. Figure 1 shows an electron micrograph obtained by Brzin[8] with sympathetic nerve fibers of frog. While the dense products due to enzyme activity are regularly present in the membranes of nonmyelinated fibers, the myelinated fibers rarely have end-product; in the one fiber in which the products are present the membrane appears detached and may be damaged. Since it appeared possible that even in a slice of 500–1000 Å thickness structural barriers may prevent or slow down the reactions between the enzyme within the membrane and the added reaction mixture, Brzin[8] applied Triton 100 X to slices of an isolated sciatic nerve fiber. After the action of the detergent he found AcCh-esterase to be present in the plasma membranes located between the myelin and the axoplasm (Fig. 2). Applying electron microscopy combined with histochemical staining, Lundin and Hellstroem[45] found AcCh-esterase to be localized in the excitable membrane of the muscle cells of plaice (*Pleuronectes platessa*) (Fig. 3). When the preparation was exposed to DFP in low concentration, 10^{-6} *M*, no enzyme activity was detectable.

Interdependence between Electrical and Enzyme Activity

As mentioned above, potent reversible inhibitors of the AcCh-esterase may reversibly block electrical activity, the irreversible ones irreversibly.

This demonstration of an essential role of the enzyme in electrical activity offers, however, serious difficulties and pitfalls. Many of them were recognized in the early phase of investigations and have been repeatedly discussed. However, during the last decade the complexity of such a phenomenon due to ultrastructural organization has become much more apparent, partly due to the information accumulated by electron microscopy, and partly by advances obtained with physiocochemical methods about the differences between enxyme reactions *in vitro* and those in the intact structure. It is today a commonplace to say that it is in many cases difficult and frequently impossible to extrapolate from findings in solution to the events in a structure. This applies particularly to structures which are apparently as well organized as the cell membranes.

Of particular importance for the problem of the essential role of AcCh–esterase in electrical activity has been the evidence accumulated in the last few years that a dependable quantitative evaluation of the AcCh-esterase in normal tissue is not obtainable with the methods available at present. This makes any estimation of enzyme activity after exposure to irreversible, or, as it is more correct to say in the light of recent information, poorly reversible, inhibitors entirely impossible. What has been nevertheless unequivocally demonstrated under proper conditions with sutiable preparations is that potent inhibitors affect electrical activity of axons in the way postulated by theory. The whole complex problem of the relationship between electrical and enzyme activity has been recently reevaluated in the light of the new information (Nachmansohn[58,59]).

Active Site of AcCh-Esterase

The discovery in 1937 of the high concentration of AcCh-esterase in electric tissue suggested the use of this material for obtaining the enzyme in solution and for its purification. The enzyme was first obtained in solution in 1938.[60] In the early 1940's a more than 400–fold purification was achieved.[72] Since the material was difficult to obtain, the amounts of protein of the highest specific activity achieved were small. However, partially purified preparations of high specific activity were suitable for kinetic studies and for analyzing the mechanism of reaction of many compounds, inhibitors and substrates, with the enzyme. AcCh-esterase was one of the first enzymes with which information about the molecular groups in the active site and about the mechanism of the hydrolytic process was obtained.[61] Effects of organophosphates and of other potent inhibitors of the enzyme widely used in pharmacology were explained in terms of their reaction with molecular groups in the active site (for summaries see, e.g.,

Nachmansohn;[50] Wilson;[84] Cohen and Oosterbaan[14]). The value of these studies is illustrated by the development of a compound, pyridine-2-aldoxime methiodide, which rapidly reactivates AcCh-esterase, phosphorylated and thereby inactivated by organophosphates, in solution,[85] and, in addition, is a most efficient antidote against poisoning by organophosphate insecticides, especially in combination with atropin.[39] The compound is even an effective antidote against some, although not all, of the potent "nerve gases."

Large-Scale Purification and Crystallization of AcCh-Esterase

The developments of protein chemistry, and macromolecular chemistry in general, during the last decade, permitting the analysis of many molecular properties including their three-dimensional structure, made it appear desirable to obtain adequate amounts of pure enzyme protein. A large-scale purification using electric tissue was recently worked out by Leuzinger and Baker.[41] From 10 kg of tissue about 70 mg of a homogeneous protein may be obtained by a procedure involving seven steps. The specific activity is about 700 mM of AcCh hydrolyzed per milligram protein per hour. The homogeneity was tested by analytical ultracentrifugation and by Canalco disc electrophoresis. From this material crystals of the enzyme were obtained in form of hexagonal prisms (Fig. 4).[42] The crystals have grown from 10 μ to 180 μ in length, with a width of 150 μ. The molecular weight is 260,000.[43] The molecule is formed by four subunits of about equal molecular weight. There are two different polypeptide chains, on the

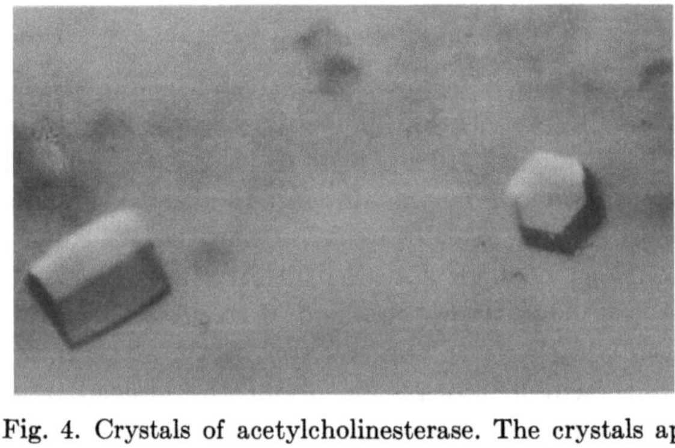

Fig. 4. Crystals of acetylcholinesterase. The crystals appear to have the form of regular hexagonal prisms. On the left side one sees a crystal from the side; on the right side, from the top. The crystals seen here are about 180 μ long and 150 μ wide.

basis of C-terminal determinations performed with two different methods, which yielded results in excellent agreement. There seem to be two active sites. Thus, the active form of the enzyme may be a dimeric hybrid.[43]

AcCh-receptor

Monocellular Electroplax Preparation

The AcCh-receptor has not been isolated as yet. A turning point in the studies of this protein has been the development of the monocellular electroplax preparation by Schoffeniels.[75–77] It is a uniquely favorable preparation for studying the properties of the receptor protein for two reasons. First, it is formed by a single cell. Therefore, the many obstacles encountered in the interpretation of reactions between ligands and their functionally associated proteins in multicellular preparations are greatly reduced. Second, the cell is highly specialized in its specific function, i.e., bioelectrogenesis. The excitable membrane is functionally the dominating feature, whereas other cell functions seem poorly developed, as indicated by the low protein and high water content and the slow rates of metabolic reactions which are not directly related to the activity of the membrane. Among other remarkable and useful characteristics may be mentioned the large size of the cell, the rectangular shape of the innervated, excitable membrane, and the presence of both synaptic and conducting parts of the membrane, which can be readily distinguished by means of electric parameters. There are thousands of synaptic junctions per cell, but their surface amounts to only 5–10% of the total surface area of the excitable face.

It may be mentioned that AcCh-esterase is evenly distributed in the membranes of the nerve terminals and in the post-synaptic and conducting membrane, as was shown by electron microscopy combined with various histochemical staining techniques by Bloom and Barrnett.[6] The presence of AcCh-recptor has also been demonstrated in the conducting as well as in the synaptic membranes. However, as in most other preparations (see next section), AcCh, curare, and other quaternary nitrogen derivatives reach the receptor in the membrane only at the synaptic junctions, while their tertiary derivatives act also on the conducting membrane. For this reason, since in most studies of the receptor AcCh and its quaternary compounds are used and also other compounds capable of reaching the receptor at the junction, but not at the conducting membrane, many of the observations on the preparations represent reactions on the level of the synaptic junctions. At present it is impossible to decide the exact site of these reactions, i.e., whether they take place in the pre- or post-synaptic membranes or in the neighborhood of the junction, which may be poorly protected, as has been described for neuromuscular junctions.

During the last decade several refinements of techniques, introduced mainly by Higman, Bartels, and Podleski, have greatly increased the usefulness and sensitivity of the preparation for studying the reactions between the receptor protein and specific ligands.[25–28] The use of electrical parameters in analyzing the reactions of the protein is comparable to the precision of other methods applied in protein studies (for recent summaries see Nachmansohn[56,58]).

Information is accumulating about the molecular groups in the active site of this protein using both receptor activators, i.e., compounds which apparently induce the conformational change and depolarize the membrane, such as AcCh, carbamylcholine, decamethonium, and related structures, and receptor inhibitors which react specifically with the active site, but do not depolarize and act as antimetabolites of AcCh, such as curare, local anesthetics, etc. The affinities to the receptor have been compared with those to the AcCh-esterase in solution.[82] Some of the molecular groups of the active site are distinctly different from those of the enzyme. SH blocking and S—S reducing agents block the action of AcCh on the receptor; washing out these agents does not restore the reaction, but chemical restoration of the SH and S—S groups does.[35] This is evidence for the protein nature of the AcCh-receptor. The same agents have no effect on AcCh-esterase activity.[33]

Essential Role of the AcCh-Receptor in Electrical Activity

Pertinent to the relationship of the receptor to its function in bioelectrogenesis has been a series of experiments performed on the electroplax with local anesthetics whose structure is closely related to that of AcCh. There are, however, two fundamental differences: (1) AcCh is a receptor activator, i.e., it reacts with the receptor and induces a conformational change, while local anesthetics apparently just block the active site and prevent AcCh from interaction with the receptor, and (2) the action of AcCh is usually limited to the membranes of the junction becasue the compound is unable to penetrate the structural barriers surrounding the excitable membranes of nerve and muscle fibers (see discussion below). In contrast, the local anesthetics penetrate the structural barriers and therefore act on the receptor in the membranes of conducting fibers as well as those of junctions. An analysis has been performed to test by which substitutions the AcCh molecule is transformed into a "local anesthetic."[3,4] The substitution of the methyl group on the carbon of the carbonyl group by a

Compound	Synaptic Junctions: Activator	Synaptic Junctions: Inhibitor	Conducting Membrane
	M concentration		
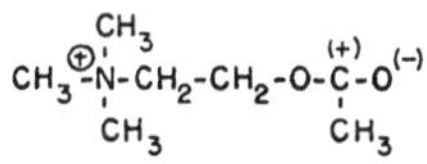 Acetylcholine	2.5×10^{-6}	0	0
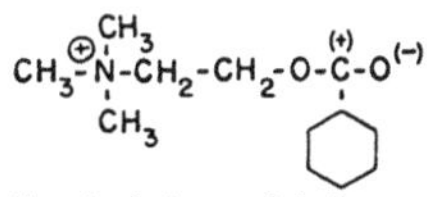 Hexahydrobenzoylcholine	5×10^{-4}	0	0
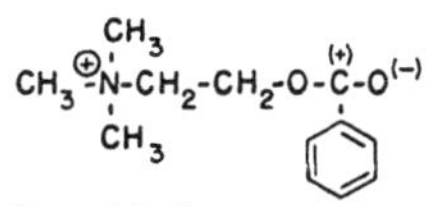 Benzoylcholine	5×10^{-4}	5×10^{-4}	1×10^{-3}
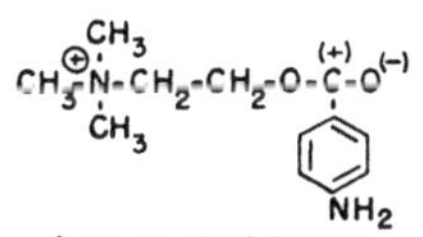 p-Aminobenzoylcholine	0	1×10^{-3}	2.5×10^{-3}
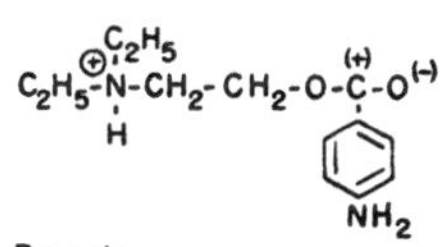 Procaine	0	2.5×10^{-4}	5×10^{-4}
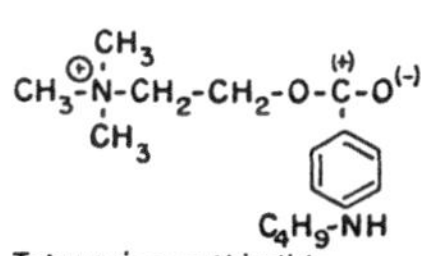 Tetracaine methiodide	0	2×10^{-5}	1×10^{-5}

Fig. 5. Substitutions of AcCh transforming the molecule from a receptor activator acting on the synaptic junctions only into a receptor inhibitor acting both on junctions and on the conducting membrane (so-called "local anesthetic"). Benzoylcholine is a transitory form both in structure and in biological activity. For details see the text.

saturated ring decreases the potency of the compound as an activator more than 100 fold. But it is still an activator and its action is limited to the junction. The substitution of the same methyl group by an unsaturated ring leads to a compound (benzoylcholine) which is an intermediary form in structure and action: depending on conditions, it may be either an activator or an inhibitor; moreover, it is capable of penetrating through structural barriers and of acting on the receptor in the conducting membrane. Addition of an amino group to the ring results in a classical "local anesthetic," blocking electrical activity of the junctions as well as of the axons. Small modifications of the structure, either of ammonium or the acyl group, may greatly increase the potency of the compound as a local anesthetic, as illustrated, e.g., by procain or tetracaine (see Fig. 5).

Recently, Changeux and Podleski applied to the electroplax membrane *p*-(trimethylammonium)benzenediazonium fluoroborate (Tdf) prepared by Wofsy.[12] This ligand has been used for affinity labeling of enzymes and antibodies by Singer[78] and his associates, and others. The diazonium group forms a covalent bond with ring amino acids. The quaternary nitrogen group is attracted by Coulombic and van der Waals' forces to the anionic group in the active site; thereby the concentration of the compound becomes high in its neighborhood and facilitates the formation of the covalent bond there. Tdf forms a covalent bond with the receptor at or near the active site. The block is irreversible. This is illustrated by Fig. 6. Another affinity labeling was achieved by Karlin in a two-step procedure; 4-(N–maleimido)phenyltrimethylammonium iodide blocks the receptor irreversibly in rather low concentrations, but only after exposure of the electroplax to dithiothreitol (DTT), which reduces S—S bridges.[36] Hexamethonium, a curarelike compound and an inhibitor of the receptor, became, after the reduction of the receptor by DTT, a receptor activator and depolarized the membrane. The same ligand thus may have different, seemingly even opposite, biological actions, dependent on the condition of the macromolecule and other factors in the membrane.

Recent observations of Changeux and Podleski[11] suggest certain similarities between the behavior of regulatory enzyme systems and the reactions of the electroplax to specific ligands acting on the receptor. The response of the electroplax to specific ligands is characterized by (1) the specific recognition of receptor activators which trigger the membrane depolarization, or of their antagonists, receptor inhibitors; and (2) the deviation of the dose-response of several activators from the Langmuir isotherm. The dose-response curve of AcCh has a sigmoid shape. The Hill coefficient (n_H) is about 2, as was first noticed by Karlin,[34] who also calculated, on the basis of the data with SH blocking and S—S reducing compounds, that after treatment with these compounds the n_H becomes approx. 1. The changes from the Langmuir isotherm were elaborated by Changeux and Podleski.[11] The sigmoid shape usually indicates that some type of cooperative action intervenes in the response to an activator. In the presence of a different activator the sigmoid shape of the curve disappears and becomes a hyperbola with an n_H of 1. This has been consistently observed

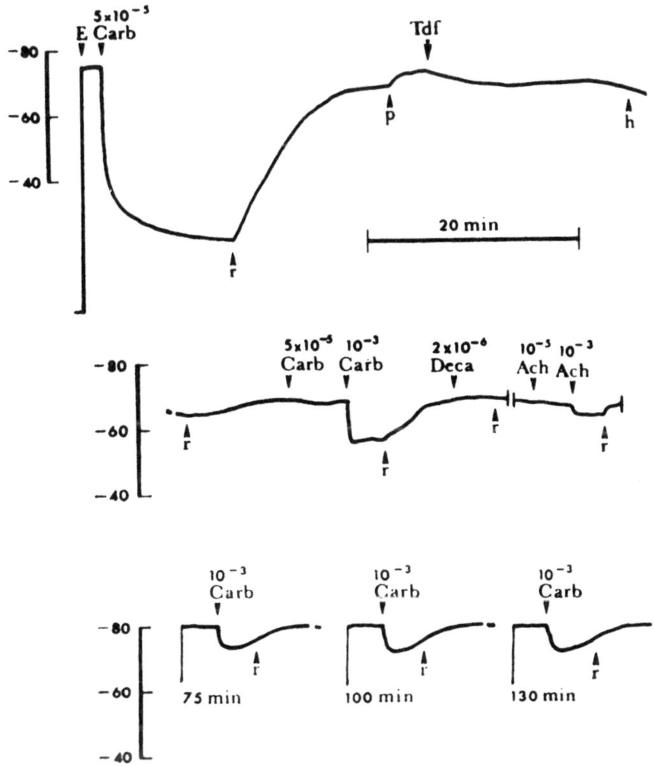

Fig. 6. Irreversible abolition of the response to a receptor activator (carbamylcholine) after a 20-min exposure to Tdf (see text), tested on the electroplax. Ordinate: membrane potentials in mV; abscissa, time. *E* impalements; *r*, Ringer's solution; *p*, Ringer's solution supplemented with 10^{-2} *M* Na-phosphate pH 5.9; Tdf, 10^{-4} *M* in *p*; *h*, 10^{-3} *M* 1-histidine in Ringer's solution to quench unreacted Tdf; Carb, carbamylcholine; Deca, decamethonium. Lower line: 10^{-3} *M* carbamylcholine is applied 75, 100, and 130 min after the end of the Tdf exposure. (From Changeux *et al.*[13])

with regulatory proteins: the cooperative interactions for the binding of a given ligand are modified by the presence of a different ligand.[13,49] Such a conversion of shape is considered as an indication that the interactions between both classes of ligands are indirect or allosteric interactions. At saturating concentrations of activators the maximal response to different activators differs independent of the ionic environment.

DIFFERENCES BETWEEN AXONAL CONDUCTION AND SYNAPTIC TRANSMISSION

In view of the progress achieved in the understanding of chemical reactions in cell membranes in general and those in excitable membranes in particular, a reevaluation of the theory of neurohumoral transmission appears appropriate.

One of the striking facts, particularly emphasized as being an indication of a mechanism of synaptic transmission different from that of conduction, is the powerful pharmacological action of AcCh on junctions in contrast to its complete failure to affect axonal conduction. Similarly, more than a century ago, Claude Bernard observed that the activity of curare blocks only the transmission of the impulse from nerve to muscle, but does not affect nerve and muscle fibers. Curare has a higher affinity to the AcCh-receptor than AcCh. It is a receptor inhibitor and therefore prevents the depolarization by AcCh released during activity and blocks the transmission at the level of the junction. Claude Bernard applied curare to the frog sciatic nerve fiber which is formed by several thousands of heavily myelinated axons and is surrounded by a sheath impervious to many chemical compounds. Dettbarn(16,18) isolated a single axon from the frog sciatic nerve. On exposure to curare he found a rapid and reversible block of axonal conduction (tested at the Ranvier nodes). A blocking action was also obtained when curare was applied to unmyelinated axons of the walking leg of lobster under appropriate experimental conditions.(17)

ACh, curare, and related compounds that act as specific receptor activators or inhibitors are quaternary nitrogen derivatives and lipid-insoluble. The excitable membranes of axons are surrounded by structural barriers usually impervious to this type of compound. Even in nonmyelinated axons, the Schwann cell surrounding the excitable membrane in most cases prevents AcCh from reaching the receptor in the membrane, although in some axons an action of AcCh and its congeners on the electrical activity of axons may be obtained, e.g., on those of the walking leg of lobster.(19)

Even if an unmyelinated axon does not react to AcCh and its congeners, chemical treatment reducing the outside barriers may permit these compounds to reach the excitable membrane and react with the receptor. In the squid giant axon, e.g., the excitable membrane is surrounded by a Schwann cell about 4000 Å thick. When these axons are exposed to AcCh, curare, or neostigmine in highest concentrations no effects are observed.(10,73) In contrast, physostigmine, a tertiary nitrogen derivative and a potent inhibitor of AcCh-esterase, about equal in strength to neostigmine, blocks electrical activity. When the axoplasm of the exposed axons was extruded physostigmine was found to have penetrated, while the other compounds had not entered the interior.[10] However, it is possible to reduce structural barriers by chemical treatment. For instance, even the whole sciatic nerve bundle, after having been desheathed, becomes sensitive to AcCh, curare, and neostigmine following the exposure to the detergent cetyltrimethylammonium.[81] In recent years Rosenberg and his associates have demonstrated that the electrical activity of squid giant axons after their exposure to certain snake venoms, and similarly to cetyltrimethylammonium, is affected by AcCh and curare in the way to be expected by their reaction with the receptor (for summaries see Rosenberg[68,69]). The venom itself, in the concentrations used and in the periods of time of exposure, had no effect on electrical activity. Examination with electron microscopy revealed marked structural changes in the Schwann cell, but none in the plasma membrane.[47] Subsequent to the venom treatment radioactive AcCh and curare were found to have penetrated in the interior.[71] The active component of the snake venom was found to be phospholipase A; its effect is fully accounted for by the formation of lysolecithin.[15,70]

Frequently, even synaptic junctions do not react to AcCh or curare. The neuromuscular junctions of lobsters, e.g., do not react to these compounds although the concentration of AcCh-esterase in these junctions is extremely high. On the other hand, lipid-soluble compounds, such as physostigmine and DFP, affect neuromuscular transmission of lobsters. Thus, the system there is functional, but protected by barriers against quaternary nitrogen derivatives. However, even when a junction reacts to AcCh, the discrepancy between the amounts of AcCh realeased in the membranes and the concentrations which must be applied externally in order to be effective raises difficult questions. In the squid giant axon the order of magnitude of AcCh released is 10^{-15} M/cm^2/impulse. This is very far from the pharmacologically active concentrations. Such a sensitive preparation as the electroplax, e.g., fails to react to 10^{-5} *M* AcCh unless physostigmine is added. This remarkable discrepancy raises the question as to the meaning of the amounts of AcCh released by stimulation at the junction and found in the extracellular perfusion fluid. No trace of AcCh appears in the perfusion fluids of synaptic junctions in the absence of physostigmine, even after prolonged stimulation. On the other hand, in the presence of physostigmine AcCh also leaks from axons, in which the surrounding barrier is not entirely impervious to AcCh; when the axons of the walking leg of lobster are kept in physostigmine AcCh appears in the outside fluid; Na^+, K^+, Ca^{++}, and Mg^{++} ions have similar effects on the release of AcCh in this preparation as those described for junctions.[20]

The AcCh-receptor, just like the AcCh-esterase, is localized in the excitable membranes of the nerve terminal and of the post-synaptic membrane; AcCh, curare, neostigmine, and related compounds act on both membranes of the junctions.[48] The nerve terminal seems to be even more sensitive than the post-synaptic membrane.[83] Since both proteins are present and functional in both membranes of the junction, it is reasonable to assume that the amplifier process takes place in the membranes of the junctions just as in those of the axon; AcCh released in the membrane of the nerve terminal acts on the receptors there and triggers off the sequence of chemical reactions resulting in the influx of Na^+ and the efflux of K^+ ions. It is known that K^+ ions release AcCh from its storage from. Thus, when the K^+ ions released from the terminal reach the post-synaptic membrane they will release AcCh there and trigger off the same amplification process. By this amplification the signal will be much more efficient than it would be if it were directly released from one cell and would act on the specific protein in a second cell. Such an assumption is all the more difficult in view of the discrepancy of the order of 10^6 to 10^7 between the amounts of AcCh released and those which are pharmacologically effective even in the presence of physostigmine. Chemical reactions in living cells are chemically and thermodynamically coupled, and most of them are structurally organized. It thus seems more plausible that the specific signal in one of the fastest cellular mechanisms known, i.e., nerve-impulse propagation, is recognized by a specific protein in the membrane where it is released and acts as a trigger of an amplifier there in a structurally organized way, than to visualize a random process in which a chemical compound released in one cell must find the specific protein in another cell.

The specific chemical forces underlying cellular mechanisms, such as, e.g., motility, vision, energy supply, etc., are remarkably similar throughout the animal kindom. It has become apparent that this also applies to the specific chemical forces underlying bioelectricity, i.e., the specific proteins controlling the changes of ion permeability in excitable membranes. The great diversity of bioelectrical phenomena and the striking differences of pharmacological effects may be explained by the great variations of cellular structure and organization: the multiformity of shape, structure, and environment of synaptic junctions is bound to modify the effects of chemical reactions in the membranes or the action of compounds applied externally, as illustrated by some of the observations discussed. Structural differences markedly affect electrical parameters even in axons: conduction velocity in different types of fibers varies from 0.1 to 100 m/sec. Since chemical composition and molecular organization of excitable membranes are at present poorly explored and unknown, it is impossible to know whether they differ in the various types of membranes. The answer to these questions requires further analyses on the molecular level.

REFERENCES

1. B. C. Abbott, A. V. Hill, and J. V. Howarth, *Proc. Roy. Soc.* (London), *B* **148,** 149 (1958).
2. R. J. Barrnett, *J. Cell Biol.* **12,** 247 (1962).
3. E. Bartels, *Biochim. Biophys. Acta* **109,** 194 (1965).
4. E. Bartels, and D. Nachmansohn, *Biochem. Z.* **342,** 359 (1965).
5. A. A. Benson, *J. Am. Oil Chemists Soc.* **43,** 265 (1966).
6. F. E. Bloom and R. J. Barrnett, *J. Cell Biol.* **29,** 475 (1966).
7. F. Brink, *Pharmacol. Rev.* **6,** 243 (1954).
8. M. Brzin, *Proc. Natl. Acad. Sci. U.S.* **56,** 1560 (1966).
9. M. Brzin, V. M. Tennyson, and P. E. Duffy, *J. Cell Biol.* **31,** 215 (1966).
10. T. H. Bullock, D. Nachmansohn, and M. A. Rothenberg, *J. Neurophysiol.* **9,** 9 (1946).
11. J.-P. Changeux and T. R. Podleski, *Proc. Natl. Acad. Sci. U.S.* **59,** 944 (1968).
12. J.-P. Changeux, T. R. Podleski, and L. Wofsy, *Proc. Natl. Acad. Sci. U.S.* **58,** 2063 (1967).
13. J.-P. Changeux, J. Thiery, Y. Tung, and C. Kittel, *Proc. Natl. Acad. Sci. U.S.* **57,** 335 (1967).
14. J. A. Cohen and R. A. Oosterbaan, *in:* "Handbuch der Experimentellen Pharmakologie: Cholinesterases and Anticholinesterase Agents" (G. B. Koelle, ed.), Vol. 15, 299, Springer–Verlag, Berlin (1963).
15. E. Condrea and P. Rosenberg, *Biochim. Biophys. Acta* **150,** 271 (1968).
16. W.-D. Dettbarn, *Nature* **186,** 891 (1960).
17. W.-D. Dettbarn, *Life Sciences* **12,** 910 (1963).
18. W.-D. Dettbarn, *in*: Symposium on cholinergic mechanism, *Ann. N.Y. Acad. Sci.* **144,** 483 (1967).
19. W.-D. Dettbarn and F. A. Davis, *Biochim. Biophys. Acta* **66,** 397 (1963).
20. W.-D. Dettbarn and P. Rosenberg, *J. Gen. Physiol.* **50,** 447 (1966).
21. P. F. Elbers, *in*: "Recent Progress of Surface Science," (J. F. Danielli, K. G. A. Pankhurst, and A. C. Riddiford, eds.), Vol. 2, p. 442, Academic Press, New York (1964).
22. J. Erlanger, *J. Neurophysiol.* **2,** 370 (1939).
23. D. E. Green and J. F. Perdue, *Proc. Natl. Acad. Sci. U.S.* **55,** 1295 (1966).
24. D. E. Green and I. Silman, *Ann. Rev. Plant Physiol.* **18,** 147 (1967).
25. H. B. Higman and E. Bartels, *Biochim. Biophys. Acta* **54,** 543 (1961).
26. H. B. Higman and E. Bartels, *Biochim. Biophys. Acta* **57,** 77 (1962).
27. H. B. Higman, T. R. Podleski, and E. Bartels, *Biochim. Biophys. Acta* **75,** 187 (1963).
28. H. B. Higman, T. R. Podleski, and E. Bartels, *Biochim. Biophys. Acta* **79,** 138 (1964).
29. A. V. Hill, *in*: "Molecular Biology," p. 17, (D. Nachmansohn, ed.), Academic Press, New York (1960).
30. A. L. Hodgkin, *Biol. Rev. Cambridge Phil. Soc.* **26,** 338 (1951).
31. A. L. Hodgkin, "The Conduction of the Nervous Impulse," p. 108, Charles C. Thomas, Springfield, Ill. (1964).
32. J. V. Howarth, R. D. Keynes, and J. M. Ritchie, *J. Physiol.* **194,** 745 (1968).
33. A. Karlin, *Biochim. Biophys. Acta* **137,** 358 (1967).
34. A. Karlin, *J. Theoret. Biol.* **16,** 306 (1967).
35. A. Karlin and E. Bartels, *Biochim. Biophys. Acta* **126,** 525 (1966).
36. A. Karlin and M. Winnik, *Proc. Natl. Acad. Sci. U.S.* **60,** 668 (1968).

37. A. Katchalsky, *in*: "Connective Tissue: Intercellular Macromolecules," Little, Brown and Co., Boston (1964); *Biophys. J.* (*Suppl.*) **4**, p. 9.
38. E. P. Kennedy, *in*: "The Neurosciences," (G. C. Quarton, T. Melnechuk, and F. O. Schmitt, eds., p. 271), The Rockefeller University Press, New York (1967).
39. H. Kewitz, I. B. Wilson, and D. Nachmansohn, *Arch. Biochem. Biophys.* **64**, 456 (1956).
40. E. D. Korn, *Science* **153**, 1491 (1966).
41. W. Leuzinger and A. L. Baker, *Proc. Natl. Acad. Sci. U.S.* **57**, 446 (1967).
42. W. Leuzinger, A. L. Baker, and E. Cauvin, *Proc. Natl. Acad. Sci. U.S.* **59**, 620 (1968).
43. W. Leuzinger, M. Goldberg, and E. Cauvin, *J. Mol. Biol.* **40**, 217 (1969).
44. W. R. Loewenstein, (ed.), Biological membranes: recent progress, *Ann. N.Y. Acad. Sci.* **137**, 403 (1966).
45. S. J. Lundin and B. Hellstroem, *Z. Zellforsch.* **85**, 264 (1968).
46. A. Marnay, *Compt. Rend. Soc. Biol.* **126**, 573 (1937).
47. R. Martin and P. Rosenberg, *J. Cell Biol.* **36**, 341 (1968).
48. R. L. Masland and R. S. Wigton, *J. Neurophysiol.* **3**, 269 (1940).
49. J. Monod, J.-P. Changeux, and F. Jacob, *J. Mol. Biol.* **6**, 306 (1963).
50. D. Nachmansohn, "Chemical and Molecular Basis of Nerve Activity," Academic Press, New York (1959).
51. D. Nachmansohn, *in*: "Handbuch der Experimentellen Pharmakologie: Cholinesterases and Anticholinesterase Agents" (G. B. Koelle, ed.), Vol. 15, p. 701, Springer–Verlag, Heidelberg (1963).
52. D. Nachmansohn, *in*: "Handbuch der Experimentellen Pharmakologie: Cholinesterases and Anticholinesterase Agents" (G. B. Koelle, ed.), Vol. 15, p. 40, Springer-Verlag, Heidelberg (1963).
53. D. Nachmansohn, *in*: "Symposium on New Perspectives in Biology," BBA Library, Vol. 4 (M. Sela, ed.). p. 176, Elsevier, Amsterdam (1964).
54. D. Nachmansohn, *in*: "Nerve as a Tissue," (K. Rodahl, ed.), McGraw-Hill, New York, p. 141 (1966).
55. D. Nachmansohn, *in*: Biological membranes: recent progress, (W. Loewenstein, ed.), *Ann. N.Y. Acad. Sci.* **137**, 877 (1966).
56. D. Nachmansohn, *in*: "Current Aspects of Biochemical Energetics: Lipmann Dedicatory Volume," (N. O. Kaplan and E. P. Kennedy, eds.), p. 145, Academic Press, New York (1966).
57. D. Nachmansohn, *Bull. Soc. Chim. Biol. Paris* **10**, 1177 (1967).
58. D. Nachmansohn, *in*: "Handbook of Receptor Physiology. Vol. I. General Principles of Receptor Biophysics and Biochemistry," (W. R. Loewenstein, ed.), Springer-Verlag, Heidelberg, in press.
59. D. Nachmansohn, *in*: Proceedings of a symposium on membrane proteins, New York Heart Assoc., *J. Gen. Physiol.*, **54**, 187s (1969).
60. D. Nachmansohn and E. Lederer, *Bull. Soc. Chim. Biol. Paris* **21**, 797 (1939).
61. D. Nachmansohn and I. B. Wilson, *in*: "Advances in Enzymology," Vol. 12, (F. F. Nord, ed.), p. 259, Interscience, New York (1951).
62. H. B. Novikoff, N. Quintana, H. Villaverde, and R. Forschirm, *J. Cell Biol.* **29**, 525 (1966).
63. E. Overton, *Pflueg. Arch. Ges. Physiol.* **92**, 346 (1902).
64. G. E. Palade, *in*: "The Scientific Endeavor, Centennial Celebration of the National Academy of Sciences," p. 179, The Rockefeller Institute Press, New York, (1963).
65. E. Racker, "Mechanisms in Bioenergetics," Academic Press, New York (1965).
66. J. D. Robertson, *Prog. Biophys. Chem.* **10**, 343 (1960).

67. J. D. Robertson, "Symposium on Molecular Biology," (D. Nachmansohn, ed.), p. 87, Academic Press, New York (1960).
68. P. Rosenberg, *Toxicon* **3,** 125 (1965).
69. P. Rosenberg, *Mem. Inst. Butantan Simp. Internac.* **33,** 477 (1966).
70. P. Rosenberg and E. Condrea, *Biochem. Pharmacol.* **17,** 2033 (1968).
71. P. Rosenberg and F. C. G. Hoskin, *J. Gen. Physiol.* **46,** 1065 (1963).
72. M. A. Rothenberg and D. Nachmansohn, *J. Biol. Chem.* **168,** 223 (1947).
73. M. A. Rothenberg, D. B. Sprinson, and D. Nachmansohn, *J. Neurophysiol.* **11,** 111 (1948).
74. W. W. Schlaepfer and R. M. Torack, *J. Histochem. Cytochem.* **14,** 369 (1966).
75. E. Schoffeniels, *Biochim. Biophys. Acta* **26,** 585 (1957).
76. E. Schoffeniels, Thèse d'agrégation, Univ. de Liège, Liège (1959).
77. E. Schoffeniels and D. Nachmansohn, *Biochim. Biophys. Acta* **26,** 1 (1957).
78. S. J. Singer, *Adv. Protein Chem.* **22,** 1 (1967).
79. F. S. Sjoestrand, *J. Ultrastruct. Red.* **9,** 340, 561 (1963).
80. R. M. Torack and R. J. Barrnett, *Exp. Neurol.* **6,** 224 (1962).
81. R. R. Walsh and S. E. Deal, *Am. J. Physiol.* **197,** 547 (1959).
82. G. D. Webb, *Biochim. Biophys. Acta* **102,** 172 (1965).
83. G. Werner and A. S. Kuperman, *in:* "Handbuch der Experimentallen Pharmakologie: Cholinesterases and Anticholinesterase Agents" (G. B. Koelle, ed.), Vol. 15, p. 570, Springer-Verlag, Heidelberg (1963).
84. I. B. Wilson, *in:* "The Enzymes," (P. D. Boyer, H. Lardy, and K. Myrbaeck, eds.), Vol. 4, p. 501, Academic Press, New York (1960).
85. I. B. Wilson and S. Ginsburg, *Biochim. Biophys. Acta* **18,** 168 (1955).

Chapter 16

The Role of Hormones in Cerebral Protein Metabolism

Stanley E. Geel and Paola S. Timiras

*Department of Physiology–Anatomy**
University of California
Berkeley, California

INTRODUCTION

Hormones can be regarded as chemical mediators which, under the control of the central nervous system (CNS), elicit responses in specific tissues necessary for the homeostasis of the organism as a whole. The last few years have witnessed an intensive search for the molecular mechanisms underlying the physiological action of hormones. Because of the complexity of the multicellular organism, the hormonal responses occurring at specific cellular sites are far removed, both spatially and temporally, from the site of hormone synthesis and release. Attempts to define the primary biochemical locus of hormonal action are therefore fraught with difficulties. Karlson[1] and Zalokar[2] formulated the concept that certain hormones involved in the functional differentiation of tissues are capable of modifying gene expression. This theory, which is inherent in the "operon" model of Jacob and Monod,[3] has had a propelling influence on recent investigations into the mode of action of hormones in regulating protein biosynthesis; such studies are providing valuable insights into the genetic regulatory mechanisms of protein synthesis.

Despite striking advances in our knowledge of the regulatory role of hormones in the protein-synthetic activity of target cells, relatively little information is available on the hormonal role in protein synthesis of the CNS. Yet the requirement for protein in terms of function is perhaps more important in the brain than in any other organ. Studies of hormonal action in protein metabolism of the developing CNS may provide practical means

* Part of the work conducted in this laboratory was supported by USPHS Grant No. GM09267.

for correcting or averting mental debilities arising from various endocrinopathies. This review does not purport to cover the literature in its entirety, but rather to focus on the present status of our knowledge concerning the role of certain hormones in protein synthesis of the developing CNS.

THYROID HORMONES

Thyroid hormones are indispensable to normal growth and development in most vertebrates.[4] The growth-promoting properties of thyroid hormones are exhibited in several organs and systems of the body, including the brain. Rats thyroidectomized at birth manifest a significant retardation of whole-body growth, with concomitant changes in brain, although, as shown in Table I, the effects are less severe on this tissue.[5] The relationship between thyroid hormones and CNS development and function has long been realized. In animals, including man, congenital hypothyroidism or experimental removal of the thyroid at an early age has a profound influence on mental function and culminates in permanent cerebral dysfunction unless replacement therapy is initiated before a critical age.

The pioneering studies of Eayrs and co-workers[6] have demonstrated that the rat CNS is highly dependent upon thyroid hormones during the critical period of morphological development. The structural changes that occur in the cerebral cortex as a result of neonatal hypothyroidism include a decrease in cell size, a hypoplastic axonal network, a reduction in the

TABLE I

Effect of Neonatal Hypothyroidism and Thyroxine Therapy on Total Body and Brain Weights and Water Content of the Cerebral Cortex of 22-Day-Old Rats[a]

Experimental group	Cerebral cortex % water	Body wt. (g)	Brain wt. (g)	Brain wt./ body wt.
Control	81.9 ± 0.06	56.7 ± 1.6	1.45 ± 0.02	0.026 ± 0.0005
Hypothyroid[b]	82.8 ± 0.09	42.1 ± 0.8	1.36 ± 0.02	0.034 ± 0.0005
	($P < 0.001$)	($P < 0.001$)	($P < 0.001$)	($P < 0.001$)
Hypothyroid + thyroxine[c]	81.9 ± 0.09	55.8 ± 1.0	1.41 ± 0.01	0.025 ± 0.0004

[a] Values are means ±S.E.M. The P values refer to significant differences between experimental and control groups.

[b] Pregnant mothers maintained on a low iodine diet and offspring injected with 100 μCi ^{131}I at one day of age.

[c] 10 μg of L-thyroxine sodium/100 g body wt. administered daily from sixth postnatal day.

TABLE II

Effect of Neonatal Hypothyroidism and Thyroxine Therapy on AChE and ChE Activity per Unit DNA in the Brain of 22-Day-Old Rats[a]

Experimental group[b]	ChE Cerebral cortex[c]	AChE[c]	
		Cerebral cortex	Hypothalamus
Control	0.26 ± 0.012	2.47 ± 0.07	5.41 ± 0.14
Hypothyroid	0.19 ± 0.009	1.59 ± 0.05	4.29 ± 0.11
	($P < 0.001$)	($P < 0.001$)	($P < 0.001$)
Hypothyroid + thyroxine	0.25 ± 0.008	2.67 ± 0.10	6.16 ± 0.26

[a] Values are means ±S.E.M. The *P* values refer to significant differences between experimental and control groups.
[b] Experimental conditions are the same as in Table I.
[c] Values are μM substrate hydrolyzed × 10^{-3}/min/μg DNA.

extent of dendritic branching, and an altered vascular pattern. Other investigators using similar experimental conditions have found the cytodifferentiation of the cerebellum to be delayed and the dendritic tree of the Purkinje cells to be hypoplastic.[7,8] These effects indicate a clear relationship between thyroid hormones and protein synthesis and may underlie alterations observed in the electroencephalogram and evoked responses[9–11] as well as the retarded patterns of innate behavior and learning capacity[12,13] observed in cretinoid animals.

Experimental studies in amphibians have shown that thyroid-hormone implants have a profound effect upon the differentiation of certain discrete neuronal units, and provided evidence that thyroid hormones are implicated in *de novo* protein synthesis.[14]

Any alterations in the type or rate of RNA and protein synthesis would be expected to be reflected in the synthesis of specific protein enzymes, and it is here that the specificity of hormonal action has been clearly demonstrated.[15] Certain enzymes, such as cholinesterase (ChE), localized principally in glial cells,[16] and acetylcholinesterase (AChE), which is associated with central synaptic activity,[17] are distinctly related to morphological differentiation and functional adaptation and exhibit markedly increased activity during critical periods of brain maturation. Radiothyroidectomy of the rat at birth has been shown to alter the activity of these enzymes (Table II),[5] and early thyroid-hormone therapy restores enzymatic activity to normal. Hamburgh and Flexner[18] had previously shown a similar decline in total cholinesterase activity under similar experimental conditions. On the assumption that the acetylcholinesterase–acetylcholine system is largely associated with central synaptic activity,

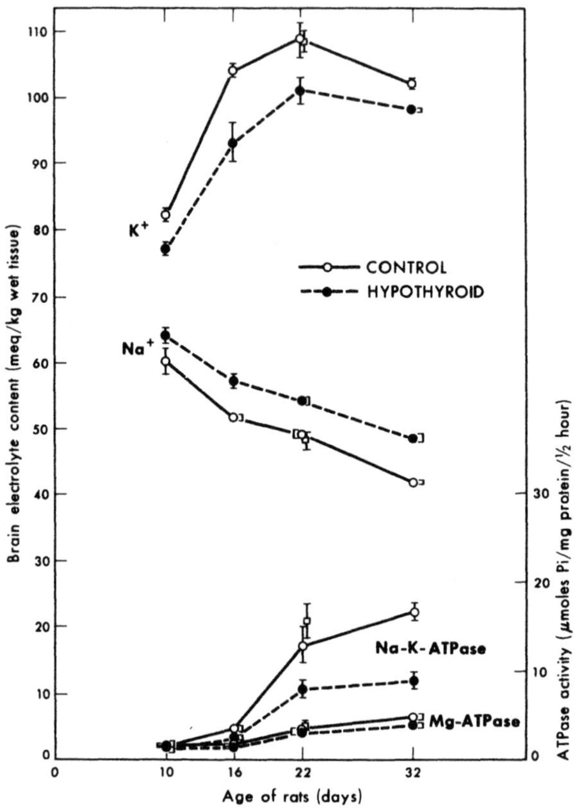

Fig. 1. Effects of neonatal hypothyroidism on Na^+ and K^+ contents, MgATPase and Na-K ATPase activity in the DOC-extractable fraction (pH 6.5) from the cerebral cortex of the rat during development. For ATPase activity each point represents mean ±S.E.M. of duplicate determinations from sixteen enzyme preparations, and for Na^+ and K^+ contents each point represents mean ±S.E.M. of eight or more samples. The squares in all parameters at day 22 represent mean values obtained from hypothyroid animals treated with thyroxine from the sixth to the 22nd day of age.

the above results, together with the anatomical findings of Eayrs,[19] form a reasonable basis on which to explain the altered learning ability of the hypothyroid rat.

Similar conclusions may be drawn from the effects of thyroid function on the Na^+- K^+-activated ATPase during CNS development. This enzyme, which is highly localized in the synaptosomal fraction of the brain,[20]

TABLE III

Effect of Neonatal Hypothyroidism and Thyroxine on Water and Electrolyte Contents in the Cerebral Cortex of the Developing Rat[a]

Experimental group[b]	Age (days)	% H_2O	Milliequivalents/kilogram wet tissue			
			Cl^-	Na^+	K^+	Mg^{++}
Control	10	85.37 ± 0.20	40.40 ± 0.61	59.87 ± 2.32	82.18 ± 0.81	13.05 ± 0.25
Hypothyroid	10	86.17 ± 0.18	42.95 ± 0.73	63.95 ± 1.59	77.05 ± 0.56	12.00 ± 0.23
		($P < 0.05$)	($P < 0.05$)	($P < 0.05$)	($P < 0.001$)	($P < 0.01$)
Control	16	82.24 ± 0.49	38.15 ± 1.96	51.13 ± 0.49	103.92 ± 1.21	13.78 ± 0.11
Hypothyroid	16	83.35 ± 0.38	45.39 ± 1.25	56.84 ± 1.08	92.83 ± 2.87	12.67 ± 0.37
		($P < 0.05$)	($P < 0.01$)	($P < 0.001$)	($P < 0.02$)	($P < 0.01$)
Control	22	81.90 ± 0.06	32.42 ± 0.48	49.33 ± 0.68	108.82 ± 2.39	14.75 ± 0.23
Hypothyroid	22	82.80 ± 0.09	34.28 ± 0.99	54.37 ± 0.75	100.71 ± 1.89	13.78 ± 0.22
		($P < 0.001$)	($P < 0.05$)	($P < 0.001$)	($P < 0.01$)	($P < 0.02$)
Hypothyroid + thyroxine	22	81.90 ± 0.09	32.00 ± 0.69	48.11 ± 1.45	109.41 ± 1.25	14.88 ± 0.19
Control	32	82.75 ± 0.96	28.02 ± 0.81	42.15 ± 0.35	102.23 ± 0.81	13.31 ± 0.12
Hypothyroid	32	82.90 ± 0.60	34.44 ± 0.94	48.39 ± 0.70	97.93 ± 0.26	12.89 ± 0.06
			($P < 0.001$)	($P < 0.001$)	($P < 0.001$)	($P < 0.01$)

[a] Values are means ±S.E.M. The P values refer to significant differences between experimental and control groups.
[b] Experimental conditions are the same as in Table I.

TABLE IV

ATPase Activity of Cerebral Microsomes in Control and Hypothyroid Rats at 22 Days of Age[a]

Group	μmoles Phosphate/mg microsomal protein/30 min		
	Total ATPase (*A*)	Mg-ATPase (*B*)	Na-K ATPase (*A* − *B*)
Control	46.03 ± 1.81	11.70 ± 0.90	34.56 ± 1.65
Hypothyroid	38.61 ± 1.94	11.25 ± 0.96	27.37 ± 1.43
	$P < 0.01$		$P < 0.01$

[a] Values are means of 13 enzyme preparations ±S.E.M.; the *P* values refer to significant differences between hypothyroid and control groups.

exhibits a rapid increase similar to that of the cholinesterases during the period of morphological growth of the brain.(21) Valcana and Timiras,(22) in a recent study, have reported that neonatal hypothyroidism in the rat significantly depresses the total specific activity of Na^+- K^+-activated ATPase (but not Mg^{++}-activated ATPase) in cerebral cortical tissue during brain development. This alteration in enzymatic activity is accompanied by a marked increase in cerebral Na^+ and a decrease in K^+ content (Fig. 1, Table III). The authors have interpreted the ionic changes as reflective of an intracellular change resulting from a depressed Na-K pump, a conclusion which confirms a hypothesis drawn from earlier studies.(23) The effects of neonatal hypothyroidism were also observed on Na^+-K^+-activated ATPase, associated with the heavy microsomal fraction in which the decrease was related to a reduced synthesis of the enzyme and/or its properties of association with cellular membranes (Table IV). None of these biochemical deviations were noted when thyroid hormone was administered to these animals from an early age.

Not only is it conceivable that the effects of thyroid hormones on ionic metabolism may underlie many of the physiological deviations attendant to hypothyroidism, but it is also possible to relate the ionic changes induced by a lack of thyroid hormones to protein biosynthesis. The characteristic sensitivity of the brain-protein synthesizing system to the ionic environment and energy-generating systems is well known and has been reviewed by Roberts and Zomzely.(24) The incorporation of amino acids into cerebral ribonucleoprotein is markedly dependent on the levels of Mg^{++}, K^+, and Na^+ in the medium and alterations in protein synthesis may be secondary to changes in ionic levels, their transport and membrane properties.

A number of other enzymes have been shown to be specifically dependent upon thyroid hormones during a critical period in CNS development. Thyroid-hormone deprivation in the rat significantly alters the brain levels of succinic dehydrogenase,[18] glutamate decarboxylase,[25] and GABA transaminase.[26] The degree to which enzyme activities are selectively dependent on thyroid function during brain development and maturation serves to emphasize the involvement of thyroid hormones in directing protein synthesis and/or their effects on the metabolic pathways and on the organelles with which these enzymes are associated.

The characteristic calorigenic effects that accompany the growth-promoting properties of thyroid hormones have led investigators in the past to identify the primary action of these hormones with processes of oxidative metabolism and mitochondrial phosphorylation. However, such effects have since been shown to be a response to an increased demand for energy as a result of the hormonal stimulation of protein synthesis. As in the testis and spleen, the mature brain possesses the unique feature of failing to respond to thyroid hormone with an increase in oxidative metabolism. In the developing brain, however, thyroxine has been found to stimulate oxygen consumption until adult respiratory rate is achieved.[7,27,28] The differential response between young and adult animals has not yet been adequately explained; however, it has stimulated further investigation of the regulatory role of thyroid hormones in cerebral protein synthesis. For example, thyroid hormone administered to intact adult rats fails to stimulate the incorporation of amino acids into protein in those tissues, including the brain, that do not respond to the hormone by an increased metabolic rate.[29] Recent studies have demonstrated that neonatal thyroidectomy depresses, whereas hormone therapy stimulates, protein synthesis in the immature brain, as measured by the incorporation of labeled leucine into cerebral cortical protein *in vivo* (Fig. 2).[23] Differences in the protein-stimulating effect of thyroid hormones in mature and immature brain preparations[30] have been ascribed to dissimilarities in the mitochondrial fractions of the respective tissues.[31] This interpretation has been questioned by Tata[32] in view of the catabolic levels of thyroxine administered.

A further unique feature of the brain is its almost exclusive dependence on blood glucose for aerobic metabolism. In addition, a considerable portion of the glucose metabolized is utilized for the production of amino acids. In the rat this process increases rapidly during a critical period of CNS development, with most of the glucose carbon appearing in the dicarboxylic amino acids.[33] Early postnatal thyroid hormone deprivation in this animal has been found to result in a decreased aerobic conversion of labeled glucose into amino acids, notably aspartic acid, in the cerebral cortex.[34] The

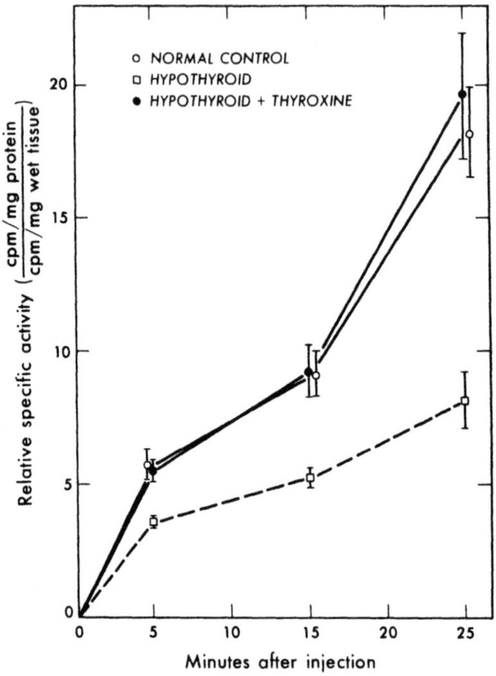

Fig. 2. Influence of neonatal hypothyroidism and thyroxine therapy on the incorporation of leucine into cerebral cortical protein in the rat at 22 days of age. Each animal was injected intravenously with 5.0 μCi *L*-(^{14}C)-leucine (0.75 μM) per 100 g body wt. Experimental conditions as for Table I.

whole brain content of N-acetyl-*L*-aspartic acid, which increases rapidly during rat brain development, is similarly depressed by propylthiouracil-induced hypothyroidism. Thyroid-hormone administration was capable of preventing this change.[35]

Tentative hypothesis can be formulated regarding the specificity of thyroid hormones in cerebral protein synthesis. It has already been speculated that hormones may exert their effects directly on target tissues or indirectly as a result of hormonal interaction. For example, changes in thyroid function are accompanied by alterations in pituitary growth-hormone content and secretion,[36–39] and the possibility that this hormone may contribute to alterations in cerebral protein synthesis cannot be excluded. The synergism between growth and thyroid hormones in promoting growth and development is well recognized.[40] Moreover, that their effects may be additive has been demonstrated in liver RNA syn-

TABLE V

Effect of Neonatal Hypothyroidism, Thyroxine, and Growth-Hormone Therapy on Nucleic Acid Content of Cerebral Cortical Fractions in the Rat at 25 Days of Age[a]

Experimental group	Fraction	RNA (μg/mg wet tissue)	DNA (μg/mg wet tissue)	RNA/DNA
Control	Nuclear	0.621 ± 0.015	0.934 ± 0.028	0.672 ± 0.009
	Cytoplasmic	1.319 ± 0.022	—	1.420 ± 0.068
Hypothyroid[b]	Nuclear	0.598 ± 0.014	1.074 ± 0.033 ($P < 0.01$)	0.569 ± 0.010 ($P < 0.001$)
	Cytoplasmic	1.210 ± 0.017 ($P < 0.01$)	—	1.165 ± 0.036 ($P < 0.01$)
Hypothyroid + thyroxine[c]	Nuclear	0.613 ± 0.027	0.969 ± 0.059	0.652 ± 0.095
	Cytoplasmic	1.337 ± 0.031	—	1.422 ± 0.080
Hypothyroid + growth hormone[d]	Nuclear	0.594 ± 0.017	1.058 ± 0.026 ($P < 0.01$)	0.585 ± 0.018 ($P < 0.001$)
	Cytoplasmic	1.191 ± 0.017 ($P < 0.01$)	—	1.131 ± 0.039 ($P < 0.01$)

[a] Values are means ±S.E.M. The P values refer to significant differences between experimental and control groups.
[b] Pregnant mothers maintained on a low iodine diet and female offspring injected with 100 μCi ^{131}I at one day of age.
[c] 5 μg *L*-thyroxine/100 g body wt. daily from sixth postnatal day.
[d] 50 μg bovine growth hormone/100 g body wt. daily from sixth postnatal day.

thesis.(41) Recently, studies were initiated to investigate the effects of thyroid hormones and of growth hormone on nucleic acid synthesis in order to further characterize the role of thyroid hormones in cerebral protein synthesis.(42) Neonatal hypothyroidism was found to produce a significant decrease in cerebral cortical nuclear and cytoplasmic RNA content (Table V), which confirms and extends earlier studies on RNA levels in unfractionated homogenates.(43, 44) Changes in RNA content were accompanied by a significant increase in DNA content, and these findings are also in agreement with previous results(43, 44) and support anatomical findings(45) demonstrating an increase in the packing density of cerebral cortical cells. Daily thyroxine therapy, initiated from the sixth postnatal day, was effective in preventing the alteration in nucleic acid levels, whereas growth hormone administered to hypothyroid rats was ineffective in this respect and, in addition, failed to correct any of the symptoms of cretinism. Even though the net RNA content of cerebral cortical fractions was re-

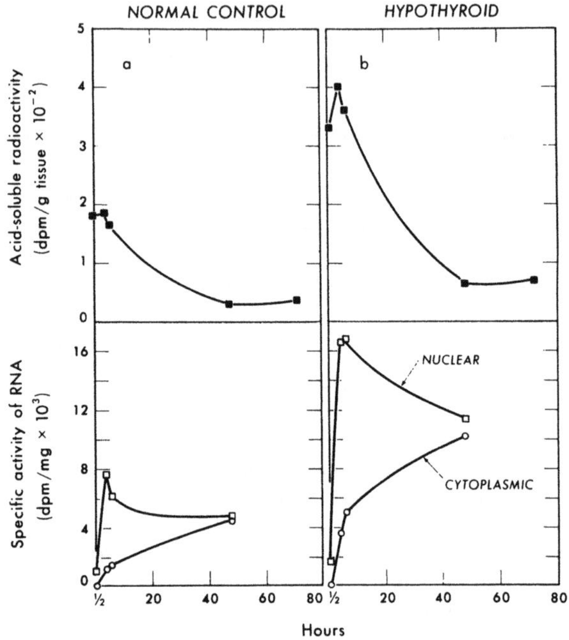

Fig. 3. Acid-soluble radioactivity (dark circles) and radioactivity incorporated into nuclear RNA (squares) and cytoplasmic RNA (open circles) following the subarachnoidal injection of (6-^{14}C)-orotic acid in hypothyroid and intact control rats at 25 days of age. Values are means of 2–3 experiments employing two animals per determination. Experimental conditions as for Table V.

duced by neonatal hypothyroidism, the rate of RNA synthesis appeared to be unaltered. The incorporation of a labeled pyrimidine precursor, (^{14}C)-orotic acid, into cerebral cortical RNA was followed at various time intervals after its injection into the subarachnoid space.[42] The time course of the incorporation of the label into RNA showed a similar pattern in intact controls and hypothyroid rats; however, the radioactivity associated with the RNA fractions was considerably greater in the case of the latter group (Fig. 3) and could not be accounted for by changes in the acid-soluble pool radioactivity. When results are expressed as a percentage of the total radioactivity incorporated, no essential difference in the rates of RNA synthesis between hypothyroid and control rats is evident, which is in agreement with recent studies conducted under similar experimental conditions.[25,46] The ratio of incorporation of the label into nuclear RNA to that for cytoplasmic RNA (Table VI) also indicates no difference in the

TABLE VI

Ratio of Incorporation of Radioactivity into Cerebral Cortical Nuclear RNA to That in Cytoplasmic RNA at Various Intervals Following Subarachnoidal Injection of (6-^{14}C)-Orotic Acid in the Rat at 25 Days of Age

Experimental group[a]	Hours post-injection				
	0.5	4	6	48	72
Control	22.0	5.6	3.5	1.4	0.9
Hypothyroid	23.3	5.2	3.9	1.3	0.9
Hypothyroid + thyroxine	—	5.1	3.6	1.2	1.1
Hypothyroid + growth hormone	21.0	6.0	3.7	1.2	1.0

[a] Experimental conditions are the same as in Table V.

rate of nuclear RNA synthesis and the appearance of newly synthesized RNA in the cytoplasm. It would seem reasonable to assume that the rate of RNA synthesis is the same in hypothyroid and control animals, but that the rate of "utilization" of RNA species is different between the two groups. On the other hand, it is also possible that the hormone affects the synthesis of pyrimidines or the pathway by which RNA precursors reach the nucleotide pool and, consequently, the size of the immediate precursor pool. It is difficult, however, to predict whether an effect on nucleotide metabolism would represent a primary effect of the hormone or merely a consequence of the increased stimulation of RNA synthesis. In this connection, supportive evidence exists that nucleotides may act as intermediaries in the stimulatory action of certain hormones on RNA and protein synthesis.(47)

All of the foregoing observations serve to substantiate the relationship between protein synthesis and the action of thyroid hormones on the developing CNS. A basically similar picture emerges from the extensive work of Tata and colleagues on the rat liver.(32) Tata's studies emphasize that the stimulatory action of thyroid hormones on cytoplasmic protein synthesis underlies the action of these hormones on growth and development.

A salient feature of all of the above studies is that the effects of thyroid-hormone deficiency can only be reversed when therapy is initiated before a critical stage of CNS maturation. The age-specificity of thyroid hormones on several tissues is well documented. However, several aspects of the thyroid–CNS relationship remain to be clarified. For example, it is important to determine whether all areas of the brain are equally affected by the hormones in terms of function and development, and whether the different cellular elements of the brain—the neurons and glial cells—respond differentially. The mental retardation characteristic of early hypothyroidism can be tentatively ascribed to defective cerebral protein synthesis, even though the studies reviewed above provide only speculative evidence regarding the precise locus or loci of hormonal action in the protein biosynthetic scheme.

Evidence for the role of other growth and developmental hormones in cerebral protein synthesis is fragmentary and largely indirect.

STEROID HORMONES

Various lines of evidence have implicated gonadal steroids in the modification of CNS activity and function.[48,49] Studies in the rat involving castration, ovarian grafts, and sex-hormone administration during critical periods of development have demonstrated that steroids have an organizing or differentiating influence (comparable to effects on genital duct differentiation) on CNS centers that mediate the control of gonadotrophin secretion and sexual behavior.[50] Testosterone administered to the female rat in the early neonatal period, or to the guinea pig or monkey prenatally, results in acyclic activity, sterility, and suppression of adult female mating behavior. Early postnatal castration of the male and consequent lack of testosterone results in the appearance of a female cyclic control of gonadotrophin secretion and an increase in female mating behavior.[51–54] The effects of electrical stimulation, ablation of certain discrete areas of the brain,[55] and local implantation of sex steroids[56, 57] have implicated defined areas of the hypothalamus in the control of gonadotrophin secretion and the mediation of sexual behavior. Recent studies involving the recording of seizure thresholds in the lateral and medial amygdala and in the hippocampus have shown that higher CNS centers are implicated in the regulation of gonadal cyclic activity in female rats.[58] In addition, these limbic areas play a significant role in regulating the onset of puberty.[59] These observations emphasize the specificity of hormonal action on discrete brain areas at critical stages of development.[60] Furthermore, specific areas of the brain selectively concentrate estradiol and the patterns of uptake and retention are similar to those of the classical target organs for this hormone.[61–64]

The manipulation of the hormonal environment in the early neonatal period affects brain excitability, as measured by the seizure response to electroshock treatment. Estradiol administered before eight days of age in the rat accelerates the development of the adult seizure pattern,[65] while castration of male rats at one day of age delays it.[66] The precocious stimulation of brain maturation by the neonatal administration of estradiol has been associated with an increased myelination of certain areas.[67,68] Additional biochemical studies are needed in order to elucidate the relationship of sex steroids to protein metabolism and the mechanisms by which sex steroids exert their effects on brain function.

Clinical observations of patients with adrenocortical insufficiency have indicated a relationship between adrenal steroids and brain function. The profound effects of adrenal steroids on CNS activity in rats and the possible metabolic events mediating these effects have been the subject of extensive reviews.[48,69] Most of the experimental evidence thus far accumulated is derived from studies employing adult animals, and consequently falls outside the context of this paper. It should be noted, however, that the brain of an adrenalectomized animal exhibits an altered electrolyte balance associated with increased excitability, decreased metabolic rate, and a decrease in total free amino acid concentration. It has been concluded that adrenocortical hormones, at least in the adult, exert a direct action on brain excitability in addition to a regulatory and a permissive action.[48]

Although relatively few such studies have been concerned with the developing CNS, there is evidence to indicate that adrenocortical hormones play a role in CNS development. The administration of cortisol to rats between the eight and fifteenth day after birth has been shown to advance the appearance of the maximal electroshock seizure pattern.[70] A single injection of cortisol immediately after birth suppresses body weight and delays the ontogeny of the spiny processes of the cerebral dendritic tree.[71] Corticosterone inhibits brain growth and depresses nucleic acid levels in mice treated from the second to seventh day after birth.[72] Parallel changes were not observed in the liver. The increase in cerebral sphingolipid and cerebroside content observed in the chick brain when cortisol is administered during embryonal development suggests that one of the mechanisms by which cortisol exerts its effects is by accelerating myelination.[73] It is relevant to note that *in vitro* studies have demonstrated that thyroid hormones have a direct effect on myelination in the developing CNS.[74] Similarly, studies on the effects of both cortisol and estradiol on nervous tissue *in vitro* indicate that these steroids are capable of regulating acetylcholinesterase activity and that this regulatory action varies according to the CNS area and its degree of maturation at the time of treatment.[75] While it is clear that steroids do participate in the control of CNS function (the physiological significance of which is clearly evident in the case of the sex steroids), the molecular basis of their stimulatory effects has not been elucidated.

The facets of steroid-hormone regulation of protein synthesis in extraneuronal tissues has been reviewed by Williams-Ashman.[76] The list of steroid hormones for which a stimulatory action on nucleic acid and protein synthesis has been demonstrated includes estrogen in the uterus,[77,78] androgens in male accessory glands,[79] and adrenal steroids in liver.[80,81]

GROWTH HORMONE

To what extent growth hormone contributes to growth during early development has been a subject of speculation for some time. The foundation of our present knowledge concerning the stimulatory effects of growth hormone on somatic growth in the rat has been established by the early studies of Evans and associates.[40,82–84] It has been shown that neonatally hypophysectomized rats continue to increase in size until 30 days of age, when a dramatic decline in body growth is observed.[40,84] Furthermore, growth hormone administered to normal but immature rats is without effect in promoting an increase in body size until 30 days of age. Plasma growth hormone has been reported to be markedly higher in children compared to adults; yet the higher levels do not appear to be related to differences in growth velocity.[85] Seckel[86] concluded that growth hormone plays a minor role, if any, in normal growth from the prenatal period to about two years of age in the human, and up to 30 days postnatally in the rat.

The requirement for growth hormone in brain development is equally contested. Hypophysectomy at an early age appears to influence skeletal development more markedly than brain growth. In fact, the chief cause of the high mortality in rats hypophysectomized at six days of age has been attributed to brain damage as a result of the disproportionate growth of the brain in relation to the cranium.[87] On the other hand, Zamenhof *et al.*[88] reported that the offspring of rats treated with high levels of growth hormone during pregnancy manifest an increase in brain DNA content and neuronal hyperplasia. Under essentially similar experimental conditions Clendinnen and Eayrs[89] had demonstrated an increase in length and number of cortical dendrites, which they interpreted as growth of an incremental nature rather than as cellular hyperplasia. An enhancement of cortical mediated behavior accompanied the structural changes induced by growth hormone. An improved learning capacity, based on the extinction of a conditioned avoidance response, has also been demonstrated in the offspring of rats receiving physiological levels of growth hormone during pregnancy.[90] None of these studies, however, takes into account that the hormonal effect may be secondary to a general protein anabolic effect on the maternal organism resulting in an improved nutritional status. A recent report, for example, shows no effect of neonatal hypophysectomy or of growth hormone therapy on cortical cell proliferation in rats.[91]

Inasmuch as changes in pituitary growth-hormone content and secretion are accompanied by alterations in thyroid function, one might expect the role of each or both of these hormones in cerebral development to assume significance. Yet, whereas growth hormone is capable of improving both body and brain weight in neonatally thyroidectomized rats, its effect in ameliorating the attendant changes in cerebral structure is limited, even when administered in high doses.[45] On the other hand, recent observations show that growth hormone is without effect on body growth and cerebral nucleic acid levels of immature hypothyroid rats.[42] It is conceivable that the disparities between the above studies may be related to differences in the level of hormone employed, its degree of purity, or the age at which the effects are measured.

In extraneuronal tissue the more obvious effects of growth hormone in the control of protein synthesis have been outlined by Korner.[92] Definitive studies are in progress in our laboratory to elucidate more clearly the regulatory influence of growth hormone in the promotion of brain development and cerebral protein synthesis.

CONCLUSIONS

In reviewing the available evidence for the regulatory role of hormones in cerebral protein synthesis a number of outstanding features are immediately apparent. One significant phenomonon is the existence of critical periods during which the CNS is differentially responsive to hormones capable of promoting growth and functional maturation.[60] Such critical periods necessarily vary according to species, functional area, and sequence of appearance of a particular adaptive response. In the rat, for example, the morphological and functional development of the cerebral cortex occurs during the early neonatal period,[93] and it is during this period that thyroid hormones exert their characteristic effects on CNS development. A similar period occurs prenatally in the guinea pig and during metamorphosis in amphibians. The appearance of the biochemical changes underlying hormonal responses in all cases supersedes the appearance of the hormone in the circulation. The events occurring during this latent period that lead to the expression of the tissue response require further investigation.

The idea that specific hormone receptor sites exist arose from observations that tissues react selectively to hormones. Thus, the interaction of a hormone with a receptor is thought to constitute the primary site of action which gives rise to a sequential series of biochemical events culminating in the physiological response. It is apparent that similar hormone-sensitive sites occur in the CNS.

Karlson's studies[1] on the effects of the insect hormone, ecdysone, on chromosomal morphology has focused attention on the possibility that hormonal interaction with specific components (repressors) of the gene-regulatory mechanism control RNA and protein synthesis. Such a hypothesis is an appealing one when applied to processes of growth and development. Thus, developmental hormones, by modifying the activity of specific genes, induce the synthesis of specific proteins involved in functional development and adaptation. However, in a recent review of the role of growth-promoting hormones in protein synthesis, Tata[94] emphasizes that recent developments do not allow one to accept the view that a control of gene transcription is the primary site of hormone action. Rather, it appears that polyribosomal formation and translation are under more direct hormonal control. The coordinated proliferation of ribosomes and membranes and the topographical segregation of protein-synthesizing units following stimulation by hormones may underlie hormonal-specific effects. According to Tata, it is possible that multiple sites of action are available to growth and developmental hormones in the control of protein biosynthesis of their respective target tissues.

Considerable advances have been made in recent years in defining the mechanism of action of hormones at the molecular level. Several lines of evidence indicate that the physiological effects of most hormones on extraneuronal tissues are mediated by an increase in RNA and protein synthesis. From the evidence reviewed a similar relationship exists between developmental hormones and CNS functional maturation. The heterogeneity of the brain—which includes differences in function and ontogenetic development of specific areas, in cytoarchitecture, and in cell types—presents obvious difficulties in the elucidation of the mechanisms of hormone action.

Great importance attaches to the understanding of the nature of the spatial-temporal sequences involved in brain development, the extent to which hormones participate in these events, and the primary site(s) of action. Such knowledge could lead to more effective prevention and correction of mental disturbances arising from endocrine imbalances.

REFERENCES

1. P. Karlson, *Perspectives Biol. Med.* **6**, 203–214 (1963).
2. M. Zalokar, *in*: "Control Mechanisms in Cellular Processes" (D. M. Bonner, ed.), pp. 87–140, Ronald Press, New York (1961).
3. F. Jacob and J. Monod, *J. Mol. Biol.* **3**, 318–356 (1961).

4. C. J. Shellabarger, *in*: "The Thyroid Gland" (R. Pitt-Rivers and W. R. Trotter, eds.), Vol. 1, pp. 187–198, Butterworth, Washington (1964).
5. S. E. Geel and P. S. Timiras, *Endocrinology* **80,** 1069–1074 (1967).
6. J. T. Eayrs, *Animal Behaviour* **7,** 1–17 (1959).
7. M. Hamburgh, E. Lynn, and E. P. Weiss, *Anat. Record* **150,** 147–161 (1964).
8. J. Legrand, *Arch. Anat. Microscop. Morphol. Exp.* **56,** 205–244 (1967).
9. P. B. Bradley, J. T. Eayrs, A. Glass, and R. W. Heath, *Electroencephalog. Clin. Neurophysiol.* **13,** 577–586 (1961).
10. R. Harris, M. Della Rovere, and P. F. Prior, *Arch. Disease Childhood* **40,** 612–617 (1965).
11. N. Hatotani and P. S. Timiras, *Neuroendocrinology* **2,** 147–156 (1967).
12. J. T. Eayrs and W. A. Lishman, *Brit. J. Animal Behaviour* **3,** 17–24 (1955).
13. J. T. Eayrs and S. Levine, *J. Endocrinol.* **25,** 505–513 (1963).
14. J. J. Kollros, *in*: "Growth of the Nervous System" Ciba Foundation Symposium, pp. 179–199, Little, Brown, Boston (1968).
15. E. C. Wolff and J. Wolff, *in*: "The Thyroid Gland" (R. Pitt-Rivers and W. R. Trotter, eds.), Vol. 1, pp. 237–282, Butterworth, Washington (1964).
16. E. Giacobini, *in*: "Morphological and Biochemical Correlates of Neural Activity" (M. M. Cohen and R. S. Snider, eds.), pp. 15–38, Harper and Row, New York (1964).
17. D. R. Curtis, R. W. Ryall, and J. C. Watkins, *in*: "Pharmacology of Cholinergic and Adrenergic Transmission" (G. B. Koelle, W. W. Douglas, and A. Carlsson, eds.), pp. 137–145, The Macmillan Co., New York (1965).
18. M. Hamburgh and L. B. Flexner, *J. Neurochem.* **1,** 279–288 (1957).
19. J. T. Eayrs, *Acta Anat.* **25,** 160–183 (1955).
20. A. A. Abdel-Latif, J. Brody, and A. Ramahi, *J. Neurochem.* **14,** 1133–1141 (1967).
21. F. E. Samson, Jr. and D. J. Quinn, *J. Neurochem.* **14,** 421–427 (1967).
22. T. Valcana and P. S. Timiras, *J. Neurochem.* **16,** 935–943 (1969).
23. S. E. Geel, T. Valcana, and P. S. Timiras, *Brain Res.* **4,** 143–150 (1967).
24. S. Roberts and C. E. Zomzely, *in*: "Protides of the Biological Fluids" (H. Peeters, ed.), Vol. 13, pp. 91–102. Elsevier Publishing Company, New York (1966).
25. R. Balázs, S. Kovacs, P. Teichgraber, W. Cocks, and J. T. Eayrs, *J. Neurochem.* **15,** 1335–1349 (1968).
26. C. A. García Argiz, J. M. Pasquini, B. Kaplún, and C. J. Goméz, *Brain Res.* **6,** 635–646 (1967).
27. J. F. Fazekas, F. B. Graves, and R. W. Alman, *Endocrinology* **48,** 169–174 (1951).
28. J. M. Reiss, M. Reiss, and A. F. Wyatt, *Proc. Soc. Exp. Biol. Med.* **93,** 19–22 (1951).
29. R. Michels, J. Cason, and L. Sokoloff, *Science* **140,** 1417–1418 (1963).
30. S. Gelber, P. L. Campbell, G. E. Deibler, and L. Sokoloff, *J. Neurochem.* **11,** 221–229 (1964).
31. C. B. Klee and L. Sokoloff, *J. Neurochem.* **11,** 709–716 (1964).
32. J. R. Tata, *in*: "Mechanisms of Hormone Action" (P. Karlson, ed.), pp. 173–184, Academic Press, New York (1965).
33. M. K. Gaitonde and D. Richter, *J. Neurochem.* **13,** 1309–1316 (1966).
34. C. J. Gómez and A. E. Ramirez de Guglielmone, *J. Neurochem.* **14,** 1119–1128 (1967).
35. E. Mussini, F. Marcucci, and S. Garattini, *J. Neurochem.* **14,** 551–554 (1967).
36. A. N. Contopoulos, M. E. Simpson, and A. A. Koneff, *Endocrinology* **63,** 642–653 (1958).
37. J. Solomon and R. O. Greep, *Endocrinology* **65,** 158–164 (1959).
38. R. A. Schooley, S. Friedkin, and E. S. Evans, *Endocrinology* **79,** 1053–1057 (1966).

39. H. Iwatsubo, K. Omori, Y. Okada, M. Fukuchi, K. Miyai, H. Abe, and Y. Kumahara, *J. Clin. Endocrinol. Metab.* **27,** 1751–1754 (1967).
40. M. E. Simpson, C. W. Asling, and H. M. Evans, *Yale J. Biol. Med.* **23,** 1–27 (1950).
41. C. C. Widnell and J. R. Tata, *Biochem. J.* **98,** 621–629 (1966).
42. S. E. Geel and P. S. Timiras, *in*: "Hormones in Development," Conference, Nottingham University, Nottingham, England, September, 1968 (M. Hamburgh and E. J. W. Barrington, eds.), National Foundation, in press.
43. S. E. Geel and P. S. Timiras, *Brain Res.* **4,** 135–142 (1967).
44. J. M. Pasquini, B. Kaplún, C. A. García Argiz, and C. J. Goméz, *Brain Res.* **6,** 621–634 (1967).
45. J. T. Eayrs, *Growth* **25,** 175–189 (1961).
46. R. Balázs, *in*: "Hormones in Development," Conference, Nottingham University, Nottingham, England, September, 1968 (M. Hamburgh and E. J. W. Barrington, eds.), National Foundation, in press.
47. O. Hechter, K. Yoshinaga, I. D. K. Halkerston, C. Cohn, and P. Dodd, *in*: "Molecular Basis of Some Aspects of Mental Activity" (O. Walaas, ed.), Vol. 1, pp. 291–341, Academic Press, New York (1966).
48. D. M. Woodbury and A. Vernadakis, *in*: "Methods in Hormone Research" (R. I. Dorfmann, ed.), Vol. 5, pp. 1–57, Academic Press, New York (1966).
49. A. Vernadakis and P. S. Timiras, *in*: "Proceedings of the Second International Congress of Hormonal Steroids," Milan, May, 1966, Excerpta Medica International Congress Series No. 125, p. 84.
50. G. W. Harris, *Endocrinology* **75,** 627–648 (1964).
51. G. W. Harris and S. Levine, *J. Physiol.* (*London*) **181,** 379–400 (1965).
52. R. A. Gorski and J. W. Wagner, *Endocrinology* **76,** 226–239 (1965).
53. H. H. Feder and R. E. Whalen, *Science* **147,** 306–307 (1965).
54. K. L. Grady, C. H. Phoenix, and W. C. Young, *J. Compar. Physiol. Psych.* **59,** 176–182 (1965).
55. C. A. Barraclough and R. A. Gorski, *Endocrinology* **68,** 68–79 (1961).
56. G. W. Harris and R. P. Michael, *J. Physiol.* (*London*) **171,** 275–301 (1964).
57. R. P. Michael, *Brit. Med. Bull.* **21,** 87–90 (1965).
58. E. Terasawa and P. S. Timiras, *Endocrinology* **83,** 207–216 (1968).
59. E. Terasawa and P. S. Timiras, *Am. J. Physiol.* **215,** 1462–1467 (1968).
60. P. S. Timiras, A. Vernadakis, and N. M. Sherwood, *in*: "Biology of Gestation" (N. S. Assali, ed.), Vol. 2, pp. 261–319, Academic Press, New York (1968).
61. J. Kato and C. A. Villee, *Endocrinology* **80,** 567–575 (1967).
62. A. J. Eisenfeld and J. Axelrod, *J. Pharmacol. Exp. Therap.* **150,** 469–475 (1965).
63. D. W. Pfaff, *Science* **161,** 1355–1356 (1968).
64. D. E. Woolley, C. F. Holinka, and P. S. Timiras, *Endocrinology* **84,** 157–161 (1969).
65. L. M. Heim and P. S. Timiras, *Endocrinology* **72,** 598–606 (1963).
66. G. L. Irvine, T. W. Ransom, W. H. Westbrook, and P. S. Timiras, *in*: "IInd International Congress on Hormonal Steroids," Milan, May, 1966, Excerpta Medica International Congress Series No. 111, Abstract No. 366, p. 211 (1966).
67. J. J. Curry and L. M. Heim, *Nature* **209,** 915–916 (1966).
68. R. Casper, A. Vernadakis, and P. S. Timiras, *Brain Res.* **5,** 524–526 (1967).
69. D. M. Woodbury, *Pharmacol. Rev.* **10,** 275–357 (1958).
70. A. Vernadakis and D. M. Woodbury, *J. Pharmacol. Exp. Therap.* **139,** 110–113 (1963).
71. S. Schapiro, *Gen. Comp. Endocrinol.* **10,** 214–228 (1968).
72. E. Howard, *J. Neurochem.* **12,** 181–191 (1965).

73. M. Granich and P. S. Timiras, *in*: "Hormones in Development," Conference, Nottingham University, Nottingham, England, September, 1968 (M. Hamburgh and E. J. W. Barrington, eds.), National Foundation, in press.
74. M. Hamburgh, *Develop. Biol.* **13,** 15–30 (1966).
75. A. Vernadakis and P. S. Timiras, *Experientia* **23,** 467 (1967).
76. H. G. Williams-Ashman, *Cancer Res.* **25,** 1096–1120 (1965).
77. G. C. Mueller, *in*: "Mechanisms of Hormone Action" (P. Karlson, ed.), pp. 228–239, Academic Press, New York (1965).
78. T. H. Hamilton, *Science* **161,** 649–661 (1968).
79. H. G. Williams-Ashman, *J. Cellular Comp. Physiol.* (*Suppl. 1*) **66,** 111–124 (1965).
80. P. Feigelson and M. Feigelson, *in*: "Mechanisms of Hormone Action" (P. Karlson, ed.), pp. 246–257, Academic Press, New York (1965).
81. F. T. Kenney, W. D. Wicks, and D. L. Greenman, *in*: Symposium on hormonal control of protein biosynthesis, *J. Cellular Comp. Physiol.* (*Suppl. 1*) **66,** 125–136 (1965).
82. R. D. Ray, M. E. Simpson, C. H. Li, C. W. Asling, and H. M. Evans, *Am. J. Anat.* **86,** 479–516 (1950).
83. D. G. Walker, M. E. Simpson, C. W. Asling, and H. M. Evans, *Anat. Record* **106,** 539–554 (1950).
84. D. G. Walker, C. W. Asling, M. E. Simpson, C. H. Li, and H. M. Evans, *Anat. Record* **114,** 19–47 (1952).
85. F. C. Greenwood, W. M. Hunter, and V. J. Marrian, *Brit. Med. J.* **1,** (5374) 25–26 (1964).
86. H. P. G. Seckel, *A.M.A. J. Diseases Children* **99,** 349–379 (1960).
87. C. W. Asling, D. G. Walker, M. E. Simpson, C. H. Li, and H. M. Evans, *Anat. Record* **114,** 49–65 (1952).
88. S. Zamenhof, J. Mosley, and E. Schuller, *Science* **152,** 1396–1397, (1966).
89. B. G. Clendinnen and J. T. Eayrs, *J. Endocrinol.* **22,** 183–193 (1961).
90. J. B. Block and W. B. Essman, *Nature* **205,** 1136–1137 (1965).
91. M. C. Diamond, *Brain Res.* **7,** 407–418 (1968).
92. A. Korner, *Recent Prog. Hormone Res.* **21,** 205–240 (1965).
93. W. A. Himwich, *Intern. Rev. Neurobiol.* **4,** 117–158 (1962).
94. J. R. Tata, *Nature* **219,** 331–337 (1968).

DISCUSSION

REISS: Investigations into the influence of hormones on brain metabolism have become very important during recent years. The elucidation of the hypothalamo-hypophyseal complex has helped us to understand why environmental and psychological variations can influence the pituitary and the glandular functions regulated by the pituitary. Changes in the peripheral glandular function can, in various ways, cause alterations in brain metabolism. There is ample evidence that various pituitary hormones can act directly upon brain metabolism, thus influencing brain functions, mentation, and behavior.

It is necessary to emphasize that the pituitary influences a multitude of enzyme systems in the brain, not only in the energy metabolism, but also in the protein metabolism, and that the nature of this influence can differ from that in other organs of the body. In our investigations on the hypophyseal effect upon energy metabolism it was found, for example, that after hypophysectomy animals showed little change in the oxygen consumption of the cortex, while that of the liver was decreased by 50%. The anaerobic glycolysis, however, increased by more than 100% in brain cortex, but very little in the liver.* The oxygen consumption in the brain of the hypophysectomized animal cannot be significantly altered by the application of growth hormone alone, but thyrotropic hormone or thyroxine can increase oxygen consumption and reduce the anaerobic glycolysis. Thyrotropic hormone is also the only substance that we have found to considerably increase the oxygen consumption of the brain of one to four day old rats. In the brain of adult intact rats it is inactive.†

After hypophysectomy the hexokinase activity and the P32 uptake of brain cortex is considerably increased.‡ Treatment with ACTH can reduce the increase of P32 in hypophysectomized brains. The increase in carbohydrate metabolism after hypophysectomy was thought to indicate that the brain was utilizing more sugar than it had at its disposal, thereby causing some psychopathological symptoms, such as the exhaustion observed in Sheenan's disease. In work concerning protein metabolism considerable increase of phenylalanine was found in the brain and serum of rats who had been kept on a phenylalanine-rich diet. After treatment with growth hormone, or anabolic steroids, this increase was significantly reduced.§ It was found that in six-week-old rats, after hypophysectomy, the incorporation of isotopically labeled phenylalanine into brain protein was diminished, but administration of growth hormone could cause a significant stimulation to near-normal levels,** demonstrating that the protein metabolism in the brain is not isolated by the blood-brain barrier from the stimulatory effect of growth hormone.

The possibility of using growth hormone in phenylketonuric patients should certainly be explored, particularly considering the extensive debates and general uncertainty at present about the advisability of treating phenylketonuric children with a phenylalanine-poor diet.

* M. Reiss, Hormones in mental disease, *in*: "Chemical pathology of the nervous system: Proceedings of the International Neurochemical Symposium, 3rd, Strasbourg, 1958," J. Folch-Pi, ed., Pergamon, New York, pp. 432–455 (1961).

† M. Reiss and L. D. Macleod, *Biochem. J.* **34**, 820 (1940). M. Reiss, F. E. Badrick, and J. M. Halkerston, *Biochem. J.* **44**, 257 (1949).

‡ M. Reiss and D. S. Rees, *Endocrinology* **41**, 437 (1947).

§ M. Reiss, M. B. Sideman, and E. S. Plichta, Influence of anabolic hormones on phenylalanine metabolism: II. Studies in animals, *J. Ment. Def. Res.* **10** (Part II), 130–140 (1966).

** S. Takahashi, N. W. Penn, A. Lajtha, and M. Reiss, Chapter 17, this volume.

Chapter 17

Influence of Growth Hormone on Phenylalanline Incorporation into Rat-Brain Protein

S. Takahashi and N. W. Penn

Willowbrook State School
Neuroendocrine Research Unit
Staten Island, New York

A. Lajtha

New York State Research Institute for Neurochemistry and Drug Addiction
Ward's Island, New York

and

M. Reiss

Willowbrook State School
Neuroendocrine Research Unit
Staten Island, New York

INTRODUCTION

Studies of the relation of endocrine glands to disturbed or retarded mentation were limited by the lack of suitable biochemical techniques and were consequently confined to behavioral observations. As in any other system in which the endocrine glands are studied, the biochemical status and competence of the target organs are of critical importance. In addition, the interrelationships among the endocrine glands themselves may alter the net metabolic effect on the receptor organs of the central nervous system, complicating any interpretation of behavioral response. During the last few years the methodology of brain chemistry has made such progress that, at least in animal experiments, one may begin to coordinate experimental results concerning endocrine function, brain metabolism, and behavior.

We have been concerned with possible treatment or amelioration of those forms of mental retardation that may be presumed to be a result of a biochemical lesion during development. Such a lesion may be genetic or may be due to infection or environmental deprivations. It has seemed possible that such aberrant patterns might in part be caused by a primary or secondary endocrine dysfunction at specific periods during the fetal stage and subsequent maturation.

Examples of hormonal action on brain metabolism and function are well established. The role of thyroid hormone in development of cretinism is known, and it has shown a possible approach to treatment of mental deficiency. The effect of hormones on the protein metabolism of the brain is discussed by Geel and Timiras,[1] and by Sokoloff.[1a]

We have studied the effect of anabolic steroids and growth hormone on retardates. In an attempt to obtain more detailed information on their effects and mode of action, these agents have also been employed in accompanying animal experiments.[2–4]

Growth hormone elicits a spectrum of effects on the major metabolic pathways of the cell. A variety of tissues has been examined, but there are few reports of the hormone's action on the adult or near-adult brain. Amino acid transport, protein synthesis, RNA synthesis, fat and carbohydrate metabolism, oxygen utilization, and electrolyte balance have been studied with respect to growth-hormone administration. It is of interest that certain biochemical functions show no response in the normal animal, but are markedly affected after hypophysectomy. Whether this is due to binding of the hormone with some receptor in the tissues, followed by gradual dissociation after hypophysectomy, or to the loss of unknown accessory factors on removal of the pituitary is a subject for future work.

In both hypophysectomized and nonhypophysectomized animals (rat, rabbit, dog) amino acid levels in plasma are significantly reduced by administration of growth hormone.[4–8] Isotopic studies indicate that in the normal rat administered growth hormone does not accelerate the subsequent incorporation of amino acid into protein.[9] The circulating levels of free fatty acids in the blood are increased shortly after growth hormone is given, in both the normal and the hypophysectomized animal.[10,11] In the intact animal it thus appears that certain actions of growth hormone associated with membrane transport may be directly stimulated by administration of the hormone.

In the *in vitro* studies of amino acid transport by the isolated rat diaphragm prior hypophysectomy is necessary for the appearance of such stimulatory effects of exogenous growth hormone, injected or added to the isolated tissue from hypophysectomized animals.[12–15]

Information on the mode of growth-hormone action has been obtained in studies of protein synthesis. The reduced incorporation of amino acid in cell-free systems from hypophysectomized rats is significantly increased by administration of growth hormone before the animals are killed.[16,17] The interrelated synthesis of RNA, and, in particular, messenger RNA, has been shown to correlate with the overall stimulation of protein synthesis by growth hormone.[18–23] It has been reported that prenatal treatment with growth hormone stimulates the proliferation of cortical neurons and improves maze performance in treated rat litters.[24]

The specific dynamic action of protein is abolished by hypophysectomy, and is restored by growth hormone.[25] It is of interest that hypophysectomy results in differential effects in liver and brain oxygen consumption. In the former, uptake is reduced about 45%, while anaerobic glycolysis rises only 20%. Oxygen utilization by cortex is barely affected, and anaerobic glycolysis is increased by over 100%.[26]

Growth hormone antagonizes the action of insulin, *in vivo* and *in vitro*. It seems to have all the qualities previously ascribed to a diabetogenic hormone. It can be shown to stimulate oxidation of fatty acids and to decrease the uptake and utilization of glucose in muscle.[27,28] In adipose tissue high levels of added growth hormone increase lipogenesis.[29,30] However, a depression of fatty acid synthesis from acetate has also been noted.[31] Growth hormone also influences the mineral metabolism in various ways.[32–43]

In previous studies it had been shown that growth hormone and anabolic steroids lower the elevated free-phenylalanine levels in the brain and liver tissue of phenylalanine-fed rats.[4] It thus appeared of interest to determine whether this reduction might be correlated with an accelerated incorporation into the brain protein of those animals in which cell division in the central nervous system had ceased.

METHODS AND MATERIALS

Animals

Male rats of the Sprague-Dawley strain, about 50 days old, weighing 140–160 gm, were divided into two groups. The rats of one group were hypophysectomized by the parapharyngeal method and the others were subjected to sham operation.

Growth Hormone and Radioactive Phenylalanine

Bovine growth hormone, distributed by the Endocrinology Study Section, N.I.H., Bethesda, Maryland, was used.

Rats of each group were injected subcutaneously with 0.5 mg of growth hormone in 0.25 ml daily for seven days, starting on the day after the operation. Control rats of each group were injected with 0.25 ml of 0.9% NaCl solution. They were fasted for 2 hr before administration of radioactive materials on the day they were killed. Each rat was injected intraperitoneally with the same amount of radioisotope (10 uCi).

Preparation of Samples

Ten, thirty, and sixty minutes after the injection of radioisotope, animals were decapitated. Blood was collected from the carotid artery in a heparinized dish, and plasma was immediately separated by centrifuging. A small piece of liver tissue was removed.

The brain was divided into two hemispheres by a sagittal incision and one of them was used for chemical analysis. All samples were promptly frozen in dry ice. When storage was required they were kept in a freezer at −20°C.

Determination of Amino Acid and Protein Activity

The frozen samples were weighed and homogenized with 2 ml of 3% perchloric acid (PCA) and centrifuged. A total of 0.5 ml of the supernatant fluid, the PCA-soluble fractions, was taken for determination of the activity of the intracellular free amino acids. The protein precipitates were then treated by standard methods.[44]

Radioactivity of the acid-soluble samples and of dried protein was determined in the liquid scintillation counter, using 500 mg of CAL-B-SIL gel powder to suspend the latter. Samples were counted to within 5% accuracy and corrected for quenching by the use of internal standards.

In order to estimate phenylalanine and tyrosine in the PCA-soluble fractions and dried protein fractions, fluorometric assays were employed.[45,46] Aliquots of the acid-soluble samples were used directly for the determination of phenylalanine and tyrosine. For protein content of these amino acids 5 mg of dried defatted protein was hydrolyzed with 0.5 ml of 6 *N* HCl in a sealed tube at 110°C for 24 hr. The hydrolyzate was dried and aliquots taken for fluorometric analysis and paper chromatography.

Paper Chromatography

A total of 0.8 ml of the PCA supernatant fluid was neutralized with 1*N* KOH and centrifuged. The supernatant solution was reserved. The

TABLE I

Effect of the Treatment of Growth Hormone (GH) and Hypophysectomy on Rat-Brain Weight

Treatment	Number of rats sacrificed	Total brain wet weight[a] (mg)	Total brain dry defatted weight[a] (mg)	% dry defatted weight
Sham operation, untreated	14	1655 ± 68	140.0 ± 17.4	8.5
Sham operation, treated with GH	14	1621 ± 91	132.1 ± 19.1	8.2
Hypophysectomy, untreated	16	1599 ± 106	144.5 ± 8.8	9.0
Hypophysectomy, treated with GH	16	1598 ± 122	147.1 ± 14.5	9.2

[a] Values are mean ± S.D.

residue was washed with 0.5 ml of cold water and the combined solutions were lyophilized. The resulting solid was dissolved in 0.1*N* HCl and aliquots were chromatographed in butanol–acetic acid–H_2O 4:1:5. The development was repeated for greater resolution. The regions of phenylalanine were counted as described for dried protein. A total of 0.2 ml of the hydrolyzates of dried protein were directly chromatographed. They were developed and counted in the same way.

Calculations

Although the same amount of radioisotope, 10 μCi, was injected in each rat, actual counts in free amino acid of blood showed large deviations among the four experimental groups because of the variation of the body weight and other effects of hypophysectomy. The comparison of experimental groups is therefore based on measurements of free phenylalanine in blood and tissue as well as the protein-bound phenylalanine.

RESULTS

Hypophysectomy alters neither the total brain weight in the six-week-old rat nor the percentage dry weight of protein (Table I). Differential results, therefore, cannot be ascribed to gross variations in brain protein content.

The incorporation of phenylalanine into brain and liver protein of sham-operated animals is unaffected by growth hormone. The protein-to-amino acid specific activity ratio in hypophysectomized animals seems to

TABLE II

Effect of Growth Hormone (GH) on Phenylalanine (Phe) Entry into Rat-Brain Acid-Soluble Fraction

Treatment	Time of sacrifice after injection (min)	Phe cpm/mg tissue	Phe μM/mg tissue $\times 10^{-5}$	SA_{Phe} [a] (cpm/μM)	$(SA_{Phe}^{B}/SA_{Phe}^{P})$ [b] $\times 10^{-3}$
Sham operated, untreated	10	5.22 ± 0.61	7.4 ± 0.70	85,300	415
	30	1.34 ± 0.07	8.7 ± 2.1	18,100	322
	60	0.45 ± 0.12	7.3 ± 1.1	5,800	966
Sham operated, treated with GH	10	4.81 ± 0.58	6.5 ± 0.50	74,000	418
	30	1.60 ± 0.46	7.4 ± 0.20	21,600	467
	60	0.51 ± 0.10	8.7 ± 0.50	6,090	985
Hypophysectomized, untreated	10	6.50 ± 1.06	7.0 ± 1.5	95,100	546
	30	4.00 ± 0.63	6.1 ± 1.5	69.100	1850
	60	1.45 ± 0.36	8.3 ± 0.90	17,400	748
Hypophysectomized, treated with GH	10	6.04 ± 1.12	8.1 ± 1.0	75,400	431
	30	4.10 ± 0.65	7.6 ± 1.4	40,200	1080
	60	1.02 ± 0.39	8.2 ± 0.70	10,600	676

[a] Specific activity phenylalanine.

[b] Ratio (specific activity phenylalanine, brain acid-soluble fraction)/(specific activity phenylalanine, plasma acid-soluble fraction).

TABLE III
Effect of Growth Hormone (GH) on Phenylalanine (Phe) Incorporation into Rat-Brain Protein

Treatment	Time of sacrifice after injection (min)	Phe cpm/mg	Phe μM/mg $\times 10^{-3}$	SA_{Phe} (cpm/μM)	$(SA_{Phe}^{Pr}/SA_{Phe}^{A})^a$ $\times 10^{-3}$
Sham operated, untreated	10	31 ± 3	262 ± 12	115	1.5
	30	78 ± 11	268 ± 12	288	15.9
	60	102 ± 16	243 ± 12	407	74.8
Sham operated, treated with GH	10	24 ± 2	252 ± 10	95	1.3
	30	77 ± 10	279 ± 11	276	13.5
	60	116 ± 22	250 ± 18	456	73.5
Hypophysectomized, untreated	10	26 ± 3	270 ± 18	98	1.0
	30	105 ± 2	257 ± 14	408	6.2
	60	185 ± 55	250 ± 14	739	45.4
Hypophysectomized, treated with GH	10	23 ± 4	272 ± 17	87	1.2
	30	123 ± 20	257 ± 16	481	11.8
	60	150 ± 26	242 ± 17	613	57.4

[a] Ratio (specific activity phenylalanine, protein fraction)/(specific activity phenylalanine, acid-soluble fraction).

be reduced in the untreated group and restored by hormone treatment. Brain and liver do not respond in quite the same manner: at shorter time (10 min) after injection of isotope there is little difference in incorporation into brain protein among the four groups, whereas in liver the difference between sham-operated and hypophysectomized groups is significant at the level P 0.01. The difference of values of specific activity ratios between the treated and untreated hypophysectomized groups increases by 30 min to a significant level, P 0.02 in both brain and liver. The restoration of activity in brain tissue is high: there is little difference between sham-operated and hypophysectomized treated groups at 30 and 60 min. This restoration is approximated in liver only at the 60-min time point. These data are summarized in Table III and V.

The specific activities of phenylalanine in protein at 30 and 60 min are higher in brain protein of the hypophysectomized group compared to the sham-operated animals. It is the accompanying high specific activity in the acid-soluble fraction (Table II) that causes the ratio of specific

TABLE IV

Effect of Growth Hormone (GH) on Phenylalanine (Phe) Entry into Rat-Liver Acid-Soluble Fractions

Treatment	Time of sacrifice after injection (min)	Phe cpm/mg tissue	Phe μM/mg tissue $\times 10^{-5}$	SA_{Phe} (cpm/μM)	$(SA_{Phe}{}^{L}/SA_{Phe}{}^{P})$[a] $\times 10^{-3}$
Sham operated, untreated	10	11.1 ± 1.11	14.3 ± 1.0	77,600	391
	30	0.94 ± 0.10	13.8 ± 2.0	7,280	129
	60	0.74 ± 0.43	14.2 ± 3.3	5,210	879
Sham operated, treated with GH	10	12.73 ± 2.86	11.9 ± 3.6	99,500	371
	30	0.92 ± 0.08	9.7 ± 0.2	9,410	142
	60	0.85 ± 0.16	12.2 ± 3.1	7,130	823
Hypophysectomized, untreated	10	27.73 ± 7.85	15.5 ± 2.4	174,500	957
	30	4.42 ± 0.59	12.6 ± 1.8	36,100	972
	60	1.72 ± 0.77	13.3 ± 2.7	14,180	686
Hypophysectomized, treated with GH	10	23.4 ± 7.10	12.4 ± 4.2	191,000	1390
	30	3.69 ± 2.20	11.5 ± 1.0	31,900	840
	60	1.42 ± 0.70	15.1 ± 2.2	9,100	497

[a] Ratio (specific activity phenylalanine, liver acid-soluble fraction)/(specific activity phenylalanine, plasma acid-soluble fraction).

TABLE V

Effect of Growth Hormone (GH) on Phenylalanine (Phe) Incorporation into Rat-Liver Protein

Treatment	Time of sacrifice after injection (min)	Phe cpm/mg protein	Phe μM/mg protein ×10⁻⁵	SA_{Phe} (cpm/μM)	$(SA_{Phe}{}^{LPr}/SA_{Phe}{}^{LA})^a$ ×10⁻³
Sham operation, untreated	10	395 ± 126	260 ± 16	1600	20.1
	30	607 ± 70	279 ± 7	2180	299
	60	478 ± 109	260 ± 18	1820	415
Sham operation, treated with GH	10	388 ± 61	265 ± 8	1460	16.3
	30	681 ± 82	278 ± 25	2440	259
	60	573 ± 77	253 ± 15	2240	370
Hypophysectomy, untreated	10	409 ± 74	281 ± 20	1310	8.4
	30	641 ± 179	278 ± 5	2310	65
	60	705 ± 183	262 ± 14	2670	216
Hypophysectomy, treated with GH	10	282 ± 66	274 ± 11	1020	5.6
	30	944 ± 175	282 ± 10	3380	146
	60	680 ± 162	256 ± 23	2640	316

[a] Ratio (specific activity phenylalanine, liver protein fraction)/(specific activity phenylalanine, liver acid-soluble fraction).

activities between protein and acid-soluble fractions to fall in the hypophysectomized animals. This pattern is not repeated in the data obtained from liver. At the 30-min point the specific activity of phenylalanine in protein is higher only in the hypophysectomized, treated group (Table V).

The ratios of protein to acid-soluble specific activity of phenylalanine indicate that at 10 and 30 min the inhibition of incorporation caused by hypophysectomy is greater in liver than in brain. At 60 min in the liver this ratio for the treated, hypophysectomized group has risen almost to the levels of the sham-operated controls. This degree of restoration does not appear in the corresponding data from brain.

It can be seen from Tables II and IV that the ratios of specific activities of phenylalanine in the cellular acid-soluble fractions to those in plasma generally show no difference between treated and untreated animals. It appears therefore that growth hormone does not act in any significant manner with respect to phenylalanine transport. Hypophysectomy itself does result in a high rate of phenylalanine entry into the liver, while no such effect can be seen in brain. Between 10 and 30 min, but not between 30 and 60 min, there is higher retention of the label in the free phenylalanine pool of both brain and liver in the hypophysectomized animals (Tables II and IV).

No consistent or significant differences due to growth hormone were found in the ratios of the specific activities of tyrosine to phenylalanine in acid-soluble fractions of either brain or liver. This problem is still under investigation.

DISCUSSION

The results are too preliminary in nature for a detailed analysis of the kinetics of protein metabolism under the experimental conditions. The analysis of considerably more time points, and of the compartments of the precursor, are planned for future study. A few tentative statements, however, can be made. The results indicate that protein metabolism of nondividing cells in the hypophysectomized rat brain is not isolated from the effects of hormones. It cannot be concluded from our data whether the observed changes are direct or indirect effects of the hormone. In short time periods in brain (Table II) the phenylalanine-specific activity in the acid-soluble fractions is greatest in the hypophysectomized, untreated animals, but it does not yield a correspondingly high level of incorporation. The specific activity of protein in these rats only approximates the values for the treated group. The ratios of specific activities of phenylalanine in protein to the corresponding values in acid-soluble fractions reveals that in hypophysectomized animals a stimulation of incorporation in hormone-treated animals does occur in shorter experimental periods (Table III).

In brain, hypophysectomy or hormone treatment does not seem to influence the entry of phenylalanine, while it increases in liver. The loss of label from the free pool is less in hypophysectomized animals in the 10–30-min period, and is somewhat increased after growth-hormone treatment (79 and 71% loss in sham-operated, 27% in hypophysectomized, and 47% in hypophysectomized and GH-treated). There is no difference in the 30–60-min period (68, 72, 75, and 74% losses, respectively).

Comparison of the estimated values for the rate of incorporation of phenylalanine into protein shows that the effect was minimal in the sham-operated animals. In the hypophysectomized animals (where incorporation was somewhat decreased in the long-term experiment) the effect of growth hormone was an increase initially toward restoration of normal rates; at longer times the rates in growth-hormone-treated hypophysectomized animals decreased again. This indicates that the action of the hormone may be different in different time periods, with high rates of formation followed by recovery periods.

Estimates of incorporation rates are more difficult in the liver, since maximal incorporation is already reached at short times. In this tissue too, however, hypophysectomy results in reduced rates of incorporation. These results confirm the observations of Korner, whose elegant studies[20] further explored the mechanism of effects of growth hormone in isolated systems.

In this paper only the effect of growth hormone on some metabolic events in the metabolism of hypophysectomized animals was investigated. It is not unexpected that with growth hormone alone a restitution of the brain metabolism to normal range is not achieved, since after hypophysectomy not only growth hormone but also TSH, ACTH, gonadotrophic hormone, and lactogenic hormone were removed. The results of treatment would have been much more complete if we would have investigated the other hormones or combinations of all the pituitary anterior lobe hormones.

The brief duration of these experiments (1 hr) indicates that growth hormone may influence protein synthesis rather than catabolic reactions. These observations suggest that in certain metabolic dysfunctions involving protein metabolism of the central nervous system (possibly phenylketonuria?) growth hormone may be considered as a possible therapeutic agent.

REFERENCES

1. S. E. Geel and P. S. Timiras, this volume, Chapter 16.
1a. L. Sokoloff, this volume, Chapter 18.
2. M. Reiss, T. Wakoh, J. C. Hillman, J. J. Pearse, J. M. Reiss, and N. Daley, *Am. J. Mental Deficiency* **70,** 4, 520 (1966).
3. M. Reiss, J. C. Hillman, J. Reiss, N. Daley, and S. Haylock, *J. Ment. Def. Res.* **10,** 116 (1966).
4. M. Reiss, M. B. Sideman, and E. S. Plichta, *J. Ment. Def. Res.* **10,** 130 (1966).
5. E. G. Frame and J. A. Russell, *Endocrinology* **39,** 420 (1946).
6. C. Griffen, J. M. Luck, V. Kulakoff, and M. Mills, *J. Biol. Chem.* **209,** 387 (1954).
7. J. A. Russell, *in*: "Hypophyseal Growth Hormone, Nature and Actions" (R. W. Smith, O. H. Gaebler, and C. N. H. Long, eds.), p. 213, McGraw-Hill, New York (1955).
8. W. D. Lotspeich, *J. Biol. Chem.* **185,** 221 (1950).
9. S. Takahashi, N. W. Penn, A. Lajtha, and M. Reiss, in preparation.
10. M. S. Raben and C. H. Hollenberg, *J. Clin. Invest.* **38,** 484 (1959).
11. H. M. Goodman and E. Knobil, *Endocrinology* **65,** 451 (1959).
12. A. Hjalmarson and K. Ahren, *Acta Endocr.* **54,** 645 (1967).
13. P. F. Brande and E. Knobil, *Proc. Soc. Exp. Biol. Med.* **110,** 5 (1962).
14. J. L. Kostyo, J. Hotchkiss, and E. Knobil, *Science* **130,** 1653 (1959).
15. J. L. Kostyo and F. L. Engel, *Endocrinology* **67,** 708 (1960).
16. A. Korner, *Biochem. J.* **73,** 61 (1959).

17. A. Korner, *Biochem. J.* **81,** 292 (1961).
18. G. P. Talwar, N. C. Panda, G. S. Sarin, and A. I. Tolani, *Biochem. J.* **82,** 173 (1962).
19. G. P. Talwar, S. L. Gupta, and F. Gros, *Biochem. J.* **91,** 565 (1964).
20. A. Korner, *Biochem. J.* **92,** 449 (1964).
21. A. Korner and A. E. Pegg, *Nature* **205,** 904 (1965).
22. K. G. Dawson, P. Patey, D. Rubinstein, and J. C. Beck, *Mol. Pharmacol.* **2,** 269 (1966).
23. C. D. Jackson and B. H. Sells, *Biochim. Biophys. Acta* **142,** 419 (1967).
24. S. Zamenhof, J. Mosley, and E. Schuller, *Science* **152,** 1396 (1966).
25. M. Reiss, *J. Endocr.* **2,** 329 (1940).
26. M. Reiss, *in*: "Chemical Pathology of the Nervous System" (J. Folch-Pi, ed.) p. 437, Pergamon, Oxford (1961).
27. J. C. Shipp, L. H. Opie, and D. Challoner, *Nature* **189,** 1018 (1961).
28. G. Bolodia and F. G. Young, *Nature* **215,** 960 (1967).
29. A. I. Winograd, W. N. Shaw, F. D. W. Lukens, and W. C. Stadie, *Am. J. Clin. Nutr.* **8,** 651 (1960).
30. A. I. Winograd, W. N. Shaw, F. D. W. Lukens, W. C. Stadie, and A. E. Renold, *J. Biol. Chem.* **234,** 1922 (1959).
31. R. D. Orth, W. D. Odell, and R. H. Williams, *Am. J. Physiol.* **198,** 640 (1960).
32. S. Ellis, J. Huble, and M. E. Simpson, *Proc. Soc. Exp. Biol. Med.* **84,** 603 (1953).
33. E. J. Collins and V. F. Baker, *Metabolism Clin. Exp.* **9,** 556 (1960).
34. C. W. Denko and D. M. Bergenstal, *Endocrinology* **57,** 76 (1955).
35. E. J. Collins, S. C. Lyster, and O. S. Carpenter, *Acta Endocr.* **36,** 51 (1961).
36. W. R. Murphy, W. H. Daughaday, and C. Hartnett, *J. Lab. Clin. Med.* **47,** 715 (1956).
37. P. H. Henneman, A. P. Forbes, M. Moldawer, E. F. Dempsey, and E. L. Carroll, *J. Clin. Invest.* **39,** 1223 (1960).
38. J. C. Beck, E. E. McGarry, I. Dyrenfurth, and E. H. Venning, *Ann. Internal Med.* **49,** 1090 (1958).
39. D. M. Bergenstal and M. B. Lipsett, *J. Clin. Endocr. Met.* **20,** 1427 (1960).
40. D. Ikkos, R. Luft, and C. A. Gemzell, *Acta Endocr.* **32,** 341 (1959).
41. D. M. Bergenstal, H. A. Lubs, L. F. Hallman, J. Patten, H. J. Levine, and C. H. Li, *J. Lab. Clin. Med.* **50,** 791 (1957).
42. H. Gershberg, *J. Clin. Endocr. Met.* **20,** 1107 (1960).
43. A. Korner, P. Randle, F. G. Young, A. C. Crooke, R. F. Fletcher, H. G. Sammons, R. Fraser, K. Ibbetson, I. D. F. Wooton, F. G. Prunty, R. R. McSwiney, R. Vaughn Jones, A. Stuart Mason, and C. E. King. *Lancet* **1,** 7 (1959).
44. A. Lajtha, S. Furst, and H. Waelsch, *Experientia* **13,** 168 (1957).
45. M. McCaman and E. Robins, *J. Lab. Clin. Med.* **59,** 885 (1962).
46. P. W. K. Wong, M. E. O'Flynn, and T. Inouye, *Clin. Chem.* **10,** 1098 (1964).
47. S. Lahiri and A. Lajtha, *J. Neurochem.* **11,** 77 (1964).

Chapter 18

The Mechanism of Action of Thyroid Hormones on Protein Synthesis and Its Relationship to the Differences in Sensitivities of Mature and Immature Brain

Louis Sokoloff

Laboratory of Cerebral Metabolism
National Institute of Mental Health
Bethesda, Maryland

INTRODUCTION

The thyroid hormones produce profound and diverse effects on the biochemical processes of most cells and tissues of the mammalian organism. These effects are nowhere more apparent then in the central nervous system, where the hormones play an essential role in the structural and functional maturation of the brain.(1) The immature brain responds to thyroid hormones not only with the characteristic augmentation of metabolic rate,(2) but also with major morphological transformations reminiscent of amphibian metamorphosis.(1) In contrast, once developed to maturity, the brain no longer exhibits any metabolic responsiveness to the thyroid hormones.(2–7)

STIMULATION OF PROTEIN SYNTHESIS BY THYROID HORMONES

Our laboratory has been engaged for the past several years in efforts to gain some insight into the biochemical basis of the lack of thyroxine-sensitivity of mature brain as contrasted with immature brain or other responsive tissues. It soon became apparent that no such insight was possible without prior understanding of the biochemical mechanism of action of the hormone. Studies in this direction were therefore undertaken, and these have led to the finding that thyroid hormones stimulate protein syn-

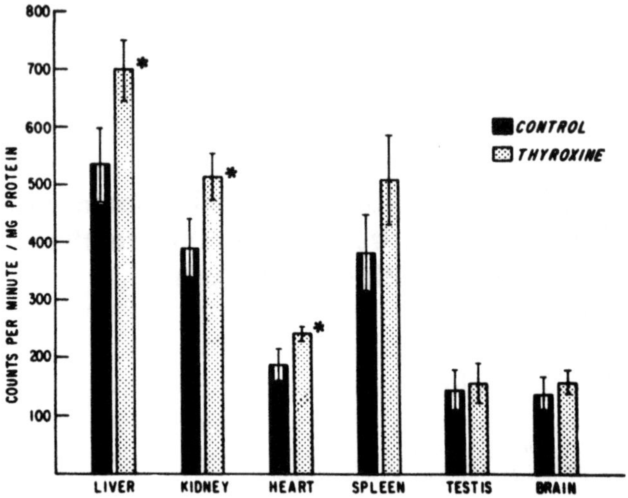

Fig. 1. Effects of thyroxine administration on ^{14}C-leucine incorporation into protein *in vivo* in various organs of the adult rat. Normal adult male albino rats were paired for age and weight. One of each pair was made hyperthyroid by the daily intraperitoneal injection of 0.5–1.0 ml of 0.01 *N* NaOH containing 100 μg of sodium *L*-thyroxine for 2–3 days; the other was given equivalent amounts of the NaOH solution alone. Protein synthesis *in vivo* was assayed simultaneously in both animals on the day following the last dose. A solution of *L*-leucine-1-^{14}C (1.25 μCi/ml in normal saline; spec. activity, 5.7 mCi/mM) was infused into the tail veins at a rate of 0.05 ml/min for 2 hr, at the end of which time the organs were removed, frozen in liquid nitrogen, and homogenized in 6% trichloroacetic acid. The precipitated protein was purified and assayed for specific activity by the method of Siekevitz.(20) There were no differences in the specific activities of the free precursor leucine pools to account for the differences in protein specific activity. (From Michels *et al.*(7))

thesis in sensitive tissues.(7–9) Furthermore, evidence was obtained to indicate that the stimulation of protein synthesis was not the consequence, but rather the prerequiste, of the increase in metabolic rate.(10,11)

It is beyond the scope of this presentation to describe again the numerous experiments which served to delineate, define, and characterize the effect on protein synthesis. For the purposes of the present discussion it is sufficient to raise and illustrate only several of the fundamental points. The stimulation of protein synthesis is observed *in vivo* in the intact animal (Fig. 1),(7) but only in tissues such as liver, kidney, and heart, which also respond to thyroid hormones with an increased oxygen consumption;(3) it

TABLE I

Effects of Hyperthyroidism and Hypothyroidism on ^{14}C-Leucine Incorporation into Protein in Cell-Free Rat-Liver Preparations[a]

Thyroid State	Protein specific activity (cpm/mg protein)		Effect of altered thyroid state	
	Control	Exper.	Δ (cpm/mg)	%
Hyperthyroid (8)	30 ± 3	41 ± 4	(+11 ± 4)[b]	+42
Hypothyroid (8)	29 ± 3	20 ± 3	(− 9 ± 4)[b]	−28

[a] From Sokoloff and Kaufman.(9) All values are means ± standard errors of the number of paired experiments indicated in parentheses. The "%" represents the mean of the individual per cent effects obtained in each experiment.

[b] Statistically significant effect; $P < 0.05$ as determined by the method of paired comparison.

is absent in known thyroxine-insensitive tissues, such as spleen, testis, and adult brain.[(3,7)] Like the effect on metabolic rate,[(2)] it is present *in vivo* in developing brain.[(12,13)]

The effect of thyroid-hormone administration *in vivo* persists in cell-free preparations from thyroxine-sensitive tissues and is mainfested as a stimulation of protein biosynthesis *in vitro* in cell-free systems containing such preparations (Table I).[(8,9)] Thyroidectomy is followed by a decrease of such activity (Table I),[(9)] which can be reversed by thyroid-hormone replacement therapy.[(14,15)]

In these experiments normal adult male albino rats were paired according to age and weight. In the studies of hyperthyroidism the experimental animal was given almost daily intraperitoneal injections of 100 μg of sodium *L*-thyroxine dissolved in 1 ml of 0.01 *N* NaOH for 6–16 days (mean, 10 days); the paired control received equivalent amounts of the NaOH solution alone. On the day following the last dose, cell-free liver preparations were prepared from both animals and assayed for protein-synthesizing activities simultaneously in parallel flasks. In the studies of hypothyroidism one rat of each pair was surgically throidectomized, and the paired control was sham-operated. Cell-free liver preparations were made and their protein-synthesizing activities assayed in paired flasks 28–41 days (mean, 32 days) after operation.

The assay system contained the following components (in μM): sucrose, 150; AMP, 5; potassium phosphate buffer (pH 7.4), 20; $MgCl_2$, 5; potassium α-ketoglutarate, 50; *DL*-leucine-1-^{14}C (spec. activity, 5.33 or 5.47 mCi/mM), 0.8. In addition, 0.45 ml of cell-free liver preparations containing mitochondria and microsomes equivalent to the yield from 200 mg

TABLE II

Effects of *L*-Thyroxine *in Vitro* on ^{14}C-Leucine Incorporation into Protein in Cell-Free Preparations from Normal Rat Liver[a]

L-Thyroxine concentration	Protein specific activity (cpm/mg protein)		*L*-Thyroxine effect	
	Control	*L*-Thyroxine	Δ (cpm/mg)[b]	%
1.3×10^{-7} *M* (10)	27.4 ± 2.7	28.5 ± 2.9	+ 1.1 ± 0.4	+ 4
6.5×10^{-7} *M* (12)	29.5 ± 2.3	30.8 ± 2.4	+ 1.3 ± 0.3	+ 5
1.3×10^{-6} *M* (16)	28.1 ± 1.9	30.5 ± 2.0	+ 2.4 ± 0.5	+ 9
1.3×10^{-5} *M* (22)	25.7 ± 1.7	30.0 ± 2.2	+ 4.3 ± 0.9	+17
6.5×10^{-5} *M* (9)	24.7 ± 3.6	34.8 ± 5.6	+10.1 ± 3.4	+42
1.3×10^{-4} *M* (13)	23.9 ± 2.1	38.3 ± 4.3	+14.4 ± 3.1	+61
3.9×10^{-4} *M* (9)	20.7 ± 2.1	36.6 ± 3.7	+15.9 ± 2.0	+77
6.5×10^{-4} *M* (5)	22.4 ± 1.5	8.4 ± 2.5	−14.0 ± 2.9	−62
1.3×10^{-3} *M* (6)	22.1 ± 1.3	3.1 ± 0.3	−19.0 ± 1.3	−85

[a] From Sokoloff and Kaufman.[9] All values presented are the means ± standard errors of the number of experiments indicated in parentheses. The "%" represents the mean of the per cent effects observed in the individual experiments.

[b] All the values in this column represent statistically significant effects; $P < 0.02$ by method of paired comparison.

and cell sap equivalent to the yield from 30 mg of fresh liver were added. The reaction mixture was brought to a final volume of 1.7 ml with water. Incubation was carried out at 37°C for 60 min. The preparation of the homogenates, incubation procedures, and the assay of the protein specific activity have been previously described.[9] The radioactivity in the labeled protein was measured by means of an end-window Geiger–Mueller tube with a counting efficiency of less than 5%, but sufficient counts were collected to achieve a coefficient of variation of less than 2%.

The addition of physiologically active thyroid hormones or their analogs directly *in vitro* to the same cell-free systems also stimulates protein synthesis.[8,9] Thyroxine, for example, stimulates in cell-free liver systems at concentrations as low as 10^{-7} *M*; graded increases in per cent stimulation are observed with progressively increasing hormone concentration until an optimum is reached at about 4×10^{-4} *M*, above which the effect on protein synthesis precipitously reverses from stimulation to inhibition (Table II; The assay conditions were the same as those described for Table I, except that the cell-free liver fractions were derived only from normal rats, and the thyroxine was added directly *in vitro*; incubation time at 37°C was 25 min).[9] The stimulation has been localized to the step in protein synthesis involving the transfer of tRNA-bound amino acid to microsomal

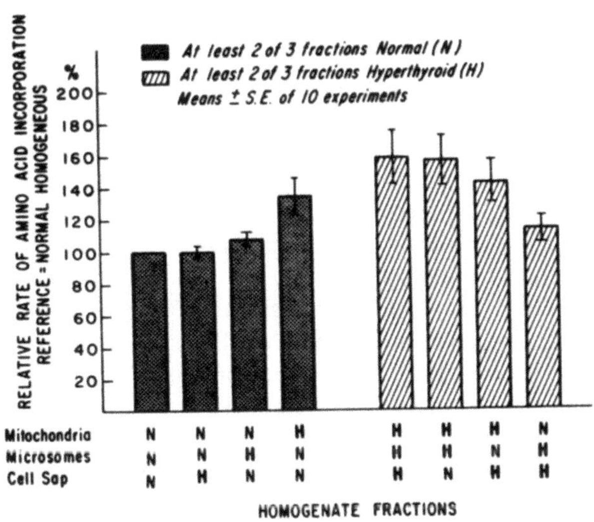

Fig. 2. Substitution experiments to determine the relative roles of the various subcellular components of liver homogenates in the stimulation of amino acid incorporation into protein by *L*-thyroxine administration *in vivo*. (From Sokoloff.[21])

or ribosomal protein,[16,17] and it reflects an enhanced activity of the ribosomal elements.[17] The thyroxine effect *in vitro* is not secondary to an effect on RNA synthesis;[18] indeed, it occurs in the absence of any new RNA synthesis[18] and is manifested on synthetic polyribonucleotide-directed amino acid incorporation into artificial polypeptides.[18] Recent studies with a rabbit reticulocyte lysate system synthesizing α and β chains of hemoglobin *in vitro* have clearly established that the thyroxine stimulation is a translational effect exerted at a late stage in the assembly of the protein molecule at the level of the elongation or completion of the nascent polypeptide chain.[19]

ROLE OF MITOCHONDRIA IN THE MECHANISM OF THE EFFECT

There is one aspect of the mechanism of the effect of thyroid hormones on protein synthesis which is of special relevance to their action in the central nervous system. When we first began these studies more than a dozen years ago, there had been very few cell-free protein-synthesis systems described in the literature. One of these was the liver system of Siekevitz,[20] which we adopted with relatively little modification and found to be quite suitable for detecting and studying the effects of thyroid hormones on protein synthesis.[8,9] Although it was mainly microsomal protein synthesis which was measured with this system, the system also contained mitochondria, which, of course, are not required for microsomal protein synthesis. This turned out to be a fortuitous circumstance in our studies. Figure 2 summarizes the results of ten paired experiments comparing the rates of protein synthesis in cell-free liver preparations from control euthyroid rats and experimental animals made hyperthyroid by thyroxine injections. Paired control and hyperthyroid rats were prepared as described for Table I, and mitochondria, microsomes, and cell sap were prepared from their livers by *Procedure B* of Sokoloff and Kaufman.[9] Amino acid incorporation into protein was assayed in the presence of all possible combinations of these three fractions derived from both types of animals. The components of the reaction mixtures and the assay conditions were the same as those described for Table I. These were standard type mixing experiments designed to identify the cell fractions responsible for the increased activity of the homogenates derived from the thyroxine-treated animals. The combination of cell fractions derived entirely from the hyperthyroid animals exhibited substantially greater activity than the corresponding mixture from the euthyroid animals. Each cell fraction of both the euthyroid and hyperthyroid mixtures was replaced, one at a time, by the corresponding cell fraction from the other type of animal. Replacement of the normal cell sap in the normal mixture with hyperthyroid cell sap made little difference in activity. Replacement of the normal microsomes with hyperthyroid microsomes caused a small but significant increase in activity, indicating that there was something different in hyperthyroid microsomes which caused increased rates of protein synthesis. The greatest difference occurred, however, when the normal mitochondria were replaced by hyperthyroid mitochondria; the activity was then increased almost to the level of the uniformly hyperthyroid mixture. Almost equivalent changes in the opposite direction were observed when the cell fractions of the hyperthyroid mixture were replaced with normal fractions; then, too, the largest change occurred with the replacement of the mitochondrial fraction. These experiments clearly demonstrated that the stimulation of protein synthesis by thyroid hormones *in vivo* was, to a small extent, the result of a change in the microsomes, presumably their ribosomes, but was mainly due to some change in the mitochondrial fraction of the cell.[9,21]

The role of mitochondria in the stimulation of protein synthesis by thyroid hormones was even more dramatically demonstrated in the completely *in vitro* system (Table III). The assay conditions for the complete system were the same as those described for Table II, except that 0.25 μM of GTP were included in the reaction mixture, and sodium *DL*-β-hydroxybutyrate was used as the oxidizable substrate in place of potassium α-keto-

TABLE III

Role of Mitochondria in the Stimulation of Microsomal Protein Synthesis by *L*-Thyroxine *in Vitro* in Cell-Free Liver Systems[a]

Protein-synthesis assay system and additions	Type of incubated mitochondrial supernatant fraction added to protein-synthesis assay system	Protein specific activity (cpm/mg protein)	Thyroxine effect	
			Δ (cpm/mg)	%
Complete (+ mitochondria, + β-hydroxybutyrate):				
Control	None	59	—	—
+ Thyroxine	None	100	+41	+69
Minus mitochondria, −β-hydroxybutyrate, + creatine phosphate, + creatine kinase,				
Control	None	49	—	—
+ Thyroxine	None	42	− 7	−14
+ Crude mitochondrial supernatant fraction	Control	26	—	—
	+ thyroxine	29	+ 3	+12
+ Boiled extract of mitochondrial supernatant fraction	Control	43	—	—
	+ thyroxine	61	+18	+42

[a] From Sokoloff.[21]

glutarate. In the assay system lacking mitochondria the mitochondria and β-hydroxybutyrate were replaced by 40 μM of creatine phosphate and 0.25 mg of creatine kinase contained in an equivalent volume. When added directly to the protein-synthesis assay system, the thyroxine concentration was 6.5×10^{-5} *M*. Incubation time at 37°C was 25 min. At the end of the incubation the reaction was terminated by the addition of 5 ml of ice-cold 0.25 *M* sucrose solution containing 1 mg of nonradioactive *DL*-leucine per milliliter. The mitochondria were then removed by centrifugation at 12,800*g* for 15 min, and the microsomal and cell-sap protein in the supernatant fraction was precipitated by the addition of an equal volume of 12% trichloroacetic acid. The precipitated protein was purified and assayed for specific radioactivity as previously described.[9] Radioactivity was measured by means of an end-window gas-flow Geiger–Mueller counter with an overall efficiency of approximately 17%. Sufficient counts were collected to achieve a coefficient of variation of less than 2%.

The mitochondrial supernatant solution added to the protein-synthesis assay system was prepared as follows: Mitochondria were incubated for 5 min at 37°C in the identical reaction mixture used for the complete protein-synthesis system, except that the GTP was absent, and the microsomes and cell sap were replaced by equivalent volumes of 0.25 *M* sucrose. The thyroxine concentration, when present, was $2.6 \times 10^{-4} M$. The incubation was terminated by chilling to 0°C in ice, and the mitochondria were then removed by centrifugation at 12,800g for 15 min. The aliquots of the supernatant fraction added to the protein-synthesis assay system were such that their thyroxine content was diluted to 6.5×10^{-5} *M*. Boiled extracts of the mitochondrial supernatant fraction were prepared by quick-boiling small aliquots of the crude fraction for 10 min, and then removing precipitated material by centrifugation at 100,000g for 1 hr.

The stimulation of amino acid incorporation into microsomal or ribosomal protein by thyroxine added directly to the cell-free system exhibited an absolute dependency on the presence of mitochondrial in the reaction mixture.[9,16,18,19] If the mitochondria were excluded from the reaction mixture and replaced by an alternative ATP generating system, the control rates of microsomal protein synthesis were the same, but there was then no stimulatioa by thyroxine (Table III).[9,16,18,19]

The increased rate of protein synthesis caused by thyroid hormones *in vitro* appears only after a lag of 5–7 min following the onset of incubation.[9,16] Studies of the events during this lag have established the occurrence of an energy-dependent interaction between the mitochondria and the thyroid hormone, leading to the formation of some product or change in the medium which is actually responsible for the stimulation of protein synthesis.[16,21] Some of the experimental evidence is represented in Table III. If mitochondria are incubated with thyroxine and an energy source for the duration of the lag and then removed by centrifugation, the addition of aliquots of the supernatant solution derived from the mitochondrial-thyroxine reaction mixture to a mitochondria-free protein-synthesizing system, in which thyroxine itself has no direct effect, results in a stimulation of protein synthesis. The stimulating activity is heat-stable, and a boiled extract of the supernatant solution, which apparently loses some inhibitory components, is even more effective (Table III).[21] The stimulating activity is associated, at least in part, if not entirely, with a small, dialyzable, heat-stable, acid labile molecule;[21,22] it does not appear to be GTP, GSH, ATP, Mg^{++}, or K^{+}, cofactors known to be required or influential in the process of polypeptide assembly.[16] The stimulation of protein synthesis by thyroid hormones *in vitro*, like the effect following their *in vivo* administration, is not a direct one on the protein biosynthetic system; it is a secondary consequence of a preceding reaction between the hormone and the mitochondrial fraction of the cell.

RELATIONSHIP BETWEEN EFFECTS ON PROTEIN AND RNA SYNTHESIS

Most hormones have now been found to cause increases in protein and RNA synthesis and to produce sometimes profound phenotypic modification and altered gene expression in their target organs. This has led to the widely accepted assumption, based mainly on premature and uncritical extrapolations, that the primary biochemical mechanism of action of all hormones is a direct transcriptional effect at the gene level.[23,24] This has not yet been clearly established for any hormone, and it is certainly not the case with the one hormone, epinephrine, the chemical action of which is presently best understood.[25]

The same issue has arisen in regard to the action of thyroid hormones. The studies described above indicated that the mitochondria were the primary locus of their biochemical action and that the stimulation of protein synthesis was cytoplasmic in origin. The *in vivo* experiments (Fig. 2) were carried out, however, at a time when it was not yet known that thyroid hormones stimulate protein synthesis, and rather large doses of thyroxine were used to minimize the possibility of missing the effect because of inadequate dosage. Tata and *et al.*[15] have questioned the significance of these *in vivo* experiments on the basis that the mitochondrial involvement in the effect on protein synthesis reflected not a physiological action of the hormone, but a toxic one due to the high doses. They found that the low rate of protein synthesis in thyroidectomized animals could be raised toward normal by a single small dose of hormone, and this effect could be observed *in vitro* in a cell-free liver system which did not contain mitochondria.[15] The increased rate of protein synthesis in this mitochondria-free, cell-free system did not appear until 30 hr after injection of the hormone. In the interim, increases in nuclear RNA turnover, nuclear RNA polymerase activity, and cytoplasmic ribosomal RNA contents occurred at approximately 4, 10, and 20 hr, respectively.[26] They concluded, therefore, that thyroid hormones act primarily in the nucleus to stimulate RNA synthesis and that the increased protein synthesis was secondary to an increased content of ribosomes and, perhaps, also messenger RNA in the cell.[26]

In our experience, however, the stimulation of protein synthesis occur-

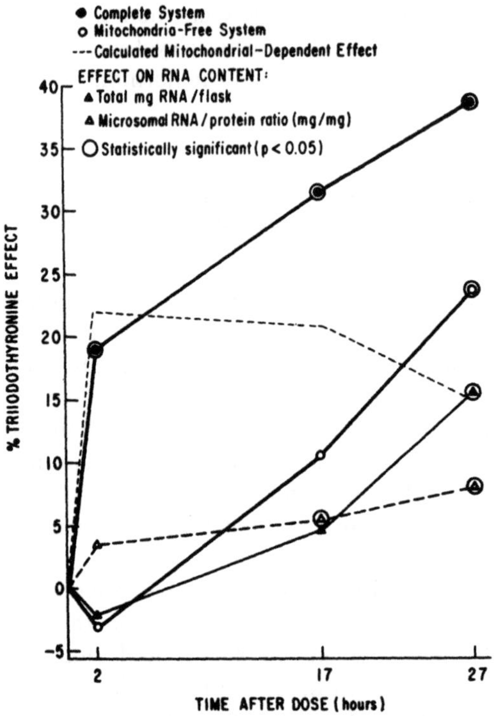

Fig. 3. Time courses of the initial mitochondria-dependent cytoplasmic stimulation of microsomal protein synthesis and the delayed, probably nuclear-mediated increases in cytoplasmic RNA and associated secondary changes in microsomal protein synthesis following a single small dose of triiodothyronine to euthyroid animals. (From Sokoloff *et al.*[21,27])

red much earlier than 30 hr after injection of the hormone. The assay systems used by Tata *et al.*[15] and by us[9] were generally quite comparable, except that ours contained mitochondria, and theirs did not. To resolve the discrepancy, we carried out the following experiments.[27] Normal adult male albino rats were paired for age and weight. One of each pair was given a single intraperitoneal injection of 60 μg of sodium triiodo-*L*-thyronine per 100 g of body weight contained in 0.01 *N* NaOH at a concentration of 75 μg/ml; the paired control animal received an equivalent volume of the NaOH solution alone. The animals were killed at the times indicated in Fig. 3, and mitochondria, microsomes, and cell sap were prepared from their livers as previously described.[27] In each experiment microsomal protein synthesis was assayed simultaneously in cell-free liver systems from both types of animals in the presence and the absence of mitochondria under assay conditions identical to those described for Table III. RNA contents were measured by the method of Fleck and Begg.[32] Each point in the figure represents the mean of 9–12 paired experiments. The dotted line (mitochondria-dependent stimulation of microsomal protein synthesis) was calculated by subtracting the per cent effect in the absence of mitochondria from the per cent effect in their presence; it is assumed that in the presence of mitochondria the total effect represents the sum of mitochondria- and nonmitochondria-dependent effects. The encircled points in the figure represent statistically significant changes ($P < 0.05$).

In the absence of mitochondria there was, indeed, as reported by Tata *et al.*,[15] a delay of approximately 27 hr before the appearance of any significant effect on protein synthesis (Fig. 3). Changes in the RNA contents occurred somewhat earlier, at about 17 hr (Fig. 3), suggesting that the increased rate of protein synthesis observed in the absence of mitochondria is secondary to changes in cytoplasmic RNA. However, when the mitochondria were present in the assay system, there was no lag in the stimulation of microsomal protein synthesis; the stimulation was observed within 2 hr after the injection of hormone, the earliest time studied, and already earlier than the changes in cytoplasmic RNA content (Fig. 3) or the reported increase in nuclear RNA turnover.[26]

There are then two separate and distinct mechanisms participating in the overall increase in the rate of protein synthesis following the administration of thyroid hormones *in vivo*. There is first an entirely cytoplasmic, mitochondria-dependent stimulation of protein synthesis which occurs with essentially no latent period and precedes any changes in cellular RNA. The time course of this early effect *in vivo* closely follows the level of the hormone in the tissues,[28] and, in fact, it is almost exactly analogous to the completely *in vitro* action of the hormone described above, which has been found to be a translational effect at a late stage in the assembly or completion of the nascent polypeptide chain.[19] *In vivo*, but not *in vitro*, this initial mitochondria-dependent effect is followed several hours later by an increase in cytoplasmic RNA and eventually a further increase in the rate of protein synthesis which is no longer dependent on the presence of mitochondria. The second delayed effect on protein synthesis is probably secondary to increased RNA synthesis in the nucleus and an elevation in the number of functional ribosomal units within the cell.[26] In contrast to the initial cytoplasmic effect, the later effects on protein synthesis and its antecedent changes in cellular RNA are completely prevented by mild fasting of the animal.[15,26] The time course and properties of the initial mitochondria-dependent effect on protein synthesis and the fact that it occurs not only *in vivo*, but can be reproduced by the hormone *in vitro*, suggest that it is closer to the true biochemical action of the thyroid hormones.[21,27] On the other hand, because the changes in RNA metabolism and their consequent additional effects on protein synthesis occur later and are manifested only *in vivo* in well-fed animals, we believe that they represent secondary biological responses or cellular adaptations to the consequences of the initial biochemical effects of the hormone.[21,27] In effect, a stimulation of the activity of the existing protein-synthesizing machinery of the cell is followed in nutritionally balanced animals by an adaptative increase in the amount of protein-synthetic machinery. Although secondary and adaptative, the changes in RNA metabolism may still comprise an important part of the overall constellation of effects of the hormone. They may, for example, be responsible for the obvious changes in gene expression manifested in amphibian metamorphosis or mammalian brain maturation. However, the initial biochemical event which triggers the entire sequence of events is a cytoplasmic action of the hormone, and the present evidence points to the mitochondria as the locus of that action.[21,27]

EFFECTS ON PROTEIN SYNTHESIS IN BRAIN

These studies of the mechanism of the effects of thyroid hormones in other tissues have led to better understanding of their anomalous actions

TABLE IV

Effects of *L*-Thyroxine *in Vitro* on ^{14}C-Leucine Incorporation into Protein in Cell-Free Preparations from Adult and Infant Rat Liver and Brain[a]

Tissue	Animal age	Protein specific activity (cpm/mg protein)		Thyroxine effect	
		Control	Thyroxine	Δ (cpm/mg)	%
Liver	Adult (3)	39	61	+22	+57
	Infant (2)	32	45	+13	+42
Brain	Adult (7)	12	9	− 3[b]	−21
	Infant (6)	44	50	+ 6[b]	+14

[a] From data of Gelber *et al.*[29] The values presented are the means of the results of the number of experiments indicated in the parentheses. The "%" represents the mean of the per cent effects obtained in the individual experiments.

[b] Statistically significant effects; $P < 0.05$ by method of paired comparison.

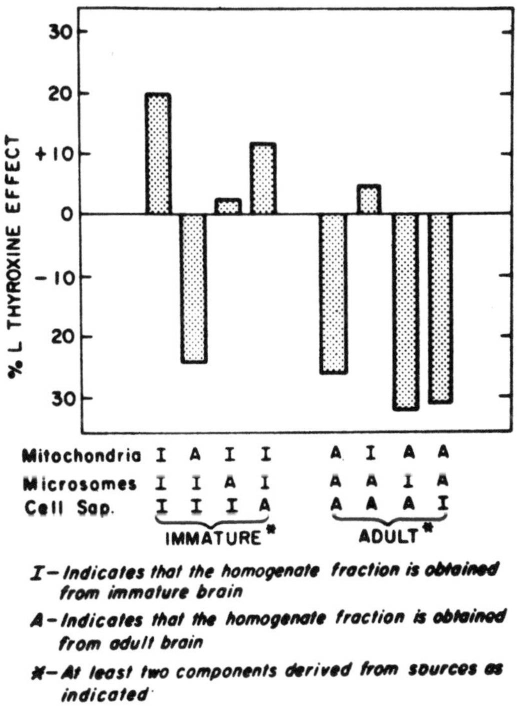

Fig. 4. Role of mitochondria in the difference in the effects of thyroxine on protein synthesis in mature and immature rat-brain cell-free preparations. Amino acid incorporation into protein was assayed exactly as described for Table IV, except that the mitochondria, microsomes, and cell sap were isolated separately and mixed together in the combinations indicated in the figure. The thyroxine concentration was 6.5×10^{-5} M. Infant and adult rats were 15–16 and 40–50 days old, respectively. (From Klee and Sokoloff.[30])

in the central nervous system. It will be recalled that the thyroid homones stimulate metabolic rate and induce maturation in developing mammalian brain, but have no comparable effects in the fully mature brain.[1–3] In parallel with these other effects, thyroid hormones stimulate protein synthesis in immature, but not in mature brain. These effects on protein synthesis are observed *in vivo*[7,12,13] as well as *in vitro*.[29,30,31] Table IV illustrates the effects *in vitro*. The assay conditions were the same as those described for the complete protein synthesis assay system in Table III.

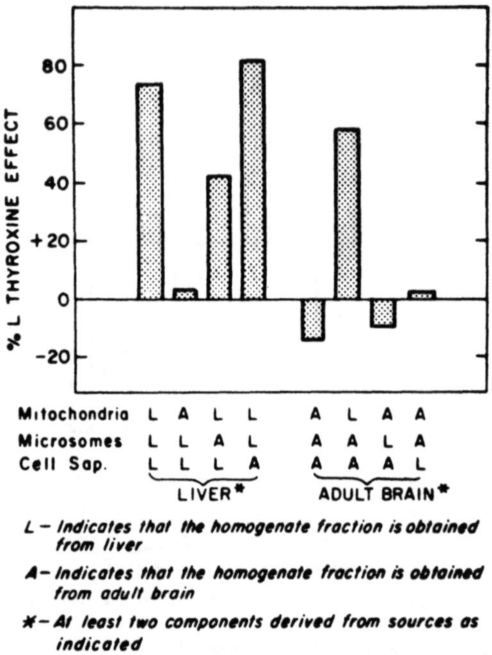

Fig. 5. Role of mitochondria in the difference in the effects of thyroxine on protein synthesis in mature rat liver and brain cell-free preparations. The assay conditions were the same as those described for Table IV, except that liver and brain cell fractions were isolated separately and mixed together as indicated in the figure. The thyroxine concentration was 6.5×10^{-5} *M*. The animals were 40–50 days old. (From Klee and Sokoloff.[30])

The liver and brain homogenates were prepared as previously described.[29] Adult rats were 40–50 days old; infant rats were 15–16 days old. Thyroxine concentration was 6.5×10^{-5} *M*. Incubation time at 37°C was 25 min. The rates of protein synthesis in cell-free preparations from the livers of infant and adult male rats are essentially the same, and both are equally well stimulated by thyroxine. In contrast, in comparable preparations from brain, protein synthesis is 3–4 times more rapid in the infant than in the adult rat and is stimulated by thyroxine, while the latter is not. The inhibition observed in the adult brain preparations in the experiments summarized in Table IV is of doubtful physiological significance; it is not observed *in vivo* (Fig. 1), and probably merely reflects the lact of a stimulation sufficiently great to overcome the effects of otherwise inhibitory conditions, such as Mg^{++}-binding by the hormone, in the *in vitro* assay.[9]

The difference in the responsivity of mature and immature brain is related to the fundamental mechanism of action of the thyroid hormones. Figure 4 illustrates the results of mixing experiments in which mitochondria, microsomes, and cell sap were isolated from both mature and immature brain, and the effect of thyroxine on protein synthesis was assayed in the presence of all combinations of these three cell fractions from both types of brain. The results indicate that the presence or absence of a thyroxine stimulation is entirely dependent on the nature of the mitochondrial fraction. Thyroxine stimulates protein synthesis in the presence of mitochondria from immature brain and fails to stimulate in the presence of adult brain mitochondria, regardless of the source of the microsomes and cell sap. The role of the mitochondria is even more clearly demonstrated by similar mixing experiments with cell fractions from adult brain and liver (Fig. 5). Furthermore, a boiled extract from the supernatant solution of an incubated liver mitochondrial-thyroxine reaction mixture, which stimulates protein synthesis in a mitochondria-free liver protein-synthesizing system (Table III), also does the same in similar adult-brain protein-synthesizing systems.

Apparently, then, immature-brain mitochondria, like those of liver and other thyroxine-sensitive tissues, contain a functional site which is the locus of the primary action of the thyroid hormones. Mature-brain mitochondria have lost this site and with it the ability to participate in the initial reaction with the hormone which triggers the chain of events characteristic of its physiological function. It is this change in the mitochondria which may be the basis of the difference in thyroid hormone sensitivity of mature brain as compared with immature brain and most other tissues.

REFERENCES

1. J. T. Eayrs, *Arch. de Biologie* (*Liège*) **75**, 529–565 (1964).
2. J. F. Fazekas, F. B. Graves, and R. W. Alman, *Endocrinology* **48**, 160–174 (1951).
3. E. S. Gordon and A. E. Heming, *Endocrinology* **34**, 353–360 (1944).
4. L. Sokoloff, R. L. Wechsler, R. Mangold, K. Balls, and S. S. Kety, *J. Clin. Invest.* **32**, 202–208 (1953).
5. P. Scheinberg, *J. Clin. Invest.* **29**, 1010–1013 (1950).
6. W. Sensenbach, L. Madison, S. Eisenberg, and L. Ochs, *J. Clin. Invest.* **33**, 1434–1440 (1954).
7. R. Michels, J. Cason, and L. Sokoloff, *Science* **140**, 1417–1418 (1963).
8. L. Sokoloff and S. Kaufman, *Science* **129**, 569–570 (1959).
9. L. Sokoloff and S. Kaufman, *J. Biol. Chem.* **236**, 795–803 (1961).
10. W. P. Weiss and L. Sokoloff, *Science* **140**, 1324–1326 (1963).

11. J. R. Tata, *Nature* **197,** 1167–1168 (1963).
12. L. Schneck, D. H. Ford, and R. Rhines, *Acta Neurol. Scand.* **40,** 285–290 (1965).
13. S. Geel, T. Valcana, and P. S. Timiras, *Brain Res.* **4,** 143–150 (1967).
14. O. Stein and J. Gross, *Proc. Soc. Exp. Biol. Med.* **109,** 817–820 (1962).
15. J. R. Tata, L. Ernster, O. Lindberg, E. Arrhenius, S. Pedersen, and R. Hedman, *Biochem. J.* **86,** 408–428 (1963).
16. L. Sokoloff, S. Kaufman, P. L. Campbell, C. M. Francis, and H. V. Gelboin, *J. Biol. Chem.* **238,** 1432–1437 (1963).
17. L. Sokoloff, P. L. Campbell, C. M. Francis, and C. B. Klee, *Biochim Biophys. Acta* **76,** 329–332 (1963).
18. L. Sokoloff, C. M. Francis, and P. L. Campbell, *Proc. Natl. Acad. Sci. U.S.* **52,** 728–736 (1964).
19. R. L. Krause and L. Sokoloff, *J. Biol. Chem.* **242,** 1431–1438 (1967).
20. P. Siekevitz, *J. Biol. Chem.* **195,** 549–565 (1952).
21. L. Sokoloff, *in;* "Regulatory Mechanisms for Protein Synthesis in Mammalian Cells" (A. Pietro, M. R. Lamborg, and F. T. Kenney, eds.), Proc. of Third Kettering Symposium, 1968, pp. 345–367, Academic Press, New York (1968).
22. L. Sokoloff, P. A. Roberts, and M. M. Januska, in preparation.
23. P. Karlson, *Perspectives Biol. Med.* **6,** 203–214 (1963).
24. A. Korner, *Acta Endocr.* (*Kbh.*) (*Suppl.* **100**) **1965,** 20–22.
25. E. W. Sutherland and G. A. Robison, *Pharmacol. Rev.* **18,** 145–161 (1966).
26. J. R. Tata and C. C. Widnell, *Biochem. J.* **98,** 604–620 (1966).
27. L. Sokoloff, P. A. Roberts, M. M. Januska, and J. E. Kline, *Proc. Natl. Acad. Sci. U.S.* **60,** 652–659 (1968).
28. N. F. Maclagan and J. H. Wilkinson, *J. Physiol.* **125,** 405–415 (1954).
29. S. Gelber, P. L. Campbell, G. E. Deibler, and L. Sokoloff, *J. Neurochem.* **11,** 221–229 (1964).
30. C. B. Klee and L. Sokoloff, *J. Neurochem.* **11,** 709–716 (1964).
31. L. Sokoloff, *Am. J. Dis. Children* **114,** 498–506 (1967).
32. A. Fleck and D. J. Begg, *Biochim. Biophys. Acta* **108,** 333–339 (1965).

Chapter 19

Glycoproteins in Neural Tissue

Eric G. Brunngraber

Illinois State Psychiatric Institute
Chicago, Illinois

INTRODUCTION

The acceleration in biochemical research that has occurred during the past decade has extended our knowledge concerning proteins that contain carbohydrates as a part of their molecular structure. An excellent review on glycoproteins has been published,[1] and a more recent review that summarizes progress in our knowledge of neural glycoproteins has appeared.[2] Glycoproteins are present in high concentrations in extracellular fluids and appear to form an important structural component of the plasma membranes of a wide variety of cell types. Apparently, these substances are synthesized within the cell in order to perform an extracellular function.[3] In some cases the secreted glycoprotein functions at locations that are remote from the site of synthesis. In other cases the secreted glycoprotein appears to function on the cell surface or in the intercellular space immediately adjacent to the parent cell, thereby playing a role in the relationship between the cell and its immediate environment. Proteolytic digestion of many purified glycoproteins releases glycopeptides the carbohydrate moieties of which generally consist of fucose, N-acetylneuraminic acid (NANA), galactose, mannose, glucosamine, and galactosamine. The purified glycopeptides are usually 60–80% carbohydrate, and in many cases may possess a relatively large molecular size. The glycopeptides from brain, for example, have molecular sizes that correspond to a molecular weight ranging from 9900 to 13,500 as judged by gel-filtration experiments using proteins and polypeptides as standards. To emphasize the predominately carbohydrate nature of these glycopeptides, they have been called "sialomucopolysaccharides".[2] It has been noted[4] that terminology has caused some confusion, and in order to distinguish these substances from the uronic acid-containing mucopolysaccharides ("glycosaminoglycuronoglycans"), we have adopted the more descriptive term *sialofucohexosaminoglycans* to identify these glycopeptides.

In addition to yielding sialofucohexosaminoglycans upon proteolytic digestion, many glycoproteins have yielded a glycopeptide that contains a shorter polysaccharide chain. The glycoproteins present in the bovine renal glomerular basement membrane[5] contains a disaccharide unit consisting of glucose and galactose, in addition to the sialofucohexosaminoglycan chains. Similarly, thyroglobulin[6] contains a smaller polysaccharide chain, consisting of five residues of mannose and one of glucosamine. A glycoprotein from the bovine aorta contains a smaller polysaccharide containing glucosamine and galactose.[7] As we shall see, brain glycoproteins also contain smaller polysaccharide chains.

N-ACETYLNEURAMINIC ACID (NANA) AND HEXOSAMINE CONTENT OF WHOLE RAT BRAIN

One gram of whole rat-brain tissue contains 1.2 mg of protein-bound carbohydrate and approximately 100 mg of protein. Since most of the well-characterized glycoproteins from various sources contain 10–33% carbohydrate, it can be calculated that 3.6 to 12% of the total brain protein may consist of glycoprotein. It was possible to obtain all of the glycopeptides from all of the glycoproteins of brain tissue by the following procedure.[8–13] Brain tissue was successively extracted with chloroform-methanol (2:1, v/v) and chloroform–methanol (1:2, v/v) in order to remove all of the gangliosides. The defatted residue was subjected to the proteolytic action of papain for 24 hr at 60°. The mixture was centrifuged and the residue was again subjected to the proteolytic action of fresh papain for an additional 24 hr. The precipitate obtained by centrifugation ("the papain-resistant fraction") contained small amounts of hexosamine and NANA. The combined supernatants obtained as a consequence of the double digestion procedure were dialyzed to yield the diffusible and nondiffusible sialofucohexosaminoglycans. The nondiffusible sialofucohexosaminoglycans were further purified by treatment with cetylpyridinium chloride, which removes by precipitation the nucleic acids and the uronic acid-containing mucopolysaccharides. The diffusible sialofucohexosaminoglycans were further purified by means of gel filtration on Sephadex G-15, which removes most of the amino acids and peptides and all of the free NANA. The diffusible sialofucohexosaminoglycans are excluded from the gel matrix. A balance sheet of the contribution of each of the fractions obtained to the total NANA and hexosamine recovered from whole brain tissue (Table I) revealed that the gangliosides

TABLE I

NANA- and Hexosamine-Containing Constituents of Rat Whole Brain

Constituent	NANA		Hexosamine	
	μM/g	%	μM/g	%
Gangliosides	2.22	67.5	0.96	25.2
Nondiffusible sialofucohexosaminoglycans	0.64	19.4	0.89	23.3
Diffusible sialofucohexosaminoglycans	0.23	7.1	0.96	25.2
Papain-resistant fraction	0.08	2.4	0.32	8.3
Free NANA	0.12	3.6	—	—
Free hexosamine	—	—	0.14	3.6
Uronic acid-containing mucopolysaccharides	—	0	0.55	14.4
TOTAL	3.29	—	3.82	—

and glycoproteins, respectively, account for 67 and 27% of the NANA and 25 and 49% of the hexosamine recovered from rat-brain tissue.

The carbohydrates derived from the glycoproteins present in 1 g of whole rat brain consists of 0.87 μM NANA, 0.44 μM of fucose, 2.71 μM of hexose, and 1.82 μM of hexosamine.

Characterization of the Nondiffusible and Diffusible Sialofucohexosaminoglycans Derived from Whole Rat-Brain Tissue

The relatively large molecular size of the nondiffusible sialofucohexosaminoglycans permitted the removal of most of the proteolytic digestion products by dialysis. Furthermore, the sialofucohexosaminoglycans are not precipitated by cetylpyridinium chloride, thereby permitting the removal of nucleic acids and uronic acid-containing mucopolysaccharides by precipitation with cetylpyridinium chloride. The nondiffusible sialofucohexosaminoglycans can be further purified by gel filtration on Sephadex G-75 or G-100,[11,14] provided water is used as eluting agent [Fig. 1(B)]. This removes impurities that absorb ultraviolet light to yield sialofucohexosaminoglycans that are 70–80% carbohydrate. In the presence of water, sialofucohexosaminoglycans aggregate and are largely excluded from the gel matrix. In the presence of buffer, sialofucohexosaminoglycans are

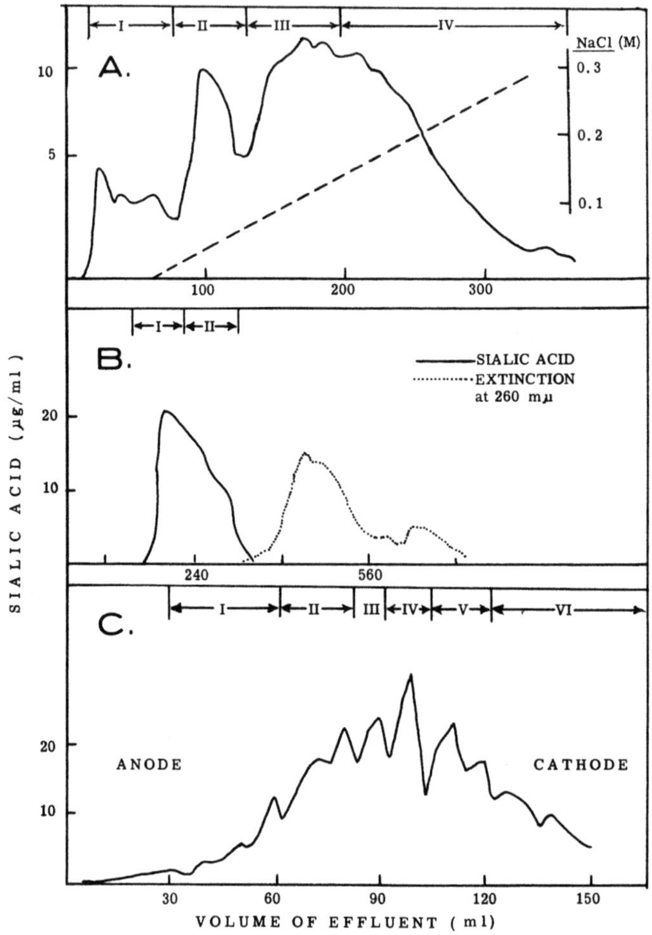

Fig. 1. Three methods that demonstrate the heterogeneity of the nondiffusible sialomucopolysaccharides isolated from brain glycoproteins. (A) Anion exchange chromatography on columns of DEAE-Sephadex A-50. (B) Gel filtration on a column of Sephadex G-200 using water as eluting agent. (C) Column electrophoresis on formalated cellulose in glycine–sodium hydroxide buffer, pH 10.3.

retarded during gel-filtration procedures and appear to possess a molecular size corresponding to a molecular weight of about 12,000. The degree of aggregation appears to be dependent upon the number of NANA residues present in the molecule. For example, analysis of the first ("high-molecular-weight") fraction and second ("low-molecular-weight") fraction to emerge from Sephadex G-200 columns using water as eluant [Fig. 1(B)] revealed that the former had a higher NANA content than the latter.

The heterogeneity of the nondiffusible sialofucohexosaminoglycans was revealed by subjecting these substances to anion-exchange chromatography on columns of Dowex-1-chloride[9] and DEAE-Sephadex[10,11] [Fig. 1 (A)], and to column electrophoresis [Fig. 1 (C)].[13] All fractions contain NANA, glucosamine, galactosamine, galactose, mannose, and fucose, but there is a variation in the molar ratios of these sugars in the different fractions. The NANA content of the sialofucohexosaminoglycans increases with the increase in electrophoretic mobility, affinity to the anion exchangers, and molecular size (Tables II and III).

After the separation of sialofucohexosaminoglycans by column electrophoresis the fractions were eluted from the column and analyzed for NANA. Inspection of the electrophoretic pattern obtained in four experiments suggested the presence of at least six major fractions. Contents of test tubes that contain the material corresponding to fractions I–VI [Fig. 1 (C)] were combined and analyzed (Table II). The ratio of NANA/fucose decreased from fraction I (highest negative charge) to fraction VI (lowest negative charge). This change is accompanied by a decrease in the ratio of NANA/hexose and NANA/hexosamine, but these changes are not as marked as the decrease observed in the ratio of NANA/fucose. On the other hand, the ratio of fucose/hexose and fucose/hexosamine increased from fraction I through VI. It is apparent that the marked decrease in the NANA/fucose ratio from fraction I through VI is a function of the decrease of NANA and an increase in fucose content relative to the hexose and hexosamine content of the fraction. The hexose/hexosamine ratio is the same for all fractions with the exception of fraction V. The data also indicates that NANA groups are not being replaced by fucose groups (i.e., the sum of fucose and NANA residues in the six fractions is not constant). These observations are consistent with the interpretation of results obtained earlier[11] by means of anion-exchange chromatography.

The carbohydrate composition of the total nondiffusible sialofucohexosaminoglycan preparation recovered from DEAE-Sephadex compares favorably with that recovered by column electrophoresis (Table III). Column electrophoresis appears to be effective in resolving fraction IV [Fig. 1(A); Table III] obtained by anion-exchange chromatography into three fractions [fractions I, II, and III in Fig. 1(C), Table III]. Both methods are effective in recovering a NANA-poor, fucose-rich (17–19% NANA and 10–11% fucose) component. This material represents 33 and 17%, respectively, of the total sialofucohexosaminoglycan preparation recovered in electrophoretic and anion-exchange fractionation procedures. In anion-exchange procedures a portion of this material is apparently not

TABLE II

Molar Ratios of Carbohydrate Constituents of the Nondiffusible Sialofucohexosaminoglycans Obtained by Column Electrophoresis[a]

Fraction	NANA: fucose	Fucose: hexose	NANA: hexose	Fucose: hexosamine	NANA: hexosamine	Hexose: hexosamine	Mannose: galactose	Glucosamine: galactosamine	Apparent mol. wt.
I	5.11	0.169	0.863	0.228	1.190	1.35	0.69	11	13,500
II	5.26	0.162	0.862	0.212	1.112	1.30	0.67	11	12,700
III	3.70	0.185	0.684	0.253	0.938	1.37	1.11	11	12,400
IV	3.18	0.215	0.685	0.272	0.864	1.28	1.42	11	11,200
V	2.11	0.216	0.450	0.398	0.837	1.89	1.05	8	10,500
VI	0.90	0.305	0.273	0.416	0.374	1.38	1.33	11	9,900
TOTAL	1.94	0.260	0.505	0.361	0.704	1.41	—	—	

[a] Analytical data obtained for fractions prepared by column electrophoresis [Fig. 1(C)]. The apparent molecular weight was estimated by determining the elution volume/void volume of the fraction by gel filtration on Sephadex G-50 using sucrose, cytochrome *c*, and bacitracin as standards.[13]

TABLE III

Carbohydrate Composition of Nondiffusible Sialofucohexosaminoglycans Obtained by Anion-Exchange Chromatography and Column Electrophoresis[a]

Experiment	Fraction	Total carbohydrate content (mg/g tissue)	Per cent of total	NANA (%)	Fucose (%)	Hexosamine (%)	Hexose (%)	NANA/fucose molar ratio
Anion exchange on DEAE-Sephadex	I	0.11	17	17	10	27	45	0.88
	II	0.10	16	26	8	26	40	1.70
	III	0.20	31	33	6	30	33	2.85
	IV	0.23	36	36	4	25	35	4.77
	TOTAL	0.64	100	30	6	27	37	—
Column electrophoresis	I	0.053	8	44	5	22	29	5.11
	II	0.093	15	43	4	23	30	5.26
	III	0.059	9	38	6	24	32	3.70
	IV	0.083	13	37	6	26	31	3.18
	V	0.139	22	30	8	22	40	2.11
	VI	0.207	33	19	11	29	41	0.90
	TOTAL	0.634	100	31	8	25	36	—

[a] Analytical data obtained for fractions prepared by anion-exchange chromatography on DEAE-Sephadex [Fig. 1(A)] and by column electrophoresis [Fig. 1(C)].

TABLE IV

Molar Ratios of Carbohydrate Constituents of the Diffusible Sialofucohexosaminoglycans and Sialohexosaminoglycan Obtained by Gel Filtration on Sephadex G-50[a]

Fraction	NANA: fucose	Fucose: hexose	NANA: hexose	Fucose: hexosamine	NANA: hexosamine	Hexose: hexosamine	Mannose: galactose	Glucosamine: galactosamine	Apparent mol. wt.
I	1.02	0.265	0.270	0.268	0.274	1.00	1.70	0.67	5900
II	1.16	0.097	0.114	0.142	0.166	1.46	5.76	0.43	3100
III	—	—	0.168	—	0.333	1.97	9.25	0.25	1350
TOTALS	1.01	0.139	0.140	0.231	0.234	1.67	—		

[a] Analytical data were obtained for fractions isolated by gel filtration on Sephadex G-50. The apparent molecular weight was estimated by determining the elution volume/void volume of the fraction by gel filtration on Sephadex G-50 using sucrose, bacitracin, and cytochrome c as standards. The molecular weight of cytochrome c and bacitracin is 12,400 and 1411, respectively.

TABLE V

Carbohydrate Composition of Diffusible Sialofucohexosaminoglycans and Sialohexosaminoglycan Obtained by Gel Filtration on Sephadex G-50[a]

Fraction	Total carbohydrate content (mg/g tissue)	Per cent of total	NANA (%)	Fucose (%)	Hexosamine (%)	Hexose (%)	NANA: fucose molar ratio
I	0.12	24	17	9	37	37	1.02
II	0.21	40	10	4	35	51	1.16
III	0.19	36	16	0	29	57	—
TOTAL	0.52	100	14	4	32	50	—

[a] Analytical data obtained for fractions prepared by gel filtration on Sephadex G-50.[13]

retained by the gel. The portion that is retained appears in fractions II and III [Fig. 1(A)]. The lower NANA content and the higher hexosamine content of fractions II and III obtained by anion-exchange chromatography as compared to fractions V and IV obtained by column electrophoresis suggests that this may be the case. It is clear that the percentage NANA of a fraction determines the sequence of elution on DEAE-Sephadex or the electrophoretic mobility. Other factors, such as the peptide portion of the molecule, may account for the differences in separation observed in the two procedures. Although it is not possible to ascertain the exact number of sialofucohexosaminoglycan components obtained by column electrophoresis or anion-exchange chromatography, the fractions obtained by either experiment are characterized by a sugar composition that is remarkably constant from experiment to experiment.

The diffusible sialofucohexosaminoglycans were separated into three fractions by gel filtration on columns of Sephadex G-50.[13] Analysis of the sugar components (Tables IV and V) revealed that one component lacked fucose. This component is therefore a sialohexosaminoglycan. The hexose/hexosamine ratios of fractions I and III (Table IV) differ markedly from those of the nondiffusible sialofucohexosaminoglycans. The major difference between the diffusible and nondiffusible sialofucohexosaminoglycans appears in the glucosamine/galactosamine and mannose/galactose ratios. The percentage of the total hexosamine that is glucosamine and the percentage of the total hexose that is galactose is much higher in the nondiffusible than in the diffusible sialofucohexosaminoglycans.

Gel filtration on Sephadex G-50 was utilized in order to determine the relative molecular sizes of the six fractions that constitute the nondiffusible sialofucohexosaminoglycans and the three fractions that constitute the diffusible sialofucohexosaminoglycans. The columns where standardized with sucrose, bacitracin, cytochrome c, and Dextran Blue. The molecular size of the six components that make up the nondiffusible sialofucohexosaminoglycans corresponded to molecular weights that ranged from 9900 to 13,500 (Table II), while the three components of the diffusible sialofucohexosaminoglycan preparation are characterized by molecular sizes corresponding to molecular weights of approximately 6000, 3000, and 1300.

SUBCELLULAR DISTRIBUTION OF GLYCOPROTEINS AND GANGLIOSIDES

The crude mitochondrial (*P*2) and microsomal (*P*3) preparations contain most of the gangliosidic NANA and glycoprotein-NANA present in the particulate matter recovered from rat-brain tissue (Table VI, Exp. 1).[9] Although mitochondrial preparations generally contain more protein-bound NANA and gangliosides than microsomal preparations, the concentration of glycoproteins and gangliosides (per milligram protein) is about two times greater in the microsomes.

The crude mitochondrial preparation was fractionated[15] into myelin-rich (*P2A*), synaptosome-rich (*P2B*), and mitochondria-rich (*P2C*) fractions by the discontinuous sucrose-density-gradient centrifugation procedure of Whittaker.[16] Analysis of the fractions (Table VI, Exp. 2) revealed that most of the glycoprotein and gangliosides present in the crude mitochondrial preparation are concentrated in the synaptosome-rich fraction, *P2B*. In general, the concentration of glycoprotein-NANA and gangliosidic NANA (per milligram protein) of the synaptosome-rich fraction is somewhat less, or equal to, that of the crude microsomal fraction.

The synaptosome-rich fraction (*P2B*) was used as the starting material for further fractionation[17] by centrifuging it through a discontinuous sucrose density gradient consisting of 0.32, 0.9, 1.0, 1.1, and 1.2 *M* sucrose to obtain five subfractions (*P2B-a*, *P2B-b*, *P2B-c*, *P2B-d*, and *P2B-e*). Analysis of the subfractions (Table VI, Exp. 3) revealed that the amount of glycoprotein-NANA (per milligram protein) was the same in all fractions, with the exception of *P2B-a*, which had a lower value. On the other hand, the amount of gangliosidic NANA (per milligram protein) decreased from a high value in the lightest fraction (*P2B-a*) to a low value for the pellet at the bottom of the tube (*P2B-e*). Myelin fragments, together with membranous fragments often seen in microsomal preparations, and a few axonal fragments were identified in *P2B-a*. Synaptosomes, many with enclosed

TABLE VI

Subcellular Distribution of Gangliosidic NANA and Glycoprotein-NANA

Exp.	Fraction	Per cent distribution		RSC		Ratio glycoprotein-NANA: gangliosidic NANA
		Gangliosidic NANA	Glycoprotein NANA	Gangliosidic NANA	Glycoprotein NANA	
1	*P*1	17	20	0.35	0.41	1.17
	*P*2	45	42	1.25	1.17	0.94
	*P*3	38	38	2.53	2.53	1.00
2	*P2A*	33	25	1.32	1.00	0.76
	P2B	63	67	1.34	1.43	1.07
	P2C	4	8	0.14	0.30	2.14
3	*P2B-a*	22	9	1.72	0.78	0.37
	P2B-b	18	13	1.54	1.10	0.60
	P2B-c	20	16	1.21	0.98	0.64
	P2B-d	19	21	0.88	0.98	0.91
	P2B-e	22	40	0.57	1.06	1.51
4	*P2A′*	16	8	0.72	0.35	0.49
	P2B′	23	12	2.36	1.25	0.53
	P2C′	31	31	1.54	1.51	0.98
	P2D′	17	20	0.92	1.05	1.25
	P2E′	13	30	0.44	1.01	2.30
5	*O*	1	10	0.31	0.37	1.19
	D	8	6	1.04	0.70	0.67
	E	22	16	3.62	2.65	0.73
	F	30	17	2.51	1.43	0.57
	G	19	28	1.53	2.14	1.40
	H	12	13	0.60	0.63	1.05
	I	7	11	0.44	1.03	2.34

[a] RSC is the per cent NANA recovered in fraction/per cent protein recovered in fraction. Description of experimental procedure and morphology of subcellular fractions is provided in the text.

mitochondria, were identified in *P2B-c*, *P2B-d*, and *P2B-e*. Free mitochondria were noted in *P2B-e*. Tubular forms, tentatively identified as axonal fragments, were noted in *P2B-b* and *P2B-c*, with lesser numbers in *P2B-a* and *P2B-d*. These experiments have revealed a difference in the subcellular distribution of gangliosides and protein-bound NANA. The amount of gangliosidic NANA (per milligram protein) is greater in those fractions that contain few or no synaptosomes, but which contain numerous tubular structures. The amount of glycoprotein-NANA (per milligram protein) is approximately the same in all fractions, suggesting that glycoproteins are associated with synaptosomes as well as with axonal fragments and other membranous fragments. As a consequence of the unequal distribution, the ratio of glycoprotein-NANA/gangliosidic NANA is smaller in the lighter, synaptosome-poor and axon-rich fractions as compared to the denser, synaptosome-rich fractions.

In another series of experiments the crude mitochondrial preparation from the cat cerebellum was fractionated[18,19] by utilizing the sucrose-density-gradient centrifugation procedure of De Robertis, *et al.*[20] The preparation was centrifuged through a sucrose density gradient consisting of 0.32, 0.8, 1.0, 1.2, and 1.3 *M* sucrose to obtain subfractions *P2A′*, *P2B′*, *P2C′*, *P2D′*, and *P2E′*. The decrease in the ratio of glycoprotein-NANA to gangliosidic NANA with increasing particle density (or size) was also observed in these experiments (Table VI, Exp. 4). Gangliosides were especially concentrated in *P2B′*, which consisted mainly of axonal fragments. Glycoproteins were most concentrated in *P2C*, which contains synaptosomes with attached axons.[18,19] A semiquantitative estimate of identifiable morphological units in these subfractions yielded the following results:

1. *P2A′* is composed predominately of myelin fragments, with some membrane fragments that appear to be of microsomal origin.
2. *P2B′* consists predominately of axonal fragments, with some membrane fragments.
3. *P2C′* consists mainly of axons with attached synaptic bulbs, a considerable number of mitochondria, and a few glial fragments.
4. *P2D′* consists largely of mitochondria, with a considerable number of axons with attached synaptic bulbs and some glial fragments.
5. *P2E′* consists mostly of mitochondria, with occasional axons with synaptic bulbs and some glial fragments.

In another series of experiments the effect of osmotic shock on the distribution of gangliosides and glycoproteins was studied using the technique described by Whittaker *et al.*[21] The osmotically shocked mitochondrial preparation (*P*2) was centrifuged at 10,000g for 20 min to remove the larger mitochondria and myelin fragments. The cloudy supernatant was subjected to centrifugation through a discontinuous sucrose density gradient (0, 0.4, 0.6, 0.8, 1.0, and 1.2 *M* sucrose) in order to obtain subfractions *O*, *D*, *E*, *F*, *G*, *H*, and *I*. Analysis of the subfractions (Table VI, Exp. 5) showed that the denser subfractions are characterized by a high glycoprotein-NANA to gangliosidic NANA ratio. These are the fractions that contain numerous synaptosome ghosts and shrunken or damaged synaptosomes. A high concentration of glycoprotein appears in fraction *E*, probably due to the presence of membranous material derived from disrupted axons, dendrites, or plasma membranes. The glycoprotein-NANA/gangliosidic NANA ratio for this fraction is low, since this fraction is also especially enriched in gangliosidic material.

Subcellular entities observed in these subfractions were as follows[21]: *O* consists largely of soluble material; *D* contains mainly synaptic vesicles and some microsomes; *E* contains microsomes, some synaptic vesicles, and occasional myelin fragments; *F* contains largely synaptosome ghosts, myelin fragments, and nonvesicular membrane fragments; *G* contains synaptic ghosts and membrane fragments; *H* contains damaged and shrunken synaptosomes, masses of small mitochondria, and synaptic vesicles that appear to be embedded in a sticky cytoplasm; and *I* largely contains small mitochondria and some shrunken synaptosomes. In earlier experiments it had been noted that exposure of the crude mitochondrial preparation to osmotic shock caused a shift of glycoproteins from more-dense to lighter fractions upon density-gradient centrifugation.[15,22]

The experimental data obtained by a variety of sucrose density-gradient centrifugation techniques led to the conclusion that synaptosomes, axons, and other unidentified membranous fragments contain gangliosides and glycoproteins as structural constituents. However, the membranous material associated with the synaptosome-rich fractions contain a higher concentration of glycoprotein relative to that of ganglioside than other portions of the neuronal cell surface.

It has been reported[23] that frog-muscle membrane preparations contain glycoproteins which yield sialofucohexosaminoglycans upon proteolytic digestion with papain.

The high content of gangliosides and glycoprotein in the crude microsomal material suggests that this particulate matter is largely made up of breakdown products derived from extensive degradation of more organized structures, such as glial processes, synaptosomes, dendrites, axons, and plasma membranes. The results of osmotic shock, which causes an increase in the appearance of glycoproteins and gangliosides in subcellular fractions that sediment into less-dense sucrose layers, lends support to this hypothesis.

Observations of subfractions derived from the cat cerebellum[18] under the electron microscope revealed the presence of fragments believed to be of glial origin. These particles sedimented into sucrose density layers that contained large numbers of synaptosomes and mitochondria. If glial processes are especially enriched in glycoprotein material, the presence of

TABLE VII

Distribution of Gangliosides and Glycoproteins in Various Parts of the Bovine Brain

Brain tissue	Gangliosidic NANA (μM/g)	Gangliosidic hexosamine (μM/g)	Glycoprotein-hexosamine (μM/g)	Nondiffusible sialofucohexosaminoglycan-NANA (μM/g)
Cerebral gray	2.77	1.14	1.87	0.62
Cerebral white	0.90	0.43	1.17	0.33
Cerebellum	1.60	0.70	1.26	0.38
Caudate nucleus globus pallidus and putamen	1.78	0.92	1.50	0.53
Corpus callosum	0.98	0.49	1.41	0.34
Thalamus	1.44	0.56	1.54	0.39
Pons	0.91	0.38	1.04	0.23
Medulla	0.96	0.37	1.06	0.28

even a small number of these fragments in the synaptosomal fraction may account for the higher glycoprotein-NANA/gangliosidic NANA ratio of this fraction.

ANATOMICAL LOCALIZATION OF GLYCOPROTEINS

Brain tissue samples that consist largely of cell bodies contain a higher concentration of glycoproteins and gangliosides[12] than areas that are enriched in myelinated fiber tracts (Table VII). However, the ratio of glycoprotein-NANA (or hexosamine) to gangliosidic NANA (or hexosamine) is higher in white matter than in gray. The values for the ratio of glycoprotein-hexosamine to gangliosidic hexosamine for various brain parts are caudate nucleus, putamen, and globus pallidus, 1.63; cerebral gray, 1.64; cerebellum, 1.80; cerebral white, 2.72; pons, 2.73; thalamus, 2.75; medulla, 2.86; and corpus callosum, 2.88. Values for sialofucohexosaminoglycan-NANA (nondiffusible)/gangliosidic NANA are: cerebral gray, 0.22; cerebellum, 0.24; pons, 0.25; thalamus, 0.27; medulla, 0.29; caudate nucleus, globus pallidus, and putamen, 0.30; corpus callosum, 0.35; and cerebral white, 0.37. Earlier, Papadopoulos[24] and Cumings *et al.*[25] had provided evidence indicating that white matter has a higher concentration of protein-bound NANA and hexosamine, relative to that of gangliosidic NANA (and hexosamine), than gray matter. The results provide a paradox: white matter and synaptosomes contain more glycoprotein per milligram ganglio-

TABLE VIII

Comparison of Nondiffusible Sialofucohexosaminoglycan Preparations Obtained from Rat and Bovine Brain

A. Molar Ratios of carbohydrate constituents of nondiffusible sialofucohexosaminoglycan preparations

	NANA: fucose	Fucose: hexose	NANA: hexose	Fucose: hexosamine	NANA: hexosamine	Hexose: hexosamine
Bovine brain	2.15	0.182	0.391	0.223	0.480	1.23
Rat brain	1.94	0.260	0.505	0.361	0.704	1.41

B. Carbohydrate composition of nondiffusible sialofucohexosaminoglycans

	Total carbohydrate (mg/g tissue)	NANA (%)	Fucose (%)	Hexosamine (%)	Hexose (%)
Bovine brain	0.47	26	6	31	37
Rat brain	0.63	31	8	25	36

side than gray matter and membrane fragments presumably derived from axons. Gray matter is enriched in synaptosomes and white matter is enriched in axons. In attempting to reconcile these findings, it must be remembered that homogenization and sucrose-density-gradient centrifugation of gray matter yields membrane fragments derived from neuronal cell membranes, dendrites, astroglia, and other structures as well as synaptosomes. Such fragments may appear in the microsomal fraction or in the less-dense particulate fractions. If these structures are especially rich in gangliosides, but poor in glycoprotein, the quantitative contribution of these structures may account for the lower glycoprotein/ganglioside ratio of gray matter. Alternatively, the higher ratio of glycoprotein/ganglioside of white matter may be due to the presence of oligodendroglial cells. Homogenization of the tissue may rupture the glia to form particulates that sediment along with synaptosomes upon sucrose-density-gradient centrifugation.[17–19] There is some experimental support for the presence of glycoprotein in glial cells (see below). In view of these considerations, the reports that glycoprotein[26] and proteins[27] are synthesized in synaptosomes should be interpreted with caution. These reports were based on the assumption that the synaptosome preparations studied were morphologically homogeneous.

It is interesting that the nondiffusible sialofucohexosaminoglycans from whole bovine brain[12] contain less fucose and NANA per mole of hexosamine and hexose than the sialofucohexosaminoglycans isolated from whole rat brain (Table VIII). The increase in brain size in a series of mammalian brains is accompanied by an increase in cell number, but the increase in intercellular areas taken up by axons is proportionately greater.[28] As a consequence, the percentage of the volume of beef brain that is taken up by axons is larger than that in rat brain. Since white matter contains less glycoprotein than gray matter (Table VII), it follows that the amount of glycoprotein per gram of *whole brain* tissue would decrease with brain size (Table VIII). Furthermore, the lower fucose and NANA content per mole of hexose and hexosamine in sialofucohexosaminoglycans isolated from whole beef brain suggests that axonal glycoproteins contain less NANA and fucose per mole hexose and hexosamine than synaptosomal glycoproteins. This concept receives some experimental support, since cerebral white matter yielded nondiffusible sialofucohexosaminoglycans with a lower fucose and NANA content per mole hexose and hexosamine than other brain areas.

HISTOCHEMISTRY

Nerve cell bodies and processes generally provide a weak or negative reaction upon staining for glycoproteins by means of the periodic acid-Schiff reaction.[29,30] However, the presence of PAS-positive granules that contain glycoproteins has been reported[31,32] and an autoradiographic reaction in the Golgi zone shortly after injection of labeled glucose was demonstrated by Droz[33] and Peterson and Leblond.[34] Schanklin and Azzam[35] noted that the incisures of Schmidt–Lantermann that occur along the length of axons at intervals between the nodes of Ranvier have a central core rich in protein and carbohydrates. Wolman[36] found that during the first stages of demyelination there is an increase in the amount of polysaccharides along the degenerating nerve fiber. Hess[29,37] studied a PAS-positive material which he believed formed an intercellular ground substance. Rambourg and Leblond[38] conducted extensive light- and electron-microscope investigations which indicated the presence of a glycoprotein material which could be traced along dendrites and axons. The cell processes were surrounded by a layer of material, the density of which increased in the synaptic cleft. Pease[39] and Bondareff[40] reported the presence of mucopolysaccharides in the synaptic clefts and on the cell surfaces. Cytochemical investigations by Shabadash[41] led to the conclusion that the isoelectric point of subcellular complexes decreased in the order: mitochrondria, microsomes, cytoplasm, axon, and synapses. Singh[42] and Lumsden[43] produced evidence that PAS-positive materials are present in the glial processes that make up a great part of the interneuronal spaces.

Schnabel[44] reported a weak PAS-positive reaction in glial fibers. Obrucnik[45] reported an increase in intracellular "neutral polysaccharide" as a consequence of cell destruction and demyelination caused by surgical severance of the cortex from the subcortex. Benetato *et al.*[46] reported that the demyelination that occurs in experimental allergic encephalitis is accompanied by a large increase in the polysaccharide content of the neuroglia and that glycoprotein accumulates in the glial cell during myelination. Many years ago Bailey and Schaltenbrand[47] and Penfield and Cone[48] reported that oligodendroglial swelling caused the cytoplasm to become swollen with fluid. The fluid was a mucin-like substance.

In general, histochemical observations suggest that neural glycoproteins are associated with nerve cell surfaces, including the synaptosomal region and the axons. Other experimental approaches have supplied support for this localization. Based on electrophysiological data, Tasaki and Singer[49] postulated that the outer surface of the axon membrane possesses a high density of negatively charged groups, possibly carboxyl in nature. Carroll and Sereda[50] have provided evidence suggesting that the smooth muscle receptors for serotonin are not gangliosides but glycoproteins that contain NANA.

FRACTIONATION OF THE INSOLUBLE GLYCOPROTEINS OF BRAIN TISSUE

Extraction of the combined mitochondrial and microsomal preparations with 0.02 *M* Tris hydrochloride buffer (pH 8.0) and water solubilizes only 10% of the protein-linked diffusible and nondiffusible sialofucohexoosaminoglycans.[2,51] The procedure solubilizes 20–25% of the total protein present in the particulates. Solutions that contain 0.5% Triton X-100 extract 60% of the glycoproteins.[52] Triton-solubilized glycoproteins can be fractionated by chromatography on calcium hydroxylapatite columns with 100% recovery, although considerable cleavage of the protein-bound NANA occurs during the procedure.[51] Approximately 20% of the applied glycoproteins are not adsorbed to the gel and are recovered with a 2.3-fold purification. A large part of the glycoproteins that adsorb to the gel fail to adsorb upon rechromatography and are recovered with a 2-to-3-fold purification. The procedure used did achieve considerable purification of several enzymes and isoenzymes present in Triton extracts,[53] which appeared in characteristic reproducible positions in the chromatograms. On the other hand, all fractions eluted from the column contain glycoprotein.[51] The proteins that failed to adsorb to the gel have the highest specific concentration (micrograms NANA per milligram protein); as the gradient of phosphate ion required to elute the protein increased, the eluted protein showed a progressive decrease in specific concentration of NANA. However, even the most tightly held protein fraction eluted at phosphate concentrations exceeding 0.2 *M* phosphate contained glycoproteins. The anomalous behavior of the glycoproteins on column chromatography on calcium hydroxylapatite may be due to aggregation and depolymerization of glycoprotein molecules,[2] or to an interaction between calcium hydroxylapatite and the lipo-glycoprotein complex present in Triton extracts.[51]

The carbohydrate content of the protein in the original Triton extract is approximately 3.2%. A threefold purification raises this value to 9.6%. It should be noted that the percentage carbohydrate in glycoproteins derived from the red blood cell membrane[54] and the kidney glomerular basement membrane[5] is about 8–10%.

Column chromatography of the Triton extracts revealed that a 2.3-fold purification of the glycoprotein that yields nondiffusible sialofucohexosaminoglycans is accompanied by a 2.3-fold purification of the glycoprotein that yields diffusible sialofucohexosaminoglycans. The ratio of nondiffusible to diffusible sialofucohexosaminoglycans was the same for all of the glycoprotein fractions recovered. These findings are consistent with the previous observations[15] that the subcellular distribution of nondiffusible and diffusible sialofucohexosaminoglycans is identical. The data suggest that diffusible and nondiffusible sialofucohexosaminoglycans are derived from the same glycoprotein or glycoproteins.

FRACTIONATION OF SOLUBLE GLYCOPROTEINS FROM BRAIN TISSUE

Drastic homogenization of whole rat brain in 0.02 *M* Tris hydrochloride buffer (pH 8.0) by means of the Serval Omni-Mixer solubilizes approximately 40% of the total protein of the tissue. Approximately 20–30% of the glycoproteins are extracted. DiBenedetta and Brunngraber[55] subjected the high-speed supernatant to electrophoresis on polyacrylamide. The resultant electropherogram was made visible by staining with amido-Black for protein and with the periodic acid-Schiff reaction for glycoprotein. A large portion of the glycoproteins migrated to form a large unresolved band in the globulin region of the electropherogram. However, five components were found to migrate faster than serum or brain albumin. Liver extracts contained only two bands that migrated faster than serum albumin, only one of which corresponded to one of the components from brain. The protein component common to liver and brain extracts was a glycoprotein.

Four of the prealbumins present in brain extracts were specific for this tissue. One of these presumably corresponded to the *S*-100 protein of Moore;[56] the three others were glycoproteins. Soluble extracts prepared from the crude nuclear, mitochondrial, and microsomal preparations from brain contained the four rapidly migrating glycoproteins. Two of the three brain-specific glycoproteins probably correspond to the brain-specific proteins identified by McEwen and Hydén,[57] one of which had a high turnover rate.

By means of column electrophoresis it was possible to separate the five prealbumins present in brain extracts from the bulk of the soluble protein. The electrophoretic pattern was divided into six fractions. The values for micrograms NANA per milligram protein were 16.3, 8.3, 2.7, 2.8, 1.7, and 1.5 for fractions I (the most negative) through VI (the least negative). Fraction I contained only the five components that migrated at a rate faster than serum albumin. The subsequent fractions contained increasing amounts of impurities corresponding to the slower moving proteins, including the bulk of the glycoproteins present in the extract.

Warecka and co-workers[58–60] have reported the extraction of a brain-specific glycoprotein from human white matter. The appearance of this brain-specific glycoprotein during ontogenesis was correlated with the differentiation of the glia during myelination, and it was suggested that this protein may play a role in the formation of myelin.

GLYCOPROTEINS IN DEVELOPING BRAIN

Suzuki[61] reported that the amount of gangliosides present in brain tissue, expressed on the basis of unit wet weight, increased slowly after birth until 6–8 days of age. This was followed by a period of rapid increase so that the ganglioside concentration reached the adult level 16 days after birth and remained constant thereafter. This finding was confirmed by Holian and Brunngraber.[62] The concentration of nondiffusible sialofucohexosaminoglycans followed the same pattern.

METABOLISM OF GLYCOPROTEINS IN THE NERVOUS SYSTEM

The metabolism of neural glycoproteins was briefly described in a recent review.[2] On the basis of the known activity of enzymes that convert hexose, hexosamine, and NANA to the nucleoside diphosphoglycosyl derivatives of the "active sugar pool" it was concluded that the metabolism of the nucleoside diphosphoglycosyl derivatives proceeds at a rate suggesting that the turnover of the brain glycoproteins is neither very rapid nor very slow. The results of a few isotopic incorporation studies supported this view. More recently, rats were injected intraperitoneally with *D*-glucosamine-1-C^{14} and the nondiffusible sialofucohexosaminoglycans were isolated and purified.[62] The time of maximal incorporation of isotope was 4–8 hr. Thereafter the amount of the incorporated isotope decreased at a rate consistent with a calculated half-life of 5.5–6 days in the 15-day-old rat and 5.2 days in the 55-day-old rat.

Brain tissue contains an N-acetylneuraminic acid-synthesizing enzyme[63] which forms N-acetylneuraminic acid 9-phosphate by condensation of N-acetyl *D*-mannosamine 6-phosphate and phosphoenolpyruvate. One gram of tissue is capable of synthesizing 0.1 μM NANA/hr. The activity of this enzyme is somewhat less in gray matter than in white matter. Cerebral gray and white matter, synthesize, respectively, 4.9–5.6 and 6.9–8.2 nM NANA/mg protein/hr. It was noted above that white matter contains considerable amounts of glycoprotein, and that the ratio of glycoprotein/gangliosides is higher in white matter than in gray matter.

Recently Hagopian and Eylar[64] reported that the basic protein encephalitogen from bovine spinal cord is a receptor for the N-acetylgalactosamine transferred from UDP-N-acetylgalactosamine by the polypeptidyl: N-acetylgalactosoamine transferase prepared from bovine submaxillary gland. Ohgushi and Yamashina[65] reported that brain tissue contains appreciable amounts of a glycopeptide-degrading enzyme. This enzyme, largely present in the lysosomes, liberated hexosamine from β-aspartylglycosylamine at a rate of approximately 0.84 μM/g/hr. Robbins *et al.*[66] reviewed the types of information involved in the biosynthesis of polysaccharides in biological systems. These include the enzymatic specificity of enzymes that catalyze the transfer of glycosyl residues from nucleoside diphosphoglycosyl derivatives, the specificity of the acceptor system, and the role of the macromolecular complex in which synthesis occurs. The authors suggest that a series of membrane-bound enzymes, which includes lipid components, form an organized synthetic system.

GLYCOPROTEINS AND NEURAL FUNCTION

Several investigations have provided hints that glycoproteins may play an active role in neural activity. Dische[67] found that the ratio of NANA/fucose of glycoproteins secreted by the dog submaxillary gland was dependent upon the nature or intensity of the stimulus. Glezer and Yakobson[68] observed an increase in aggressive behavior and pain sensitivity following administration of Triphtazine to rats. The behavioral change was accompanied by an increase in the polysaccharide content of the cortex and some subcortical regions. This increase appeared in structures that surrounded the neuron, rather than in the nerve cell body itself. Taylor *et al.*[69] reported that the administration of reserpine to rats caused an increase in glycoprotein accumulation in the denervated submaxillary gland, but could not find similar changes in brain extracts. These authors suggested that changes in glycoprotein structure associated with cell membranes might explain the long-lasting effects of reserpine on the ability to store biogenic amines. Humphreys[70] and Moscona[71] obtained a species-specific glycoprotein that promoted selective reaggregation of dissociated cells of marine sponges. Margoliash *et al.*,[72] who demonstrated the presence of fucose, glucosamine, mannose, galactose, and glucose, suggested that this glycoprotein is involved in contact selectivities and grouping properties of cells. Kramer[73] found that puromycin appears to inhibit the incorporation of glucosamine into glycoproteins present on cell surfaces of the Chinese hamster ovary cells by preventing the production of an intracellular precursor of surface sialoproteins. This observation is of interest in view of the reported inhibitory effect of puromycin on protein synthesis and memory.

A recent report by Bogoch[74] on the effect of training on glycoprotein levels in pigeon brain is unconvincing. Soluble pigeon-brain extracts were reported to contain 0.66 mg protein-bound hexose per gram brain at rest. This value increased to an average value of 6 mg per gram brain tissue 10 min after training. If this increase is due to *de novo* synthesis, as implied by Bogoch, brain tissue appears to be capable of incorporating hexose into glycoprotein at a rate of 180 μM/g/hr. Enzymes in brain tissue that synthesize nucleoside diphosphoglycosyl derivatives for the "active sugar pool" proceed at rates that fall in the range of 0.1–5 μM of substrate converted per gram tissue per hour. Utilization of glucose by the brain *in vivo* proceeds normally at a rate of 18 μM/g/hr, and brain tissue contains 2.6 μM of glucose per gram of tissue.[75] This is hardly sufficient to supply the hexose required to maintain an incorporation rate of 180 μM hexose/g/hr. The available glucose would be depleted within 1 min after training. Furthermore, in order to incorporate hexoses into glycoproteins at this rate, a minimum of 360 μM of ATP per gram tissue per hour must be made available during the same period. This represents approximately 50% of the ATP that is made available during the same time by the oxidative processes of the tissue. In order to form a complete glycoprotein molecule, the incorporation of one molecule of hexose must be accompanied by the incorporation of approximately 1.2 molecules of NANA, fucose, and hexosamine. This would require the utilization 100% of the ATP produced (per gram tissue per hour) during the time of maximal glycoprotein synthesis induced by training. Clearly, the increase in protein-bound hexose reported by

Bogoch cannot be due to *de novo* synthesis. Since Bogoch confined his studies to the amount of glycoprotein present in soluble extracts, it may be postulated that training causes a change in the membrane-bound glycoproteins so that these become more amenable to extraction. Soluble pigeon-brain extracts were reported by Bogoch to contain 6 mg protein-bound hexose (average value) per gram brain tissue after training. This is ten times the value known to be present in one gram of whole rat-brain tissue. The content of glycoprotein in pigeon brain is in fact less than that found in rat brain.(76) Even if all of the "insoluble" glycoprotein becomes "solubilized" by training, the amount of "insoluble" glycoprotein in the tissue is insufficient to account for the large increase of protein-bound hexose in the soluble extracts reported by Bogoch. The extracts prepared by Bogoch contain several carbohydrate-containing constituents which might be expected to react with the method used for the determination of hexose. These include uronic acid-containing mucopolysaccharides, glycoproteins, glycolipids, nucleic acids, and possibly some polyunsaturated lipids that produce chromogens with concentrated sulfuric acid. Bogoch provides data showing that his extracts contain ribose, xylose, and glucose. None of these sugars have been found to be present in brain glycoproteins. Although it would appear that Bogoch is measuring some chemical change that accompanies the learning process, the identity of the component responsible for this change remains to be determined.

It is reasonable to suppose that glycoproteins play a role in information storage.(67,74,77) These substances are located at cell surfaces and at sites of contact between neurons, especially the synaptic region. There is evidence suggesting that NANA-containing proteins may act as receptors for transmitter action. Glycoproteins bound to the surface of the cell may be the carriers of the information that facilitates the establishment of the correct interneuronal connections during the development of the nervous system or during nerve regeneration.(67) Once this connection is established these compounds may influence the impulse that passes from cell to cell. During this process an impulse may alter the configuration or structure of the surface glycoprotein molecules that are secreted by the cell. This alteration may be such that the cell membrane is somewhat different from what it was before, and this difference will be reflected in the subsequent behavior of the cell. It appears to be well established that nerve function is accompanied by alterations in nucleic acid structure and increased protein synthesis. Conceivably, all or a part of this newly synthesized protein material is glycoprotein or a glycoprotein precursor. Deposition of this newly synthesized material at the cell surface could have the practical effect of altering the transmission capabilities of the synaptic regions.

The variety of polysaccharide chains available can provide for an almost infinite number of glycoprotein molecules that differ only in the number and type of polysaccharide chains attached to the protein structure. The large number of glycoprotein types may serve as the biological road signs that direct the exceedingly complicated process of the prenatal development of the organism on the one hand, and the postnatal development of the mental process on the other.

REFERENCES

1. A. Gottschalk (ed.), "Glycoproteins, Their Composition, Structure, and Function," Elsevier Publishing Co., Amsterdam (1966).
2. E. G. Brunngraber, *in*, "Handbook of Neurochemistry" (A. Lajtha, ed.), Vol. 1, Plenum Press, New York 1969.
3. E. D. Eylar, *J. Theoret. Biol.* **10,** 89–113 (1965).
4. R. U. Margolis, *J. Histochem. Cytochem.* **16,** 486 (1968).
5. R. G. Spiro, *J. Biol. Chem.* **242,** 1915–22, 1923–32 (1967).
6. R. G. Spiro, *J. Biol. Chem.* **240,** 1603–10 (1965).
7. B. Radhakrishnamurthy and G. S. Berenson, *J. Biol. Chem.* **241,** 2106–12 (1966).
8. E. G. Brunngraber and B. D. Brown, *Biochim. Biophys. Acta* **69,** 581–2 (1963).
9. E. G. Brunngraber and B. D. Brown, *J. Neurochem.* **11,** 449–459 (1964).
10. E. G. Brunngraber and B. D. Brown, *Biochim. Biophys. Acta* **83,** 357–60 (1964).
11. E. G. Brunngraber and B. D. Brown, *Biochem. J.* **103,** 65–72 (1967).
12. E. G. Brunngraber, B. D. Brown, and V. Aguilar, *J. Neurochem.*, **16,** 1059 (1969).
13. C. Di Benedetta, E. G. Brunngraber, G. Whitney, B. D. Brown, and A. Aro, *Arch. Biochem. Biophys.*, **131,** 404 (1969).
14. E. G. Brunngraber and G. Whitney, *J. Chromatog.* **32,** 749–50 (1968).
15. E. G. Brunngraber, H. Dekirmenjian, and B. D. Brown, *Biochem. J.* **103,** 73–78 (1967).
16. V. P. Whittaker, *Biochem. J.* **72,** 694–706 (1959).
17. H. Dekirmenjian and E. G. Brunngraber, *Biochim. Biophys. Acta* **177,** 1–10 (1969).
18. L. N. Johnston and L. M. H. Larramendi, *Exp. Brain Res.* **5,** 326–340 (1968).
19. H. Dekirmenjian, E. G. Brunngraber, L. N. Johnston, and L. M. H. Larramendi, *Exp. Brain Res.*, **8,** 97 (1969).
20. E. De Robertis, A. Pellegrino De Iraldi, G. Rodriquez de Lores Anaiz, and L. Salganicoff, *J. Neurochem.* **9,** 23–35 (1962).
21. V. P. Whittaker, I. A. Michaelson, and R. J. A. Kirkland, *Biochem. J.* **90,** 293–303 (1964).
22. E. G. Brunngraber, V. A. Ziboh, and W. G. Occomy, *in.* "Protides of the Biological Fluids" (H. Peeters, ed.), Vol. 13, 207–10, Elsevier Publishing Co., Amsterdam (1966).
23. L. G. Abood, K. Kurahasi, E. G. Brunngraber, and K. Koketsu, *Biochim. Biophys. Acta* **112,** 330–39 (1966).
24. N. T. Papadopoulos, *Analyt. Biochem.* **1,** 486 (1960).
25. J. N. Cumings, H. Goodwin, and G. Curzon, *J. Neurochem.* **4,** 234–237 (1959).
26. S. H. Barondes, *J. Neurochem.* **15,** 699–706 (1968).
27. K. von Hungen, H. R. Mahler, and W. J. Moore, *J. Biol. Chem.* **243,** 1415–23 (1968).
28. D. B. Tower, *J. Comp. Neurol.* **101,** 19–52 (1954).
29. A. Hess, *J. Comp. Neurol.* **98,** 69–91 (1953).

30. I. J. Young and L. G. Abood, *J. Neurochem.* **6,** 89–94 (1960).
31. N. M. Sulkin, *J. Neurochem.* **5,** 231–235 (1960).
32. L. Chouinard, *Compt. Rend.* **250,** 3061 (1960).
33. B. Droz, *J. Microsc.* **6,** 419 (1967).
34. M. Peterson and C. P. Leblond, *J. Cell Biol.* **21,** 143–148 (1964).
35. W. M. Schanklin and N. A. Azzam, *J. Comp. Neurol.* **123,** 5–9 (1964).
36. M. Wolman, *J. Neurochem.* **1,** 370–376 (1956–7).
37. A. Hess, *J. Anat.* (*London*) **92,** 298–303 (1958).
38. A. Rambourg and C. P. Leblond, *J. Cell Biol.* **32,** 27–53 (1967).
39. D. C. Pease, *J. Ultrastruct. Res.* **15,** 555–588 (1966).
40. W. Bondareff, *Anat. Rec.* **157,** 527–535 (1967).
41. A. L. Shabadash, "Tret'ya Vses. Konf. po Biokhim. Nervnoi Sistemy," p. 283 Akad Nauk ArmSSR, Inst. Biokhim., Sb. Dokl., Erevan 1962 (Pub. 1963).
42. R. Singh, *J. Histochem. Cytochem.* **12,** 712–13 (1964).
43. C. E. Lumsden, *in:* "Biology of Neuroglia" (W. F. Windle, ed.), pp. 141–61, C. C. Thomas, Springfield, Illinois (1958).
44. R. Schnabel, *Experimentia* **17,** 28 (1961).
45. M. Obrucnik, *Acta Histochem.* **11,** 289 (1961).
46. G. Benetato, E. Gabrielescu, L. Parteni, A. Bordeianu, and I. Boros, *Fiziol. Normala Patol.* (*Bucharest*) **7,** 73 (1961).
47. P. Bailey and G. Schaltenbrand, *Deut. Z. Nervenheilk.* **97,** 231 (1927).
48. W. Penfield and W. Cone, *Arch. Neurol. Psychiat.* **20,** 1065 (1928).
49. I. Tasaki and I. Singer, *Ann. N.Y. Acad. Sci.* **137,** 792–805 (1966).
50. P. M. Carroll and D. D. Sereda, *Nature* **217,** 667–8 (1968).
51. E. G. Brunngraber, V. Aguilar, and A. Aro, *Arch. Biochem. Biophys.*, **129,** 131 (1969).
52. E. G. Brunngraber and E. A. Bejnarowicz, *in:* "Protides of the Biological Fluids" (H. Peeters, ed.), Vol. 13, pp. 201–5, Elsevier Publishing Co., Amsterdam (1966).
53. E. G. Brunngraber and V. Aguilar, *J. Neurochem.* **9,** 451–61 (1962).
54. A. H. Maddy, *Biochim. Biophys. Acta* **117,** 193–200 (1966).
55. C. Di Benedetta, and E. G. Brunngraber, submitted for publication.
56. B. W. Moore, *Biochem. Biophys. Res. Commun.* **19,** 739 (1965).
57. B. S. McEwen and H. Hydén, *J. Neurochem.* **13,** 823–33 (1966).
58. K. Warecka, *Life Sci.* **6,** 1999 (1967).
59. K. Warecka and D. Müller, *J. Neurol. Sci.*, **8,** 329–46 (1969).
60. K. Warecka and H. Bauer, *Deut. Z. Nervenheilk.*, **194,** 66 (1968).
61. K. Suzuki, J. Neurochem. **12,** 969–79 (1965).
62. O. Holian and E. G. Brunngraber, to be submitted for publication.
63. R. Joseph and B. K. Bachhawat, *J. Neurochem.* **11,** 517–26 (1964).
64. A. Hagopian and E. H. Eylar, *Arch. Biochem. Biophys.* **126,** 785–94 (1968).
65. T. Ohgushi and I. Yamashina, *Biochim. Biophys. Acta* **156,** 417–19 (1968).
66. P. W. Robbins, A. Wright, and M. Dankert, *J. Gen. Physiol.* **49** (*Suppl. Part 2*), 331–45 (1966).
67. Z. Dische, *in:* "Protides of the Biological Fluids" (H. Peeters, ed.), Vol. 13, pp. 1–20, Elsevier Publishing Co., Amsterdam (1966).
68. I. I. Glezer and I. S. Yakobson, *Zh. Nervopatol. Psikhiat.* **67,** 576 (1967).
69. P. W. Taylor, Jr., K. C. Richardson, P. M. Roddy, and E. Titus, *J. Pharm. Exp. Therap.* **156,** 483–491 (1967).
70. T. Humphreys, *Devel. Biol.* **8,** 27–47 (1963).
71. A. A. Moscona, *Proc. Natl. Acad. Sci. U.S.* **49,** 742 (1963).

72. E. Margoliash, J. R. Schenck, M. P. Hargie, S. Burokas, W. R. Richter, and G. H. Barlow, *Biochem. Biophys. Res. Commun.* **20,** 383–88 (1965).
73. P. M. Kramer, *J. Cell. Physiol.* **69,** 199–207 (1967).
74. S. Bogoch, "The Biochemistry of Memory," Oxford University Press, New York (1968).
75. H. McIlwain, "Biochemistry and the Central Nervous System," Little, Brown and Co., Boston (1966).
76. N. T. Eldredge, G. Read, and W. Cutting, *Med. Exp.* **8,** 265 (1963).
77. E. G. Brunngraber, *Perspect. Biol. Med.,* **12,** 467 (1969).

Chapter 20

Development of Enzymes Involved in AMP Metabolism in Chick Embryo Brain

P. L. Ipata

Institute of Biological Chemistry
Free University of Chieti
Chieti, Italy

and

F. A. Manzoli and I. Wegelin

Institute of Histology and General Embryology
University of Bologna
Bologna, Italy

INTRODUCTION

The role exerted by AMP as "modulator" of several biosynthetic enzymic pathways has long been recognized. The biochemical features of AMP metabolism during differentiation, however, have so far received scant attention.

Two main pathways for AMP breakdown have been described in several tissues, both yielding NH_3 and inorganic phosphate. The first involves deamination of AMP, followed by dephosphorylation of IMP ("direct pathway"), the second dephosphorylation of AMP, followed by deamination of adenosine ("indirect" pathway).[1] In mature brain ATP has a double effect on AMP metabolism, being at the same time a positive allosteric effector of adenylate deaminase, and a negative allosteric effector of adenylate phosphatase.[6,10] The rate of adenosine deamination in CNS of adult animals is unaffected by ATP and other nucleoside triphosphates.[5]

In view of the importance of AMP in purine interconversion in brain, we thought it interesting to study the developmental patterns of enzymes catalyzing its breakdown in chick embryo brain: adenylate deaminase, adenylate and inosinate phosphatase, and adenosine deaminase. The activity of nucleoside phosphorylase is also reported.

MATERIALS AND METHODS

White Leghorn eggs (57 ± 2 g average weight) were purchased from a local agricultural station. They were incubated at 39°C at 60% relative humidity and turned once a day. The chick embryos and chicks were sacrificed at various developmental stages (evaluated according to Hamburger and Hamilton,[3] ranging from six days of incubation to four days after hatching. Brains were dissected and weighed, and the dry weight determined.[11] A 20% tissue homogenate (w/v) was made in cold 0.25 *M* sucrose and centrifuged at 1.000g for 15 min. The supernatant fluid was directly used for enzyme assays.

Nucleotides and nucleosides were purchased either from Sigma Chem. Co. St. Louis, or from Boheringer und Sohne, Mannheim, Germany.

The assays for adenylate and inosinate phosphatases contained 0.8 ml 0.2 *M* Tris-acetate buffer pH 7, 0.1 ml 12 m*M* AMP or IMP, 0.1 ml of brain extract, and water to a final volume of 1.03 ml. Tubes were incubated at 37°C for 20 min, and the reaction was stopped by addition of 0.5 ml 35% perchloric acid. The precipitated protein was removed by centrifugation, and orthophosphate was estimated in the supernatant fluid by the method of Fiske and Subbarow.[2] All values have been corrected for zero time controls. The effect of ATP or Mg^{++} was studied by including in the standard reaction mixtures 0.030 ml of 0.5 m*M* ATP or 0.030 ml of 400 m*M* $MgCl_2$.

The assays for adenylate deaminase activity contained 0.2 ml 0.2 *M* Tris-acetate buffer, pH 7, 0.2 ml 9 m*M* AMP, 0.1 ml of brain extract, and water to a final volume of 0.7 ml. Tubes were incubated at 37°C for 40 min, and the reaction was stopped by addition of 0.030 ml of 30% trichloroacetic acid. The precipitated protein was removed by centrifugation, and ammonia was estimated in the supernatant fluid by the Nessler procedure. The effect of ATP was studied by including in the standard reaction mixture 0.2 ml of 3.5 m*M* ATP. All values have been corrected for zero time controls.

The activity of adenosine deaminase was determined spectrophotometrically according to Kalckar.[7] The activity of nucleoside phosphorylase was assayed spectrophotometrically, with guanosine as substrate, by measuring the increase in absorption at 290 nm at alkaline pH, where there is a great difference in the extinction coefficients of guanosine and guanine.

The assays contained 0.3 ml potassium phosphate buffer, pH 6.8, 0.2 ml of 9 m*M* guanosine, 0.150 ml of brain extract, and 0.2 ml of water. At zero time and after 10 min and 20 min incubation at 37°C, 0.2-ml aliquots of the reaction mixture were added to 2 ml of 0.3 *N* NaOH, tubes were centrifuged, and the optical density of the supernatant fluid was measured at 290 nm. The increase in absorption is a measure of the guanine formed.

RESULTS

Figure 1 shows the developmental pattern of adenylate deaminase of chick embryo brain as measured in the absence and in the presence of ATP. It can be seen that during the whole incubation time the addition of ATP results in a marked activation of the enzyme activity.

Figure 2 shows the developmental pattern of chick embryo adenylate phosphatase and inosinate phosphatase. It can be seen that both in the presence and in the absence of Mg^{++} ions there is a peak of activity at the 15th day of incubation. The activation by Mg^{++} ions, however, is maximal during the period of maximal activity, and is practically absent at the hatching. Inhibition of adenylate phosphatase by ATP in the absence of Mg^{++} ions was always observed, and was about 75–85% during the period considered.

The developmental pattern of nucleoside phosphorylase in chick embryo brain shows an increase of the enzyme activity from the sixth until the hatching day (Fig. 3).

The levels of adenosine deaminase activity in chick embryo brain are given in Fig. 4. The developmental pattern of the enzyme is markedly different from that of adenylate deaminase, the activity being absent from the 13th until the 19th day, and rising abruptly the day before hatching. No change in activity could be detected by addition of ATP.

In order to investigate the possible role of adenosine deaminase in CNS, an extract of sheep brain containing high levels of adenosine de-

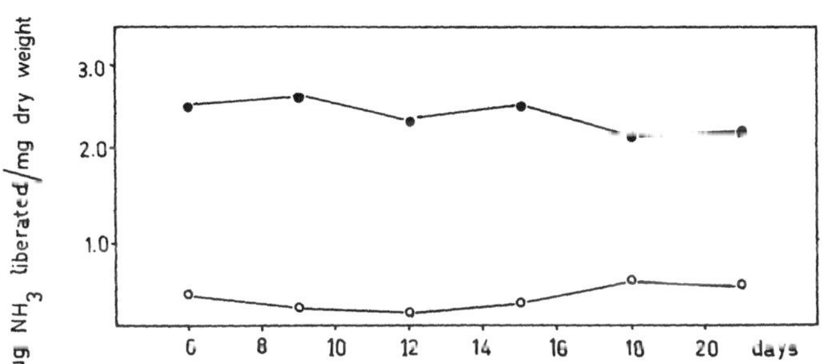

Fig. 1. Adenylate deaminase activity in chick embryo brains, in the absence of ATP (○), and in the presence of ATP (●). Experimental conditions are described in the text.

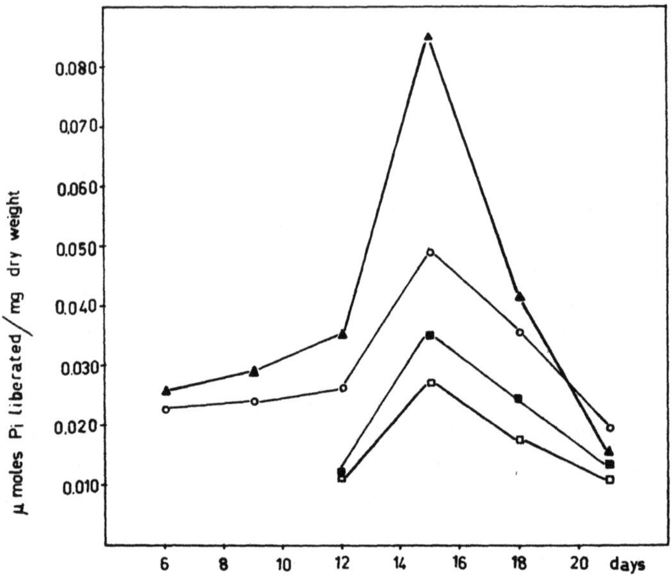

Fig. 2. Developmental pattern of adenylate and inosinate phosphatase in chick embryo brain. Adenylate phosphatase (○) in the absence of Mg^{++} ions; adenylate phosphatase in the presence of Mg^{++} ions (▲); inosinate phosphatase in the absence of Mg^{++} ions (□); inosinate phosphatase in the presence of Mg^{++} ions (■).

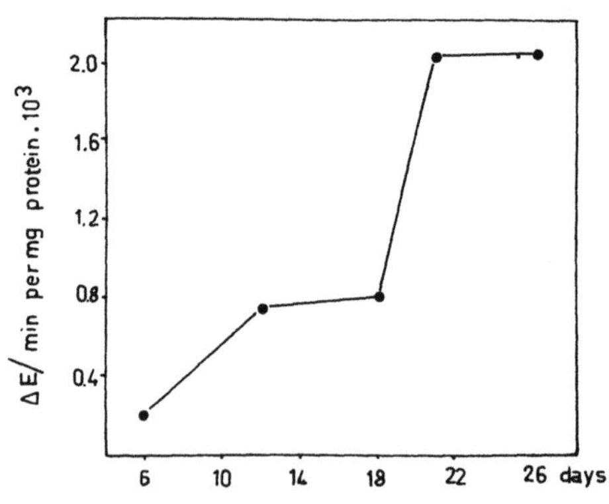

Fig. 3. Developmental pattern of nucleoside phosphorylase in chick embryo brain. The experimental conditions are described in the text.

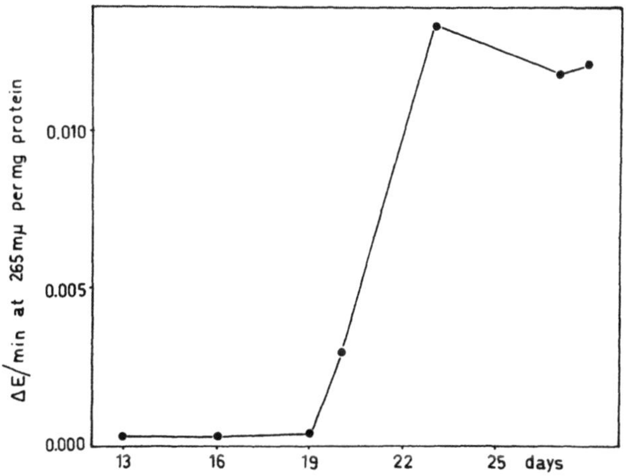

Fig. 4. Developmental pattern of adenosine deaminase in brain of chick embryo. The activity was assayed according to Kalckar.[7]

aminase[5] was incubated with adenosine in the absence and in the presence of 5-phosphorybosyl-1-pyrophosphate.

The reaction mixtures contained 0.6 ml of a 0.2% solution of adenosine, 0.4 ml of 0.05 *M* potassium phosphate buffer, pH 7.4, and either 0.6 ml of a 0.2% solution of 5-phosphorybosyl-1-pyrophosphate (P-R-PP), or 0.6 ml of water; 0.2 ml of the supernatant fluid obtained after centrifugation of a sheep-brain homogenate (20% w/v) in 0.9% NaCl were added, and incubation was carried out at 37°C. At the times indicated below (Table I) 0.2-ml aliquots were heated at 100°C for 3 min and centrifuged. An

TABLE I

Column *n* (left to right)	Time of incubation (min)
1 Adenosine + P-R-PP	0
2 Adenosine + P-R-PP	2
3 Adenosine + P-R-PP	5
4 Adenosine + P-R-PP	10
5 Adenosine + P-R-PP	20
6 Adenosine + P-R-PP	30
7 Adenosine − P-R-PP	30
8 A standard of IMP	—
9 Standards of IMP, inosine hypoxanthine, and adenosine (from top to bottom)	—

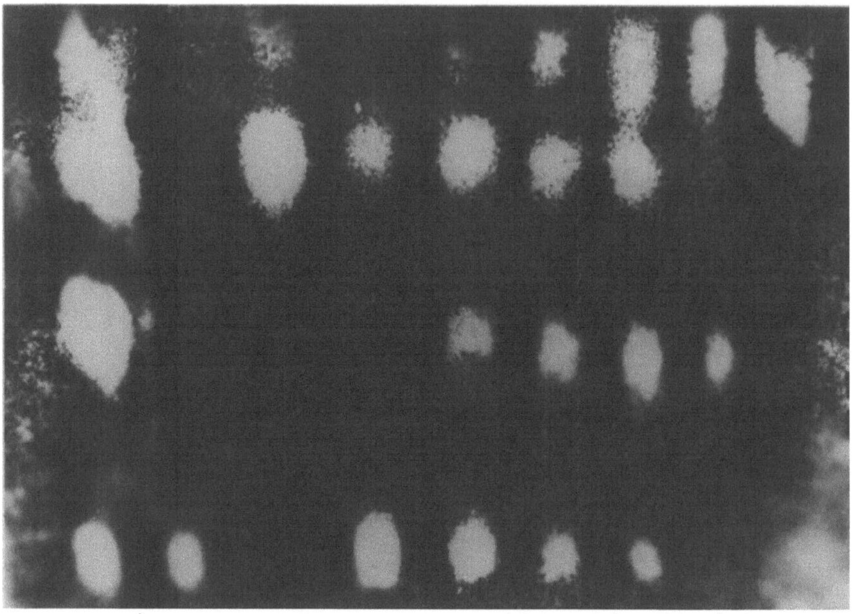

Fig. 5. Utilization of adenosine by sheep brain extracts.

amount of 0.050 ml of the supernatant fluid was chromatographed on paper using *n*-propanol:water:100% trichloroacetic acid:22% NH_4OH (75:20:5:0.3). The chromatogram was then photographed with the UV light according to the method of Markham and Smith.[9]

Figure 5 shows that after 30 min incubation adenosine was converted to IMP, probably through the successive action of adenosine deminase, nucleoside phosphorylase, and IMP pyrophosphorylase. When 5-phosphorybosyl-1-pyrophosphate was omitted adenosine was converted to hypoxanthine.

DISCUSSION

In a recent review on AMP metabolism in heart and lung extracts of adult animals Burger and Lowenstein[1] have shown that in the presence of ATP the inosinate (or "direct") pathway is the preferred one. Thus, in the presence of ATP the rate of the pathway involving AMP deamination is 3.1 times higher than that involving dephosphorylation to adenosine. In the absence of ATP the rate of the pathway involving AMP dephosphorylation is 16 times higher than the direct pathway. These results have

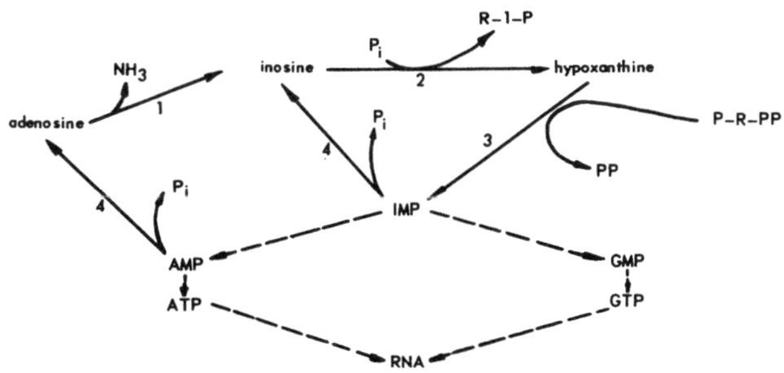

Fig. 6. Role of adenosine deaminase in nucleotide metabolism (1) adenosine deaminase; (2) nucleoside phosphorylase; (3) IMP pyrophosphorylase; (4) adenylate or inosinate phosphatase.

been attributed to the simultaneous activation of adenylate deaminase and inhibition of adenylate phosphatase exerted by ATP.

Our results suggest that during development of the chick embryo AMP catabolism in brain might occur through the "direct" pathway. This view is substantiated by (1) the marked activation of adenylate deaminase by ATP during the whole incubation period, (2) the absence of brain adenosine deaminase activity from the 13th until the 19th day of incubation, and (3) by the inhibition exerted *in vitro* by ATP on adenylate phosphatase. It must be emphasized that the intracellular concentrations of ATP in chick embryo brains are sufficiently high to cause the observed inhibition. The physiological significance of the regulation of AMP metabolism would be to convey more IMP for guanyl-compound biosynthesis. The concentration of guanyl compounds is in fact considerably higher in CNS than in other organs.[4,8]

The increase of adenosine deaminase activity during the period of hatching suggests that the "indirect" pathway may also contribute to AMP metabolism in mature brain. The metabolic significance of adenosine deaminase is still obscure. Its possible role in the so-called "salvage pathway" of purine biosynthesis is illustrated in Fig. 5, showing that adenosine can be converted to IMP, the common precursor of AMP and GMP, in adult sheep brain. Figure 6 illustrates the role of adenosine deaminase in nucleotide metabolism in brain.

It is interesting to note that the activity of nucleoside phosphorylase, another enzyme involved in the "salvage pathway" of purine biosynthesis, also shows a rapid increase before hatching.

Our results suggest that the ability to synthesize nucleotides *via* pathways different from the *de novo* synthesis is acquired by chick embryo in the period of hatching.

SUMMARY

The developmental patterns of enzyme activities related to AMP metabolism have been investigated in chick embryo brain.

Adenylate deaminase was found to be ATP-activated during the whole incubation period, without significant changes of its specific activity. Conversely, adenylate phosphatase shows a peak of activity at the 15th day of incubation, and appears to be inhibited by ATP.

Adenosine deaminase is absent until the 19th day of incubation and increases abruptly before hatching. The conversion of adenosine to IMP in the presence of P-R-PP in CNS is shown.

The results suggest that AMP breakdown in chick embryo brains occurs through deamination of AMP, followed by dephosphorylation of IMP, and that the adenosine deaminase activity of mature brain may be involved in the so-called "salvage pathway" of purine biosynthesis.

REFERENCES

1. R. Burger and J. M. Lowenstein, Adenylate deaminase. III Regulation of deamination pathways in extracts of the heart and lung, *J. Biol. Chem.* **242,** 5281–5288 (1967).
2. C. H. Fiske and Y. Subbarow, The colorimetric determination of phosphorus, *J. Biol. Chem.* **66,** 375–400 (1925).
3. F. R. Lillie and H. L. Hamilton, "Lillie's development of the chick. Introduction to embryology," Holt, Rinehart, and Winston, New York (1952).
4. P. J. Heald, ed., "Phosphorus Metabolism of Brain," pp. 1–15, Pergamon, New York, (1960).
5. P. L. Ipata, Studies on the inhibition by nucleoside triphosphates of sheep-brain 5-nucleotidase, *Biochem. Biophys. Res. Commun.* **27,** 337–343 (1967).
6. P. L. Ipata, Sheep-brain 5-nucleotidase. Some enzymic properties and allosteric inhibition by nucleoside triphosphates, *Biochemistry* **7,** 507–515 (1968).
7. H. M. Kalckar, Differential spectrophotometry of purine compounds by means of specific enzymes. II Determination of adenine compounds, *J. Biol. Chem.* **167,** 445–459 (1947).
8. F. A. Manzoli and I. Wegelin, I nucleotidi liberi del fegato embrionale, *Arch. Ital. Anat. Embriol.*, in press.
9. R. Markham and J. D. Smith, Chromatographic studies of nucleic acids. 1. A technique for the identification and estimation of purine and pyrimidine bases, nucleosides and related substances, *Biochem. J.* **45,** 294–298 (1949).
10. J. Mendicino and J. A. Muntz, The activating effect of adenosine triphosphate on brain adenylic deaminase, *J. Biol. Chem.* **233,** 178–183 (1958).
11. I. Wegelin and F. A. Manzoli, Free nucleotides in the chick embryo brain during development, *J. Neurochem.* **14,** 1161–1165 (1967).

III

Alterations of Metabolism

Chapter 21

Macromolecular Aspects of the Nerve Growth Factor Proteins*

Eric M. Shooter

Departments of Genetics and Biochemistry, and
Lt. Joseph P. Kennedy, Jr. Laboratories for Molecular Medicine
Stanford University School of Medicine
Stanford, California

and

Silvio Varon

Department of Biology
University of California at San Diego
La Jolla, California

INTRODUCTION

A number of reviews have appeared recently[1–3] which describe the discovery and the biological properties of the nerve growth factor (NGF) in detail. Consequently, this paper will focus more on the macromolecular properties of NGF. This is particularly pertinent, since several apparently divergent reports on this subject have been published in the past two years.

The term nerve growth factor was given to specific proteins which produce hyperplasia and hypertrophy of chick embryonic sensory and sympathetic ganglia *in vivo*.[4] The mediodorsal population of chick sensory

* Research supported by funds from the Lt. Joseph P. Kennedy, Jr. Foundation, by U.S. Public Health Service research grants from the National Institute of Neurological Diseases and Stroke (NB 04270 and 07606), and by National Science Foundation Grants GB 4430 and 6878. We wish to acknowledge the contributions made by H. Ronald Fisk, Lloyd A. Greene, Junichi Nomura, Charles Raiborn, and Andrew P. Smith to the work from our laboratories discussed in this review. We are indebted to Valerie MacPhee and Elaine S. Shooter for the experiments shown in Figs. 1–3.

ganglionic neurons shows increased mitotic activity between the sixth and eleventh days of incubation, accelerated differentiation of early neuroblasts, and hypertrophy of differentiated neurons.[5] The response of the sympathetic cells is greater than that of sensory neurons with growth seen even in cells at advanced stages of differentiation.[6,7] A similar *in vivo* response to NGF is obtained with sympathetic neurons of mice, hyperplastic effects being noted between three and nine days of postnatal age and hypertrophic effects alone after this age.[1,8,9] In contrast, sensory neurons are receptive only during a restricted period of their embryonic development and show no response in newborn or adult animals. When explanted seven to nine day old chick embryonic sensory and sympathetic ganglia are exposed to optimum concentrations of NGF in culture a uniform halo of nerve fibers grows out from the ganglionic membrane, a phenomenon which forms the basis for the bioassay of NGF activity.[10] For several of the NGF species these optimum concentrations are around 10–20 ng/ml.

Antisera to NGF when injected daily into newborn mice for 3–5 days results in the almost total destruction of the developing sympathetic nervous system.[11] These observations, together with the demonstration of NGF activity in extracts of freshly dissected sympathetic ganglia of various species[1] and the finding that fluorescent antibody to NGF localizes within the cytoplasm of the receptive sympathetic neurons,[11a] suggest that NGF is normally present in these nerve cells and is essential for their growth and differentiation. Levi-Montalcini and Angeletti[12] have also found that trypsin-isolated chick-embryo ganglionic neuroblasts do not grow in synthetic media in the absence of NGF. Hoffman and McDougall,[13] however, have observed that a proportion of sensory ganglionic cells can grow in synthetic media and suggest that this may be due to bound NGF carried by such cells.

METABOLIC AND MORPHOLOGICAL CHANGES INDUCED BY NGF

The mechanism by which NGF stimulates growth in the responsive neuroblasts is unknown. It results in an overall increase in the synthetic activities of the receptive cells, but the initial event which catalyzes these changes has not been defined. Oxidation of C-1 of glucose is significantly increased,[14,15] as is the incorporation of labeled acetate into lipid, all lipids being labeled to the same relative extent.[16,17] Ganglioside synthesis also appears enhanced without any gross change in pattern.[17a] The incorporation of labeled amino acids into protein is stimulated by NGF.[14,18] The amino acid label appears preferentially in the more acidic of the water-soluble proteins of the ganglia,[19] as well as in membrane protein insoluble in nonionic detergents.[20] The addition of puromycin suppresses incorporation into both control and NGF-treated embryonic sensory ganglia,[1] while addition of Actinomycin D only suppresses the NGF-inducible increase.[18] RNA synthesis is also stimulated by NGF.[14] The NGF stimulation of RNA synthesis is observed, in embryonic sensory ganglia, before that of protein synthesis and is not affected by puromycin.[18,21] The density-gradient profiles of the RNA labeled in the presence of NGF were indistinguishable from those of the control.[21] Using ^{32}P-orthophosphate Toschi *et al.*[22] report that this label in the presence or absence of NGF appears in a high-molecular-weight RNA species with a DNA-like base composition. Other studies show that the amount of RNA in isolated neuroblasts from embryonic sensory ganglia is increased by exposure to NGF, although no change in base ratio occurs.[23] The stimulation of RNA synthesis appears therefore to be a primary event, but what brings this about is not known.

These metabolic events are accompanied by morphological changes in the receptive cells. Crain *et al.*[24] noted the development of an ordered paracrystalline structure in the cytoplasm of the neuroblasts. More recently Levi-Montalcini *et al.*[25] found that the nuclei of such cells develop irregular contours and have dense and clumped chromatin after 4 hr exposure to NGF. The Golgi apparatus becomes dilated and some cells show a considerable increase in the number of ribosomes associated with membranous endoplasmic reticulum. After 12 hr a very large increase in neurofibrils and neurotubules was observed. Both metabolic and morphological changes are those anticipated for an actively growing and differentiating cell.

BIOLOGICAL SOURCES OF NGF

Discovered originally in two mouse sarcomas, NGF is now obtained primarily from either snake venom or the adult male mouse submaxillary gland. The latter organ has two morphologically and functionally distinct components, the acini, where the mucopolysaccharides and amylases are concentrated, and the convoluted tubules, in which are found the proteolytic and hydrolytic enzymes and NGF.[2] The evidence for the location of NGF in the convoluted tubules comes directly from the use of fluorescent antibody to NGF,[26,27] from the parallelism between development of the tubules and the NGF increase in the gland, and from the sexually dimorphic character common to both the tubules and NGF.[28,29] Despite the fact that removal of the submaxillary gland has no influence on the sympathetic nervous system in mice,[6] that there are large differences in terms of NGF levels between male and female submaxillary glands, and that the high content of NGF appears to be restricted to the gland of the mouse as contrasted with other mammalian species, there is considerable evidence that NGF is actually synthesized in the gland. Burdman and Goldstein[30] argued that if NGF was being transported to the submaxillary gland from some other site of synthesis, then intravenous injection of antibody to NGF should interfere with the accumulation of NGF in the gland. Under a wide variety of conditions and with animals at different stages of development they were unable to demonstrate any such effect, nor could they show any accumulation of fluorescent antibody to NGF in the submaxillary gland when large quantities of this antibody were injected *in vivo*. More directly, Levi-Montalcini and Angeletti,[31] using slices of submaxillary gland, showed that the incorporation of labeled amino acids into NGF paralleled that of incorporation into other gland proteins. This would not be the result anticipated if the gland merely stored NGF.

MACROMOLECULAR STRUCTURE OF NGF

NGF from Snake Venoms

The Original Cohen Preparation and Multiple Forms Based on a Common Subunit

In the original preparation from venom of *Angkistrodon piscivorus* Cohen[14] achieved a 40-fold purification in the isolation of a protein which sediments as a single peak with an $s_{20,w}$ of 2.2*S*. The protein contains 1.6% hexose and its biological activity, optimal at about 100–200 ng/ml, is destroyed either by proteolytic enzymes or by antiserum to snake venom. The preparation showed some proteolytic activity as measured on denatured hemoglobin, but since this activity is destroyed by exposure to alkali while the NGF activity remains unaffected, it appears that the two activities are not associated. NGF activity was first extracted from the dried venom by dissolving it in approximately 8 *M* urea in 0.01*N* NaOH. After taking appropriate ammonium sulfate cuts to obtain an active fraction the purification continued with a series of ion-exchange chromatography steps. Since this NGF activity is not adsorbed to DEAE-cellulose at pH 6.0 and requires reasonably high salt concentrations for elution from CM-cellulose, it is apparently associated with a relatively basic protein. However, it should be noted that the initial extraction step at high pH in urea would disaggregate most subunit-containing proteins.

When dilute buffers at neutral pH are used to extract the venom, biological activity is obtained, as judged by gel filtration, in species whose molecular weights range from 20–60,000 depending on the species of snake used.[32,32a] This activity is not adsorbed onto DEAE-cellulose at neutral pH. There is considerable confusion in the literature on the properties of the DEAE and subsequent NGF fractions. Angeletti *et al.*[33] report that both gel filtration and sedimentation give it a molecular weight of about 40,000, that it can be resolved by acrylamide gel electrophoresis at pH 6.6 into two, basic, biologically active fractions, each of molecular weight about 24,000. In other words, electrophoresis under these conditions causes disaggregation. On the other hand, in more recent reports[2,3] the same DEAE fraction is described as having a molecular weight of 20,000. Perhaps the most detailed studies are those reported by Angeletti.[32a] After removing esterase activity from the DEAE fraction derived from the venom of *Crotalus adamanteus* by gel filtration, further chromatography on CM-cellulose at pH 5.0 resolves a number of biologically inactive proteins from two basic biologically active components which are only eluted from the resin at high salt concentration. The major part of the activity resides in the last fraction, and on subsequent gel filtration at pH 5.0 this can be partially resolved into a relatively homogenous component of $s_{20,w}$-2.5S and molecular weight 20–30,000 and a more heterogeneous component of $s_{20,w}$-1.56S and molecular weight 12,000. The latter can be further resolved by gel filtration on Sephadex G-25 at pH 5.0 into the 12,000 molecular weight species and a second species of equal specific biological activity but of molecular weight around 5000. If the fraction from the CM-cellulose column is first brought to neutral pH before gel filtration, three NGF species are found. The first has a molecular weight in excess of 100,000, the second is the 20–30,000 species noted above, and the third has a molecular weight significantly less than 12,000, but electrophoretically is not identical to the 5000 molecular weight species. Essentially the same results were obtained for NGF from the venom of *Bothrops jararaca*. Thus, five NGF species of different molecular weight have been identified. These are a form of molecular weight >100,000, species of molecular weight 20–30,000 and 12,000, respectively, which show the same two protein zones on electrophoresis, a 5000 molecular weight species which also shows two components on electrophoresis, one being identical to one of the components in the 12,000 molecular weight species, and a fifth component of unspecified low molecular weight. On the basis of these data, Angeletti suggests that these are interconvertible multiple-molecular-weight forms of NGF all based on the fundamental 5000 molecular weight NGF monomer. Supporting data for this hypothesis is apparently given by Levi Montalcini and Angeletti.[2] They state that the two components obtained from CM-cellulose chromatography of molecular weights 12,000 and 22,000, respectively, have the same amino acid composition and give superimposable complement fixation curves. Also, the specific biological activity of the larger component is twice that of the smaller. However, it is not clear to which of the two Angeletti fractions these last data pertain. The Cohen preparation may be identical to the larger of the above two components, especially since their specific activities are similar (see Angeletti *et al.*[3]). The method described above for isolating NGF from snake venom follows closely that previously described by Varon *et al.*[34] for the isolation of a high-molecular-weight form of NGF from mouse submaxillary glands. It will therefore be discussed further in a following section.

A Highly Active NGF from Venom of Angkistrodon piscivorus

Banks *et al.*[35] have recently described an alternative procedure for isolating a protein with properties very similar to those of the Cohen preparation, including its level of optimum activity (100 ng/ml). They also describe the isolation of other basic proteins from the venom of *Vipera russelli* which are optimally active at extremely low concentrations of 10^{-5} ng/ml. Their procedure is thus worth discussing in some detail. A solution of the crude venom was passed first through Sephadex G-150 and the active fractions (active at 50 ng/ml), after concentration, filtered again on Sephadex G-75. The active fractions from this column (active at 10 ng/ml and therefore already some 10 times more active than the regular snake NGF) were chromatographed on an anion-exchange resin at pH 10.8. The first peak from this column contained a basic protein, homogeneous by sedimentation and diffusion criteria, of molecular weight around 40,000 and active at 10^{-5} ng/ml. Since the ion-exchange chromatography achieved only a 30-fold purification in terms of protein but 10^6 fold in terms of activity, it must either be removing a potent NGF inhibitor or altering the properties of the basic protein in such a way as to drastically increase its specific activity. This highly active form of NGF rapidly loses activity on storage in solution, but stabilizes at a level where the optimal response is at 10^{-1} ng/ml. Differences in the electrophoretic patterns of a number of preparations led these authors to conclude that distinct but closely related types of NGF may be present in the various venoms, or that this form is susceptible to conformational changes. Because of its high activity they also conclude that they have isolated the active NGF molecule.

The molecular weight of this NGF species is in the size range described by Angeletti[32a] for NGF in the crude preparations from a number of venoms, although it does not appear to undergo the aggregation–disaggregation reactions reported for the latter. Both the normal and the highly active NGF species from snake venom are basic proteins. The methods of isolation have steps in common, but Banks *et al.*[35] employ an anion exchange at alkaline pH for the final purification step, while Angeletti *et al.*[3] use a cation exchanger at mild acid pH. The crucial difference between the two methods in terms of the level of activity of the product must be in this final chromatographic step.

NGF from the Mouse Submaxillary Gland

The Early Preparations

The original method for isolating NGF from this gland was also devised by Cohen.[29] The final protein has an $s_{20,w}$ of 4.3*S* and is optimally active at 15 ng/ml. The isolation procedure included precipitation by streptomycin sulfate, alcohol, and ammonium sulfate and three ion-exchange chromatographic steps. The behavior of the biologically active fraction on the first CM-cellulose column and probably also on the second DEAE-cellulose column is not that of a basic protein, but rather of a protein whose isoelectric point is less than pH 7.4. The third chromatographic step on CM-cellulose is initiated at pH 4.38, and at this stage the NGF activity requires high salt concentrations for elution and is probably now associated with a basic protein. This change in the protein character of the biological activity during the isolation procedure can be explained by the more recent results of Varon *et al.*,[34,36,37] which will be discussed shortly.

Schenkein and Bueker[38] used a similar isolation procedure to that of Cohen up to and including the DEAE-cellulose chromatography, with similar results. When the active fraction from this column, after dialysis against water, was subjected to paper electrophoresis at pH 7.4 three inactive proteins were resolved. A combination of two of them, protein A and the acidic polypeptide C of molecular weight 3500, gave activity in the usual bioassay. Protein A showed two components on sedimentation with $s_{20,w}$ of 2.4 and 4.2*S*, respectively. These authors concluded that the association of the smaller polypeptide C with the larger protein A was essential for activity.

A High-Molecular-Weight Form, 7S NGF

An alternative method for isolating NGF activity from the submaxillary gland has been worked out by Varon *et al.*[34] The three stages, gel filtration on G-100, chromatography on DEAE-cellulose, and gel filtration on G-150, are all carried out at neutral pH and the yield of final product is much higher in terms of protein than by the Cohen method, although the specific activity is the same—optimal response occurring at a concentration around 13 ng/ml. The major differences between the two products are in size and electrochemical properties. The alternative method produces a protein, between 95 and 98% of which sediments in one peak with an $s_{20,w}$ of 7.1S (designated 7S NGF) and whose isoelectric point is less than pH 7.0. The molecular weight of the 7S NGF species is thus of the order of 140,000 and it is not a basic protein. This same species can be identified as the only one with NGF activity at any of the intermediate purification stages, including the original homogenate of the submaxillary gland, and thus appears to be the NGF species first liberated from the gland.[34] Because of its availability in relatively large quantities (it accounts for some 2% of the soluble protein of the gland) the 7S NGF species has been extensively characterized. It is obtained from preparation to preparation as a protein with well-defined, reproducible physical properties. Besides eliciting the typical *in vivo* and *in vitro* NGF effects, 7S NGF preparations also exhibit enzymatic properties. The latter is evident as an esterase activity which can be assayed by the hydrolysis of simple argininyl esters.[40,40a] The kinetics of the hydrolysis by 7S NGF preparations exhibit a characteristic lag phase before reaching maximum velocity. The 7S molecule therefore carries at least two activities. The 7S NGF is a stable protein provided it is stored in solutions or freeze-dried from solutions whose pH's lie between 5 and 8. Outside of these pH limits it undergoes a reversible dissociation into smaller subunits with a concomitant drop in biological activity.[37] The subunits, although about the same in size, are of three widely differing types which can be easily separated by ion-exchange chromatography on CM-cellulose (Fig. 1).

Only one of them, the β subunit, elicits the characteristic response in the usual *in vitro* NGF bioassay. The β subunit is a basic protein with an $s_{20,w}$ around 2.5S and, as presently prepared, has approximately the same specific biological activity, 50–100 BU/μg protein, as the 7S NGF species. It is therefore optimally active in the same concentration range of 10–20 ng/ml.

Since a preparation of the β subunit can be obtained from the penultimate step in the 7S NGF purification which has slightly higher biological activity, it is possible that these levels of activity are lower than the intrinsic activity of a β subunit which has suffered no inactivation during isolation. Since the β subunit comprises about one-fifth of the total protein of 7S NGF,[39,41] only a similar fraction of the total biological activity is obtained after dissociation of the 7S NGF species. The relative inaccuracy

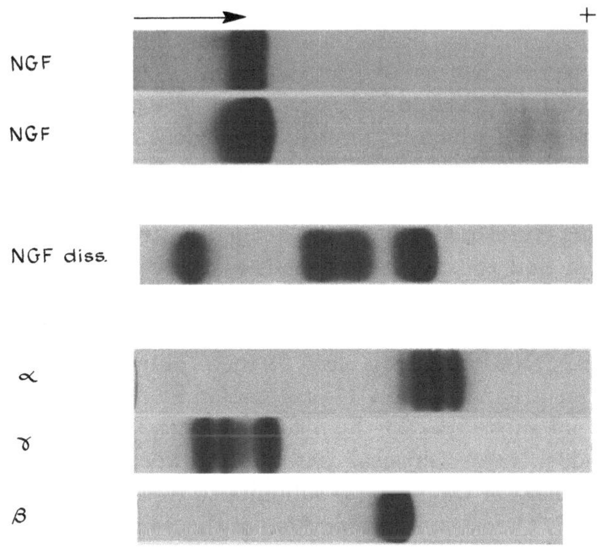

Fig. 1. 7*S* NGF and its subunits. The top two lines are analyses of 100 μg 7*S* NGF in the bistris-tes system at pH 7.55 and the tris tricinate system pH 8.3. The third line shows an analysis of 7*S* NGF in sodium glycinate buffer at pH 10.3, where it is completely dissociated into, from left to right, the β, γ, and α subunits. The bottom three lines contain analyses of the isolated subunits, the first two in tris tricinate and the last in an acetate system at pH 3.8. Migration in this last gel is toward the cathode. All gels were 7½% acrylamide.

of the bioassay procedure should be borne in mind when considering these figures. The esterase activity exhibited by 7*S* NGF preparations resides only in the γ subunits.[40] In contrast to 7*S* NGF, the γ enzyme shows no lag phase, but hydrolyzes substrate immediately at a maximal and linear rate. The specific activity of the γ enzyme is 6–10 times higher than that exhibited by 7*S* NGF. Since the γ subunits contribute about 45% of the 7*S* protein,[39] their enzymatic activity is decreased 2.5-to-4-fold in the 7*S* complex relative to the activity in the free form. Four individual species of the γ subunits are recognizable electrophoretically (Fig. 1); at least the three major ones have the same specific activities, identical to that of the γ enzyme. The third type of subunit obtained upon dissociation of 7*S* NGF, the α subunits, also comprise a group of four closely related, relatively acidic proteins (Fig. 1). No distinct activity has yet been recognized for the α subunits.

Regeneration of 7S NGF from Its Subunits

A characteristic feature of the isolated α, γ, and β subunits is their ability to recombine when mixed together at neutral pH to regenerate a larger protein species which by physical, biological, and enzymatic criteria is identical to 7*S* NGF (Fig. 2).[37,39–41] This recombination process is not just an aggregation of the β subunits with two other indifferent proteins, but the specific regeneration of a well-defined protein. It occurs optimally with particular and constant proportions of the three subunits, excess of one or two subunits in the mixture being readily detected by appropriate physical methods (Fig. 2). The relative proportions of the subunits needed to regenerate 7*S* NGF are very close to those obtained in the initial dissociation of an original 7*S* NGF preparation.[39] Moreover, other basic proteins will not substitute for the β subunit in the recombination process (Fig. 3). Besides these physical criteria, the recombination selectively alters the activities of the γ and β subunits to reproduce those characteristic of 7*S* NGF. Thus, the recombination of the β subunit with the α and γ subunits increases the total biological activity of the system by from 3-to-5-fold, as would be expected because the β subunit and 7*S* NGF have the same specific activity per unit of protein. This increase is seen to advantage in experiments analyzed in sucrose density gradients, since the activity of the β subunit, but not that of 7*S* NGF, is relatively unstable under these

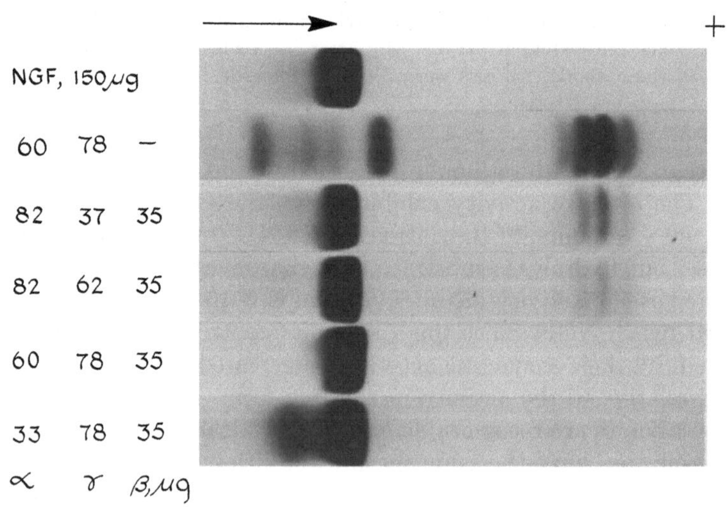

Fig. 2. The recombination of the 7*S* NGF subunits. Various proportions of the three subunits were mixed as indicated in 0.05 *M* Tris HCl buffer pH 7.4 and analyzed in 7½% acrylamide gels in the bistris-tes system at pH 7.55.

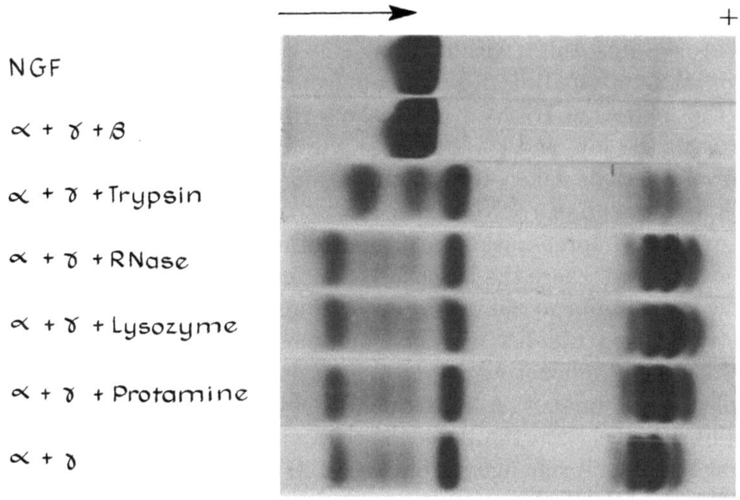

Fig. 3. The attempted recombination of α and γ subunits with a number of basic proteins. In each instance 60 μg of α and 78 μg of γ subunits were mixed with either 35 μg of β subunit or 150 μg of the other basic proteins and analyzed in 7½% acrylamide gels in the bistris-tes system at pH 7.55. Trypsin effectively digests γ^1 and the α subunits under these conditions.

conditions.[37] Again the specific enzymatic activity of the γ subunits decreases during recombination to approach that of 7*S* NGF, and at the same time the lag-phase kinetics seen only with 7*S* NGF reappears.[40,40a] Thus, the activity of the β subunits is specifically modified by their interaction with α and γ subunits, and that of the γ subunits by their interaction with α and β subunits. No pair of subunits can reproduce this specific recombination process. Although α or γ subunits react to some extent with β subunits, neither the biological activity nor the enzymatic activity of these intermediate complexes appears to differ from those of the corresponding free subunits.

The α and γ subunits contain more than one protein species. These individual subunits can be isolated by chromatography and each one shown to recombine with a complementary pair of subunits to produce a 7*S* NGF species containing only single types of α or γ or of both subunits.[41] Dissociation of such species produces only those individual subunits used in their formation. The original NGF preparation therefore consists of multiple forms of 7*S* NGF, each form being composed of α, γ, and β subunits but differing in the type of α or γ subunits they contain, and although these cannot be readily isolated, a gradient of such species can be demonstrated in the migrating zone of 7*S* NGF. This is possible because the mobility of a given 7S NGF species is determined by the type of α subunit it contains, those with α^1 being slower than those with α^2 subunits and so on. Subunit analysis of serial cuts from the low- to the high-mobility side of the migrating 7S NGF zone show that α^1 containing NGF's predominate on the low and α^4 NGF's on the high mobility side. NGF prepared from one lobe taken from a single mouse shows the same distribution of subunits and of 7S NGF species.

The 7S NGF preparation is not a static system, but, like hemoglobin,[42] is a system where the larger $\alpha\beta\gamma$ species are in dynamic equilibrium with free α subunits at all pH's, even around neutrality. This may be demonstrated in a number of ways, but most directly by observing the incorporation of radioactivity into 7S NGF when it is incubated with ^{125}I-labeled α subunits.[43] A consequence of this dynamic system is that the equilibrium between the parent species and subunits can be displaced by passage over ion-exchange resins. If 7S NGF is chromatographed slowly enough on DEAE-cellulose, some α subunits remain bound to the column and the corresponding excess of γ and β subunits are eluted as a $\beta\gamma$ complex.[44] This behavior has practical implications for the isolation of NGF by ion-exchange chromatography and may explain some of the apparently conflicting data in the literature.

NGF of Lower Molecular Weight from the Submaxillary Gland

An active NGF species comparable in presumed molecular size to the β subunit of 7S NGF has more recently also been reported by Zanini *et al.*[48] These authors describe a method for isolating NGF which follows very closely that of Varon *et al.*[34] using virtually identical gel filtration and DEAE-cellulose chromatographic procedures as the first two purification steps. They agree that NGF activity as extracted from the gland is in a high-molecular-weight form. For their subsequent and final fractionation step they use chromatography on CM-cellulose at pH 5.0 under conditions very similar to those employed by Varon *et al.*[37] for the separation of the 7S NGF subunits. It is not surprising, therefore, that the NGF activity is eluted last in association with a basic protein which appears to have the physical properties of the β subunit, i.e., it has an $s_{20,w}$ of 2.3S and gives a single band on acrylamide gel electrophoresis.[2] The corresponding molecular weight of 28,000[46] is at variance with the values of 40–50,000 previously quoted from the same laboratory[32,33] for similar fractions. The elution characteristics of the two other biologically inactive fractions from the CM-cellulose column leave little doubt that they contain the α and γ subunits. These authors therefore confirm the original findings of Varon *et al.*[34,36] that the high-molecular-weight form of NGF (7*S* NGF) can be dissociated at acid pH and the biological activity subsequently isolated as a single, smaller basic protein. While agreeing that this basic protein can aggregate with other proteins, such aggregates accounting for the NGF species in the initial gland extract, Zanini *et al.*[46] state that the other proteins separated by CM-cellulose chromatography (the α and γ subunits) do not contribute to the biological activity of the NGF.

The basic NGF protein separated by Zanini *et al.* has been assigned a molecular weight of 28,000 by gel filtration and apparently disaggregates further on standing in dilute solution into half-molecules of molecular weight 14,000. These two forms give identical immunodiffusion and immunoelectrophoretic patterns and complement fixation curves and are therefore considered to be different aggregates of a common subunit. Since species of higher molecular weight than the 2.3*S* or 28,000 molecular weight component are described, i.e., a 4.4*S* species[2] and a 40,000 or 44,000 molecular weight species,[3,33] higher states of polymerization of the common subunit are postulated. This hypothesis is similar to one put forward for NGF from snake venom.[32a]

The data of Zanini *et al.*[46] differ from that of Cohen[28] and of Varon *et al.*[37] in at least one significant detail. The concentration required for optimum *in vitro* response for the preparation of Zanini *et al.* is 1 ng/ml, while for the other two preparations it is 10–20 ng/ml. However, Zanini *et al.* now report that the activity of their NGF preparation from snake venom is optimally active at 1 ng/ml, in contrast to their previous agreement[3] with the 100–200 ng/ml level for the Cohen[14] or the Banks *et al.*[35] preparations. The newer levels of NGF required for optimum response can further be lowered by a factor of 10 if the dilutions are made in albumin solution.[46] Antibodies elicited against the mouse NGF cross-react to some extent with snake-venom NGF, but about 50 times more antiserum is required with the snake-venom NGF to produce a complement fixation curve equal to that obtained with mouse NGF.

NGF of Very High Activity from the Submaxillary Gland

An adaptation of their own method for preparing NGF from snake venom[38] and of Cohen's method[28] has led Schenkein *et al.*[45] to isolate an NGF preparation claimed to be optimally active at both 30 ng/ml and 10^{-6} ng/ml. This form is separated on DEAE-Sephadex at pH 7.5 using as starting material that part of the active fraction obtained after streptomycin sulfate, alcohol, and ammonium sulfate precipitation of the homogenate of the gland, which is not retained on CM-cellulose at neutral pH. This is a similar fraction to the one from which Cohen isolated his original

NGF by chromatography on CM-cellulose at acid pH. The highly active NGF is eluted from the DEAE column after a fraction rich in argininyl esterase(s), and Schenkein *et al.* suggest that these enzymes are responsible for the inactivation of their new NGF species. However, if this is so, it is not clear why simple separation of biological and enzymatic activities allows the very high biological activity to be expressed. Although these authors interpret their results as evidence for two different NGF's, NGFA and NGFB, the latter being highly active, an alternative conclusion would be that both activities reside in one type of molecule. It may be noted that the enzyme-rich fraction gives an electrophoretic pattern similar to that of the γ subunits, while the NGFB fraction contains at least zones from both α and γ subunits. It follows also that NGFB is relatively acidic in nature, in direct contrast to the basic character of the β subunit and also of the highly active form isolated by Banks *et al.*[35]

NGF Detected by Culture on Acrylamide Gel after Electrophoresis

Hoffman and McDougall[13] have described some very interesting results obtained by culturing chick-embryo sensory ganglia directly on top of an acrylamide gel through which submaxillary gland extracts have been electrophoresed. In these cultures cells migrate out from the ganglionic mass over the gel surface. After electrophoresis of the extract three positions on the gel show a higher migration and differentiation of cells from the ganglia indicative of three zones of NGF-like activity. Two of these zones are close together and electrophoretically resemble 7*S* NGF, but the third is associated with more-acidic proteins and is not interpretable in terms of the subunit properties of 7*S* NGF. An even more fascinating result was their finding that thrombin could mimic the biological action of NGF in their system. Because of the differences in assay conditions, it is perhaps unreasonable to directly compare these results with those obtained by the conventional bioassay, but the approach opens up new possibilities for the detection of NGF and other biological activities.

DISCUSSION

The Mouse Submaxillary Gland NGF

Since the 7*S* NGF preparation from the mouse submaxillary gland is the NGF which has been characterized most extensively, an attempt will be made to compare the other preparations with it. Starting with the other submaxillary gland preparations, it is possible, on the basis of the data obtained on the subunit structure of 7*S* NGF, to rationalize, in part, the Cohen preparation.[28] In the latter, NGF activity remains associated with a protein of isoelectric point less than 7.4, presumably the 7*S* NGF, until it is subjected to ion-exchange chromatography at acid pH. At this stage the activity is recovered with a more-basic protein. Since a similar step results in the dissociation of 7*S* NGF, the separation of its constituent subunits, and the recovery of NGF activity solely in the basic β subunits, it is reasonable to conclude that the Cohen method is achieving essentially the same result. The differences in the size of the final products, $s_{20,w}$ of 4.3*S* for the Cohen preparation and 2.5*S* for the β subunit, remains to be explained, but the higher $s_{20,w}$ of the former may result from aggregation—either under the conditions of the column or simply because they are eluted together—of the β subunit with one of the biologically inactive subunits. Alternatively, along the lines suggested by Levi-Montalcini and Angeletti,[2] the differences may be explained by polymerization of the β subunit. The stability of the biological activity compared to the proteolytic activity of the Cohen preparation at high pH can be understood in terms of the known stability of the β subunit and the presumed instability of the γ subunit or other proteases under these conditions.

The method of Schenkein and Bueker,[38] being similar to that of Cohen, presumably also isolates 7*S* NGF at least up through the chromatographic steps. However, the subsequent isolation of three inactive components and of the recovery of activity by the addition of the peptide fraction C to fraction A is not readily explained in terms of the known properties of 7*S* NGF. More recently, Schenkein *et al.*[45] presented data suggesting that a highly active NGF is uncovered in parallel with the removal of argininyl esterase. Separation of the γ subunits of 7*S* NGF, which have potent esterase activity against arginine esters, from the biologically active β subunits does not result in such a dramatic increase in NGF activity. The observations of Schenkein *et al.*,[45] therefore, must reflect some other, as yet unsuspected, phenomenon.

Accepting the physicochemical data of Zanini *et al.*[46] and of Levi-Montalcini and Angeletti,[2] there seems little reason to doubt that since their preparation follows so closely that of Varon *et al.*,[34] it also results in the isolation of the 7*S* NGF β subunit. The new levels of activity reported for their preparation are, however, some tenfold higher than those previously reported from the same laboratories[2] and those obtained by Varon *et al.*,[34] and the reasons for this have not yet been given. This may be the reflection, as noted earlier, of choosing conditions which cause less inactivation of the biological activity of the β subunit. Zanini *et al.* infer that the biologically inactive proteins separated from the β subunit by CM-cellulose chromatography at acid pH (the α and γ subunits) are just indifferent proteins with which the β subunit happens to be nonspecifically aggregated in the homogenate of the gland. This viewpoint ignores the detailed evidence presented by Varon *et al.*[34,36] and substantiated by later papers from this group [37,40,41] that the NGF activity in the homogenate is associated with a stable, well-characterized, subunit-containing protein, 7*S* NGF. Such evidence has already been summarized in an earlier section, but to emphasize just one point, it is difficult to conceive how the changes in the specific enzymatic activity and kinetics of the γ subunit which occur upon association with α and β subunits could result from anything other than a very specific interaction between subunits from a parent protein. The molecular heterogeneity of NGF activity can therefore be explained at one level by the existence of the $\alpha\beta\gamma$ complex or 7*S* NGF, the free β subunit, and the other complexes that can form between this subunit and the α or γ subunits.[34] Polymerization of the β subunit by itself has not been demonstrated and cannot at present be invoked to explain, at least for mouse NGF, the existence of aggregates of molecular weight greater than 30,000. The reports that both mouse and snake-venom NGF activity persists in species smaller than the β subunit[2,32a] are of considerable interest in terms of the possible substructure of the latter. Another level of molecular heterogeneity of NGF exists because of the occurrence of multiple forms of the 7*S* $\alpha\beta\gamma$ complex[41] (and also of the $\alpha\beta$ complex), but these forms are the same size and differ only in the type of individual subunits they contain. Whether the 7*S* NGF is the form in which NGF exists in the actual gland has not yet been determined. There remains the possibility, although unlikely, that the subunits are synthesized in different regions and only specifically recombine when released from the gland.

What relationship, if any, the α and γ subunits have to the biological activity of NGF has been discussed by Varon *et al.*[34,37] in terms of a number of alternatives, none of which could be emphasized on the basis of available data. The possibilities range from fine regulation of biological activity to its mere stabilization, to, as Zanini *et al.* also state, no relationship at all. It is pertinent that 7*S* NGF calls forth *in vivo* the same dramatic effects seen with the earlier preparations, which presumably do not contain the inactive subunits.[34] On the other hand, the β subunit is active alone, and antibodies made against NGF probably comprising only this subunit are sufficient to inhibit *in vitro* fiber growth in the bioassay and to produce *in vivo* the reported selective and permanent destruction of the sympathetic nerve cells.[11,11a] The question, therefore, remains an open one.

Snake-Venom NGF

The conditions used by Cohen[14] for extracting NGF activity from venom would dissociate a 7*S*-like NGF (if this were present in venom) and result in the subsequent isolation of a β-like subunit. The properties of Cohen's preparation, an $s_{20,w}$ of 2.2 and a basic protein, are indeed close to those of the β subunit. The gel filtration studies of Angeletti[32,32a] suggest, however, that activity in venoms is associated with smaller species than 7*S* NGF even though on DEAE-cellulose chromatography the activity behaves identically to 7*S* NGF (cf. Fig. 1 of Angeletti[32a] and Fig. 1 of Zanini *et al.*[46]) The final step in some of the methods of isolation of NGF from snake venom is similar to that used to separate the 7*S* NGF subunits and the elution patterns from the column do have areas in common. Major venom NGF activity is associated with the last and most basic protein(s) to be eluted, as is the activity from the mouse gland in the form of the β subunit. It is clear, therefore, that the final snake-venom NGF preparation may have many similarities to the active β subunit of 7*S* NGF, and it will be of considerable interest to make further detailed comparisons. This is emphasized by the finding[32a] that NGF activity from the venom is associated, as is that from the mouse, with proteins of different molecular weight, one at least being $>$100,000 in size. Of equal importance is the clarification of what sorts of physical or chemical changes in the NGF protein from either source result in the expression of an extremely high biological activity.[35,45] The characterization of the proteins which exhibit NGF activity has made real progress over the past few years, and this will obviously be continued. There is also hope that with this increased knowledge some further progress can be made in understanding the events which follow from the interaction of these proteins with the responsive neurons.

REFERENCES

1. R. Levi-Montalcini, *Harvey Lecture Ser.* **60,** 217–259 (1966).
2. R. Levi-Montalcini and P. Angeletti, *Physiol. Rev.* **48,** 534–569 (1968).
3. P. Angeletti, R. Levi-Montalcini, and P. Calissano, *Adv. Enzymol.* **31,** 51–75 (1968).
4. S. Cohen, R. Levi-Montalcini, and V. Hamburger, *Proc. Natl. Acad. Sci. U.S.* **40,** 1014–1018 (1954).
5. R. Levi-Montalcini, *in*: "Chemical Basis of Development" (W. D. McElroy and B. Glass, ed.), pp. 646–664, Johns Hopkins Press (1958).
6. R. Levi-Montalcini and S. Cohen, *Proc. Natl. Acad. Sci. U.S.* **42,** 695–699 (1956).
7. R. Levi-Montalcini and S. Cohen, *Ann. N.Y. Acad. Sci.* **85,** 324–341 (1960).
8. R. Levi-Montalcini and B. Booker, *Proc. Natl. Acad. Sci. U.S.* **42,** 373–384 (1960).
9. R. Levi-Montalcini, *in*: "Experimental Biology and Medicine, Morphological and Biochemical Aspects of Cytodifferentiation" (E. Hagen, W. Wechsler, and P. Zilliken, eds.), Vol. 1, pp. 170–182, Karger, Basel (1967).
10. R. Levi-Montalcini, H. Meyer, and V. Hamburger, *Cancer Res.* **14,** 49–57 (1954).
11. R. Levi-Montalcini and B. Booker, *Proc. Natl. Acad. Sci. U.S.* **42,** 384–391 (1960).
11a. R. Levi-Montalcini and P. U. Angeletti, *Quart. Rev. Biol.* **36,** 99–108 (1961).
12. R. Levi-Montalcini and P. U. Angeletti, *Develop. Biol.* **7,** 653–569 (1963).

13. H. Hoffman and J. McDougall, *Exp. Cell Res.* **51**, 485–503 (1968).
14. S. Cohen, *J. Biol. Chem.* **234**, 1129–1137 (1959).
15. P. Angeletti, A. Liuzzi, R. Levi-Montalcini, and D. G. Gandini–Attardi, *Biochim. Biophys. Acta* **90**, 445–450 (1964).
16. P. Angeletti, A. Liuzzi, and R. Levi-Montalcini, *Biochim. Biophys. Acta* **84**, 778–781 (1964).
17. A. Liuzzi, P. Angelletti, and R. Levi-Montalcini, *J. Neurochem.* **12**, 705–708 (1965).
17a. M. Graves, S. Varon, and G. McKhann, *J. Neuropath. Exp. Neurol.*, in press.
18. P. Angeletti, D. Gandini-Attardi, G. Toschi, M. Salvi, and R. Levi-Montalcini, *Biochim. Biophys. Acta* **95**, 111–120 (1965).
19. D. Gandini-Attardi, P. Calissano, and P. U. Angeletti, *Brain Res.* **6**, 367–370 (1967).
20. J. A. Burdman and A. R. Dravid, *Brain Res.* **6**, 355–358 (1967).
21. G. Toschi, E. Dore, P. U. Angeletti, and R. Levi-Montalcini, *J. Neurochem.* **13**, 539–544 (1965).
22. G. Toschi, E. Dore, and O. Sapera, *in*: "1st Intern. Meeting Neurochemistry," Strasbourg, Abstr. of Communications (1967).
23. J. A. Burdman, *J. Neurochem.* **14**, 367–371 (1967).
24. S. Crain, H. Benitez, and A. E. Vatter, *Ann. N.Y. Acad. Sci.* **118**, 206–231 (1964).
25. R. Levi-Montalcini, F. Caramia, S. A. Luse, and P. U. Angeletti, *Brain Res.* **8**, 347–362 (1968).
26. M. N. Goldstein and J. A. Burdman, *Anat. Record* **151**, 199–208 (1965).
27. R. Levi-Montalcini and P. U. Angeletti, *Quart. Rev. Biol.* **36**, 99–108 (1961).
28. S. Cohen, *Proc. Natl. Acad. Sci. U.S.* **46**, 302–311 (1960).
29. R. Levi-Montalcini and P. U. Angeletti, *in*: "Regional Neurochemistry," Proc. 4th Intern. Neurochemical Symp. (S. S. Kety and J. Elkes, eds.), pp. 362–376, Pergamon, New York (1961).
30. J. A. Burdman and M. N. Goldstein, *J. Exp. Zool.* **160**, 183–188 (1965).
31. R. Levi-Montalcini and P. U. Angeletti, *in*: "Ciba Found. Symp. Growth of the Nervous System" (G. E. W. Wolstenholme and M. O'Connor, eds.), pp. 126–147, Churchill, London (1968).
32. R. A. H. Angeletti, *J. Chromatog.* **36**, 535–537 (1968).
32a. R. A. H. Angeletti, *J. Chromatog.* **37**, 62–69 (1968).
33. P. Angeletti, P. Calissano, J. S. Chen, and R. Levi-Montalcini, *Biochim. Biophys. Acta* **147**, 180–182 (1967).
34. S. Varon, J. Nomura, and E. M. Shooter, *Biochemistry* **6**, 2202–2209 (1967).
35. B. E. C. Banks, D. V. Banthorpe, A. R. Berry, H. S. Davies, S. Doonan, D. Margaret Lamont, R. Shipolini, and C. A. Vernon, *Biochem. J.* **108**, 157–158 (1968).
36. S. Varon, J. Nomura, and E. M. Shooter, *Proc. Natl. Acad. Sci. U.S.* **57**, 1782–1789 (1967).
37. S. Varon, J. Nomura, and E. M. Shooter, *Biochemistry* **7**, 1296–1303 (1968).
38. I. Schenkein and E. D. Bueker, *Ann. N.Y. Acad. Sci.* **118**, 171–182 (1964).
39. A. Smith, S. Varon, and E. M. Shooter, unpublished data.
40. L. A. Greene, E. M. Shooter, and S. Varon, *Proc. Natl. Acad. Sci. U.S.* **60**, 1383–1388 (1968).
40a. L. A. Greene, E. M. Shooter, and S. Varon, *Biochemistry* **8**, 3735–3741 (1969).
41. A. P. Smith, S. Varon, and E. M. Shooter, *Biochemistry* **7**, 3259–3268 (1968).
42. G. Guidotti, W. Konigsberg, and L. C. Craig, *Proc. Natl. Acad. Sci. U.S.* **50**, 774–782 (1963).
43. A. Smith, L. Greene, H. R. Fisk, S. Varon, and E. M. Shooter, *Biochemistry*, in press.

44. E. M. Shooter, S. Varon and J. Nomura, *Chimia* **22,** 143–144 (1968).
45. I. Schenkein, M. Levy, E. D. Bueker, and E. Tokarsky, *Science* **159,** 640–643 (1968).
46. A. Zanini, P. Angeletti, and R. Levi-Montalcini, *Proc. Natl. Acad. Sci. U.S.* **61,** 835–842 (1968).

DISCUSSION

SCHENKEIN: Would you care to comment on what groups are responsible for the stability of the 7*S* NGF, in view of the fact that it contains both acidic and basic subunits? Secondly, the finding of a lag phase in the kinetics of hydrolysis of substrate by the esteropeptidase is very interesting. Over the past year we have studied this or a similar enzyme intensively, and we are puzzled by this lag period. Our feeling now is that the lag period is related to the production of benzoyl arginine from the ester substrate, the amino acid derivative then being an activator of the enzyme. Do you have any comments on this?

SHOOTER: On your first point, I think the facts that (1) 7*S* NGF is stable between pH 5 and 8, but not outside this pH range, and (2) that two of the subunits, α and β, will have significant net charge at neutral pH, suggest that ionic interactions between oppositely charged groups play perhaps a major role in stabilizing the complex. That dissociation occurs below pH 5 would appear to implicate aspartate and glutamate residues, and, at the alkaline end, histidine, α-amino, or even lysine residues. Concerning the lag period, your suggestion is an interesting possibility. We suspect, however, that it has something to do with dissociation of the 7*S* NGF complex, perhaps involving substrate.

PORCELLATI: Is the esteropeptidase activity inhibited by organophosphate compounds?

SHOOTER: Yes, but even at many-fold excess the inhibition is not complete.

PORCELLATI: Have you made any studies on NGF during regeneration or demyelination?

SHOOTER: No, we haven't; I don't recall any that have been done.

FOLCH-PI: Could Drs. Shooter and Schenkein comment on the marked differences between the amounts of their two preparations required for optimal response in the bioassay. For 7*S* NGF this is about 10^{-8} g/ml and for Dr. Schenkein's preparation 10^{-14} g/ml. For a biological activity the requirement of 10^{-8} g/ml is not particularly good, and it may be that in spite of your carefully controlled isolation procedure and detailed characterization you are simply carrying a much more active component on a larger amount of inert protein.

SHOOTER: The 7*S* NGF preparation only elicits growth in the usual bioassay system at concentrations around 10^{-8} g/ml, and we have as yet not observed growth response at the much lower concentrations used by Dr. Schenkein and his colleagues.

SCHENKEIN: Dr. Folch-Pi is possibly quite right. The activity which we have observed by diluting out our preparation to these incredibly low protein concentrations immediately led us to suspect that none of us is really dealing with the true NGF molecule. The phenomenon we observe is that there is a peak of biological activity in our preparations at 10^{-8} g/ml. However, if you keep on diluting the solution the activity never entirely disappears, and, indeed, a second peak is observed at 10^{-14} g/ml. This is difficult to accept unless you actually see it. The question we are now asking is whether we are getting a different phenomenon at very high dilution which appears to have the same final biological effect as NGF.

FOLCH-PI: With respect to the activity at very high dilutions, you must bear in mind that even here you still have billions of molecules which could easily saturate all possible sites on the ganglia. Tetanus toxin, for example, exerts its potent biological effect at much smaller concentrations still.

SHOOTER: The experiments I have discussed certainly offer no clue at this stage as to whether or not the NGF activity will finally turn out to be associated with a much smaller entity associated with 7*S* NGF or its β subunit. However, even if it is, one must still be impressed by the specificity of this association. The β subunit, for example, is isolated using a number of different physicochemical procedures, it can be recombined with the other subunits to reform 7*S* NGF, and the latter can be isolated and redissociated to yield the same biologically active subunit. In other words, if there is a small peptide attached to this subunit, it is bound sufficiently strongly to survive a wide variety of treatments, and this suggests a high degree of specificity.

RAPPOPORT: I just wondered if you have any evidence to indicate whether a protein similar to 7*S* NGF is responsible for the activity of snake venom. I think if you do find a similar protein in a completely independent source it would be strong evidence in favor of your model.

SHOOTER: The evidence given by Levi-Montalcini and her colleagues suggests that NGF activity in snake venom is associated not only with a basic β-like protein, but also with higher-molecular-weight forms. Whether or not the latter will be found to contain α and γ-like proteins remains to be seen.

BOGOCH: Do you have any structural analysis on any of these fractions? Are any of them conjugated proteins?

SHOOTER: Yes, Levi-Montalcini and her colleagues showed in 1965 that the Cohen NGF preparation is a glycoprotein. We find carbohydrate attached to the β subunit.

KIES: Do you know the isoelectric point of the basic β subunit?

SHOOTER: It has an isoelectric point between pH 9 and 10.

KIES: The myelin basic protein has an isoelectric point we think in excess of 10, that is, even more basic than the β subunit. We also have evidence that the myelin basic protein can complex with other, rather acidic proteins in such a manner that the complex survives the conditions of extraction, electrophoresis, etc. Do you think this is happening with your β subunit?

SHOOTER: It is true that one must always be concerned about possible artifacts in the sense of complex formation, particularly between proteins of opposite charge. The critical point about the interaction of the β subunit, as I have outlined, is that it is with two other subunits, not just one. Moreover, these interactions are characterized by specific changes in the biological and enzymatic properties of the interacting subunits. This is one major reason we regard 7*S* NGF as a specific multisubunit protein.

Chapter 22

Regulation by Amino Acids of Protein Synthesis in a Cell-Free System from Immature Rat Brain: Stimulatory Effect of γ-Aminobutyric Acid and Glycine*

Claude F. Baxter

Neurochemistry Laboratories
Veterans Administration Hospital
Sepulveda, California,
Division of Neurosciences
City of Hope Medical Center
Duarte, California
and
Department of Physiology
UCLA School of Medicine
Los Angeles, California

and

Sujata Tewari

Neurochemistry Laboratories
Veterans Administration Hospital
Sepulveda, California

INTRODUCTION

There are several steps along the metabolic pathways of protein synthesis and degradation which can be regulated, directly or indirectly, by substances of low molecular weight. These include reactions at the operon level, where transcription of DNA into RNA may be affected,[1] and reactions at the substrate level, where the overabundance or short supply

* The expert technical help of Mrs. B. Manely during the course of the experiments and the diligent secretarial help of Miss L. Eaton during the preparation of this manuscript are gratefully acknowledged. This project is supported by NIH Grant No. NB-03743.

of a specific amino acid may act as a regulatory factor.[2–4] Recent data which hint at an interrelationship between neurochemical transmitter substances and protein metabolism[5, 6] are of particular physiological interest. If acetylcholine, catecholamines, γ-aminobutyric acid (GABA), or any other potential neurochemical transmitter were to modify protein metabolism at or near nerve endings, one could envision numerous mechanisms by which the repeated use of a specific neuronal pathway would lead initially to changes in the local rate of protein synthesis, and, subsequently, perhaps, to an alteration in the "connectivity" or conductivity of these or adjacent neuronal pathways. The mechanisms would be similar to some that have been proposed to explain certain aspects of learning and habituation.[7]

Motivated by these considerations and encouraged by a preliminary report (published later in detail[6]), we initiated studies to determine the effect and mechanisms of action of amino acids in general, and GABA in particular, upon the incorporation of radioactive amino acids into the proteins of a ribosomal system from immature rat brain. This particular *in vitro* system was chosen in order to avoid (as far as possible) major membrane barriers which could impede the passage of amino acid substrates to the sites of protein synthesis.

It was possible to show with this system that in the presence of Na^+ and low levels of amino acid substrates, both GABA and glycine (at physiologically reasonable levels) stimulate the incorporation of [^{14}C]-labeled amino acids into protein. The stimulating effect of GABA was demonstrable in brain tissue and, to a lesser extent, in kidney, but not in cell-free systems from liver, heart, or reticulocytes. One step in protein synthesis, which is accelerated in the presence of GABA, was localized within the reaction sequence which leads from the [^{14}C]amino acid substrate to the formation of [^{14}C]aminoacyl-tRNA.

METHODS AND MATERIALS

The details of the experimental procedures are described fully elsewhere.[8] Briefly: Ribosomes and a fraction containing the pH-5 enzymes were prepared by established methods from brains and other organs of 21-day-old male Sprague-Dawley rats. Both ribosomes and pH-5 enzymes were suspended in a medium containing 0.25 *M* sucrose, 0.02 *M* Trizma buffer pH 7.6, 0.01 *M* MgAc:, 0.04 *M* NaCl, 0.10 *M* KCl, and 0.006 *M* mercaptoethanol.[9] A normal incubation mixture of 1 ml contained 0.03 ml ribosomal and 0.05 ml of the pH-5 enzymes fraction. This corresponded to 250–300 μg ribosomal protein and 750–900 μg pH-5 enzymes protein per milliliter. In the incubation mixture the relative contribution made by protein from the ribosomal and pH-5 enzymes fraction varied by a ratio of from 1:3.0 to 1:3.5. Final concentrations of ATP, GTP, and U[^{14}C]*L*-leucine (or alternative amino acid) in the incubation system were 2.5 m*M*, 0.10 m*M*, and 0.5 μCi, respectively. GABA and other amino acids were added as indicated. Additions were made while keeping all solutions and incubation tubes at 0–4°C. Conditions for incubations lacking ribosomes or the pH-5 enzyme fraction were the same as for the whole system. All samples were run in duplicate and all experiments repeated at least once.

The incubation at 37°C for a given period of time were followed by inactivation of aliquots of the incubation mixture with ice-cold TCA. For the determination of radioactivity incorporated into protein the aminoacyl-tRNA was solubilized and removed with hot TCA. The proteins were precipitated on glass filter-disks and washed extensively.[8] Radioactivity incorporated into aminoacyl-tRNA was determined by measuring the difference in counts between pellets extracted with hot TCA (protein) and pellets extracted with ice-cold TCA (aminoacyl-tRNA and protein). When necessary the [^{14}C]leucyl-tRNA containing pH-5 fraction was isolated,[10] analyzed for radioactivity, and occasionally used for further incubations.

All radioactivity was counted in a scintillation counter. Unless otherwise stated, results are expressed as $\mu\mu$moles amino acid incorporated per milligram ribosomal protein/hour. Determinations of protein and RNA levels were made by conventional methods.[11, 12]

RESULTS

The basic characteristics of the cell-free system have been reported previously.[6,8] ATP, GTP, Mg^{++}, and the pH-5 fraction were at optimal levels as determined by actual test and as described for similar systems in the literature.[9,13,14] In confirmation of the earlier report,[6] an addition of GABA to this system stimulated the incorporation of U[^{14}C]*L*-leucine into protein. The rate of incorporation was linear for more than 60 min, both in the absence and presence of GABA [Fig. 1(A)]. Close to optimal stimulation of leucine incorporation was achieved with concentrations of GABA which reflect the actual levels found in different areas of rat brain [Fig. 1(B)].[15]

Under all conditions more than 90% of the incorporated counts were found in the particulate ribosomal protein fraction. This result appears similar to that obtained with a preparation from adult rat brain.[16] GABA stimulation of [^{14}C]leucine incorporation was most pronounced into the particulate protein fraction.

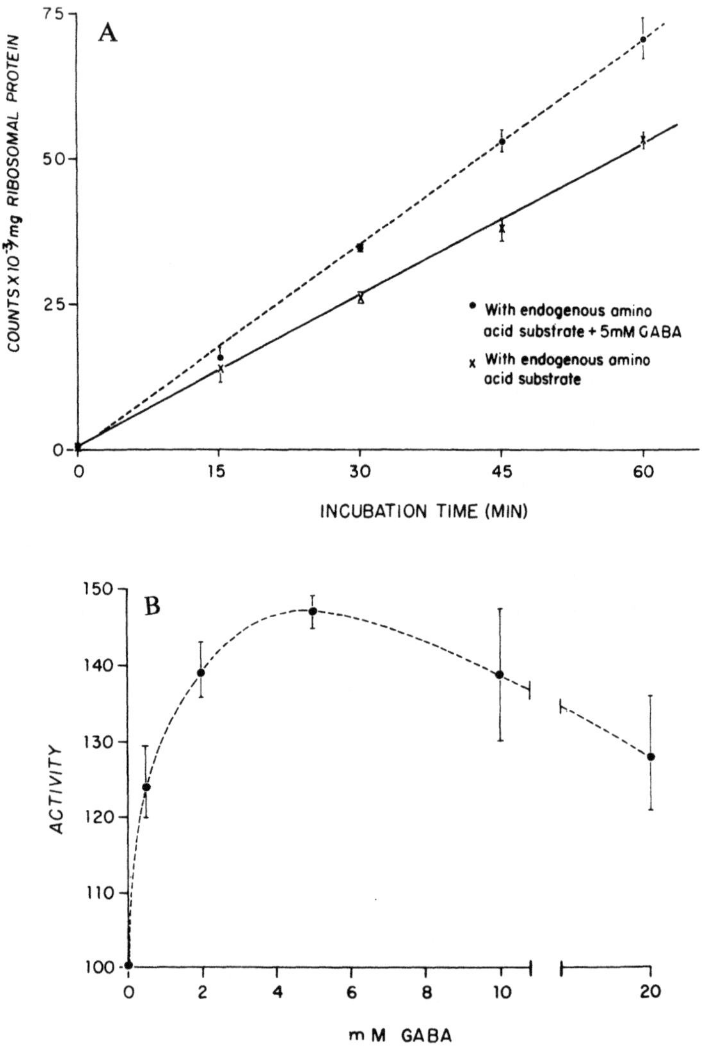

Fig. 1. U[^{14}C]leucine incorporation into ribosomal proteins of immature rat brain. (A) Time course: Vertical bars indicate the range of values obtained. (B) Effect of various concentrations of GABA. Activity is expressed as a percentage of control values. 100% = 213 $\mu\mu$moles leucine incorporated/mg ribosomal protein/hr. Vertical bars indicate range.

Effect of Na^+ in Mitochondrial and Ribosomal Systems

In earlier experiments the rates of leucine incorporation into protein of both mitochondrial and ribosomal systems were measured. GABA did *not* stimulate leucine incorporation into the mitochondrial system (S. Tewari and C. F. Baxter, unpublished). Upon closer examination it was found that the incubation medium used[6] for the study with the mitochondrial system was very low in Na^+ as compared to the Na^+ in the ribosomal medium. By adding Na^+ or by replacing K^+ with Na^+ in the mitochondrial medium, it was possible to demonstrate a stimulation of U[^{14}C]leucine incorporation by GABA (Table I). Conversely, when potas-

TABLE I

Effect of Na^+ and K^+ Upon Incorporation of U[^{14}C]L-Leucine into Protein of Ribosomal and Mitochondrial System from Immature Rat Brain[a]

System	Ionic substrate concentrations		Amino acid substrate: Endogenous [^{14}C]-leucine incorporation		Endogenous + 5m*M* GABA [^{14}C]-leucine incorporation	
	Na^+ (mM)	K^{+b} (mM)	μμmoles/mg prot/hr	Rel. activity (%)	μμmoles/mg prot/hr	Rel. activity[c] (%)
Normal mitochondrial	0.8	26.1	11.7	100 ± 3	9.36	82 ± 2
	0.8	56.1	11.5	98 ± 5	10.8	92 ± 7
Modified mitochondrial	20.8	26.1	9.5	81 ± 7	14.3	122 ± 9
	20.8	6.1	9.4	80 ± 5	14.4	123 ± 9
Normal ribosomal	40	100	275[d] (244–306)	100 ± 10	382	139 ± 4
Modified ribosomal	—	140	264	96 ± 4	231	84 ± 0

[a] Mitochondria were isolated[18] and incubated in a medium containing the following components: radioactive leucine and mitochondria; mannitol 0.3 *M*; KCl 20 m*M*; $MgCl_2$ 10 m*M*; Trizma pH 7.6 10 m*M*; α-ketoglutarate (neutralized with KOH) 5 m*M*; ADP 2 m*M*; K_2HPO_4 10 m*M*; EDTA (Na salt) 0.2 m*M*; and GTP 0.1 m*M*. The ionic concentration of this medium is calculated (Table I) with the assumption that K_2HPO_4 is 11% dissociated at pH 7.6. The 0.8 m*M* Na^+ concentration is derived from the EDTA addition. The composition of the medium described in the preceding section was used to calculate the ionic components in the normal ribosomal system.

[b] The contribution made by KOH, which was used to neutralize amino acids and Tris ATP, in no case increased the K^+ in the various systems by more than 1% of the milimolar values shown.

[c] The relative activity is expressed as a function of leucine incorporation in the normal system without the addition of GABA.

[d] This value was quite variable with different preparations of ribosomes.

sium ions were substituted for sodium ions in the incubation medium of the ribosomal system, stimulation by GABA of leucine incorporation into protein could no longer be observed (Table I).

Effect of Glycine

A number of compounds structurally related to GABA were tested to see if they too would stimulate leucine incorporation into the protein of the ribosomal system (Table II). Only glycine produced an enhancement of leucine incorporation greater than that observed with GABA. Homocarnosine had a slight stimulatory effect. No other GABA analog significantly stimulated the isotope incorporation into protein; long-chain ω-amino acids actually inhibited U[^{14}C]*L*-leucine incorporation by about 50%.

TABLE II

U[^{14}C]*L*-Leucine Incorporation by Ribosomal System in the Presence of Compounds Structurally Related to GABA

Amino acid substrate addition, 5 *mM* [a]	"Relative" leucine incorporation activity (%)
Endogenous	100[b]
+ Glycine	157
+ P-Alanine	92
+ GABA	145
+ δ-Amino valeric acid	53
+ ε-Amino caproic acid	51
+ α-Aminobutyric acid	98
+ α-γ-Diaminobutyric acid	93
+ N-Acetyl GABA	68
+ γ-Guanidinobutyric acid	98
+ Homocarnosine	115

[a] All compounds were neutralized (if necessary) with KOH or HCl to a final pH between 6.0 and 7.0 before addition.

[b] 100% activity in the system containing only endogenous amino acid substrate represents a net incorporation of 369 μμM of U[^{14}C]*L*-leucine into protein/mg ribosomal protein/hr (or 115,313 cpm incorporated/mg ribosomal protein/hr). All systems with additions to the substrate are compared to this value. In three separate experiments variability of results was usually no greater than ±3% except in the case of GABA, ε-aminocaproic acid, α-γ-diaminobutyric acid, and homocarnosine, where variability of up to 6% was found.

TABLE III

Effect of GABA upon U[^{14}C]*L*-Leucine Incorporation into Ribosomal System Protein, in the Presence of Different Amino Acid Substrates[a]

Test system[a,b]	(Ref.)	Exogenous amino acid substrate: GA	GME	ASP	ASGN	ALA	GLY	SER	VAL	THRE	LYS	PROL	ILEU	HIST	METH	TYR	ARG	φ-ALA	TRYP	CYSH	CYSSCY	HEXAM	U[^{14}C]*L*-Leucine incorporation, activity[c] (%): Without GABA	+5 m*M* GABA
1	(9)	—	—	—	—	—	—	—	—	—	—	—	—	—	—	—	—	—	—	—	—	—	100	140
2	(9)	—	—	—	—	—	—	—	—	—	—	—	—	—	—	—	—	—	—	—	—	—	82	104
3	(18)	0.05	0.05	0.05	0.05	0.05	0.05	0.05	0.05	0.05	0.05	0.05	0.05	0.05	0.05	0.05	0.05	0.05	0.05	0.05	0.05	0.05	122	144
4	(24)	0.59	0.22	0.50	—	0.39	0.52	0.72	0.61	0.76	0.32	0.52	0.23	0.08	0.06	0.13	0.21	0.21	0.06	0.04	—	0.22	100	134
5	(16)	11.4	2.23	4.29	2.23	1.0	1.0	2.0	0.14	1.0	0.36	0.14	0.05	0.11	0.06	0.10	0.14	0.06	0.14	—	0.14	—	134	122
6	(16)	2.85	0.551	1.07	0.551	0.25	0.25	0.50	0.035	0.25	0.09	0.035	0.013	0.028	0.015	0.025	0.035	0.015	0.035	—	0.035	—	120	143

[a] The endogenous source of amino acids in all incubation systems was from ribosomes and pH-5 enzymes with the exception of test system 2 in which ribosomes and cell sap was used. In this latter system, each 1 ml incubation system contained 0.208 mg ribosomal protein and 0.210 mg unprocessed cell sap protein (105,000g supernatant). In test system 4 the exogenous amino acid levels are 6.8% lower than those reported in the literature.[24]

[b] 100% Activity – 05,500 cpm/mg ribosomal protein/hr or 210 μμmoles leucine in corporated/mg ribosomal protein/hr.

Effect of Amino Acid Substrate

The choice of amino acid substrate conditions for the study of protein synthesis in ribosomal systems varies greatly from one laboratory to another, and effects observed in one system cannot always be duplicated in another. In fact, the inclusion of amino acid mixtures in brain ribosomal systems have been reported variously to stimulate or inhibit [^{14}C]amino acid incorporation.[16,17–19] In a comparative study we tested the effect of GABA upon leucine incorporation in the presence of six substrates which (with the exception of cell sap) differed only in their amino acid composition (Table III). Substrates 1 and 2 contained only endogenous amino acids,[9,20–22] either from the ribosomal fraction or from the cell sap.[9]

The third substrate, containing 0.05 μM of 20 amino acids, was originally used to study protein synthesis in bacterial systems.[23] It has since been used extensively for the study of protein synthesis in rat-brain ribosomes.[18]

The fourth amino acid substrate of low molar concentration was used to study protein synthesis in rabbit-brain ribosomes.[24] Its *relative* amino acid composition is based upon the concentrations reported to occur in brain extracts. However, the *actual* amino acid concentrations are considerably lower than those found in extracts of rabbit brain.

The fifth substrate, containing high levels of amino acids, was originally tested in a study of amino acid incorporation into mature rat-brain cortex ribosomal protein.[16] The concentration and composition of this mixture is based upon the "free" amino acid levels of mature rat-brain cortex extracts. It was tested both at full and at one-quarter strength (substrate 6).

The results in Table III show that inclusion of most of the amino acid mixtures enhanced the incorporation of U[^{14}C]*L*-leucine into proteins of the immature rat-brain ribosomal system. This finding differs from observations with preparations from mature rat brain in which no stimulation of leucine incorporation could be elicited by the addition of exogenous amino acids.[16] GABA stimulated the incorporation of leucine in the presence of cell sap and in conjunction with all but the highest level of added amino acid substrate.

Effect of Individual Amino Acid Addition to Endogenous Substrate

The foregoing observations led to a study of the effect of individual amino acids upon leucine incorporation and the possible interaction of single amino acids with GABA in the ribosomal system. Results are summarized in Table IV.

TABLE IV

Interaction of Individual Amino Acids at Tissue Extract Concentrations, With GABA, As Measured by U[^{14}C]*L*-Leucine Incorporation into Ribosomal Protein[a]

Group	Amino acid substrate addition	(m*M*)	U[^{14}C]*L*-Leucine incorporation, activity (%) Without GABA	+5 m*M* GABA	GABA effect in the presence of amino acid Δ (%)
	Endogenous only		100	140	+40
A	+ Threonine	1.00	98	131	+34
	+ Lysine	0.36	70	87	+24
	+ Proline	0.14	93	114	+23
B	+ Methionine	0.06	66	75	+14
	+ Phenylalanine	0.06	75	84	+12
	+ Cysteine	0.05	86	95	+11
	+ Histidine	0.11	98	107	+ 9
C	+ Glutamic acid	11.14	104	111	+ 7
	+ Aspartic acid	4.23	108	114	+ 6
	+ Glutamine	2.23	77	81	+ 5
	+ Tryptophan	0.14	76	79	+ 4
	+ Asparagine	2.23	82	84	+ 2
	+ Isoleucine	0.05	76	75	− 1
	+ Valine	0.14	72	69	− 4
	+ Arginine	0.14	75	71	− 5
	+ Serine	2.00	73	68	− 7
	+ Cystine	0.14	66	61	− 8
D	+ Alanine	1.00	84	76	− 9
	+ Tyrosine	0.10	86	73	−15
	+ Glycine	1.00	147	112	−35

[a] The tissue extract concentrations are those determined for mature rat brain[(16)] except for cysteine, which was tested at the concentration used in another system.[(16)] Amino acids are listed in an order which reflects their interference, at tissue extract concentrations, with the stimulatory effect of GABA upon leucine incorporation. Threonine is the least inhibitory and glycine the most inhibitory. The numerical results are based upon three separate sets of experiments for each amino acid. For comparative purposes leucine incorporation into protein, in the presence of endogenous amino acids only, was designated as 100% activity. This represents an average value of 210 $\mu\mu$moles of leucine incorporated/milligram ribosomal protein/hour. Absolute incorporation rates differed slightly from one preparation of ribosomes to another. The percentage stimulation or inhibition of leucine incorporation after the addition of GABA to the system was calculated as $[(5)/(4) \times 100] - 100$ and was found to be reproducible within a range of ±4% of the values shown in the last column.

Of 20 amino acids at tissue extract levels only glycine stimulated leucine incorporation into protein to a greater extent than did GABA. Eleven amino acids by themselves inhibited leucine incorporation (Table IV). Contrary to results with tissue slices,[25] glutamic and aspartic acid stimulated leucine incorporation mildly in some of the cell-free preparations. At a concentration of 5 m*M*, histidine stimulated leucine incorporation by a few per cent.

Amino Acid–GABA Interaction

The interaction of individual amino acids with GABA as measured by U[C^{14}C]leucine incorporation into protein is shown in the fifth column of Table IV. No amino acid, at levels found in brain extracts, potentiated the GABA stimulation of U[^{14}C]*L*-leucine incorporation. Virtually unaltered stimulation of leucine incorporation by GABA was observed only in the presence of threonine. When glycine, tyrosine, or alanine (group D) were present the addition of GABA resulted in a substantial inhibition of leucine incorporation.

Use of Different "Marker" Amino Acids

The incorporation of [^{14}C]*L*-leucine, [^{14}C]*L*-phenylalanine, or [^{14}C]*L*-lysine into protein has been used widely as a criterion of protein synthesis in various organisms and tissues, including brain.[20,26–28] The stimulatory effects of GABA and glycine upon the incorporation of these three isotopically labeled amino acids into the proteins of the brain ribosomal system are compared in Table V. Overall rates of incorporation of the three amino acids varied, but incorporation of all was stimulated, to a larger or smaller degree, in response to the addition of GABA or glycine.

The addition to the incubation mixture of 5 m*M* GABA as well as glycine resulted in an inhibited incorporation of both U[^{14}C]*L*-leucine and U[^{14}C]*L*-lysine into protein. However, under identical conditions U[^{14}C]*L*-phenylalanine incorporation was stimulated by an additional 75%.

Conversely, when a low-level amino acid mixture was added to the endogenous substrate both [^{14}C]leucine and [^{14}C]lysine incorporation was stimulated significantly, while [^{14}C]phenylalanine incorporation was inhibited. Addition of GABA further stimulated incorporation of U[^{14}C]-leucine and U[^{14}C]lysine, but had no significant effect upon the incorporation of U[^{14}C]phenylalanine.

These complex results emphasize the fact that studies of "protein synthesis" in immature rat-brain ribosomes, as measured by amino acid

TABLE V

Effect of GABA, Glycine, and an Amino Acid Mixture upon Incorporation into Protein of Three U[^{14}C]-Labeled Amino Acids in Immature Rat-Brain Ribosomal System

Amino acid substrate	Radioactive amino acid used[a]					
	U[^{14}C]*L*-Leucine		*U*[^{14}C]*L*-Lysine		U[^{14}C]*L*-Phenylalanine	
	$\mu\mu$moles/mg/hr	%	$\mu\mu$moles/mg/hr	%	$\mu\mu$moles/mg/hr	%
Endogenous	296	100	172	100	80	100
+ 5 m*M* GABA	417	141	186	108	98	123
+ 1 m*M* Glycine	456	154	239	139	101	126
+ 5 m*M* GABA + 1 m*M* glycine	332	112	153	89	162	203
+ 11 m*M* Glutamic acid	308	104	144	84	69	87
+ 5 m*M* GABA + 11 m*M* glutamic acid	332	112	165	96	80	100
+ 0.05 m*M* amino acids[b]	382	129	222	129	53	66
+ 5 m*M* GABA + 0.05 m*M* amino acids[b]	444	150	256	149	54	68

[a] Specific activity of isotopically labeled amino acids was U[^{14}C]*L*-leucine, 200 μCi/μM; U[^{14}C]*L*-lysine, 198 μCi/μM; U[^{14}C]*L*-phenylalanine, 355 μCi/μM. For every 1 ml of incubation mixture 0.5 μCi of amino acid was used.

[b] The exogenous substrate contained 20 amino acids[18] (see Table III), except that in each case the radioactive amino acid being tested was omitted from the mixture.

incorporation, will yield answers which depend heavily upon substrate conditions and upon the labeled amino acid used as the marker for protein synthesis or turnover.

Relative Organ Specificity

A comparison was made between the effect of GABA upon cell-free protein-synthesizing systems from brain, liver, kidney, heart, and reticulocytes of immature rats. Ribosomes and the pH-5 enzyme fractions were prepared from each tissue by very similar procedures, except that heart ribosomes were first isolated in a medium low in Na^+ and K^+,[29] and reticulocyte preparations were not subjected to desoxycholate treatment.[30] The purity of the ribosomal preparations was checked by electron micro-

Ribosomes \ pH 5 enzymes	BRAIN	LIVER	KIDNEY	HEART	RETICULOS
BRAIN	++	−	−	−	−
LIVER	++	−			
KIDNEY	−*		+		
HEART	+			−	
RETICULOS	+				−

Fig. 2. Effect of GABA upon U[^{14}C]-leucine incorporation into ribosomal proteins of different rat organs. Methods for the preparation of the various subcellular fractions and incubation conditions are described in the text. (+): Systems in which the addition of 5 m*M* GABA did (+) or did not (−) increase U[^{14}C]-leucine incorporation into protein significantly. The incubation of liver, heart, kidney, or reticulocyte ribosomes with brain pH-5 enzymes resulted in an increase of 50–100% above normal in the rate of leucine incorporation into ribosomal protein. Despite this increase, GABA stimulated protein synthesis in all of these systems except kidney. In the case of kidney ribosomes, the rate of U[^{14}C]leucine incorporation was 300% greater with brain pH-5 enzymes than with kidney pH-5 enzymes.

scopy. Only the ribosomes from the reticulocyte preparation were contaminated with appreciable amounts of membranous material. In cell-free preparations from liver, heart, and reticulocytes the addition of GABA had no stimulatory effect upon the incorporation of U[^{14}C]leucine into protein. The GABA effect was confined to preparations from brain and, to a lesser extent, to kidney (Fig. 2). It is of interest to note also that it is only in these two tissues that substantial amounts of endogenous GABA have been found.(31)

The relative degree of organ specificity for the GABA effect appears to reside in the pH-5 enzyme fraction of brain tissue. This was shown by experiments in which brain ribosomes were incubated with the pH-5 enzymes of other tissues, or ribosomes from liver, heart, kidney, and reticulocytes were incubated with the pH-5 enzymes from brain tissues. Whenever brain ribosomes were used with pH-5 fractions from other organs the incorporation of U[^{14}C]leucine into protein was not enhanced by GABA. However, in all systems in which the pH-5 enzyme fraction of brain were used together with ribosomes from other organs the addition of GABA enhanced the rate of U[^{14}C]lysine incorporation into protein. Kidney ribosomes were the one notable exception (Fig. 2).

Stimulation of [^{14}C]leucyl-tRNA Formation

The time course of GABA-stimulated U[^{14}C]leucine incorporation into the pH-5 enzyme fraction and, presumably, [^{14}C]leucyl-tRNA is shown in Fig. 3. Incorporation of radioactivity in the RNA fraction reached a maximum at about 15 min and thereafter declined. At no time period was there any radioactivity associated with the proteins of this ribosome-free system (see also Table VI).

TABLE VI

Effect of GABA on Formation of [^{14}C]Leucyl-tRNA from U[^{14}C]Leucine, and [^{14}C]-Labeled Ribosomal Protein from [^{14}C]Leucyl-tRNA[a]

System	Incubation time (min)	Activity in [^{14}C]leucyl-tRNA[b]	
		counts/min/ml mix.	counts/min/mg RNA
U[^{14}C]Leucine,	0	20	16
pH-5 enzymes	15	5340	4243
(control)			
	Δ15	+5320	+4227
U[^{14}C]Leucine,	0	23	18
pH-5 Enzymes	15	9652	7425
+5 m*M* GABA			
	Δ15	+9629	+7407
GABA stimulation	—	81%	75%
		Activity (counts/min/ml incubation mixture) in	
		[^{14}C]leucyl-tRNA	[^{14}C]ribosomal protein
[^{14}C]leucyl-tRNA,	0	4289	85
transferases, and	15	1976	2341
ribosomes			
(control)	Δ15	−2313	+2256
[^{14}C]leucyl-tRNA,	0	4281	89
transferases,	15	1889	2379
ribosomes, and			
5 m*M* GABA	Δ15	−2392	+2290
GABA stimulation	—	n.s.	n.s.

[a] Incubations of the 1-ml ribosomal system consisted of a 1.8 mg aliquot of the labeled pH 5 enzyme fraction containing the [^{14}C]leucyl-tRNA and transferase complex, 180 μg ribosomes, and 0.1 m*M* GTP. The incubation medium was the same as above, as were the extraction methods. Incubation with the pH-5 enzyme fraction was with 9.5 mg pH-5 enzyme protein, 2.5 m*M* ATP, and 5.0 μCi U[^{14}C]leucine per 10 ml incubation medium at pH 7.6. The incubation medium itself has been described previously.(9) At the end of the incubation period the pH-5 enzymes and aminoacyl-tRNA were reprecipitated out of a cold, diluted aliquot of the incubation mixture by adjusting the pH to 5.2 with *N* acetic acid. Further purification of these fractions by TCA precipitation, ethanol—ether extraction, etc., have been described previously.(9)

[b] All counts were soluble in hot TCA, indicating no incorporation into protein or into leucyl-adenylate complex.(32)

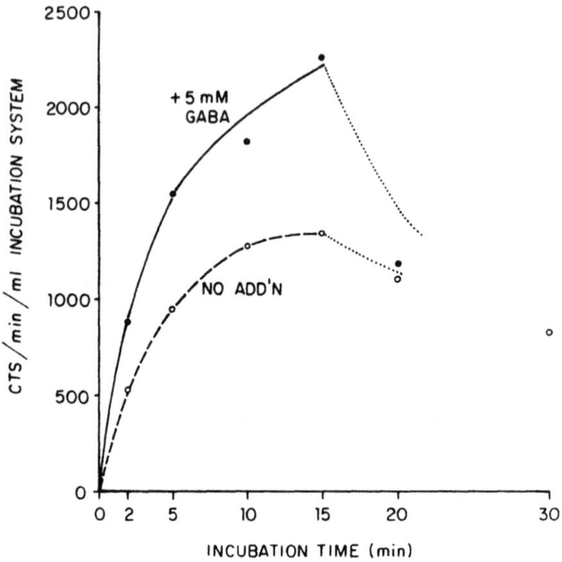

Fig. 3. Time study of GABA stimulation upon U[^{14}C]-leucine incorporation into [^{14}C]leucyl-tRNA of the pH-5 enzyme fraction from immature rat brain.

To test the effect of GABA upon the reaction sequence from leucyl-tRNA to ribosomal protein, the [^{14}C]leucyl-tRNA containing pH-5 enzyme fraction(10) was incubated with an unlabeled brain ribosomal preparation (Table VI).

The transfer of label from [^{14}C]leucyl-tRNA to ribosomal proteins was not enhanced by the addition of GABA to the system (Table VI). These results indicate that in the cell-free system from immature rat brain GABA stimulates a rate-limiting step of protein synthesis which is located somewhere between the free amino acid pool and the formation of the aminoacyl-tRNA.

DISCUSSION

In the foregoing studies the term "protein synthesis" has been used to designate the incorporation of radioactively labeled amino acids into a fraction of a cell-free ribosomal system which is precipitated with hot TCA. On the basis of this definition GABA, glycine, and various amino acid mixtures appear to stimulate protein synthesis or turnover in immature rat brain. The apparent requirement for Na^+ to demonstrate the

GABA effect may relate to the transport or binding of GABA to a specific subcellular site. The involvement of Na^+ in such functions, both with GABA and glycine, is well documented.[33–36]

Stimulatory and inhibitory effects of specific amino acid substrates upon protein-synthesizing systems of brain tissue have been reported from several laboratories.[3,25,28,37] The effect of GABA on protein synthesis is unusual only insofar as it involves an amino acid which is not known to be incorporated into protein. Furthermore, the effect appears to have some organ specificity. Both glycine and GABA have in common, not only the stimulatory effect on protein synthesis *in vitro*, but also the potential to simulate the action of an inhibitory transmitter substance at specific synapses in the nervous system.[38–41] There is evidence of limited protein synthesis and turnover in the synaptic regions of brain tissues,[42–44] and several authors have suggested an altered synaptic transmission as a consequence of an alteration in regional protein metabolism.[45,46] Thus, the dual effect of GABA and glycine suggests a regulatory function of inhibitory neurotransmitter agents upon protein synthesis.

The mechanism of such regulation might occur at various levels of the protein-synthetic process. At the substrate level the interaction of GABA with isolated amino acids, as well as with various amino acid mixtures, revealed an extremely complex set of results which, at this point, cannot be interpreted in a satisfactory manner. Although the interaction of glycine, alanine, or tyrosine with GABA (Table IV, group D) does not exclude competition by GABA for sites on specific activating enzymes, such a mechanism cannot be invoked to explain stimulation rather than inhibition of leucine incorporation. Competition for common carriers across membrane barriers also appears unlikely, both because ribosomes contain few well-defined membranes[47] and because the interaction of amino acids with GABA does not seem to follow the transport pattern of acidic, basic, and neutral amino acid groups.[36,48,49] No correlation was observed, either, with carbon chain length (see Table II).[50]

GABA in the presence of a number of substrates had differing effects upon the incorporation into protein of three [^{14}C]-labeled amino acids. The pattern of results may indicate differences in size of the endogenous pools of these amino acids, or differences in their respective activating enzymes, aminoacyl-tRNA or RNAse. It might also reflect stimulation in the synthesis or turnover of specific proteins whose amino acid composition differs from the average amino acid composition of the total ribosomal protein pool.

In many studies of protein synthesis, conclusions have been drawn on the basis of the incorporation rate of only one radioactive amino acid. The possible errors and limitations of such interpretations, in cases when more than just one specific protein is turning over or is being synthesized, are amply illustrated by the foregoing results.

Aside from the regulation at the substrate level, amino acids are known to affect polysomal aggregations,[51] RNA synthesis,[52] and a variety of other factors involved in protein synthesis. It is well to remember also that it is usually impossible to distinguish between a stimulation of [^{14}C]-amino acid incorporation into a protein and an inhibition in the rate of [^{14}C]amino acid excorporation out of the protein. Our recent data suggest that GABA is neither inhibiting proteinase activity, affecting RNAse activity, nor contributing to the energy metabolism of the *in vitro* system (S. Tewari and C. F. Baxter, unpublished).

The results presented in Fig. 3 and Table VI have focused attention upon the early metabolic steps in protein synthesis, and it is in this area that we are presently attempting to study the mechanism by which GABA exerts its stimulatory effect.

SUMMARY

1. The incorporation of [^{14}C]-labeled leucine, lysine, and phenylalanine into the proteins of a ribosomal system from immature rat brain has been studied *in vitro*.

2. With Na^+ and low levels of exogenous amino acids in the incubation medium GABA significantly stimulated the incorporation of these amino acids into brain proteins. A 40–50% stimulation of U[^{14}C]leucine incorporation was obtained with a 4 m*M* GABA concentration. Of all amino acids and GABA analogs tested, only glycine had a similar effect.

3. The mechanism by which GABA stimulates protein synthesis has been investigated. The interaction of GABA with other amino acids is complex. Both amino acid substrates and the [^{14}C]amino acid used as a marker profoundly affected the results obtained.

4. The stimulatory effect of GABA upon protein synthesis was most pronounced in cell-free systems from brain tissue, but was also evident, to a smaller degree, in a cell-free system from kidney. Organ specificity appears to be a function of the pH-5 enzyme fraction in the incubation system.

5. GABA stimulated the rate of [^{14}C]leucyl-tRNA formation from U[^{14}C]leucine by 75% but had no effect upon the subsequent incorporation of the [^{14}C]leucine into the ribosomal system.

6. It is suggested that GABA and glycine may have a regulatory function upon protein metabolism at or near nerve endings and thus may indirectly promote alterations in synaptic transmission.

REFERENCES

1. N. R. Cohen, *Biol. Rev.* **41**, 503–506 (1966).
2. H. Borsook, E. H. Fischer, and G. Keighley, *J. Biol. Chem.* **229**, 1059–1070 (1957).
3. B. M. Hanking and S. Roberts, *Biochim. Biophys. Acta* **104**, 427–438 (1965).
4. A. Fleck, J. Shepherd, and H. N. Munro, *Science* **150**, 628–629 (1965).
5. D. B. Peakall, *Proc. Soc. Exp. Biol. Med.* **126**, 198–201 (1967).
6. M. K. Campbell, H. R. Mahler, W. J. Moore, and S. Tewari, *Biochmistry* **5**, 1174–1184 (1966).
7. E. R. Kandel and W. A. Spencer, *Physiol. Rev.* **48**, 65–134 (1968).
8. S. Tewari and C. F. Baxter, *J. Neurochem.*, **16**, 171–180 (1969).
9. A. J. Munro, R. J. Jackson, and A. Korner, *Biochem. J.* **92**, 289–299 (1964).
10. M. B. Hoagland, M. L. Stephenson, J. F. Scott, L. I. Hecht, and P. C. Jamicnik, *J. Biol. Chem.* **231**, 241–257 (1958).
11. O. H. Lowry, N. J. Rosebrough, A. L. Farr, and R. J. Randall, *J. Biol. Chem.* **193**, 265–275 (1951).
12. P. N. Campbell and J. R. Sargent, *in*: "Techniques in Protein Biosynthesis" (P. N. Campbell and J. R. Sargent, eds.), Vol. 1, pp. 301–303, Academic Press, New York (1907).
13. S. Løvtrup and T. Zelander, *Exp. Cell Res.* **27**, 468–473 (1962).
14. M. R. V. Murthy and D. A. Rappoport, *Biochim. Biophys. Acta* **95**, 121–131 (1965).
15. C. F. Baxter and E. Roberts, *Proc. Soc. Exp. Biol. Med.* **101**, 811–815 (1959).
16. C. E. Zomzely, S. Roberts, and D. Rapaport, *J. Neurochem.* **11**, 567–582 (1964).
17. M. Satake, Y. Takahashi, K. Mase, and K. Ogata, *J. Biochem.* (*Tokyo*) **56**, 504–511 (1964).
18. M. R. V. Murthy and D. A. Rappoport, *Biochim. Biophys. Acta* **95**, 132–145 (1965).
19. A. T. Campagnoni and H. R. Mahler, *Biochemistry* **6**, 956–967 (1967).
20. D. H. Clouet, M. Ratner, and N. Williams, *Biochim. Biophys. Acta* **123**, 142–151 (1966).
21. Y. Takahashi, K. Mase, and S. Abe, *J. Biochem.* (*Tokyo*) **60**, 363–371 (1965).
22. D. H. Adams and L. Lim, *Biochem. J.* **99**, 261–265 (1966).
23. J. H. Mathaei and M. W. Nirenberg, *Proc. Natl. Acad. Sci. U.S.* **47**, 1580–1588 (1961).
24. K. H. Stenzel, R. F. Aronson, and A. L. Rubin, *Biochemistry* **5**, 930–936 (1966).
25. F. Orrego and F. Lipmann, *J. Biol. Chem.* **242**, 665–671 (1967).
26. R. S. Piha, M. Cuenod, and H. Waelsch, *J. Biol. Chem.* **241**, 2397–2404 (1966).
27. S. Yamagami, R. Friz, and D. A. Rappoport, *Biochim. Biophys. Acta* **129**, 532–547 (1966).
28. M. R. V. Murthy, *Biochim. Biophys. Acta* **119**, 586–598 (1966).
29. D. C. N. Earl and A. Korner, *Biochem. J.* **94**, 721–734 (1965).
30. H. R. V. Arnstein, R. A. Cox, and J. A. Hunt, *Biochem. J.* **92**, 648–661 (1964).
31. M. Zachmann, P. Tocci, and W. L. Nyhan, *J. Biol. Chem.* **241**, 1355–1358 (1966).
32. P. Rouget and F. Chapeville, *European J. Biochem.* **4**, 310–314, (1968).
33. K. A. C. Elliott and N. M. Van Gelder, *J. Physiol.* (*London*) **153**, 423–432 (1960).
34. K. Sano and E. Roberts, *Biochem. Pharmacol.* **12**, 489–502 (1963).
35. S. Varon, H. Weinstein, C. F. Baxter, and E. Roberts, *Biochem. Pharmacol.* **14**, 1755–1764 (1965).
36. P. N. Abadom and P. G. Scholefield, *Can. J. Biochem. Physiol.* **40**, 1575–1590 (1962).
37. S. Roberts and B. S. Morelos, *J. Neurochem.* **12**, 373–387 (1965).
38. D. R. Curtis and J. C. Watkins, *J. Neurochem.* **6**, 117–141 (1960).

39. A. Takeuchi and N. Takeuchi, *J. Physiol.* **183,** 418–432 (1966).
40. M. Otsuka, E. A. Kravitz, and D. D. Potter, *J. Neurophysiol.* **30,** 725–752 (1967).
41. R. Werman, R. A. Davidoff, and M. H. Aprison, *Nature* **214,** 681–683 (1967).
42. I. G. Morgan and L. Austin, *J. Neurochem.* **15,** 41–51 (1968).
43. K. Von Hungen, H. R. Mahler, and W. J. Moore, *J. Biol. Chem.* **243,** 1415–1423 (1968).
44. M. W. Gordon and G. G. Deanin, *J. Biol. Chem.* **243,** 4222–4226 (1968).
45. R. Elul, *Nature* **210,** 1127–1131 (1966).
46. I. B. Gartside, *Nature* **220,** 383–384 (1968).
47. E. D. P. De Robertis, W. W. Nowinski, and F. A. Saez, *in*: "Cell Biology," 4th Ed., p. 137, W. B. Saunders Co., Philadelphia (1965).
48. R. Blasberg and A. Lajtha, *Arch. Biochem. Biophys.* **112,** 361–377 (1965).
49. G. Levi, R. Blasberg, and A. Lajtha, *Arch. Biochem. Biophys.* **114,** 339–351 (1966).
50. K. D. Neame, *J. Neurochem.* **11,** 67–76 (1964).
51. B. S. Baliga, A. W. Pronczuk, and H. N. Munro, *J. Mol. Biol.* **34,** 199–218 (1968).
52. D. P. Nierlich, *Proc. Natl. Acad. Sci. U.S.* **60,** 1345–1352 (1968).

DISCUSSION

MAHLER: We have recently tried to repeat and extend our earlier observations with GABA using the more highly resolved polysomal system that we have described. Initially, we were very disappointed in finding that the results appear to be quite irreproducible; we could, however, make them completely reproducible and conform to the pattern that was so neatly presented by Dr. Baxter by working in an all-Tris system—by isolating the polysomes in Tris, and by omitting the amino acid mixture. We reproducibly find a stimulation of 40%, and when the amino acid mixture is added back the effect is wiped out. The effect is completely specific for GABA, none of the other GABA analogs work, and when glycine and the other neutral amino acids (we didn't use glycine alone) are added the effect is wiped out also.

FOLCH: You are using synthetic amino acid mixtures. Do those amino acid mixtures contain GABA?

BAXTER: No, they do not.

GOODWIN: I was very interested in the effect on the incorporation starting with the leucyl RNA. I think that some of the confusion on amino acid effects, effects of amines, drugs, etc., might be considerably reduced if these systems would be run in parallel while studying the activating enzymes which apparently seem to be much more specific for amino acids. We've studied a series of catecholamines, indolamines, amino acid conditons on the transferase step alone, and it's not nearly so confusing as when one studies the specific activating enzymes.

MURTHY: You seem to get similar effects either by adding the amino acids or by adding GABA. Have you measured the concentration of free amino acids in a reaction mixture after the addition of GABA? Does GABA affect the proteases or, in some manner, the release of amino acids?

BAXTER: We have not yet checked whether GABA affects proteolysis. Of course, when you get an increase of incorporation it can be either increased protein synthesis or a decrease in breakdown, and we are very much concerned to find out which it is. We do know that GABA has no effect on the latent endogenous RNAse of the system.

LAJTHA: You said that GABA affects a specific fraction and the phenylalanine incorporation patterns seem to be different from other stimulations. Do you have any evidence that the effects concern a specific protein fraction?

BAXTER: We have not fractionated the protein that we obtained with the incorporated phenylalanine. However, this is something that is technically very feasible and will be done eventually.

ROBERTS: On the basis of a specific stimulatory effect of GABA, how could you actually explan the observation that GABA stimulated the incorporation of certain amino acids, such as leucine and lysine, and inhibited phenylalanine. All proteins would have all of these amino acids in rather high concentration.

BAXTER: It might be unwise at this stage of our knowledge to speculate about the differential effects we obtained with different amino acids. But please note that when only endogenous amino acids were present in the test system GABA stimulated the incorporation of all of the radioactive amino acids tested, but to different degrees. Whether this reflects selective stimulation in the synthesis of proteins low in lysine or high in leucine, or indicates selective stimulation in the breakdown of proteins high in lysine (or any other number of possibilities) I do not know. The inhibition of phenylalanine incorporation in the presence of an exogenously added amino acid mixture (0.05 *M* final concentration) is indeed noteworthy. It has also been observed by Dr. Murthy. In his system cystine turned out to be the offending amino acid, but the reason for this has not been established.

MANDEL: If the effect is on leucyl *t*RNA synthesis, since your phenomena depends on sodium, this would also be sodium dependent.

BAXTER: Tris works as well as sodium ion. I'm not claiming that the GABA effect is on the synthesis of leucyl tRNA exclusively, because when the pH-5 fraction is precipitated and washed many transferases and other things may have been removed. When you reincubate the ribosomes with the pH-5 fraction there is a very good possibility that some essential factors are missing from the system. All we can say at this time is that GABA does stimulate the reaction sequence from amino acid to leucyl tRNA, but that in our system we cannot demonstrate stimulation beyond this point.

PORCELLATI: There are effects on other systems also. For instance, Tris seems to be a competitive inhibitor of the incorporation into phospholypids.

Chapter 23

Changes in RNA and Proteins Induced by Stimulation*

Donald A. Rappoport and Hatim F. Daginawala†

Division of Molecular Biology
Department of Pediatrics
and
The Marine Biomedical Institute
The University of Texas Medical Branch
Galveston, Texas

INTRODUCTION

One of the oldest and most intriguing problems in neurobiology is the elucidation of brain function. It has been apparent for a long time that the brain is essentially an electric organ which is capable of accrual, retention, and retrieval of information. In the past quarter of a century neurophysiologists made extensive studies of evoked electroencephalograms (EEG's) from the brain under a variety of conditions, but none of these observations revealed any clues as to the mechanism underlying brain function. Ablation experiments did show that the information storage was not uniquely localized[1] and that, excluding evoked potentials due to arousal, no correlations were noted in EEG's with conditioning or learning.

The chemists' efforts in brain research were initially analytical, such as the early work of Thudichum.[2] Due to the complexity of chemical constituents as well as the heterogeneity of cellular components in the brain, this did not yield any meaningful results which were relevant to brain function. Thus, up to 1950 little, if any, significant information was accrued from the chemical studies toward clarification of this problem.

* This work was supported by a James W. McLaughlin Fellowship awarded to Hatim F. Daginawala during the period January 1, 1963 through December 31, 1965, a grant from the Medical Research Foundation of Texas, three National Institutes of Health grants, MH 10053, 1S01 FR-05427, and DHEW 1 R01 NB 07707, and a National Science Foundation grant, GU 484-CO-3.

† Present address: Department of Biochemistry, Nagpur University, Nagpur, India.

The continuous cognizance and apperception of the physical environment immediately surrounding each one of the higher organisms is a consequence of a variety of sensory information received and integrated in the brain. Although each sensor receives information from the environment by a unique mechanism, all sensors transduce this information to an encoded electrical impulse conveyed to the brain as an afferent signal. Precisely how the individual sensors transduce information, how this is conveyed to the brain, how the afferent impulse is coded, and how the brain reads and retains this information is unknown, although fragmentary concepts and explanations for each of these processes are available.

Historically, the neurophysiologists conceptualized the accrual process of sensory information as an electrical phenomenon. Sir Charles Sherrington[3] postulated that a group of neurons, associated by means of their dendrites, set up a circuit in which one neuron, stimulated by an impulse transmitted from a sensor *via* the axon, initiates a series of discharges. These discharges excite a group of neurons, thus setting up a circuit of evoked consecutive discharges which he designated as a "reverberatory circuit." Lashley[1] indicated that this, or a similar electrical phenomenon, represented the electrical "engram" of newly acquired information, and so, additional bits of information would also set up new reverberating circuits. This, in essence, was the conceptualized electrical model representing the form in which information is retained in the brain. Hebb[4] further expanded this concept by developing a two-process theory of memory. In the initial process of learning a trace is established in the brain, essentially a reverberatory circuit, which is evanescent and unstable. This, in turn, if reinforced by additional conditioning, will cause growth of the synaptic ends, resulting in a stable engram of memory. This concept was based upon the evidence of Duncan,[5] who showed that animals receiving electroconvulsive shock (ECS) immediately after short-term conditioning did not retain the learned skill. But, if a long interval of time elapsed between conditioning and ECS, the learned skill was retained in spite of numerous ECS treatments. The concept of synaptic growth as an engram of memory storage was also supported by Weiss and Hiscoe,[6] who observed axoplasmic flow of protein toward the synapses during their growth. None of the existing evidence directly proved Hebb's concept. Sir John Eccles[7] found that repeated conditioning of an identical skill facilitated synaptic transmission of the evoked potentials *via* the same synaptic pathways, thus suggesting that the information acquired during conditioning affected some form of coding along a particular synaptic route, which, after acquiring this engram in the initial phase of conditioning, facilitated transmission of identical coded signals through this coded pathway. The concept of "facilitated synaptic routes," which represents coded pathways for specific types of information, suggested that memory is an engram residing solely in the synaptic junctions.

Currently neurobiochemists have become interested in the mechanism of information accrual based on the findings of Hydén and co-workers[8–11] that following conditioning of rats there were changes in brain nuclear RNA. Their findings demonstrated that Deiter cell nuclei showed both an increase in RNA and a change in base ratio due to specific vestibular stimulation. In similar experiments they also demonstrated base ratio changes in glial RNA immediately surrounding Deiter neurons. In another series of experiments the Swedish investigators found changes in RNA content and base ratios in rat cortical neurons during training in transfer of handedness. They trained rats who are normally right-handed to reach for their food using their left paws. In this experiment changes were noted in neurons of that hemisphere associated with left-handedness compared to the control neurons which were available in the contralateral side of the brain. Their results showed that an increase in RNA occurred only after training. The RNA produced during forced conditioning also reflected an increase in (adenine + guanine)/(cytosine + uracil), $(A + G)/(C + U)$, ratio. They concluded from these studies that an acute learning situation, without prior precedant to the animal, acts as a stimulant which results in the production of RNA with an altered base ratio in the nuclei of directly involved neurons.

These findings led Hydén to propose a mechanism of information accrual in the brain in which neuronal nuclear RNA was designated as the memory trace. He stated that the coded afferent impulses arising from the sensors, consisting of frequency modulations, were responsible for the changes in base composition in the RNA. He further implied that since DNA-dependent RNA-polymerase was responsible for the RNA memory trace, in some unknown fashion the DNA, and, consequently, the RNA-polymerase, were affected by the afferent impulse so that a specific RNA was formed transcribing the information coded in the afferent impulse. The incongruous aspect of this mechanism is that it does not show how an afferent impulse can alter template properties of DNA so as to permit synthesis of unique RNA for each bit of information transmitted by the impulse.

Talwar *et al.*[12] also studied the effect of stimulation on brain RNA. They investigated the changes in rat-brain RNA during convulsion induced by pentamethylene tetrazol (Metrazole or Cardiazole). They also studied the influence of light stimulus on the synthesis of RNA in the occipital cortex. Their results showed that a portion of the RNA (10–12% of total RNA) undergoes a turnover in response to stimuli. Moderate stimuli tend to increase the content of RNA in the cells, while intensive sustained convulsive stress caused a decrease in RNA. They concluded that this decrease in RNA reflected the inability of the synthesis to keep pace with utilization, where utilization may consist of either (1) a breakdown of RNA, or (2) a transfer of RNA to other compartments.

Similar experiments have also been conducted by Zemp *et al.*[13] They showed that 15 min of conditioned avoidance training in rats increased the incorporation of radioactive uridine into brain RNA in contrast to unconditioned controls. The increased incorporation occurred in RNA that sediments in a sucrose gradient like mRNA from liver and other tissues. Although the increase in radioactivity occurs throughout the gradient, the greatest increase in radioactivity occurs in RNA which is isolated from brain ribosomes and which sediments at about 18 *S*. This corresponds to fractions of liver RNA reported to stimulate amino acid incorporation into acid-insoluble polypeptides. Jarlstedt[14] has studied RNA changes in rat cerebellar Purkinje cells after proprioceptive, exteroceptive, and vestibular stimulation, respectively, of rats. His results show that Purkinje cells increased their RNA content following stimulation.

Kogan and Zaguskin[15] have correlated changes in ribonucleic acid with electrical activity in crayfish single stretch receptor neurons during excitation and inhibition of nerve activity. Moderate excitation of the neuron caused disorganization of the fusiform arrangement of RNA in concentric layers and an increase in the total RNA. Strong excitation brings uniform distribution of RNA in the neuron besides an increase in total quantity of RNA. Short periods of inhibitory nerve activity restores the arrangement of RNA in concentric rows and increases the total quantity of RNA.

Recently Pevzner[16] summarized the results obtained by different groups of Russian investigators concerning the changes in brain RNA metabolism during excitation and inhibition of nervous activity. These results indicate that there was an increase in RNA synthesis during excitation of nerve cells and a preponderance of RNA breakdown during the inhibition of nerve cells.

Some workers have reported that electrical stimulation did not induce any changes in RNA. Edstrom and Grampp[17] studied the effect of electrical stimulation on the RNA metabolism in the nerve cell and its surrounding capsular tissue of the lobster stretch receptor organ. They showed that the amount and the base composition of RNA in nerve cells after continuous stimulation for about 8 hr at a frequency of about 5–10 sec^{-1} was not significantly different from those of unstimulated control cells. They also found that the uptake of ^{32}P into RNA of nerve cells or capsular tissue of the receptor organ did not change on stimulation. Addition of actinomycin D (2 μg/ml) completely inhibited the synthesis of RNA, and this had no effect on electrophysiologically measurable membrane properties, nor did it interfere with the continuous generation of action potentials of receptor organ for 12–24 hr.

Other investigators have studied the effect of RNA and protein inhibitors on animal behavior. Dingman and Sporn[18] injected intracisternally an analog of purine, 8-azaguanine-2-^{14}C, into rat brains, and this was incorporated into the RNA of the brain. In the presence of this analog the rats required more time to learn a new maze, although it did not affect their ability to perform a task and to recall a previously well-learned maze.

Flexner *et al.*[19–21] have evaluated the role of proteins in the maintenance of memory by using several inhibitors of protein synthesis, such as puromycin and acetoxycycloheximide. They found marked effects on the learning ability of mice following brain injection of puromycin. Using avoidance conditioning as a means of training the animals, mice were trained in a Y-maze with a grid floor that could be electrified. The mouse was placed at the stem of the Y; to avoid a shock he had to move within 5 sec into the correct arm of the maze. When puromycin was injected bilaterally into the hippocampi and adjacent temporal cortices of the mice short-term memory (one day after learning) was practically abolished. Injections into the other regions of the brain were essentially without effect. If mice were injected 11–43 days after training, injections in the hippocampal and temporal regions were also without effect. Only a combined bilateral, frontal, and ventricular injection destroyed memory in these animals as measured by their inability to perform learned tasks. Similarly, acetoxycycloheximide caused inhibition of protein synthesis in all cerebral areas equivalent to that produced by puromycin. However, in contrast to puromycin, the heximide was without effect on either recent or longer-term retention of learned task. Furthermore, acetoxycycloheximide protected memory against the destructive effects of puromycin, although mixture of the two drugs produced profound inhibition of protein synthesis.

Agranoff and Davis[22, 23] have studied the effect of puromycin on memory fixation in goldfish. Goldfish were placed in individual shuttle boxes in which light stimulus was paired with repetitive electric shock. To avoid the shock, fish had to swim over a hurdle from the light to the dark end of the box. The trial cycle was 20 sec of light alone, then 20 sec of light paired with shock, followed by 20 sec of darkness. A correct response was scored when the fish swam over the hurdle before the onset of the shock. All fish were given 20 trials in a 40-min session on the first day of an experiment and 10 retraining trials in a 20-min session 1–72 hr following the training session. Memory was inferred from an increase in the number of correct responses by the goldfish in blocks of 10 trials. Puromycin injected intracranially into goldfish immediately following avoidance training resulted in a gradual loss of all memory of the avoidance response within 2–3 days. Injection of puromycin before the training did not appear to cause a more rapid decay of memory than did injections immediately after the training. Their previous experiments showed that memory formation was not affected by puromycin 1 hr after training.

Barondes and Cohen[24] have also studied the effects of puromycin and actinomycin D on successive phases of memory fixation. Mice injected bitemporally with puromycin 5 hr before training learned to escape or to avoid shock by choosing the correct limb of a Y-maze. When retested 15 min after training they had normal retention. In the ensuing $2\frac{3}{4}$ hr the animals injected with puromycin, unlike the controls, showed a progressive decrease in conditioned response to less than 7%. They concluded that there is an initial phase of memory storage which is independent of protein synthesis in the temporal lobe, and that this overlaps a second phase which is dependent on protein synthesis in this region of the brain. Cohen and Barondes,[25] in earlier experiments, had found that injections of actinomycin D, which inhibits cerebral RNA synthesis 94–96%, did not interfere with retention of a solution to a Y-maze within 4 hr after training. This suggests that if protein synthesis is indeed required for the second phase of memory storage, such protein synthesis is directed by a stable messenger RNA which was synthesized before and independent of the time of acquisition, hence they concluded that RNA is not concerned with information accrual. They further concluded that their experiments, and those of Flexner *et al.*,[20] suggest that there are at least three phases of memory storage in the shock-motivated maze learning: an initial phase, uninfluenced by puromycin, which extends for a number of minutes after learning; a second phase which is inhibited by temporal injections of puromycin and which may extend for several days; and a third phase of memory which can be interfered with only by more extensive intracerebral injections of puromycin.

Talwar *et al.*[26] recently reported some data on the protein content and the rate of protein biosynthesis in the sensory cortical area (occipital cortex) of rabbits in response to light stimulus. From their experiments they concluded that the residual protein content per unit DNA rises markedly in the occipital cortex, cerebral cortex, tectum, and hippocampus of the developing rabbit in the period immediately following the opening of the eyes. The greatest change was in the cerebellar cortex during this period. Prior blinding of the animals significantly diminished the increase in protein concentration of the occipital cortex. Blinding did not significantly affect the increase in proteins of other areas of the brain. Exposure of adult animals to light increased the incorporation of (1-^{14}C)-valine and (U-^{14}C)-glucose, respectively, into proteins of the occipital cortex, in contrast to the incorporation of each of these precursors into similar brain areas of animals kept in the dark.

On the other hand, Metzger *et al.*[27] have found that visual stimulation did not affect protein synthesis in monkeys with split brains. After injection of tritiated water, split-brain monkeys were conditioned using a combination of light and shock. The relative rate of protein synthesis in nuclei, synaptic endings, mitochondria, and microsomes from each occipital and temporal cortex were determined. With one eye covered during conditioning the relative rates of protein synthesis in the corresponding brain region remained unchanged.

In an attempt to summarize the recent biochemical studies relating RNA and protein synthesis, it appears that the accrual of conditioned information in the brain is probably a two-step process. The first step is the transduction of the information from coded afferent impulse into nuclear RNA. This must be a rapid phenomenon occurring within the time at which afferent impulses reach the neurons in the brain. The resulting RNA probably contains the information of the afferent impulse in the form of base sequences (at least similar, if not identical, to the genetic code) and it appears from the reported data that this RNA most likely represents the immediate and the temporary information generally referred to as "short-term memory." As a corollary, it appears that this RNA can then serve as a template for the synthesis of protein. The second step in the two-step process occurs at much slower rate, requiring reinforcement of afferent information in order to obtain sufficient protein synthesis. Conceptually, it is suggested that protein represents the long-term storage of information ("long-term memory"). The information encoded in the protein is in the form of sequential linkage of specific amino acids, each amino acid arising as a product of the triplet code in RNA presumably identical to genetic code.

Although the above concept has some similarity to the mechanism proposed by Hydén, it is inconceivable that the synthesis of RNA induced by the coded afferent impulse should in any way be dependent on DNA. The present understanding of DNA as an invariant genetic component implies that even electrical impulse could not induce conformational changes within DNA to permit alterations in the nature of DNA-dependent synthesis of new RNA as a transcript of information from the afferent impulse.

In order to maintain a clear concept which would permit experimental testing of the mechanisms involved in the accrual of short- and long-term memory of information transmitted by afferent impulses from the sensors, the following concepts may be deemed as a working hypothesis for information accrual by the brain. First, it is conceivable that sensory information is stored essentially in the nucleus of the brain cells, since it is a stable environment in which genetic information is retained for the life span of the organism. Second, irrespective of the accepted mechanism responsible for the transfer of coded impulses along the axons and dendrites of the neurons, it is conceived here that the impulse is also transmitted through the cytoplasm in the brain cells, possibly *via* the endoplasmic reticulum, thus reaching the nucleus. In the nucleus the impulse initiates *de novo* synthesis of RNA, thus transcribing the information from the impulse into the sequence of bases within this polymer. The latter subsequently acts as a template to transcribe this information into a specific sequence of amino acids in a newly formed protein. This mechanism specifies that an RNA-synthesizing enzyme is present which does not depend upon DNA as a template, hence differs from RNA polymerase. The only known enzyme which can synthesize RNA, and has been found in animal tissue,[28–30] is polynucleotide phosphorylase,[31] which acts without requiring either DNA or RNA as templates and which uses nucleoside diphosphates as substrates. It is possible that some other enzyme yet to be discovered may perform this function. However, designation of polynucleotide phosphorylase for this role avoids presumptive assignments to an imaginary entity and allows experimental approach to the concept.

The above ideas were the basis for the formulation of a working concept designating RNA as a "short-term memory" and protein as a "permanent memory" in the process of information accrual and storage in the brain. This permitted design of experiments to examine the proposed mechanism in the laboratory. The first definitive experimental design was concerned with the qualitative and quantitative changes in brain nuclear RNA during and following sensory stimulation.

The schematic diagram in Fig. 1 illustrates the concept of information processing in the brain cell nuclei. It is assumed that information originating from the sensory organs is transmitted to the neuronal nuclei in the form of an afferent impulse which contains the information as frequency-modulated signals spelled out in the diagram as *DEF*. When this coded afferent impulse reaches the neuronal nuclei a particulate component of the nuclei receives this signal and this activates an enzyme within this particle capable of synthesizing an RNA in the absence of a template or primer. Such an enzyme is polynucleotide phosphorylase complexed in a transducer particle which, in response to the frequency modulation of the afferent impulse, can select one specific nucleotide diphosphate from the surrounding pool, and hence synthesize an RNA whose base sequence is determined by the frequency components in afferent impulse. In Fig. 1 this particle is designated as the transducer, since it converts the coded afferent impulse into a coded RNA. This transduction occurs rapidly, at

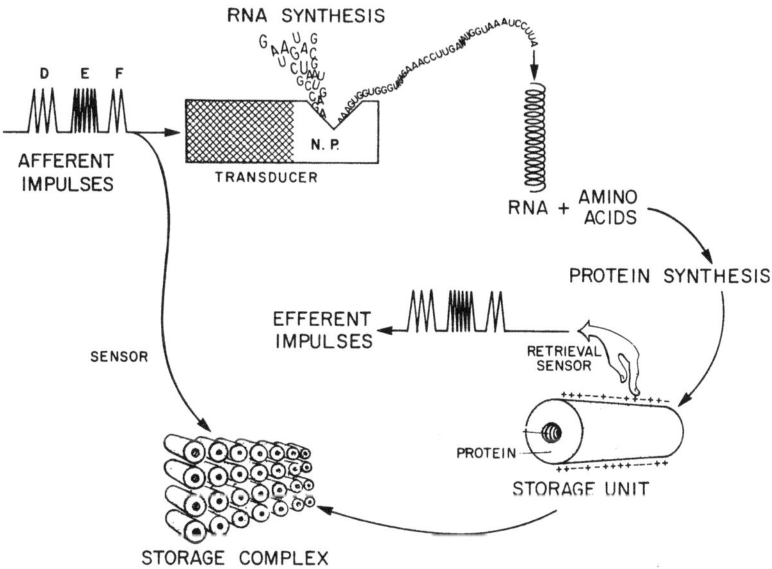

Fig. 1. Scheme of the working concept on information processing in the neuronal nuclei.

speeds equal to or greater than the speed with which afferent impulses enter the neuronal nuclei.

The resulting coded RNA then acts as a template for the biosynthesis of a protein[32] through the action of nuclear ribosomes and adjuvent enzymes, which will contain the information in the encoded RNA. However, information represented by the sequence of amino acids in this newly synthesized protein is important from the standpoint of charge distribution on the surface of the protein due to the amino acid side-chains. Thus, the nature of the charged and the uncharged side groups on the surface of the protein will represent the code originally presented to the nuclei as frequency components in the afferent impulse. It is assumed that the newly synthesized protein forms a tight helix constituting the storage unit, as indicated in Fig. 1. It is visualized that the surface charges and their distribution on this protein particle will be a permanent coded memory unit from which the information can be retrieved by an electrical sensor (indicated as a finger in the diagram) whose identity, at present, cannot be depicted, and thus an efferent impulse can be generated at any time when retrieval is necessary. A group of these storage units representing particulate proteins fixed in relative position to each other and in a temporal sequence dependent on the time of information accrual, represent stored memory. These units can be imagined as permanent entities which are destroyed only when the neurons are disrupted by physical injury.

It is conceived in this hypothesis that the afferent impulse entering the neuronal nuclei is initially bifurcated, with one branch of the impulse immediately scanning the storage complex and rapidly noting whether the new information in this impulse has previously been experienced by the presence or absence of a corresponding storage unit. If such information has been accrued and duly stored, then that component of the impulse, which would normally pass through the transducer, will not pass through the transducer and no new RNA will be formed. This is essentially a conservation mechanism. Further examination of this concept indicates that both the transducer and the storage unit are postulated to be insoluble particles. This concept also presupposes that the transducer will show enhanced polynucleotide phosphorylase activity only under the influence of an electrical impulse; thus, presumably, some fraction of the insoluble particles isolated from neuronal nuclei should exhibit pronounced polynucleotide phosphorylase activity under direct current stimulation.

BACKGROUND CONSIDERATIONS LEADING TO THIS CONCEPT

In order to justify the unique molecular changes proposed in this concept of neuronal information processing the speculations underlying the proposed scheme are summarized below (scheme is illustrated in Fig. 1).

Since the brain does receive and emit information in the form of electrical impulses, and RNA has been implicated as a chemical depository of information in neuronal nuclei,[9] then it follows that a transduction of coded electrical impulses into a coded RNA must occur. This change must be mediated by a discrete particle which is capable of accepting the impulse, here assumed to be a frequency-modulated waveform, and this then affects the active site on an associated enzyme which synthesizes RNA. It is presumed that this enzyme is polynucleotide phosphorylase, since each packet of information introduced by the electrical impulse to the transducer must dictate the formation of a coded RNA independent of any template or primer. Polynucleotide phosphorylase has been reported to be present in liver nuclei by Hilmore and Heppel.[28] It is presumed that the transducer is a defined structure capable of accepting afferent impulses and simultaneously converting the coded information into a corresponding nucleotide sequence in RNA. The afferent impulse enters the transducer and affects the ionized side-chains on the active site of the polynucleotide phosphorylase in such a way that a specific nucleotide diphosphate combines at this site preferentially from a pool of purine and pyrimidine nucleotide diphosphates in the immediate environment. Essentially, it is presumed that the waveform of the impulse directly affects both the ionization and hydrogen bonding in the active enzyme site, so that binding of a specific nucleotide diphosphate is dependent, first, on whether it is a purine or pyrimidine, then on the absence or presence of OH or amine side groups and their positions.

Examination of any RNA molecule clearly shows that in spite of the information coded in the base sequences, the electrical properties of the molecule are essentially that of a secondary phosphate ester anion. To permit electrical readout of this information, it was necessary to involve the synthesis of protein whose sequence of amino acids would correspond to the nucleotide sequence in RNA; hence, the RNA produced by the transducer is here considered to act as a template for ribosomes in the nuclei,[32] thus to produce a protein in which the side-chains of the amino acids exhibit charge distributions, both in kind and location, which correspond to the nucleotide sequence in the RNA. To maintain an error-free charge distribution, it was necessary to assume that both RNA and the protein disassociated from the ribosome, and formed a single, tight, helical complex where the outer protein shell of the complex was free of random coiling. This complex was designated as a memory-storage unit particle whose charges accurately reflect information similar to that of a Morse code, dot, dash, space, but in the form of plus, minus, space.

In Fig. 1 the storage units are shown to form a storage complex which in essence represents memory, and no attempt is made here to explain how a retrieval wave (retrieval sensor) can select information from a specific group of storage units within the storage complex. It can only be mentioned that the retrieval sensor is imagined as an electrical system which is either attracted by positive charges or repelled by negative charges, or remains unaffected by neutral sites, resulting in an efferent impulse. The mechanism for retrieval is completely unknown.

In order to avoid misconceptions due to large deviations between observations in each experiment and from individual organisms in identical experiments, it was necessary that each organism serve as its own control. Second, it was considered necessary that observations be made both *in vitro* as well as *in vivo* in order to ascertain that *in vitro* observations represent phenomena actually operating in the living organism. A search of the literature showed that Adrian and Ludwig[33] reported that exposed brain in isolated catfish heads evoked what appeared to be normal EEG's from surface electrodes for approximately 1 hr after decapitation. Also, these authors reported that olfactory stimulation by lavage of the nares with aqueous extracts elicited changes in EEG for approximately 45 min after decapitation. This was duplicated in this laboratory using salt water catfish (*Galeichthys felis*) which are indigenous to the Gulf Coast.

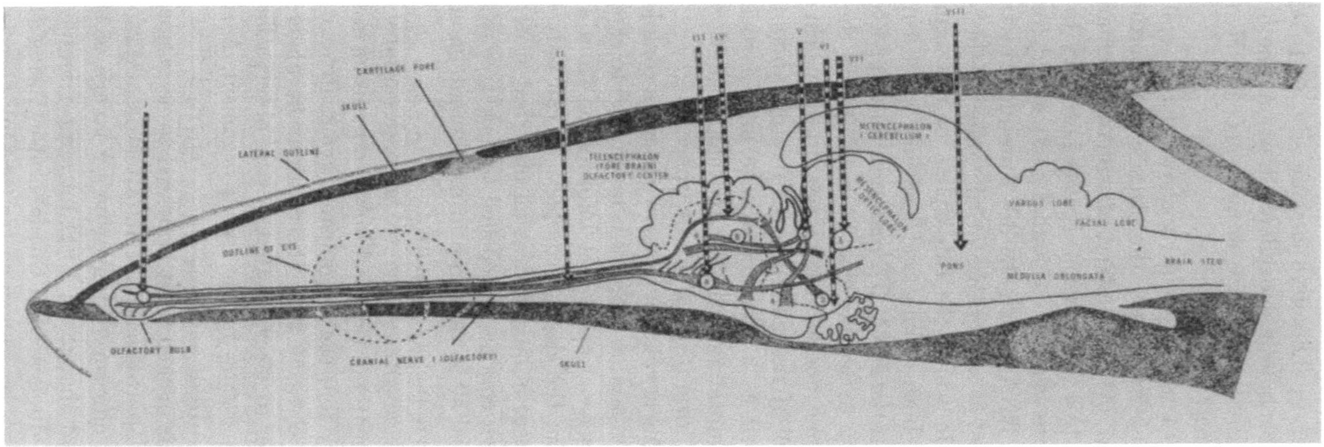

Fig. 2. A diagrammatic illustration of a sagittal section of a catfish head and brain showing: (I) olfactory bulb; (II) olfactory tract; (III) olfactory nuclei; (IV) central white matter of telencephalon; (V) habenular nuclear complex; (VI) preoptic area; (VII) dorsal nucleus of thalamus; and (VIII) pons. The.brain centers are designated as: (A) olfactory nuclei; (B) anterior commissure; (C) habenular nuclei; (D) ventral nucleus of thalamus; and (E) dorsal nucleus of thalamus. The numbers on the nerve tracts represent: (1) dorsal forebrain; (2) medial forebrain; (3) olfactohabenular; (4) preopticohavenular; (5) striothalamic; (6) hippocampothalamic; and (7) striohypothalamic. (Based on the histological evidence reported by Kappers *et al.*[61])

The use of a poikilothermic animal such as the catfish in brain studies had certain advantages in contrast to the mammalian brain. The catfish brain, although highly developed, is anatomically much more primitive than the mammalian brain. As can be seen in Fig. 2, the teleost brain has no cortex, and the top of the brain is an extension of the cerebellum; on the sides are the optic lobes, and the forebrain represents a highly developed telencephalon. The rhinencephalic region in the teleost is rudimentary in comparison to the mammalian brain, and this suggests that there are fewer feedback circuits due to emotionality resulting from sensory stimulation. Thus, the catfish represents a much less complex central nervous system then any encountered in mammals, and hence these preparations offer an opportunity to make interpretative observations in a simplified system.

According to Allison,[34] the morphology of the vertebrate olfactory system has undergone little change during the process of evolution. The olfactory system of the fish is, in neuroanatomical respects, identical to that of mammals. By observing the behavior of fish, Parker[35] demonstrated that the fish have high sensitivity to odorous substances. Most odoriferous substances which play a meaningful biological role in the lives of fishes are of organic nature. They are detected through the olfactory epithelium connected through the olfactory nerve in the nasal sac, although some of them may be perceived through the taste buds. However, the sense of smell in most fish is more acute than that of taste and permits detection of stimulants present in minute concentrations. Olmstead[36] observed that blinded catfish found foods as readily as did normal fish. Currently, many investigators[37–41] have confirmed the observation that decapitated-fish-brain preparations have an olfactory system which evokes EEG's during stimulation.

In order to clarify the role of RNA in accrual of information arising from sensors and transmitted to the brain through a coded afferent impulse, studies were undertaken to correlate changes in the synthesis of brain nuclear RNA during stimulation of the olfactory system in catfish, which represents a highly developed system, but phylogenetically a simpler one than that of the mammal.

MATERIALS AND METHODS[42]

Catfish

In the experimental studies seawater catfish (*Galeichthys felis*) indigenous to the surrounding water of the Gulf of Mexico were used. They were caught alive and stored in a large tank with running seawater. This

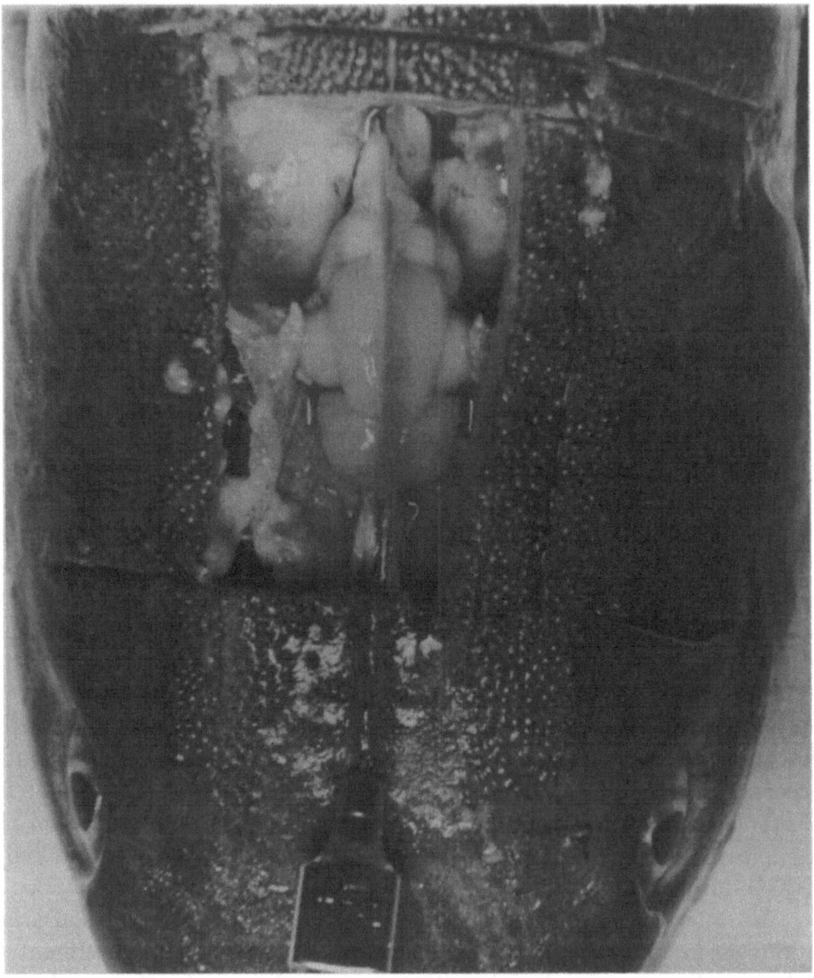

Fig. 3. The exposed split catfish brain, with Parafilm inserted into the incision. The position of the hypodermic needle for injection of labeled precursors and drugs is also illustrated.

facility was provided by the US Bureau of Commercial Fisheries, Biological Laboratories, through the courtesy of Mr. M. Lindner, Director, in Galveston. The sharp barbs were trimmed before the fish were transferred into the tank to prevent injury among them during storage. For each experiment fish of approximately the same size and weight were selected. Intracranial injections of radioisotopes or actinomycin D were

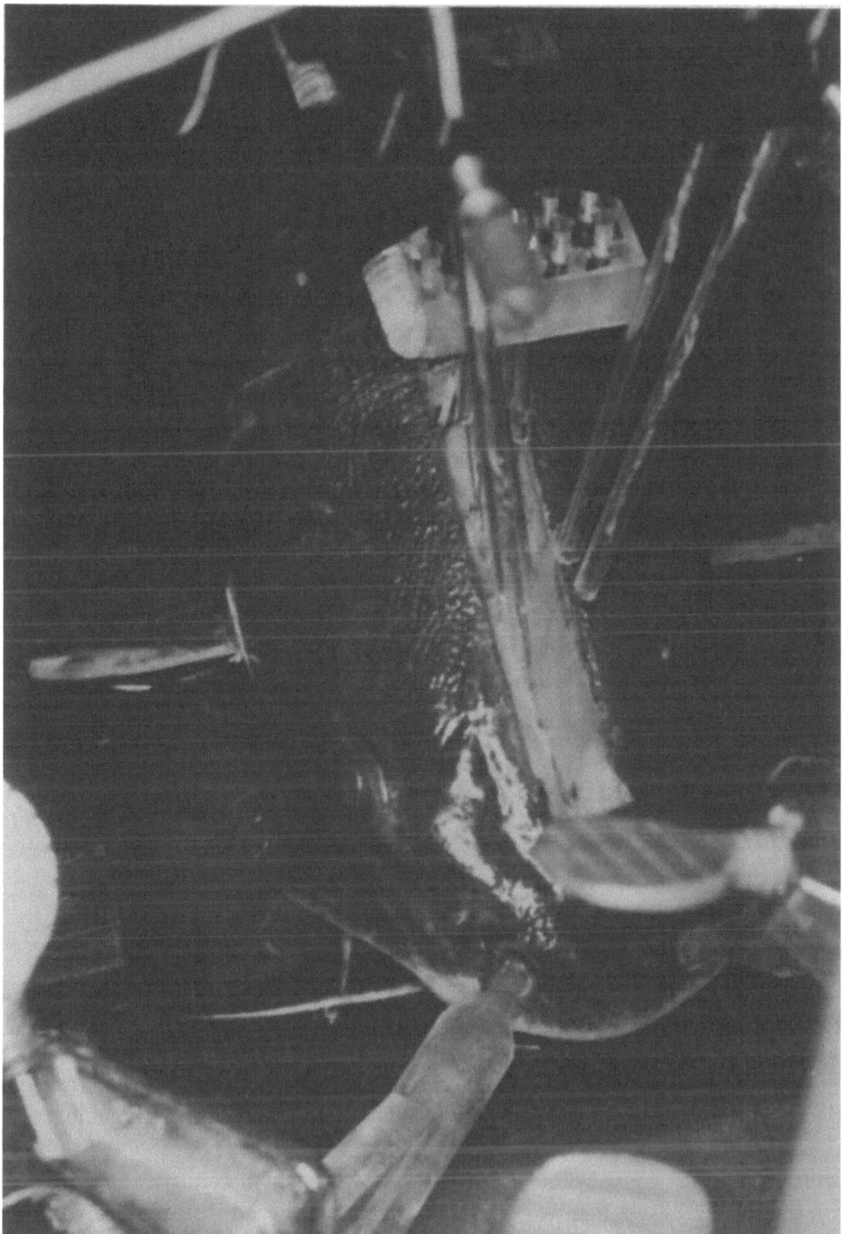

Fig. 4. The position of electrodes in the exposed catfish brain and the cannulae leading to the nares for the introduction of fluids. The Parafilm separating the split-brain is also shown.

made in live fish with a hypodermic needle through a cartilaginous pore in the head of the fish as illustrated in Fig. 3. During experiments seawater in the tank was aerated by a fritted-glass gas-dispersion tube using compressed air.

Split-Brain Preparation

The skull of the catfish was cut by means of a small electric-powered saw, and the brain was exposed as shown in Fig. 3. A split-brain preparation of the decapitated catfish was made by a longitudinal midline incision with a sharp scalpel, the line of incision passing between the forebrain and extending through the brain stem. A Parafilm was inserted into the incision, thus separating the two halves of the brain (Fig. 3). For olfactory stimulation two cannulae were fitted to the nares (Fig. 4) to direct the flow of seawater. One naris was lavaged with seawater as a control, while the other naris was lavaged with seawater containing an odorant. Surface bipolar macroelectrodes were placed on each of the olfactory tracts and on each telencephalon (Fig. 4), which allowed EEG recordings from each of the two intact olfactory systems in the split-brain preparation. In all experiments charcoal-treated and filtered seawater was used and each odorant was dissolved in seawater to a final concentration of 10^{-6} *M*.

RESULTS

Changes in Nuclear RNA

Studies on Live Catfish

The live catfish were placed in seawater containing morpholine in a final concentration of 10^{-4} *M* and allowed to swim (with aeration) for various intervals of time. For stimulation of live catfish seawater containing 10^{-4} *M* morpholine was used. Since the mucous surface of these fish adsorbed morpholine, the actual concentration was probably closer to 10^{-6} *M* when six to ten live fish were kept in 12–15 liters of seawater containing 10^{-4} *M* morpholine. The brains were then removed and nuclei isolated. When live fish were placed in seawater containing morpholine (10^{-4} *M*) there was a consistent increase in nuclear RNA content, based upon unit weight of DNA in brain nuclei, as well as an increase in guanine and uracil and a concomitant decrease in adenine and cytosine content, in contrast to nuclear RNA from control fish swimming in seawater (Table I).

TABLE I

Effect of Morpholine on the Composition of RNA in Brain Nuclei from Free, Swimming Catfish[a]

Fish No.	Base composition (% mole fraction)				RNA/DNA	A/U	G/C
	Adenine	Guanine	Cytosine	Uracil			
Control							
1	20.8	3.05	25.5	23.2	0.151	0.89	1.20
2	20.6	30.8	25.4	23.3	0.151	0.88	1.21
3	19.5	32.6	25.5	22.4	0.140	0.87	1.28
Mean ± S.D.	20.3 ± 0.7	31.3 ± 1.1	25.5 ± 0.05	23.0 ± 0.5	0.147 ± 0.006	0.88 ± 0.01	1.23 ± 0.04
Stimulated (morpholine)							
4	18.4	33.7	20.2	27.7	0.190	0.66	1.67
5	16.8	35.0	22.2	26.0	0.200	0.65	1.58
6	15.4	34.2	22.8	27.6	0.203	0.56	1.50
7	17.1	34.3	22.8	25.8	0.195	0.66	1.50
Mean = S.D.	16.9 ± 1.2	34.3 ± 0.5	22.0 ± 1.0	26.8 ± 1.0	0.197 ± 0.005	0.63 ± 0.05	1.56 ± 0.08

[a] All fish were allowed to swim in aerated seawater for 1 hr. Subsequently, some of the fish were kept in 10^{-4} M morpholine–seawater for 1 hr before sacrificing.

TABLE II

Reversal Experiments of the Morpholine Effect on RNA Composition of Brain Nuclei in Free, Swimming Catfish[a]

Bases	Controls (5)	Base composition (% mole fraction) at various times of sacrificing				
		Immediately (5)	3 hr (2)	24 hr (2)	48 hr (2)	72 hr (2)
Adenine	20.4	17.1	18.0	20.9	21.2	21.9
Guanine	31.4	35.7	35.8	31.9	31.2	30.9
Cytosine	25.6	23.3	22.7	25.6	25.6	25.4
Uracil	22.5	23.8	23.4	21.6	22.1	22.2
RNA/DNA	0.159	0.239	0.229	—	0.134	0.140
A/U	0.90	0.72	0.77	0.96	0.96	0.98
G/C	1.22	1.53	1.57	1.24	1.22	1.22

[a] Fish were allowed to swim in aerated seawater with 10^{-4} M morpholine for 1 hr, and then were placed in running seawater for various time intervals as indicated before sacrificing. Numbers in parentheses indicates the number of fish used for the analyses.

In another set of experiments a group of catfish was placed in seawater containing morpholine (10^{-4} M), and after 1 hr in this environment all the fish were removed and placed in fresh-running seawater. Two of the fish were killed immediately and the nuclear RNA isolated and analyzed. From the fresh-running seawater groups of previously stimulated fish were removed, respectively, at 3, 24, 48, and 72 hr and their nuclear RNA isolated and analyzed. Analysis of the isolated nuclear RNA from each fish brain showed that the changes in content and composition of nuclear RNA following stimulation persisted for 3 hr. However, within 24 hr the RNA changes almost completely reverted to the values of nuclear RNA in the unstimulated controls (Table II).

Studies on Isolated Catfish Heads

In another group of experiments the catfish were decapitated and the brain was exposed (Fig. 3). Two cannulae were attached to the nares (Fig. 4) to allow the flow of seawater into one naris as a control, while the other naris was lavaged with seawater containing 10^{-6} M morpholine. This arrangement allowed stimulation of one-half of the brain while the other half was kept as a control; each half innervated to an intact olfactory system. Stimulation was continued for 30 min and then the brain halves

TABLE III

Effect of Morpholine on the Composition of RNA in Catfish-Brain Nuclei Using Split-Brain Preparation[a]

Fish No.	Base composition (% mole fraction)				RNA/DNA	A/U	G/C
	Adenine	Guanine	Cytosine	Uracil			
Control half							
1	20.7	31.8	24.9	22.6	0.217	0.91	1.28
2	19.5	31.6	25.3	23.5	0.210	0.83	1.25
3	21.0	32.0	25.3	21.6	0.228	0.97	1.26
4	21.4	31.0	24.6	23.0	0.210	0.93	1.26
Mean ± S.D.	20.6 ± 0.8	31.6 ± 0.4	25.0 ± 0.3	22.6 ± 0.8	0.216 ± 0.008	0.91 ± 0.06	1.26 ± 0.01
Stimulated half							
1	15.8	36.3	23.1	23.0	0.291	0.69	1.57
2	15.8	33.8	24.1	26.2	0.255	0.60	1.40
3	18.2	33.7	23.5	24.6	0.261	0.74	1.43
4	18.2	33.9	23.6	24.3	0.247	0.75	1.44
Mean ± S.D.	17.0 ± 1.4	34.4 ± 1.2	23.5 ± 0.4	24.5 ± 1.3	0.263 ± 0.019	0.69 ± 0.07	1.46 ± 0.07

[a] Conditions: The live fish were placed in ice. The brain was exposed, then cut in two by a midline incision. The brain halves were separated by a strip of Parafilm. One naris was lavaged with seawater while morpholine (10^{-3} M)-seawater was lavaged through the other naris for 30 min before sacrificing.

TABLE IV

Effect of Amyl Acetate on the Composition of Catfish-Brain Nuclear RNA in Split-Brain Preparations[a]

Fish No.	Amyl acetate (M)	Base composition (% mole fraction)				RNA/DNA	A/U	G/C
		Adenine	Guanine	Cytosine	Uracil			
Control half								
1	0	21.0	32.7	24.7	21.6	0.205	0.97	1.32
2	0	21.4	32.2	25.0	21.4	0.194	1.00	1.29
3	0	20.9	31.8	25.9	21.4	0.208	0.98	1.23
4	0	21.2	30.7	25.9	22.2	0.211	0.95	1.19
Mean ± S.D.	—	21.1 ± 0.2	31.9 ± 0.9	25.4 ± 0.6	21.6 ± 0.4	0.204 ± 0.007	0.97 ± 0.02	1.25 ± 0.06
Stimulated half								
1	10^{-6}	23.1	35.7	20.3	20.9	0.243	1.10	1.76
2	10^{-6}	23.0	35.3	21.1	20.6	0.237	1.12	1.67
3	10^{-4}	24.4	36.7	18.4	20.5	0.233	1.19	1.99
4	10^{-4}	23.2	36.5	18.1	22.2	0.248	1.05	2.00

[a] Conditions used in this experiment were identical to those described in Table III.

were removed and homogenized separately, the nuclei isolated, and the RNA composition and content determined (Table III). The changes in the nuclear RNA base ratios in split-brain preparations following olfactory stimulation were identical to those found in live fish swimming in seawater containing morpholine (Tables I and II), and, concomitantly, there was an increase in total RNA content in stimulated brain halves in contrast to the controls. When the content of brain nuclear RNA in the split-brain preparation was compared with such data from live fish, the former showed a higher RNA content based on unit weight of DNA. This is probably due to the disturbances in the brain incurred by decapitation and brain-splitting procedures.

In order to determine whether brain cytoplasmic RNA undergoes any changes during olfactory stimulation, cytoplasmic RNA was isolated from split-brain preparations following morpholine (10^{-6} M) stimulation (15 min) under exactly the same experimental conditions as those described above. Total cytoplasmic RNA from catfish brain did not exhibit any changes following olfactory stimulation.

Using split-brain preparations in isolated catfish heads, amyl acetate in seawater was tested as an olfactory stimulant in concentrations of 10^{-6} and 10^{-4} M, respectively. The resultant changes in brain nuclear RNA are shown in Table IV. Both levels of amyl acetate caused an increase of equal magnitude in nuclear RNA and an identical change in RNA base ratio. Comparing the base ratio changes induced by morpholine (Table III), where the content of adenine and cytosine decrease while guanine and uracil increase, amyl acetate induced an increase in adenine and guanine and a decrease in cytosine and uracil.

Camphor and menthol in seawater (10^{-6} M) were tested as odorants using split-brain preparation in isolated catfish heads as described above. The resulting analyses of brain nuclear RNA (Table V) showed that neither compound caused a significant change in RNA base ratios; however, camphor did induce an increase in total nuclear RNA, while menthol decreased the content of RNA.

Since in the above experiments the odorants used were novel to the catfish, it was of some interest to determine the effect of natural and familiar odorants to the catfish. In one set of experiments live shrimp were homogenized in charcoal-treated seawater (1 g fresh shrimp free of exoskeleton in 50 ml of charcoal-treated seawater). The homogenate was centrifuged at 2000g for 15 min and the clear supernatant was used to lavage one naris of a split-brain catfish preparation for 15 min, while the other naris was lavaged with seawater. In another experiment shrimp were allowed to remain in seawater (1 g shrimp per 50 ml charcoal-treated seawater) for 24 hr at room temperature to permit some putrefaction to

TABLE V

Effect of Camphor and Menthol[a] on the Composition of Catfish-Brain Nuclear RNA in Split-Brain Preparations[b]

Fish No.	Stimulant	Base composition (% mole fraction)				RNA/DNA	A/U	G/C
		Adenine	Guanine	Cytosine	Uracil			
Control half								
1	None	21.3	30.3	26.8	21.6	0.200	0.98	1.13
2	None	22.2	29.8	25.2	22.8	0.190	0.97	1.18
3	None	20.7	31.4	26.5	21.4	0.195	0.96	1.18
4	None	21.8	30.2	25.0	22.9	0.205	0.95	1.20
Mean ± S.D.	—	21.5 ± 0.6	30.4 ± 0.7	25.9 ± 0.9	22.2 ± 0.8	0.197 ± 0.007	0.97 ± 0.02	1.17 ± 0.04
Stimulated half								
1	Camphor	21.5	31.1	25.1	22.3	0.242	0.96	1.23
2	Camphor	21.2	31.9	25.1	21.8	0.224	0.97	1.27
3	Menthol	21.3	32.1	24.8	21.8	0.173	0.97	1.29
4	Menthol	21.5	30.8	26.1	21.6	0.188	0.99	1.18

[a] Each at 10^{-6} M.

[b] Conditions used in this experiment were identical to those described in Table III.

TABLE VI

Effect of Shrimp Extracts as an Odorant on RNA in Catfish-Brain Nuclei Using Split-Brain Preparations

Fish No.	Base composition (% mole fraction)				RNA/DNA	A/U	G/C
	Adenine	Guanine	Cytosine	Uracil			
Control half							
1	20.3	32.7	26.2	20.8	0.194	0.97	1.25
2	20.3	32.5	26.3	20.9	0.198	0.97	1.24
3	19.2	32.9	26.9	21.0	0.207	0.91	1.22
4	20.9	32.6	24.8	21.5	0.202	0.97	1.31
Mean ± S.D.	20.2 ± 0.7	32.7 ± 0.2	26.0 ± 0.9	21.0 ± 0.3	0.200 ± 0.006	0.95 ± 0.03	1.25 ± 0.04
1[a]	22.8	32.9	23.3	21.0	0.203	1.09	1.41
2[a]	23.6	32.4	22.3	21.7	0.192	1.09	1.45
3[b]	24.5	31.2	22.5	21.8	0.211	1.12	1.39
4[b]	23.3	32.3	23.4	21.0	0.218	1.11	1.38

[a] Dead shrimp were allowed to remain in charcoal-treated, filtered seawater (1 g shrimp per 50 ml seawater) for 24 hr at room temperature (for putrefaction). Then the liquid was filtered and this was used for 15-min lavage of nares for olfactory stimulation.

[b] Live shrimp were homogenized in charcoal-treated, filtered seawater (1 g fresh wt. shrimp free of exoskeleton in 50 ml of seawater). The homogenate was centrifuged at 2000 g for 15 min and the clear supernatant was used to lavage the nares for 15 min of olfactory stimulation.

TABLE VII

Effect of Red-Fish Skin Extracts on Brain Nuclear RNA in Catfish Split-Brain Preparations[a]

	Base composition (% mole fraction)						
Fish No.	Adenine	Guanine	Cytosine	Uracil	RNA/DNA	*A/U*	*G/C*
Control half							
1	22.5	30.3	24.6	22.6	0.196	0.99	1.23
2	21.8	30.1	26.2	21.9	0.190	0.99	1.15
Stimulated half							
1	24.7	30.7	24.5	20.1	0.236	1.23	1.25
2	24.0	30.2	25.0	20.8	0.217	1.15	1.20

[a] 20 g of red fish (*Sciaenops ocellatus*) skin was homogenized in charcoal-treated, filtered seawater. The homogenate was centrifuged at 2000 g for 15 min and clear supernatant was used to lavage the nares for 15 min of olfactory stimulation.

occur. The aqueous phase was then filtered, and the filtrate was used to lavage one naris of a split-brain catfish preparation for 15 min, while the other naris was lavaged with seawater. The nuclear RNA from each half of the brain was isolated and analyzed. The results from these studies (Table VI) show that both the live-shrimp extract, as well as the decayed extract from frozen shrimp, induced a slight increase in RNA content and an identical change in base ratios of RNA in the stimulated portion of the catfish-brain nuclei. These extracts induced an increase in adenine and a decrease in cytosine without affecting guanine and uracil in the nuclear RNA (Table VI).

Extracts from red fish (*Sciaenops ocellatus*) skin were prepared by homogenizing 20 g of skin in one liter of charcoal-treated seawater. The homogenate was then centrifuged at 2000g for 15 min and the clear supernatant was used as an odorant under conditions similar to those described for the above studies with shrimp extract. The nuclear RNA from each half of the brain was isolated and analyzed (Table VII). Red-fish skin extract induced an increase in total content of RNA with an increase in adenine and a decrease in uracil in the stimulated halves of the catfish brain. Table VIII summarizes the changes induced in catfish-brain nuclear RNA by different odorants tested.

TABLE VIII

A Summary of Changes Induced by Odorants in Catfish-Brain Nuclear RNA

Exp.[a]	Condition	Brain injection	Odorant	Conc. of odorant (M)	RNA/DNA	A/U	G/C
1	Control	—	—	—	0.147	0.88	1.23
	Stimulated	—	Morpholine	10^{-4}	0.197	0.63	1.56
2	Control	—	—	—	0.216	0.91	1.26
	Stimulated	—	Morpholine	10^{-6}	0.263	0.69	1.46
3	Control	2-C^{14}-uridine	—	—	0.202	0.93	1.19
	Stimulated	2-C^{14}-uridine	Morpholine	10^{-6}	0.250	0.81	1.43
4	Control	Actinomycin D	—	—	0.201	0.98	1.26
	Stimulated	Actinomycin D	Morpholine	10^{-6}	0.225	0.74	1.40
5	Control	—	—	—	0.200	0.98	1.31
	Stimulated	—	Amyl acetate	10^{-6}	0.240	1.11	1.72
6	Control	—	—		0.210	0.97	1.21
	Stimulated	—	Amyl acetate	10^{-4}	0.240	1.12	2.00
7	Control	—	—	—	0.195	0.98	1.16
	Stimulated	—	Camphor	10^{-6}	0.233	0.97	1.25
8	Control	—	—	—	0.200	0.96	1.20
	Stimulated	—	Menthol	10^{-6}	0.181	0.99	1.23
9	Control	—	—	—	0.196	0.97	1.24
	Stimulated	—	Dead shrimp extract	—	0.197	1.08	1.43
10	Control	—	—	—	0.204	0.94	1.26
	Stimulated	—	Live shrimp extract	—	0.214	1.11	1.38
11	Control	—	—	—	0.193	0.99	1.19
	Stimulated	—	Red fish skin homogenate	—	0.227	1.19	1.23

[a] All experiments were split-brain, except for Exp. 1, which was with live fish.

DISCUSSION

The present studies showed that olfactory stimulation induced changes in brain nuclear RNA in both intact and isolated heads of the catfish. The RNA isolated from brain nuclei was found to be uniform and consistent among various size catfish, based on the diversity in total brain DNA content. When live catfish were placed in seawater containing morpholine there was a consistent increase of brain nuclear RNA as well as a change in its base composition (Tables I and II). These changes in brain nuclear RNA induced by morpholine in live catfish were not altered when fish were transferred to fresh seawater and kept there for 3 hr. However, within 24 hr, the RNA changes were almost completely reversed to the levels of nuclear RNA in unstimulated controls (Table II). These results demonstrate that the nuclear RNA changes induced in the catfish brain were not permanent and disappeared within 24 hr.

Using catfish split-brain preparations with morpholine as the olfactory stimulant, there was a change in RNA base ratio in the stimulated half of the brain (Table III) identical to that found in morpholine-stimulated living fish (Table I). Concomitantly, there was also an increase in RNA/DNA ratio in stimulated halves of the split-brain preparation similar to that found in free, swimming fish (Tables I and III). When the increase in RNA between the free, swimming fish and the isolated-head preparations were compared the latter showed a higher RNA content. This is probably due to the electrical disturbances incurred during decapitation and brain-splitting procedures.

Using amyl acetate (10^{-6} M) as an odorant, the amount of brain nuclear RNA increased similar to that noted with morpholine, but the base ratio change was different (A/U from 0.97 to 1.12 and G/C from 1.25 to 1.70; Table IV) from that found with morpholine stimulation (A/U from 0.92 to 0.69 and G/C from 1.26 to 1.46; Table III).

In contrast to the effect of synthetic organic odorants discussed above, a natural food for catfish, such as shrimp, elicits both an increase in RNA and a change in RNA base ratio distinctly different from the base-ratio changes observed with the novel synthetic odorants (Table VI). The red fish, which represents a cohabitant of the catfish, emits a substance from the skin which is an odorant and probably serves as the means for the catfish to recognize specific cohabitants. This substance(s) elicits a slight increase in RNA and a definite change in RNA base ratio when tested in a catfish split-brain preparation (Table VII).

On the other hand, both camphor and menthol, natural compounds which are odorant to humans, did not cause a significant change in RNA base ratio; however, camphor did induce an increase in total RNA, while menthol decreased total RNA (Table V). From these results it appears that neither camphor or menthol acts as true odorants for catfish, but they do affect the olfactory epithelium by nonspecific activation or irritation.

Although the observations reported here are solely with the changes noted in nuclear RNA, total cytoplasmic RNA from catfish brain did not exhibit any changes following olfactory stimulation using decapitated split-brain preparation.

These results suggest that novel odorants, as well as natural stimulants, elicited coded afferent impulses from the olfactory epithelium which induced changes in brain nuclear RNA (Table VIII). The observation that each odorant elicits a unique change in the sequence of bases in the brain nuclear RNA appears to suggest that the molecular structure of the odorant is responsible for the synthesis of a specific RNA; however, such an interpretation at this time is premature.

In preliminary tests, when actinomycin D was injected into the catfish brain it was found that it did decrease nuclear RNA by 7–12% when compared to fish brain without actinomycin D. However, the presence of actinomycin D did not interfere either with the synthesis of or the base-ratio change in brain nuclear RNA synthesized during olfactory stimulation by morpholine. Since actinomycin is known to be a specific inhibitor of DNA-dependent RNA polymerase,[(43,44)] the above observation suggests that RNA polymerase is probably not involved in the synthesis of RNA in brain nuclei induced by olfactory stimulation. The possibility of DNA-dependent RNA polymerase being responsible for the formation of RNA in brain nuclei during information accrual has been suggested by Hydén.[(11,45)] A number of investigators have used actinomycin D as an indicator of DNA participation in the synthesis of RNA during conditioning of animals.[(25,46–50)] The results from studies of actinomycin D injected into the brain of various animals before and after training have been equivocal, in that in rats[(46,47)] and mice[(25,51)] it has been found not to interfere with training, but it does interfere with learning in goldfish[(48)] and rats.[(52)] Since the findings of Hyden and co-workers[(8–11)] and present studies implicate brain nuclear RNA synthesis during brain function, it was necessary to determine what alternate enzyme, other than RNA polymerase, is responsible for RNA synthesis. It is suggested that polynucleotide phosphorylase may be the enzyme responsible for brain nuclear RNA synthesis during olfactory stimulation. It is now well known that polynucleotides formed by the action of this enzyme *in vitro* resemble natural RNA's. These products are of high molecular weight, and are attacked by ribonucleases. Such polynucleotides have been used as messenger RNA's under certain conditions for the incorporation of amino acids into protein.[(53)] One of the reasons for suggesting polynucleotide phosphorylase as the enzyme responsible for brain nuclear RNA synthesis during olfactory stimulation is the fact that it does not require a template for the synthesis, and hence it is not inhibited by actinomycin. Thus, it is conceived that a coded afferent impulse emitted from an olfactory epithelium or any other sensor, in some manner affects the activity of polynucleotide phosphorylase in brain nuclei so that the newly formed RNA has a base composition corresponding to the coded information in the afferent impulse.

Macromolecular changes in the brain such as induced by stimulation have been interpreted in different ways by various groups of investigators.

Some authors believe that during the transmission of an afferent impulse a group of neurons set up a circuit or an electrical "engram" which consists of a set of facilitated electrical connections between sensory and motor neurons. The formation of such an engram is governed by the increased synthesis of one or more specific enzymes in certain neurons.[54] These enzymes catalyze the formation or the destruction of neural transmitter substances. Acetylcholine is considered by some to be of primary importance in synaptic transmission between neurons.[55] It has been postulated that the release of transmitter substances during sensory stimulation derepresses a particular gene, which results in an increase in RNA synthesis. Derepression may be induced by small molecules, such as acetylcholine, or by some inhibitor released as a part of the cycle of synaptic transmission, such as γ-aminobutyric acid, as suggested by Roberts.[56] Derepression may be directly induced in a neuron by the electrical impulse, as suggested by Bonner.[57] Pribram[58] has proposed that RNA itself can initiate or induce derepression of a gene.

Hechter and Halkerston[59] have suggested that all significant sensory inputs involve activation of certain specific neurons, designated as *A* neurons, which contain biogenic amines. The *A* neurons are interconnected with *B* neurons such that when *A* neurons "fire" *B* neurons are activated by acetylcholine and biogenic amines released by *A*. This activation in *B* neurons catalyzes formation of cyclic 3′,5′-adenylic acid (cyclic AMP). Hechter and Halkerston assume that increased biosynthesis of RNA and protein in stimulated neurons may result from increased cyclic AMP formation. The *B* neurons are also connected to other *C*, neurons such that when *B* neurons are "fired" a network of firing neuron is established involving *A*, *B*, and *C* neurons which represent short-term memory. The maintenance of this specific neuronal network (or short-term memory) for a duration of approximately 30 min is dependent upon intracellular generation of cyclic AMP in the *B* neurons during this time interval.

Landauer[60] has suggested that the basic event of conditioning or learning is the transfer of RNA molecules from surrounding glial cells into conducting neurons and subsequent participation of this RNA as a messenger RNA for protein synthesis in the neurons. The transfer of RNA across membranes of activated neurons during impulse conduction is presumed to be due to the removal of a barrier consisting of an unfavorable potential gradient which is represented by membrane impermeability.

The above investigators have made certain assumptions in their interpretation of information accrual in the brain. A few have implicated some role to DNA in the process, mainly as a template for RNA synthesis. Since DNA is a very stable structure, it is difficult to imagine how configuration changes in DNA can be affected by an afferent impulse so that it can act as a template for RNA synthesis and yield an RNA whose base sequence reflects the information in the coded impulse. Others have implicated the role of biogenic amines in the initiation and maintenance of specific "firing" network of neurons for approximately 30 min, representing short-term memory induced by afferent impulses. How the specificities of various afferent impulses are encoded in the neuronal network is difficult to imagine. Similarly, how the short-term memory in the neuronal network is converted into long-term memory, in growing synaptic connections or in macromolecular formation, is not clearly understood.

From the existing reports in the literature and from the results obtained in the present studies the conceptual process of information accrual in the brain is visualized as follows: The coded afferent impulse is transmitted along the axons and dendrites of the neurons, and thus a fair number of neurons and glia receive the same information. Simultaneously, this signal passes through the cytoplasm to the nucleus within the brain cells. As the afferent impulse enters the nuclei, it is transduced in some unknown manner so that RNA synthesis occurs in such a manner that the sequence of bases corresponds to the coded information in the impulse. Thus, the coded afferent impulse induces the synthesis of RNA whose base sequence contains the information transmitted by the impulse. From the data presented here the newly synthesized RNA represents short-term "memory." It is further conceptualized that this RNA subsequently acts as a messenger RNA which directs the biosynthesis of a protein whose amino acid sequence is determined by the base sequence in this RNA. Thus, this newly synthesized protein represents the long-term information storage unit or engram (Fig. 1).

REFERENCES

1. K. S. Lashley, "Physiological Mechanisms in Animal Behavior," Academic Press, New York (1950).
2. D. L. Drabkin, "Thudichum," University of Pennsylvania Press, Philadelphia (1958).
3. C. Sherrington, "The Integrative Action of Nervous System," Yale University Press, New Haven (1947).
4. D. O. Hebb, "The Organization of Behavior," John Wiley and Sons, New York (1949).
5. C. P. Duncan, *J. Comp. Physiol. Psychol.* **42,** 32–44 (1949).
6. P. Weiss and H. B. Hiscoe, *J. Exp. Zool.* **107,** 315–395 (1948).
7. J. C. Eccles, "The Neurophysiological Basis of Mind," Clarendon Press, Oxford (1953).
8. H. Hydén and E. Egyhazi, *Proc. Natl. Acad. Sci. U.S.* **48,** 1366–1373 (1962).
9. H. Hydén and E. Egyhazi, *Proc. Natl. Acad. Sci. U.S.* **49,** 618–624 (1963).
10. H. Hydén and E. Egyhazi, *Proc. Natl. Acad. Sci. U.S.* **52,** 1030–1035 (1964).

11. H. Hydén and P. W. Lange, *Proc. Natl. Acad. Sci. U.S.* **53,** 946–952 (1965).
12. G. P. Talwar, S. P. Chopra, B. K. Goel, and B. D'Monte, *in*: "Macromolecules and Behaviour" (J. Gaito, ed.), pp. 71–88, York University Press, Toronto (1966)
13. J. W. Zemp, J. E. Wilson, K. Schlesinger, W. D. Boggan, and E. Glassman, *Proc. Natl. Acad. Sci. U.S.* **55,** 1423–1431 (1966).
14. J. Jarlstedt, *Acta Physiol. Scand.* **67,** 243–252 (1966).
15. A. B. Kogan and S. L. Zaguskin, *Fed. Proc.* **25,** T191–T195 (1966).
16. L. Z. Pevzner, *in*: "Macromolecules and Behaviour" (J. Gaito, ed.), pp. 43–70. York University Press, Toronto (1966).
17. J. E. Edstrom and W. Grampp, *J. Neurochem.* **12,** 735–741 (1965).
18. W. Dingman and M. B. Sporn, *J. Psychiat. Res.* **1,** 1–11 (1961).
19. L. B. Flexner, J. B. Flexner, R. B. Roberts, and G. de la Haba, *J. Neurochem.* **12,** 535–541 (1965).
20. L. B. Flexner, J. B. Flexner, R. B. Roberts, *Proc. Natl. Acad. Sci. U.S.A.* **56,** 730–735 (1966).
21. L. B. Flexner and J. B. Flexner, *Proc. Natl. Acad. Sci. U.S.* **55,** 369–374 (1966).
22. B. W. Agranoff, R. E. Davis, and J. J. Brink, *Proc. Natl. Acad. Sci. U.S.* **54,** 788–793 (1965).
23. R. E. Davis and B. W. Agranoff, *Proc. Natl. Acad. Sci. U.S.* **55,** 555–559 (1966).
24. S. H. Barondes and H. D. Cohen, *Science* **151,** 594–595 (1966).
25. H. D. Cohen and S. H. Barondes, *J. Neurochem.* **13,** 207–211 (1966).
26. G. P. Talwar, S. P. Chopra, B. K. Goel, and B. D'Monte, *J. Neurochem.* **13,** 109–116 (1966).
27. H. P. Metzger, M. Cuenod, A. Grynbaum, and H. Waelsch, *Life Sciences* **5,** 1115–1120 (1966).
28. R. J. Hilmore and L. A. Heppel, *J. Am. Chem. Soc.* **79,** 4810–4811 (1957).
29. A. A. Hakim, *Enzymologia* **21,** 81–95 (1960).
30. K. Yagi, T. Ozana, and H. Konogi, *Nature* **184,** 1939 (1959).
31. M. Grunberg-Manago and S. Ochoa, *J. Am. Chem. Soc.* **77,** 3165–3166 (1955).
32. J. H. Frenster, V. G. Allfrey, and A. E. Mirskey, *Biochim. Biophys. Acta* **47,** 130–137 (1961).
33. E. D. Adrain and C. Ludwig, *J. Physiology* **94,** 441–460 (1938).
34. A. C. Allison, *Biol. Rev.* **28,** 195–244 (1953).
35. G. H. Parker, *J. Exp. Zool.* **10,** 1–6 (1911).
36. J. M. D. Olmstead, *Am. J. Physiol.* **46,** 443–458 (1918).
37. J. P. Schade and I. J. Weiler, *J. Exp. Biol.* **36,** 435–452 (1959).
38. T. Shibuya, *Jap. J. Physiol.* **10,** 317–326 (1960).
39. J. Boudreau, *Jap. J. Physiol.* **12,** 272–278 (1962).
40. J. Seegar, *Prog. Brain Res.* **14,** 143–231 (1965).
41. R. Nanba, B. Djahanparwar, and R. von Baumgarten, *Pflugers Archiv.* **288,** 134–150 (1966).
42. D. A. Rappoport and H. F. Daginawala, *J. Neurochem.* **15,** 991–1006 (1968).
43. I. H. Goldberg, M. Rabinowitz, and E. Reich, *Proc. Natl. Acad. Sci. U.S.* **48,** 2094–2101 (1962).
44. I. H. Goldberg, M. Rabinowitz, and E. Reich, *Proc. Natl. Acad. Sci. U.S.* **49,** 226–229 (1963).
45. H. Hydén, *in*: "The Cell" (J. Brachet and A. E. Mirsky, eds.), Vol. 4, pp. 215–323, Academic Press, New York (1960).
46. L. J. Goldsmith, *J. Comp. Physiol. Psychol.* **63,** 126–132 (1967).
47. S. H. Barondes and M. E. Jarvik, *J. Neurochem.* **11,** 187–195 (1964).

48. B. W. Agranoff, R. E. Davis, L. Casola, and R. Lim, *Science* **158,** 1600–1601 (1967).
49. J. J. Brink, R. E. Davis, and B. W. Agranoff, *J. Neurochem.* **13,** 889–896 (1966).
50. B. W. Agranoff and R. E. Davis, *Science* **158,** 523 (1967).
51. S. H. Barondes and H. D. Cohen, *Proc. Natl. Acad. Sci. U.S.* **58,** 157–164 (1967).
52. S. H. Appel, *Nature* **213,** 1253–1254 (1967).
53. M. W. Nirenberg and H. J. Matthaei, *Proc. Natl. Acad. Sci. U.S.* **47,** 1588–1602 (1961).
54. M. H. Briggs and G. B. Kitto, *Psychol. Rev.* **69,** 537–541 (1962).
55. J. C. Eccles, *Science* **145,** 1140–1147 (1964).
56. E. Roberts, *Brain Res.* **1,** 117–166 (1966).
57. J. Bonner, *in*: "Macromolecules and Behaviour" (J. Gaito, ed.), pp. 158–164, York University Press, Toronto (1966).
58. K. H. Pribram, *in*: "Macromolecules and Behaviour" (J. Gaito, ed.), pp. 165–187, York University Press, Toronto (1966).
59. O. Hechter and I. D. K. Halkerston, *Perspect. Biol. Med.* **7,** 183–198 (1964).
60. T. K. Landauer, *Psychol. Rev.* **71,** 167–179 (1964).
61. C. V. A. Kappers, G. C. Huber, and E. C. Crosby, *in*: "The Comparative Anatomy of Nervous System of Vertebrates," Vol. 2, pp. 1260–1292, Macmillan, New York (1936).

Chapter 24

Effects of Actinomycin D on RNA Transcription, Protein Synthesis, and Nuclear Structure*

Harold Koenig, Chung Y. Lu, Stanley Jacobson, Panna Sanghavi, and Rajinder Nayyar

Neurology Service, VA Research Hospital
and
Department of Neurology
Northwestern University Medical School
Chicago, Illinois

INTRODUCTION

Neurons and glia are sites of active turnover of RNA and protein.[1,2] These metabolic activities are of fundamental importance for the biosynthetic renewal of a wide assortment of structural, soluble, enzyme, and other proteins necessary for maintenance of normal neural structure and function. However, the relation of RNA and protein synthesis to neurophysiological functions such as excitation, inhibition, nerve impulse conduction, and synaptic transmission has not been clearly defined. Earlier studies[3–6] from this laboratory showed that the pyrimidine analogs, 5-fluoroorotic acid and 5-fluorouridine, after a latent period of several days cause severe neurophysiological disturbances and neuropathological lesions in cat spinal cord concomitant with impaired synthesis of RNA and protein. Edström and Grampp[7] found that a crustacean stretch receptor neuron could generate action potentials *in vitro* for 12–24 hr after RNA synthesis was blocked by actinomycin D. Cohen and Barondes[8] were able to inhibit RNA synthesis in rat brain by intracerebral injection of AMD without adversely affecting learning in short-term experiments.

* Supported in part by grants from the NIH (NB 01456, NB 05509, and NB 04493), and the National Multiple Sclerosis Society (No. 372). The excellent technical assistance of the following persons is acknowledged: B. Newson, C. Hughes, T. Konga, D. Thomas, S. Koenig, L. Koenig, S. Silverman, M. D. Gunther, L. Simons, S. Boegner, and E. Washington.

Over the past several years we have used actinomycin D (AMD), a potent inhibitor of RNA transcription *in vivo*, and of RNA polymerase activity *in vitro* (reviewed by Reich and Goldberg[9]), to explore in detail the relation of RNA and protein metabolism to function in cat spinal cord. AMD acts by forming a stable complex with DNA. Several other antibiotic inhibitors which have a similar mechanism of action, including mithramycin, olivomycin, and chromomycin A_3, were also employed with essentially the same results. When injected intrathecally in cat AMD causes a distinctive neurological disorder after a latent period. During the latent period well-defined structural and biochemical lesions appear: (1) diffuse chromatin in nuclei of neurons and glia condenses and subsequently coalesces into large, dense masses[10–14]; (2) RNA synthesis, as measured by ^{14}C-adenine or ^{14}C-orotic acid uptake, is simultaneously inhibited[11,14]; and (3) protein synthesis, demonstrated by ^{14}C-lysine incorporation, is depressed somewhat later.[11,13,14] These investigations have elicited several distinctive features with regard to the biological activity of AMD in neural tissue, and have elucidated the mechanism for the repression of RNA transcription.

MATERIALS AND METHODS

Biochemical Experiments

AMD (a generous gift of Merck, Sharp and Dohme) or ^{3}H-AMD (3.8 Ci/mM, Schwarz Bioresearch, Inc.) diluted with cold AMD (5 μg/8.74 μCi, or 1 mg/50 μCi, in 0.5 ml of saline) was injected into the lumbar subarachnoid space of adult cat following surgical laminectomy at the L-7 level performed under nembutal anesthesia. To measure RNA synthesis, ^{14}C-adenine (28.9 mCi/mM, Nuclear Chicago) (3 μCi/0.2 ml/kg) was given by the same route 30, 60, or 120 min before sacrifice. In experiments on rat brain, ^{3}H-AMD (10 μg/17.8 μCi/0.02 ml) was injected intracerebrally, followed by ^{14}C-adenine (2 μCi/0.02 ml) 30 or 50 min before sacrifice. Cat cerebrum was used for separating white and gray matter; ^{3}H-AMD (20 μg/6.94 μCi/0.01 ml) was injected into three sites in each hemisphere through burr holes made with a dental drill. ^{14}C-Uridine (59.7 mCi/mM, Nuclear Chicago), (1 μCi/0.01 ml) was injected into the same sites 15 min before sacrifice. White and gray matter were dissected and processed separately.

Spinal cord was removed under anesthesia to about the T-6 segment and homogenized in six volumes of 0.25 *M* sucrose with 1 m*M* EDTA. Rat brain was homogenized in nine volumes of medium. Homogenates were fractionated into crude nuclear, cytoplasmic, and soluble fractions by spinning at 1000g for 10 min and 105,000g for 90 min, respectively. The nuclear pellet was resuspended in 1.5 *M* sucrose and centrifuged at 105,000g for 90 min to deposit a purified nuclear pellet. All manipulations were at 4°C. For *in vitro* experiments slices of rat cerebrum cut at 0.23 mm with a tissue chopper[15] were incubated in a physiological salt medium (128 m*M* NaCl, 6.3 m*M* KCl, 1.3 m*M* Mg_2SO_4, 2.7 m*M* $CaCl_2$, 0.2 m*M* Na_2HPO_4, 10 m*M* MgCl, 2.5 m*M* Tris adjusted to pH 7.35 with HCl) at 37° under atmospheric air in a Dubnoff shaker. ^{14}C-Adenine (27.7 mCi/mM, Nuclear Chicago) (5 μCi) was introduced 20 min before stopping the reaction with cold TCA (final concentration 17%), and the entire tissue slices processed. RNA and DNA were extracted from acid- and lipid-extracted tissue samples and measured by ultraviolet spectrophotometry.[16] In some experiments RNA was also measured by an orcinol method,[17] and DNA by a modified diphenylamine reaction.[18] Doubly labeled samples of acid- and lipid-soluble and RNA fractions were assayed for ^{14}C and ^{3}H radioactivity in a liquid scintillation counter (Nuclear Chicago 725) by a channel ratio technique.[19] In two experiments the ^{3}H-AMD complex was isolated by zonal equilibrium centrifugation over a $CsCl_2$ density gradient.[20]

Morphological and Cytochemical Studies

In biochemical experiments samples of cat spinal cord and cerebrum were fixed by immersion in 4% cold neutral formaldehyde, or by vascular perfusion with phosphate-buffered 4% formaldehyde or cacodylate-buffered 4% glutaraldehyde. For electron microscopy, glutaraldehyde-fixed tissue slices were postfixed for 1–2 hr in 1% buffered OsO_4, and imbedded in araldite. Ultrathin sections were stained with uranyl acetate or lead hydroxide and examined in an RCA EMU3F electron microscope. Thick (1 μ) sections of araldite-imbedded tissue and paraffin sections were stained for DNA by the Feulgen method, by thionin after RNase digestion, and for histone by fast green at pH 8, as well as by routine neuropathological stains. H^3-AMD was demonstrated in radioautographs prepared by a slide-dipping procedure with Kodak NTB or NTB 3 nuclear emulsion.

RESULTS

Biochemical Effects of AMD

RNA Synthesis

Spinal Cord. ^{14}C-Adenine, when injected intrathecally in control cat, is incorporated into nuclear RNA five times more rapidly than into cyto-

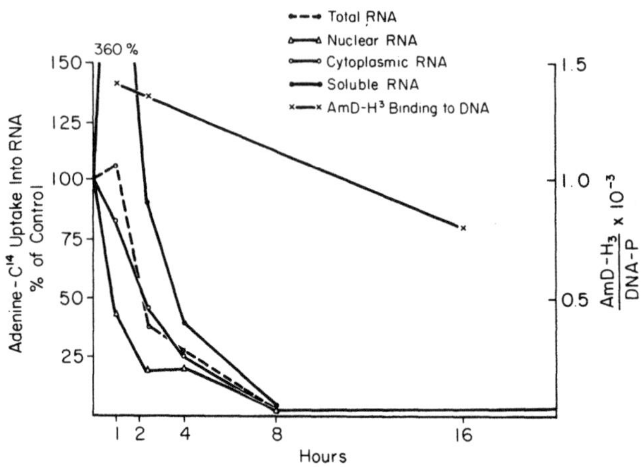

Fig. 1. Effect of H^3-AMD (1 mg) on ^{14}C-adenine uptake into RNA in subcellular fractions of cat spinal cord as a function of time. At various times after injection of ^3H-AMD into the lumbar subarachnoid space ^{14}C-adenine was injected *via* the same route, and the caudal half of the spinal cord removed under anesthesia. The tissue was fractionated and processed as described in the text. ^{14}C-Adenine uptake into RNA was expressed as the *RSA*, (*SA* of RNA)/(*SA* of acid-soluble fraction), and is given here as per cent of the control value. The amount of ^3H-AMD firmly bound to DNA was computed from the tritium radioactivity of the acid- and lipid-insoluble alkaline hydrolyzate of purified nuclear pellets and the known *SA* of the ^3H-AMD. Further experimental details are given in the section on materials and methods. (From Koenig *et al.*[12])

plasmic RNA, and 12 times more rapidly than into soluble RNA. Uptake into the several subcellular fractions is linear over a 2-hr period. RNA synthesis in cat spinal cord, as measured by the RSA (relative specific activity) of ^{14}C-adenine in total RNA, was initially augmented and subsequently inhibited by AMD. After a large (1-mg) dose the *RSA* of total RNA was 106% of the control at 1 hr, and 3% by 8 hr (Fig. 1). With a small (5-μg) dose the *RSA* of total RNA was 135% of the control at 1 hr, and 40% by 6 hr (Fig. 2). ^{14}C-Adenine incorporation into nuclear and cytoplasmic RNA was invariably inhibited by AMD, while ^{14}C-adenine incorporation into soluble RNA was at first markedly elevated, and only later depressed. After a 5-μg dose the *RSA* of nuclear and cytoplasmic RNA declined gradually to 30% of the control by 6 hr, while the *RSA* of soluble RNA rose to 360% at 1 hr and subsequently declined (Fig. 2).

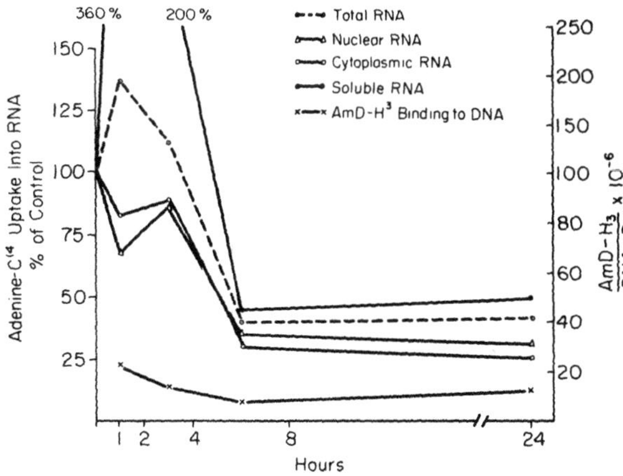

Fig. 2. Effect of H^3-AMD (5 μg) on C^{14}-adenine uptake into RNA in subcellular fractions of cat spinal cord as a function of time. Procedures were as described in legend to Fig. 1. (From Koenig *et al.*[12])

After a 1-mg dose the *RSA* of nuclear RNA declined to 44% of control at 1 hr, and 3% by 8 hr, while the *RSA* of cytoplasmic RNA fell more gradually to 3% by 8 hr. The *RSA* of the soluble RNA rose to 360% of the control at 1 hr and subsequently declined to 5% by 8 hr (Fig. 1).

A smaller dose of AMD (3 μg), given intrathecally, stimulated the incorporation of ^{14}C-adenine into RNA in all subcellular fractions of cat spinal cord at two and seven days after injection (Fig. 3). This enhancement

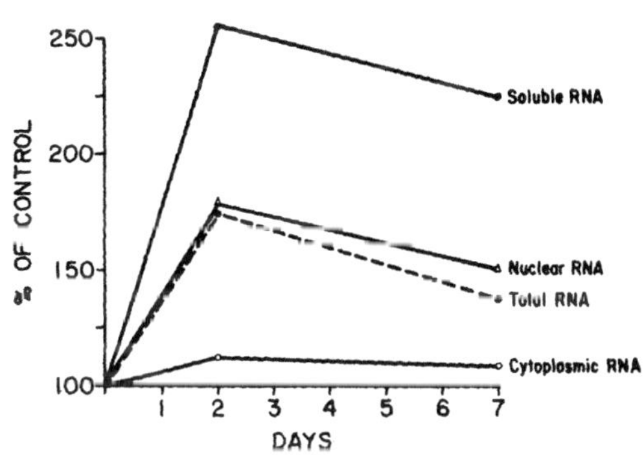

Fig. 3. Effect of AMD (3 μg) on ^{14}C-adenine incorporation into RNA in subcellular fractions of cat spinal cord as a function of time. Procedure as described in legend to Fig. 1.

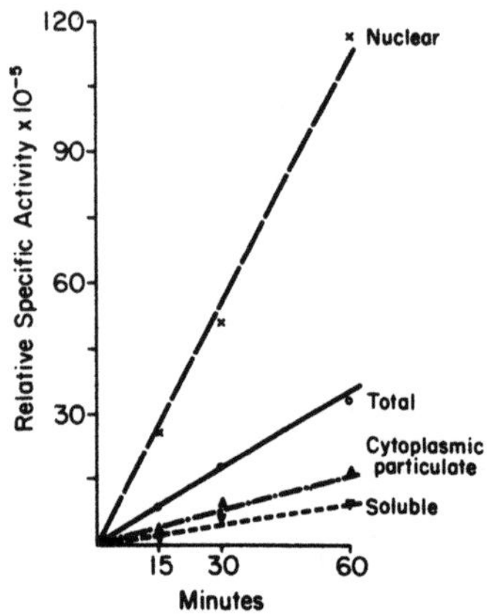

Fig. 4. Incorporation of ^{14}C-adenine into RNA in subcellular fractions of normal rat brain as a function of time. See legend to Fig. 1 for experimental details.

was most marked in the soluble RNA, the *RSA* of which was 225–250% of the control.

Cerebrum. ^{14}C-Adenine, injected intracerebrally in rat, is incorporated linearly into RNA in the several subcellular fractions for 2 hr (Fig. 4). The labeling of nuclear RNA was 6–7 times more rapid than cytoplasmic RNA, and 8–13 times faster than soluble RNA. After intracerebral injection of AMD (5 μg per hemisphere), the *RSA* of total RNA fell to 55% of the

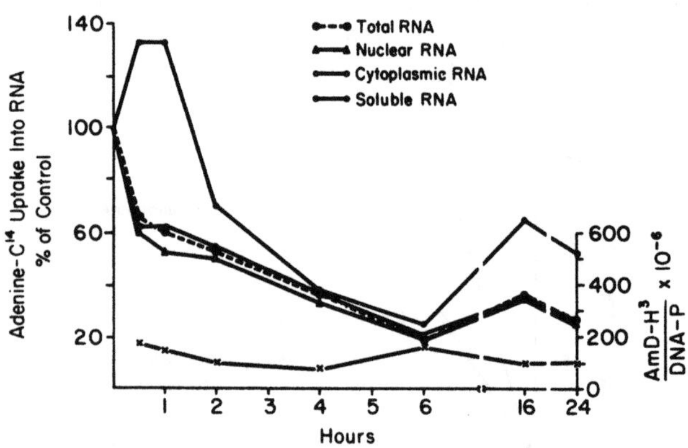

Fig. 5. Effect of ^{3}H-AMD (10 μg) on ^{14}C-adenine uptake into RNA in subcellular fractions of rat brain as a function of time. At various time intervals after injection of 5 μg of ^{3}H-AMD into each cerebral hemisphere ^{14}C-adenine was injected *via* the same route 30 or 50 min before killing. Further processing was as described in the legend for Fig. 1. (From Koenig *et al.*[12])

Fig. 6. Effect of ^{3}H-AMD (120 μg) on ^{14}C-uridine uptake into total RNA of cortex and subcortex of cat cerebrum as a function of time. At various times after injection of 20 μg of ^{3}H-AMD into three separate sites in each cerebral hemisphere ^{14}C-uridine was injected into the same sites 15 min before sacrifice. White and gray matter were dissected and processed separately, as described in the legend for Fig. 1. (From Koenig *et al.*[12])

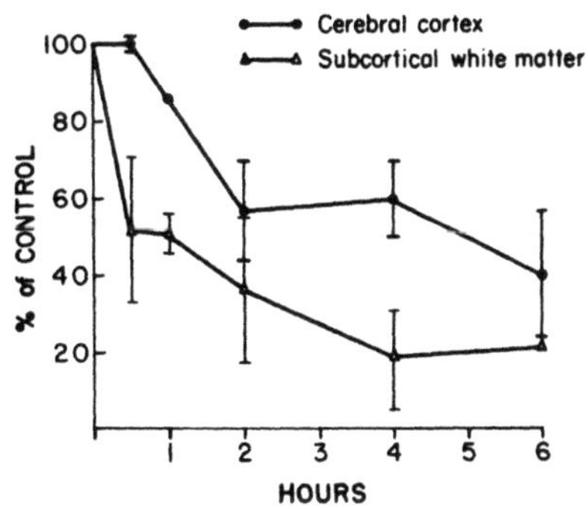

control at ½ hr, and 19% by 6 hr (Fig. 5). The *RSA* of nuclear and cytoplasmic RNA was inhibited to a similar degree, but the *RSA* of soluble RNA increased initially and subsequently declined.

In cat cerebrum ^{14}C-adenine uptake into total RNA declined more rapidly in subcortical white matter than in cortex after AMD injection (20 μg in each of six sites) (Fig. 6). The *RSA* of total RNA in white matter was 54% of control at 30 min, and 20% by 4 hr, while the *RSA* of total RNA in the overlying gray matter was 100% of the control at 30 min, and 60% by 4 hr. Subcellular fractionation was not carried out. A similar effect has been seen in rat cerebrum.[21]

Inhibition of RNA Synthesis in Vitro. After a 2-hr preincubation period with ^{3}H-AMD, ^{14}C-adenine incorporation into total RNA in rat-brain slices was inhibited 30–47% over a wide range of AMD concentration. Prolongation of preincubation time to 4 hr increased the inhibition to 78% (Fig. 7).

Fig. 7. Effect of varying time of preincubation with ^{3}H-AMD on *in vitro* ^{14}C-adenine uptake into total RNA of rat brain. Rat-brain slices were preincubated without and with ^{3}H-AMD (5 μg, 2 μCi) for various time intervals in a physiological salt medium in a shaking incubator at 37°C under air. ^{14}C-Adenine (5 μCi) was added to the reaction mixture 20 min before stopping the reaction with cold TCA. Procedures were as described in the legend for Fig. 1 and the section on materials and methods. (From Koenig *et al.*[12])

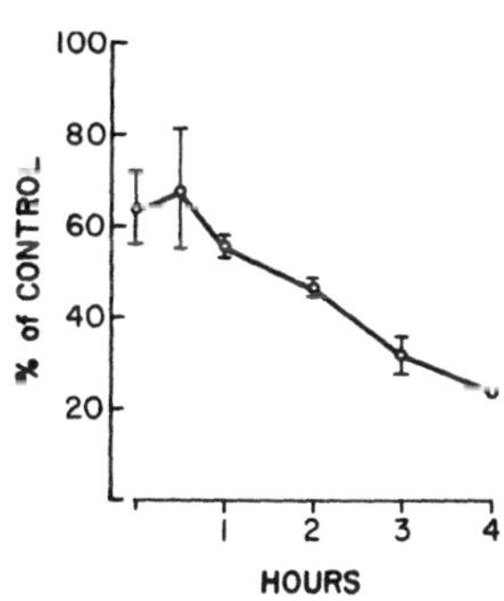

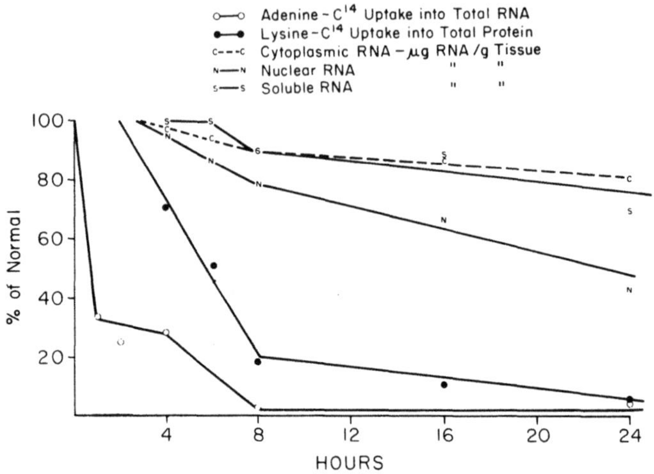

Fig. 8. Effect of AMD (1 mg) on isotope incorporation and chemical composition of cat spinal cord as a function of time. ^{14}C-Lysine (150 mCi/mM) (1 μCi/kg) was injected subarachnoidally 1 hr before sacrifice. Total protein was measured by the method of Lowry *et al.*[48] and uptake expressed as the *RSA*, (*SA* of protein)/(*SA* of acid-soluble fraction). Methods are similar to legend in Fig. 1. (From Koenig.[14])

Protein Synthesis

Cat Spinal Cord. ^{14}C-Lysine uptake, expressed as the *RSA* of total neural protein, served as a measure of protein synthesis. The effect of a large dose (1 mg) and a small dose (5 μg) of AMD on protein synthesis in feline spinal cord is shown in Figs. 8 and 9, respectively. Protein synthesis

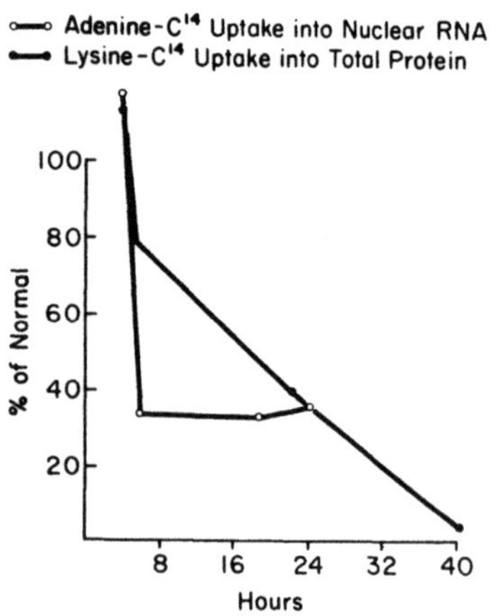

Fig. 9. Effect of AMD (5 μg) on isotope incorporation into total RNA and total protein in cat spinal cord as a function of time. See legends to Figs. 1 and 8 for experimental methods. (From Koenig.[14])

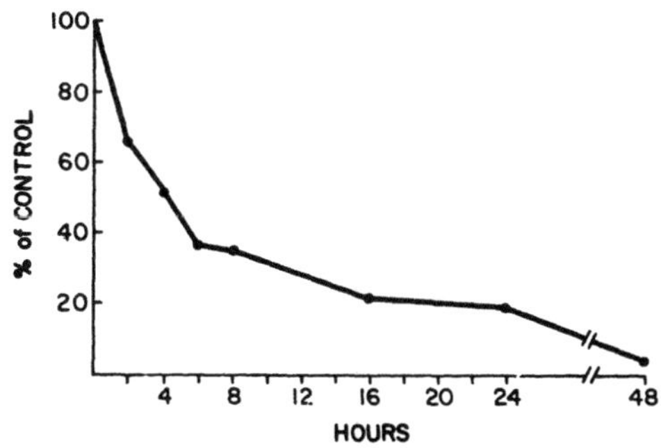

Fig. 10. Effect of AMD injected intracerebrally (10 μg into each hemisphere) on ^{14}C-lysine uptake into total protein in rat brain as a function of time. See legends to Figs. 1 and 8 for experimental methods.

declines more slowly than RNA synthesis. A significant depression (29%) was already evident 4 hr after AMD (1 mg), when the pool of cytoplasmic RNA was barely affected. Protein synthesis was severely inhibited (81%) at 8 hr when cytoplasmic RNA was diminished by only 11%. After a 5-μg dose protein synthesis was 39% of normal at 19 hr, when cytoplasmic RNA was 91% of normal.

Rat Brain. After AMD (20 μg), injected intracerebrally, ^{14}C-lysine uptake into total brain protein declined to 50% of control at 4 hr, 20% at 24 hr, and 3% by 48 hr (Fig. 10).

Fig. 11. Effect of AMD (5 μg) on RNA concentration in subcellular fractions of cat spinal cord as a function of time. Methods are as in legend to Fig. 1.

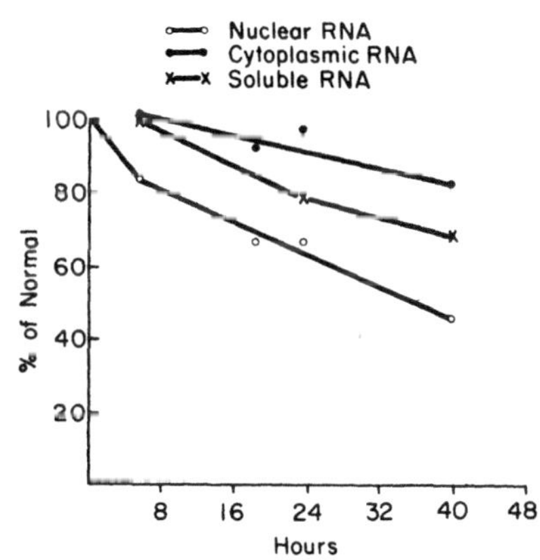

TABLE I

Concentration of RNA, DNA, and Protein in Subcellular Fractions of Spinal Cord of Normal Cat

	Concentration[a]			
	Total	Nuclear	Cytoplasmic	Soluble
RNA	45.9 ± 0.94	7.12 ± 0.27	30.28 ± 0.96	9.45 ± 0.51
DNA	26.1 ± 0.56			
Protein	58.9 ± 1.42	8.38 ± 0.16	38.08 ± 0.63	14.12 ± 0.47

[a] RNA and DNA concentrations are given as μg nucleotide-P per g tissue ± S.E. (N = 8); protein concentration is given as mg protein per g tissue ± S.E. (N = 8).

Concentration of RNA, DNA, and Protein

RNA. Control values in cat spinal cord are given in Table I. The effect of a large dose (1 mg) of AMD on the concentration of RNA in the primary fractions of feline spinal cord is shown in Fig. 8. Nuclear RNA declined most rapidly, reaching 43% of normal by 24 hr. Cytoplasmic RNA was 80%, and soluble RNA 70%, of control by 24 hr. After a small (5-μg) dose of AMD the RNA concentration fell at about half this rate (Fig. 11). Nuclear RNA was 45% of control, cytoplasmic RNA 82%, and soluble RNA 68%, of control at 40 hr. When a smaller dose (3 μg) of AMD was given nuclear RNA was 55% at two days, and normal at seven days. Cytoplasmic RNA was unaltered, while soluble RNA increased to 150% of the control by seven days (Fig. 12).

DNA. The DNA concentration in feline spinal cord was unaltered 24 hr after AMD (1 mg and 5 μg). However, 40 hr after a 5-μg dose DNA

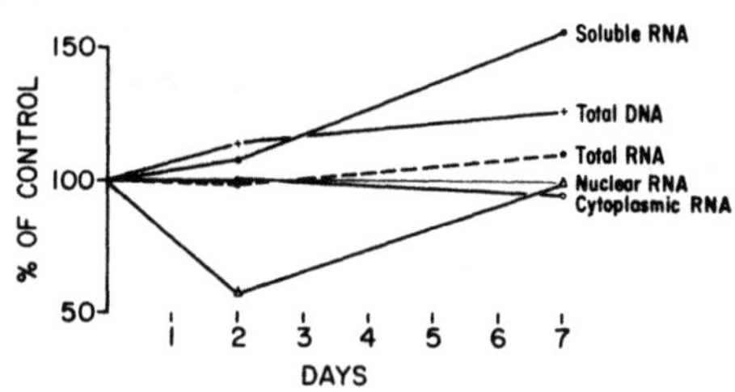

Fig. 12. Effect of AMD (3 μg) on the chemical composition of cat spinal cord as a function of time. Methods are as in legend to Fig. 1.

concentration declined to 70% of control in some experiments, suggesting a loss of cells in treated spinal cord. After a 1- or 3-μg dose the DNA concentration in spinal cord rose to 125–135% of control by 7–9 days, suggesting an increased cellularity in spinal cord.

Protein. The protein in the nuclear fraction declined to 60–70% of control 24 hr after a big (1-mg) dose. A comparable fall in nuclear protein obtained 40 hr after a small (5-μg) dose of AMD. The cytoplasmic and soluble protein were unaffected during the first 24 hr (1 mg and 5 μg), but these decreased to about 75% of control 40 hr after AMD (5 μg). With a very small dose (1 and 3 μg) of AMD, nuclear protein rose along with the DNA to about 150%, and cytoplasmic and soluble protein to 115%, of control by 7–9 days after injection.

Binding of ^{3}H-AMD

Subcellular Distribution and Solvent Extraction. In cat spinal cord the bulk of ^{3}H-AMD, 89–93% of the total, occurred as acid- and/or lipid-labile complexes (Table II). In rat brain 75–80% of the total ^{3}H-AMD occurred in this state (Table III). The residual tritium radioactivity, 7–11% in feline spinal cord and 20–25% in rat brain, was largely recovered in the RNA fraction (acidified alkaline hydrolysate). This fraction represents ^{3}H-AMD bound firmly to DNA, as it occurred mainly in the purified

TABLE II

Solubility and Distribution of ^{3}H-AMD in Subcellular Fractions of Cat Spinal Cord[a]

Solubility	Per cent of total ^{3}H-AMD							
	Homogenate		Nuclear		Cytoplasmic		Soluble	
	6 hr	24 hr	6 hr	24 hr	6 hr	24 hr	6 hr	24 hr
TCA-soluble	16.0	20.0	2.9	2.9	3.5	4.2	9.6	12.9
Lipid-soluble	77.4	69.0	19.5	21.9	46.5	32.4	11.4	14.7
Alkali-soluble	6.6	11.0	5.9	10.0	0.5	0.8	0.2	0.2
TOTAL	100.0	100.0	28.3	34.8	50.5	37.4	21.2	27.8

[a] ^{3}H-AMD (5 μg, 8.74 μCi, 0.5 ml) was injected into the lumbar subarachnoid space 6 and 24 hr before sacrifice. The homogenate was fractionated, the resulting fractions extracted serially with acid and lipid solvents, and the residue hydrolyzed in alkali. The pooled extracts and the alkaline hydrolysate were radioassayed for tritium as described in the section on materials and methods. (From Koenig *et al.*[12])

TABLE III

Solubility and Distribution of H^3-AMD in Subcellular Fractions of Rat Brain[a]

Solubility	Per cent of total ^{3}H-AMD							
	Homogenate		Nuclear		Cytoplasmic		Soluble	
	1 hr	6 hr	1 hr	6 hr	1 hr	6 hr	1 hr	6 hr
TCA-soluble	15.9	10.1	3.8	3.3	5.6	2.8	6.5	4.0
Lipid-soluble	64.1	64.9	35.6	36.4	22.9	16.8	5.6	11.7
Alkali-soluble	20.0	25.0	18.9	23.8	1.0	1.0	0.1	0.2
TOTAL	100.0	100.0	58.3	63.5	29.5	20.7	12.2	15.9

[a] ^{3}H-AMD (10 μg, 5 μCi, 0.02 ml), was injected intracerebrally 1 and 6 hr before sacrifice. Tissue was processed as in Table II. (From Koenig *et al.*[12])

nuclear pellet. AMD is linked to DNA by hydrogen bonds[9] which would not be susceptible to acid or lipid solvents.

Isolation of the AMD–DNA Complex. The validity of the solvent extraction procedure for measuring firmly bound ^{3}H-AMD was corroborated by direct isolation of AMD–DNA complex formed *in vivo*. Rats were given 10 μg of ^{3}H-AMD intracerebrally, killed 2 hr later, and a portion of the purified nuclear pellet processed by zonal centrifugation over a $CsCl_2$ gradient.[20] This gave 80.5 μM of ^{3}H-AMD bound per mole of DNA-P. The portion that was processed by solvent extraction yielded a value of 80 μM of ^{3}H-AMD per mole of DNA-P. A part of the ^{3}H-AMD was associated with nuclear proteins poor in DNA in the top of the $CsCl_2$ gradient.

Stoichiometry of the AMD-DNA Complex. The stoichiometry of the AMD–DNA complex varied with the dose, ranging from 8 to 1400 μM of ^{3}H-AMD per mole DNA-P in cat spinal cord (Figs. 1, 2), and 80 to 187 μM of ^{3}H-AMD per mole of DNA-P in rat brain (Fig. 5). In cat cerebrum, gray matter contained 26–50 μM of AMD, and white matter 11–40 μM of AMD, per mole of DNA-P. High molar ratios were present at the earliest intervals tested. *In vitro* studies[22] with purified brain nuclei indicate that firm binding of AMD to DNA occurs directly after addition of the drug. At a given time interval the inhibitory effect of AMD varies approximately as the logarithm of the bound AMD (Fig. 13).

Radioautography. Radioautographs of paraffin sections of cat spinal cord made 1–8 days after ^{3}H-AMD (1–2 μg/3.04–6.08 μCi/0.5 ml) revealed radioactivity, mainly nuclear in location, in glia, neurons, and other neural

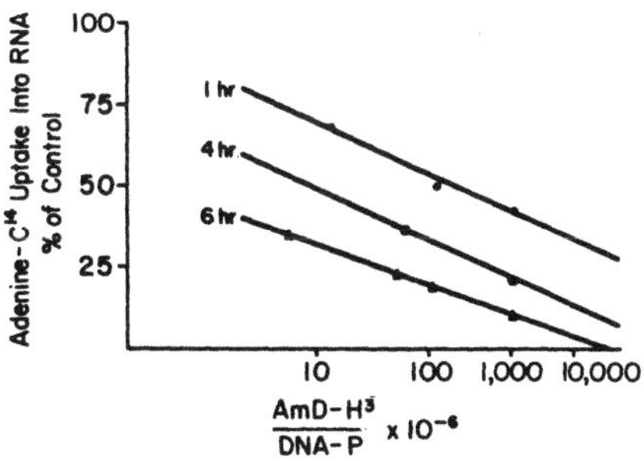

Fig. 13. ^{14}C-Adenine incorporation into nuclear RNA of cat spinal cord and rat brain as related to the stoichiometry of the AMD–DNA complex at several time intervals. The abscissa is on a logarithmic scale.

cells [Fig. 15(A–D)]. No activity was found in myelinated axons. In some neurons radioactivity also occurred in the cytoplasm, where it was presumably bound to mitochondrial DNA. Neuronal nuclei generally exhibited much more radioactivity than glial nuclei [Fig. 15(A–D)]. This radioactivity represents ^3H-AMD which is firmly bound to DNA, as the acid- and lipid-labile component would have been removed during processing.

Ultrastructural and Cytochemical Effects of AMD

The ultrastructural and cytochemical changes were largely restricted to the nuclei of neurons and glia during the first 24 hr after AMD injection. In general, glia and small neurons were more rapidly affected than large neurons.

Electron Microscopy

In spinal cord of control cats large motoneurons contain sparse small clumps of dense chromatin; small neurons contain abundant coarse chromatin masses, and medium-size neurons are intermediate in this respect. The nucleoplasm features innumerable osmiophilic granules from 30 to 250 Å in diameter, both scattered and in cloudlike aggregates up to 2 μ which seem to be partly lipoprotein in nature (see below). Nucleoli contain a granular zone with numerous 150-Å particles, a fibrous nucleonema, and amorphous areas. Glial nuclei possess abundant coarse chromatin masses, nucleoli, and osmiophilic granules [Fig. 14(A, E)].

After AMD (1 mg) fine-structural changes become visible in many neuronal nuclei at $\frac{1}{2}$ hr, and reach major proportions by 8 hr [Fig. 14(B D)]. Small clumps of dense chromatin appear throughout the nucleoplasm. These fuse to form coarse, elongated or rounded masses resembling chromosomes of mitotic cells. The nucleoplasmic granules are

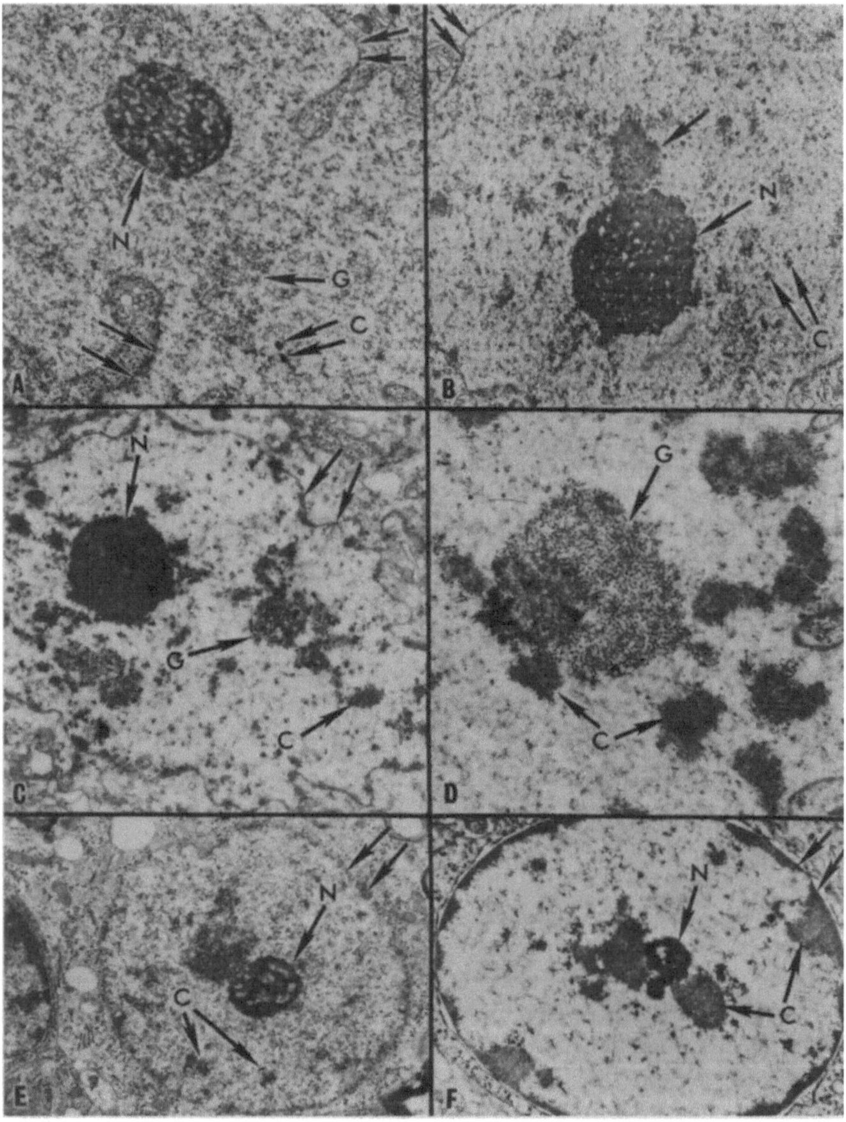

Fig. 14. Electron micrographs showing nuclear changes produced in cat spinal cord by AMD (1 mg). C: chromatin; G: granular aggregates; N: nucleolus; unlabeled arrows: nuclear membrane. (A–D) Motoneurons: (A) Control (×5640); (B) ½ hr after AMD; single arrow indicates heterochromatic sex chromatin next to the nucleolus (×5640); (C) 4 hr after AMD (×5640); (D) 4 hr after AMD (×16,800). (E–F) Glia (×3150): (E) Control; (F) ½ hr. (Partly from Koenig *et al.*[12])

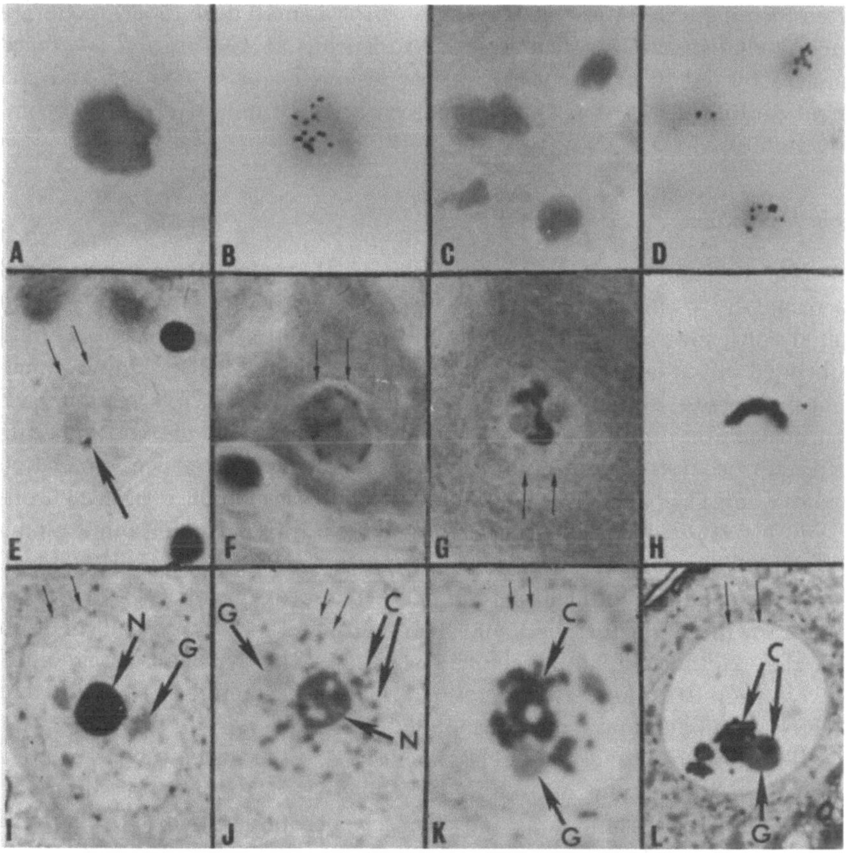

Fig. 15. Light micrographs of spinal motoneurons. (A–D) Radioautographs showing the nuclear radioactivity in spinal motoneuron (A, B) and spinal glia (C, D) due to ^{3}H-AMD bound to DNA. (E–H) Feulgen stain for DNA (×938): (E) Control showing sex chromatin body of Barr (single arrow); (F) 8 hr after AMD (1 mg); (G) 24 hr after AMD (1 mg); (H) 40 hr after AMD (5 μg). (I–L) 1-μ sections of glutaraldehyde–OsO_4-fixed, araldite-imbedded spinal cord stained with toluidine blue (×1500): (I) Control; (J) ½ hr after AMD (1 mg); (K) 4 hr after AMD (1 mg); (L) 24 hr after AMD (1 mg). (Partly from Koenig *et al.*[12])

gathered into large compact aggregates, and the component granules enlarge up to 300 Å, possibly by fusion. The intervening nucleoplasm becomes rarefied. Advanced nuclear changes appear in glia by 30–60 min (Fig. 14F), in small neurons by 1–2 hr, and in large motoneurons by 4–8 hr after AMD (1 mg). Nucleoli of neurons and glia exhibit a loss of 150-Å granules, fragmentation and collapse of the nucleonema, and segregation of these components [Fig. 14(B–D)]. Chromatin condenses about the nucleolus, often forming a perinucleolar envelope. After a small dose (5 μg) moderate nuclear changes are discernible in large neurons at 1–3 hr, and advanced changes by 24–40 hr. Identical nuclear lesions occur in neurons and glia of rat and cat cerebrum, appearing as early as 10 min after intracerebral injection of AMD.

Light Microscopy

The morphology of DNA is similar to that of nuclear histone, and corresponds to chromatin seen in electron micrographs. In the control spinal cord, motoneuronal nuclei contain sparse, fine chromatin granules imbedded in a faintly colored nucleoplasm [Fig. 15(E)]. Small spinal neurons possess abundant coarse chromatin particles, while medium-size neurons exhibit an intermediate chromatin pattern. Glial chromatin occurs as discrete coarse masses, and as diffusely stained nucleoplasm. In treated tissues (1 mg) a coarsening of chromatin granules is evident in large neurons within 1–2 hr, and in glia within 30 min. The diffuse nucleoplasmic staining for DNA, representing euchromatin, rapidly disappears. By 16–24 hr all the chromatin of motoneurons is compacted into one or several masses that stain intensely for DNA and histone [Fig. 15(F–H)]. Small neurons and glia exhibit comparable changes within 1–2 hr. Nucleoli change more slowly; they are moderately smaller and less basophilic at 4–8 hr, and severely shrunken and feebly basophilic by 16–24 hr.

Two kinds of abnormal nuclear inclusions, often intimately associated, are seen in 1-μ sections of araldite-imbedded tissues: (1) dense chromatin clumps which are osmiophilic and intensely basophilic; and (2) spheroidal masses corresponding to the granular aggregates in electron micrographs that are osmiophilic and moderately basophilic [Fig. 15(I–L)]. The latter structures seem to contain lipoproteins, as they stain with Sudan black, as well as with basic and acidic dyes. Moreover, their osmiophilia is curtailed by prior delipidation or bromination. These nuclear lesions develop more slowly after a small (5-μg) dose of AMD. Identical nuclear changes occur in AMD-treated rat and cat cerebrum. In these loci, as in feline spinal cord, there was a close correlation between the severity of the nuclear lesions and the inhibition of RNA synthesis.

DISCUSSION

These investigations have brought to light several significant features with regard to the biological activity of AMD in neural tissue which are inconsistent with the steric hindrance theory (reviewed by Reich and

Goldberg[9]). First, the kinetics of the AMD-induced repression of RNA transcription differ sharply from the kinetics of AMD binding. When administered locally ^{3}H-AMD penetrates into neural tissue and a portion, approximately 20–25% of the total in rat brain and 7–11% in cat spinal cord, is rapidly bound in an acid- and lipid-stable linkage to nuclear DNA. However, inhibition of RNA synthesis develops relatively slowly *in vivo* and *in vitro*, 6–8 hr elapsing before a maximum effect is attained. A time-dependent increase in the inhibitory action of AMD is discernible in other biological systems, e.g., fibroblasts[23] and developing retina,[24] but its significance has not hitherto received attention. According to the steric hindrance theory, the inhibition of RNA synthesis should attain a maximum value coincident with the formation of the AMD–DNA complex. Our results clearly indicate that the repression of RNA transcription is not a direct consequence of the binding of AMD, but is due to a secondary, time-dependent process which is initiated by the formation of the AMD–DNA complex.

A second feature is the remarkable effectiveness of AMD in repressing RNA synthesis at a low saturation level of DNA. Thus, as little as 10^{-5} M of AMD bound per mole of DNA-P (Fig. 2) inhibited RNA synthesis in cat spinal cord by 70% after 6 hr. Under these conditions DNA strands 10^5 nucleotide units in length between deoxyguanine sites occupied by AMD should still be available to prime the RNA polymerase reaction. Thus, our stoichiometric findings are also incompatible with the steric hindrance theory.

The differential susceptibility of gray matter and white matter to AMD is noteworthy. Despite binding less AMD, subcortical white matter of cat cerebrum exhibited an earlier and more severe inhibition of RNA synthesis than the cortex, suggesting that glia are more sensitive to the inhibitory effects of AMD than neurons. Appel[21] made similar observations in rat cerebrum, but did not measure AMD binding. It should also be noted that repression of RNA transcription developed more slowly in cat spinal cord than in rat brain. This was correlated, in part, with the greater stimulation of ^{14}C-adenine uptake into soluble RNA observed in spinal cord, and may reflect a response pattern of the large neurons which abound in this tissue. The enhanced synthesis of soluble RNA, not hitherto described, calls attention to the need for detailed study of the types of RNA formed in nervous tissue in the presence of AMD. Of particular interest is the finding that the augmented labeling of soluble RNA may overshadow the inhibitory effect of AMD on isotope incorporation into the other RNA's

Protein synthesis declines more slowly than RNA synthesis in AMD-treated nervous tissue. In cat spinal cord, ^{14}C-lysine uptake into protein was significantly depressed (−29%) 4 hr after AMD (1 mg), when the concentration of cytoplasmic RNA was unaffected, and severely inhibited (−81%) at 8 hr, when cytoplasmic RNA was diminished by only 11%. These results are consistent with the view that protein synthesis in neural tissue is dependent upon a small pool of rapidly turning-over messenger RNA. From these experiments it may be inferred that all messenger RNA's in nervous tissue have a half-life measured in hours, since protein synthesis in spinal cord is virtually abolished (−96%) 24 hr after AMD (1 mg). It is of interest that neuronal and glial polysomes in electron micrographs of AMD-treated tissues are ultrastructurally intact when protein synthesis was severely inhibited. It would appear, therefore, that the integrity of the polysome *in situ*, in contrast to its behavior *in vitro*,[25] is not dependent upon messenger RNA alone, as this component is presumably exhausted at this time.

Appel[26] has reported that protein synthesis declines more rapidly in subcortex than in cortex of AMD-treated rat cerebrum. He concluded that in addition to unstable messenger RNA's, there may also be a population of stable messenger RNA's in cerebral cortex. However, two features of AMD introduce uncertainties in its use for investigating the half-life of messenger RNA's. First, the repression of RNA transcription effected by AMD in neural tissue attains a maximum value only after some hours, and may be incomplete. Second, AMD is known to accelerate the breakdown of nuclear RNA in liver and other tissues,[27,28] and may possibly effect the degradation of cytoplasmic RNA. Our findings seem to support the latter possibility. After a large dose (1 mg) of AMD the concentration of nuclear RNA in cat spinal cord fell most rapidly, with an estimated half-life of 20 hr. Cytoplasmic RNA, mainly ribosomal, declined at one-third the rate, with a half life of 60 hr; soluble RNA was unaffected for a few hours and then fell more rapidly, with a calculated half-life of 40 hr. The half-life of ribosomal RNA as estimated with AMD, 60 hr, is much shorter than that calculated from isotope uptake studies, six days in rat brain,[29] suggesting that there may indeed be an augmented breakdown of cytoplasmic RNA's in AMD-treated tissues. This may be partly due to an increased neurophysiological activity in AMD-poisoned nerve cells.

Our correlative studies have shown that AMD produces characteristic changes in the nuclear morphology of neural cells which parallel the establishment of the repressed state of RNA transcription. The most important alterations are; (1) the formation of clumps of dense chromatin through compaction of diffuse chromatin; (2) an agglomeration and enlargement of osmiophilic granules; (3) a rarefaction of intervening nucleoplasm; and (4) nucleolar changes leading to atrophy and loss of RNA-containing components.

Morphological changes in nuclear chromatin following AMD treatment have been previously noted in HeLa cells,[30] dipteran salivary gland,[31] frog embryonal cells,[32] lampbrush chromosomes of amphibian oocytes,[33] rat exocrine pancreas,[34] rat salivary gland,[35] baboon kidney cells,[36] and neurons of mouse facial nucleus.[37] These changes have been variously attributed to an apparent increase in DNA,[30,36] inhibition of RNA synthesis,[33] and an effect of binding of AMD to DNA.[31,32] However, they have not been hitherto correlated with the metabolic effects of AMD, nor have they been invoked as a basis for the mechanism of action of AMD. Our studies indicate that the nuclear DNA content of AMD-treated nervous tissue is unaltered for at least 24 hr, although the concentration of nucleus protein was reduced about 30% 24 hr after AMD (1 mg). Moreover, the nuclear lesions are well developed prior to the abolition of RNA synthesis, at which time the nuclear RNA content is reduced only 5–10%. Thus, the nuclear alterations, at least during the early hours, reflect a rearrangement of nuclear components which is induced directly by AMD, rather than a change in chemical composition.

We postulate that AMD, upon binding to the guanine groups of DNA, cross-links the extended fibrils of diffuse chromatin, converting these into a contracted, impervious mass of dense chromatin which is incapable of priming the RNA polymerase reaction. AMD can cross-link DNA molecules *in vitro*,[38,39] possibly by dimerization of bound AMD molecules through hydrophobic bonds between the cyclic peptide side chains.[40] The presence of substantial lipid-soluble H^2-AMD in neural nuclei and its association with buoyant nuclear proteins in the $CsCl_2$ gradient suggest that AMD also interacts with nuclear lipoprotein. Recent experiments have corroborated this inference.[41,42] This interaction may account for the agglomeration of nucleoplasmic lipoprotein granules, and may also contribute to the condensation and repression of euchromatin. Phospholipids are abundant in euchromatin,[43,44] and are believed to play an important role in maintaining its expanded state and high metabolic activity. These phospholipids are associated with dense lipoproteins that are strongly acidic, and constitute a significant portion of nonhistone protein in nuclei of rat brain and liver.[41,42]

In contrast to the binding of AMD, which is essentially instantaneous, the cross-linking of euchromatin microfibrils *in situ* is envisaged as a time-dependent process whose rate would vary inversely with the mean distance between the extended microfibrils. The euchromatin of glia and small neurons is contained within a small nuclear volume, and is hence more readily cross-linked by AMD. The euchromatin of motoneurons, on the other hand, is dispersed in a large nuclear volume, and is consequently more resistant to cross-linkage. This would explain the greater sensitivity of the former cell types to AMD. Similarly, the restriction of the chromatin template for transcription of ribosomal RNA to the nucleolus would facilitate its cross-linkage by AMD, and thus account for the great sensitivity of ribosomal RNA synthesis to AMD.[45]

Recent studies in this laboratory indicate that a variety of agents which bind to DNA *in vitro* modify RNA synthesis in cat spinal cord and simultaneously produce changes in nuclear chromatin.[14,46] The chromopeptides, chromomycin A_2, mithramycin, and olivomycin, given in a small dose (10–1000 μg), resemble AMD, in that they repress RNA transcription and produce a coarse clumping of chromatin together with an aggregation of granular material.[14] Other drugs which complex with DNA *in vitro* are considerably less potent inhibitors in cat spinal cord (indeed, they may stimulate isotope uptake into RNA). These include: the anthracyclines, nogalomycin, daunomycin, and cinerubin[14]; the basic dyes, proflavine, acridine orange, neutral red[47]; ethidium bromide and 4-nitroquinoline-N-oxide.[46] When given in sufficiently large doses, however, these agents also elicit a heterochromatization in neural nuclei and coincidentally inhibit RNA synthesis. Basic dyes, in addition to their putative complexing with DNA, bind to acidic nuclear lipoproteins *in vivo*,[41,42] and may exert their biological effects partly in this manner. These findings suggest that an induced heterochromatization may be a general mode of action of agents which repress RNA transcription by interacting with nuclear DNA and/or nuclear lipoproteins.

SUMMARY

We have investigated the biochemical and metabolic effects of actinomycin D (AMD) in nervous tissue in relation to the binding of AMD and the production of nuclear changes, with the following results.

1. Actinomycin D (AMD), given into the lumbar theca of cat or intracerebrally in cat and rat, inhibits RNA synthesis, as measured by ^{14}C-adenine incorporation in spinal cord and brain, respectively. A kinetic study has shown that this inhibition develops gradually, reaching a maximum by 6–8 hr. ^{14}C-Adenine uptake into nuclear and cytoplasmic RNA is decreased from the start, but its uptake into soluble RNA is initially stimulated, and subsequently inhibited. In cat spinal cord the latter effect is particularly marked, and may mask the inhibitory effect of AMD when uptake into total RNA is measured. AMD blocks RNA synthesis more rapidly in white matter than in gray matter. AMD produces a time-dependent inhibition of RNA synthesis *in vitro* in rat-brain slices comparable to that *in vivo*.

2. Protein synthesis, measured by ^{14}C-lysine uptake, is depressed more slowly than RNA synthesis, but is nearly abolished (−97%) in cat spinal cord 24 hr after 1 mg, and in rat brain (−97%) by 48 hr after a 20-μg dose.

3. Twenty four hours after AMD (1 mg) the concentration of RNA in cat spinal cord falls 52% in the nuclear fraction, 30% in the soluble fraction, and 20% in the cytoplasmic fraction. DNA is unaltered, but the protein concentration of the nuclear fraction declines by 30–40% at 24 hr.

4. ^{3}H-AMD binds to all subcellular fractions in cat spinal cord and rat brain. Acid- and lipid-extractible complexes account for the bulk of tissue ^{3}H-AMD, the remainder occurring mainly in the nuclear fraction in an alkali-labile state. Direct isolation of the ^{3}H-AMD–DNA complex by zonal centrifugation over a $CsCl_2$ gradient confirms the inference that the latter represents AMD firmly bound to DNA. ^{3}H-AMD binds very rapidly to DNA *in vivo* and *in vitro*. Stoichiometric measurements reveal that the inhibitory effect varies logarithmically with the amount of AMD bound to DNA. As few as 10 molecules of AMD bound per 10^6 molecules of DNA-P inhibited RNA synthesis in spinal cord by 70% at 6 hr.

5. Light- and electron-microscope investigations have demonstrated striking changes in nuclear morphology of neurons and glia. These include: (a) a conversion of diffuse chromatin to a dense chromatin; (b) an aggregation of granular material; (c) a rarefaction of nucleoplasm; and (d) nucleolar changes leading to loss of RNA components and atrophy. These nuclear changes, designated nuclear hyperchromatization, advance more rapidly in glia and small neurons than large neurons. The evolution of the nuclear lesions closely parallels the development of the inhibition of RNA synthesis. It is proposed that an AMD-induced heterochromatization of neural nuclei is responsible for the repression of RNA transcription.

REFERENCES

1. H. Koenig, *J. Biochem. Biophys. Cytol.* **4**, 785–792 (1958).
2. H. Koenig, *in*: "Morphological and Biochemical Correlates of Neural Activity" (M. M. Cohen and R. S. Snider, eds.), 39–56, Hoeber, New York (1964).
3. H. Koenig, *Science* **127**, 1238–1239 (1958).
4. H. Koenig, *Arch. Neurol.* **2**, 463–475 (1960).
5. H. Koenig, *in*: "Response of the Nervous System to Ionizing Radiation" (T. Haley and R. Snider, eds.), pp. 109–121, Academic Press, New York (1962).
6. H. Koenig, *in*: "International Review of Neurobiology" (C. C. Pfeiffer and J. R. Smythies, eds.), Vol. 10, 199–230, Academic Press, New York (1967).
7. J.-E. Edström and W. Grampp, *J. Neurochem.* **12**, 735–742 (1965).
8. H. D. Cohen and S. Barondes, *J. Neurochem.* **13**, 207–212 (1966).

9. E. Reich and I. H. Goldberg, *in*: "Progress in Nucleic Acid Research and Molecular Biology" (J. N. Davidson and W. E. Cohn, eds.), Vol. 3, pp. 184–235, Academic Press, New York (1964).
10. H. Koenig and S. Jacobson, *J. Cell Biol.* **31**, 61A–62A (1966).
11. H. Koenig, C. Y. Lu, and S. Jacobson, *Fed. Proc.* **26**, 291 (1967).
12. H. Koenig, C. Y. Lu, S. Jacobson, P. Sanghavi, and R. Nayyar, submitted for publication.
13. H. Koenig and C. Y. Lu, *Trans. Am. Neurol. Assoc.* **92**, 250–252 (1967).
14. H. Koenig, *in*: "Motor Neuron Disease: Research on Amyotrophic Lateral Sclerosis and Related Disorders" (F. H. Norris, Jr., and L. T. Kurland, eds.), pp. 347–368, Grune and Stratton, New York (1968).
15. R. E. Smith and M. G. Farquhar, *Nature* **200**, 691 (1963).
16. W. C. Hutchinson and H. N. Monro, *The Analyst* **86**, 768–815 (1961).
17. G. Ceriotti, *J. Biol. Chem.* **214**, 59–70 (1955).
18. K. W. Giles and A. Myers, *Nature* **206**, 93 (1965).
19. R. W. Hendler, *Anal. Biochem.* **7**, 110–120 (1964).
20. C. W. Dingman and M. B. Sporn, *Science* **149**, 2052–2054 (1965).
21. S. H. Appel, *Nature* **207**, 1163 (1965).
22. P. Sanghavi and H. Koenig, unpublished observations.
23. E. Reich, R. M. Franklin, A. J. Shatkin, and E. L. Tatum, *Proc. Natl. Acad. Sci. U.S.* **48**, 1238–1245 (1962).
24. D. L. Kirk, *Proc. Natl. Acad. Sci. U.S.* **54**, 1345–1353 (1965).
25. T. Staehelin, F. O. Wettstein, and H. Noll, *Science* **140**, 180 (1963).
26. S. H. Appel, *Nature* **213**, 1253–1254 (1967).
27. R. Wiesner, G. Acs, E. Reich, and A. Shafiq, *J. Cell Biol.* **27**, 47 (1965).
28. G. A. Stewart and E. Farber, *J. Biol. Chem.* **243**, 4479–4485 (1968).
29. D. M. Dawson, *J. Neurochem.* **14**, 939–946 (1967).
30. M. N. Goldstein, I. J. Slotnick, and L. J. Journey, *Ann. N. Y. Acad. Sci.* **89**, 474–483 (1960).
31. B. J. Stevens, *J. Ultrastruct. Res.* **11**, 329–353 (1964).
32. K. W. Jones and T. R. Elsdale, *J. Cell Biol.* **21**, 245–252 (1964).
33. M. Izawa, V. S. Allfrey, and A. E. Mirsky, *Proc. Natl. Acad. Sci. U.S.* **49**, 544–551 (1963).
34. A.-M. Jézéquel and W. Bernhard, *J. Microscopie* **3**, 279–296 (1964).
35. H. T. Jhee, S. S. Han, and J. K. Avery, *Am. J. Anat.* **116**, 631–651 (1965).
36. G. E. Schoefl, *J. Ultrastruct. Res.* **10**, 224–243 (1964).
37. A. Torvik and A. Heding, *Acta Neuropath.* **9**, 146–157 (1967).
38. R. Haselkorn, *Science* **143**, 682 (1964).
39. W. Kersten, H. Kersten, and W. Szybalski, *Biochemistry* **5**, 236–244 (1966).
40. D. M. Crothers, S. L. Sabol, D. I. Ratner, and W. Müller, *Biochemistry* **7**, 1817–1823 (1968).
41. H. Koenig and C. Y. Lu, *J. Histochem. Cytochem.*, **17**, 199–200 (1969).
42. H. Koenig and C. Y. Lu, unpublished observations.
43. J. Frenster, *Nature* **206**, 680–681 (1965).
44. F. Dolbeare and H. Koenig, *J. Cell Biol.* **39**, 35a (1968).
45. R. P. Perry, *Exp. Cell Res.* **29**, 400–406 (1963).
46. H. Koenig, unpublished observations.
47. R. Nayyar and H. Koenig, *J. Cell Biol.* **49**, 99a (1968 .
48. D. H. Lowry, N. J. Rosebrough, A. L. Farr, and R. J. Randall, *J. Biol. Chem.* **193**, 265 (1951).

DISCUSSION

McEWEN: Have you looked at the distribution of the inhibitor in the spinal cord? Is it concentrated near the point of injection?

KOENIG: Yes. There is a definite concentration gradient of AMD, or other agents, injected into the lumbar subarachnoid space, so that higher uptake occurs into the lumbosacral spinal cord than into the ceroical cord.

ROBERTS: You mentioned that your evidence suggested that there was no stable messenger RNA in brain comparable to that present in liver, but, of course, this was on the basis of the fact that protein synthesis was abolished completely in 24 hr. On the basis of some of our studies with the turnover of hybidizible RNA in the adult brain at least, there are components which turn over with a half-life of about four days, which is quite stable, of course. I wonder whether the appearance of intact polysomes in the nerve cell after actinomycin D might not represent the presence of this apparently stable messenger.

KOENIG: As indicated, neural polysomes *in situ* in AMD-treated tissue seem ultra structurally intact after protein synthesis is essentially abolished. Hence, this criterion is of little value in judging the presence of intact messenger. Several objections to the use of AMD in studies on the turnover of messenger RNA's are indicated in my paper.

OCHS: We found with puromycin that at some level it seemed like there was a specific action of puromycin at protein synthesis, and that at another level it was destructive in general. I just wonder if you have such a dose relationship, so that at a higher level you may go on to chromatolysis, and then see some general signs of degeneration.

KOENIG: The cytoplasm of neurons and glia is quite normal for the first day. It takes about 24–36 hr for necrotizing changes to become evident in cat spinal cord after a big dose of AMD (1 mg). AMD and other agents which inhibit RNA transcription affect polysome structure late. Puromycin and other inhibitors of polypeptide synthesis act at the level of translation at the polysomal sites, and hence may disrupt the polysomal aggregates early. Nissl bodies lose their cripsness 24 hr after a big dose of AMD (1 mg). A typical chromatolytic picture develops subsequently. After a very small dose of AMD (1–5 μg) chromatolysis is delayed for several days. Chromatolysis regularly commences in the periphery of the neuron, suggesting that the Nissl substance, i.e., parallel stacks of rough endoplasmic reticulum, farthest from the nucleus is the first to undergo disorganization when RNA transcription is arrested. Theoretically, this may be due to impaired synthesis of messenger, ribosomal, or soluble RNA, since the polysomal structure requires the combined presence of messenger RNA and the amino-acylated soluble RNA at the ribosomal sites. When a very small dose of AMD (1–2 μg) is used ribosomal RNA synthesis seems to be selectively blocked in neurons, as judged from radioautographs showing relatively normal uptake of ^{14}C-adenine into spinal motoneurons with nucleolar atrophy and peripheral chromatolysis (Koenig, unpublished observations).

RAPPOPORT: Since actinomycin D does combine rather firmly, is it physically possible to extract the bound and unbound actinomycin and make some logical deduction from these findings?

KOENIG: To the best of my knowledge, the neurotoxicity of actinomycin D is due largely, if not exclusively, to its repressive effect on RNA synthesis. Some authors have suggested that AMD has other biological actions. However, in our experiments the primary effect of AMD seems to be a repression of RNA transcription.

WOLFGRAM: To what extent are these necrotic and to what extent are they reversible processes? Have you allowed any of the animals to survive and observed the reformation of normal chromatin?

KOENIG: Extensive clumping of neuronal and glial chromatin in AMD-treated tissue inevitably leads to necrobiotic changes. I have had 30–60-day survivals after a very small dose of AMD (½–1 μg). In these instances the clumped chromatin granules are fine and dispersed throughout the nucleoplasm. I have no information on the question of reversibility once AMD-induced clumping of chromatin occurs.

MANDEL: We find a strong binding of actinomycin to DNA.

KOENIG: That is certainly in accord with our experience. We consider ^{3}H-AMD to be firmly bound to DNA when tritium radioactivity resists extraction by cold TCA (10–20%) and lipid solvents, but is solubilized by alkaline hydrolysis or hot TCA. There is little question that the deoxyguanine residues of DNA are the binding sites for AMD. However, I must reiterate that as few as 10 AMD molecules bound per million DNA nucleotides suffices to repress RNA transcription in cat spinal cord after 6–8 hr. This finding is difficult to reconcile with the steric hinderance theory, as long stretches of DNA polymer should still be available to prime the RNA polymerase reaction. It an, however, be explained by our hypothesis that AMD, on binding to guanine sites n diffuse chromatin, cross-links this chromatin to form a dense, inactive chromatin. With regard to the effect of AMD on soluble RNA, this is a complex problem. I don't believe that turnover of the terminal CCA group can account for the increased labeling that we see. Stimulation of precursor uptake into soluble RNA is short-lived and is promptly followed by a marked inhibition after a large dose of AMD. With very low doses (1–3 μg), however, an increased labeling of soluble RNA may persist for days and is accompanied by an augmentation in pool size of soluble RNA. I must emphasize that the term "soluble RNA" is strictly operational. We don't know the sedimentation coefficients of these RNA s, or whether they have amino acid receptor or transfer function. We are currently characterizing the soluble RNA s which are formed in AMD-treated nervous tissue. In any event, it is important to recognize that labeling of soluble RNA s can be enhanced by AMD, and by many other drugs with similar action (Koenig, unpublished observations), for a period of time. During this period the increased labeling of the soluble RNA may overshadow the inhibitory effect on other RNA species. Therefore, it may be important in some experiments to study the incorporation of RNA precursors in the various subcellular fractions in order to fully understand what the inhibitor is doing.

BOGOCH: I am interested in the fact that is disparity between the inhibition of the RNA synthesis and protein synthesis and the preservation of reflex activity. In the axon, actinomycin-D doesn't seem to have any effect on amino acid incorporation. I wonder whether this could have any relevance to your observation of the differences in the preservation of activity and the inhibition of protein synthesis as a whole.

KOENIG: In our experiments we did not study structural units, and do not know what is responsible for the early deterioration in spinal cord function. However, I believe that the first stage of deterioration results from a breakdown in the function of spinal interneurons. Paralysis regularly develops at a time when muscle stretch reflexes can still be obtained and may even be hyperactive, indicating that motoneurons can respond to muscle stretch receptor input. It would appear, therefore, that the supersegmental impulses are not effectively relayed to the motoneurons through the spinal interneurons. We have not identified the specific protein molecules whose deficiency may be responsible for the loss of interneuronal function. These may conceivably be enzyme proteins with a short half-life involved in synthesis of transmitter substances.

BOGOCH: Since the chromatin change is visible on the light microscope, I wonder whether those who are studying RNA in relation to functional activity of cells shouldn't be looking at chromatin stains as control rather than EEG's. I wonder whether the chromatin changes are early enough and reliable enough to indicate a toxic agent.

KOENIG: Well, that is an interesting thought. Although my interest in the nervous system spans 25 years, my interest in chromatin is only three years old. Like most neurobiologists, I used the chromatin pattern to identify the various types of neurons and glia, and measured DNA content to estimate cell numbers, but otherwise ignored it. We are now becoming wiser. Modern molecular biology teaches that the morphology of chromatin is an important determinant of its template function in RNA transcription. Pathologists have long regarded nuclear pyknosis and other changes in nuclear structure as nonspecific changes of moribund or dead cells. This simplistic view is no longer tenable. The role of nuclear alterations in disease states and experimental and drug effects needs reassessment. There is some evidence that a clumping of neuronal chromatin is an early event in the evolution of certain neuronal atrophies.* It is possible that such nuclear changes and a resultant repression of template function are initiators of the atrophic process.

* Koenig.[14] Also see S. A. Luse, *in*: "Pathology of the Nervous System" (J. Minckler, ed.), Vol. 1, p. 564, McGraw-Hill, New York (1968).

Chapter 25

The Breakdown of Polysomes and the Stimulation of Protein Synthesis as Cerebral Mechanisms of Defense against Seizures*

Otto Z. Sellinger and Julio M. Azcurra

Mental Health Research Institute
The University of Michigan Medical Center
Ann Arbor, Michigan

INTRODUCTION

The recent advances in our understanding of the relationships among nucleic acids and the ribosome on the one hand and protein synthesis, its control and modulation on the other[1–5] have revealed that in neural tissue the mutual interactions of these and other macromolecular systems with one another and with the environment must be very complex if, by virtue of their integration, a correct specification of cellular function and cellular behavior is to be achieved.[6–8] That is why the intricate network of capabilities that this tissue must have available for the specification and the programming of the myriad of its complex supramolecular functions has been barely formulated, and, when so, only speculatively.[9–11] Hence, the question: "Do the cellular and the organismic responses of neural tissue originate in molecular events, and, if so, how are these transduced into specific modes of expression, termed behavior?" must elicit, at present, a guarded "We don't know, but give us more time to find out" for an answer.

For the past three years our laboratory has been engaged in attempts to formulate answers to this and related questions. The principal areas of research interest have been and continue to be (1) the molecular anatomy of brain tissue, with focus on (a) metabolic differences between the two types of endoplasmic reticulum, the ribosome-studded and the smooth,[12–15] (b) the regional microheterogeneity of brain lysosomes,[16,17] and (c)

* This research was supported by Grant No. NB-06294 from the United States Public Health Service.

the nature and properties of the synaptic membrane[18,19]; (2) protein synthesis in the immature brain, where we have been guided by the thought that the process, as it occurs simultaneously on free and endoplasmic-reticulum attached ribosomes, may be qualitatively different[20–22]; (3) the mechanisms of defense against seizures, where we have high-lighted the role of methionine in protecting animals from methionine sulfoximine seizures[23,24]; and (4) the intracerebral disposition of drugs, where we have concentrated on exploring the fate and the regulatory functions of the cholinergic blocking agent hemicholinium-3 in the synthesis of brain lipids.[25,26]

In the present report we shall describe a study of the interactions between the convulsant methionine sulfoximine and the RNA- and protein-synthesizing machinery of the immature rat brain. Before doing this, however, a brief survey of certain aspects of drug–macromolecule interactions in neural tissue is deemed in order.

Effects on Nucleic Acids*

The studies of Hydén and associates[27–30] have revealed that the amount of RNA per nerve cell (rabbit Deiters nerve cell) can be increased and the normal base ratio of the total RNA changed significantly 1 hr after the administration of 20 mg/kg of tricyano-aminopropene. A more detailed analysis of this effect revealed that the base-ratio changes were not characteristic of the RNA whose formation had been "induced" by the drug and which was ribosomal in type, but rather that they were the result of a drug-elicited loss (16%) from the nucleus of the affected nerve cells (analyzed to contain only 3.5% of the cell RNA) of a special RNA with an unusually low base ratio $(C + G)/(A + U)$. Hyden and co-workers have also shown that the administration of the monamine oxidase inhibitor inhibitor phenylcyclopropylamine results in neuronal RNA gains and glial RNA losses.[31] The existence of drug-susceptible transport mechanisms conveying RNA from the glia into the neurons has been invoked to account for these findings.

Another, rather controversial drug, said to affect cerebral RNA, is magnesium pemoline.[32] An enhancement of RNA synthesis in the brain, as a result of RNA polymerase activation[33] or by virtue of a decreased RNA breakdown[34] (inhibition of RNA-ase to which pemoline may bind), has been reported, but the original findings have been challenged,[35,36] and a failure to detect any stimulation of RNA polymerase of rat brain, either

* Excluding actinomycin D, cytosine arabinoside, and similar inhibitory agents.

in vitro or *in vivo*, has been reported. Recently, however, a reconfirmation of the stimulatory effect of the drug on RNA synthesis has appeared from the laboratory in which the effect was originally described.(33)

Of somewhat more cogent relevance to our own studies are the reports of the effects of convulsant agents or conditions on cerebral RNA.(37) Geiger *et al.*(38) found that short-lasting afferent stimulation of the brain was followed, within 1 min, by a decrease of cortical nucleic acids, while Mihajlovic *et al.*(39) observed a decrease of RNA in various brain parts of the cat 1 min after electroshock. Talwar *et al.*(40) reported on a fall of the RNA content in rat brain during the tonic phase of metrazol convulsions, while Noach *et al.*(41) described a significant drop in the RNA of cerebral cortical slices isolated from animals which had been electroshocked for 30 sec and which was not seen in animals receiving phenobarbital before the electroshock.

Pechan and Marko(42) have reported that the administration of metrazol leads to a significant decrease in the concentration of cerebral ATP, GTP, and UTP which could not be entirely accounted for by increases in the concentrations of the corresponding di- or mononucleotides.

The above cited, and admittedly rather randomly collected, analytical findings have been complemented in recent years by attempts to examine the effects of convulsant conditions and agents on some metabolic parameters which, in the long run, may govern RNA turnover and hence determine its functional capability. These studies, which have included an investigation of the changes in cerebral glycolysis and of its regulation during convulsions,(43) have revealed that the specific radioactivity of cerebral RNA following the intracisternal administration of short pulses of ^{32}P-orthophosphate increases during the tonic (preconvulsive) phase of the metrazol-elicited seizure, falls during the clonic (convulsive) phase, and returns to control levels after seizure termination.(44) On the basis of these findings it has been suggested that "there is no irreversible damage suffered by the tissue in this type of convulsion" and that "the brain after convulsion would be synthesizing RNA at an accelerated pace to regain the normal levels."(44) In subsequent reports the (nonspecific) interaction of a number of convulsant agents with the RNA of cerebral ribosomes was examined.(45–47) One of the findings was of the loss of hydrogen-bonded secondary structure and, generally, of stability of ribosomal RNA following its *in vitro* treatment with drugs, especially with strychnine. Ghosh(47) has recently reviewed this area of study, high-lighting the biological implications of the findings.

In related experiments designed to assess the effects of the seizure state on the relative distribution of cerebral RNA species Talwar *et al.*(37) extracted total tissue RNA by a proceudre using phenol and sodium dodecyl sulfate[50] and subjected the mixture of the RNA species present to chromatographic analysis on methylated albumin-kieselguhr columns. The profiles obtained from the brains of convulsing animals (metrazol) were found to be in no way different from those obtained using control brains.

The effect of electroshock-elicited seizures on the polysomal aggregate size in rabbit brain has also been studied,[51] and an appreciable (about 20%) disruption of the highly aggregated polysomes to monomeric ribosomes noted. It has been suggested "that the shift in (the) polysome-ribosome equilibrium could be simply interpreted in terms of a relative decrease in the rate of ribosome attachment to mRNA in comparison with the readout rate."

Effects on Cerebral Protein Synthesis*

Many reports describing effects of drugs on cerebral protein synthesis have appeared, but because of space limitations only a few of these, and then only if the drug was given *in vivo*, will be mentioned here. (For the review of the literature prior to 1961 see Waelsch and Lajtha.[52])

The effect of chlorpromazine and of lysergic acid diethylamide (LSD) on the regional turnover of protein was examined by Piha *et al.*,[53] who noted a partly opposite mode of action for the two drugs. Protein synthesis was found to be inhibited by chlorpromazine, maximally in the basal ganglia and least in the cerebral cortex, while LSD, although its action was less pronounced, tended to exert an enhancing effect on protein turnover.

The work of Clouet and Ratner[54] has revealed that the incorporation of ^{14}C-leucine by cell-free preparations of brain tissue isolated from morphine-treated rats proceeds at a lower than normal rate if tested more than 1 hr after the administration of morphine, but at a slightly higher than normal rate if tested within 45 min of morphine administration. No conclusions regarding the biological significance of these findings have been suggested but further results dealing with this important facet of morphine action are presented by Clouet in this volume (Chapter 38).

The ability of the convulsant methionine sulfoximine (MSO) to inhibit the *in vivo* incorporation of ^{35}S-methionine into brain protein was briefly noted long ago,[56] yet subsequent *in vitro* experiments appeared to be in conflict with some of the results obtained *in vivo*, for they failed to confirm the interference by MSO of the conversion of ^{35}S-methionine to protein in a slice system.[57]

* Excluding puromycin, cycloheximide, chloramphenicol, and similar inhibitory agents.

Effects of MSO on Polysomes and Protein Synthesis

Recently, as part of our continuing study of the biochemical and molecular correlates of the MSO seizure phenomenon, we have examined the effect of this agent on an enzyme system of rat brain, glutamine synthetase-transferase, which, as reported elsewhere,[23,24] has a plurimembranous, intracellular localization in this organ. Selective effects of MSO, particularly when given in combination with methionine (molar ratio 5:1), on the transferase component located, respectively, in the ribosome-studded, as opposed to that located in the ribosome-devoid, membranes of the endoplasmic reticulum (ER) were noted.[24] The further exploration of this interaction of MSO + methionine with cell elements, presumably involved in the defense against the MSO-induced seizures, is described in the present report. A preliminary account of some of the findings has appeared.[58]

Briefly, we describe a transient, but readily reproducible, breakdown to mono- and disomes of a portion of the polysomal population of rat cerebral cortex following the administration of the convulsant MSO in combination with the protective agent methionine. The relative specificity of this effect and its modulation by actinomycin D are also briefly documented. Moreover, results are presented which show that the observed disaggregation of polysomes is coupled to an increased rate of protein synthesis which sets in in synchrony with the disaggregation.

EXPERIMENTAL PROCEDURE

Materials and Methods

L-Methionine: Nutritional Biochemicals; *L*-methionine-*DL*-sulfoximine (MSO): Calbiochem or Pierce; *L*-methionine amide, α-methyl-*DL*-methionine, and *D*-methionine: Cyclo; *L*-valine, *L*-alanine, bovine serum albumin, and enzyme grade sucrose: Mann. The ^{14}C-*L*-amino acids (U.L.) were from International Chemical and Nuclear Corp. (specific activity >200 mCi/mM). Actinomycin D was a generous gift from Dr. B. W. Agranoff of this Institute.

Animals

Rats between 16 and 18 days of age were used throughout. This age was chosen for reasons of convenience and economy, and also because polysomes of the cerebral cortex of animals this age have been extensively studied and the techniques for their isolation amply described.[59]

Injections

These were intraperitoneal (i.p.) and intrathecal (i.t.). MSO, methionine, and all of the other substances tested in lieu of methionine were given i.p. No ill effects were noted as a result of the administration of the MSO + methionine combination, regardless of the dose of each. The i.t. route was used for the administration of ^{14}C-*L* amino acids (see legends of Fig. 2 and Table III for details).

Radioactivity

A Packard liquid scintillation spectrometer, Model 3302, and, in later experiments, a Nuclear Chicago Unilux II spectrometer were used for the measurement of radioactivity. Total and soluble radioactivity were determined in a cocktail of xylene–ethylene glycol monoethyl ether and dioxane (1:1:3) (XDC) containing (per liter) 30 g of naphthalene, 1 g of PPO (2,5-diphenyloxazole), and 0.35 g of dimethyl-POPOP [1,4-bis(4-methyl-5-phenyloxazolyl) benzene]. In some experiments Biosolv No. 3 (Beckman Co.) was used to solubilize particulate samples. This allowed counting in a toluene cocktail (see below). The radioactivity, precipitated by 2 ml of cold 10% (w/v) trichloroacetic acid (TCA), was adsorbed on Millipore filters (pore size: 0.45 μ); 100 μg of bovine serum albumin were used as carrier when gradient fractions were to be counted. The filters were treated successively with 1 and 0.5 ml of 5% cold TCA, 3 and 2 ml of 95% EtOH, and 5 ml of ether. They were then transferred to counting vials and dried under an IR lamp. Toluene (10 ml) containing 4 g of PPO and 0.1 g of dimethyl POPOP/liter was used as the fluor solution.

Centrifugal Methods

Free polysomes were isolated from cerebral cortex homogenates of 15–18-day-old rats prepared in a solution containing 5 m*M* Mg^{++}, 20 m*M* Tris buffer (pH 7.2), and 25 m*M* K^+.[59] After centrifugation (15 min, 20,000g) the supernatant (4 ml) was immediately layered on 3.5 ml of 0.5 *M* sucrose, which rested on 4 ml of 2.0 *M* sucrose, both sucrose solutions containing 10 m*M* Mg^{++}, 20 m*M* Tris, and 25 m*M* K^+. Centrifugation was for 160 min at 160,000g. The pellet (FP) was suspended in 1 ml of a solution containing 1 m*M* Mg^{++}, 20 m*M* Tris, and 25 m*M* K^+, and 0.5–0.7 ml were layered on a linear sucrose gradient (limits as specified) which was centrifuged in a Spinco SW 50 rotor for 60 min at 180,000g. The ER-associated polysomes (BP) were collected from the 0.5–2.0 *M* sucrose interface.

For gradient analysis they were freed of membranes by treatment with 0.2% sodium deoxycholate (final concentration) and were then isolated in pellet form by centrifugation of the clear suspension, as above for the free polysomes.

The sucrose gradients were pumped through a hole in the bottom of the centrifuge tube into a rectangular flow-through cell (Uvicord, LKB Industries Co., 0.4 cm in diameter) and were collected in the drop mode (11 drops/tube). The absorption of the effluent was recorded continuously.

In an attempt to quantitate the extent of polysomal breakdown, the area under the monosome + disome peaks was measured with a planimeter; it is expressed as per cent of the total area. While there was no major run-to-run variation in the appearance of the polysomal profile obtained

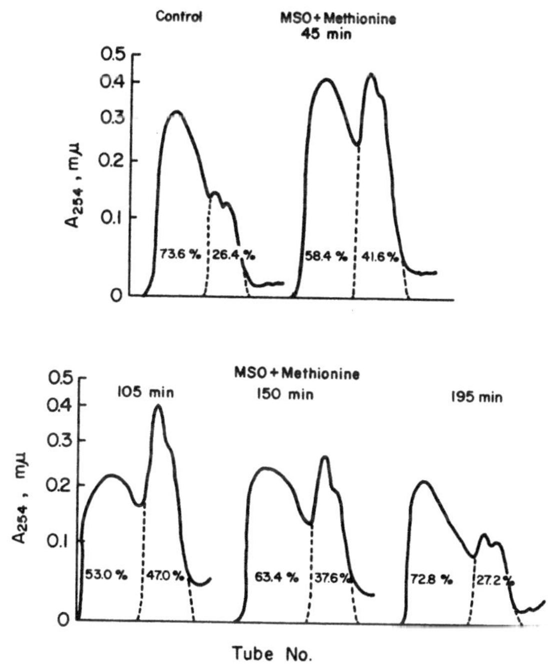

Fig. 1. The disaggregation of the free polysomes of rat cerebral cortex into mono- and disomes following the administration of MSO (0.94 mM/kg) and *L*-methionine (4.7 mM/kg). The free polysomes were isolated from animals killed at the times indicated. The areas under the respective A_{254} peaks were delimited by the dashed lines and were then integrated with a planimeter. They are expressed as per cent of the area under the total A_{254} profile.

from control (water-injected) animals, several factors which could only be partly controlled influenced the run-to-run contribution of the monosome + disome peaks to the total profile in the experimental group, the main variable being individual variability. This was reflected by fluctuations in the monosome + disome area values seen in different animals at a given time after injection of MSO + methionine.

RESULTS

Effects on Polysomes

The increase above normal values of the contribution of mono- and disomes to the free polysome population following the administration of MSO–methionine is shown in Fig. 1. It may be seen that the onset of the disaggregating process is rather rapid, that the disaggregation at its peak involves no more than about 20% of the polysomes, and that the return to normal aggregate size is complete within 4 hr. It could be shown in separate experiments, the details of which will be reported elsewhere,[60] that the disaggregation is identical if, in lieu of the free polysomal population, the post-mitochondrial supernatant, which contains virtually all of the poly-

TABLE I
The Specificity of the Polysomal Disaggregation[a]

Compound injected	No. of animals	Molar ratio to MSO	Difference in % of disaggregation[b]
MSO	3	—	0.8
L-methionine	3	5	5.8
MSO + *L*-methionine	10	5	12.7
MSO + *L*-methionine	1	1	8.4
MSO + *D*-methionine	1	5	4.1
MSO + *D*-methionine, 2×	1	5	6.1
MSO + α-methyl-*DL*-methionine	1	5	1.3
MSO + α-methyl-*DL*-methionine, 2×	1	5	0.5
MSO + *L*-alanine, 2×	1	5	− 2.1
MSO + *L*-valine	1	5	2.1
MSO + *L*-methionine amide	2	5	7.6

[a] Rats between 16 and 21 days of age received 0.94 mM/kg of MSO together with the compound indicated. A booster injection given 30 min after the first one is shown by 2×. All animals were sacrificed 60 min after the first injection.

[b] Denotes the difference between the monomer + dimer areas (expressed as % of the total area) of the treated and control brains. The mean value is given.

TABLE II

Delay Caused by Actinomycin D in the Reaggregation of Polysomes

Compounds administered	Time of actinomycin D action (min)	Monomer + dimer[a] (%)	Reaggregation delay[b] (%)
MSO + methionine[c]	—	27–34	—
Actinomycin D[d]	240	36.7	—
MSO + methionine +actinomycin D			
10 min later	230	46	9.3
45 min later	195	46.3	9.6
120 min later	120	34	−2.7

[a] Determined by measuring the area under the monomer + dimer peaks with a planimeter.

[b] By comparison with the value of 36.7% monomer + dimer, which represents the maximum of disaggregation after 240 min of actinomycin D.

[c] MSO: 0.94 mM/kg; *L*-methionine: 4.7 mM/kg. Sacrifice: 240 min later.

[d] Administered intrathecally, at a dose of 1 μg/g of body weight.

somes of the cell, is examined. On the other hand, no disaggregation was seen using any hepatic polysome-containing fraction.(60) When the cortical homogenate of a normal rat was made 1 m*M* in MSO and 5 m*M* in *L*-methionine the free polysomes isolated therefrom showed intact aggregate size profiles.

The specificity of the disaggregating effect is illustrated in Table I. The admittedly preliminary survey reveals the necessity of the joint presence of MSO and methionine, for either compound alone produced only negligible alterations of the polysome profile. Replacement of *L*-methionine by α-methyl-*DL*-methionine, valine, or alanine was ineffective, yet the substitution of the *L*-isomer by the *D*-isomer or by its amide resulted in slight but consistent disaggregation.

One of the interesting features of the MSO + methionine effect was its transient nature and the return of the polysomes to intact aggregate size within about 2 hr after reaching a maximum of disaggregation. An inquiry into the process of reaggregation on the premise that if it depended on synthesis of new mRNA, it could also be delayed by actinomycin D was therefore attempted, and the results of this experiment are shown in Table II.

When actinomycin D was allowed to act for 240 min, a period of time entirely sufficient for RNA synthesis to be virtually arrested,(61) the aggregation state of the polysomes was comparable to that observed after

240 min of MSO + methionine. Yet, disaggregation as measured 240 min after MSO + methionine was higher if actinomycin D was administered either at the inception of the disaggregation process (10 min after MSO + methionine) or near its peak (45 min after MSO + methionine). Injecting actinomycin D as late as 120 min after MSO + methionine, i.e., during the reaggregation process, and allowing only 120 min for it to block RNA synthesis, produced no delay in the reaggregation of the polysomes, presumably because the block of mRNA synthesis was incomplete.[62]

Effects on Protein Synthesis

The results of experiments in which ^{14}C-valine was administered to animals pretreated with MSO + methionine (Fig. 2) indicate that the

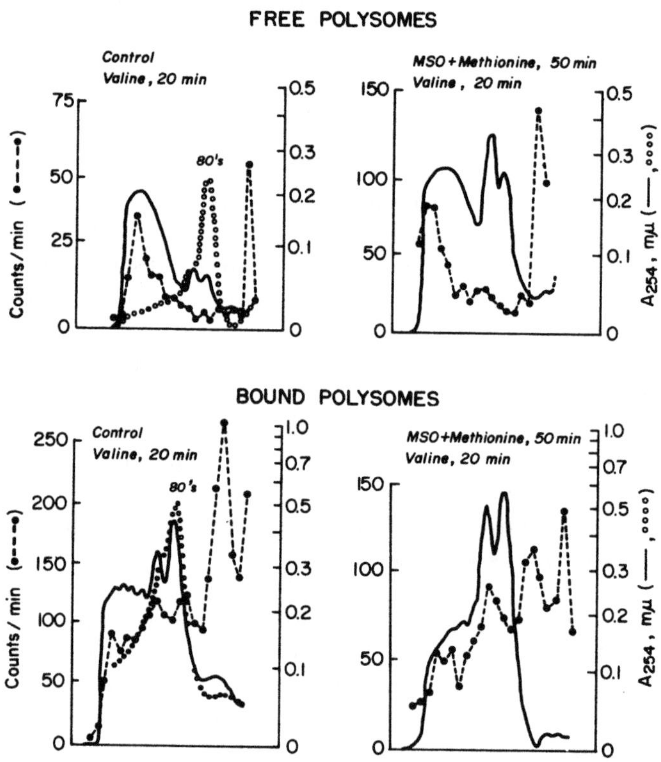

Fig. 2. Pulse-labeling of nascent protein by the free and the bound polysomes of cerebral cortex: effect of MSO + methionine. Darkened circles, dashed curve: counts/minute. Solid curve: A_{254}. The A_{254} profile indicated by the open circles serves as a positional marker for the single (80-S) ribosome.

TABLE III

Effect of MSO + Methionine on the Specific Radioactivity of Brain Cortex Homogenate, the Post-Mitochondrial Fraction, and the Free Polysomes

Fraction	Specific activity for various pulse times[a]								
	2 min			5 min			7 min		
	Control	Treated	T/C	Control	Treated	T/C	Control	Treated	T/C
Homogenate	791	1802	2.28	1752	3693	2.10	2202	5213	2.37
Post-mitochondrial supernatant	363	740	2.06	1553	4087	2.62	2915	6548	2.25
Free polysomes	2200	3430	1.56	2150	4120	1.92	2610	6190	2.37

[a] 6.5 μCi of ^{14}C-valine (U.L., SA: 190 mCi/mM) given intrathecally to each rat. The brains of two control and two treated animals were pooled in each experiment. The specific radioactivity is expressed as counts/min of TCA-insoluble radioactivity/mg of protein. Treated: with 0.94 mM/kg of MSO and 4.7 mM/kg of *L*-methionine 1 hr before sacrifice; controls received equal volumes of distilled water. T/C = treated/control.

TABLE IV

The Effect of MSO + Methionine on the Conversion of C^{14} Leucine into Protein[a]

Fraction	Effect for various pulse times[a]					
	2 min		5 min		7 min	
	Control	Treated	Control	Treated	Control	Treated
Homogenate	69	238	150	260	74	245
Post-mitochondrial supernatant	39	81	67	135	39	137
Free polysomes	—	—	484	825	191	828

[a] Values are expressed as: [(counts/min/TCA-insoluble)/(counts/min/total)] × 1000; all other conditions are as in footnote to Table III. The total radioactivity was determined as described in the section on experimental procedure.

qualitative distribution of the polysome-associated radioactivity (nascent protein) was the same in the control and the treated groups. Animals were injected with MSO (0.94 mM/kg) and *L*-methionine (4.7 mM/kg). Thirty minutes later they were given an intrathecal injection of 6.5 μCi (SA:250 mCi/mM) of U.L. C^{14}-valine and were killed 20 min thereafter. The free polysomes were layered on a 15–40% linear sucrose gradient which was centrifuged for 60 min at 180,000g in a SW-50 Spinco rotor. The gradient was drop-collected, as described in the section on experimental procedure, and the contents of each tube were analyzed for radioactivity. Cold TCA (10%, 2 ml) was added to each tube together with 100 μg of bovine serum albumin as carrier. The suspension was passed through a Millipore filter, washed with 5 ml of 5% cold TCA, 5 ml of 95% ethanol, and 5 ml of ether. The discs were counted in a toluene cocktail. The bound polysomes were isolated from the 0.5–2.0 *M* sucrose interface after treatment with deoxycholate (0.2%, w/v) for 30 min at 0°. They were collected by centrifugation for 45 min at 269,000g in the R 65 angle-head Spinco rotor. The gradient was collected as above. The free polysomes were treated with 0.01 μg/ml of pancreatic ribonuclease for 30 min at 0° and the resulting suspension was gradient-centrifuged, as described above. The bound-polysome A_{254} profile shows that the detergent treatment failed to prevent the expression of the MSO + methionine effect, which may be seen to superimpose on the deleterious effect of the detergent. When the TCA-insoluble radioactivity present in the cortex of short-pulsed (C^{14}-leucine) animals was compared it was noted that the MSO + methionine-treated group, whether injected 60 min (Table III) or 30 min (not shown) before death, incorporated the label at about twice the rate of the control animals. Expressing the results by taking into account differences in the total tissue pool radioactivity confirmed this finding (Table IV).

As a further control, it was established that the increased TCA-insoluble radioactivity in the MSO + methionine-treated brains is not of the leucyl-tRNA type.(63)

DISCUSSION

It is generally presumed(64) that the extent of polysomal aggregation in a mammalian cell mirrors its synthetic competence, and that, conversely, the disaggregation of polysomes brought about by *in vivo* intervention denotes a loss, albeit sometimes only a temporary one,(65) of such competence. Although cases of *in vivo*-induced polysome disaggregation have been frequently observed in liver tissue,(65–67) cerebral polysomes have been shown to disaggregate *in vivo* only in one instance, i.e., as a result of acute, intensive, and repeated electroshocks.(51)

The present results thus seem to be the first recorded instance of polysome disaggregation in brain which is not the result of exogenously applied trauma. Furthermore, as we show here, the affected cells appear to exhibit an increased rate of protein synthesis, at the time when the disaggregation is in progress, as indicated by the increased utilization of C^{14}-leucine on the part of the polysomal assemblies left available for synthesis (Table III).*

A striking feature of the disaggregation phenomenon is that it involves only a portion of the polysomes. A possible interpretation of this finding assumes that, like liver,(68) immature brain has two classes of polysomes with widely differing half-lives and that only the shorter-lived of the two populations is called upon to respond to the MSO + methionine trigger. This view appears to be consistent with the recent reports of multiple turnover rates for,(69) and of the extensive size(70) and functional(71–73) heterogeneity among, cerebral mRNA's. Some experimental support for it is provided by the actinomycin results (Table II), which indicate that the administration of MSO + methionine causes the depletion of selected mRNA species whose resupply, for purposes of polysome reassembly, cannot be maintained by drawing from the preexisting pools of mRNA (nuclear and/or cytoplasmic) for longer than 2 hr, after which time the system acquires actinomycin sensitivity, and reassembly is delayed. This interpretation implies the presence of short-, rather than long-term, reserves of free and stable mRNA in neural tissue cytoplasm, a conclusion which

* The virtually unlabeled mono- and disome peaks are, by virtue of this fact, considered nonfunctional and noncontributory to the synthetic process.

stands in contrast to the situation in hepatic cytoplasm,[74] but which agrees fully with the view recently expressed by Bondy and Roberts[71] that: "cerebral cells may have the potential for adapting rapidly to changing physiological conditions by alterations in protein synthesis made possible by the existence of a relatively large population of unstable mRNA molecules."

Finally, the intriguing question: "How does the transient and partial disaggregation of polysomes elicited by MSO + methionine and the parallel increase in the rate of cerebral protein synthesis relate to and/or mediate resistance to seizures?" One possible answer may be formulated as follows: If the administration of MSO + methionine were indeed to trigger an exceptional demand on some of the presumed, short-lived cytoplasmic mRNA's, as suggested above, and this, in turn, led to the specific disaggregation of polysomal assemblies with a rapid turnover, then, among the replacement mRNA's emanating from the nucleus and emerging *de novo* in the cytoplasm there may be some which are breakdown products of nuclear polycistronic RNA[71,75] and thus contain a newly transcribed "emergency" code for the synthesis of "seizure defense protein." It is conceivable that because of the "crash" nature of the process the latter would occur at higher than normal rates.

REFERENCES

1. J. R. Warner and R. Soeiro, *in*: "Macromolecular Synthesis and Growth" (R. Malt, ed.), pp. 31–60, Little, Brown & Co., Boston (1967).
2. S. Penman, *in*: "Macromolecular Synthesis and Growth" (R. Malt, ed.), pp. 121–148, Little, Brown & Co., Boston (1967).
3. R. P. Perry and D. E. Kelley, *J. Mol. Biol.* **35**, 37–59 (1968).
4. H. Harris, *in*: "Nucleus and Cytoplasm," Clarendon Press, Oxford (1968).
5. B. S. Baliga, A. W. Pronczuk, and H. N. Munro, *J. Mol. Biol.* **34**, 199–218 (1968).
6. S. L. Palay, *in*: "The Neurosciences, A Study Program" (G. C. Quarton, T. Melnechuk, and F. O. Schmitt, eds.), pp. 24–31, Rockefeller University Press, New York (1967).
7. M. V. Edds, Jr., *in*: "The Neurosciences, A Study Program" (G. C. Quarton, T. Melnechuk, and F. O. Schmitt, eds.), pp. 230–240, Rockefeller University Press, New York (1967).
8. J. D. Ebert, *in*: "The Neurosciences, A Study Program" (G. C. Quarton, T. Melnechuk, and F. O. Schmitt, eds.), pp. 241–247, Rockefeller University Press, New York (1967).
9. E. R. Kandel, *in*: "The Neurosciences, A Study Program" (G. C. Quarton, T. Melnechuk, and F. O. Schmitt, eds.), pp. 666–690, Rockefeller University Press, New York (1967).
10. J. Altman, *in*: "The Neurosciences, A Study Program" (G. C. Quarton, T. Melnechuk, and F. O. Schmitt, eds.), pp. 723–743, Rockefeller University Press, New York (1967).

11. B. W. Agranoff, *in*: "The Neurosciences, A Study Program" (G. C. Quarton, T. Melnechuk, and F. O. Schmitt, eds.), pp. 756–765, Rockefeller University Press, New York (1968).
12. O. Z. Sellinger, F. de Balbian Verster, R. J. Sullivan, and C. Lamar, Jr., *J. Neurochem.* **13**, 501–513 (1966).
13. J. M. Azcurra and O. Z. Sellinger, *Brain Res.* **6**, 359–362 (1967).
14. O. Z. Sellinger and J. M. Azcurra, *in*: "Macromolecules and the Function of the Neuron" (Z. Lodin and S. P. R. Rose, eds.), pp. 256–268, Excerpta Medica Press, Amsterdam (1968).
15. O. Z. Sellinger and W. G. Ohlsson, *J. Cell Biol.* **39**, 121a (1968).
16. O. Z. Sellinger and R. A. Hiatt, *Brain Res.* **7**, 191–200 (1968).
17. O. Z. Sellinger and L. M. Nordrum, *J. Neurochem.*, **16**, 1219 (1969).
18. O. Z. Sellinger and R. N. Borens, *Biochim. Biophys. Acta* **173**, 176–184 (1969).
19. O. Z. Sellinger, R. N. Borens, and L. M. Nordrum, *Biochim. Biophys. Acta* **173**, 185–191 (1969).
20. J. G. Nievel and J. N. Cumings, *Nature (London)* **214**, 1123–1124 (1967).
21. T. Hallinan, C. N. Murty, and J. H. Grant, *Arch. Biochem. Biophys.* **125**, 715–720 (1968).
22. O. Z. Sellinger and W. G. Ohlsson, *Fed. Proc.* **28**, 896 (1969).
23. N. E. Ghittoni and O. Z. Sellinger, *Fed. Proc.* **27**, 837 (1968).
24. O. Z. Sellinger, J. M. Azcurra, and W. G. Ohlsson, *J. Pharmacol. Exp. Therap.* **164**, 212–222 (1968).
25. O. Z. Sellinger, E. F. Domino, V. B. Haarstad, and M. E. Mohrmann, *J. Pharmacol. Exp. Therap.* **167**, 63–76 (1969).
26. M. V. Gomez, E. F. Domino, and O. Z. Sellinger, *Fed. Proc.* **28**, 292 (1969).
27. E. Egyhazi and H. Hydén, *J. Biophys. Biochem. Cytol.* **10**, 403–410 (1961).
28. H. Hydén and E. Egyhazi, *J. Cell Biol.* **15**, 37–44 (1962).
29. H. Hydén, *in*: "The Neuron" (H. Hydén ed.), pp. 179–219, Elsevier Publishing Co., Amsterdam (1967).
30. H. Hydén, *in*: "The Neurosciences, A Study Program" (G. C. Quarton, T. Melnechuk, and F. O. Schmitt, eds.), pp. 248–266, Rockefeller University Press, New York (1967).
31. E. Egyhazi and H. Hydén, quoted by Hydén.[29,30]
32. A. J. Glasky and L. N. Simon, *Science* **151**, 702–703 (1966).
33. L. N. Simon and A. J. Glasky, *Life Sciences* **7**, 197–202 (1968).
34. P. Puig-Muset, J. Ramia, and J. Martin-Esteve, *Nature (London)* **215**, 522–523 (1967).
35. N. R. Morris, G. K. Aghajanian, and F. E. Bloom, *Science* **155**, 1125–1127 (1967).
36. H. Stein and T. Yellin, *Science* **157**, 96–98 (1967).
37. G. P. Talwar, B. K. Goel, S. P. Chopra, and B. D'Monte, *in*: "Macromolecules and Behavior" (J. Gaito, ed.), pp. 71–88, Appleton, Century and Crofts, New York (1966).
38. A. Geiger, S. Yamasaki, and R. Lyons, *Am. J. Physiol.* **184**, 239–247 (1956).
39. Lj. Mihajlović, B. D. Janković, M. Petković, and K. Isaković, *Experientia (Basel)* **14**, 144–145 (1958).
40. G. P. Talwar, B. Sadasivudu, and V. S. Chitre, *Nature (London)* **191**, 1007–1008 (1961).
41. E. L. Noach, J. Joosting Bunk, and A. Wijling, *Acta Physiol. Pharmacol. Neerlandica* **11**, 54–69 (1962).
42. T. Pechan and P. Marko, *Physiol. Bohemoslov.* **12**, 458–462 (1963).

43. B. Sacktor, J. E. Wilson, and C. G. Tiekert, *J. Biol. Chem.* **241**, 5071–5075 (1966).
44. V. S. Chitre, S. P. Chopra, and G. P. Talwar, *J. Neurochem.* **11**, 439–448 (1964).
45. R. K. Datta and J. J. Ghosh, *J. Neurochem.* **11**, 357–366 (1964).
46. J. J. Ghosh, R. K. Datta, and K. C. Bhattacharya, *Can. J. Biochem.* **43**, 959–975 (1965).
47. J. J. Ghosh, *J. Indian Chem. Soc.* **44**, 872–886 (1967).
48. C. Vesco and A. Giuditta, *Biochim. Biophys. Acta* **142**, 385–402 (1967).
49. I. D. Herriman and G. D. Hunter, *J. Neurochem.* **12**, 937–948 (1965).
50. R. P. Perry, *Proc. Natl. Acad. Sci. U.S.* **48**, 2179–2185 (1962).
51. C. Vesco and A. Giuditta, *J. Neurochem.* **15**, 81–85 (1968).
52. H. Waelsch and A. Lajtha, *Physiol. Rev.* **41**, 709–736 (1961).
53. R. S. Piha, R. M. Bergstrom, L. Bergstrom, A. J. Uusitalo, and S. S. Oja, *Ann. Med. Exp. Fenn.* **41**, 498–515 (1963).
54. D. H. Clouet and M. Ratner, *J. Neurochem.* **15**, 17–25 (1968).
55. D. H. Clouet, this volume, Chapter 38.
56. Z. Lodin and J. Kolousek, *Physiol. Bohemoslov.* **5**, 43–48 (1956).
57. J. Folbergrova, *J. Neurochem.* **10**, 775–782 (1963).
58. O. Z. Sellinger and J. M. Azcurra, *Fed. Proc.* **27**, 756 (1968).
59. A. T. Campagnoni and H. R. Mahler, *Biochemistry* **6**, 956–967 (1967).
60. J. M. Azcurra and O. Z. Sellinger, manuscript in preparation.
61. H. D. Cohen and S. H. Barondes, *J. Neurochem.* **13**, 207–211 (1966).
62. S. H. Barondes and M. E. Jarvik, *J. Neurochem.* **11**, 187–196 (1964).
63. A. von der Decken, *in*: "Techniques in Protein Biosynthesis" (P. N. Campbell and J. R. Sargent, eds.), Vol. 1, pp. 61–131, Academic Press, New York (1967).
64. T. Staehelin, F. O. Wettstein, and H. Noll, *Science* **140**, 180–183 (1963).
65. S. W. Kwan and T. E. Webb, *J. Biol. Chem.* **242**, 5542–5548 (1967).
66. H. Sidransky, D. S. P. Sarma, M. Bongiorno, and E. Verney, *J. Biol. Chem.* **243**, 1123–1132 (1968).
67. P. G. W. Plagemann, *J. Biol. Chem.* **243**, 3029–3037 (1968).
68. S. H. Wilson and M. B. Hoagland, *Biochem. J. (London)* **103**, 556–566 (1967).
69. S. H. Appel, *Nature (London)* **213**, 1253–1254 (1967).
70. S. C. Bondy and S. Roberts, *Biochem. J. (London)* **105**, 111–118 (1967).
71. S. C. Bondy and S. Roberts, *Biochem. J. (London)* **109**, 533–541 (1968).
72. M. R. V. Murthy, *Biochim. Biophys. Acta* **166**, 115–123 (1968).
73. C. E. Zomzely, S. Roberts, C. P. Gruber, and D. M. Brown, *J. Biol. Chem.* **243**, 5396–5409 (1968).
74. G. A. Stewart and E. Farber, *Science* **157**, 67–69 (1967).
75. B. Mach and P. Vassalli, *Proc. Natl. Acad. Sci. U.S.* **54**, 975–982 (1965).

Chapter 26

Protein Synthesis and Memory Formation

Bernard W. Agranoff

Mental Health Research Institute
and Department of Biological Chemistry
University of Michigan, Ann Arbor

INTRODUCTION

A role for macromolecules in the storage of behavioral information has long been championed.[1–5] This postulated role remains at present highly inferential, even though many investigators have attempted to confirm or to disprove it. Experiments have thus far produced too few constraints for those who would theorize on the biological basis of learning and memory. While we are clearly at a very speculative stage, the ranks of experimenters are growing, and with them the firm hope that an assault on this least-known aspect of living systems is underway.

Several experiments have been reported which suggest a change in mobility of an electrophoretic band of protein[6] or in the amount of nucleic acid labeling[7] in whole brain extracts as a result of training. The excitement of training or the motor activity involved in responding are reportedly ruled out by appropriate control-animal experiments. We do not know *a priori* what amount of change in protein or RNA we could reasonably expect were there a change in a specific macromolecular species or in a relatively abundant macromolecule in a specific anatomical locus as a result of learning or memory formation. It is clear, however, that large changes in brain protein or nucleic acid composition on labeling reflect at best a concomitant process rather than one unique to specific sensory motor aspects of the behavioral experience. Otherwise, we would soon find that the brain does not have the capacity for the theoretically large number of possible behavioral changes characteristic of the nervous system.

Even reported changes in single neurons, such as in the Deiter's cells (in which total RNA as well as base ratios are reported altered) or hippocampal pyramidal cells (in which protein labeling and electrophoretic mobilities vary) would seem to reflect general, rather than specific, changes in behavioral states when we consider the multiple functions of neurons.

The major consideration of the present paper is another approach to the biological study of learning and memory: What happens behaviorally when a macromolecular synthetic process is severely inhibited in part or all of the brain at various times during or after a training session? In such experiments total inhibition of DNA, RNA, or protein synthesis is not possible. Hence, inability to block learning or memory with an inhibitor would not conflict with the biochemical changes described above. Alternatively, present-day techniques might not detect a change in protein or nucleic acid composition which could nevertheless be suppressed by a blocking agent. We see, then, from the onset that neither the search for biochemical changes characteristic of learning and memory, nor the use of antibiotic blocking agents can ever prove the other approach invalid. Nor can we say that positive findings by the two approaches confirm a common basic phenomenon. While the various inhibitory studies described below do not demonstrate unequivocally that protein synthesis is required for memory formation, this explanation remains at present the simplest working hypothesis which is consistent with results of studies in our laboratory with the goldfish and most of the studies of others with mice.

Many studies have employed puromycin, a naturally occurring antibiotic, consisting of a nucleoside and amino acid in amide linkage. It mimics the terminus of an acyl-transfer RNA and blocks protein synthesis by combining with growing peptide chains on the polysome and releasing them prematurely as peptidyl-puromycin.[8,9] The antibiotic blocks protein synthesis without necessarily slowing ribosomal travel along m-RNA. Another class of protein inhibitors, the cycloheximides, block protein synthesis by slowing ribosomal travel and peptide formation.[10] They therefore also prevent peptidyl-puromycin formation.

EXPERIMENTS WITH MICE

In 1963 Flexner *et al.*[11] reported the amnestic effects of intracerebral injections of puromycin on mice previously trained in a single session to turn right or left to avoid a punishing electrical shock. The antibiotic (90 μg in 12 μl) was injected into a right and left temporal site *via* a stereotaxic apparatus and inhibited protein synthesis in the hippocampus and entorhinal cortex $>80\%$ for 8–10 hr. Injections given 1–3 days following training produced amnesia, while saline injections into the temporal sites or bilateral ventricular or frontal injections of puromycin had no effect on memory. Bilateral injections of hydrolyzed puromycin or puromycin analogs had no effect. It was tentatively concluded that intact protein synthesis was required following training to form memory. Chloramphenicol had no effect, and this was initially attributed to an insufficient inhibition of protein synthesis.[12] An interesting aspect of this finding is a recent report on the effect of various agents on brain protein synthesis. While chloramphenicol, but not acetoxycycloheximide (AXM), blocks liver mitochondrial protein synthesis, the reverse is reported for brain mitochondria.[13]

In experiments employing bilateral injections of 60–120 μg of AXM the Flexners[14] found no impairment of memory, although protein synthesis was blocked more intensively than by puromycin. Further, a mixture of 120 μg of AXM and puromycin did not appear to block memory as tested several days after the injection. AXM blocked the amnestic effects of puromycin.

These findings gave rise to two hypotheses: The first is that puromycin allows m-RNA to degrade, while AXM, by preserving polysomal aggregation, preserves m-RNA until the antibiotic effect wears off.[14] Flexner *et al.*[15] further suggested that m-RNA is produced during learning and codes for a protein which by "self-induction" assures further m-RNA formation.

The second hypothesis is that peptidyl puromycin is the amnestic agent.[16] AXM and the combination AXM and puromycin would not be expected to produce a behavioral effect. Supporting evidence for the role of peptidyl-puromycin in memory blocking is as follows:

1. Gambetti *et al.*[17] find that puromycin, like polypeptides, produces mitochondrial damage, demonstrated by electron microscopy of mouse brain following intracerebral injection.

2. Injection of saline, even two months after puromycin, appears to relieve the memory loss.[18] Correlative studies with ^{3}H-puromycin gives suggestive evidence of residual radioactivity in the brain, possibly peptidyl-puromycin.[16] The saline somehow "washes out" the peptidyl-puromycin.

Studies by Barondes and Cohen[19] verified that puromycin, but not cycloheximides or puromycin + cycloheximide, produced amnesia in mice. Cohen *et al.*[20,21] gave further evidence that the mechanism of puromycin (presumably peptidyl-puromycin) amnesia was *via* the production of occult seizures.

3. Puromycin (100 μg in 10 μl) injected into the lateral sites causes loss of the characteristic electrical activity of the mouse hippocampus.[20] Cycloheximide (100 μg) or saline produces no effect. The combination puromycin + cycloheximide does not give reliable results.

4. In subsequent experiments similar amounts of puromycin and cycloheximide were injected intracerebrally, and 1 hr later the convulsant

drug pentylenetetrazol (PT, metrazol) 50 mg/kg was injected subcutaneously.[21] While this amount does not produce PTC (pentylenetetrazol convulsions), pretreatment with intracerebral injections of puromycin causes the mice to convulse. Cycloheximide blocks this effect of puromycin, while pretreatment with AXM (10 μg in each site) does not increase convulsive activity. The experiments on electrical activity and PTC potentiation were taken to mean that puromycin gives rise to peptidyl-puromycin, which is a convulsant agent, and that the occult convulsions, rather than a block of protein synthesis, may cause the amnesia. Contrary to their earlier studies, Barondes and Cohen[22] later found that AXM (10 μg in 10 μl) injected into each temporal site blocks memory, and hence they proposed that the latter drug acted by inhibition of protein synthesis, while puromycin might have at least two modes of action: block of protein synthesis *and* a convulsant action, presumably *via* peptidyl-puromycin production.

PREVIOUS EXPERIMENTS WITH FISH

Our laboratory has been engaged for the past five years on the study of memory of shock-avoidance coupled to a light stimulus in the goldfish. This work has been summarized elsewhere,[23–28] and only relevant aspects will be stated here.

1. Both puromycin (170 μg) and AXM (0.2 μg) produce amnesia if given before or just following initial training. AXM does not block the puromycin amnestic effect. Note that while the dose of puromycin is comparable to that used in the mouse, the dose of AXM used is 1/100–1/1000 that used in the mouse.

2. The unanesthetized fish are injected through the cranium into the intracranial space *over* the brain. No detectable neurologic effects are noted. While our results seem to preclude the possibility that peptidyl-puromycin causes memory loss in the goldfish, it remained possible that puromycin produces occult convulsions and that these convulsions are one of the means by which puromycin produces amnesia.

THE EFFECT OF ANTIMETABOLITE ANTIBIOTIC PRETREATMENT ON PENTYLENETETRAZOLE CONVULSIONS IN THE GOLDFISH

We first investigated the effect of fish condition and dose on convulsions. Goldfish are shipped weekly from Stoutland, Missouri *via* rail in sealed plastic bags containing 100 fish in two gallons of water. While there are generally few deaths (<1%) in transit, we ordinarily store 200–300

TABLE I

Effect of Time after Arrival on Pentylenetetrazol Convulsions (PTC)

Days	PTC/20 Fish
0	18
1	11
2	5
5	4
7[a]	2, 4, 2, 0, 1, 2

[a] Control groups from experiment 3. Each fish received PT (50 mg/kg).

fish/200-gallon tank of water for 1–2 weeks before using them for behavioral experiments. We have noted that fish placed in shuttle-boxes 1–2 days after arrival are occasionally stunned by the mild punishing shock used in the training task. It was therefore not unexpected that the number of convulsions/group were a function of time after arrival (Table I).

Experiment 1

Groups of 20 goldfish were weighed and set out in individual plastic containers (home tanks) containing 1500 ml of aged water. After 15–60 min they were injected intraperitoneally with pentylenetetrazol (50 mg/kg) and were observed for 10 min. All animals that convulsed did so within 10 min, almost all of them within 5 min. Some recovered and convulsed again for longer periods of time. Convulsions were often preceded by hyperactivity, banging into the walls and lid of the plastic container, and finally by loss of vertical orientation, usually accompanied by clonic tremors of the pelvic and pectoral fins.

Using animals stored > one week, we established 50 mg/kg as our standard intraperitoneal dose. The following number of convulsions were seen in each group of 10 fish: 50/kg, 0; 75 mg/kg, 1; 100 mg/kg, 7; 125 mg/kg, 7; 150 mg/kg, 9; 175 mg/kg, 10. Pretreatment with diphenylhydantoin (35 mg/kg) was ineffective in protecting against PTC, as has been observed in mammals.[29]

Twenty-one groups of 20 fish each were each injected intracranially with varying amounts of puromycin in 10 μl of saline and given 50 mg of PT intraperitoneally 1 hr later (Table II). It can be seen that 170 μg of puromycin injected intracranially in the fish is just enough to produce a near-maximal effect on PTC.

TABLE II

Potentiation of PTC by Intracranial Injections of Puromycin

Puromycin di-HCl (μg)	PTC/20 Fish
0	2, 4, 2, 0, 1, 2
25	1, 4, 5
50	9, 5, 8
85	13, 12, 13, 7, 10, 8
170	15, 19, 15

Experiment 2

Subsequent experiments investigated the effects of AXM, of AXM in puromycin-potentiated PTC, and of actinomycin D, an antibiotic which blocks RNA synthesis and memory in the goldfish.[25] Table III shows that AXM, an amnestic agent in goldfish, is not a convulsant agent. Further, unlike the studies in the mouse, AXM does not protect against puromycin-potentiated PTC. On the other hand, puromycin aminonucleoside, the nucleoside moiety of puromycin, appears to be a convulsant agent, even though it does not produce memory loss.[24] Since puromycin contains 2 equivalents of HCl, the effect of additional HCl as well as of HCl itself was examined.

TABLE III

Effect of Various Agents on PTC

Agent and Dose	PTC/20 Fish
AXM (0.2 μg)	1
AXM (0.1 μg) + puromycin (85 μg)	10
AXM (0.2 μg)[a] + puromycin (170 μg)[b]	16
Puromycin aminonucleoside (92 μg)	8, 7, 9
Puromycin aminonucleoside · 2 HCl (92 μg + 0.63 μM HCl)	12, 10
HCl (0.63 μM)	3, 3, 7
Actinomycin D (2.0 μg)	0

[a] 2 hr before PT.
[b] 1 hr before PT.

TABLE IV

Effect of Diphenylhydantoin on PTC and Puromycin-Potentiated PTC

Diphenylhydantoin (mg/kg) (intraperitoneal)	Puromycin (g) (intracranial)	PT (mg/kg) (intraperitoneal)	PTC/20 Fish
35[a]	170[a]	50	16
70[a]	—	—	19
70[b]	—	100	18

[a] Given 1 hr before PT.
[b] Given 4 hr before PT.

Experiment 3

While diphenylhydantoin did not block PTC, it was possible that it blocked puromycin-potentiated PTC. This was of interest, since diphenylhydantoin is reported to block puromycin-mediated amnesia in mice.[21] In the goldfish we found no protection under any conditions (Table IV).

The results support the conclusion that puromycin potentiates PTC in the goldfish, as has been reported in the mouse. This action does not appear to be directly related to the antibiotic's amnestic action.

The convulsive activity is probably not related to the formation of peptidyl-puromycin, but may be related to a property conferred by the nucleoside structure it shares with puromycin aminonucleoside. Since AXM is amnestic in the fish and since it does not protect against puromycin-induced amnesia,[30] we presently interpret our results to associate memory block with inhibition of protein synthesis, actions that the two disparate molecules, puromycin and AXM, share. In other studies we have suggested that RNA synthesis,[25] but not DNA synthesis,[31] may block memory.

The results with mice are at present difficult to explain simply. Under varying conditions cycloheximide can be shown to block,[22] not to block,[19] and to temporarily block,[32] memory. Barondes *et al.* suggested in earlier studies that overtraining overcame the action of AXM. It is also possible that animals were retested before amnesia could develop.[33] The intracranial dose of AXM which has been reported to block memory in the mouse (20 μg) is much lower than the dose reported not to block memory in the mouse. It is possible that AXM delays the decay of short-term memory, particularly in high doses.

SUMMARY

Studies on the effects of antibiotic blocking agents on learning and memory have been reviewed. Arguments supporting the role of peptidyl-puromycin formation as part of the amnestic mechanism have been reviewed. Experiments are presented which suggest that in the goldfish the convulsive action of puromycin is not mediated *via* peptidyl-puromycin, nor does this action appear to be related to the amnestic action of puromycin.

REFERENCES

1. J. J. Katz and W. C. Halstead, Protein organization and mental function, *Comp. Psych. Monographs* **20,** 1–38 (1950).
2. R. W. Gerard, The fixation of experience, *in*: "Brain Mechanisms and Learning" (J. F. Delafresnaye, A. Fessard, R. W. Gerard, and J. Konorski, eds.), pp. 21–32, Blackwell, Oxford (1961).
3. H. Hydén, The neuron, *in*: "The Cell: Biochemistry, Physiology and Morphology" (J. Brachet and A. E. Mirsky, eds.), Vol. 4, pp. 215–323, Academic Press, New York (1960).
4. W. Dingman and M. B. Sporn, Molecular theories of memory, *Science* **144,** 26–28 (1964).
5. F. O. Schmitt, Molecules and memory, *New Scientist* **23,** 643–645 (1964).
6. H. Hydén and P. W. Lange, Protein synthesis in the hippocampal pyramidal cells of rats during a behavioral test, *Science* **159,** 1370–1373 (1968).
7. J. W. Zemp, J. E. Wilson, K. Schlesinger, W. O. Boggan, and E. Glassman, Brain function and macromolecules, I. Incorporation of uridine into RNA of mouse brain during short-term training experience, *Proc. Natl. Acad. Sci. U.S.* **55,** 1423–1431 (1966).
8. M. B. Yarmolinsky and G. L. de la Haba, Inhibition by puromycin of amino acid incorporation into protein, *Proc. Natl. Acad. Sci. U.S.* **45,** 1721–1729 (1959).
9. D. Nathans, Puromycin inhibition of protein synthesis: incorporation of puromycin into peptide chains, *Proc. Natl. Acad. Sci. U.S.* **51,** 585–592 (1964).
10. B. Colombo, L. Felicetti, and C. Baglioni, Inhibition of protein synthesis by cycloheximide in rabbit reticulocytes, *Biochem. Biophys. Res. Commun.* **18,** 389–395 (1965).
11. J. B. Flexner, L. B. Flexner, and E. Stellar, Memory in mice as affected by intracerebral puromycin, *Science* **141,** 57–59 (1963).
12. L. B. Flexner, Loss of memory in mice as related to regional inhibition of cerebral protein synthesis, *Texas Rep. Biol. Med.* **24,** 3–19 (1966).
13. M. W. Gordon and G. C. Deanin, Protein synthesis by isolated rat brain mitochondria and synaptosomes, *J. Biol. Chem.* **243,** 4222–4226 (1968).
14. L. B. Flexner and J. B. Flexner, Effect of acetoxycycloheximide and of an acetoxycycloheximide–puromycin mixture on cerebral protein synthesis and memory in mice, *Proc. Natl. Acad. Sci. U.S.* **55,** 369–374 (1966).
15. L. B. Flexner, J. B. Flexner, and R. B. Roberts, Memory in mice analyzed with antibiotics, *Science* **155,** 1377–1383 (1967).

16. L. B. Flexner and J. B. Flexner, Studies on memory: the long survival of peptidyl-puromycin in mouse brain, *Proc. Natl. Acad. Sci. U.S.* **60**, 923–927 (1968).
17. P. Gambetti, N. K. Gonatas, and L. B. Flexner, The fine structure of puromycin-induced changes in mouse entorhinal cortex, *J. Cell Biol.* **36**, 379–390 (1968).
18. J. B. Flexner and L. B. Flexner, Restoration of expression of memory lost after treatment with puromycin, *Proc. Natl. Acad. Sci. U.S.* **57**, 1651–1654 (1967).
19. S. H. Barondes and H. D. Cohen, Comparative effects of cycloheximide and puromycin on cerebral protein synthesis and consolidation of memory in mice, *Brain Res.* **4**, 44–51 (1967).
20. H. D. Cohen, F. Ervin, and S. H. Barondes, Puromycin and cycloheximide: different effects on hippocampal electrical activity, *Science* **154**, 1557–1558 (1966).
21. H. D. Cohen and S. H. Barondes, Puromycin effect on memory may be due to occult seizures, *Science* **157**, 333–334 (1967).
22. S. H. Barondes and H. D. Cohen, Delayed and sustained effect of acetoxycycloheximide on memory in mice, *Proc. Natl. Acad. Sci. U.S.* **58**, 157–164 (1967).
23. B. W. Agranoff, R. E. Davis, and J. J. Brink, Memory fixation in the goldfish, *Proc. Natl. Acad. Sci. U.S.* **54**, 788–793 (1965).
24. B. W. Agranoff, R. Davis, and J. J. Brink, Chemcal studies on memory fixation in goldfish, *Brain Res.* **1**, 303–309 (1966).
25. B. W. Agranoff, R. E. Davis L. Casola, and R. Lim, Actinomycin D blocks formation of memory of shock avoidance in the goldfish, *Science* **158**, 1600–1601 (1967).
26. B. W. Agranoff, Agents that block memory, *in*: "The Neurosciences: A Study Program" (G. C. Quarton, T. Melnechuk, and F. O. Schmitt, eds.), pp. 756–764, The Rockefeller University Press, New York (1967).
27. J. J. Brink, R. E. Davis, and B. W. Agranoff, Effects of puromycin, acetoxycycloheximide, and actinomycin D on protein synthesis in goldfish brain, *J. Neurochem.* **13**, 889–896 (1966).
28. R. E. Davis and B. W. Agranoff, Stages of memory formation in goldfish: evidence for an environmental trigger, *Proc. Natl. Acad. Sci. U.S.* **55**, 555–559 (1966).
29. J. E. P. Toman, Drugs effective in convulsive disorders, *in*: "The Pharmacological Basis of Therapeutics" (L. S. Goodman and A. Gilman, eds.), 3rd ed., pp. 215–236, MacMillan, New York (1965).
30. B. W. Agranoff, unpublished.
31. L. Casola, R. Lim, R. E. Davis, and B. W. Agranoff, Behavioral and biochemical effects of intracranial injection of cytosine arabinoside in goldfish, *Proc. Natl. Acad. Sci. U.S.* **60**, 1389–1395 (1968).
32. L. B. Flexner, J. B. Flexner, and R. B. Roberts, Stages of memory in mice treated with acetoxycycloheximide before or immediately after learning, *Proc. Natl. Acad. Sci. U.S.* **56**, 730–735 (1966).
33. B. W. Agranoff, Recent studies on the stages of memory formation in goldfish, *in:* "Molecular Approaches to Learning and Memory" (W. L. Bryne, ed.), pp. 35–39, Academic Press, New York (1969).

DISCUSSION

UNGAR: Have you tried washing out the puromycin, as Flexner did in mice, with saline and binder to reestablish memory?

AGRANOFF: We don't make a hole in the brain, so we can't do the appropriate experiment, but Dr. Barondes replicated Flexner's finding in the mouse.

BARONDES: The puromycin effect that Flexner finds is that when he gives the drug 1–3 days after training he finds that it causes amnesia, whereas if he gives it seven days later, it does not. We replicated the experiment, and, indeed, if you give saline one day after injecting puromycin the day after training, it does cause recovery of the memory. None of those experiments are relevant to the question "is protein synthesis required for memory storage?" I think that the action of protein synthesis in memory occurs much before the drug is administered in these kinds of experiments. They are interesting phenomena, but they are probably germane to a different thing.

SELLINGER: Under certain conditions one can impair memory with ECS and with KCl; but then, in terms of the occult seizures, which can be potentiated by puromycin or on which puromycin has an effect, these seem not to have a very important part in memory retention. My interest is simply in whether or not one can make the generalization that a hyperexcited state will or will not facilitate any of these processes.

AGRANOFF: It appears that it is not the convulsion-potentiating action of these drugs in the fish that correlates with memory loss. I would like to summarize in what the mouse experiments have in common with the fish experiments. In both species, in the one case using an intracranial injection, and in the other an intracerebral or subcutaneous injection, two inhibitors of protein synthesis each produce a block in memory formation. There is partial training in each instance, and a short-term decay which does not particularly correlate with susceptibility to the antimetabolite. Learning does not seem to be blocked, while the formation of long-term memory is. At the present time there is a great deal that these two preparations have in common. What remains to be explained is what puromycin does in the mouse brain in addition to blocking memory.

FOLCH-PI: What is the proof that puromycin and all the various inhibitors do not do many other things besides directly and indirectly inhibiting protein synthesis? There have been reports, by Kanfer, that puromycin does in fact alter the biosynthesis of ganglioside. What is the proof that you are not giving the animals a gigantic hangover? Can you get the same results with alcohol? The basic point is, what is the proof that you are not just interfering with brain function in general, instead of dealing with a specific interference?

AGRANOFF: I would like to make two comments. First, from our behavioral data, one can see that animals who received drug immediately following training had a memory deficit three days later. Animals who received the same amount of drug 3 hr after training (the first group was retrained 72 hr after puromycin, the others 69 hr after puromycin) should have had the same "hangover." This is a kind of control for nonspecific sickness. Also, we measure both avoidance and escape, and the ability to escape is not affected by the drugs. This would seem to be an argument against the "hangover." Next, I agree with you that when we talk about protein synthesis we don't know whether it is *the* phenomenon or some epiphenomenon of the inhibition that blocks memory. In order to make ganglioside, Kanfer has apparently shown that you must simultaneously make protein. In order to make membranes, you certainly must make protein. Such a large number of biochemical processes involve protein synthesis that what is surprising is that these animals are able to show normal acquisition at a time when their brains are not making protein normally.

AGRANOFF: I would like to ask Dr. Barondes about the actinomycin study he published some time ago.

BARONDES: Well, the problem with actinomycin is that in order to effect, say, 95% inhibition of cerebral RNA synthesis in the mouse, you have to give enormous doses.

These enormous doses make the mice untestable after a certain number of hours, so that we have to do our tests within 3 hr after training. We know from the acetoxycycloheximide experiments that 3 hr after training is just not long enough. The short-term memory still survives at this time. If we give a smaller dose, which inhibits 75% of brain RNA synthesis, we can test them for a longer period of time, but we find that while they remember, now we probably haven't inhibited at all, or may not have inhibited enough, because we know from the acetoxycycloheximide experiments that if we only inhibit 75%, we get no effect at all. Thus, the current status is that the phenomena that we reported are correct, namely, that learning can go on in the presence of 95% inhibition of cerebral RNA synthesis and it can last for about 3 hr. We have never been able to find out what happens beyond that time under a condition of appropriately severe inhibition.

AGRANOFF: Do you think then that the RNA is not involved?

BARONDES: I don't know. The limiting thing in protein synthesis is not always RNA synthesis. That is, you can activate RNA that is already present in the cells, and this is true in differentiation, and might be true in memory.

AGRANOFF: In the three-out-of-four criterion that you use, an animal will randomly give two out of four positive responses. Won't some naive animals look as if they are well trained simply by chance? What numbers of animals are you using?

BARONDES: We use about 15 or 20 animals for a time point and we get very good statistically significant differences at the level of 0.01; furthermore, with the light–dark discrimination, which is a more delicate task, we can use the 9-out-of-10 criterion and get the same phenomenon.

MC EWEN: I would like to ask Dr. Agranoff to comment on Bitterman's experiments, and perhaps generally on the question of task, because Bitterman shows a more or less specific effect of fear as opposed to a discrimination.

AGRANOFF: Yes, Dr. Bitterman has essentially replicated the shock-avoidance result that we published using puromycin. Superimposing on that a discrimination test, he finds that puromycin does not knock this out. The animals were injected daily, and his would be rather hard to compare with other experiments. While there are psychologists who do not feel that he demonstrated this very convincingly, if he has, then there is a difference between the mouse and the fish. We have modified our task in several ways. We have used light-to-dark, and dark-to-light, and we have put a plastic gate over the hurdle, which the fish must deflect. We have the impression that modifications which make the task more dfficult for the fish produce memories which are susceptible to a smaller amount of drug. In other words, with a more difficult task, we can use smaller amounts of amnestic agents. One more point. We use about the same amount of puromycin as is used in the mouse, 170–180 mg. The amount of acetoxycycloheximide we use is, however, much less, 0.2 μg. This is less than 1/1000 of the amount used in some of the mouse intracerebral injections. When 0.10 μg of acetoxycycloheximide is given intraperitoneally it does not have amnestic effects.

MANDEL: I think we can believe from all these experiments that proteins are involved in memory. Do you think that a new protein is synthesized during learning?

AGRANOFF: If you mean a unique primary sequence, I do not believe that there is any evidence or precedent for a hypothesis involving such a molecule which would characteristically be produced as a result of behavior. I refer to an earlier paper.[26]

Chapter 27

Is the Amnesic Effect of Cycloheximide Due to Specific Interference with a Process in Memory Storage?*

Samuel H. Barondes

Departments of Psychiatry and Molecular Biology
Albert Einstein College of Medicine
Bronx, New York

INTRODUCTION

Because of its participation in so many important cellular regulatory processes, it is likely that cerebral protein synthesis is required for the cellular alterations which characterize "long-term" memory.[1] The proteins whose synthesis might play a role in facilitating the formation of new functional synaptic relationships in the brain could fall into a variety of classes. They could be enzymes which regulate the synthesis of neurotransmitters, receptor proteins, structural proteins, or proteins which direct some specific type of intercellular recognition. Since little is known about all such proteins in the brain, and since the possible candidates which could be critical for the formation of new functional synapses are so great even within each of these classes, a "shot-gun" approach has been taken as a first step in implicating cerebral protein synthesis in the process of formation of long-term memory. This approach is based on the use of inhibitors of cerebral protein synthesis, particularly the cycloheximide class, with which we will be primarily concerned. Although it has not as yet helped to identify the existence of specific proteins or classes of proteins whose synthesis is critical for long-term memory storage, it has been useful in supporting the hypothesis that cerebral protein synthesis is indeed required for this process, and has provided some indication of the time relative to training when this process occurs. Other data which support this hypothesis are reviewed by Agranoff in Chapter 26.

* This work was supported by Career Development Award K3-MH-18232 and by Grants MH 12773 and MH 06118 from the National Institute of Mental Health.

Both the usefulness and a major potential limitation of this approach are due to the fact that *general* inhibition of cerebral protein synthesis appears to be achieved with cycloheximide.[2,3] When inhibition of cerebral protein synthesis is established in a memory experiment in an attempt to block the synthesis of cerebral protein required for long-term memory storage other ongoing cerebral protein synthesis is also blocked. This ongoing synthesis is presumably directed toward replacement of proteins which are being metabolized and also toward other cellular regulatory processes which may be proceeding independently at the time of the experiment. Interference with these processes would ultimately impair brain function. However, the half-life of most brain proteins tends to be in the range of days or weeks.[4,5] Therefore, inhibition of cerebral protein synthesis for a number of hours might have no significant effect on general ongoing brain function. Nevertheless, to support a *specific* effect on memory, one would like to impair memory with as little inhibition of cerebral protein synthesis as possible, and one would also like to establish this inhibition for only as brief a time as possible. In this way there would be least suspicion of interference with other ongoing processes which might be dependent on cerebral protein synthesis.

In our initial studies of this problem we produced intense and prolonged inhibition of cerebral protein synthesis and studied its effects on memory after prolonged training in a conventional learning task. The purpose of this report is to indicate that using the cycloheximide class of inhibitors, it is possible to achieve effects on formation of long-term memory even with relatively short periods of inhibition of cerebral protein synthesis. This is becoming possible through studies of memory after relatively brief training of easily learned tasks. Although even relatively marked and prolonged inhibition of cerebral protein synthesis appears to have no significant effect on general brain function (gross behavior, cerebral electrical activity[6,7]), less-drastic inhibition appears to be effective in impairing the formation of long-term memory. As these experiments accumulate, it becomes increasingly likely that cycloheximide and acetoxycycloheximide impair the formation of long-term memory by inhibiting cerebral protein synthesis specifically required for this process.

THE BASIC EXPERIMENT

The basic experimental design that we use is to administer an inhibitor of cerebral protein synthesis shortly before training, so maximal inhibition of cerebral protein synthesis is established during the time of training and for minutes to hours thereafter. When mice are trained to escape shock by choosing the left limb or the lighted limb of a T-maze (to a criterion of 3 out of 4 consecutive correct responses or 5 out of 6 consecutive correct responses, respectively) at a time when 90–95% of cerebral protein synthesis is inhibited by injection of cycloheximide or acetoxycycloheximide before training, the mice learn normally.[8–11] They also remember normally when tested 3 hr after training, but have markedly impaired memory when tested 6 hr after training or thereafter.[8–11] This marked impairment is also observed when testing is conducted seven days after training or six weeks after training.[8] These results suggest that cerebral protein synthesis is required for "long-term" memory (memory for six or more hours after training), but is not required for learning or for "short-term" memory (memory for up to three or more hours after training). The fact that learning and short-term memory are normal in mice whose cerebral protein synthesis is markedly inhibited indicates that general brain function is not grossly impaired by the inhibitors in these experiments.

The time of administration of the inhibitor is critical. This too argues against a nonspecific effect of inhibition of cerebral protein synthesis on brain function. If acetoxycycloheximide was administered *long before* training, so that maximal inhibition was achieved before training, and inhibition at the time of training had decreased to about 70%, the animals learned and remembered normally in the behavioral situations studied.[8] If inhibition was established immediately after training was completed, there was a relatively small amnesic effect,[11] and if it was established ½ hr after training, in some situations there was no amnesic effect.[11] Since establishment of marked inhibition ½ hr after training in certain situations has absolutely no effect on long-term memory, it appears that the critical protein synthesis required for long-term memory storage occurs during training or within minutes after training, or both. In other situations it may continue, to some extent, for at least 30 min after training.[12] However, its absence is not detectable until the short-term process, which appears to be independent of cerebral protein synthesis, has decayed.[8–11]

IMPORTANCE OF DEGREE OF TRAINING AND OF TASK

There is considerable evidence, dating back to the classic ablation experiments of Lashley,[13] that neurons all over the brain are involved in the memory of the solution of maze tasks. This is important for the design of experiments to test the possible participation of protein synthesis in memory. Each trial in a multiple-trial training situation is presumed to be somewhat different from the preceding one. The animal may perceive things somewhat differently and behave somewhat differently during

TABLE I

Effect of Acetoxycycloheximide or Cycloheximide on Memory after Different Training Procedures[a]

Task	Training criterion	Savings (%)
Position discrimination (shock escape)	3/4	35
	9/10	76
Light–dark discrimination (shock escape)	5/6	34
	9/10	40
	15/16	63
Position discrimination (water reward)	3/4	6

[a] Mice were injected intracerebrally or subcutaneously with acetoxycycloheximide or cycloheximide before training, so that 90–95% of cerebral protein synthesis was inhibited during training. They were trained to choose the correct limb of a T-maze until they had made the indicated number of correct responses (e.g., 3 out of 4 consecutive correct responses). Retention was determined seven days later by comparing the number of trials required to reach criterion at that time with the number of trials initially required to reach criterion. This is expressed in Table I as savings (%).[8] Saline-injected mice had 70–80% savings in all these tasks when tested one week after training. The mice trained in the position discrimination to a criterion of 9 out of 10 consecutive responses had savings which were indistinguishable from those of controls. In all the other tasks the drug-treated mice had markedly impaired savings. (For details see works cited.[8–10])

each training trial, although the net measured result, the final performance of the correct response, is scored identically. Presumably, some different neurons are involved in each trial, so that with an increasing number of trials larger and larger numbers of neurons may begin to participate in the memory-storage process. In addition, the critical process for memory storage (presumably, cerebral protein synthesis) may be augmented in participating neurons with each trial. Therefore, one would expect that with increasing training greater and greater inhibition of cerebral protein synthesis would have to be established in order to prevent the formation of measurable long-term memory. Results might vary with the task used since the "information" obtained per trial might vary.

Some experiments are summarized in Table I which demonstrate that both the degree of training and the nature of the task are important variables. When marked inhibition of cerebral protein synthesis is established during brief training (3 out of 4 criterion) in a position discrimination, long term memory is markedly impaired. However, with prolonged training (9 out of 10 criterion) it is indistinguishable from the control level.

TABLE II

Effect of Degree of Inhibition of Cerebral Protein Synthesis during Training on "Long-term" Memory[a]

Acetoxycycloheximide (μg)	Inhibition (%)	Savings (%)
0	—	81
0.4	76	78
2	86	59
10	92	37
20	94	31
120	96	38

[a] Mice were injected intracerebrally with the indicated amount of acetoxycycloheximide (for details see Cohen and Barondes[(8)]). Inhibition of cerebral protein synthesis was estimated in mice at the time that training was conducted by determining incorporation of subcutaneously administered ^{14}C-leucine into cerebral protein.[(8)] The degree of inhibition is overestimated somewhat by the method of calculation (see [(11)]). Mice were trained to escape shock by choosing the left limb of a T-maze to a criterion of 3 out of 4 consecutive correct responses. Savings was determined seven days later. Inhibition of approximately 76% of cerebral protein synthesis during training had no effect on memory in this situation.

Position discriminations are apparently stored more efficiently than light-dark discriminations in mice since marked inhibition of cerebral protein synthesis interferes with long-term memory only of the latter, when training is given to an identical criterion (Table I). Therefore, care must be taken to choose the task and degree of training which are used in experiments of this type. Presumably this is of particular importance since total inhibition of cerebral protein synthesis is not achieved in these studies.

IMPORTANCE OF DEGREE AND DURATION OF INHIBITION

In our early experiments, in which we sought to determine the dependence of long-term memory storage on inhibition of cerebral protein synthesis, we attempted to achieve maximal and sustained inhibition of cerebral protein synthesis. This was achieved with intracerebral injections of acetoxycycloheximide. The importance of extensive inhibition of cerebral protein synthesis in producing a demonstrable effect on memory in the behavioral situation studied in these experiments is shown in Table II. Indeed, if less than about 80% of cerebral protein synthesis was inhibited during training in this particular situation, there was absolutely no measurable effect on long-term memory. This finding in itself underscores the relative resistance of general brain function to marked inhibition of cerebral protein synthesis. It is noteworthy that some retention was found even

TABLE III
Inhibition of Cerebral Protein Synthesis by Subcutaneous Injection of Cycloheximide[a]

Time after injection (min)	Inhibition (%)		
	120 mg/kg	30 mg/kg	8 mg/kg
30–60	95	91	74
60–90	77	72	60
120–150	52	40	39
180–210	19	—	—
210–240	9	—	—

[a] Mice were injected subcutaneously with the indicated dose of cycloheximide. ^{3}H-leucine was injected subcutaneously at the indicated time thereafter, the mice were sacrificed 30 min later, their cerebral hemispheres were removed, and TCA-precipitable and TCA-soluble radioactivity were determined.[8] Incorporation of ^{3}H-leucine was compared with that in saline-injected controls. The TCA-soluble radioactivity estimates the specific activity of the leucine pool at the time of sacrifice, but overestimates the specific activity of this pool for the entire 30-min period after the injection of tritiated leucine.[11] This is particularly true in the mice injected with inhibitor. To simplify these studies, we took the TCA-soluble radioactivity at the time of sacrifice as a measure of the specific activity of the leucine pool for the entire 30-min period. The degree of inhibition of cerebral protein synthesis indicated is, however, an overestimate. The overestimate is relatively slight when the calculated inhibition is great, but becomes more significant when the calculated inhibition is smaller.

when very large amounts of drug were given. It is not known whether the small amount of residual protein synthesizing capacity or some other process is responsible.

Since intracerebral injections even of relatively small doses of acetoxycycloheximide produced very sustained inhibition (considerable inhibition was even observed 24 hr after injection of the drug[8]), attempts were made to produce more transient inhibition. This was achieved by subcutaneous administration of the inhibitor and by using cycloheximide, a shorter-acting derivative of acetoxycycloheximide. Subcutaneous administration of cycloheximide produced prompt onset of inhibition of cerebral protein synthesis, and, depending on dose, the inhibition was relatively short-lived. The effect of several doses of subcutaneously administered cycloheximide on cerebral protein synthesis at various times after injection of the drug are shown in Table III. We have found that doses of cycloheximide which produce a maximum of about 90–95% inhibition of cerebral protein synthesis, but whose action is virtually completely over within 4 hr of the injection, produce marked impairment of long-term memory when administered 30 min before training.[14] Therefore, decreasing the total duration of inhibition does not diminish the amnesic effect.

In addition to work with more-transient inhibition of cerebral protein synthesis, we have also been engaged in studies of effects of less-intense cerebral protein synthesis inhibition, using even smaller amounts of training than we have in the past. We have recently studied the effect of cycloheximide on memory after training to a criterion of one correct response in a T-maze in which the correct side is marked both by the presence of light and a position cue. Mice which have been trained to escape shock until they have made one correct response show significant retention for at least one week. Impairment of this retention can be achieved with a subcutaneous injection of 30 mg/kg of cycloheximide 30 min prior to training,[15] which inhibits 90% of cerebral protein synthesis for only a short time (Table III). Reduction of the dose to 8 mg/kg had a very slight and inconstant effect. Therefore marked inhibition must be established to interfere with memory even if very little training is given.

OTHER CONSIDERATIONS

In addition to concern about the nonspecific effects of general inhibition of cerebral protein synthesis on behavior, other possible effects of the drugs must be considered in evaluating our results. Since inhibition of protein synthesis is also produced in nonneural tissues when these drugs are injected either intracerebrally or subcutaneously, the possibility must be considered that this general abnormality might result in memory impairment. This has been examined by comparing the relative effect of subcutaneous or intracerebral administration of 20 μg of acetoxycycloheximide. This dose impairs cerebral protein synthesis more effectively if given intracerebrally, but has equal or greater effects on other tissues if given subcutaneously. It was effective in impairing long-term memory only if given intracerebrally.[8] Therefore, the relevant action of the drug is on the brain, rather than other tissues. The use of cycloheximide, which has a much shorter duration of action than acetoxycycloheximide, has also helped to diminish the likelihood that systemic illness is involved in the amnesia. Doses of cycloheximide administered subcutaneously, which are quite effective in impairing long-term memory formation, produce little or no evidence of systemic illness, in contrast with acetoxycycloheximide, which, in the doses used in earlier experiments, produced diarrhea and dehydration.

Another problem in interpreting our studies arises from the possibility that the drugs may impair memory by some action on the brain unrelated to their effect on cerebral protein synthesis. This possibility was underscored by our finding[4,5] that puromycin, another potent inhibitor of protein synthesis, produces occult seizures and marked cerebral electrical abnormalities. However, cycloheximide and acetoxycycloheximide do not produce these abnormalities.[4,5] The possibility that some side effect of the cycloheximide group of drugs may be responsible for their amnesic effect can also be considered in the light of the relative effectiveness of cycloheximide and acetoxycycloheximide on memory. Using these related drugs, which differ markedly in their potency in inhibiting cerebral protein synthesis, the effective dose of the drug is highly correlated with the degree of inhibition which is achieved. A dose of cycloheximide which inhibits cerebral protein synthesis far less than the equivalent dose of acetoxycycloheximide may be ineffective in impairing the formation of long-term memory, whereas the acetoxycycloheximide is markedly effective. Therefore, the effects of the drugs are correlated with the degree of inhibition of cerebral protein synthesis and not with the absolute amount of drug administered. Were there other effects of the cycloheximide series of drugs which account for their amnesic effect, such a correlation need not have been found.

CONCLUSION

Our experiments indicate that administration of cycloheximide or acetoxycycloheximide prior to training, so that marked inhibition of cerebral protein synthesis is established during training, impairs the formation of long-term memory. Recent improvements in our experimental technique include: (1) use of very little training and of different tasks; (2) administration of drugs subcutaneously, which produces more rapid onset of generalized inhibition of cerebral protein synthesis than localized intracerebral injections; and (3) use of cycloheximide, which has a shorter duration of action than acetoxycycloheximide. These improvements have increased the likelihood that our findings are due to inhibition of cerebral protein synthesis specifically involved in the long-term memory storage process.

REFERENCES

1. S. H. Barondes, *Nature* **205,** 18–21 (1965).
2. H. D. Sisler and M. R. Siegel, *in*: "Antibiotics. I. Mechanism of Action" (D. Gottlieb and P. D. Shaw, eds.), pp. 283–308, Springer-Verlag, New York (1967).
3. M. B. Gordon and G. G. Deanin, *J. Biol. Chem.* **243,** 4222–4226 (1968).
4. A. Lajtha, *J. Neurochem.* **3,** 358–365 (1959).
5. A. Lajtha, *Int. Rev. Neurobiol.* **6,** 1–98 (1964).
6. H. D. Cohen, F. Ervin, and S. H. Barondes, *Science* **154,** 1557–1558 (1966).

7. H. D. Cohen and S. H. Barondes, *Science* **157**, 333–334 (1967).
8. S. H. Barondes and H. D. Cohen, *Proc. Natl. Acad. Sci. U.S.* **58**, 157–164 (1967).
9. H. D. Cohen and S. H. Barondes, *Nature* **218**, 271–273 (1968).
10. H. D. Cohen and S. H. Barondes, *Commun. Behav. Biol.* **1** (Part *A*), 337–340 (1968).
11. S. H. Barondes and H. D. Cohen, *Science* **160**, 556–557 (1968).
12. A. Geller, F. Robustelli, S. H. Barondes, H. D. Cohen, and M. E. Jarvik, *Psychopharmacologia* **14**, 371–376 (1969).
13. K. Lashley, *in*: "The Neuropsychology of Lashley" (F. A. Beach, D. O. Hebb, C. T. Morgan, and H. W. Nissen, eds.), pp. 478–506, McGraw-Hill, New York (1960).
14. S. H. Barondes and H. D. Cohen, *Proc. Natl. Acad. Sci. U.S.* **61**, 923–929 (1968).
15. L. Squire, A. DiMele, H. D. Cohen, and S. H. Barondes, unpublished.

Chapter 28

Glycoproteins of the Brain of the Training Pigeon

Samuel Bogoch

Foundation for Research on the Nervous System
and Boston University School of Medicine
Boston, Massachusetts

INTRODUCTION

Recent interest in the molecular events which are associated with various brain dysfunctions has drawn attention to our fundamental ignorance of normal neurochemistry.[35] In much the same way, recent interest in the biochemistry of learning and memory has pointed up our ignorance of the biochemical differences between quite different states of activity of the nervous system, all occurring for varying periods of time in each day for each normal animal.

In the case of certain secretory glands different functional states of the secretory cells range from immediate post-depletion, through relative rest, to build-up of secretory material, to actual discharge of secretory material. All of these are associated with biochemical changes which are so gross as to be visible under the light microscope. Even in this case, while most of the cells act in concert, at least four phases can be distinguished.

In the case of the nervous system, rather than a collection of fairly uniform cells, a collection of subsystems of different phylogenetic ages operate together in the most complex structure in biology. Curiously, the neurochemical literature has rarely noted the fact that a nervous system asleep, one awake at rest, another awake but active, another awake and training but not learning, still another awake, training, and learning, may all need to be considered as very different biochemical states. Certainly, all of the previous data of this laboratory,[4] as well as that to be presented here, support the need for these distinctions.

The terms training, learning, and memory are used independently in our studies to indicate different levels of certainty about the nature of an operational laboratory event, and indicate different levels of complexity of concept. Each of these three is different from the general term "activity" in that a higher degree of ordered activity of the nervous system is implied. Each of the three is also different from the other two. Thus, when we state that an animal is "training," or, even more carefully, "in the training situation," we imply nothing about learning or memory except that these may or may not occur. When the term "learning" is used it requires evidence of a changed response to a stimulus, with elaboration of concomitant variables, as in several useful definitions recently cited.[4] And when the term "memory" is used it implies a much more comprehensive, indeed, higher heirarchical level of organizational change in the nervous system. Learning is a *process* which utilizes old memories and constructs new ones; memory is a *state* which contains learning in its past. Further, in learning and memory biochemical changes need to be considered for a variety of the at least partially discrete operations of sensory reception, transmission, encoding, storage, retrieval, association, and discharge, together with their separate supporting reactions.[4] We have proposed[4] that the brain mucoids are in some way involved in transmission functions relevant to training, learning, and memory. Some of the recent evidence relevant to this hypothesis will be considered here.

Regardless of their ultimate function, quantitative analytic data on the structure and localization of the three brain glycoproteins 10*B*I, 10*B*II, and 11*A* will be presented.

METHODS: EXTRACTION AND PURIFICATION OF BRAIN GLYCOPROTEINS

The methods developed in our laboratory over the past six years for the extraction and separation of brain glycoproteins,[2–6] now included in the "modified procedure,"[3,4] include the following steps, the rationale and evidence for which has been previously discussed, and all of which have been found to be critical for obtaining both the quantitative and the qualitative yield of brain glycoproteins herein described:

1. Homogenization: repeated 6–14 times; i.e., residue is exhaustively extracted with 0.005 *M* phosphate buffer, pH 7, as extractant until no further appreciable protein is obtained.
2. Centrifugation at high speed.
3. Concentration of soluble supernate by perevaportion.
4. Cellex–D (DEAE) chromatography with stepwise rather than gradient elution with solvents of decreasing pH. Nonelutable protein removed from column at end of run only, with Triton X-100.

5. Tube-to-tube monitoring in all instances of protein content by Folin–Lowry quantitative determination (optical density readings of effluants not adequate). Tube-to-tube quantitative determination of hexose, hexosamine, and neuraminic acid where detailed carbohydrate information required for preliminary separation.

6. Standard grouping of effluants by tube number to produce 13 major fractions, and with *A*, *B*, etc., subfractions, and I, II, etc. sub-subfractions, 16 fractions are currently used for brain-to-brain, regional, subcellular, and other comparisons.

7. The use of disc-gel electrophoresis and immunological data obtained through the preparation of specific antisera[29,30] to identify subcomponents of each of the above fractions.

8. Subcomponents further purified by further column chromatography as well as by other means, and their discreteness or heterogeneity tested by specific amino acid content and specific carbohydrate content (before and after sequential stepwise hydrolysis and identification and quantitation of each liberated sugar).

9. Finally, the functional state of the animal, i.e., sleep, rest, training, etc., is as carefully defined as possible in each instance.

By means of the above methods we are able to quantify and compare from whole brain, dissected areas, or subcellular fractions:

1. Total glycoprotein from brain (extracted by this method).

2. Individual reproducible groups of glycoproteins such as 10*B*I *versus* 10*B*II or 11*A*.

3. Individual glycoprotein species purified from any of the above groups.

Furthermore, all of the above information is on *intact* nonhydrolyzed glycoproteins. After quantitative hydrolysis the constituent carbohydrate groups can be traced back to their glycoprotein of origin, as distinct from other approaches to these substances.[36]

A single method of extraction may not provide all of the information which is required on the brain glycoproteins, so that it is important to understand clearly in addition to both the advantages and the limitations of any given method, the specific kind of information a given method can supply. Therefore, in contrast to the methods of our laboratory, it may be instructive to compare those of another laboratory:[36]

1. Extraction. Instead of starting with a mild nondenaturing buffer such as neutral phosphate, in this method brain tissue is extracted with chloroform–methanol.[36] This initial step has the advantage of removing

lipids, but unfortunately has the double disadvantage of (a) removing appreciable hexose- and hexosamine-rich glycoproteins (in addition to classical ganglioside, which is protein-bound),[9] and (b) denaturing some proteins which remain. That is, a large number of polysaccharides which are not classical gangliosides are removed by this procedure. In fact, in distinction from the brain gangliosides, these substances have a much higher percentage of hexosamine than of neuraminic acid. They form a heterogenous group of aminoglycolipids, and many of them are *bound to proteins*. Since all of them do not contain neuraminic acid, but do contain many of the other carbohydrates, it is clear that they are being discarded in the chloroform–methanol extract, and that the remaining water-soluble carbohydrate macromolecules represent only a part of the glycoproteins of the brain. Thus the use of the terms "soluble" or "insoluble" must refer to the solvent used and other parameters of the extraction in each instance.

2. Papain digestion. The entire "defatted residue" is subjected to the proteolytic action of papain not once, but twice, for 24 hr, in order to produce as complete hydrolysis as possible. It should be noted that from here on this method permits *only* the study of hydrolytic fragments. There is no opportunity to study discrete intact brain glycoproteins.

3. Treatment with cetylpyridinium chloride in an attempt to remove uronic acid-containing mucopolysaccarides. This method, previously used effectively by Wardi *et al.*,[32] also on hydrolytic split fragments of macromolecular carbohydrate from brain, has not yet been shown to be a quantitative means of exclusively separating uronic acid-containing from other carbohydrate fragments.

4. Excluding the protein-bound sugars initially discarded in the chloroform–methanol extract as lost, the papain–digested material cannot give total concentration of carbohydrate in brain. Indeed, since there is no dialysis step in the above procedure, even the inclusion of the chloroform–methanol-extracted carbohydrate will not permit the distinction between protein-bound and only lipid-bound carbohydrate, both of which can be nondialyzable, and neither of them from small-molecular-weight dialyzable, free, or bound carbohydrate.

5. The analysis of individual sugars in the split fragments, since it cannot be related back to the original intact glycoproteins in this method, can provide only a clue as to which split fragments are relatively rich in one or another sugar constitutent. However, this can be useful when the intact glycoproteins are studied for key constituents by which their microheterogeneity can be further resolved. Certainly nothing can be said after digestion about the molecular weight or any other physicochemical property of of the *pooled* glycoproteins of origin from which these fragments have been cleaved.

This rather detailed discussion of methodology is included because at the early stages of the study of any group of substances, such as the brain glycoproteins, much confusion will be avoided in the literature, and subsequent investigations appreciably aided, if the actual steps of any method are the subject of constant attention and short-cut terms such as "soluble glycoproteins" are avoided.

CELLULAR AND SUBCELLULAR FRACTIONATION

There is now considerable evidence that 10*B* is predominantly or exclusively glial glycoprotein.[2–6] There is some preliminary data on the subcellular localization of 10*B* and 11*A*, but this work has not progressed sufficiently to adequately distinguish fine-structure localization. Thus, for example, while 11*A* is prominent in synaptosome preparations,[2–6] it is uncertain whether it is part of pre- or post-synaptic membranes, vesicles, or associated mitochondria, or associated with glial fragments.

IMMUNOCHEMICAL DATA

Specific antisera have been prepared in rabbits to protein 10*B* of both normal human and Tay–Sachs' disease[30] human brain, as well as from normal pigeon brain.[29] Disc gel electrophoresis of the antigen always has shown it to contain between two and three distinct components, depending on the functional state of the brain. The immunochemical data correspond, in that one major and two minor bands are demonstrated by Ochterlony plate reactions. In pigeon brain, antisera to 10*B*I, 10*B*II, and 11*A* have been prepared.[29] Preliminary evidence shows all three to share at least one antigenic determinant. The nature and concentration of carbohydrate moieties glycosidically bound to these three pigeon-brain proteins also differs.[2–6] The correlation of these quantitative chemical observations on bound sugars with the physical properties on column chromatography and gel electrophoresis and with the immunochemical data mentioned above thus provides a four-dimensional analysis of discreteness or heterogeneity of each purified fraction.

DEVELOPMENTAL DATA

Group 2 brain proteins have been shown to increase in concentration temporarily in the training situation, but, in contrast to 10*B* and 11*A* changes, the change in 2 does not persist.[4] The development of these brain proteins is therefore of interest.

There is a suggestion of a "critical period" between 15 and 20 gestational weeks in which the protein of groups 1 and 2 increases markedly. In contrast, the stability of other groups during this period is noteworthy.[4] The change in groups 1 and 2 correlate in time with the massive appearance of new dendritic cell processes, and with the rapid growth of the cortical gray mantle, beginning in the insular area and spreading over both occipital and frontal poles of the cerebral cortex. Discrete specimens from the insular region as distinct from the poles, in a 20-week-old human brain, showed the insular area to contain 1.3 mg/g of groups 1 and 2, while the poles contained only 0.1 mg/g. This temporal relationship of concentration of specific proteins to nerve cell process growth has led to the use of the term dendritic proliferation protein for the relevant proteins of groups 1 and 2.

While groups 1 and 2 clearly represent a group of proteins, and there is as yet no direct evidence that they are actually located in the dendritic processes, the abrupt increase in the concentration of these proteins has a developmental phase specificity and is distinct from the minimal changes in total protein and other protein groups during this period.[4]

CHANGES IN THE PROTEIN MOIETIES OF 10B AND 11A DURING TRAINING: HYDROLYSIS METHODS

Our previous work has demonstrated that the brain glycoproteins may be considered in terms of various possible information processing roles, because of their heterogeneity,[1,3–5,7–9] location in membranes and at synapses,[3–5,10,33] development,[4] pathology,[4,9,11] recognition functions,[12–15] biosynthesis,[4,5] variation in different behavioral states,[16–22] and concentration in different training states.[2,4,5]

If glycoproteins serve recognition and contact functions in the nervous system, it may be possible to detect structural differences in the carbohydrate components of the brain glycoproteins as a function of nervous system activity in general, of training more specifically, and, eventually, specifically in relation to learning. We now report the comparison of the carbohydrate components of three brain glycoproteins, which have previously been shown to vary in the concentration of their protein components in the training as compared with the resting state. In passing, it may be noted that this new series of studies on training pigeons confirmed the previous observation[4] that, determined through their protein moieties, in training there is a decrease in the concentration of 10*B*, an increase in the concentration of 11*A*, and a consequent change in the ratio of 11*A* to 10*B*. The stepwise acidic hydrolytic cleavage and determination of the carbohydrate units of these three glycoproteins indicates considerable structural differences, especially in the composition of their most readily hydrolyzable carbohydrate units.

Male white Carneaux pigeons were trained in the standard operant-conditioning apparatus as previously described[4] in a nonstressful, non-aversive food-reward paradigm, first to peck each of two keys for food reward, and then to perform discriminations between two keys, performance on one only being rewarded. The programmed instruction and automaticity of recording of all responses are advantages in this learning system, as is the ability to state quantitatively at any given time in the training process the amount of the training task which has been accomplished. All pigeons had the same number of 1-hr training sessions, 13 for the preexperimental training, where both keys were equally reinforced, and 21 for the experimental sessions, where only one key was reinforced. The total training period extended over a three-month period with rest days interspersed irregularly. In the present experiment (*B* series), where a random distribution between right and left keys would be represented by a score of 50 (50% of pecks on each key), the "amount learned" could be expressed by the deviation from random, i.e., preference of the correct key over the incorrect key, expressed as deviation from 50. Thus, complete "learning" of the task would be represented by pecking only on the correct key (a score of 50).

A double-blind procedure was again maintained, whereby those performing the training of pigeons had no information of the chemical studies being undertaken, and *vice versa*. Pigeons were placed in rank order according to the degree to which the correct preference had been acquired,

TABLE I

Carbohydrate Constituents of Pigeon-Brain 10*B*I, 10*B*II, and 11*A*[a]

Hydrolyzed fraction	Hexose			Hexosamine			Neuraminic Acid		
	10*B*I	10*B*II	11*A*	10*B*I	10*B*II	11*A*	10*B*I	10*B*II	11*A*
Dialyzates									
0.1*N*	1.0	7.5	1.4	0	0	0	0.64	0.97	0
2*N*	6.8	5.7	6.6	14.6	16.2	3.3	—	—	—
4*N*	2.0	2.8	2.0	1.0	1.1	1.3	—	—	—
Nondialyzable	0.4	0.7	0.9	0.6	0.6	0	—	—	—
TOTAL	11.1	16.7	10.9	16.2	17.9	4.6	0.64	0.97	0

[a] As per cent of protein.

and their last day score, as described above, was compared with the absolute concentration of 10*B* and 11*A* glycoproteins, expressed in mg/g of whole brain.

The pigeons were sacrificed immediately after the last session by dipping the head in dry ice–acetone, and the brain extracted for glycoproteins by the "modified procedure," all as previously described.[4] In addition to the concurrent freezing of the brain at sacrifice, the repeated homogenizations of brain, high-speed centrifugation, concentration, stepwise elution from purified DEAE-cellulose (Cellex-D), and Folin–Lowry determination of proteins in each of 350 tubes of eluate are all important components of the method and all as previously described.[4] The total protein from each individual pigeon brain was thereby separated by combination of elution tubes into the 16 groups previously referred to.[4] 10*B*I and 10*B*II are separable by these methods into two quite sharp, discrete peaks, whereas 11*A*, occasionally a discrete sharp peak, is more often continuous with the last portion of the 10*B*II peak. These glycoproteins were collected separately from each brain, then combined, concentrated by perevaporation to 5–10 cc volume, and dialyzed exhaustively in the cold against 0.005 M phosphate buffer, pH 7.

Each glycoprotein was then subjected to stepwise acidic hydrolysis and dialysis under the following conditions: (1) 0.1*N* HCl hydrolysis, 80°C, 1 hr, followed by dialysis; (2) hydrolysis of the residue from the 0.1*N* HCl hydrolysis with 2*N* HCl hydrolysis, $4\frac{1}{2}$ hr at 100°C, followed by dialysis; (3) hydrolysis of the nondialyzable residue remaining after the 2*N* hydrolysis with 4*N* HCl, 100°C, $4\frac{1}{2}$ hr, followed by dialysis. This resulted in three dialyzates, representing the hydrolytic split products which became dialyzable after increasing acidic hydrolysis, and the final nondialyzable residue remaining after 4*N* HCl hydrolysis. The analysis of each of these fractions in terms of hexose, hexosamine, and neuraminic acid, expressed as per cent of protein in the original unhydrolyzed fraction, is shown in Table I. Protein was determined according to the Folin–Lowry method;[24] the hexose determinations were the orcinol–H_2SO_4 method of Sorensen and Haugaard[25] with 50% glactose:50% mannose as standard; the hexosamine determination was a modified Elson–Morgan method[26] with galactosamine as standard. Neuraminic acid was determined two ways, according to the thiobarbituric acid method[27] and by means of the Bial's orcinol reaction.[28] All determinations were repeated on all fractions between two and five times. On the 0.1*N* dialyzate there was close agreement between Bial's orcinol and thiobarbituric acid methods for neuraminic acid, the former reading slightly higher, as expected (for example, 0.69 and 0.59%, respectively, and 1.02 and 0.92%, respectively). The true values probably lie between those obtained with each method.

The data from a previous series (EPII and CP, Table XVI of Bogoch[4]) have been fully discussed elsewhere.[4] While the experimental conditions for the training of pigeons were different in that situation, and the pigeons were allowed to rest for three to eleven months after training had been completed before they were sacrified and the glycoproteins determined, the relationship of 10*B* to 11*A* in untrained pigeons (CP) as compared to those who have been trained (EPII) was demonstrated: the concentration of 10*B* was decreased, of 11*A* increased, and the ratio of 11*A* to 10*B* increased in the EPII group. The present (*B*) series again showed the relationship of the concentration of 10*B* and 11*A*, in that there was an increase in 11*A* concentration, a decrease in 10*B* concentration, and an increase in the ratio of 11*A* to 10*B*. These changes accompanied training which was comparable in many ways but not identical to that observed in the previous experiment.

THE CARBOHYDRATE COMPOSITION OF PIGEON 10BI, 10BII AND 11A

The data in Table I demonstrate the composition of the carbohydrate units determined in isolated 10*B*I, 10*B*II and 11*A* from series *B*. Some comparative indication is given of the ease or difficulty with which these different constituents can be hydrolytically released. With 0.1*N* HCl hydrolysis at 80°C, as expected, the neuraminic acid which is present is liberated, together with some hexose components, but no hexosamine. With 2*N* HCl more of the hexose components and the bulk of hexosamine components are liberated. With 4*N* HCl most of the remainder of the hexose and hexosamine components are liberated. 10*B*I, 10*B*II, and 11*A* are seen to differ in terms of the total concentration of hexose: 10*B*I and 11*A* being close, whereas 10*B*II contains some 50% more hexose than the other two. With 0.1*N* HCl hydrolysis the readily hydrolyzable carbohydrate units are released. 10*B*II has 50% more neuraminic acid than 10 *B*I, but 11*A* has no demonstrable neuraminic acid. In addition, 10*B*II has four times the amount of hexose in this (0.1*N*) fraction than is derived from 10*B*I, and some five times the amount seen in 11*A*. The total amount of hexose in the combined, less readily hydrolyzed fractions, that is, those released by 2*N* and 4*N* HCl hydrolysis, together with the small amount in the nondialyzable remaining residue, are seen to be very similar in all three glycoproteins, 9.2%, 9.2%, and 9.5%, respectively, for 10*B*I, 10*B*II and 11*A*. There is slightly more hexosamine in 10*B*II than in 10*B*I, but much less hexosamine in 11*A* than in the 10*B* glycoproteins.

10*B*I and 10*B*II can therefore be distinguished from each other by the above quantitative analytical data as two discrete glycoproteins. Furthermore, each of these is distinct from 11*A*, another glycoprotein with considerably less hexosamine and no neuraminic acid. The identification of the individual sugars in each of these glycoproteins has been partially reported[4] and in hydrolytic fractions are now in progress in each of these. Antisera prepared in rabbits[29] separately to 10*B*I and 10*B*II, and to 11*A*, suggest the presence of one shared antigenic component in all three, and the possibility of additional individual antigenic sites for each. Antisera prepared earlier in this laboratory[30] to Tay–Sachs' disease whole 10*B* fraction have given at high dilution immunofluorescent straining of astrocytic glia only.[31]

Both the protein data and the carbohydrate data of this series of experiments again support the conclusion that these glycoproteins are involved in biochemical reactions occurring during the training process.[2–6] With the definition of these distinguishing carbohydrate compositions as shown in Table I added to the chromatographic separation of these substances, and with the additional evidence of 10*B* decrease and 11*A* increase accompanying the training procedure, the glycoproteins involved have been named Trainein I and II (10*B*I and 10*B*II, respectively) and Learnein (11*A*). While the change in concentration of Trainein and Learnein is further evidence that these glycoproteins are in some way related to training and learning processes, their exact function, and whether of a primary or secondary nature,[4] remain to be elucidated. The biosynthetic data for the incorporation of ^{14}C-glucose into these glycoproteins[3–6] support an extremely rapid and sizable synthetic potential commensurate with their proposed function. The fact that both Trainein and Learnein may contain even greater amounts of glycosidically bound carbohydrate constituents during the earlier phases (indeed minutes) of the training process (see EPI in Bogoch[2]) taken together with the possibility of microheterogeneity within each of these glycoproteins, points to considerable remaining complexity in the resolution of their exact functions.

Nonetheless, an important point has now been clarified. If the carbohydrate constituents of each of these glycoproteins showed no appreciable structural differences, there would be reason to question the validity of the recognition aspects of the mucoid–memory hypothesis.[2] With the shift from Trainein to Learnein supported, and the present data indicating marked distinctions between them in terms of the structure of their carbohydrate components, the possibility that carbohydrate specificity underlies specificity of transmission at junctures between cells is further supported. The recent electron-microscope evidence by Droz and associates[10,33] that glycoproteins are heavily concentrated at synaptic junctures is certainly relevant to our earlier subcellular fractionation studies.[4,5] The demonstration of Kanfer and Richards[34] that puromycin blocks ganglioside

TABLE II

Amino Acid Composition of Pigeon-Brain Proteins 10*B*I and 10*B*II[a]

Amino acid	10*B*I	10*B*II
Glutamic	129.5	132.9
Aspartic	103.2	98.7
Leucine	87.7	85.2
Alanine	87.0	92.9
Glycine	78.9	76.8
Lysine	72.2	80.7
Serine	66.1	55.5
Valine	61.4	63.2
Threonine	52.6	51.0
Arginine	48.6	59.4
Isoleucine	43.8	43.2
Proline	41.8	43.2
Phenylalanine	33.0	31.6
Tyrosine	26.3	22.6
Histidine	25.0	24.5
Cysteic	23.6	25.8
Methionine (+cystine)	19.6	12.9

[a] Residues per 1000 residues.

synthesis is also relevant. Thus it remains plausible that a glycoprotein "sign-post" mechanism[(2)] represents the (or one) chemical basis of the establishment of the cluster of specific circuits in the nervous system which could constitute a memory record. Some ways in which both genetic and experiential influences might operate under these conditions, together with additional theoretical considerations, have been recently discussed.[(4)]

AMINO ACID COMPOSITION OF 10*B*

Following 21-hr hydrolysis with 6*N* HCl and at 105°C the amino acid composition of 10*B* given as residues per 1000 residues is as shown in Table II. These figures can only be considered a first approximation in view of the lack of definitive data on the true molecular weight and the elimination of heterogeneity.

SUMMARY OF PROPERTIES OF 10*B*

Brain protein 10*B* was first isolated in 1964 from human brain glial tumors,[(37)] where it was found to be 6–9 times its concentration in normal human brain specimens.[4] It was then shown to be increased sevenfold in Tay–Sachs' disease brain (sphingolipidosis)[11] and referred to as Tay–Sachs' disease protein. It has also been isolated from tissue culture of glial tumor cells, from membranous "ghosts" of these lysed cells,[4] from a brain cell tumor of the ependymoma type grown subcutaneously in mouse,[4] from normal cat hippocampus and caudate nuclei,[4] frozen normal mouse brain, from normal pigeon brain,[4] and from microsomal, mitochondrial, and synaptosome subcellular fractions of guinea pig brain.[4] Because of its increased concentration in glial tumors, in Tay–Sachs' disease, where there is a marked gliosis, and in isolated tissue culture glial cells and their membranes, it was assumed to be predominantly or totally a glial constituent.[4] Antiserum to 10*B* from each of normal human brain, glial tumor, both human and mouse, mouse brain, and pigeon brain have revealed two antigenic components in all cases other than human, in which there appears to be only one. These do not cross-react with material isolated in the same way from liver, kidney, and blood. On immunofluorescence with suitable controls and at high dilution of Tay–Sachs' 10*B* antiserum, only the astrocytic glia fluoresce.[31] There is no cross-reaction with *S*-100 protein. The high carbohydrate content of 10*B* and the amino acid content further distinguish 10*B* from *S*-100. ^{14}C-Glucose is rapidly (within 10–20 min) and markedly incorporated into the carbohydrates of 10*B* during training.[4] Changes in the concentration of both the amino acid and the carbohydrate portion of 10*B* in pigeon brain during training are of interest. In pigeon brain 10*B* is separable into two distinct components, Trainein I and Trainein II.

REFERENCES

1. S. Bogoch, *Nature* **190,** 153 (1960).
2. S. Bogoch, *Neurosci. Res. Prog. Bull.* **3,** 38 (1965).
3. A. Quamina and S. Bogoch, *in*: "Proc. XIIIth Colloq., Protides of Biological Fluids, Brugge," (H. Peeters, ed.), p. 211, Amsterdam, Elsevier (1966).
4. S. Bogoch, "The Biochemistry of Memory: With an Inquiry into the Function of the Brain Mucoids," Oxford University Press, New York (1968).
5. S. Bogoch, P. C. Belval, W. H. Sweet, W. Sacks, and G. Korsh, *in*: "Proc. XVth Collog., Brugge, Protides of Biological Fluids," (H. Peeters, ed.), p. 129, Amsterdam, Elsevier (1968).
6. S. Bogoch, Nervous system proteins, *in*: "Handbook of Neurochemistry," Vol. I (A. Lajtha, ed.), Plenum Press, New York, 1969.
7. S. Bogoch, *J. Biol. Chem.* **235,** 16 (1960).
8. S. Bogoch, *Nature* **185,** 392 (1960).
9. S. Bogoch, "Cerebral Sphingolipidoses," (S. M. Aronson and B. W. Volk, eds.), p. 249, Academic Press, New York (1962).
10. A. Rambourg and C. P. Leblond, *J. Cell Biol.* **32,** 41 (1967).

11. S. Bogoch and P. C. Belval, *in*: "Inborn Disorders of Sphingolipid Metabolism," (S. M. Aronson and B. W. Volk, eds.), p. 273, Pergamon Press, New York (1966).
12. S. Bogoch, *Virology* **1,** 458 (1957).
13. S. Bogoch, P. Lynch, and A. S. Levine, *Virology* **7,** 161 (1959).
14. S. Bogoch and E. S. Bogoch, *Nature* **183,** 53 (1959).
15. S. Bogoch, M. K. Paasonen, and U. Trendelenburg, *Brit. J. Pharmacol.* **18,** 325 (1962).
16. S. Bogoch, *Am. J. Psychiat.* **114,** 122 (1957).
17. S. Bogoch, *AMA Arch. Neurol. Psychiat.* **80,** 221 (1958).
18. S. Bogoch, *Am. J. Psychiat.* **114,** 1028 (1958).
19. S. Bogoch, K. T. Dussik, and P. Conran, *New Eng. J. Med.* **264,** 521 (1961).
20. S. Bogoch, K. T. Dussik, C. Fender, and P. Conran, *Am. J. Psychiat.* **117,** 409 (1960).
21. S. Bogoch, *in*: "Biological Treatment of Mental Illness," (M. Rinkel, ed.), p. 406, Farrar, Straus and Giroux, New York (1966).
22. R. Campbell, S. Bogoch, N. J. Scolaro, and P. C. Belval, *Am. J. Psychiat.* **123,** 952 (1967).
23. H. Nikaido, *in*: "The Specificity of Cell Surfaces," (B. D. Davis and L. Warren, eds.), p. 3, Prentice-Hall, Englewood Cliffs, New Jersey (1967).
24. O. H. Lowry, N. J. Rosebrough, L. Farr, and R. J. Randall, *J. Biol. Chem.* **193,** 265 (1951).
25. M. Sorensen and G. Haugaard, *Biochem. Z.* **206,** 247 (1933).
26. L. A. Elson and W. T. Morgan, *J. Biochem.* **28,** 988 (1934); as modified in S. Bogoch, *J. Biochem.* **68,** 319 (1958).
27. L. Warren, *J. Biol. Chem.* **234,** 1971 (1959).
28. S. Bogoch, *Biochem. J.* **68,** 319 (1958).
29. S. Bogoch and B. R. Das, unpublished data.
30. S. Bogoch and P. C. Rajam, unpublished data.
31. P. F. Benda, K. Mori, and W. H. Sweet, personal communication.
32. A. H. Wardi, W. S. Allen, D. L. Turner, and Z. Stary, *Arch. Biochem. Biophys.* **117,** 44 (1966).
33. B. Droz and H. L. Koenig, this volume, Chapter 4.
34. J. N. Kanfer and R. L. Richards, *J. Neurochem.* **14,** 513 (1967).
35. S. Bogoch (ed.), "Future of the Brain Sciences," Plenum Press, New York (1969).
36. E. Brungrabber, this volume, Chapter 19.
37. S. Bogoch, *in*: "Proc. 8th International Congress of Neurological Sciences, Vienna," Excerpta Medica, Amsterdam (1965).

DISCUSSION

SHNEOUR: I think that learning in macromolecules raises a problem which may or may be not genuine, but is worth considering. The work of the Bell Telephone Laboratories and Ackerman and others has shown that the total rate at which information can be acquired through the five senses of a human being approaches 33 bits/sec. Now, if we assume that this is a low number and use a higher figure for simplicity, and set 10^2 or 100 bits/sec, this enables us to find a total number of bits that a person can acquire during an average life-time. If we now assume a redundancy factor, according to the work of Lashley, each bit is recorded 100 times, and it makes no difference whether an error of one order of magnitude is made. If we now say that each bit is recorded in a molecule of 1000 molecular weight, relatively large, this calculation gives us a total of 10^{12}. If we divide by 10^{23}, the order of the Avogadro number, this gives us 10^{-11}. If we assume an average molecular weight of 1000, this means that the total mass of all the information-storing systems has a total weight in the human being, assuming that the calculations are right, of 10^{-8} g. Even if I have made a mistake of three orders of magnitude or more, this calculation points out the possibility that the macromolecules that are involved in memory storage may be extremely small in the total mass. In addition, if they are distributed among the 10^{10} or 10^{11} neurons, I think it raises a very substantive issue of what we are looking at.

BOGOCH: If the storage of information in the brain is thought of as occurring in bits, and if each bit has a molecular structure for a counterpart, then Dr. Shneour's calculations are indeed of interest for the storage aspects of information handling in the nervous system. However, as I have pointed out elsewhere,(4) the storage aspect is but one of many components of the total number of processes which have behavioral and physiological identity, and which must be considered to have biochemical equivalents as well. Thus, the required processes for the handling of experiential information in the nervous system include mechanisms for sensory reception, for transduction (the conversion of primary sensory energy modes of pressure, heat, light, etc., to the electrochemical equivalents of nerve cell transmission), for coding for transmission, for storage (are the mechanisms for short- and long-term storage the same?), for retrieval or nonretrieval, for association, for restorage and reretrieval, for abstraction, and for discharge in cognitive, neuroendocrine, or motor pathways. All of the above processes, though perhaps overlapping to some degree, probably require quite discrete molecular mechanisms for performance. Finally, all of the above processes may be thought of as "primary mechanisms," each of which requires a host of secondary supportive chemical reactions for realization. For this is, after all, a biological rather than a nonbiological mechanism. Some of the confusion which has arisen is semantic or conceptual in nature, as well discussed by Von Foerster (see Bogoch(35)). He also points out the difference between the bit of information stored and the complex processes necessary for its retrieval and acquisition of meaning. The calculations which we will need to make will, I believe require a much more detailed knowledge of all of the subprocesses involved in the molecular processing of information in the brain.

FOLCH: One point bothers me: you get much more proteins, about 20% more from the trained animals than from the naive animals. However, at the same time you insist, quite properly, that you have to homogenize over and over and over again. Are you homogenizing the same number of times in each case? Are you sure that that in one case you are not extracting proteins that in the other case may remain behind? Have you studied the residue that is left behind to see with regard to those two very interesting fractions that you get out, whether you are dealing with a difference in extractability?

BOGOCH: The increased yield of protein reflects extractable protein. The increased yield on extraction does not necessarily mean increased synthesis, but may only reflect some physical change with training which influences extractability. For each pigeon brain we extract only four times. We have found that we get essentially no more after four extractions. The more complex the brain, the more often we have to homogenize. A human adult brain needs 9–13 times to get optimal yield; fetal human brain only six. The number of homogenizations appear to directly relate to the complexity of the brain structure.

FOLCH: Do you have a protein pattern in the successive extractions? Is it there throughout?

BOGOCH: Yes, $10B$ and $11A$ are there, and throughout.

FOLCH: But you have not examined the insoluble residue.

BOGOCH: We have not as yet studied the insoluble residue. All of these data are only on some 60% of the total brain protein. We have no idea of what is occurring in the other 40%.

MARKS: Do you think it would be worthwhile to try the effect of puromycin on the appearance of your $10B$ and $11A$?

BOGOCH: We have held back from the use of any pharmacological agent. We wanted to get solid base lines for the past five years on the normal concentrations at rest, with activity, and with training. The puromycin effect on memory storage in other species has certainly been well studied, and we look forward to examining the effect of puromycin on the concentration of $10B$ and $11A$.

SHOOTER: I think you get much more reproducible results on extracting water-soluble proteins if you use sonic disruption rather than repeated homogenization. I think what repeated homogenizations is doing is simply disrupting small amounts of membrane.

BOGOCH: Your comment that the probable basis of the increased yield on repeated homogenizations is the disruption of membranes is in agreement with our own conclusion, which we reached several years ago on the basis of getting even higher yields when subcellular fractions were first prepared, then separately extracted.[3] The total yields which you report with sonication are close to those we obtain with repeated homogenization. We have not yet tried sonication, but it would be necessary to show with sonication that not only the total yield, but also the yield of each protein fraction, such as $10B$ and $11A$, is comparable to that which we obtain by repeated homogenization. I understand that denaturation is not minimal with sonication, although I expect that there is some denaturation in whichever method is used.

UNGAR: I was struck by the fact of the temporary increase in $10B$ concentration followed by a decrease. Now, in transfer experiments it is well established that you can only transfer recently acquired information. So it seems that there is a rapid synthesis of the material and then it decays, and this agrees with our experiments. What do you think the carbohydrate part plays in the actual coding in these molecules?

BOGOCH: I have dealt with this question at length in a recent monograph,[4] so I wouldn't like to say much about the theoretical implications of a change in concentration of carbohydrate. The thing that attracted me to the idea that the macromolecular carbohydrates might have recognition functions in nervous system is the fact that they do have recognition functions in bacteria, and this I would like to underline. There is one way to consider recognition function which might have an application to coding in the nervous system. That is, rather than thinking of coding of bits of information, if neuron A will or will not fire B depending on the degree to which neuron B contacts it and the degree of contact is a function of the carbohydrate end-groups at the synapse, then we have a "sign-post mechanism" for the routing of impulses and the development of circuitry. This theory has been more fully developed both in terms of genetic and experimental aspects in the above-mentioned monograph.[4]

Chapter 29

Chemical Transfer of Learned Information*

Georges Ungar

Baylor University College of Medicine
Houston, Texas

INTRODUCTION

The inclusion of problems related to memory and learning in a volume devoted to protein metabolism in the nervous system is a sign of progress. It shows a readiness to consider the possibility that information may be handled in the nervous system on a chemical principle. This possibility is still meeting considerable, even if steadily declining, opposition, and is in need of some spectacular breakthrough to gain full acceptance.

The hypothesis favoring a molecular mechanism in learning and memory reposes on three lines of evidence: (1) existence of chemical changes correlated with learning, (2) impairment of memory fixation by inhibitors of RNA and protein synthesis, and (3) chemical transfer of learned behavior.

Since the first two types of evidence are covered by other contributors to this volume, I shall deal here only with the experiments of chemical transfer which should represent the most definite evidence for a molecular mechanism in learning and memory. If information can be transferred from one individual to another by means of a brain extract, the information must be stored in terms of molecular structure. The transfer experiments are based on the same principle as the bioassay methods which gave the first demonstration of the chemical mediation of synaptic transmission. One has to realize, of course, that transfer of a behavioral pattern is a considerably more complex situation than that of the vagal effect on the heart, but in both cases a neural effect produced in a donor preparation is transferred to a receptor preparation by purely chemical means.

The first transfer experiments were done in planarian worms in 1962.[1]

* Supported by Grant No. MH-13361 from the United States Public Health Service.

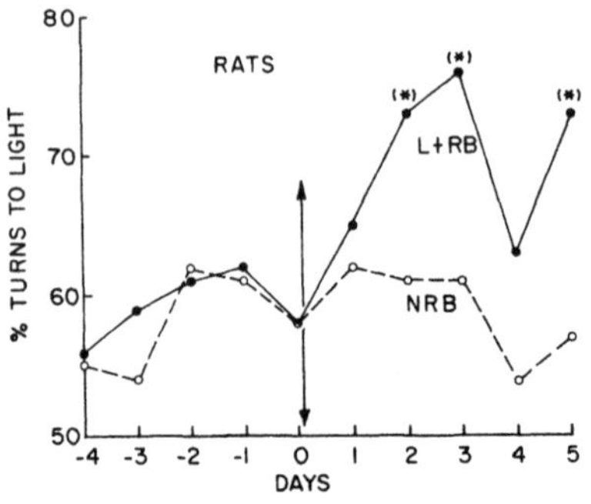

Fig. 1. Transfer of light-dark discrimination to unreinforced recipients. Abscissa: days before and after injection of brain extracts (equivalent to 1 g of rat brain) from donors trained to escape into the lighted arm of the maze (L+RB) and untrained rats (NRB). The arrow indicates the time of injection. Asterisks show statistically significant differences between experimental and control values (U-test). Each point represents the mean of eight animals.

Although they were replicated in a number of laboratories, the results remained controversial and are still not universally accepted. The first mammalian transfer experiments appeared in 1965,[2–5] but several other laboratories failed to confirm them.[6–10] In the last three years, however, at least nine other laboratories published positive results.[11–19] Further work was also published by three of the original groups, and I should like to give a few examples of the experiments done in my laboratory.

EXAMPLES OF TRANSFER EXPERIMENTS

I was encouraged to attempt transfer of learned behavior by the results of experiments in which we were able to induce morphine tolerance in animals without previous experience of the drug by means of material extracted from brain of tolerant animals.[20] Drug tolerance can be conceived of as a form of learning, since it develops as a result of experience. Our first paradigm was habituation, which is a sort of tolerance acquired to physical stimuli. Rats were submitted to a sound which originally elicited a startle response. By repeating the stimulus, the donors' responses were decreased to 10% or less. When extracts of brain taken from these animals were injected into naive recipients, these responded significantly less (below 50%) to the sound than those that were given brain from untrained rats.[4] These experiments were followed by successful attempts at transferring conditioned avoidance and light–dark discrimination. The interpretation of these experiments was, however, somewhat obscured by the fact that the recipients were reinforced in the same way as the donors, so that the effect of the extracts might be construed as a simple acceleration of the learning process.[21]

In subsequent experiments, therefore, the recipients were never reinforced, so that their behavior during successive tests could have been influenced only by whatever information was supplied to them in the brain

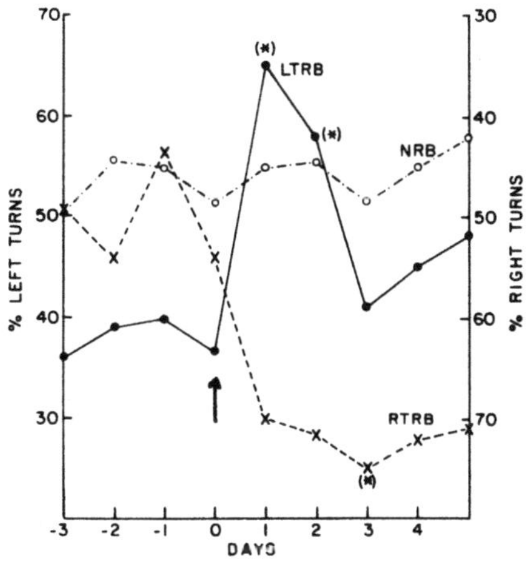

Fig. 2. Transfer of spatial discrimination to unreinforced recipients. Abscissa: days before and after injection of brain extracts (equivalent to 0.75 g of rat brain) from donors trained to run into the left arm (LTRB), into the right arm (RTRB) of the maze, and untrained rats (NRB). Arrow shows the time of injection. Asterisks indicate statistically significant differences from preinjection means (U-test). Each point represents the mean of six animals for LTRB and RTRB and 10 animals for NRB.[23]

extracts. In the first series of experiments donors were trained in a Y-maze to escape shock by running into the lighted arm. The light was switched on randomly in one or the other arm of the maze. Extracts of brain taken from the donors trained to a 90% criterion or, in most experiments, trained several days beyond criterion, were injected to recipients. These had been tested for a few days before injection for their behavior in the maze so that their pre- and post-injection runs could be compared. One such experiment is summarized in Fig. 1, which shows that after injection the recipients of trained brain ran more often into the lighted arm than before, while the animals treated with untrained brain did not change their behavior. In my laboratory these experiments showed a high degree of reproducibility, and we were able to determine the optimal conditions of donor training and dosage.[22,23]

Another paradigm based on spatial discrimination gave some excellent results, but its reproducibility was poor. In these experiments the donors were trained to escape in the left or right arm of the maze. In the successful experiments the recipients of left-trained brain increased their turns to the left and those which received right-trained brain ran more often to the right (Fig. 2). The results were, however, strongly influenced by the innate or acquired right or left biases of the donors and the recipients. Some of the factors of the transfer were analyzed,[22,23] with the general conclusion that the most effective message transmitted to the recipients was the command to overcome their bias.

The largest number of experiments was done with a passive avoidance paradigm first used successfully for chemical transfer by Gay and Raphelson.[16] This situation is based on the innate preference of rodents for dark enclosures. The donors were trained to reverse this preference by

TABLE I

Frequency Distribution of Extracts According to the Time Spent in the Dark Box (DBT) by the Animals Treated with Them[a]

DBT (sec)	Experimental		Control	
	N	%	*N*	%
21–30	3	3.6	0	0
31–40	14	16.7	0	0
41–50	13	15.5	0	0
51–60	15	17.8	0	0
61–70	11	13.1	1	4.5
71–80	15	17.8	1	4.5
81–90	6	7.2	0	0
91–100	5	5.9	1	4.5
101–110	2	2.4	4	18.3
111–120	0	0	5	22.9
121–130	0	0	3	13.5
131–140	0	0	3	13.5
141–150	0	0	4	18.3

[a] Each extract represents a pool of brains taken from groups of identically trained recipients, varying in size from 20 to 150 rats. Each extract was tested after intraperitoneal injection into 6–8 mice. DBT is the mean value obtained in each group of recipients during three post-injection days. *N*: number of extracts with DBT falling within the range indicated.

being given an electric shock when they took refuge in the dark. They were subjected to this treatment for five periods of 5 sec every day for at least five days. Recipients which had been given extracts of brain taken from these donors showed avoidance of the dark. Recipients of control brains taken from untrained or pseudotrained donors continued to show dark preference.[24] Up to date 84 extracts taken from a total of 2810 rats were tested in 638 recipient mice, and the results are shown in Table 1. These results were obtained with a uniform dose of the equivalent of 1 g of donor brain per mouse. The effect was shown to be proportional to the dose in the range of 0.5–1.5 g; the latter dose giving usually a maximal effect. The result also depended on the length of training given to the donors; the effect increased up to about eight days, but decreased when training was further prolonged.

Another passive avoidance paradigm used more recently is based on the fact that rodents placed on a narrow platform tend to step down onto the surrounding wider area. This preference can be abolished if the animals receive an electric shock when they step down. Donors were trained for step-down avoidance on a schedule similar to the one used in dark avoid-

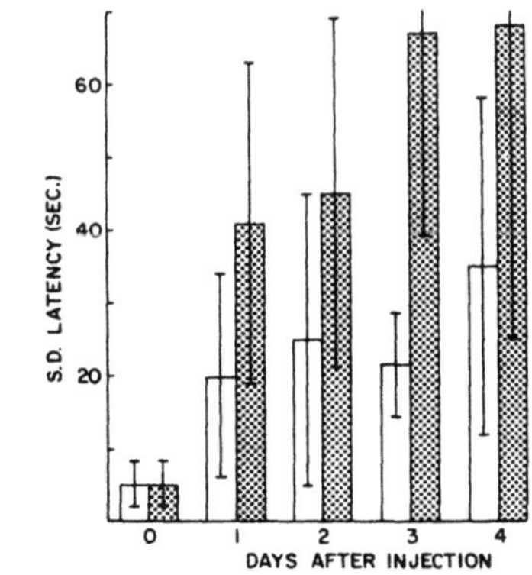

Fig. 3. Transfer of step-down avoidance. Mean latency of step-down (±S.D.) in recipient mice injected with brain extracts from trained donor rats (dotted columns) and from untrained rats (blank columns). Preinjection values (0) and mean latencies at four consecutive post-injection days. The difference between experimental and control groups was significant (t-test) on day 3 and with the pooled values. Means of two measurements every day in 12 animals in each group.

ance. The recipients injected with extract of brain taken from these donors were tested for latency of step-down. The results of one of the experiments are summarized in Fig. 3. In most recent experiments we recorded not only the latency of step-down, but also the total time spent on the platform.

The foregoing are only a few examples selected among the experiments done in my laboratory and do not do justice to the variety of behavioral situations which have been successfully transferred. Many of them involved positive reinforcement[2,3,5,11,14,17,19] in animals deprived of food or water, and their complexity went from simple avoidance[18] to complex instrumental conditioning.[11–14] Recently, Dr. Fjerdingstad in my laboratory was able to transfer the tendency to press alternately the right or left levers to obtain reward. This is probably the most complex paradigm transferred up to now.

REPRODUCIBILITY AND SPECIFICITY

Unsuccessful attempts at replicating[6–10] one of the early experiments[5] created some doubts about the reality of the transfer phenomenon. Now, three years later, in the light of the positive results published in this interval it is obvious that transfer experiments are too complex and involve too many variables to be dismissed by doing one or two experiments with a few leftover animals. After having surveyed all the published work, I was able to point to some of the critical factors,[25] but there are certainly others which remain to be tracked down.

The first important factors concern the training of the donors. Irrespective of how fast they learn the task to be transferred, the training should be continued for a period of 6–12 days. Donors should be able to learn the task to a criterion of 90%, and those that cannot reach this criterion within a reasonable interval should not be used.

The selection of suitable recipients represents another set of factors. Before injecting them with the brain extracts one must ascertain that the species, the strain, and the age group used is capable of learning the task to be transferred by the usual training methods. The recipients should also be screened individually so as to eliminate those subjects that do not respond normally. For example, in the transfer of dark avoidance, the recipients should, before injection, show a distinct preference for the dark.

The dose of brain extract to be administered is one of the most commonly neglected factors. As a rule, amounts equivalent to two to four recipient brains are necessary for significant results. There are, however, great variations from one situation to the other, so that the optimum dose has to be determined by trial and error.

The schedule of testing has also proved to be a critical factor. In most of the negative experiments the recipients were tested only once or twice at arbitrarily chosen intervals after injection. Since the optimum interval seems to vary with each situation, it is essential that the recipients be tested repeatedly for five or six days after administration of the brain extract.

When a new paradigm is explored all the conditions should be varied until the best procedure is found. This is, at least, the procedure recommended for those who want to obtain positive results. There is an infinite number of ways for disproving a hypothesis, but only a very few can produce the positive evidence.

This leads us to the interpretation of the results. In at least one of the series of experiments which were considered negative[9] most of the results were in the direction anticipated by the transfer hypothesis, although they fell short of the usual criteria of statistical significance. In a case like this the reasonable attitude is to try to improve the conditions of the experiment rather than to reach hasty negative conclusions.

As seen above (Table I), even under the best conditions available at present one cannot expect 100% positive and significant results in the transfer experiments. It is by pooling the results of several experiments and computing the probability of their being due to chance that one can test the reality of the findings. Table II gives a summary of all the experiments done in my laboratory. Even with the inclusion of the completely negative results, the overall probability of the phenomena being due to chance is negligible.

If we add to this the other published results, there can hardly be any doubt that *administration of brain extracts taken from trained animals can*

TABLE II

Behavioral Changes in Animals Injected with Extract of Brain Taken from Trained (TRB) and Untrained (NRB) Donor Rats

Training	Number of recipients[a]				P^c
	TRB		NRB[b]		
	+	0 or −	+	0 or −	
Habituation	41	1	0	22	<0.001
Conditioned avoidance	10	5	2	11	<0.01
Left or right escape	110	67	2	28	<0.001
Escape to light					
Reinforced	15	9	5	15	<0.02
Unreinforced	80	12	2	38	<0.001
Passive avoidance	451	187	48	156	<0.001
TOTAL	697	281	50	270	<0.001
%	71	29	18	82	—
N	978		329		—

[a] +: Change in the direction expected from the training of donors. 0: No change; −: Change in opposite direction.

[b] In control animals the direction of the change was counted as in the corresponding experimental animals.

[c] By the χ^2 method.

modify the behavior of the recipients in the direction expected from the training of the donors. This conclusion, however, leaves open the problem of the specificity of the behavioral changes induced by the brain extracts.

A first assumption was that the brain of trained animals contains some factor that has a general stimulating effect on learning and that this factor is absent from untrained brain. This is a legitimate objection to those experiments in which the recipients are subjected to training with reinforcement and where the transfer is assessed in terms of learning rates. In most of the recent transfer experiments, including ours, however, the recipients are not reinforced, and since they do not learn anything during testing, the results cannot be due to stimulation of learning.

Another objection was that the brain extracts transfer the stress which is always associated with learning, especially under negative reinforcement. This objection was particularly applicable to our passive avoidance experiments, and we tried to answer it. We submitted control donors to the same amount of electric shock, following the same schedule as the trained donors, but either in the lighted box or entirely outside the context of the

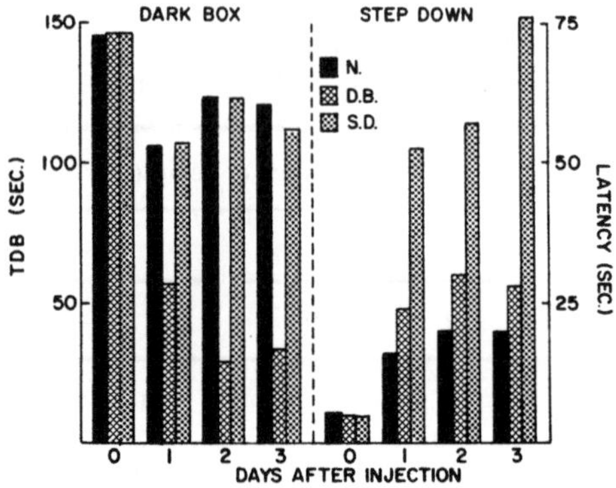

Fig. 4. Failure to produce cross-transfer between dark avoidance and step-down avoidance. At left: time spent in dark box (TDB) by recipients injected with brain extracts from untrained rats (N), from rats trained to avoid the dark box (D.B.), and from rats trained to avoid step-down from platform (S.D.). At right: latency of step-down in group in three identically treated groups of recipients. Each group consisted of six mice.

avoidance situation. In all cases the recipients of brain taken from these "pseudotrained" donors behaved like the controls when they were tested for dark avoidance.

The most reliable evidence for specificity was given by the cross-transfer experiments. To test the specificity of the transfer of habituation,[21] we trained two groups of donors; one group was habituated to a sound, the other to an air-puff. Brain extracts from each group were administered to two groups of recipients. One of these two groups was tested for sound habituation, the other for air-puff habituation. It was found that the recipients of sound-habituated brain were habituated only to sound but not to air-puff and *vice versa*.

Cross-transfer was also attempted between the two passive avoidance paradigms mentioned above: dark avoidance and step-down avoidance. Figure 4 shows that the recipients of brain from donors trained to avoid the dark did not differ significantly from the controls in the latency of step-down. Conversely, the recipients of step-down avoidance brain failed to show avoidance of the dark.

To find out how far this specificity goes would require experiments with sound or light discrimination using different frequencies or intensities.

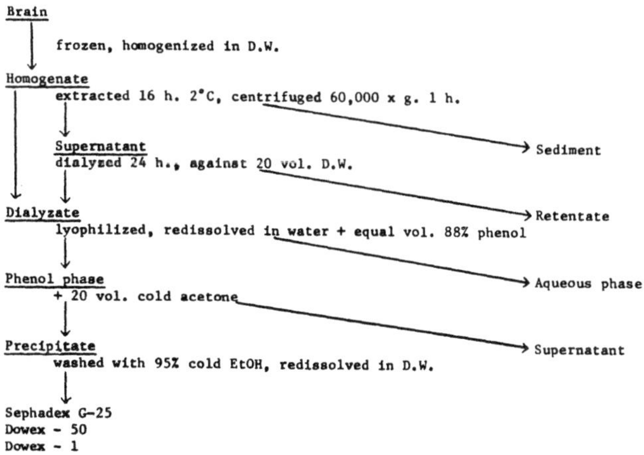

Fig. 5. Outline of the purification procedure for dialyzable transfer materials extracted from brain.

CHEMISTRY OF THE TRANSFER FACTORS

In two of the four original publications[2,4] the initial experiments were done with crude brain extracts and, at least in my laboratory, subsequent purification was guided by assays of activity. Figure 5 shows the main steps by which some of the transfer factors were partially purified. They were all dialyzable, with the exception of the factor which transfers dark–light discrimination in the Y-maze. It is important to note that throughout the procedures the brain preparations were kept either frozen or at ice-bath temperatures.

The dialyzable factors were partitioned between water and phenol, and the activity was always found in the phenol phase, suggesting that the active material is not a nucleic acid sequence, but a peptide. This was confirmed by the fact that treatment with ribonuclease was without effect on the activity of the material. Proteases, on the other hand, seem to destroy the behavioral effect of the extract. The factor responsible for the transfer of sound habituation is destroyed by chymotrypsin, the material which transfers dark avoidance is inactivated by trypsin, and the transfer factor of step-down avoidance is hydrolyzed by both enzymes.

These differences in the chemical properties of the transfer factors give further support to the specificity of the phenomenon. It is evident that if the brain extracts produce their behavioral effects by imparting specific information to the recipients, each of the effects must be produced by a distinct molecular entity. The information content of even comparatively short amino acid sequences is more than enough to account for the coding of memory.

Gel-filtration on Sephadex G-25 indicates a chain length of 8–12 amino acid residues for all three factors. Two of them were found to behave on ion-exchange resins as basic substances, the third one has not yet been investigated. The transfer factor of dark avoidance, whose purification is the most advanced, has to be administered at the dose of 30 μg per mouse for a discernible effect. Judging by the potency of some of the known pharmacologically active peptides, our material still requires at least a thousand-fold purification.

Rosenblatt *et al.*[(11)] also concluded that the transfer factor in their experiments was a peptide. Several laboratories, however, started out with the assumption that the active material was an RNA sequence.[(3,5,15–17)] Their "RNA" extracts, however, still contained enough protein to account for the effect. Recent work[(26)] claims that the transfer factor is found in the ribosomal RNA fraction.

Some of our recent studies with more concentrated solutions of the transfer factors suggest that they can form complexes with acidic material, as basic peptides are known to do. This suggests that the transfer peptides may combine with ribosomal RNA. Experiments are being conducted to verify this assumption, which would explain the presence of the peptides in RNA preparations.

It also remains possible that RNA sequences can act as transfer factors, especially in the planarian experiments and when the extract is administered to mammals by the intracisternal route.[(3)] However, in most experiments the extracts were injected intraperitoneally, and it is unlikely that RNA could go through the blood-brain barrier. Peptides such as vasopressin, MSH, ACTH, substance P, on the other hand, are known to penetrate into the brain.

MECHANISM OF TRANSFER

One of the main reasons for the opposition to the molecular hypothesis in general and to chemical transfer in particular is the lack of a theoretical framework which could integrate them with the established concepts. There is a widespread misconception according to which the molecular interpretation is incompatible with the highly organized structure of the nervous system and the usually accepted "connectionist" view of the learning mechanism.

It seems, however, that the molecular hypothesis is not only compatible with the extreme degree of structural differentiation of the nervous system, but can also explain the organization of the pathways by a mecha-

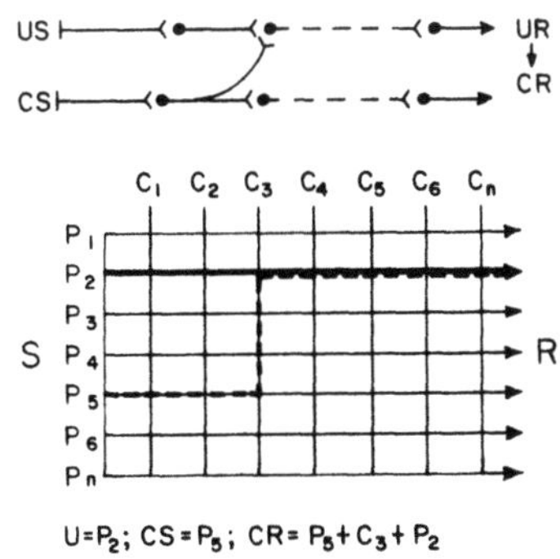

Fig. 6. Diagrammatic representation of the pathways and connectors involved in Pavlovian conditioning. Top: pathway of unconditioned stimulus (US) and response (UR) and conditioned stimulus (CS); when the synapse indicated between the two becomes functional, CS elicits CR. Bottom: Neural network. Stimuli acting on innate pathways (P_1 − P_n) elicit innate responses (for example, $P_2 \rightarrow R_2$). When a stimulus acts on a receptor of P_3 in a definite temporal relationship with the firing of P_2 a connection is established through C_5, so that firing of P_5 now will elicit response R_2. The connector hypothesis assumes that this is brought about by the combination of the recognition molecules between the pathways involved, as summarized in the bottom line.

nism of chemical recognition between neurons.[27] This process represents an extension and refinement of the well-known adhesion between cells of the same category caused by genetically determined specific coating substances. The mechanism could operate in the differentiation of innate pathways and also in the creation of new neural connections, which is the central phenomenon in learning and memory.

Such a hypothesis would supply a single explanation for innate as well as for learned behavior. In the absence of any learning, animals can respond to certain stimuli in a stereotyped fashion which utilizes the inborn neural channels. Learning results in the creation of new patterns of behavior by establishing new connections between existing pathways, as shown in an extremely simplified way in Fig. 6. The new connections are probably created by the simultaneous or almost simultaneous firing of neurons belonging to two different pathways. By increasing the permeability of the pre- and post-synaptic structures, this firing may result in an interchange and combination of the "recognition molecules" of the two pathways, and thus creates a new synaptic connection. One can also assume that training results in a greatly increased synthesis of these molecular "connectors." An extract of the brain of trained donors should contain, therefore, a comparatively large amount of the particular substance which represents the newly acquired connections.

We have to assume that the connectors have a high affinity for the structures from which they originated. When they are injected into recipients they probably bind to these sites and are capable of establishing synaptic connections similar to those that were created in the donors by training. It is probable that transfer is possible only when the donor brain contains a large excess of the transfer factors, since only a small fraction of the injected material reaches the site of action. There is some evidence that normally the storage of memory traces requires extremely minute quantities of the connectors. Only during the early periods of intensive training does the brain contain the amounts required for transfer, and it is unlikely that innate or long-established behavior patterns are transferable.

The relationship of these phenomena with protein and RNA metabolism of the neuron has been discussed elsewhere.[(27,28)] Let us just note here that the combination of connectors creates only a temporary synaptic facilitation. Permanent fixation of the memory trace requires the synthesis of the connector by the neurons which acquired the new connections. This must involve activation of DNA sequences which until then were repressed, and the formation of messenger RNA new to the cell. According to recent observations, the transfer effect can be suppressed by administration of actinomycin D or puromycin to the recipients.[(29)] This finding suggests that the transfer factor acts by initiating the synthesis of some protein in the recipients.

According to the connector hypothesis[27] the molecular code of acquired information represents a system of road signs or signposts which directs the flow of impulses along the organized channels of the brain across the junctions created by learning. These coded molecules have meaning only in terms of the neural network, just as a road map can be used only in reference to an existing highway system.

A "memory trace" or "engram" is probably represented in the brain by a neural circuit whose neurons are able to recognize each other because the synapses linking them are marked by the same connector molecule. The significant point that distinguishes the brain from an electronic computer is the chemical nature of the connection between the functional unit. The computer is no doubt the best model that has yet been proposed for the nervous system, provided one recognizes that the brain is not an electronic, but a chemical computer.

REFERENCES

1. J. V. McConnell, *J. Neuropsychiat.* **3,** 1–42 (1962).
2. S. Reinis, *Activ. Nerv. Sup.* **7,** 167 (1965).
3. E. J. Fjerdingstad, T. Nissen, and H. H. Røigaard-Petersen, *Scand. J. Psychol.* **6,** 1–5 (1965).
4. G. Ungar and C. Oceguera-Navarro, *Nature* **207,** 301–302 (1965).

5. F. R. Babich, A. L. Jacobson, S. Bubash, and A. Jacobson, *Science* **149,** 656–657 (1965).
6. M. W. Gordon, G. G. Deanin, H. L. Leonhardt, and R. H. Gwynn, *Am. J. Psychiat.* **122,** 1174–1178 (1966).
7. C. G. Gross and F. M. Carey, *Science* **150,** 1749 (1965).
8. R. J. Kimble and D. P. Kimble, *Worm Runner's Digest* **8,** 32–36 (1966).
9. M. Luttges, T. Johnson, C. Buck, J. Holland, and J. McGaugh, *Science* **151,** 834–837 (1966).
10. W. L. Byrne, D. Samuel, E. L. Bennett, M. R. Rosenzweig, E. Wasserman, A. R. Wagner, R. Gardner, R. Galambos, B. D. Berger, D. L. Margules, R. L. Fenichel, L. Stein, J. A. Corson, H. E. Enesco, S. L. Chorover, C. E. Holt, III, P. H. Schiller, L. Chiappetta, M. E. Jarvik, R. C. Leaf, J. D. Dutcher, Z. P. Horovitz, and P. L. Carlson, *Science* **153,** 658–659 (1966).
11. F. Rosenblatt, J. T. Farrow, and W. F. Herblin, *Nature* **209,** 46–48 (1966).
12. J. A. Dyal, A. M. Golub, and R. L. Marrone, *Nature* **214,** 720–721 (1967).
13. W. L. Byrne and A. Hughes, *Fed. Proc.* **26,** 676 (1967).
14. O. Wolthuis, *in*: "Molecular Approaches to Learning and Memory" (W. L. Byrne, ed.), AAAS Symposium held in New York, December, 1967, Academic Press, New York, in press.
15. G. Adam and J. Faiszt, *Nature* **216,** 198–200 (1967).
16. R. Gay and A. Raphelson, *Psychon. Sci.* **8,** 369–370 (1967).
17. K. Weiss, *in*: "Molecular Approaches to Learning and Memory" (W. L. Byrne ed.), AAAS Symposium held in New York, December, 1967, Academic Press, New York, in press.
18. G. Chapouthier and A. Ungerer, *Comp. Rend.* **267,** 769–771 (1968).
19. E. Rosenthal and S. B. Sparber, *The Pharmacologist* **10,** 168 (1968).
20. G. Ungar and M. Cohen, *Int. J. Neuropharmacol.* **5,** 183–192 (1966).
21. G. Ungar, *in*: "Proceedings of the Vth International Congress of the Collegium Internationale Neuropsychopharmacologicum," pp. 169–175, Excerpta Medica, Amsterdam (1967).
22. G. Ungar, *J. Biol. Psychol.* **9,** 12–27 (1967).
23. G. Ungar and L. N. Irwin, *Nature* **214,** 453–455 (1967).
24. G. Ungar, L. Galvan, and R. H. Clark, *Nature* **217,** 1259–1261 (1968).
25. G. Ungar, *in*: "Methods in Pharmacology" (A. Schwartz, ed.), Appleton-Century-Croft, New York, in press.
26. J. Faiszt and G. Adam, *Nature* **220,** 367–368 (1968).
27. G. Ungar, *Perspect. Biol. Med.* **11,** 217–232 (1968).
28. G. Ungar, *in*: "Protides of the Biological Fluids" (H. Peeters, ed.), pp. 151–162, Elsevier Publishing Company, Amsterdam (1966).
29. S. Reinis, *Nature* **220,** 177–178 (1968).

DISCUSSION

McEWEN: Have you tried the effects of substances such as ACTH or α-MSH, which, according to DeWied's experiments, have some nonspecific behavioral effects?

UNGAR: No, I have not. DeWied's experiments deal with the effects of these peptides on the extinction of conditioned avoidance responses. The whole problem could be approached by trying out all the available peptides and studying their behavioral effects, but at present it would be like looking for a needle in a haystack. With luck it could pay off handsomely, but otherwise it would probably take many years before any pattern would emerge.

McEWEN: I think we are in substantial agreement, but the use of ACTH and α-MSH could help to rule out some nonspecific effects.

MARKS: I have two brief questions: one concerns the route of administration and the other the penetration of macromolecules into the brain.

UNGAR: In all our experiments the extracts were injected intraperitoneally; some other workers gave them intravenously, and at least one group utilized the intracisternal route. I do not believe that RNA injected intraperitoneally can reach the brain. This does not mean that RNA does not play a role in the processing of learned information, it means merely that it is unlikely to be the actual transfer factor.

MARKS: What other macromolecules do you have in mind that can get into the brain after intraperitoneal injection?

UNGAR: All the evidence suggests that the transfer factors are peptides. They are destroyed by proteases and not by RNase, and most of them are dialyzable. Many peptides are known to go through the blood-brain barrier; for example, vasopressin, oxytocin, and angiotensin have been shown to have central effects; substance P and the peptides just mentioned by Dr. McEwen have distinct behavioral effects after systemic administration. There is even evidence that I^{135}-labeled γ-globulin can be found in the brain after intravenous injection.

LAJTHA: You mentioned that the factors are dialyzable. I assume that the nondialyzable part of the homogenate is inactive, but can you retain some activity if you submit it to digestion?

UNGAR: We have not done this, but it may be worth trying.

BAXTER: How soon after the intraperitoneal injection do you see the effect?

UNGAR: It depends on the situation transferred. In dark avoidance it can be seen in 3 or 4 hr. It is probably a diphasic effect: it appears in a few hours, decreases somewhat after 24 hr and increases in 48 hr. We have been speculating that the early effect is due to the direct action of the peptide, while the late one is caused by the synthesis of the transfer factor by the recipient.

COHEN: What happens when you inject recipients with extracts from two groups of donors trained for different tasks? For example, if some recipients received extracts from donors trained to turn left and from donors trained to turn right, would the recipient behave like the controls, or would there be a greater latency in his choice because it could not decide which way to go?

UNGAR: This type of conflict is used in producing experimental neuroses in animals. We have not tried any of this; the closest we came to it was to test recipients for a behavioral task different from the training of the donors.

BOGOCH: I noticed that all your experiments use aversive situations. Have you or others transferred nonaversive learning?

UNGAR: I have always used negative reinforcement, but most of the other published transfer experiments used food or water reward in the training of the donors. I mentioned Dr. Fjerdingstad's experiments, in which the donors were trained to prefer the white box by positive reinforcement. The recipients, however, were tested in exactly the same way as in the experiments I described, without any reinforcement.

MANDEL: My first question concerns the consistency of the results. In fig. 3, showing the results of the transfer of step-down avoidance, only one out of four experiments was positive.

UNGAR: The figure shows the same experiment on four consecutive days; it was positive on all four days, but statistical significance was reached on only one day. In any case, not all the individual experiments are significant. The real test is to pool together the results of several experiments, good and bad, and calculate the probability of their being due to chance.

MANDEL: My second point is the labeling of your peptide to see if it really goes to the brain after intraperitoneal injection.

UNGAR: We plan on doing this as soon as one of the transfer factors is synthesized. We have a whole list of things we want to do when, and if, we reach that point.

IV

Metabolism Related to Pathology

Chapter 30

Hydrosoluble Proteins of Human Nervous Tissue

D. Karcher, L. Thiry, W. Zeman, and A. Lowenthal

Department of Neurochemistry, Born-Bunge Foundation Berchem-Antwerp and Institute Pasteur, Brussels

INTRODUCTION

Relatively little is known about the proteins of nervous tissues, especially in regard to the human central nervous system. No preferred method of extraction has been developed and no accepted classification of CNS proteins exists. Nevertheless, some progress has been made in this field, largely because of improvements in investigative techniques such as electrophoresis and column separation.

The present paper is concerned with certain hydrosoluble proteins, which were studied by the following methods:

1. Agar gel electrophoresis after Wieme;[1,2] acrylamide electrophoresis after Ornstein and Davis;[3] starch gel electrophoresis after Smithies[4] and Poulik;[5] and immunoelectrophoresis after Scheidegger[6] and Grabar and Burtin.[7]
2. Column separation by gel filtration and by ion-exchange chromatography after Karcher and Lowenthal.[8]
3. Immunodiffusion after Ouchterlony[9] and Hartmann and Toilliez.[10]
4. Determination of measles antibodies after Norrby[11]: Hemagglutination inhibition with measles virus, dissociated with tween 80 and ether.

The results presented here fall into two categories: (1) Proteins with rapid migration in agar gel, which occur in the alpha globulin field; and (2) proteins with slow migration in agar gel, which occur in the gamma globulin field.

FAST-MIGRATING PROTEINS

In previous publications[12,13] attention was drawn to the fact that the concentration of fast-migrating proteins, probably albumins, depends upon the topography of the sample from which the extracts are obtained. In specimens which are considered as controls, because of the absence of a demonstrable pathomorphologic and functional alteration, fast-migrating proteins occur only in the spinal cord and in the peripheral nerves. On agar gel electrophoresis those extractable from the cord migrate in the alpha globulin field, whereas proteins eluted from peripheral nerve invariably exhibit the same mobility as albumin. Detailed topographical studies are underway which will provide more precise results.[14] Suffice it to state that the presence of fast-moving hydrosoluble proteins, slightly slower than albumin, but somewhat faster than serum $alpha_1$ globulin, has been demonstrated by agar gel electrophoresis in 66% of the control spinal cords, as defined above, but not in the corresponding brains. In 17% of the controls a protein with the mobility of albumin was observed, but in the remaining 17% no fast-migrating proteins were present. To date it has been impossible to establish any correlation between the presence of this protein and its variations of electrophoretic mobility on the one hand and the age of the patient and his underlying disease on the other. Accordingly, no speculations as to its significance are offered.

This fast-migrating protein cannot be stained with Sudan Black B or by the PAS reaction, which rules out the possibility of a lipo- or a glycoprotein. Since the relative mobility of this protein corresponds to that of the first isozyme of lactate dehydrogenase, it is important to state that the protein possesses no corresponding enzyme activity.[15]

TABLE I

Relation between Brain Weight and Presence of an Albuminlike Protein in Tay–Sachs Brains

Weight (g)	Albuminlike protein	Cases from
1180	−	Volk
1205	−	Zeman
1420	−	Volk
1600	−	Tourtellotte
1620	+	Volk
1620	+	Volk
1630	+	Volk
2150	+++	Inst. Bunge

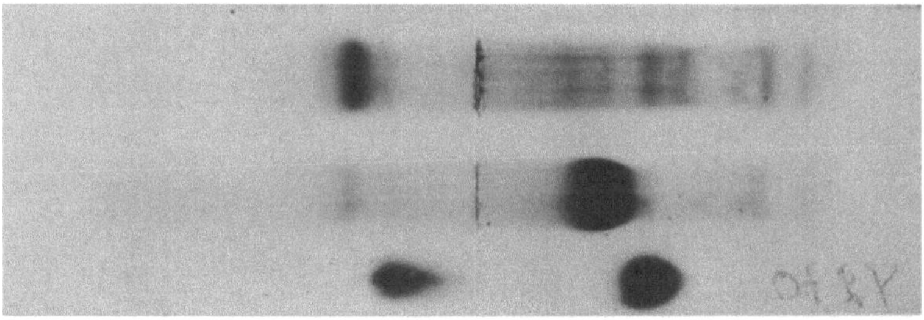

Fig. 1. Agar gel electrophoresis of hydrosoluble proteins of spinal cord, showing a strong fraction which migrates in the alpha globulin field. Control at bottom. The large indicator spot represents serum albumin, the other transferrin.

In addition to a fraction which migrates in the albumin field, present in abnormally high quantities,[8] a protein of slightly lower mobility such as found in spinal cord extracts is also observed in edematous brain tissue. More recent studies revealed the presence of a protein with the same alpha globulin mobility as in the spinal cord extracts in the brains of patients suffering from Tay–Sachs' disease with megalencephaly (Table I), in a cerebral biopsy from a patient with Schilder's encephaloclastic cerebral sclerosis, in a brain biopsy of a patient with cerebral necrosis secondary to infantile hypoxic encephalopathy, and, finally, in one biopsy- and in two autopsy-derived brains from patients with subacute sclerosing panencephalitis.

It follows that the protein under discussion can be extracted from autopsy material as well as from biopsy-derived tissues. The protein localizes predominantly in the white matter, but is also present in the gray. On agar gel electrophoresis this protein has always the same relative mobility, which we estimate at about 0.900 (Fig. 1); in other words, it migrates to a position between serum albumin and serum alpha$_1$ globulin. On electrophoresis in 6% acrylamide the protein remains close to the cathode and displays a mobility comparable to that of an albumin octomer (Fig. 2), and in starch gel it moves in the alpha globulin field (Fig. 3). On immunodiffusion the protein under question behaves like albumin, as it interacts with antialbumin just like serum albumin. Although it forms a continuous precipitation line with albumin, its diffusion velocity is slower than that of serum albumin, probably the result of a higher molecular weight (Fig. 4). On immunoelectrophoresis the protein is precipitated by human serum antialbumin (Fig. 5). By determination of the void volume of a Sephadex G-200 column, a molecular weight of about 600,000 was calculated for this protein.

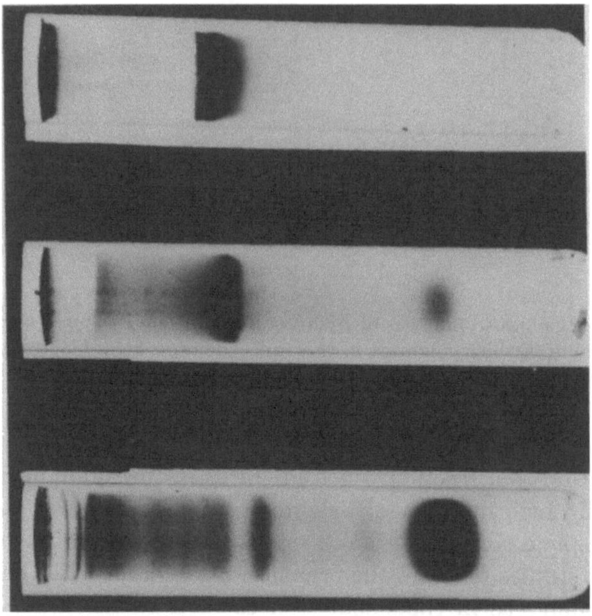

Fig. 2. Acrylamide electrophoresis of hydrosoluble proteins from nervous tissues. Left: Normal serum. Center: Optic nerve, displaying little albumin, but considerable quantities of the predominating protein. Right: Mixture of extracts from brain with Tay-Sachs' disease and another with SSPE exhibiting the predominating protein in the same position.

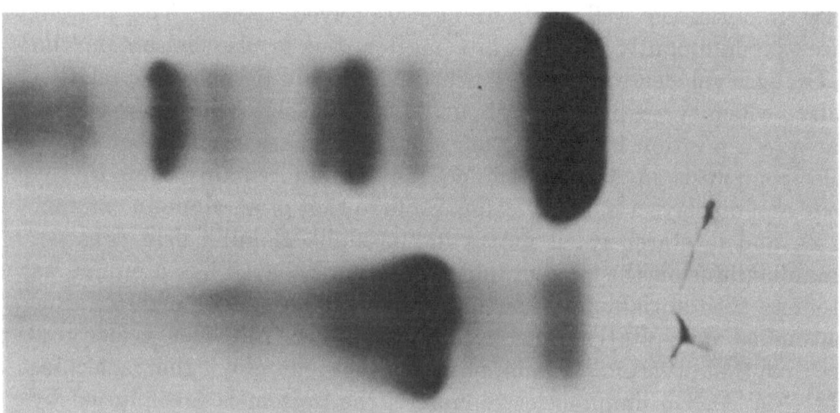

Fig. 3. Starch gel electrophoresis of hydrosoluble proteins from the spinal cord with a predominating fraction migrating at the level of alpha globulins. Bottom: normal serum.

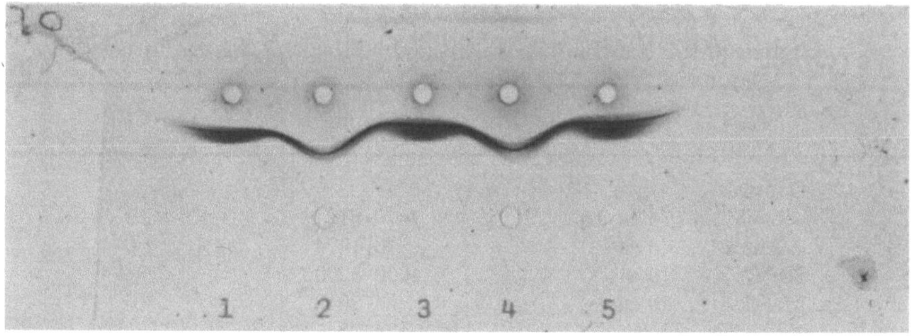

Fig. 4. Immunodiffusion showing the relationship between serum albumin and the predominating proteins isolated from hydrosoluble extracts of brain and spinal cord. The upper holes contain antialbumin. Lower holes: (1) Purified serum albumin from patient with Tay-Sachs' disease; (2) purified predominating protein from brain of the same patient; (3) commercial serum albumin (Calbiochem); (4) purified predominating protein from normal spinal cord; (5) purified serum albumin from normal individual.

In summary, a protein which is immunologically related to albumin, but which possesses a slower electrophoretic mobility, can be isolated from clinically and morphologically unaltered spinal cord and from brain tissues affected by certain pathological conditions. The molecular weight of this protein as determined by different methods seems to support the assumption that the protein represents either a polymer or an aggregate. Our present working hypothesis stipulates that this macromolecular protein shares an immunologic parentage with serum albumin. To our knowledge, this protein has not been previously described in nervous tissue or in any other organ.

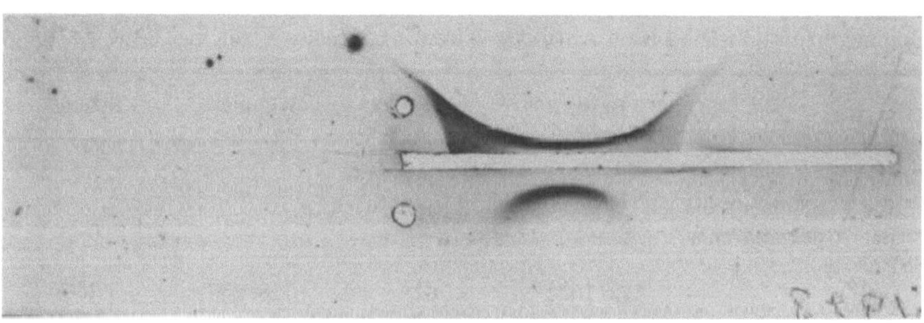

Fig. 5. Immunoelectrophoresis showing precipitation of normal serum albumin (above) and of the predominating protein from hydrosoluble extracts of the spinal cord by antialbumin.

TABLE II

Hemagglutination-Inhibition Antibody Titers to Measles Virus[a]

		Serum	CSF[a]
Controls			
Children	(4 cases)	0–512	0
Adults	(5 cases)	0–1024	0
SSPE[b]	(6 cases)	4000–32000	8–64
Measles	(9 cases)	64–2048	0

[a] Cerebrospinal fluid.
[b] Subacute sclerosing panencephalitis.

SLOW-MIGRATING PROTEINS

Our studies on the gamma globulins of the central nervous tissue progressed less rapidly than the work on albumins. Previous publications were concerned with the demonstration of gamma globulins in water-soluble extracts from central nervous tissue in cases of multiple sclerosis[13] and subacute sclerosing panencephalitis (SSPE).[16] In a detailed study Tourtellotte[17] demonstrated that these proteins are actually immunoglobulins. In SSPE these gamma immunoglobulins are present in the serum, the cerebrospinal fluid, as well as in the brain tissue, and for serum Tourtellotte[17] and Kolar[18] have established their identity as immunoglobulins. These observations are in agreement with the more recent, uncontestable observation that the titer of measles antibodies in the serum and in the spinal fluid of patients with SSPE is significantly elevated (Table II).

TABLE III

Hemagglutination-Inhibition Antibody Titers to Measles Virus in Tissue Extrects

		Brain	Liver	Kidney	Spleen
Controls[a]					
Ker	Schilder's disease	—			
Ste	Oligophrenia	—	—	—	—
SSPE[b]					
Def		512, 1024	64	128–256	256
Mes		64	—	—	—
Boy		512			

[a] Three-letter designations refer to names of patients.
[b] Subacute sclerosing panencephalitis.

Quantitative determinations showed that the measles antibody titers in the biological fluids are higher than in the central nervous tissue (Table III). Nevertheless, comparative studies on different organs from patients with SSPE make it probable that the measles antibodies present in the central nervous tissues are not blood-borne, but are locally produced.[19] These antibodies have never been observed in control material or in central nervous tissues affected by diseases other than SSPE.

SUMMARY

1. The majority of human spinal cords not affected by clinical or pathomorphologic alterations harbor a protein which migrates between the serum albumin and $alpha_1$ globulin fields. A similar protein is found in the human brain subject to certain pathological conditions. This protein appears to be a macromolecular complex, either a polymer or an aggregate, which shares an immunologic parentage with human serum albumin. The significance of its presence and the mode of formation remain obscure.

2. In subacute sclerosing panencephalitis, a disease probably produced by the measles virus, measles antibodies are found in the serum and the spinal fluid as well as in aqueous extracts of the nervous tissue and of other organs. In addition to the presence of these antibodies, agar gel electrophoresis demonstrates qualitative changes in the appearance of the gamma globulins in serum and in spinal fluid.

REFERENCES

1. R. J. Wieme, "Studies on Agar Gel Electrophoresis. Techniques. Applications," Arscia, Brussels (1959).
2. R. J. Wieme, "Agar Gel Electrophoresis," Elsevier, Amsterdam (1965).
3. L. Ornstein and B. J. Davis, "Disc Electrophoresis," Distillation Products Industries, Division of Eastman Kodak Company.
4. O. Smithies, Zone electrophoresis in starch gel; group variations in the serum proteins of normal human adults, *Biochem. J.* **61**, 629–641 (1955).
5. M. D. Poulik, Starch gel electrophoresis in a discontinuous system of buffers, *Nature* **180**, 1477 (1957).
6. J. J. Scheidegger, Une micromethode de l'immuno-électrophorèse, *Intern. Arch. Allergy Appl. Immunol.* **7**, 103–110 (1950).
7. P. Grabar and P. Burtin, "Analyse Immuno-électrophorétique," Masson, Paris (1950).
8. D. Karcher and A. Lowenthal, Hydrosoluble proteins of edematous human nervous tissue, *in*: "Brain Edema" (I. Klatzo and F. Seitelberger, eds.), pp. 195–201, Springer-Verlag, Wien (1967).
9. O. Ouchterlony, Antigen-antibody reactions in gels, *Acta Path. Microbiol. Scand.* **26**, 507–515 (1949).

10. L. Hartmann and M. Toilliez, Microméthode d'étude en gélose de la réaction antigène anticorps (variante du procédé d'Ouchterlony), *Rev. Franc. Etudes Clin. Biol.* **2,** 197–199 (1957).
11. E. Norrby, Hemagglutination by measles virus. 4. A simple procedure for production of high-potency antigen for hemagglutination-inhibition (HI) tests, *Proc. Soc. Exper. Biol. Med.* **111,** 814–818 (1962).
12. A. V. Palladin and N. M. Polyakova, *in*: "Biochemistry of the Central Nervous System" (F. Brücke, ed.), p. 185, Pergamon, London, (1959).
13. M. van Sande, "Contribution a l'Étude des Proteines Cérébrales et de Deux Enzymes (la Déhydrogénase de l'Acide Lactique et la Déhydrogénase de l'Acide Malique du Système Nerveux Central)," Acta Medica Belgica, Bruxelles (1962).
14. D. Karcher, W. Zeman, N. Chamoles, and A. Lowenthal, Alpha-Albumin in normal and pathologic nervous tissue, I. Biochemical Investigations (*Brain Research*, in press).
15. M. van Sande, D. Karcher, and A. Lowenthal, Caractérisation électrophorétique des protéines spinales, *in*: "Protides of the Biological Fluids" (H. Peeters, ed.), p. 253, Elsevier, Amsterdam (1964).
16. A. Lowenthal, "Agar Gel Electrophoresis in Neurology," Elsevier Publishing Co., Amsterdam (1964).
17. W. W. Tourtellotte, Y. A. Parker, R. M. Herndon, and C. V. Cuadros, Subacute sclerosing panencephalitis brain immunoglobulin G, measles antibody and albumin, *Neurology* **18** (No. 1, Part 1), 117–121 (1968).
19. W. W. Tourtellotte, in preparation.

DISCUSSION

BOGOCH: The albuminlike brain protein fraction described here can be distinguished in several respects from the brain proteins I described in Chapter 28 of this volume, especially 10*B* and 11*A*. Figure 6 shows three column chromatograms of human-brain proteins obtained using the same methods described in Chapter 28 for pigeon-brain proteins. The bottom chromatogram is of proteins extracted from normal 2½-year-old human gray matter. The uppermost chromatogram in Fig. 6 is from a case of Tay–Sachs' disease. In the later stage of Tay–Sachs' disease the small brain contains a large number of broken-down nerve cells and a large number of glia. Thus, it represents a natural experiment of nature in which one has both glial enrichment, so to speak, and neuronal depletion. The normal mean value of 10*B* is 0.64 mg/g in normal gray matter, and it is 7 mg/g in this Tay–Sachs' situation. This protein (10*B*) was referred to as Tay–Sachs' disease protein. There is quite a sharp peak, and only one component is shown on disc gel electrophoresis. A distinguishing point is that it is a very fast-moving band, not the slow-moving band described here. The second point is that it occurs in small brains too, whereas the protein described here is characteristic of large (edematous) brains. The third point is that antisera which we have prepared to 10*B* do not cross-react with serum albumin or with any other serum component, nor with kidney or liver protein prepared the same way. It does react only with 10*B* from pigeon brain, human brain, guinea pig brain, and rat brain. 10*B* is therefore apparently a native brain protein which is organspecific, but not species specific, which is increased in Tay–Sachs' disease for some reason, possibly gliosis, and in glial brain tumors. It has been isolated from glia grown in tissue culture and from outer glial membranes derived from these cells. It is likely, therefore, that it is totally or predominantly a glial protein. Figure 7 shows

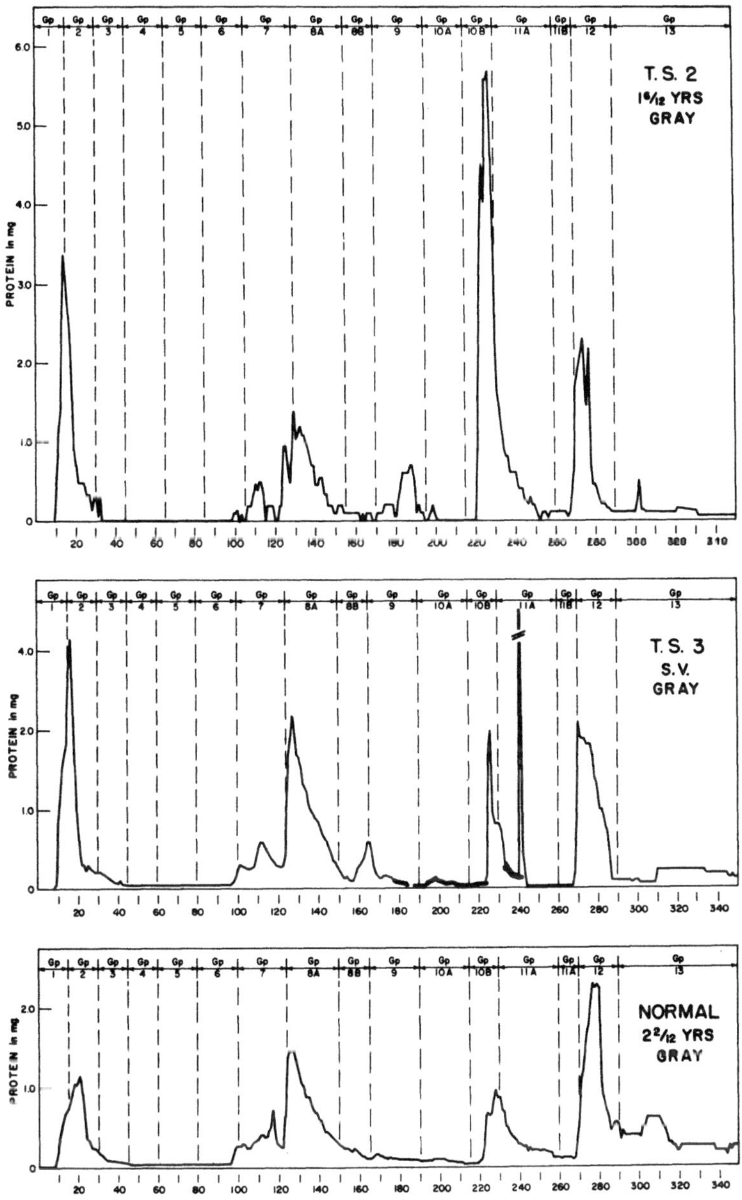

Fig. 6. Column chromatography (DEAE cellulose) of proteins extracted from Tay–Sachs' disease brain (upper), Spielmayer–Vogt disease brain (middle), and normal 2½-year-old brain (lower) from Bogoch and Belval, *in*. "Inborn Disorders of Sphingolipid Metabolism," Pergamon Press, New York (1966).

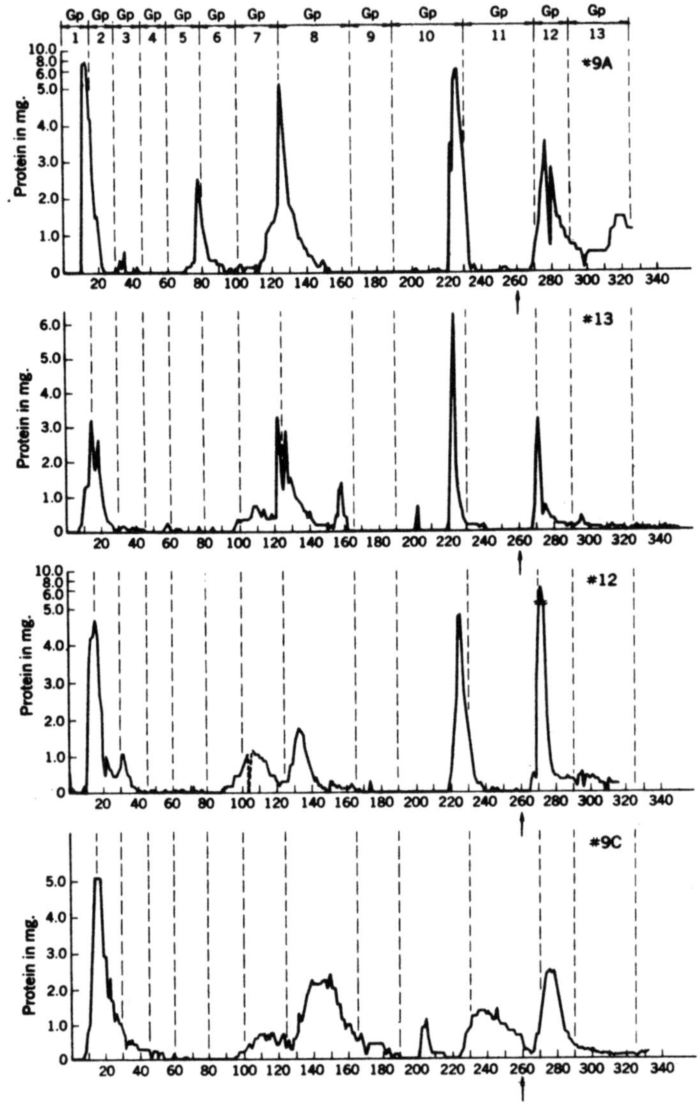

Fig. 7. Column chromatography (DEAE cellulose) of proteins extracted from three brain glial tumors (9A, 13, and 12) and from normal human grey matter (9C). Note increase in 10*B* and decrease in 11*A* in tumor tissue when compared with normal values.

the presence of 10*B* in human glial tumors—the glioblastomas and gliomas. In the three upper (tumor) chromatograms (9A, 13, and 12) one sees a sharpening of the whole curve for the proteins of the glial tumors when compared with the pattern of normal adult gray matter (9C) on the bottom. 10*B* is quantitatively increased 3–6-fold in glial tumors.

Here, as in Tay–Sachs' disease, there is an enrichment of 10*B*, suggesting that it is a glial cell constituent. The mean for the normal human adult values of 10*B* is 0.64 mg/g, but the glial tumors are markedly elevated in fraction 10*B*, 2–4 mg/g. The fourth point is that immunoflourescence with antisera which we prepared to 10*B* does not visualize brain vessels or neurons, but demonstrates in dramatic and high detail the astrocytic glia (K. Mori and W. H. Sweet, personal communication). The fifth important difference is that whereas 10*B* is a glycoprotein, with high concentration of sugars, the protein described here has no constituent carbohydrates.

RAPPORT: Dr. Lowenthal, your protein shares with albumin one common determinant, insofar as you have been able to pick up by cross-reacting the protein with antialbumin serum. This could have something very interesting to do with the way the protein is put together. However, have you made antibodies against this protein to see if you could detect very substantial antigenic differences between the protein and albumin by cross-reacting this antiserum with albumin as antigen?

LOWENTHAL: No antiserum has so far been prepared. Isolated albumin from edematous brain tissue reacts with human serum antialbumin. Precipitating lines of edematous albumin and serum albumin with human serum antialbumin show partial identity between both substances.

RAPPORT: There was a recent report on the high-density lipoprotein of rat plasma, which appears to share a common antigenic determinant with albumin, but which is also quite different in its other physical characteristics.

LOWENTHAL: In this protein there are no lipids, nor any glucose moiety.

MANDEL: If you have many proteins in these fractions which have a similar charge and a similar molecular weight, how can you be sure your fraction is really homogeneous? In addition, I think it would be possible to see if your fraction is similar to serum albumin in its "fingerprint," for instance. Finally, I would note that one has to be very careful when estimating the molecular weight by gel-filtration.

LOWENTHAL: This protein is homogeneous, according to the different tests we carried out. Elution volume on Sephadex column indicates a high molecular weight, close to 600,000. No other methods were used so far to confirm it.

KIES: What Sephadex gel were you using?

LOWENTHAL: Sephadex G-200 with phosphate buffer.

KIES: From your disc gel electrophoresis have you evidence about size? Since you had to use 5% acrylamide gel, that in itself would indicate that it is large and the problem here is not retention on Sephadex.

LOWENTHAL: According to our investigation, we are led to believe that this protein is a macromolecule, but we cannot demonstrate it so far.

KIES: But the acrylamide gel of different concentrations is a very sensitive indication of size, too. Dr. Bogoch, is your 10*B* fraction nonspecies specific? Did you say that you got cross-reaction with the pigeon brain and human brain?

BOGOCH. The data that we have so far on 10*B* indicates that it is not species specific, but it is organ specific. It does not cross-react with serum, with kidney or liver. The fraction of Dr. Rajam, of our laboratories, is also different. This fraction is another acidic glycoprotein which is called AP, for acidic peak. AP does not cross-react with 10*B*I or 10*B*II subfractions, or whole 10*B* or with 11*A*. Our 10*B* and 11*A* also do not cross-react with Moore's S-100 or with AP. AP has not yet been tested with Moore's S-100, but it appears that all four are different. The only thing they appear to share is that they are all fast-moving acidic proteins. The other big difference is that Moore's protein does not have sugar in it, whereas $10B$ and $11A$ do. Thus, ours are glycoproteins (see Chapter 28), while Moore's S-100 is not a glycoprotein.

GAULL: Dr. Lowenthal, how or why does your protein occur in the spinal cord, and how do you reconcile the occurrence in the spinal cord with your theories about its appearance in pathological tissues, because of presumable vascular problems.

LOWENTHAL: Cerebrospinal fluid is richer in albumin in the lower spaces and in prealbumin in the ventricular regions. We do not exclude the possibility that a similar protein distribution exists in the central nervous system. Human spinal cord can only be studied postmortem. Therefore, we need an experimental animal in order to investigate the pathophysiology of α-albumin. We are now searching for such material.

KIES: You mentioned Dr. Tourtellotte's work on leucoencephalitis. Have you made any comparisons of this protein in the situations in which he finds the extremely elevated globulins in the brain, and where are the lesions?

LOWENTHAL: In multiple sclerosis and subacute sclerosing panencephalitis the gamma globulin content of hydrosoluble nervous tissue proteins are increased, and, as was shown by Dr. Tourtellotte, those gamma globulins are predominantly IgG.

KIES: Is there a corresponding increase in this protein? Have you ever done any collaboration on this? I know that Tourtellotte tries to prove that his gamma globulin can't come from serum, because he doesn't have an increase in serum albumin, but I wondered about an increase in this protein.

LOWENTHAL: Two subacute sclerosing panencephalitis brains mentioned here were also studied by Dr. Tourtellotte. On this material IgG increase was demonstrated. Measles-antibody determinations were similar to ours in this material.

LAJTHA: Did you measure any enzymatic activity in these fractions? Do you think that the major difference between your fraction and Dr. Bogoch's is that perhaps yours is exogenous and his is endogenous?

LOWENTHAL: LDH isozymes were measured. No activity was located on α-albumin. This fraction could be exogenous, perhaps modified serum albumin. This is only a working hypothesis.

Chapter 31

Studies on Proteinase Enzymes during Wallerian Degeneration

Giuseppe Porcellati

Department of Biological Chemistry
University of Pavia
Pavia, Italy

INTRODUCTION

A Review of Previous Work

Recent summaries on protein metabolism in Wallerian degeneration (WD) have been reported elsewhere.(1–3) In these paragraphs we shall review only the most pertinent results on proteolytic enzymes.

An active proteolysis takes place at neutral pH within the peripheral nervous system (PNS) of the hen during some stages of WD;(4–6) this result has also been reported by other workers,(7–10) who have also presented data on peptidase activity. The same, though less evident, finding has been observed following administration of demyelinating organophosphorus compounds;(11) tetraethylpyrophosphate, a nondemyelinating agent, does not produce the proteinase increase.(12)

Neutral proteinase activity is unchanged in the proximal stump of sectioned cat nerve,(1) unlike acidic cathepsin. This activity increases in both proximal and distal stumps of a sectioned peripheral nerve (PN) at an early stage of WD,(1) whereas the increase of the neutral proteinase activity in the distal stump occurs later. The activity of acidic and neutral proteinase has also been examined in various segments of the degenerating cat sciatic nerve, more precisely, in the upper, medium, and lower portions of the distal stump;(13) increases are quantitatively similar throughout the whole length of the PN, only the most distal part of the sectioned nerve displaying an unusually higher acidic catheptic activity. During nerve regeneration the neutral proteinase activity is reduced in both the proximal and distal reinnervating stumps to about 50% of normal.(1)

The enzymatic changes of the acidic and neutral proteinase in the central white tracts after transection appear to be less dramatic and more protracted than those reported for the sectioned PN[13]: the increases appear at about the 15th and 30th day from chordotomy, respectively.

Present Concepts

The results outlined above indicate that proteinases are involved in secondary demyelinations. The increase of their activities is supposed to be due, not to increased cellular secretion, but to *in vivo* "liberation" or unmasking of enzyme preexisting in or around the normal nerve.[2,7] It has been indicated that a basic trypsin-digestible myelin protein, which is present in both the CNS and PNS,[2,14,15] may constitute the real substrate for the early increase of proteolytic activities in the degenerating PN.[16] Moreover, it has been inferred that myelin lipid becomes dissociated from the protein of the sheath during the preliminary stages of demyelination.[7,17] However, how does the lipid–protein bond dissociate in Wallerian processes? And how do the early biochemical changes in the lipoprotein layers of the myelin sheath accompany the early physical alterations of the axoplasm in the damaged axon?

Current ideas favor the view that in WD some kinds of injury, represented by the sudden lesion between axon and myelin, may cause a greater instability of the lipoprotein layers in the well-ordered structure of the myelin sheath.[1,2,7,13,16] This may depend on early detectable biochemical changes[1] which take place at the beginning of the physical fragmentation of the myelin.[1–3,7,13,17,18] Among these biochemical changes, the increase of proteinase activities certainly takes place at early stages of WD, and has been well documented, as shown above. However, some uncertainty exists whether the enzymatic change depends on conformational alterations of the multilayered lipoprotein architecture of the PN, or constitutes the primary cause of the alteration itself. Present concepts favor the opinion that the early and partial alteration of the myelin lipoprotein structure is due to disruption of lipid bound to trypsin-digestible protein,[2,7,17] which is caused by "liberation" or activation of the proteases within the myelin sheath. Proteinase activity therefore appears to constitute a primary factor involved in demyelinating processes.[2,7,19] The enzymatic system is mainly a constituent of the myelin sheath itself,[1,2,20] and is associated with the lipid and proteolipid complexes of the nerve.[20]

The acidic proteinase activity, which is probably located in predominant amounts in the mitochondrial and possibly lysosomal fractions of nervous tissues, increases in proximal and distal stumps of PN at an early stage of the WD.[1,13] The increase might be related to the initial accumulation and/or damage of neuronal and axonal mitochondria, which is indicated as one of the earliest signs of Wallerian changes.[21–24]* The later increase of neutral proteinase activity,[4–6,10] on the other hand, might signify a chemical response due to cellular proliferation.[1] According to Adams and Tuqan,[7] neutral proteinase activity also increases in PN in the very early periods of WD.

Present Studies

In our work attempts have been made to obtain separation of intracellular fractions of PN by centrifuging homogenates of normal nerves and of distal segments of degenerating cat PN, in order to find out whether our previous ideas on the behavior of proteinase enzymes during WD were correct. Procedures have been developed for the separation of the nerve subfractions from sucrose homogenates by taking into account previous work on properties of brain and nerve tissues. It does not seem very meaningful, at present, to work out nerve subfraction preparations, since great difficulty may be encountered in preparing the material, and a great deal of caution is necessary for assessing its morphology. We started, however, to prepare the more reproducible and less sophisticated fractions, with the aim of developing more reliable and precise techniques, which may enable us to obtain additional results in the future.

EXPERIMENTAL

Nerve Injury and Nerve Treatment

Experiments were carried out on adult male cats of the same age kept together with controls under the same nutritional, environmental, and hygienic conditions. Sectioning of the sciatic nerve was done as reported elsewhere.[5,26] Proximal and distal segments of the same length were excised at the appropriate time, together with segments of the contralateral normal nerve taken at the same level. The nerves were carefully freed of connective and adipose tissues, and the perineural sheath dissected away from each individual fascicle. They were successively cut into smaller fragments and immediately frozen under −25°C. Each estimation was carried out at a given interval on a pool of sciatic nerve derived from two or three similarly treated animals.

* Additional references in this connection have been quoted by Fischer and Malík.[25]

Homogenate and Subfractions

The frozen fragments were weighed on a torsion balance, minced with scissors, and immersed, in small beakers, in eight volumes of 0.25 *M* sucrose, 0.005 *M* EDTA, and 0.006 *M* β-mercaptoethanol for 20 min at +2°C. They were successively homogenized with this ice-cold solution in a glass–glass Potter–Elvehjem homogenizer (clearance 0.18 mm) at 800 rpm during a total time of 120 sec, with eight excursions in each direction. The homogenate always left a small nondispersed residue, which constituted 10% of the original fresh weight. After filtration through cheesecloth, homogenates were made up to the original volume with the above mentioned ice-cold solution.

The material was successively centrifuged at 500g for 10 min to remove debris, and the corresponding supernatant (S) brought to 1200g for 15 min in the Servall SS 34 rotor. A pellet and a cloudy supernatant fraction (S_1) were obtained. The pellet was washed once with 3–4 volumes of 0.25 *M* sucrose and recentrifuged to yield fraction P_1, which was not further processed. The supernatant S_1, together with the washing, was centrifuged at 11,000g_{av} for 20 min (SS 34 rotor) to yield a precipitate (P_2), a floating white layer (P_3), and a supernatant (S_2). The layer P_3 was brought down as a pellet by resuspension in small volumes of 0.32 *M* sucrose, and centrifuged at 110,000g_{av} (40,000 rpm) in the Spinco No. 40 rotor for 30 min, to yield fraction P_3. The precipitate P_2 was taken up in small volumes of 0.32 *M* sucrose, layered over 0.8 *M* and 1.2 *M* ice-cold sucrose,[27] and centrifuged at 95,000g_{av} (39,000 rpm) for 90 min in the No. 40 rotor. In this way, three distinct subfractions were obtained, *A*, *B*, and *C*, from top to bottom, respectively. Layers *B* and *C* were accurately removed with fine-tipped Pasteur pipettes, adjusted to 0.32 *M* ice-cold sucrose concentration, and brought down as individual pellets by centrifuging at 110,000g_{av} (40,000 rpm) for 30 min in the No. 40 rotor. The layer *A* was diluted to 0.32 *M* sucrose solution, layered over a 0.656 *M* sucrose, and centrifuged for 30 min at 47,000g_{av} (21,500 rpm) in the Spinco SW 25.1 rotor. A fraction which resembled myelin concentrated as a white layer at the interface between the two sucrose solutions; it was aspirated off, diluted to 0.32 *M* sucrose, and brought down (fraction *A*) as above. The original supernatant S_2, together with washings, was centrifuged at 105,000g_{av} for 90 min in the Spinco No. 40 head to yield a small pellet (P_4) and a final supernatant (S_3), neither of which was further purified. Poor results were obtained on trying to purify fraction P_1, or to better purify fraction P_2 according to the procedure of de Robertis *et al.*,[28] as described elsewhere.[29]

Each washed pellet was resuspended in 0.25 *M* sucrose, in order to give a final concentration of 1 ml/150 mg fresh tissue, which was used soon after preparation.

Biochemical Tests and Provisional Identification of Subfractions

For phosphoinositide, cerebroside, and cholesterol estimation each PN subfraction was normally lyophylized and dried. Mono- and polyphosphoinositides were detected and estimated as described previously,[30] total cerebroside was analyzed according to Svennerholm,[31] and cholesterol as reported by Lefler and McDougald.[32] Succinic dehydrogenase (1.3.99.1), cytochrome oxidase (1.9.3.1), glucose-6-phosphate phosphohydrolase (3.1.3.9), acid phosphatase (orthophosphoric monoester phosphohydrolase, 3.1.3.2), and aryl sulfatase (aryl-sulphate sulphohydrolase, 3.1.6.1) were determined according to Laatsch *et al.*,[33] Cooperstein and Lazarow,[34] Hübscher and West,[35] Gianetto and de Duve,[36] and Roy,[37] respectively. Released orthophosphate (P_i) was measured by the method of Lowry *et al.*[38] It was not possible to assay all the fractions enzymatically, because the methods often gave unreliable results and the subfractions were not present in sufficient amounts.

Phase-contrast-microscope observations or electron-microscope analyses of the subfractions were not performed because we were not sufficiently confident of the morphological homogeneity of the respective populations.

TABLE I

Chemical and Enzymic Analyses on Homogenate and Subfractions from Cat Peripheral Nerve[a]

Fraction[b]	PI[c]	CHOL[c]	CER[d]	SDH[e]	G-6Pase[f]	APase[f]
Homogenate	0.021	0.49	29.2	1.47	2.10	0.28
P_3	0.035	0.86	64.9	Negligible	0	0
A	0.041	1.01	81.7	Negligible	0	0
B	0	0.11	0	3.60	>0.10	0.10
C	0	—	> 1	7.90	4.21	1.96
P_4	>0.01	—	—	Negligible	5.90	0.80

[a] Abbreviations: PI, phosphoinositides; Chol, cholesterol; Cer, cerebrosides; SDH, succinic dehydrogenase activity; G-6Pase, glucose-6-phosphatase activity; APase, acid phosphatase activity.

[b] For the denomination of the fractions see the text.

[c] Results expressed as μmoles/mg protein.

[d] Results expressed as μg/mg protein.

[e] Activity expressed as mmoles of succinate oxidized/mg protein/hr.

[f] Activity expressed as mmoles P_i released from substrate/mg protein/30 min.

However, useful indications have arisen from studies with chemical and enzymatic markers. The results, reported in Table I, show that fractions P_3 and A are depleted of the enzymatic activities examined, and are enriched with phosphoinositides, cholesterol, and cerebrosides; these lipids were not detected, or were observed only in very low amounts, in other subfractions. These two fractions were thought to be similar, and were equated to myelin. Unlike other fractions, subfraction C shows the highest SDH activity (Table I), and some cytochrome oxidase activity, and contains cardiolipin.[39] For these reasons C was tentatively equated to a mitochondrial-rich material. Fraction C also displays the highest acid phosphohydrolase activity (Table I), together with noticeable aryl-sulphate sulphohydrolase activity; this fraction could therefore have some relationship with lysosomal-like material. High glucose-6-phosphate phosphohydrolase activity is present, although not consistently, in subfraction P_4, probably containing membranes of the endoplasmic reticulum. Fraction B, which shows some SDH activity, was considered as an intermediate premitochondrial subfraction; it can probably be equated to a membrane-rich fraction, possibly comparable to a similar fraction obtained from brain tissue.[29]

The provisional identification of the various PN subfractions made by us on the basis of biochemical analyses compares satisfactorily well with the results obtained with the brain tissue.[29] PN subfractions A, B, C, and P_4 may, in fact, be compared, respectively, with fractions AP_2, BP_2, CP_2, and P_3 of the brain-tissue homogenates; in addition, the floating layer (P_3) has also been observed in PN structures by other workers,[26] who have similarly equated it to myelin. A correct identification of the PN subcellular fraction is certainly a prerequisite for an adequate study of biochemical activities during demyelination. The provisional comparison with the brain particles reported in Table II has been worked out, but only on the basis of analyses with biochemical markers, and of analogy with studies on proteinase enzymes in brain homogenate subfractions.[29,42] These criteria of identification have been adopted by us, and will be used throughout this work.

Methods for Estimating Proteinase Activity

Neutral and acidic proteinase activities were determined at pH 7.4 and 3.8, respectively[1,5] with freshly prepared, acid-denatured hemoglobin (Hb) as substrate.[29] With 4 mg Hb/ml and 0.3–2.0 mg of particulate protein/ml, zero-order kinetic reactions were obtained. Release of amino groups, determined through ninhydrin procedures,[1,5,29,40] was linear with respect to enzyme concentrations up to 2.0 mg protein/ml, and was corrected for the release of ninhydrin-positive material that occurred in the absence of added substrate. Estimations for each experimental point were always done in duplicate.

Analytical Procedures, and Reagents

Protein was determined according to Lowry *et al.*,[41] with crystalline bovine serum albumin as standard; its N content was determined before use by the micro-Kjeldhal method. Blank and standard tubes were prepared with the same concentrations of the components of the sucrose solution present in the various subfractions. Turbidity developed only occasionally, and was eliminated by extraction with equal volumes of distilled diethyl ether, and centrifugation of the colored aqueous solution at 2000g for few minutes. The P_i was estimated by the method of Lowry *et al.*[38]

Calculations

The proteinase-activity values of normal nerves are reported as μmoles of NH_2 N/g of original fresh tissue/hr, or as μmoles of NH_2 N/g of protein/hr. In order to obtain a more accurate estimation of the proteinase activity during demyelination, the results for degenerating nerves are expressed as the percentage change in the sectioned nerve as compared with the control-nerve values of the same cat, calculated as μmoles amino nitrogen released per gram of protein. Since during the period of nerve degeneration examined a noticeable increase of the wet wt of the tissue occurs (from about 70 mg/cm to 120 mg/cm in the damaged distal nerve), results have also been calculated on the basis of the activity per cm of nerve and per mg of fresh nerve.

RESULTS

Normal Peripheral Nerves

The acidic and neutral proteinase activities in PN homogenates are similar when estimated throughout the whole length of the PN, or assayed on the left and right sciatic nerve preparations. The distribution of both activities in the PN subfractions is reported in Table II. The values are approximate, since the per cent of recovery of each fraction is not known. However, 1 mg of protein homogenate, after filtration through cheesecloth, yielded 0.3, 0.12, 0.04, 0.06, 0.03, 0.14, and 0.15 mg of protein in fractions

TABLE II

The Distribution of Proteinase Activity in Subcellular Fractions of Cat Sciatic Nerve[a]

Fractions	Provisional comparison with brain fractions	Acid proteinase		Neutral proteinase	
		(a)	(b)	(a)	(b)
Whole homogenate	—	8.1 ± 1(7)	61 ± 8(7)	21.6 ± 3.5(8)	162 ± 24(8)
P_1	Crude nuclear fraction	2.1 ± 0.1(4)	52 ± 2.6(4)	3.36 ± 0.3(5)	84 ± 7(5)
P_2	Crude mitochondrial fraction	3.1 ± 0.2(6)	179 ± 12(6)	5.11 ± 0.6(4)	295 ± 30(4)
A	Myelin	0.21 ± 0.02(5)	40 ± 3(5)	1.28 ± 0.1(4)	242 ± 21(4)
B	Nerve endings, vesicles, "membranes"	0.26 ± 0.01(4)	32 ± 2.4(4)	3.05(3)	381(3)
C	Mitochondria, lysosomes	3.1 ± 0.3(4)	773 ± 51(4)	0.75 ± 0.07(4)	184 ± 17(4)
P_3	Myelin	0	0	3.1(2)	192(2)
P_4	Crude microsomal fraction	0.80 ± 0.05(5)	43 ± 2(5)	1.21 ± 0.1(5)	65 ± 5(5)
S_3	Final supernatant	0.19 ± 0.02(4)	9.5 ± 0.8(4)	0.91 ± 0.1(4)	45 ± 4(4)

[a] Experimental conditions: the reaction mixture (1 ml), incubated for 1 hr at 37° under N_2, contained 10 mg of homogenate protein or 0.3–2.0 mg of corresponding particulate protein (about 75 mg of fresh tissue or corresponding fraction); 25 mM Tris-HCl buffer, pH 7.4, (or 40 mM acetate buffer, pH 3.80); and 4 mg of denatured hemoglobin. Results are corrected for ninhydrin-positive material released in the absence of substrate, and are expressed either as (a) μmoles of NH_2 N/g of original fresh tissue/hr, or (b) μmoles of NH_2 N/g of protein/hr. Values are given ±S.D., with numbers of estimations in parentheses.

P_1, P_3, A, B, C, P_4, and S_3, respectively, with a total amount of 0.84 mg of protein.

Although results are not clear-cut, fraction P_2 appears to possess the highest activity under both neutral and acidic conditions; acidic proteinase activity is, however, more concentrated in this fraction than the neutral enzyme, which, on the contrary, is much more concentrated in the myelin-containing fraction, P_3. On further purification of the crude P_2 fraction (sedimenting between $1200g_{av}$ for 15 min, and $11{,}000g_{av}$ for 20 min) into subfractions A, B, and C, the myelin-containing subfraction (fraction A), together with other membranous structures (fraction B), becomes enriched with the neutral proteinase enzyme, whereas the highest acidic proteinase activity is displayed by the heavier subfraction C. Fraction P_4 is very poor in both enzymatic activities.* With some exceptions, these results do not differ from those obtained on brain proteinases.[29,42]

Proteinase activity of the supernatant is low if compared with that of the whole homogenate (Table II). However, acidic activity of the supernatant increases significantly if the PN homogenate is left for 60–90 min in the cold and then fractionated. On the contrary, neutral proteinase activity of the supernatant decreases on standing, owing to a general decrease of activity of the whole homogenate; the neutral enzyme is really more labile than the acidic proteinase, and its activity varies noticeably among the different experiments.

Degenerating Peripheral Nerves

Acidic and neutral proteinases of cat sciatic nerve subfractions have been examined during WD for periods up to 32 days from transection. The results, reported in Figs. 1 and 2, are expressed as the percentage change in the sectioned nerve as compared with the control-nerve values of the same animal, determined as μmoles NH_2 N released/g of protein/hr. Similar relationships between normal and damaged nerves have also been obtained by determining the results as μmoles/mg of fresh nerve/hr or as μmoles/cm of nerve/hr. The results indicate that the acidic and neutral proteinase activities are really higher than normal in the sectioned nerve during the time of observation, and that the changes do not merely reflect variations in the weight or protein content of the nerve.

The changes in the acidic proteinase activity (Fig. 1) of subfractions B and C from the injured nerve follow those in the whole nerve homoge-

* It has not been possible to further fractionate subfraction P_4 following deoxycholate treatment. Results were often unreliable.

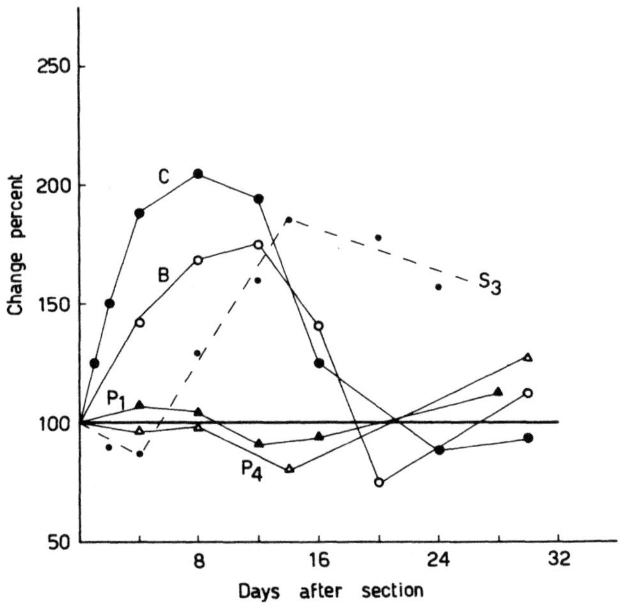

Fig. 1. Acidic proteinase activity in subfractions of normal and degenerating distal cat sciatic nerves at various times after sectioning. Mean values of four estimations each.

nate.[1,13] The activity of subfraction C, the richest source of the acidic proteinase system (Table II), increases by more than 50% of the control nerve at the first days from sectioning, and a further increase is observed at the eighth day; similar, although less evident, changes are displayed by subfraction B. Fractions P_1 and P_4 do not show increase of the acidic proteinase activity, at least for the first 16–20 days from sectioning. The results may signify that material sedimenting with fractions P_1 and P_4 is not responsible for the early increase of acidic proteinase activity during demyelination. The activity of supernatant S_3 increases noticeably up to the 12th–14th day from sectioning, but it displays a latency time before increasing: at the 12th–14th day from transection the acidic proteinase activity of the subfractions B and C is well below the normal levels (Fig. 1).

Neutral proteinase activity of degenerating PN subfractions (Fig. 2) also increases, as in the whole homogenate.[1,4,5,13] The increase is much more evident in the subfractions A and B, which have been equated (by analogy with the brain-tissue experiments) to myelin and other membranous structures, than it is in fraction C (mitochondria, lysosomes?). Moreover, little increase of the enzymatic activity appears in the S_3

fraction of the injured nerves; this result is at variance with the findings reported for the acidic proteinase system (Fig. 1).

DISCUSSION

Present Results

The isolation of pure PN subfractions is a prerequisite for an adequate investigation of proteinase activities during WD. We cannot certainly state that the various subfractions separated with the techniques described in this work are pure. On the contrary, fractions P_1 and P_4, which have not been further fractionated, are presumably contaminated with both lighter and heavier material. Moreover, contamination of these fractions with myelin elements is possible. The other fractions may present more homogeneity, at least as judged by analyses with enzymatic markers. Myelin fractions A and P_3 may be contaminated with portions of adjacent structures, such as axolemma and perinuclear plasma membrane of the

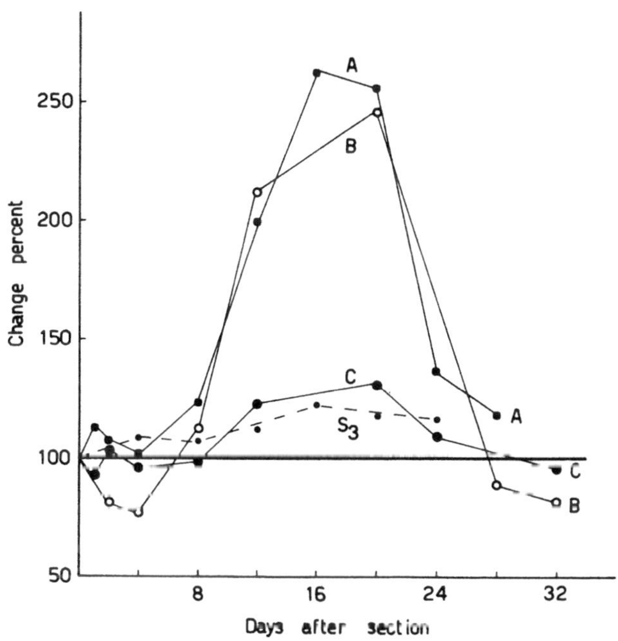

Fig. 2. Neutral proteinase activity in subfractions of normal and degenerating distal cat sciatic nerves at various times after sectioning (mean values of four estimations for each point).

Schwann cells, and, unfortunately, at the present time there are no methods available to solve this problem.

The results of the present contribution show that the acidic proteinase system of normal PN is more concentrated in the mitochondrial (and possibly lysosomal) structures, whereas the neutral proteinase is located preponderantly in myelin and other membranous structures. However, these results have to be considered approximate if related to *in vivo* conditions; in the *in vitro* experiments we may not be using the proper substrates on which the enzymes act *in vivo*. Moreover, the PN protein of the different subfractions may differ deeply in nature, and therefore may interact differently with the proteinase enzymes.

The changes shown by the acidic and neutral proteinases in the subfractions of the degenerating nerve resemble those observed in the whole homogenates,[1,4,5,13] although this applies only to some subfractions. The early enzymatic changes of the acidic proteinase which take place in fractions *B* and *C* and in the supernatant may reflect a characteristic physiochemical and probably morphological change of some of the PN particulate subfractions, rather than a response or a property of the enzymatic system itself. In this last case the change would probably have uniformly affected all the PN subfractions. A general change in the organization of all the nerve lipoprotein membrane systems is unlikely to occur simultaneously in a degenerating PN at an early time. If this would be the case, changes in proteinase activity of all of the particulate fractions and soluble enzyme would have taken place. It is therefore possible that only a selective damage of some of the PN subfractions takes place at the axonal level at early periods from transection.

The acidic proteinase activity of the supernatant S_3 of the degenerating nerve displays a latency time of 5–6 days before increasing, compared to other fractions; at the time this activity begins to increase, that of fractions *B* and *C* is already decreasing. It is worth saying, in this connection, that a similar increase of the acidic proteinase activity was obtained with the supernatant of intact normal nerves if homogenates were left for long periods at room temperature before processing or if they were prepared by other methods. All these results may provide evidence for postulating that some nerve organelles (mitochondria or lysosomes?) become more fragile during WD for a sort of physicochemical conformational change of their lipoprotein architecture, and may more easily release proteinase activity into the supernatant during manipulations or even *in vivo*. This consideration also takes into account some interesting results reported by Adams *et al.*[2,7,16] in this connection. Neutral proteinase activity also increases in the subfractions of a degenerating PN, as in the whole homogenate, but only at later periods from transection. The increase is, however, much more evident in those subfractions which have been equated to peripheral myelin than in other subcellular structures. Unlike acidic proteinase, little increase of the neutral enzymatic system takes place in the supernatant S_3 of the injured nerves. The increase of this enzymatic activity in the PN could be related more to membrane transformation or cellular proliferation than to rupture of subcellular elements or damage with subsequent release of enzymes.

General Discussion

The increase of the acidic proteinase, which represents an early biochemical phenomenon occurring in WD, might be related to the initial damage to axonal mitochondria, which is indicated as one of the earliest signs of Wallerian changes and of other demyelinations,[21–24] and to axonal lysosomes. On the contrary, the late increase of the neutral proteinase activity represents more a chemical response to cellular proliferation or myelin transformation than properly a sign of damage to subcellular particles. The increase seems to represent an effect of many demyelinating processes, being, in this connection, less intense when cellular proliferation is less evident, as during poisoning by organophosphates.[11,12]

The changes of acidic proteinase activity might have some relationship to alterations of PN lysosomes. At the present time little is known about PNS lysosomes, which have, however, been definitely detected in axons.[43] Evidence has been accumulating in these last years from various types of experimental approaches that lysosomes may play an interesting part in demyelination processes, probably acting as an important initiating factor, and that some enzymatic changes which occur at the early stages of the WD might be considered as lysosomal changes. Variation in the membrane permeability of the lysosomes has been reported to occur in injured nerve cell of the motoneurons of the cervical spinal cord and of the hypoglossal nucleus during axon reaction;[44] it has been found that the number of lysosomes also increases during this situation.[44,45] Comparable findings have been reported for the rat lumbar spinal cord after section of the sciatic nerve.[46] Holtzman and Novikoff[47] have recently described morphological alteration of lysosomes in the rat sciatic nerves following injury, and have reported the presence of various types of lysosomes within the altered PN. Comparable results have been reported for transected sciatic nerves in the rabbit.[48] Moreover, increases in lysosomes in the Schwann cytoplasm have been reported by Gould and Holt during WD in the rat,[49] and increased production of primary lysosomes has been reported in the neurons of the superior cervical ganglia of the rabbit after axon reaction.[50]

It has been suggested that these primary lysosomes behave in a phagocytic manner, engulfing small regions of cytoplasm, mitochondria, and endoplasmic reticulum, and digesting them,[50] probably by means of catheptic activity.

Lysosomes therefore are involved in demyelinating processes. Together with mitochondria, they may become more fragile during WD for a sort of physicochemical conformational change of their lipoprotein architecture, their limiting membrane may be easily disrupted, and the organelles may more easily release enzymatic activity *in vivo*, such as the proteinase enzymes.[1] It is known, in this connection, that in the lysosomal group of enzymes cathepsins[29,42,51–55] are considered the enzymes most involved in degradation and in digestive processes,[1,52] as well as in the hydrolysis of the tissue proteins.[55] The change of acidic proteinase activity in the preliminary stages of the process due to release of enzyme from the damaged axon may therefore be an important initiating factor in demyelinations. However, how does this liberation occur? And how does the mitochondrial or lysosomal matrix become damaged during the early stages of the degeneration process, when no detectable signs of histological demyelination occur?

It is known that the lipoprotein membrane of lysosomes is resistant to the lytic agents outside and the hydrolytic enzymes contained within. To bring lysosomal enzymes and exogenous material or structures together, fusion of lysosomes with cell structures has to occur.* Some change in property of the lysosomal membrane system may therefore occur during WD, allowing diffusion of enzymes into the enclosed regions, or the membrane may dissolve. Lysosomes have been indicated recently by Mellors *et al.*[57] to be capable of catalyzing a rapid disruption of liver and heart mitochondria, with an irreversible membrane damage. The alteration closely parallels that produced by the free fatty acids, which are believed to be the principal factors in initiating the swelling of the mitochondria,[57] and to promote the initiation of the digestive phenomena. An acid phospholipase activity has, in fact, been found in the liver lysosomes, which liberates chiefly free fatty acids from endogenous lipids;[58] some workers have also shown that amounts of lysocompounds are formed.[59] Lysocompounds are known to act upon brain homogenates,[60] to injure membranes,[61,62] and to release enzymes from their normal sites within brain.[61] It follows that if free fatty acids and lysocompounds are simultaneously produced in degenerating materials, then a potentiating effect of both toxic compounds on cell damage and disruption may be visualized. Recent

* It has been reported recently that lysosomal hydrolases are able to react with substrate within intact lysosomes *in vitro*.[56]

evidence has indicated that acid phospholipase activity may also be present in lysosomes of nervous tissues,[63] having the character of a phospholipase A_1 and producing both fatty acid and the 2-acyl isomer of lysolecithin. There is considerable evidence, therefore, indicating that a relationship exists between lysosomes and the metabolism of complex lipids.

Free fatty acids and lysocompounds may not only affect cell membranes and their integrity, but also induce the release of proteinases from the damaged membranes of the lysosomes and presumably of the mitochondria, and these enzymes may further produce membrane alterations. In other words, lysocompounds may liberate the enzymes from the myelin (neutral proteinase) or from nerve organelles (acidic proteinase). We have recently obtained experimental evidence[64] which indicates that preparations of lysolecithin (1-acyl isomer), added to the crude P_2 fraction of normal cat sciatic nerve preparations brings about a twofold increase of the acidic proteinase activity, whereas the neutral proteinase enzyme is unaffected. Lysolecithin may act by producing a release of proteinase activity from mitochondria and possibly lysosomes, with subsequent extraction of more enzyme. Recently a detectable increase of the amounts of lysophosphatidylcholine and lysophosphatidylethanolamine has been found in hen, cat, and rabbit PN during the first days from transection,[65–67] thus suggesting that increasing amounts of lysocompounds and free fatty acids may be formed in the PN structures during early periods of demyelination.

In conclusion, a working hypothesis put forward following our experimental results suggests that lysosomal-like particles of PN structures can, for the sudden primary lesion occurring in WD between axon and myelin, catalyze the release of free fatty acids and/or other surface-active lipids from the membranes, which labilize mitochondrial and lysosomal structures and produce early "liberation" of the proteinase enzymes. This enzymatic system may then be one of the causes of the conformational change in the multilayered lipoprotein architecture of the PN structures, with all the subsequent modifications.

Current Research Problems

High-resolution studies of WD are necessary, particularly of the first structural changes that start the drastic destruction of myelin, in order to examine early alterations in the lysosomal and mitochondrial systems. These changes need to be quantitatively documented. These early alterations need to be compared with parallel metabolic events in any PN constituents, myelin sheath, axon, Schwann cells, and surrounding membranes.

Most of the biochemical changes which occur during WD are purely secondary phenomena, with the exception of the change in proteinase activity, which might have a more direct connection with the primary causative mechanisms of the Wallerian process. In addition to the increase in the content of lysophosphatide, the increase of proteolytic activity is an early biochemical phenomenon which occurs in WD. It is still difficult, however, to prove whether the change in activity results in demyelination or is merely a response to the demyelinating process. It is therefore necessary with further work to relate this early increase to the morphological changes of the injured PN and to locate with greater precision the enzymatic activity within the different subfractions of PN. Similarly, further research is necessary to clearly differentiate "neutral" proteinase from other similar enzymes.[68]

The importance of the role of the lysosomes in neuronal degeneration is still uncertain. Their changes may constitute a relatively late occurrence in the sequence of degradative events, rather than a partial cause of the lesion itself; this point needs to be further clarified. On the other hand, there is considerable experimental evidence which indicates a relationship between lysosomes and metabolism of complex lipids. Free fatty acids and lysocompounds, which can be liberated from phospholipids and complex lipids by lysosomal acid hydrolases,[58,69,70] may alter the integrity of lysosomes and mitochondria.[57,69] Following reactions catalyzed by the acid hydrolases, lysosomes may be converted into myelin body residues with membrane arrays of lipid material (see Weinreb *et al.*[69]). If these consideration can be proved for nervous tissues during early phases of WD, then a more understandable picture of the first biochemical events of WD will be obtained, and the lysosomal phospholipases will be considered as important initiators of catabolism and digestion in this connection. Indications about interesting relationships existing among lysocompound action, proteinase "liberation," and alteration of the mitochondrial lipoprotein organization[71,72] also are of value for fully understanding the biochemical findings and mechanisms of nerve degeneration.

REFERENCES

1. G. Porcellati, *in*: "Protides of the Biological Fluids" (H. Peeters, ed.), Vol. 13, pp. 115–126, Elsevier, Amsterdam (1966).
2. C. W. M. Adams (ed.), "Neurohistochemistry," Elsevier, Amsterdam (1965).
3. G. Porcellati, *in*: "Handbook of Neurochemistry" (A. Lajtha, ed.), Vol. VII, Plenum Press, New York (1969).
4. G. Porcellati and B. Curti, *in* "IVth International Congress of Biochemistry, Vienna, 1958," pp. 13–50, International Union of Biochemistry, Vienna (1958).
5. G. Porcellati and B. Curti, *J. Neurochem.* **5**, 277–282 (1960).

6. G. Porcellati and R. H. S. Thompson, *J. Neurochem.* **1,** 340–347 (1957).
8. R. E. McCaman, *in*: "Ultrastructure and Metabolism of the Nervous System" (S. R. Korey, A. Pope, and E. Robins, eds.), pp. 169–181, Williams and Wilkins, Baltimore (1962).
9. R. E. McCaman and E. Robins, *J. Neurochem.* **5,** 32–42 (1959).
10. J. J. Bubis and M. Wolman, *Israel J. Med. Sci.* **1,** 410–414 (1965).
11. G. Porcellati, A. Millo, and I. Manocchio, *J. Neurochem.* **7,** 317–320 (1961).
12. G. Porcellati, *in*: "Proc. 7th Intern. Congr. Neurology, Neurochemistry Symposium, Rome, 13th September, 1961," pp. 815–822, Societa Grafica Romana, Rome (1961).
13. G. Porcellati, *Il Farmaco* **20,** 586–605 (1965).
14. F. Wolfgram, *Ann. N.Y. Acad. Sci.* **122,** 104–115 (1965).
15. G. Porcellati, *in*: "Handbook of Neurochemistry" (A. Lajtha, ed.), Vol. II, Plenum Press, New York (1969).
16. C. W. M. Adams and O. B. Bayliss, *J. Histochem. Cytochem.* **16,** 110–114 (1968).
17. M. Wolman, *in*: "Chemical Pathology of the Nervous System" (J. Folch-Pi, ed.), pp. 254–259, Pergamon Press, London (1961).
18. M. L. Karnovsky and G. Majno, *in*: "Chemical Pathology of the Nervous System" (J. Folch-Pi, ed.), pp. 261–266, Pergamon Press, London (1961).
19. M. Wolman, *Biochim. Biophys. Acta* **102,** 261–268 (1965).
20. C. W. M. Adams, A. N. Davison, and N. A. Gregson, *J. Neurochzm.* **10,** 383–395 (1963).
21. H. de F. Webster, *J. Cell Biol.* **12,** 361–383 (1962).
22. G. Kreutzberg and W. Wechsler, *Acta Neuropath.* **2,** 349–361 (1963).
23. S. Blümcke, H. R. Niedorf, and J. Rode, *Acta Neuropath.* **7,** 44–61 (1966).
24. W. E. Watson, *J. Neurochem.* **13,** 849–856 (1966).
25. J. Fischer and V. Malík, *Acta Histochem.* **19,** 369–376 (1964).
26. H. S. Bachelard and G. D. Silva, *Arch. Biochem. Biophys.* **117,** 98–105 (1966).
27. V. P. Whittaker, *Biochem. J.* **72,** 694–706 (1959).
28. E. de Robertis, A. Pellegrino de Iraldi, G. Rodriguez de Lores Arnaiz, and L. Salganicoff, *J. Neurochem.* **9,** 28–42 (1962).
29. N. Marks and A. Lajtha, *Biochem. J.* **89,** 438–447 (1963).
30. G. Porcellati, *Acta Neurol.* **22,** 168–172 (1967).
31. L. Svennerholm, *J. Neurochem.* **1,** 42–53 (1956).
32. H. H. Lefler and C. H. McDougald, *Amer. J. Clin. Pathol.* **39,** 311–316 (1963).
33. R. H. Laatsch, M. Kies, S. Gordan, and E. C. Alvord, *J. Exp. Med.* **115,** 777–784 (1962).
34. S. J. Cooperstein and A. Lazarow, *J. Biol. Chem.* **189,** 665–674 (1951).
35. G. Hübscher and G. R. West, *Nature* **205,** 799–800 (1965).
36. R. Gianetto and C. De Duve, *Biochem. J.* **59,** 433–438 (1955).
37. A. B. Roy, *Biochem. J.* **53,** 12–15 (1953).
38. O. H. Lowry, N. R. Roberts, M. L. Wu, W. S. Hixon, and E. J. Crawford, *J. Biol. Chem.* **207,** 10–37 (1954).
39. R. M. C. Dawson, N. Hemington, and J. B. Davenport, *Biochem. J.* **84,** 497–501 (1962).
40. N. Marks and A. Lajtha, *Biochem. J.* **97,** 74–83 (1965).
41. O. H. Lowry, N. J. Rosebrough, A. L. Farr, and R. J. Randall, *J. Biol. Chem.* **193,** 265–274 (1951).
42. A. Lajtha and N. Marks, *in*: "Protides of the Biological Fluids" (H. Peeters, ed.), Vol. 13, pp. 103–114, Elsevier, Amsterdam (1966).
43. D. R. Sharma and H. A. Matzke, *Anat. Rec.* **151,** 59–68 (1965).

44. K. D. Barron and S. Sklar, *Neurology* **11,** 866–879 (1961).
45. K. D. Barron and T. O. Tuncbay, *Amer. J. Path.* **40,** 637–652 (1962).
46. R. P. Gould and S. J. Holt, *J. Physiol.* (*London*) **184,** 2-3P (1966).
47. E. Holtzman and A. B. Novikoff, *J. Cell Biol.* **27,** 651–669 (1965).
48. S. Blümcke, H. R. Niedorf, and J. Rode, *Acta Neuropath.* **7,** 44–61 (1966).
49. R. P. Gould and S. J. Holt, *in*: "Cytology of Nervous Tissue," Supplement, Symposium of Anatomical Society of Great Britain and Ireland, p. 45, London (1961).
50. J. S. Dixon, *Nature* **215,** 657–658 (1967).
51. A. J. Barrett, *Biochem. J.* **104,** 601–608 (1967).
52. P. L. Sawant, I. D. Desai, and A. L. Tappel, *Biochim. Biophys. Acta,* **85,** 93–102 (1964).
53. K. F. Firfarova, "5th Meeting of the Federation of European Biochemical Societies, Prague, 1968," Commun. n. 1038, Prague University Press (1968).
54. J. M. W. Bouma and M. Gruber, *Biochim. Biophys. Acta* **113,** 350–358 (1966).
55. J. W. Coffey and C. De Duve, *J. Biol. Chem.* **243,** 3255–3263 (1968).
56. M. Rosenberg and A. Janoff, *Biochem. J.* **108,** 889–891 (1968).
57. A. Mellors, A. L. Tappel, P. L. Sawant, and I. D. Desai, *Biochim. Biophys. Acta* **143,** 299–309 (1967).
58. A. Mellors and A. L. Tappel, *J. Lip. Res.* **8,** 479–485 (1967).
59. W. Stoffel and H. Greten, *Hoppe-Seyl. Z.* **348,** 1145–1150 (1967).
60. G. R. Webster, *Nature* **180,** 660–661 (1957).
61. E. A. Marples, R. H. S. Thompson, and G. R. Webster, *J. Neurochem.* **4,** 62–70 (1959).
62. R. H. S. Thompson, *in*: "Metabolism and Physiological Significance of Lipids" (R. M. C. Dawson and D. N. Rhodes, eds.), pp. 541–552, J. Wiley and Sons, London (1964).
63. S. Gatt, *Biochim. Biophys. Acta* **159,** 304–316 (1968).
64. G. Porcellati and I. Montanini, unpublished observations.
65. J. Domonkos and L. Heiner, *J. Neurochem.* **15,** 87–91 (1968).
66. J. F. Berry, W. H. Cevallos, and R. R. Wade, *J. Am. Oil Chem. Soc.* **42,** 492–500 (1965).
67. G. Porcellati, *Acta Neurol.* **24,** 119–124 (1969).
68. P. J. Riekkinen and U. K. Rinne, *Brain Res.* **9,** 126–135 (1968).
69. N. J. Weinreb, R. O. Brady, and A. L. Tappel, *Biochim. Biophys. Acta* **159,** 141–146 (1968).
70. S. Mahadevan and A. L. Tappel, *J. Biol. Chem.* **243,** 2849–2862 (1968).
71. E. C. Weinbach and J. Garbus, *Biochim. Biophys. Acta* **162,** 500–505 (1968).
72. I. Honjo and K. Ozawa, *Biochim. Biophys. Acta* **162,** 624–627 (1968).

DISCUSSION

RICHTER: You relate the increase in activity of the neutral proteinase after nerve section to membrane damage and/or cellular proliferation; I wonder whether stimulation may also play any part in this change.

PORCELLATI: You are right; stimulation may also give the same type of change. The increase of the neutral proteinase activity in our experiments is, however, related I think, to histologically detectable demyelination. It does not occur if we inject non-demyelinating organophosphorus compounds, and it increases when histological demyelination starts; therefore, I believe it is an effect of the demyelination, and probably more related to membrane changes. By the 8th–10th day after sectioning myelin really changes, probably in conformational structure, shape, and composition.

MANDEL: Is the increase of proteinase activity due to increase of enzymes, that is, to a *de novo* synthesis, or to increase of activity caused by activation phenomena, or is it due to disappearance of inhibitors? Under all these conditions you can get increases of enzymatic activity. In order to know if, for instance, there are compounds which increase activity, you should do cross-incubation experiments, taking half of the nerve preparation from one section and half from another. If you have production of some stimulating or releasing substances, you may have activity which is higher than normal, and if you have a decrease, this result may indicate presence of inhibitory compounds.

PORCELLATI: I am sure these experiments can be carried out from this point of view. I want to say that substrate concentration and structure can also play a role in enzyme activation. It may be that the proteins are changing in conformation and structure during nerve damage, and may become more available to enzymes, although their concentration does not vary quantitatively from controls, at least during the first periods after sectioning.

COHEN: There have been some studies on phosphatases in cut peripheral nerves, and a certain number of the results obtained might be of some interest. The acid phosphatase activity not only increases during nerve degeneration, but, if you have a nerve crush with ultimate regeneration, you have an increase in the distal portion, not only upon degeneration, but also upon regeneration. You have higher levels well after a year, so that these studies are well worth carrying on for a long period of time. Thereafter phosphatase activity goes down, but after you get return of function apparently you still have higher levels than normal. There is also some evidence, using detergents, that for the acid phosphatase there is a release of structural bound enzyme during nerve degeneration together with production of acid phosphatase in the activated Schwann cells. In addition, there was also a study done 10 years ago by Samorajski, who cut sciatic nerves of rabbits and, besides using the uncut nerve on the opposite side as a control, also assayed nerves on unoperated animals. He claimed that for both acid phosphatase and alkaline phosphatase there was a marked rise on the unoperated side (the control side), compared to control animals which had not been operated, which suggests that at least some precaution should be taken to see whether controls in these conditions really are controls.

PORCELLATI: I have never had for the proteinase enzymes the type of result you mention. The only thing I may say in regard to phosphatases is that in the degenerating nerves the acid phosphatase activity increases first, and the alkaline enzyme increases much later.

SELLINGER: I would like to comment on this very interesting model in which lysolecithin would activate lysosomes so as to release some of the enzymes. Actually, I think lysolecithin may have been among the many compounds that have been tested to see whether or not they act as lysosomes activators *in vitro*. Do you know whether there is a system where this particular compound has been shown to exert this kind of effect on purified lysosomes? If so, this would be a good correlation to make.

PORCELLATI: I don't think that there have been any studies on the direct action of lysolecithin in lysosomal rupture and damage, and in the release of enzymes, although results have been reported on mitochondrial damage and alterations of its metabolic systems. The only thing I ever did was to try to find out whether acid or neutral proteinases were activated in the presence of lysolecithin, and we got results of activation, or, say, of "liberation," of acid proteinase activity, but not of the neutral enzyme.

LAJTHA: We were looking at the differences in acid proteinases and have some evidence that there is more than one, but couldn't really find a specific one which we could specify as lysosomal. Can you tell us anything about the specificity of this particular enzyme?

PORCELLATI: We wanted to do this; however, we have no clear answer in regard to this problem.

MARKS: As to the source of these enzymes, I think it is agreed that during Wallerian degeneration there is an infiltration of many nonaxonal elements into the lesion, so that it is possible that some of these proteinases may arise from macrophage lysosomes, and so forth.

PORCELLATI: I agree completely. What really struck me is that the activity of proteinases increases in the very early periods after sectioning, when, certainly, macrophages will not be invading myelin structure. I think macrophagocytosis has been described as taking place during demyelination at the sixth or seventh day after transection, histologically speaking.

Chapter 32

Inhibition of Brain Protein Synthesis*

Stanley H. Appel

Division of Neurology
Duke University Medical Center
Durham, North Carolina

An understanding of the molecular basis of normal and altered neurological function represents one of the significant challenges of modern neurology. In many neurologic disorders we can specify the enzymatic defect and the resultant alterations in precursors and products. But even in conditions so defined it remains unclear what altered brain functions produce the devastating neurologic disorder. The amino acidurias serve as an excellent example of disorders whose enzymatic and neuropathologic defects are reasonably well delineated, but whose brain dysfunction cannot be defined biochemically.[1] Alterations in neurohormones and in mitochondrial function have been implicated, but not proven.[2–6] For our present purposes the most intriguing hypotheses relate to alterations in protein synthesis which may occur as the result of amino acid imbalances.[7,8] In this report we review some of our experimental data which confirm reported alterations in amino acid transport produced by amino acid imbalances,[9–12] and suggest that both by inhibition of transport and by interference with amino acid activation, a significant inhibition of protein synthesis may occur *in vitro*. Assuming that protein synthesis is a rate-limiting reaction in brain function, studies of amino acid inhibition *in vitro* may represent a meaningful approach to the understanding of neurologic dysfunction at a molecular level.

Our initial studies direct attention to the role of branched-chain amino acids and their α-keto analogs in inhibiting protein synthesis in brain slices.[8] Slices were obtained from cerebral cortices of 80-g male Wistar rats, and incubations were performed in Krebs–Ringer phosphate

* This work was supported by the William Krug Memorial Grant, No. 558-A-2, from the National Multiple Sclerosis Society; and Grant No. NB-07872 from the National Institutes of Health, United States Public Health Service.

TABLE I

Effect of Amino Acid Excess on ^{14}C-Amino Acid Uptake and Incorporation into Protein in Brain Tissue Slices

^{14}C amino acid	^{12}C amino acid (2.0 mM)	Tissue conc. (% of control)	Protein (% of control)
Valine ^{14}C	None	100	100 (2486 cpm/100 mg)
	Leucine	38	36
	Isoleucine	24	42
	α-Ketoisocaproic	63	64
	α-Ketoisovaleric	96	42
Leucine ^{14}C	None	100	100 (5636 cpm/100 mg)
	Isoleucine	31	40
	Valine	35	38
Isoleucine ^{14}C	None	100	100 (5336 cpm/100 mg)
	Leucine	29	9
	Valine	29	22
Arginine ^{14}C	None	100	100 (3119 cpm/100 mg)
	Leucine	63	67
	Isoleucine	75	62
	α-Ketoisocaproic	96	119
	α-Ketoisovaleric	88	126

buffer with 12 mM dextrose at 37° for varying time periods. The appearance of radioactive amino acids in the acid-soluble phase reached a steady state within 5–10 min. At steady state conditions ^{14}C-valine had accumulated intracellularly 1.75-fold compared to the incubation medium. ^{14}C-Leucine had accumulated 1.6-fold; isoleucine, 1.8-fold; arginine, 0.98-fold; and ^{14}C-tyrosine, 2.2-fold. The amino acid accumulation was inhibited by substances which interfere with maintenance of energy metabolism, such as 2,4-dinitrophenol, oligomycin, and potassium cyanide.

In these studies no attempt was made to separate the steady-state accumulation into the separable processes of influx and efflux, both of which would contribute to the steady-state level.[12]

After completing these baseline studies ^{12}C-amino acids were added to the incubation medium to assess the extent of inhibition of steady-state accumulation of ^{14}C-amino acids. More specifically, with valine as the radioactive amino acid, ^{12}C-leucine, isoleucine, α-ketoisocaproic acid, and α-ketoisovaleric acid were all able to inhibit radioactive valine accumulation (Table I). ^{12}C-Leucine was inhibitory for most of the ^{14}C-amino acids employed. Detailed kinetics of these inhibitions were not performed. However, the results were in accord with the observations of other investigators[12] in which leucine was found to block uptake and to enhance efflux of other amino acids. Thus, a low influx and high efflux would lead to lower amino acid levels than might occur with altered influx alone.

Having demonstrated the pattern of inhibition of amino acid transport, we turned our attention to the consequences of such inhibition for protein synthesis in brain slices. In contrast to the rapid equilibrium achieved with amino acid accumulation, the incorporation of amino acid into protein was linear for at least 60 min. Under these conditions, as noted also in Table I, ^{12}C-amino acids inhibited the incorporation of the ^{14}C-amino acids into protein as extensively as they inhibited the steady-state accumulation of these compounds. The correspondence between the extent of inhibition of amino acid accumulation and inhibition of its incorporation into protein was striking. At first approximation, these data suggested that alteration of transport of the radioactive amino acid was responsible for inhibition of radioactivity incorporation into protein. However, the fact that leucine inhibited incorporation of isoleucine into protein far more than it inhibited isoleucine accumulation into the slice stood out, and sug-

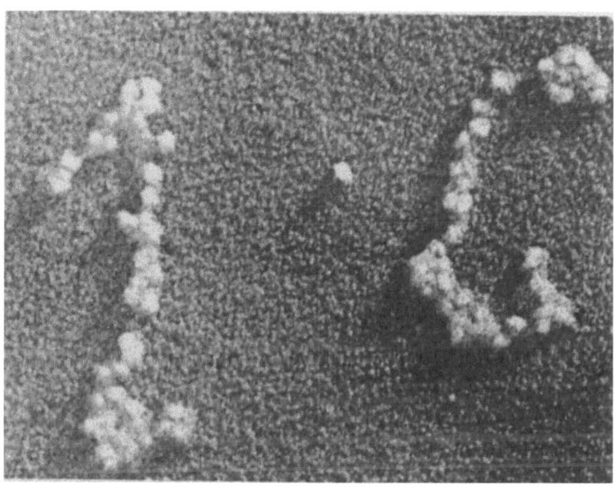

Fig. 1. Rat-brain polysomes. Polysomes were isolated in a continuous sucrose gradient as previously described.[13] A sample was selected from the heavy region of the gradient, air-dried, shadowed with a carbon–platinum pellet, and examined in a Phillips EM-200 microscope. (×100,000.)

TABLE II

Effect of Excess Amino Acids on Cell-Free Protein Synthesis

Branched-chain amino acids (^{14}C amino acid added)	^{12}C amino acid (2.0 mM)	Per cent of control
Valine ^{14}C (0.25 μCi /0.010 μM)	None	100
	Leucine	74
	Isoleucine	77
	α-Ketoisocaproic	68
	α-Ketoisovaleric	46
Leucine ^{14}C (0.25 μCi /0.010 μM)	None	100
	Isoleucine	36
	Valine	95
	α-Ketoisocaproic	63
	α-Ketoisovaleric	108
Isoleucine ^{14}C (0.25 μCi /0.010 μM)	None	100
	Valine	72
	Leucine	34
	α-Ketoisocaproic	63
	α-Ketoisovaleric	100
Arginine ^{14}C (0.25 μCi /0.010 μM)	None	100
	Leucine	91
	Isoleucine	93
	α-Ketoisocaproic	108
	α-Ketoisovaleric	162
Tyrosine ^{14}C (0.25 μCi /0.005 μM)	None	100
	Phenylalanine	39
	Phenylalanine	73
	Phenylalanine	90

gested a possible effect on protein synthesis which may be independent of the transport into the cell.

A cell-free protein-synthesizing system was prepared from rat brain which was dependent upon the presence of ribosomes, brain supernatant pH-5 fraction, ATP, GTP, and an energy-generating system.[13] When the ribosomes were centrifuged in a 15–50% continuous sucrose density gradient, they migrated as polysomal aggregates. Figure 1 represents a of platinum–carbon shadowing of such ribosomes examined by electron microscope (100,000×). The polysomes can be disaggregated into single 80-S monosomes in the presence of ribonuclease; however, neither desoxyribonuclease, nor lipolytic or proteolytic enzymes could disaggregate these polysomes. With this system, incorporation of radioactive valine into protein was linear for at least 30 min and was sensitive to ribonuclease and puromycin, but not to desoxyribonuclease nor actinomycin D. With this *in vitro* system we noted a definite reproducible inhibition by the 19 ^{12}C-amino acids (minus valine) on the incorporation of ^{14}C-valine into protein. On further investigation the inhibition of valine incorporation could be reproduced by two of the amino acids, isoleucine and leucine, as well as their α-keto analogs (Table II). ^{14}C-Leucine incorporation is similarly inhibited by isoleucine and valine, and ^{14}C-isoleucine by valine and leucine. A similar inhibition of ^{14}C-tyrosine into protein was demonstrated with phenylalanine or some of its closely related derivatives.

Although kinetic studies cannot be given full significance in such a

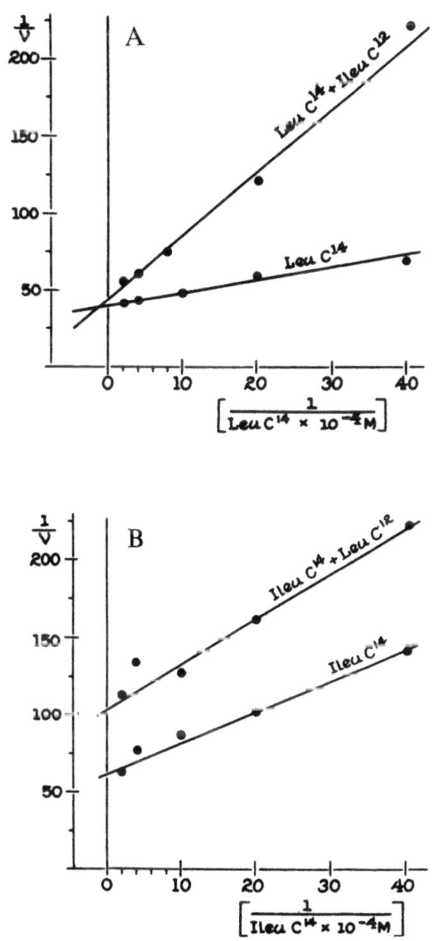

Fig. 2. Kinetics of amino acid incorporation into brain-slice protein. Incubations were performed with varying concentrations of ^{14}C-amino acid and 2.0 m*M* ^{12}C-amino acid for 30 min with 100 mg rat cortical slices in 3.0 ml of Krebs–Ringer phosphate medium with 12 m*M* glucose. Proteins were precipitated by sonication in 5% trichloracetic acid and then processed as previously described.[13] Results are expressed as double reciprocals.

multicomponent system, it was of interest that the effect of ^{12}C-isoleucine on the incorporation of ^{14}C-leucine into protein had the appearance of a competitive interaction, whereas the inhibitory effect of ^{12}C-leucine on isoleucine incorporation into protein in the cell-free system was apparently noncompetitive or uncompetitive [Figs. 2(A) and (B)].

In the cell-free system we have no transport barriers, and yet an inhibition similar to that described in brain slices can be demonstrated. Cell-free systems can be separated into two distinct phases: (1) the enzymatic activation of an amino acid and subsequent reaction with its specific transfer RNA, and (2) the formation of the polypeptide chain from aminoacyl transfer RNA's assembled on ribosomes containing messenger RNA template. Although radioactive aminoacyl sRNA can be readily incorporated into acid-insoluble material, with the ribosomal system preferred, we have been unable to demonstrate an inhibition of this incorporation by leucine or isoleucine. Thus, it seems unlikely that our observed inhibition occurs at the level of amino acid assembly on ribosomes.

Amino acid activation and formation of aminoacyl sRNA can be readily demonstrated either with brain pH-5 fraction or supernatant. We employed the assay of Loftfield and Eigner[(14)] in which the formation of ^{14}C-aminoacylhydroxamates is used to assess the presence of amino acid-activating enzymes (Table III,A). In this system a significant fraction of the amino acid-activating enzymes is isolated with the brain microsomal fraction and cannot be removed even with extensive washing. However, the majority of activity is present in the soluble fraction. With either particulate or soluble enzymes isoleucine will inhibit the formation of ^{14}C-leucine hydroxamate approximately 40% at low substrate concentrations and 10% at higher substrate concentrations (Table III,B). Leucine will inhibit the formation of ^{14}C-isoleucine hydroxamate approximately 30% at both low and high substrate concentrations (Table III,C). These results are in accord with those noted in the overall cell-free system, in which leucine's effect on isoleucine incorporation appeared to be essentially noncompetitive or mixed-competitive, whereas isoleucine's effect on leucine incorporation was clearly competitive. These data suggest that the activation of the amino acid to form the aminoacyl–AMP enzyme complex may be inhibited prior to the reaction with the specific transfer RNA.

The inhibition of an activating enzyme by a related amino acid recalls the experiments of Bergman *et al.*[(15)] in which a purified isoleucine activating enzyme from *E. coli* was shown to be inhibited by large concentrations of valine. However, from *E. coli* the activated valine could not be donated either to a specific valine or leucine sRNA. If the mechanism were similar in brain, high levels of one amino acid could effectively compete with lower levels of a related compound for a specific amino acid-activating en-

TABLE III

Assay of Amino Acid Activation[a]

			cpm/200 μg protein
A. Complete System (^{14}C-Leu., 28 × $10^{-5}M$)			8310
−ATP-Mg			264
−Brain enzyme			464
−Hydroxylamine			186
B. ^{14}C-Amino acid	Control	^{12}C-Ileu. (2.8 mM)	Per cent of control
^{14}C-Leu. (2.8 × 10^{-5} M)	2307	1380	60
^{14}C-Leu. (28 × 10^{-5} M)	8007	7201	90
	Control	^{12}C-Leu. (2.8 mM)	Per cent of control
C. ^{14}C-Ileu. (2.8 × 10^{-5} M)	201	143	71
^{14}C-Ileu. (28 × 10^{-5} M)	1509	963	64

[a] Incubations were at 37° for 20 min in 0.14 ml containing 4 μM ATP; 10 μM $MgCl_2$; 0.5 mM hydroxylamine (pH 7.0); and 0.6–0.8 mg brain 100,000g supernatant protein. Samples (0.05 ml) were processed according to Loftfield and Eigner.[14] 0.02 μM ^{14}C-leucine = 98,140 cpm; 0.02 μM ^{14}C-isoleucine = 125,240 cpm.

zyme and impair the overall process of protein synthesis. By analogy with the bacterial system, the consequence would be primarily a change in the rate of protein synthesis rather than the nature of the synthesized protein.

We might then postulate that in the amino acidurias, such as maple syrup urine disease, the critical defect is a lowered rate of protein synthesis secondary to a block in transport and activation of amino acids. The recent reports of Prensky and Moser,[16] analyzing the brains of two patients with maple syrup urine disease, may be particularly relevant. Proteolipids were found to be decreased to approximately $\frac{1}{3}$ of control values, while their amino acid composition and that of residual proteins were essentially normal. Although there are several possible explanations of this result, a prominent possibility is the impairment of the rate of protein synthesis by toxic levels of leucine or its α-keto derivatives.

The critical question is whether impairment of protein synthesis in the developing brain will cause long-term effects on information processing, storage, and upon neural activity. The experiments of Flexner *et al.*,[17] Agranoff *et al.*[18] and Barondes and Cohen[19] are relevant in their demonstration that inhibition of protein synthesis and alterations of macromolecular metabolism can affect information storage in brain function in mice and goldfish. The studies demonstrated above would suggest that alterations in transfer and competition at the amino acid-activating enzyme level may inhibit the rate of protein synthesis without affecting the specific amino acid composition of the proteins synthesized.

The question still remains, however, as to what particular structure or what particular function is most immediately affected by this inhibition of protein synthesis. Since the time of Cajal, the point of approximation of two communicating cells, the synapse, has often been implicated as the site where transient or permanent changes may influence information transfer. Recent studies from our laboratory have been concerned with defining the biochemical properties of synapses *in vitro*.[20,21] These experiments demonstrated that nerve endings isolated from rat-brain cortex incorporate amino acids into membrane protein *in vitro*.[20] The incorporation was linear for 20 min and continued for approximately 50 min. Such synthesis was inhibited 70–80% by puromycin and cycloheximide and 25% by chloramphenicol. The protein synthesis was not dependent upon exogenous substrates or energy supplies, including ATP, ADP, AMP, or GTP. The ionic content of the medium appeared critical: 100 mM sodium and 10 mM potassium resulted in a fourfold stimulation over control incubations without any ions.[21] Ouabain (5×10^{-4} M) produced a 50% inhibition, and dinitrophenol, oligomycin, and potassium cyanide were all potent inhibitors. The contributions of bacteria, extrasynaptosomal ribosomes, microsomes, or mitochondria were ruled out by the electron-microscope monitoring, the ionic activation and antibiotic inhibition, and the kinetics of incorporation.

The uptake of amino acids into synaptosomes possessed many characteristics of the brain-slice system. It was dependent upon sodium concentration, upon endogenous energy availability, and upon sodium-potassium-ATPase activity.[22] The apparent K_m for leucine uptake was 3.0 mM in the absence of sodium and 0.7 mM in its presence. In addition, in preliminary experiments radioactive amino acid uptake was inhibited by closely related ^{12}C-amino acids in a fashion similar to the results reported with whole slices.

It thus becomes possible to suggest that an altered pattern of amino acids may slow the rate of synaptic protein synthesis as well as that of neuronal and glial synthesis as previously demonstrated in brain slice or cell-free ribosomal systems. Inhibition of any or all of these processes may be critical in determining the devastating effects noted in the amino acidurias. Previous investigators have stressed the apparent alterations in myelin as being the most critical event in the pathogenesis of behavioral dysfunction. However, there is no evidence that such alterations are associated with altered axonal conduction or impaired neural activity. During the critical periods of differentiation in the first year of life not only myelin formation, but also neuronal growth and development and synapse formation are proceeding at a rapid rate. We suggest that the pattern of synaptic connections between cells and the efficiency of such synapses in interneuronal communication may represent the key to the organization and function of the nervous system. Thus, in developing organisms such synapses may be extremely sensitive to alterations in their macromolecular metabolism or membrane constituents which could be produced by the amino acidurias. Such synaptic alterations and dysfunction would not be expected to show up on conventional light microscopy. The effects may well be devastating on the growth of new synapses and the efficiency of those already established, and may explain why disturbances during early synapse formation may be more catastrophic than perturbations of a more stable differentiated adult system.

Implicating the synapse as the target of the dysfunction in metabolic disorders affecting the nervous system has the further advantage that it may explain both short-term reversible and long-term irreversible deficits. The rapid neurological deterioration noted by Snyderman *et al.*[23] following administration of leucine may be explained by short-term effects at the synapse interfering with transport and resulting in depolarization as noted in pancreas with excess leucine.[24] Sustained alteration in transport and amino acid activation may impair macromolecular metabolism and severely compromise newly forming synapses and intercellular communication being established in the developing brain. Further work is obviously necessary to validate these hypotheses.

REFERENCES

1. J. B. Stanbury, J. B. Wyngaarden, and D. S. Fredrickson, "The Metabolic Basis of Inherited Disease," 2nd ed., McGraw-Hill, New York (1966).
2. J. H. Fellman, *Proc. Soc. Exp. Biol. Med.* **93,** 413 (1956).
3. A. N. Davison and M. Sandler, *Nature* **181,** 186 (1958).
4. R. E. Tashian, *Metabolism* **10,** 393 (1961).
5. R. K. Howell and M. Lee, *Proc. Soc. Exp. Biol. Med.* **113,** 660 (1963).
6. H. E. Himwich and J. F. Fazekes, *Arch. Neurol. Psychiat.* **51,** 73 (1944).
7. I. Bickis, J. P. Kennedy, and J. H. Quastel, *Nature* **179,** 1124 (1957).
8. S. H. Appel, *N.Y. Acad. Sci. (Trans.)* **29,** 63 (1966).
9. G. Guroff and S. Udenfriend, *J. Biol. Chem.* **237,** 803 (1962).
10. G. Guroff, W. King, and S. Udenfriend, *J. Biol. Chem.* **236,** 1773 (1961).
11. K. D. Neame, *J. Neurochem.* **11,** 67 (1964).
12. A. Lajtha, G. Levi, and R. Blasberg, *in*: "Brain Edema" (I. Klatzo and F. Seitelberger, eds.), pp. 367–381, Springer-Verlag, New York (1967).
13. S. H. Appel, W. Davis, and S. Scott, *Science* **157,** 836 (1967).
14. J. B. Loftfield and E. A. Eigner, *Biochem. Biophys. Acta* **72,** 372 (1963).
15. F. H. Bergman, P. Berg, and M. H. Dieckman, *J. Biol. Chem.* **236,** 173 (1961).

16. A. L. Prensky and H. W. Moser, *J. Neurochem.* **11,** 77 (1964).
17. J. B. Flexner, L. B. Flexner, and E. Stellar, *Science* **141,** 57 (1963).
18. B. W. Agranoff, R. E. Davis, and J. J. Brink, *Proc. Natl. Acad. Sci. U.S.* **54,** 788 (1965).
19. S. Barondes and H. D. Cohen, *Science* **151,** 594 (1966).
20. L. Autilio, S. H. Appel, P. Pettis, and P. L. Gambetti, *Biochemistry* **7,** 2615 (1968).
21. A. V. Escueta and S. H. Appel, *Biochemistry,* **8,** 725 (1969).
22. S. H. Appel, L. Autilio, B. W. Festoff, and A. V. Escueta, "Biochemical studies of synapses in vitro. III. Ionic activation of protein synthesis," *J. Biol. Chem.* **244,** 3166 (1969).
23. S. Snyderman, P. M. Norton, E. Roitman, and L. Holt, Jr., *Pediatrics* **34,** 454 (1964).
24. P. Dean and E. Mathews, *Nature* **219,** 389 (1968).

Chapter 33

Studies on a Tissue-Specific Histone from Pig Brain*

Steven Edward Kornguth

Departments of Neurology and Physiological Chemistry
University of Wisconsin Medical Center
Madison, Wisconsin

INTRODUCTION

I would like to describe some of the biochemical properties and the localization by immunochemical techniques of a tissue-specific basic protein isolated from pig brain. Much of this study is the result of efforts by Mr. Lawrence Tomasi of my laboratory. The protein was isolated as previously described.[1]

The pig brains were repeatedly extracted with a chloroform–methanol solution (2:1) by homogenization. The residue was dried with acetone, washed with water and with 10% sodium chloride, and then repeatedly extracted with 0.01 *N* HCl. The material extracted between pH 2.4 and pH 2.6 was further purified as follows: the pH was adjusted to 5.0 and the supernatant solution was treated with ammonium sulfate. The fraction precipitating from solution between 25 and 40% saturation with ammonium sulfate was purified further by ion-exchange chromatography on IRC-50 using a concentration gradient of guanidinium chloride. The basic protein, which was eluted from the column as a discrete peak at a concentration of 9% guanidinium chloride, was utilized for further experiments.

The protein was shown to be homogeneous by the criteria of electrophoresis, ultracentrifugation, N-terminal analysis, and gel filtration on Sephadex G-100. The protein migrated toward the cathode as a single band at pH 4.3 (Fig. 1) and at pH 8.6. Amino acid analysis of the acid hydrolyzates[1] indicated that the basic amino acids (lysine, arginine, and histidine) comprise 24 mole % and the dicarboxylic amino acids

* This study was supported by Grant No. NB-05631 from the National Institutes of Health.

TABLE I
Amino Acid Analysis of Purified Basic Protein from Brain

Amino acid	Pig histone (moles %)[a]	Guinea pig myelin protein
Lysine	8.1	7.7
Histidine	5.9	2.6
Arginine	9.9	4.5
Aspartic acid	6.2	8.3
Threonine	4.3	4.6
Serine	8.3	5.8
Glutamic acid	6.6	11.5
Proline	7.8	4.8
Glycine	15.1	6.9
Alanine	9.2	8.1
Half-cystine	0	0.7
Valine	1.7	5.5
Methionine	1.3	1.9
Isoleucine	1.9	3.5
Leucine	5.8	7.8
Tyrosine	2.9	2.4
Phenylalanine	4.9	3.5
Ammonia	9.9	9.8

[a] Values are reported as moles of amino acid per 100 moles of all amino acids recovered after 24-hr hydrolysis in 6 *N* HCl at 110°. No correction has been made for the hydrolytic losses of amino acids.

(aspartic and glutamic) 12 mole % of the protein (Table I). These data are consistent with the cationic properties of the molecule. Other unusual characteristics of this protein are the high glycine content (15 mole %) and high proline content (8 mole %). Neither cysteine nor the oxidized products of cysteine was found. Glycine was shown to be the only N-

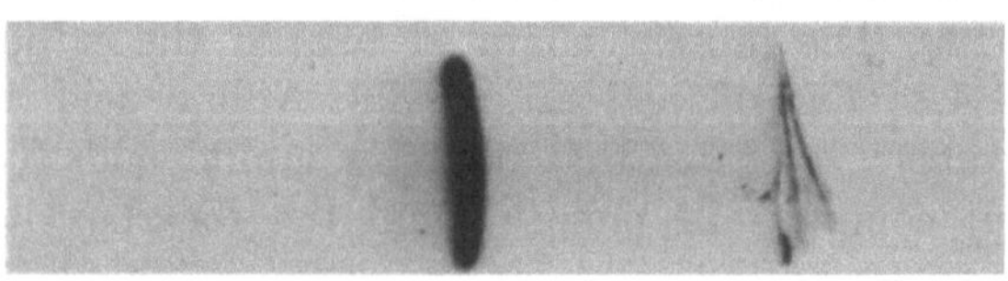

Fig. 1. Electrophoretic pattern on 15% polyacrylamide gel of the chromatographic component eluted from amberlite 1 RC-50 with 8–10% guanidinium chloride. Electrophoresis was performed at pH 4.3, 3 mA per gel for 3 hr.

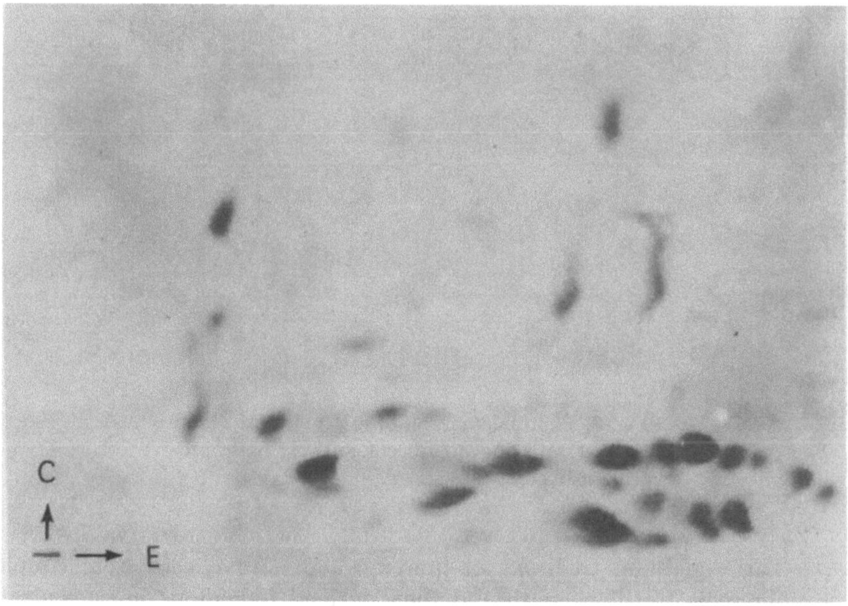

Fig. 2. Peptide map of the tryptic digest of the pig-brain basic protein. Descending chromatography was run for 26 hr with the upper phase of *n*-butyl alcohol–acetic acid–water (4:1:5) as the developing solvent. Electrophoresis was performed in pyridine-acetate buffer, pH 3.7, at 60 V/cm for 1 hr. Peptides were visualized with the cadmium-ninhydrin reagent. Here *E* represents the electrophoretic direction and *C* the chromatographic direction.

terminal amino acid, by the phenylisothiocyanate method. The protein sedimented as a single symmetrical peak with a $S_{20,w}$ of 1.5*S* (phosphate buffer pH 7.4; 0.2 ionic strength). Fingerprints of the tryptic digests of this protein indicated that the molecular weight of the pig-brain protein is 27,000 (Fig. 2).[2]

IMMUNOHISTOCHEMICAL STUDIES

Antibodies were prepared against this protein and characterized by the Ouchterlony and immunoelectrophoretic techniques.[2] Only one precipitin line was seen by both of these methods, and after electrophoresis of the antigen the precipitin line was in the region of migration of the cationic protein.

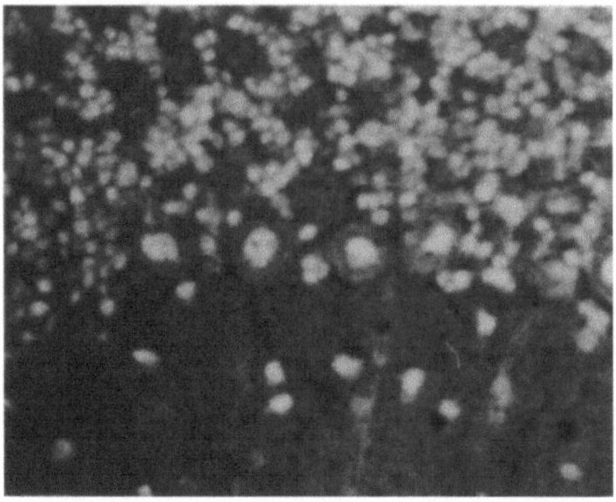

Fig. 3. Photomicrograph of rat cerebellar corter reacted with the fluorescent anti-pig basic protein antibody

The γ-globulin fractions of immune and normal sera were coupled to fluorescein isothiocyanate and then purified to remove nonspecifically reacting material. When cerebellum was reacted with the fluorescent immune globulin the nuclei of Purkinje cells and of granule cells were the primary sites of binding (Fig. 3).[2] If the fluorescent immune γ-globulin was first absorbed with the specific antigen, or if fluorescent normal γ-globulin was used, no specific nuclear reaction was seen. When brain stem, cervical spinal cord, or dorsal root ganglia (Fig. 4) was reacted with the fluorescent immune γ-globulin the nuclei of the neuronal cell populations

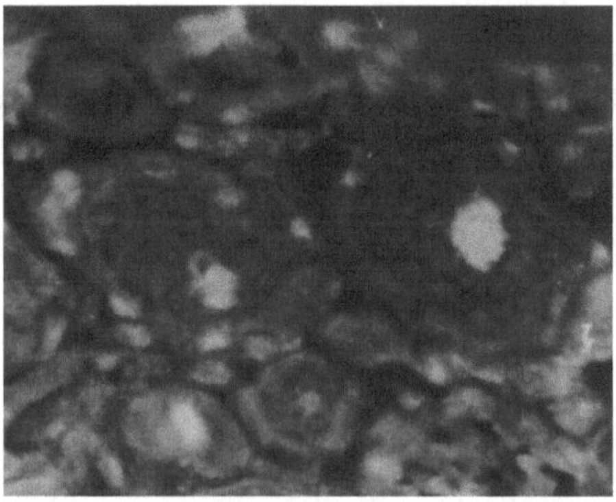

Fig. 4. Photomicrograph of rat dorsal root ganglia (cervical region) reacted with the fluorescent anti-pig basic protein antibody.

were the primary sites of reactivity.[2] The nuclei of adrenal medullary cells also reacted with the fluorescent immune γ-globulin; these cells, however, are derived from sympathetic ganglion neurons during embryological development. The nuclei of the glial cells did not react with fluorescent immune globulin, indicating that the localization of the antigen was restricted to neurons.

Of the various other tissues examined including the liver, kidney, spleen, ovary, and testis, only the testis showed a positive reaction with fluorescent immune globulin, and that reaction was restricted to the nuclei of spermatogonia.[2] Tissues obtained from animals other than pig (including frog, rat, guinea pig, cow, and monkey) showed the presence of a protein immunologically identical to, and similarly distributed as, the original antigen. Therefore, since this protein is basic and is restricted in localization to the nuclei of cells, it can be described as a histone. This histone is tissue-specific, since it is found in nervous tissue and in testis, but not in liver, kidney, spleen, or ovary. An immunologically identical protein is found in a wide variety of animals.

The following changes were observed in the localization of this protein within the nuclei of dorsal root ganglion neurons during ontogenic development.[3] At the earliest stages examined (20-day fetus and 2-day old rat) the protein was restricted to the nucleoli. At 10 days after birth threads of fluorescence were seen to join the nucleoli or to emerge from the nucleoli into the nucleoplasm (Fig. 5). At 40 days post-natum the distribution of

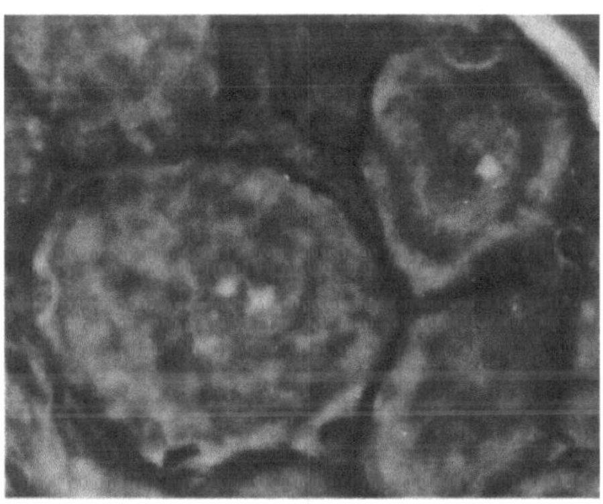

Fig. 5. Photomicrograph of dorsal root ganglion (cervical region) obtained from a 10-day-old rat. The section was reacted with the fluorescent anti-pig basic protein antibody.

the fluorescence was the same as that observed in adult cells. The localization of this histone within the cell nucleus therefore changes during neuronal maturation.

DISCUSSION

This immunohistochemical study demonstrates both the existence of a tissue-specific histone and the changes in its distribution during ontogenesis. The major arguments against the existence of tissue-specific histones have been the similar electrophoretic migration, amino acid composition, and the N-terminal amino acid analysis of histone fractions obtained from different organs. These similarities, which have been based on the analysis of heterogeneous histone fractions, have led several investigators[4–7] to conclude that histones cannot have the specificity required of molecules which can intrinsically recognize distinct regions along the DNA molecules, and therefore cannot act as gene regulators. If similarities in electrophoretic migration, amino acid analysis, and N-terminal amino acids were sufficient evidence of molecular identity for proteins in general, we would have to conclude that another class of proteins, such as the immunoglobulins, are identical and that their specificity is not a function of primary structure. However, recent studies have shown that the immunoglobulins represent a heterogeneous population of proteins, as demonstrated by immunological and chemical techniques.[8,9] The specificity of the immunoglobulins is now thought to be a function of their primary structure.

It is the feeling of this investigator that as more studies concerning the antigenic properties and primary structure of histones appear, microheterogeneity within this protein class will also be found. When these properties of histones become more clearly elucidated it will be possible to discuss more meaningfully the question of whether histones do possess the specificity required of gene regulators.

REFERENCES

1. L. G. Tomasi and S. E. Kornguth, *J. Biol. Chem.* **242,** 4933–4938 (1967).
2. L. G. Tomasi and S. E. Kornguth, *J. Biol. Chem.* **243,** 2507–2513 (1968).
3. S. E. Kornguth and L. G. Tomasi, *J. Cell Biol.* **38,** 515–522 (1968).
4. J. Bonner, M. E. Dahmus, D. Fambrough, R. C. Huang, K. Marushige, and D. Y. H. Tuan, *Science* **159,** 47–56 (1968).
5. A. Neidle and H. Waelsch, *Science* **145,** 1059–1061 (1964).
6. M. G. Ord and L. A. Stocken, *Biochem. J.* **98,** 888–897 (1966).
7. R. M. Gutierrez and L. S. Hnilica, *Science* **157,** 1324–1325 (1967).

8. J. Killander (ed.), "Gamma Globulins—Structure and Control of Biosynthesis, Nobel Symposium 3," Interscience Publishers, New York (1967).
9. L. Frisch (ed.), Cold Spring Harbor Symp. Quant. Biol., Vol. 32, Cold Spring Harbor Publishers, New York (1967).

DISCUSSION

MANDEL: What is the test according to which you say that the proteins are histones?

KORNGUTH: A basic protein that is cationic at pH 9 and localized essentially to the nucleus of cells can be described as a histone.

MANDEL: If you want to be sure that they are histones, you have to isolate chromatin, or to isolate this compound from chromatin. This is because there are many basic proteins in the nuclei which are not histones.

KORNGUTH: Our isolated basic protein precipitates DNA, RNA, and other polyanions. Basic proteins combine electrostatically with polyanions, and, as a consequence, the separate components of chromatin isolated from a cellular homogenate may not be identical with components of chromatin as it exists within the nuclear compartment. For this reason, and in view of the immunohistochemical findings, I believe that the initial isolation of chromatin would not be particularly helpful for classification purposes. Our protein is similar in some respects to the histone rich in glycine and arginine that was isolated from calf thymus by H. Busch and co-workers. The high proline content of our basic protein from pig brain is a feature not seen in the thymus protein.

KIES: There are several amino acids besides proline which have significantly different molar ratios in these two proteins: histidine, serine, valine, isoleucine, and phenylalanine. It is more realistic to say that they are similar only in that both are basic and have high glycine contents. The molecular weight of glycine–arginine-rich histone is reported to be 10,000, whereas Tomasi and Kornguth[1] estimate thei pig-brain basic protein to be 20,000.

KORNGUTH: I would like to indicate that the molecular weight of the pig brain histone was placed at 27,000[2] on the basis of all available information. Asymmetric molecules give rise to aberrantly low molecular weight estimates by ultracentrifugation and high molecular weight estimates by gel filtration. This has been discussed previously.[1,2]

BOGOCH: Dr. Rajam in our laboratory has isolated a protein and purified it. It is an acidic protein. It contains sialic acid as an end-group, and by removing the sialic acid, we can convert it to "basic protein." By putting sialic acid back on, it should become an acidic protein. The immune properties change markedly; the antigenic group is dependent upon whether sialic acid is present or not. Dr. Rajam produces very nice fluorescent antibody pictures in cell nuclei, very similar to yours. I think it would be a problem for the future to attempt to separate out his acidic protein from this "basic protein."

EINSTEIN: The basic protein with encephalitogenic property has no sialic acid in the molecule. We have tested it for sialic acid, but did not find it even in the crude preparations. A glycoprotein which contains sialic acid is extracted with glycine at high pH, but the protein moiety is not the same as the basic protein which we are discussing. The basic proteins—the encephalitogen and histones, and probably ribosomal basic proteins—react with acidic mucoproteins. The basic protein on which we and others are working is a constituent of the CNS myelin, and sialic acid has never been demonstrated in the myelin. The encephalitogenic protein is combined with lipids in the form of a proteolipid complex. The proof is indirect, based on solubility in organic solvents. We were interested in a combination of encephalitogen with other types of compounds, namely with RNA (DNA). We joined forces with Dr. Huang of Johns Hopkins University in an effort to find out whether the basic protein is bound to RNA. It was our aim to find some physiological function for the encephalitogenic protein. Dr. Huang, during her tenure at the California Institute of Technology, together with Professor Bonner isolated a histone combined with RNA and concluded that histone might play a role in the regulation of protein synthesis. We also had some indication that this might be the case in the brain, because the electrophoretic pattern of the acid extracts of the developing brain indicated that at the time the encephalitogen is synthesized some other protein "disappears." In order to explore the possibility that the encephalitogen might play a role in protein synthesis, various preparations, including purified myelin, were sent to Dr. Huang. She employed the technique of cesium chloride banded centrifugation. The outcome of the experiments was that there is no RNA associated with the myelin. We may conclude that the basic protein is not attached either to carbohydrates (specifically the one mentioned by Dr. Bogoch to sialic acid) or to RNA or DNA. The only indirect evidence we have is that the basic protein(s) of the myelin is combined with lipids in the form of a proteolipid complex. As far as the comparison between the encephalitogenic protein and histones is concerned, both proteins are high in basic amino acids and both preparations move toward the negative electrode; however, the position of the encephalitogen in the acrylamide gel electrophoretic pattern is different from that of the histones. Histone(s), however (and this applies when they are prepared from neural tissue), is not encephalitogenic. There is some difference in the immunogenic behavior between salt and acid extracted encephalitogens, as Dr. Helene Rauch found. The one extracted with neutral salt gave precipitating antibodies; the other one, extracted with acid, did not, unless the encephalitogen was complexed with acidified albumin.

KIES: There are several criteria for differentiating CNS histones from myelin basic protein which clearly indicate to us that Dr. Kornguth's basic protein extracted from CNS at pH 2.5 is not a histone, but is in fact myelin basic protein. First, it was obtained in far too large an amount to be derived from nuclei; second, it was obtained from tissue previously extracted with chloroform–methanol at a pH which removes almost all of the myelin basic protein, but does not extract more than traces of histones; third, the amino acid composition of Dr. Kornguth's protein is almost identical to that of other myelin basic proteins, whereas it differs markedly (as he himself has pointed out) from thymus histones. The high degree of homology among histones of all higher creatures, plants, and animals has been noted and discussed repeatedly.* It seems very unlikely that one species or one organ would be atypical.

KORNGUTH: In regard to Dr. Bogoch's comment, we have not been concerned with the sugar content of the unfractionated basic proteins, but after your paper and Dr.

* D. M. Fambrough and J. Bonner, *J. Biol. Chem.* **243,** 4434 (1968); D. M. Fambrough and J. Bonner, *J. Biochem.* **5,** 2563 (1966); L. S. Hnilica, *Biochim. Biophys. Acta* **117,** 163 (1966).

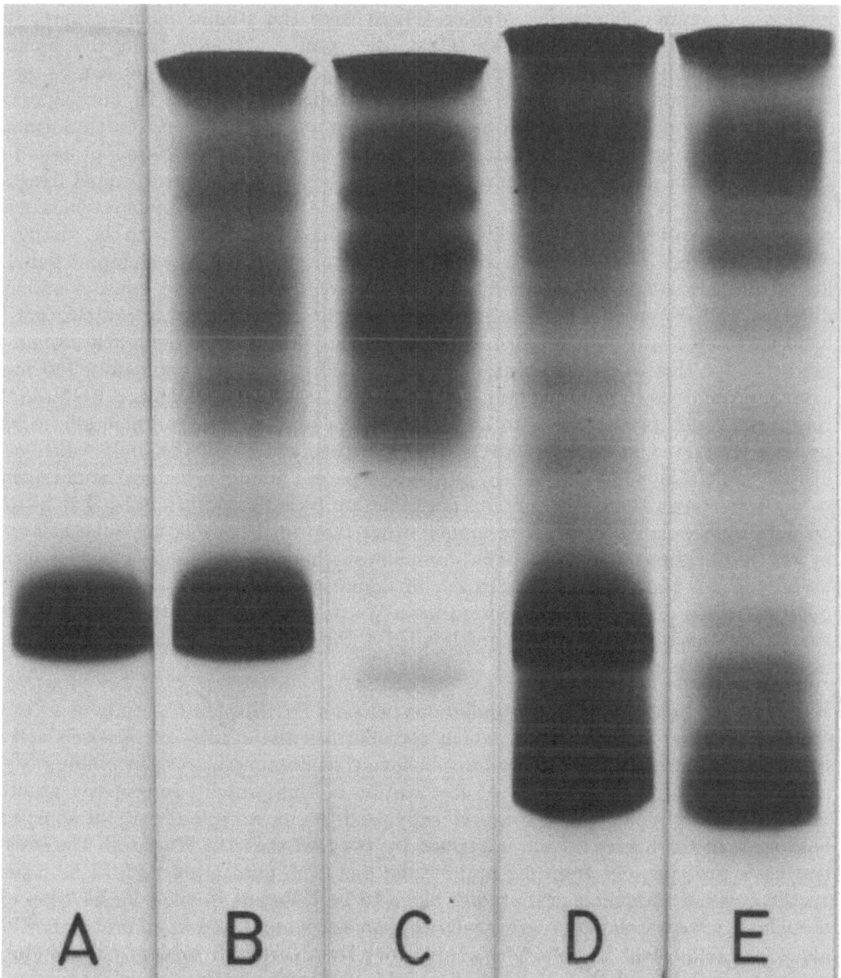

Fig. 6. Basic proteins of guinea pig brain and liver. Electrophoresis, pH 7.0, 5% gel. Origin at top of figure; cathode at bottom. A. Encephalitogen prepared from isolated myelin [R. H. Laatsch, M. W. Kies, S. Gordon, and E. C. Alvord, Jr., J. Exp. Med. **115,** 777 (1062)]. B. 0.03 N HCl extract of brain (pH 1.9). C. 0.03 N HCl extract of liver (pH 2.3). D. 0.1 N HCl extract of brain (pH 1.2). E. 0.1 N HCl extract of liver (pH 1.2). (Martenson, Deibler, and Kies, unpublished.)

Brunngraber's I certainly think this would be a rewarding area for investigation. We did not isolate nuclei prior to extraction. Our initial goal was to isolate large quantities of a homogeneous basic protein from brain. After convincing ourselves of the homogeneity of our protein, we thought it was appropriate to prepare antiserum and initiate our immunofluorescent studies. These were the studies reported here. With regard to the question of dependability, the results obtained from the immunofluorescent technique, as well as from any other experimental technique, can be viewed only in the appropriate context involving appropriate controls. In our experience the immunofluorescent technique has provided interesting and consistent information with respect to two basic proteins from brain: the myelin-associated protein from guinea pig brain and the basic protein from pig brain. When we extracted liver and other organs in a manner similar to that used for the pig-brain preparation a major component was obtained that was similar in electrophoretic mobility with the component we have isolated. The proteins from the different organs have very different immunological properties and their solubilities in $(NH_4)_2SO_4$, ethanol, and aqueous ammonia differ as well. From this I would say that these similarly migrating proteins are not identical in structure. With regard to the statement that there is too much protein per pig brain for this to be a histone, it should be noted that approximately 140 mg of protein is obtained from 14 pig brains. It is difficult to see on what basis it is stated that this is too much protein for it to be a histone. As to the types and amounts of basic proteins extracted at different pH's—this question will be settled only when other histone populations are sufficiently purified to permit immunochemical studies. Since it has been shown that basic proteins are taken up by isolated nuclei and that histones are lost from isolated nuclei, a technique other than nuclear isolation is necessary for *in situ* localization of histones. The immunohistochemical technique provides a very useful tool for such studies. The extent of homology among histones is precisely the question raised by the studies reported here. We do not anticipate that this is the only specific histone, but we feel that more detailed studies will indicate marked heterogeneity in this protein population.

KIES: In the light of all the contradictions between Dr. Kornguth's study of a "brain-specific" histone and our work, which corroborates the studies of Waelsch and his co-workers on brain histones,* I do not believe that immunofluorescent binding can be accepted as a primary proof of the localization of this protein. Improperly absorbed immune sera frequently bind nonspecifically, and the most probable site of nonspecific binding is in the nucleus. I am impressed by the fact that Dr. Kornguth has isolated two basic proteins, one from pig brain[(1)] and one from guinea pig,† which he believes on the basis of immunological studies alone to be different in their localization even though they were isolated by essentially the same technique and in all probability have identical amino acid compositions. How does Dr. Kornguth explain the differences between his data and that of Waelsch and his co-workers? Figure 6, prepared by my associates, Dr. Martenson and Mrs. Deibler, indicates the relative ease with which basic protein specific to myelin can be distinguished from the group of basic proteins common to both brain and liver (presumably nuclear histones). These patterns are derived from extracts of guinea pig CNS. Similar comparisons of extracts prepared from

* A. Neidle and H. Waelsch, *Science* **145,** 1059 (1964); R. S. Piha, M. Cuenod, and H. Waelsch, *J. Biol. Chem.* **241,** 2397 (1966).

† S. E. Kornguth and H. G. Thompson, Jr., *Arch. Biochem. Biophys.* **105,** 308 (1964).

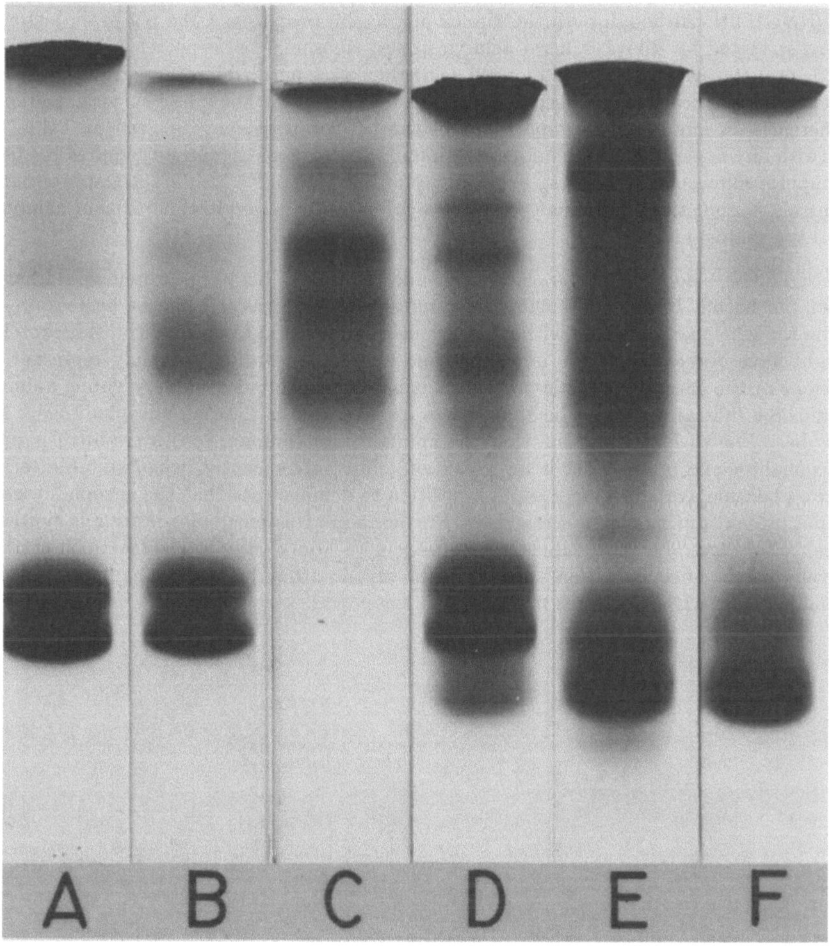

Fig. 7. Basic proteins of rat brain and liver. Electrophoresis, pH 7.0, 5% gel. Origin at top of figure; cathode at bottom. A. 0.1 *N* HCl extract of isolated myelin. B. pH 3 (titrated with *N* HCl) extract of brain. C. pH 3 (titrated with *N* HCl) extract of liver. D. 0.1 *N* HCl extract of brain. E. 0.1 *N* HCl extract of liver. F. 0.1 *N* HCl extract of isolated brain nuclei (Martenson, Deibler, and Kies, unpublished)

rat myelin and brain nuclei lead to the same conclusion (see Fig. 7). Martenson and Gaitonde* have also made analogous extractions of bovine CNS in which they found that the only basic protein extracted in significant quantities at pH values of 2 or above was derived from myelin.

* R. E. Martenson and M. K. Gaitonde, *J. Neurochem.*, in press.

KORNGUTH: Although both the guinea pig myelin protein and the pig brain histone were extracted by dilute mineral acid (histones, ribosomal basic proteins, and myelin basic proteins are all extracted by mild acid) they were purified in different ways as has been reported previously[1] (see Kornguth and Thompson above). The very marked differences in amino acid compositions of the two proteins may be seen in Table I. As with any experimental technique, the immunohistochemical method requires the use of appropriate control studies. From our published work[1–3] using such appropriate controls, we must conclude that our histone protein is localized in the nuclei of neurons and spermatogonia.

EINSTEIN: I don't think Dr. Kornguth's description of his basic protein could have been put better. It was not isolated from nuclei. His aim was to isolate a basic protein from the whole neural tissue without previous fractionation. Consequently, this protein might have derived from the nuclei or from myelin. Accordingly, it may represent a histone or the encephalitogenic protein. If it has been isolated from very young animal brain, the indication would be that it is a histone, rather than an encephalitogen. It has been known for some time that the animal brain (human, too) after birth is not encephalitogenic (guinea pig is an exception), but only recently, when suitable techniques became available, were we in a position to demonstrate that the encephalitogen is not present during the first few days. Furthermore, this particular protein is synthesized later than the proteolipids, the other, more abundant protein fractions of the myelin. If the encephalitogen appears in the myelin after the "tolerance period," this would explain why it behaves like a foreign compound.

Chapter 34

Problems Related to the Protein-Eliciting Experimental Allergic Encephalomyelitis*

Elizabeth Roboz Einstein

Department of Neurology
and
Department of Biochemistry and Biophysics
University of California School of Medicine
San Francisco, California

and

Li-Pen Chao

Department of Neurology
University of California School of Medicine
San Francisco, California

INTRODUCTION

Experimental allergic encephalomyelitis (EAE) is an autoimmune disease with central nervous system (CNS) involvement. The compound responsible for the condition is a basic protein and derives from the CNS myelin. This has been established by localization of the encephalitogen through immunofluorescent studies of Rauch and Raffel[41] and by actual isolation of the encephalitogenic protein(s) from the myelin.[13,18,27,29]

The two most crucial questions related to the problem of EAE are: (1) Is there more than one encephalitogen in the myelin? (2) Why is this particular myelin protein encephalitogenic? An attempt will be made to answer these questions.

* The studies were supported by USPHS grant NB-02545, the National Multiple Sclerosis Society grant No. 457, and, in part, by the Hartford Foundation. The collaboration of Judit Csejtey in these studies is gratefully acknowledged.

IS THERE MORE THAN ONE ENCEPHALITOGEN IN THE MYELIN?

The idea of multiple encephalitogens is based on reports that encephalitogenic proteins have been isolated with molecular sizes ranging from 3000 to 50,000.[3,5,6,11,23,27,30,36,37]

The divergence in reported molecular size of the encephalitogenic protein may be attributed to several factors:

1. To the mode of preparation, especially the initial steps consisting of selection of organic solvents for removal of the lipids and extraction of the encephalitogen with salt or acid of the lipid-free residue. The pH of the acid used for the extraction also plays a role.
2. To proteolysis due to the action of an endogenous enzyme resulting in an active fragment(s).
3. To selection of particular methods used for measuring the molecular weight.
4. Notwithstanding the validity of points 1–3, we do not exclude the possibility that there is more than one encephalitogenic protein present in the neural tissue.

Influence of the Mode of Preparation

Prior to extraction of the encephalitogen the lipids are removed with organic solvents. Acetone and petrol ether were used by Einstein *et al.*;[11] Nakao *et al.*;[36] Kibler and Shapira;[23] Lumsden *et al.*;[30] and Carnegie *et al.*[3] A chloroform–methanol mixture was first employed by Folch and Lees[21] for the extraction of proteolipids. This solvent mixture was also used very successfully for the removal of lipids by Kies[25] and by others including ourselves. For extraction of the encephalitogen two types of chemicals have been used: salts, such as potassium chloride, by Einstein *et al.*,[10,11] citrate by Roboz (Einstein) and Henderson[42] and Kibler and Shapira,[23] and acetate by Eng *et al.*[18] Acid extraction with dilute of H_2SO_4 or HCl, a method started by Kies,[25] at present is used by a majority of investigators. It should be mentioned that it is not necessary to remove the lipids before extraction of the encephalitogen. The two operations may be carried out together.[13,49]

We have found differences in the immunological response to the encephalitogens, depending on whether initially the neural tissue was extracted with acid or salt. Although these encephalitogens are very similar in their physicochemical properties and both are highly encephalitogenic, the one prepared by salt extraction elicits a precipitating antibody, while the one prepared by acid extraction does not. Partial amino acid sequence has been carried out on the acid-extracted encephalitogen[19] and on the salt-extracted encephalitogen,[24] but results are not complete enough yet to make comparisons between the two encephalitogens. When the acid-extracted encephalitogen was complexed with acetylated, succinylated, or phosphorylated serum albumin and rabbits were immunized with any of these preparations a precipitating antibody to the encephalitogen was obtained.[40] Antibodies apparently are not required for the development of EAE,[12] as tested by the passive cutaneous anaphylaxis (PCA) reaction. However, Kibler and Barnes[22] found a close relationship between the pathology of EAE and antibodies when the Farr[20] technique was employed.

Since there is a certain correlation between the conditions outlined in 1 and 2 above, the two will be discussed together in the following paragraph.

Proteolysis Due to the Action of Endogenous Enzymes

For extraction of the encephalitogen we have used H_2SO_4 and HCl with various pH.[13,35,36] The acetone petrol ether preextraction, followed by acid extraction of pH 2.8–3.2, resulted in three active proteins with different mobilities and molecular weights.[36] The question arose whether all three encephalitogens are present endogenously in the spinal cord, or whether the two smaller ones are enzymatic breakdown products of the larger one. There was some indication that the lower-molecular-weight proteins are breakdown products of the larger one, through enzymatic hydrolysis. When a preparation obtained by extraction with a pH of 1.8 and representing essentially one major encephalitogen was incubated with a preparation of pH 3.2 this was degraded.[37] This became clear to us through the report of Marks and Lajtha,[32] who isolated an acid proteinase from brain which had an optimal activity at pH 3.2, the very same pH we used for extraction of the encephalitogen. On the other hand, the one prepared at pH 1.8, although preextracted with acetone and petrol ether, consisted of one major protein with only some minor contaminants. Marks and Lajtha[32] have shown that at this pH the acid proteinase is not, or is only slightly, active

Using the purified acid proteinase, it was established that after 5 hr incubation encephalitogen G was completely degraded, with the formation of three smaller proteins (polypeptides), as demonstrated by acrylamide gel electrophoresis (Fig. 1). The degradation product was proved to be encephalitogenic; therefore, isolation of the active component was undertaken. Using repeated gel filtration on Sephadex G-75 and chromatography on carboxymethyl cellulose, an active fragment was isolated which appeared

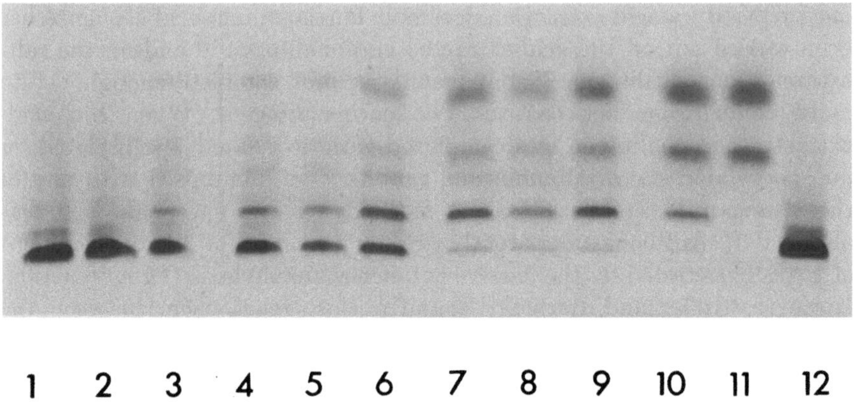

Fig. 1. Acid proteinase activity on encephalitogenic protein G. (1) Control; (2) 10-min incubation, no enzyme added; (3) 10-min incubation; (4) 20-min incubation; (5) 30-min incubation; (6) 45-min incubation; (7) 60-min incubation; (8) 2-hr incubation; (9) 3-hr incubation; (10) 5-hr incubation; (11) 24-hr incubation; (12) 24-hr incubation, no enzyme added.

to be homogeneous by the criteria of ultracentrifugation, acrylamide gel electrophoresis, and amino- and carboxyl-terminal analyses. It has been established that the active fragment derived entirely from the carboxyl end of the encephalitogen G (Table I).[6]

Dependence of the Molecular Weight on the Method Used for the Determination

Using the gel permeability technique for estimation of the molecular size, according to the technique of Andrews,[1] a molecular weight of 35,000 was obtained whereas, by ultracentrifugation, molecular weights of 12,500

TABLE I

Some Properties of the Encephalitogen and of the Active Fragment

	Encephalitogenic protein	Active fragment
N-Terminal groups	Blocked	Phenylalanine
C-Terminal groups	-Gly-Ala-Arg	-Gly-Ala-Arg
Biological activity	Tested at the same level. Both active	

TABLE II

Molecular Weight of the Encephalitogenic Protein

Method of Determination	Molecular weight
Yphantis[53]	17,800
Mandelkern and Flory[31]	20,200
Tanford *et al.*[45]	17,900

and 15,800 were found, respectively, based on the number of peptides.[37] Because of the divergence in molecular weight when measuring the same protein, and to other anomalous properties of the protein, it was subjected to various physicochemical measurements, such as sedimentation velocity and viscosity in the presence and absence of 6 *M* guandine hydrochloride, and optical rotatory measurements (ORD) and molecular weight as determined by sedimentation equilibrium. These measurements resulted in a molecular weight of approximately 17,800 (see Table II). The ORD measurements indicated that the encephalitogen has no alpha helix.[7]

It should be mentioned that the encephalitogenic protein gave a single symmetrical peak during the sedimentation-velocity studies both in the presence and the absence of 6 *M* guanidine hydrochloride. This measurement, in addition to the acrylamide gel electrophoretic analysis, which resulted in one band, indicated that the preparation is not only homogeneous, but also that there was no deaggregation in the presence of the denaturing agent, guanidine hydrochloride. The same applies for measurements of the intrinsic viscosity.[7]

The hydrodynamic properties of the encephalitogenic protein G are summarized in Table III. The data clearly indicate that there is no difference in size measured in the presence or absence of guanidine hydrochloride. The results from the intrinsic viscosity measurements showed that this protein is in random or disordered conformation for both conditions (see Table III). For most globular proteins the magnitude of the intrinsic viscosity is between 3 and 4 ml/g. Much higher values are reported for denatured, random, or fibrous proteins.[52] The viscosity measurements indicate that the encephalitogen is in random conformation.

The results of the ORD measurements indicate that the encephalitogen is a flexible, disordered protein without any alpha helix. It behaves as a polyelectrolyte in either dilute salt solution or in aqueous solution. Globular proteins are not suitable for measurement of the molecular weight of the disordered-protein encephalitogen (Chao and Einstein).[7a] For estimation of the molecular weight by sedimentation-equilibrium technique

TABLE III
The Hydrodynamic Properties of the Encephalitogenic Protein

Condition	Intrinsic viscosity $[\eta]$ (ml/g)	Sedimentation const. $S^{\circ}_{20,w}$ in S
0.05 *M* sodium acetate– 0.1 *M* NaCl, pH 4.6		
without 6 *M* guanidine HCl	16.9	1.33
with 6 *M* guanidine HCl	20.8	1.67

high pH or salt concentrations have to be used to overcome the strong cationic charge effect of this protein.[7,46]

Is There More Than One Encephalitogenic Protein Present in the Myelin?

Evidence points to the fact that the lower-molecular-weight protein(s) is a breakdown product of the larger, due to the action of endogenous proteolytic enzyme(s).[26,36] The possibility of the encephalitogen *in vivo* being multiple should not, however, be excluded at this time, although some of our experimental data speak against it. Myelin was solubilized with a mixture of phenol, acetic acid, and water (14:3:3) and electrophoresis done with the same solvent; the isolated and purified encephalitogen and proteolipid preparations were used as a control. (Details of the procedure are given in the next section.) The three classes of proteins in the myelin—namely, the proteolipids, the acidic proteins, and the encephalitogen(s)—were clearly separated (Fig. 2). The proteolipid fraction, kindly supplied by Dr. Lees, and prepared according to Folch and Lees,[21] was also compared in another run with our proteolipid preparation. The two gave identical patterns. The Wolfgram and Kotorii[50] acidic protein, although not yet localized in acrylamide gel electrophoresis, but moving close to the position of the proteolipids, is anodal. Therefore, a distinct separation of these proteins from the encephalitogen could be obtained. Only one wide band appeared in the position of the encephalitogen when the myelin was electrophoresed. The mixture of phenol and acetic acid is considered not only a suitable deaggregation agent, but at the same time it inactivates the proteolytic enzymes;[2] therefore, if several bands should appear, it would indicate multiple encephalitogens *in vivo*, not proteolytic breakdown. However, only a very minor faster-moving band appeared; but since this was not present in other preparations, the multiplicity was not indicated.

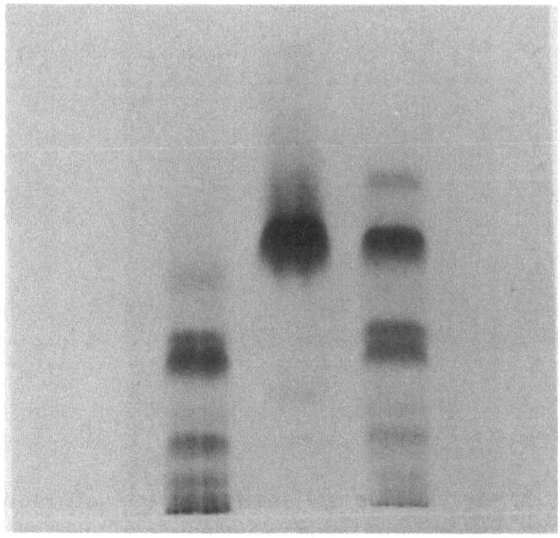

Fig. 2. Acrylamide electrophoresis on neural proteins using 10% gel and a buffer of phenol: acetic acid:water (2:1:1). (1) Proteolipids; (2) encephalitogenic preparation; (3) myelin.

The electrophoretic patterns of the basic proteins from human, bovine, and rabbit myelin are very similar; however, the rat presents a different picture. Joan Kauttu, working in our laboratory on the myelin formation in developing brain, found two equally intense bands moving far apart in the myelin of Long-Evans rats. Whether these two bands with different mobilities are both encephalitogenic has not yet been established. Eng *et al.*[18] also showed two basic proteins in the electrophoretic pattern of rat myelin.

It should be mentioned here that the number of bands increased in the electrophoretic pattern of the acid extract of brain when urea was used in the buffer. This technique has been used for histones and adopted for the basic proteins of neural tissue.[4] It should be mentioned that after the encephalitogen is purified it presents only one band with this technique.

WHY IS THIS PARTICULAR PROTEIN ENCEPHALITOGENIC?

This is the most enigmatic problem in the field of experimental allergic encephalomyelitis. The answer, most likely, rests not in a unique structure, if such will be proved, but in the underlying immunological reaction.

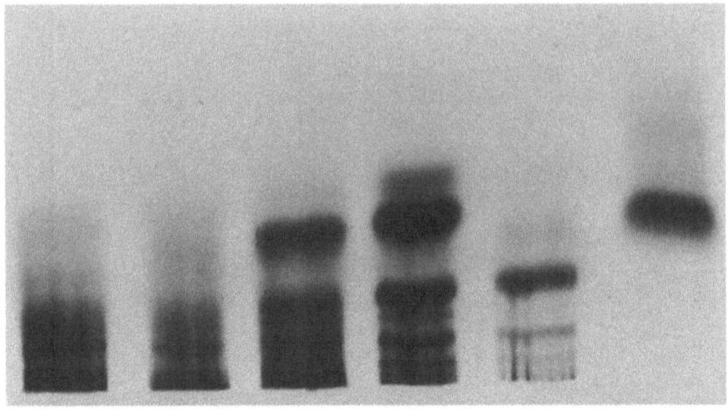

Fig. 3. Acrylamide gel electrophoresis of rabbit-brain myelin preparations using 10% gel and a buffer of phenol: acetic acid:water (2:1:1). (1) One day; (2) 5 days; (3) 9 days; (4) adult; (5) proteolipids; (6) encephalitogenic preparation.

Toward this end, two explanations may be offered. One, the sequestered nature of the encephalitogen, and two, its absence during the early development of the brain. These are not mutually exclusive. It has been known for some time that the fetal brain and the brain shortly after birth are not encephalitogenic; however, no proof has yet been offered on a molecular level.

With the development of techniques for the complete solubilization and electrophoresis of all the proteins present in the myelin, we were able to follow the appearance of the encephalitogenic protein during the course of brain development. The solvent which fulfills both requirements—namely, dissolving the myelin and at the same time being suitable for electrophoresis—is the mixture of phenol, acetic acid, and water. It was first used for solubilization of leaf proteins,[2] and later for solubilization and electrophoresis of ribosomal proteins.[51] This solvent mixture was employed for electrophoresis of the myelin,[9,15,16,34] and also for other membrane proteins.[8] The phenol–acetic acid extract was used directly in some cases, while in other cases it was freeze-dried first, then dissolved and electrophoresed. Since commercially available electrophoretic tanks made from plastic cannot be used, because of the corrosive nature of the phenol, the tank was constructed from glass in our shop. Dimensions of the flat glass dish were 44 × 28 × 6 cm and it had removable platinum electrodes. Ten per cent acrylamide gel was prepared and soaked in phenol, acetic acid, and water (14:3:3) overnight. The gel slab was removed, and the solubilized myelin sample inserted in the slot of the gel. Buffer was made up of a mixture of phenol, acetic acid, and water in a ratio of 2:1:1; it was run at 275 V in the cold room or in a refrigerator at 4°C for 24 hr. After completion of the electrophoresis the proteins were stained with 0.5% amido black. Destaining of the gel, when made up as a slab (block), is a somewhat lengthy procedure, but the time may be reduced with electrophoretic destaining; however, since we observed the disappearance of some of the minor bands with electrophoretic destaining, we now place the slab in a flat glass dish (obtainable commercially as a baking dish) and remove the excess dye with a mixture of acetic acid, methanol, and water (1:4:5) using a magnetic stirrer and exchanging the solvent several times a day.

It should be pointed out that by this technique all the proteins are solubilized, separated, and visualized in the electrophoretic pattern. The encephalitogenic protein and a proteolipid preparation were used as control and compared with myelin extracts from various ages of brain. The electrophoretic distribution of these are presented in Fig. 3. It can be seen from the electrophoretic pattern that the proteolipids are definitely formed before the encephalitogenic proteins. These are missing in the first few days after birth.

It should be mentioned here that myelin prepared by ultracentrifugation from the brains of one-, five-, and seven-day-old rabbits has a different consistency from the one prepared from mature brain. In order to exclude the possibility that the ultracentrifugation introduced an artifact, we also carried out electrophoretic studies by extracting the encephalitogenic and other basic proteins directly from brain.

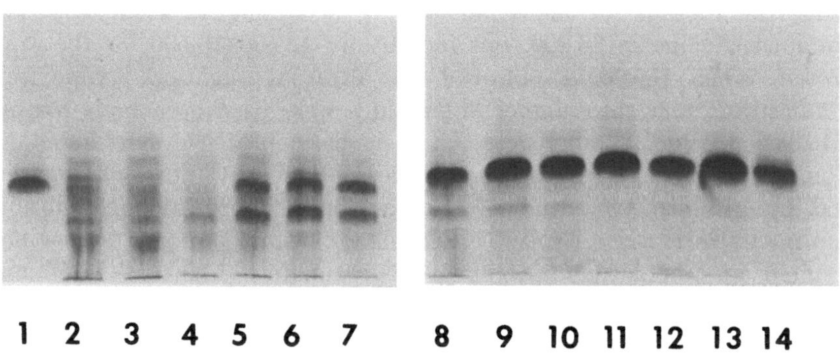

Fig. 4. Electrophoretic distribution of basic proteins in developing rabbit brain from one day to 170 days. (1) Encephalitogenic protein; (2) one day; (3) 5 days; (4) 7 days; (5) 9 days; (6) 12 days; (7) 14 days; (8) 18 days; (9) 30 days; (10) 60 days; (11) 90 days; (12) 120 days; (13) 170 days; (14) encephalitogenic protein.

The extract made from brain contains other basic proteins besides the encephalitogen. Most prominent among these are the histones. However, earlier electrophoretic studies with brain histones indicated that these migrate differently from the encephalitogen. Consequently, this technique may be used for study of the appearance of the encephalitogenic protein, as related to development. In a study where 1-, 5-, 7-, 9-, 12-, 14-, 18-, 30-, 60-, 80-, 120-, and 170-day-old rabbit brain acid extracts were included (Fig. 4), the patterns clearly showed that the encephalitogen is not present up to the first nine days. On the ninth day two distinct bands appear, both apparently in the same concentration. The second band has not been identified at the present time, but it should be pointed out that this slower-moving protein fades later, when the animal matures. The distance between the two bands and the intensity of the color in this immature brain are very similar to that observed in mature rat-brain myelin by Joan Kauttu in our laboratory.

DISCUSSION

Several aspects of this complex problem of experimental allergic encephalomyelitis have been discussed. It is recognized that the complexity stems partly from the nature of the encephalitogen. We refer here to its highly basic composition and to its anomalous physical properties. Studies in our laboratory on optical rotatory dispersion (ORD) and viscosity suggest that this protein is in disordered conformation and has no alpha helix.[7] For this type of protein, which is a flexible, disordered molecule, the method of molecular-weight determination by gel filtration is not applicable when globular proteins are used for comparison (Chao and Einstein).[7a] When using the method of molecular sieving according to Andrews,[1] we obtained a value of 36,000, but found only 12,500–16,500 for the same protein when the data obtained by ultracentrifugation, amino acid composition, and the number of peptides were used as a basis for our calculation. Similar differences for their basic proteins were found by Tomasi and Kornguth[47] between values obtained by ultracentrifugation and by gel filtration. Studies on data from sedimentation equilibrium and on intrinsic viscosity have changed the values obtained earlier and resulted in a molecular weight of approximately 17,800. Kies[26] and Caspary and Field[5] reported a molecular weight of 50,000 when measured by gel filtration. This figure most likely would have been much lower if the same protein had been subjected to ultracentrifugal measurements.

It is evident, however, that the difference in molecular weights is not due solely to methodology, since encephalitogens with different molecular sizes have been obtained from the same extract measured by the same technique. The possibility of an enzymatic breakdown of the encephalitogen is certainly greater when acetone and petrol ether are used instead of the chloroform–methanol mixture, since the former "preserves" the enzymes.[4,26] However, in our experience this applies only when the pH of the extracting media is close to the optimal pH (3.2) of the acid proteinase. When the pH was lowered to 1.8–2.0, where there is hardly any activity of this enzyme,[32] the purified encephalitogenic protein gave about the same molecular weight as the one preextracted with chloroform methanol.[6,36,37]

No divergence was observed in the electrophoretic mobility between encephalitogens obtained from various species, except rats, and even from different laboratories. Encephalitogens isolated from human and bovine cord moved exactly in the same position, as did the encephalitogens sent to us from other laboratories—from Dr. Eylar of the Salk Institute (bovine) and from Dr. Eng of the VA Hospital at Palo Alto (human source). We have also studied encephalitogens obtained from human, rabbit, sheep, and rat. These, however, represent only crude acid extracts of the myelin, except the human.[13] Among these, only the rat myelin was found to be different in its electrophoretic pattern, according to Joan Kauttu in our laboratory and Eng *et al.*[18]

TABLE IV

Amino Acid Composition of the Encephalitogen

Amino acid (mole %)	The present study	Eng *et al.*[18]
Lysine	8.2	8.1
Histidine	5.9	5.8
Arginine	11.2	9.8
Aspartic	6.8	7.0
Threonine	4.3	4.4
Serine	9.6	9.9
Glutamic	6.4	6.6
Proline	6.8	7.3
Glycine	15.6	15.0
Alanine	8.6	8.3
Valine	0.8	1.8
Half-cystine	Trace	0.0
Methionine	1.2	1.1
Isoleucine	1.2	1.9
Leucine	6.0	6.0
Tyrosine	2.6	2.3
Phenylalanine	4.9	4.9

The amino acid composition from two different sources, one from human[18] and one from bovine[36,37] are nearly identical (see Table IV).

As far as the existence of more than one basic protein present endogenously in the myelin is concerned, no definite answer can be given. No one has yet isolated and well characterized two or more encephalitogens and proved that they do not originate from enzymatic breakdown.

As far as the difference in the electrophoretic pattern is concerned, there are indeed differences depending on the conditions and the buffer applied. The number of lines increased when urea was used in the buffer for the acid extract. We have not separated the bands representing basic proteins after electrophoresis with urea; consequently, we have not established if they represent proteins with different mobilities and different molecular entities, or even if they are encephalitogenic. Martenson and Kies reported at the 1968 Federation conference (also see Chapter 35 of this volume) that at high pH (10.6) the basic proteins of the myelin form multiple bands. Caspary and Field[5] reported that the basic protein tends to undergo hydrolysis at high pH: "Six or more major components were demonstrable by chromatography following dialysis against a carbonate buffer at pH 9.0." Since the antigen was exposed for an unreasonably long time to this high pH (24 hr to six weeks), no conclusion can be drawn as to the nature of the changes caused by the high pH.

Depending on the mode of preparation, we found no differences in the chemical, but did in the immunogenic, properties between the two encephalitogens. According to the technique employed for their extraction, they reacted differently. The protein extracted with neutral salt gave precipitating antibodies; the one extracted with acid at pH 1.8 did not, although both were equally encephalitogenic.[40] The question is, did the acid affect the antigenic part of the molecule? Or did we obtain two encephalitogens with subtle differences?

It has been known for some time that the brain before and shortly after birth is not encephalitogenic. This was based on experiments of Paterson,[38] who injected immature brain into experimental animals and found that the animals did not develop EAE. Waksman,[48] discussing the Paterson paper, suggested that the antigen-paralytogenic agent is absent during the tolerance period. Others[44] also reported that the very young brain is not encephalitogenic. An explanation for these findings is presented by our studies on the molecular level. First, we studied human brain (from autopsy) in the various stages of development, starting with six weeks premature. The electrophoretic patterns indicated that the encephalitogen is absent at six and four weeks premature, at seven weeks and probably eight months.[14,17]

In the present experiments our attention has been focused on the appearance of the encephalitogenic protein in comparison with other proteins in myelin of the rabbit. The electrophoretic pattern indicates that there is no band corresponding to the encephalitogen up to seven days. In contrast, the proteolipids have already appeared in the myelin of one-day-old rabbits. It is possible that the proteolipid synthesis occurs during the time when the immunocompetent cells are formed. On the other hand, the particular basic protein (the encephalitogen) is absent during this time.

The other possible explanation for the encephalogenicity, namely, its sequestered nature, i.e., the fact that it is kept apart from the systemic circulation by tissue barriers, does not stand up in the light of the reports of Patterson,[39] Sherwin *et al.*,[43] and Klatzo *et al.*[28] According to these reports, compounds from the brain may pass into the systemic circulation. If the sequestered nature is the decisive factor, then the other major constituent of the myelin, the proteolipids, would be "equally protected," and, if injected with them the animal should react similarly as to the encephalitogen. The time difference in the appearance between the encephalitogen and proteolipids in the myelin offers a more plausible explanation, provided the sequence is correct; namely, that the immunocompetent cells are formed first, and synthesis of the encephalitogenic protein is formed afterward. This would explain why this protein is not recognized as "its self," and when injected results in encephalomyelitis.

SUMMARY

Problems concerned with the protein inducing experimental allergic encephalomyelitis have been discussed. Some of the factors which influence the status of questions concerning the multiplicity of the encephalitogen were traced back to the mode of preparation and to the techniques used in the determination of the molecular weight. The discrepancy between the various values obtained here has been explained by the physical properties of the encephalitogen. The likelihood that the lower-molecular-weight encephalitogen is an enzymatic breakdown product has been strengthened by the actual isolation of an active degradation product from the digest of encephalitogen G, hydrolyzed with neural proteinase.

New data have been presented on the sequential appearance of the protein correlated to age, using the technique of complete solubilization and electrophoresis. The electrophoretic pattern clearly demonstrates that the encephalitogen appears later than the other proteins—notably, the major protein components of the myelin, the proteolipids. The finding is interpreted by the hypothesis that the proteolipids are recognized by the immunocompetent cells whereas the encephalitogen, as a late-comer, is not.

This offers a possible explanation for the encephalitogenic property of this myelin protein.

REFERENCES

1. P. Andrews, *Biochem. J.* **91,** 222 (1964).
2. M. Bagdasarian, N. A. Matheson, R. L.M . Synge, and M. A. Youngson, *Biochem. J.* **91,** 91 (1964).
3. P. R. Carnegie, B. Bencina, and G. Lamoureux, *Biochem. J.* **105,** 559 (1967).
4. P. R. Carnegie, G. Lamoureux, and B. Bencina, *Nature* **214,** 407 (1967).
5. E. A. Caspary and E. J. Field, *Ann. N.Y. Acad. Sci.* **122,** 182 (1965).
6. L.-P. Chao and E. R. Einstein, *J. Biol. Chem.* **243,** 6050 (1968).
7. L.-P. Chao and E. R. Einstein, *Fed. Proc.* **28,** 842 (1969).
7a. L.-P. Chao and E. R. Einstein, *J. Chromatag.*, **42,** 480 (1969).
8. C. W. Cotman and H. R. Mahler, *Arch. Biochem. Biophys.* **120,** 384 (1968).
9. J. Csejtey, W. J. Davis, and E. R. Einstein, Pacific Conference of Chemistry and Spectroscopy, Abstract No. 133 (1967).
10. E. R. Einstein, D. M. Robertson, and J. DiCaprio, *Fed. Proc.* **19,** 331 (1960).
11. E. R. Einstein, D. M. Robertson, J. M DiCaprio, and W. Moore, *J. Neurochem.* **9,** 353 (1962).
12. E. R. Einstein, A. Nakao, J. DiCaprio, and W. D. Moore, "Proc. VII Int. Congr. Neurol. (Rome)" Vol. 2, p. 748 (1964).
13. E. R. Einstein, J. Csejtey, W. Davis, and H. Rauch, "Proc. VIII Int. Congr. Neurol. (Vienna)" Vol. 4, p. 137 (1965).
14. E. R. Einstein and J. Csejtey, *Trans. Amer. Neurol. A.* **91,** 218 (1966).
15. E. R. Einstein and J. Csejtey, Int. Soc. Neurochem. (Strasbourg) Abstract 61 (1967).
16. E. R. Einstein, J. Csejtey, and N. Marks, *FEBS Letters* **1,** 191 (1968).
17. E. R. Einstein, J. Csejtey, W. J. Davis, and C. H. Rauch, *Immunochem.* **5,** 567 (1968).
18. L. F. Eng, F. C. Chao, B. Gerstl, D. Pratt, and M. G. Tavastsjerna, *Biochemistry* **7,** 4455 (1968).
19. E. H. Eylar and G. A. Hashim, *Proc. Natl. Acad. Sci. U.S.* **61,** 644 (1968).
20. R. S. Farr, *J. Infect. Dis.* **103,** 239 (1958).
21. J. Folch and M. Lees, *J. Biol. Chem.* **191,** 807 (1951).
22. R. F. Kibler and A. E. Barnes, *J. Exp. Med.* **116,** 807 (1962).
23. R. F. Kibler and R. Shapira, *J. Biol. Chem.* **243,** 281 (1968).
24. R. F. Kibler and R. Shapira, *Fed. Proc.* **27,** 464 (1968).
25. M. W. Kies, *in*: "Allergic Encephalomyelitis" (M. W. Kies and E. C. Alvord, eds.) pp. 293–299, Thomas, Springfield, Ill. (1959).
26. M. W. Kies, VI Int. Congr. Neuropath., Zurich, 1965, Abstracts, p. 27.
27. M. W. Kies, E. B. Thompson, and E. C. Alvord, Jr., *Ann. N.Y. Acad. Sci.* **122,** 148 (1965).
28. I. Klatzo, J. Miquel, P. J. Ferris, J. D. Prokop, and D. E. Smith, *J. Neuropath. Exp. Neurol.* **23,** 18 (1964).
29. R. H. Laatsch, M. W. Kies, S. Gordon, and E. C. Alvord, Jr., *J. Exp. Med.* **115,** 777 (1962).
30. C. E. Lumsden, D. M. Robertson, and R. Blight, *J. Neurochem.* **13,** 127 (1966).
31. L. Mandelkern and P. J. Flory, *J. Chem. Phys.* **20,** 212 (1952).

32. N. Marks and A. Lajtha, *Biochem. J.* **97,** 74 (1965).
33. R. E. Martenson, G. E. Deibler, and M. W. Kies, *Fed. Proc.* **27,** 619 (1968).
34. E. Mehl, Proc. Ist Int. Soc. Neurochem., Strasbourg, 1967, Abstracts, p. 154.
35. A. Nakao and E. R. Einstein, *Ann. N.Y. Acad. Sci.* **122,** 171 (1965).
36. A. Nakao, W. Davis, and E. R. Einstein, *Biochim. Biophys. Acta* **130,** 163 (1966).
37. A. Nakao, W. Davis, and E. R. Einstein, *Biochim. Biophys. Acta* **130,** 171 (1966).
38. P. Y. Paterson, *Ann. N.Y. Acad. Sci.* **73,** 811 (1958).
39. P. Y. Paterson, *Adv. Immunol.* **5,** 131 (1966).
40. H. C. Rauch and E. R. Einstein, *Int. Arch. Allergy,* in press.
41. H. C. Rauch and S. J. Raffel, *J. Immunol.* **92,** 452 (1964).
42. E. Roboz (Einstein) and N. Henderson, *in*: "Allergic Encephalomyelitis" (M. W. Kies and E. C. Alvord, eds.), pp. 281–292, Thomas, Springfield, Ill. (1959).
43. A. L. Sherwin, M. Richter, J. B. R. Cosgrove, and B. Rose, *Neurology* **13,** 113 (1963).
44. G. J. Svet-Moldavsky, O. G. Andjaparidze, S. S. Unanov, *et al.*, *Bull. World Health Org.* **32,** 47 (1965).
45. C. Tanford, K. Kawahara, and S. Lapanje, *J. Biol. Chem.* **241,** 1921 (1966).
46. D. C. Teller, J. M. Kinkade, and R. D. Cole, *Biochem. Biophys. Res. Commun.* **20,** 739 (1965).
47. L. G. Tomasi and S. E. Kornguth, *J. Biol. Chem.* **242,** 4933 (1967).
48. B. H. Waksman, *Ann. N.Y. Acad. Sci.* **73,** 817 (1958).
49. F. Wolfgram, *Ann. N.Y. Acad. Sci.* **122,** 104 (1965).
50. F. Wolfgram and K. Kotorii, *J. Neurochem.* **15,** 1281 (1968).
51. T. S. Work, *J. Mol. Biol.* **10,** 544 (1964).
52. J. J. Yang, *Adv. Protein Chem.* **16,** 323 (1961).
53. D. A. Yphantis, *Biochemistry* **3,** 297 (1964).

Chapter 35

Specificity of Myelin Basic Proteins*

Marian W. Kies

Section on Myelin Chemistry
Laboratory of Cerebral Metabolism
National Institute of Mental Health
Bethesda, Maryland

A number of studies have appeared in the last ten years on the isolation from central nervous system (CNS) of certain highly basic proteins which are capable of inducing experimental allergic encephalomyelitis (EAE).[1-10] These studies have depended largely on tests of biological activity to monitor the fractionation procedures. Such tests are still an important part of our investigative procedures, even though the myelin basic proteins[2] are probably the best characterized of the proteins isolated from the CNS. They have been isolated from many species, and the amino acid compositions of several have been reported.[7-13] Considerable information is also available regarding amino acid sequence of certain parts of the molecule.[8,14]

Because encephalitogenic activity was never detected in any organ other than the CNS, the encephalitogens were assumed to be "organ-specific." On the other hand, they were originally thought to lack species-specificity because tests of CNS from many species revealed a remarkable ubiquity in the distribution of encephalitogenic activity. Whole CNS tissue of almost any mammalian species was encephalitogenic in practically every species investigated (Table I). Qualitative tests were gradually replaced by bioassay procedures,[16] but it was not until purified protein fractions were available that differences between fractions derived from different species could be evaluated.

It was not surprising that isolation and purification of myelin basic proteins required revision of some of the earlier concepts regarding their

* I am indebted to the many colleagues who have collaborated with me in these studies, particularly Drs. Alvord, Lisak, and Martenson, and Mrs. Deibler.

immunologic specificity. What was originally considered to be lack of species-specificity, owing to the marked similarities in biological activity of the CNS of several species, must be modified to the extent that bioassay and immunoassay as well as certain physicochemical procedures detect differences among the proteins isolated from various species. On the other hand, the organ-specific nature of the encephalitogen remains a valid concept.[17,18]

Myelin basic proteins have certain immunogenic characteristics which may or may not be related to their encephalitogenicity. They are not recognized immunologically as "self" constituents and are therefore capable of inducing their immunologic reactions even in the autologous situation. Kabat *et al.*[19] have shown that a fragment of CNS removed from a monkey will induce EAE when injected into the same animal. Encephalitogenic proteins induce both immediate[20–23] and delayed hypersensitivity[24] in homologous as well as isologous animals,[25] and they can be used to analyze these immunologic phenomena in humans with CNS pathology.[26,27] Several aspects of their immunogenicity, e.g., the function of adjuvant,[28] passive induction of EAE and delayed hypersensitivity,[25,29] suppression of EAE,[20,29] etc., have been studied in detail.

Purification of the encephalitogenic basic proteins permitted a more careful examination of the question of species-related differences, one of the first indications of such differences being obtained in studies on the rat. This animal is susceptible to EAE when sensitized with rat or guinea

TABLE I

Species Distribution of Susceptibility to Experimental Allergic Encephalomyelitis and the Encephalitogenic Activity of CNS Tissue[15]

Susceptible species	Source of encephalitogen
Human[a]	Human
Beef	Beef
Monkey	Monkey
Sheep	Sheep
Guinea pig	Guinea pig
Rat	Rat
Rabbit	Rabbit
Mouse	Mouse
Goat	Pig
Cat	
Dog	

[a] Rabies post-vaccinal encephalitis, post-infectious encephalitides.

TABLE II

Induction of EAE and Antibody to Homologous BP in Lewis Rats

Sensitization				EAE		Antibody incidence[b] (%)
Species BP	Wt. (Mcg)	Adjuvant		Clinical (%)	DO_{50}[a]	
Guinea pig	25	CFA + Pertussis[c]		100	10	0
	25	CFA	—	100	13	0
	200	CFA	—	100	11	100
	200	—	Pertussis	0	—	0
Beef	200	CFA + Pertussis		0	—	100
Rat	100	CFA	—	88	15	0

[a] Median day of onset.
[b] Specific binding of I^{125}-labeled BP, detected by radioimmunodiffusion.[21]
[c] Complete Freund's Adjuvant plus Pertussis.

pig myelin basic protein (BP), but is much less susceptible to EAE when injected with human, monkey, rabbit, or bovine BP's.[30,31] Thus, the rat clearly detects chemical (structural?) differences among myelin BP's of various species which cannot be detected by the guinea pig.

In spite of their resistance to its encephalitogenic activity, rats readily produce antibody to bovine BP, suggesting that antibody *may* not (or *need* not) be related to pathogenesis of EAE (Table II). The antibody induced in the rat by bovine myelin BP cross-reacts with similar proteins isolated from several other species (Table III), some of which are relatively non-encephalitogenic in the rat.[31] Amino acid-sequence studies are not suffi-

TABLE III

Cross-Reactivity of Anti-BP Antibody Induced in Lewis Rats by Heterologous Myelin Basic Protein

Species of sensitizing antigen (in CFA)	EAE	Species of test antigen					
		Rat	Guinea pig	Beef	Monkey	Rabbit	Human
Guinea pig, 25γ	+	0/5	3/5	0/5	—	—	—
Guinea pig, 200γ	+	4/5	4/5	4/5	4/5	4/5	4/5
Beef, 200γ	−	5/5	5/5	5/5	5/5	5/5	5/5

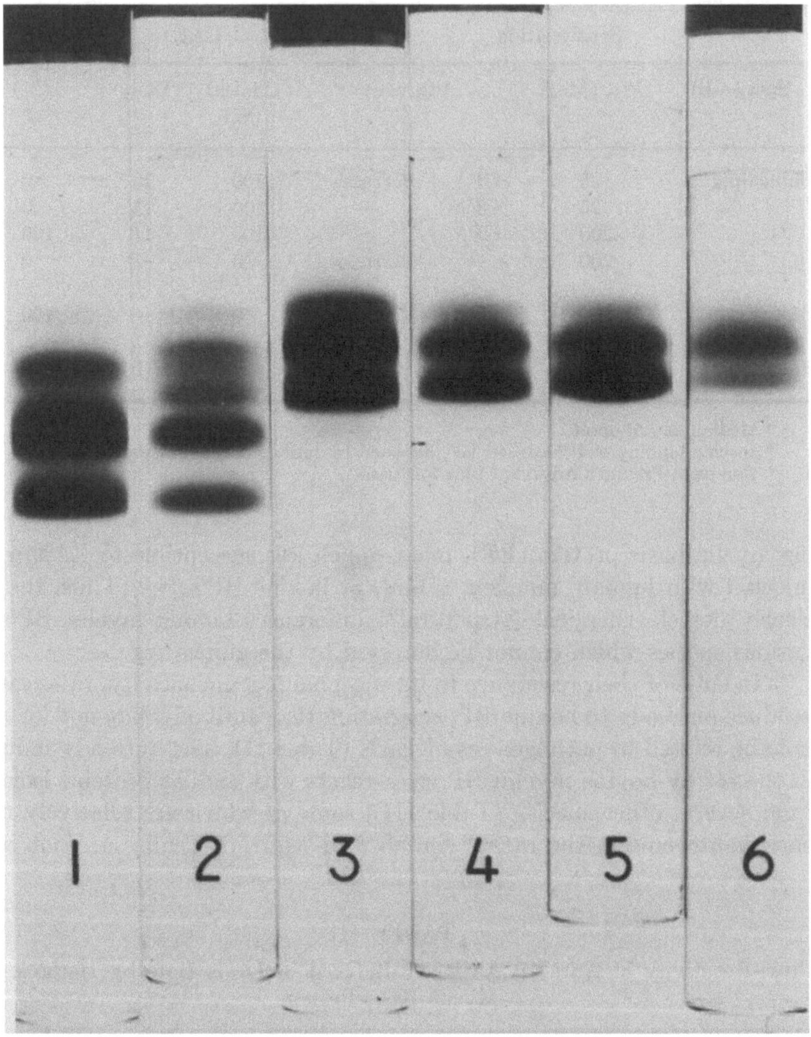

Fig. 1. Comparison of rat myelin basic proteins with those of other species. (1) "Small"-molecular-size fraction obtained from (2); (2) rat total myelin basic proteins; (3) "large"-molecular-size fraction obtained from (2); (4) guinea pig, (5) rabbit, and (6) beef total myelin basic proteins. Electrophoresis in 5% acrylamide gel, pH 10.6; origin at top; direction of migration toward the cathode (Martenson *et al.*[34]).

ciently complete at the present time to define the encephalitogenic site(s?), much less the antigenic sites, and it remains to be seen how the various portions of the molecules participate in these two biologic activities.

Another study which predicted chemical or structural differences among myelin basic proteins of different species was that of Alvord,[32] in which antisera produced in rabbits with purified bovine myelin basic protein were found to produce five precipitin lines on immunodiffusion against the inducing antigen. Cross-reactivity was demonstrated with human, guinea pig, rabbit, pigeon, and rat encephalitogenic preparations, but not with sheep. However, not all five precipitin lines were demonstrable with all of these encephalitogens, suggesting again that myelin basic proteins from different species differed with respect to certain antigenic sites in the molecules.

With regard to the organ-specific character of these proteins, it is obvious that myelin is specific to the nervous system, but it does not necessarily follow that its constituent proteins are limited to this tissue. However, extensive studies in our laboratory with both electrophoretic and immunologic techniques have failed to yield any evidence for identical proteins in other organs.[17,33] Electrophoretic comparisons of extracts of brain and other organs (adrenal gland and liver) have also demonstrated that the basic proteins of myelin can be clearly differentiated from the histones of brain as well as those of other organs.[34]

The original concept of "the encephalitogen" was that of a single CNS constituent. This idea of its homogeneity *in situ* has influenced much of the study on the isolation and purification of myelin basic protein. Purified "homogeneous" preparations have been characterized by chromatography, electrophoresis at neutral or acid pH, and ultracentrifuge data. Recently Martenson and Gaitonde[35] reported that acrylamide gel electrophoresis at high pH can distinguish multiple forms in purified bovine myelin basic protein found to be homogeneous by other criteria.[10] In continuing these studies, Martenson *et al.*[34] have shown that the heterogeneity detected in the purified protein is not an artifact of purification. Comparable heterogeneity can be demonstrated in guinea pig myelin basic proteins isolated either from whole CNS by a procedure designed to minimize chemical or enzymatic changes, or from myelin by extraction with 0.1 *N* HCl.

The basic proteins of rat myelin are unique, in that their heterogeneity involves more than the differences in charge observed in myelin basic proteins of other species.[34–36] It is possible to separate rat myelin basic proteins into two groups of different molecular size (Fig. 1), whereas the myelin basic proteins of each of the other species appear to consist of proteins of the same size differing only with respect to charge. The striking similarity of the electrophoretic patterns in Fig. 2 suggests a fundamental

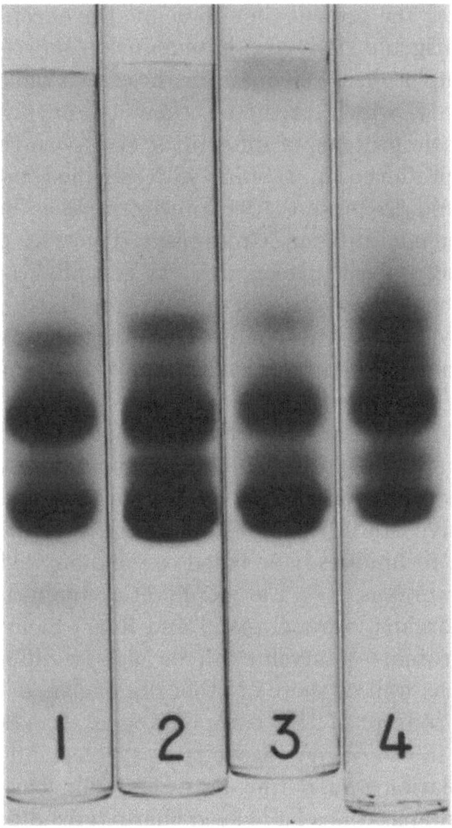

Fig. 2. Electrophoresis of myelin basic proteins in pH gradient (pH 8.6–10.6). (1) Rat "large" fraction, (2) guinea pig, (3) rabbit, (4) beef. Direction of migration toward the cathode (Martenson *et al.*[34]).

homology between the "large" fraction of rat myelin basic proteins and the total myelin basic proteins of three other species.

Gel filtration data indicate they are all essentially identical in molecular weight (~17,000) and that the electrophoretically distinct "small" rat myelin basic proteins have a molecular weight of about 14,000, a difference in chain length of approximately 25 amino acids. Whether all members of the group(s) are equally encephalitogenic or antigenic cannot be determined until further fractionation has been achieved. Studies designed to answer this question are underway at the present time.

In summary, studies on purified myelin basic proteins have confirmed the organ-specificity of these proteins. These studies have also shown that myelin basic proteins from different species possess subtle differences which can be demonstrated either by bioassay in the rat, by immunologic cross-reactions, or by electrophoretic analysis. The latter, combined with gel filtration data, indicate that the myelin basic proteins in most species consist of a group of proteins of uniform size, differing only slightly from one another in charge. Rat myelin contains basic proteins which can be separated into two sizes—the larger-size group resembling the myelin basic proteins of other species, and the smaller one characteristic of the rat. If the various components within a species are found to differ significantly in amino acid composition and immunologic activity, present concepts of antigenic and encephalitogenic determinants will have to be revised.

REFERENCES

1. M. W. Kies and E. C. Alvord, Jr., *in*: " 'Allergic' Encephalomyelitis" (M. W. Kies and E. C. Alvord, Jr., eds.), pp. 293–299, Charles C. Thomas, Springfield, Ill. (1959).
2. R. H. Laatsch, M. W. Kies, S. Gordon and E. C. Alvord, Jr., *J. Exp. Med.* **115,** 77–88 (1962).
3. E. Roboz-Einstein, D. M. Robertson, J. M. DiCaprio, and W. Moore, *J. Neurochem.* **9,** 353–361 (1962).
4. R. F. Kibler, R. H. Fox, and R. Shapira, *Nature* **204,** 1273–1275 (1964).
5. M. W. Kies, *Ann. N.Y. Acad. Sci.* **122,** 161–170 (1965).
6. A. Nakao and E. Roboz-Einstein, *Ann. N.Y. Acad. Sci.* **122,** 171–181 (1965).
7. E. A. Caspary and E. J. Field, *Ann. N.Y. Acad. Sci.* **122,** 182–198 (1965).
8. R. F. Kibler and R. Shapira, *J. Biol. Chem.* **243,** 281–286 (1968).
9. P. R. Carnegie, B. Bencina, and G. Lamoureux, *Biochem. J.* **105,** 559–568 (1967).
10. R. E. Martenson and F. N. LeBaron, *J. Neurochem.* **13,** 1469–1479 (1966).
11. A. Nakao, W. J. Davis, and E. Roboz-Einstein, *Biochim. Biophys. Acta* **130,** 163–170 (1966).
12. E. Roboz-Einstein, A. Nakao, J. DiCaprio, and W. D. Moore, *in*: ' Proc. VII Int. Congr. Neurol.," Vol. 2, pp. 748–763, Soc. Graf. Romana (1964).
13. S. Hruby, C. M. Shaw, E. C. Alvord, Jr., and M. W. Kies, *Int. Arch. Allergy,* in press.
14. E. H. Eylar and G. A. Hashim, *Proc. Natl. Acad. Sci. U.S.* **61,** 644–650 (1968).
15. E. C. Alvord, Jr., *in*: " 'Allergic' Encephalomyelitis" (M. W. Kies and E. C. Alvord, Jr., eds.), pp. 518–557, Charles C. Thomas, Springfield, Ill. (1959).
16. E. C. Alvord, Jr. and M. W. Kies, *J. Neuropath. Exp. Neurol.* **18,** 447–457 (1959).
17. C. S. August, M. W. Kies, S. Hruby, and C. M. Shaw, *Fed. Proc.* **25,** 356 (1966).
18. E. C. Alvord, Jr., S. Hruby, K. Hughes, C. M. Shaw, and M. W. Kies, *Int. Arch. Allergy,* in press.
19. E. A. Kabat, A. Wolf, and A. E. Bezer, *J. Exp. Med.* **89,** 395–398 (1949).
20. E. C. Alvord, Jr., C. M. Shaw, S. Hruby, and M. W. Kies, *Ann. N.Y. Acad. Sci.* **122,** 333–345 (1965).
21. R. P. Lisak, R. G. Heinze, M. W. Kies, and E. C. Alvord, Jr., *Proc. Soc. Exp Biol. Med.* **130,** 814–818 (1969).

22. G. A. Falk, R. G. Heinze, M. W. Kies, and E. C. Alvord, Jr., *J. Immunol.* **100**(2), 321–328 (1968).
23. R. F. Kibler and A. E. Barnes, *J. Exp. Med.* **116**, 807–825 (1962).
24. C. M. Shaw, E. C. Alvord, Jr., J. Kaku, and M. W. Kies, *Ann. N.Y. Acad. Sci.* **122**, 318–331 (1965).
25. G. A. Falk, M. W. Kies, and E. C. Alvord, Jr., *J. Immunol.* **101**(4), 638–644 (1968).
26. P. O. Behan, N. Geschwind, J. B. Lamarche, R. P. Lisak, and M. W. Kies, *Lancet* **II**, 1009–1012 (1968).
27. R. P. Lisak, R. G. Heinze, G. A. Falk, and M. W. Kies, *Neurology* **18**(2), 122–128 (1968).
28. E. C. Alvord, Jr., C. M. Shaw, W. J. Fahlberg, and M. W. Kies, *Z. Immun.-Forsch.* **126**, 193–206 (1962).
29. S. Levine, E. M. Hoenig, and M. W. Kies, *Science* **161**, 1155–1156 (1968).
30. S. Levine and E. J. Wenk, *Ann. N.Y. Acad. Sci.* **122**, 209–224 (1965).
31. R. P. Lisak, unpublished.
32. E. C. Alvord, Jr., *in*: "The Central Nervous System," Int. Academy of Pathology Monograph No. 9, pp. 52–70, The Williams and Wilkins Company, Baltimore (1968).
33. R. E. Martenson, G. E. Deibler, and M. W. Kies, *Fed. Proc.* **27**, 619 (1968).
34. R. E. Martenson, G. E. Deibler, and M. W. Kies, *J. Biol. Chem.* **244**, 4261–4267 (1969).
35. R. E. Martenson and M. Gaitonde, *J. Neurochem.* **16**, 333 (1969).
36. E. Mehl, Abstracts 1st Int. Congr. Neurochem., Strasbourg, 1967, p. 154

DISCUSSION

LAJTHA: Don't you think that if you consider autoimmunity to brain protein as a mechanism, then brain proteins should get to the antibody-forming site? Making them smaller would make it possible for them to get out of the brain, and therefore liberation of a hydrolytic enzyme, which makes a protein smaller, enables them to become an antigen in a sense.

KIES: This isn't a very big protein; even before proteolytic digestion its molecular weight is about 17,000. This chapter has really dealt with immunochemistry, and really had very little to do with disease. We hope that studies of the antibodies will help us understand the antigen, but sensitized cells, and not antibodies, are thought to induce the disease. It is very easy to speculate that when the sensitized cells get to the target organ some "immunologic" event takes place which liberates lysosomal enzymes, or perhaps involves the complement system. The details have not been worked out. I don't think one need postulate that an intrinsic proteinase is involved in this, although that doesn't mean that it isn't.

RAPPORT: Is there any enzyme activity directly on myelin that would cause release of the small fragments?

KIES: Adams *et al.** discuss this in their paper on enzymatic activity in myelin; an amino peptidase has also been noted by Beck *et al.*†

* C. W. M. Adams, A. N. Davison, and N. A. Gregson, *J. Neurochem.* **10**, 383 (1963).

† C. S. Beck, C. W. Hasinoff, and M. E. Smith, *J. Neurochem.* **15**, 1297 (1968).

LAJTHA: However, it is obvious that myelin as such is not an encephalitic enzyme. It is possible that damage to the membrane would liberate proteolytic enzyme.

RAPPORT: Now, if you could just disrupt the myelin a little bit in some way, you could perpetuate the mechanism by releasing these small fragments.

BOGOCH: Kabat reported, I think in the forties, that in polio studies in monkeys he found antibodies to polio in spinal fluids at a time when they were not present in serum, and suggested at that time that there might be an antibody-forming site within nervous systems, discrete from those outside. Has there been any further evidence, pro or con, in terms of the ability of the nervous system to produce antibodies by itself?

KIES: Scheinberg *et al.** have done some experiments, also Tourtellote, *et al.*†

FOLCH: Would you care to comment on Lumsden's report of encephalitogenic activity in proteolipids?

KIES: He found encephalitogenic activity of proteolipids when he tested 14 mg, and that weight is more than 2000 times the weight of BP required to induce EAE. It is very easy to see how one might have that much contamination.

FOLCH: Would that also explain the fact that the proteolipids are encephalitogenic in rabbits?

KIES: I don't know whether that's the explanation or not. Rabbits are very different in their immunologic reactivity from guinea pigs. This is one of the problems which has always bothered us regarding proteolipids and the basic protein. I know that both proteolipids and basic protein occur in myelin, so I think that it is highly probable that there was some encephalitogenic protein in the proteolipids tested

FOLCH: Have you been worried by the fact that your result may be due to a very slight contaminant?

KIES: My only answer is that no material has been tested which possessed higher specific activity than myelin basic protein or a peptide derived from it.

SELLINGER: If this protein has been buried in some way within the myelin so that it isn't available to proteolytic attack *in vivo*, then the problem of its synthesis arises. Now, we have a membrane (myelin), and within it synthetic processes would have to occur. The alternative would be that at some time prefabricated protein is transferred into myelin.

ENG: We've been looking at purified myelin from bovine and human white matter, and when we prepare myelin by a continuous and discontinuous medium we can extract the basic protein fraction of myelin with $\frac{1}{2}M$ salt without removal of the lipids, and by disc-electrophoresis this basic protein gives one main band. This protein elicits precipitating antibodies in rabbits, it is highly encephalitogenic in amounts of about 5 gamma.

KIES: Dr. Eng, I have to take issue with your use of electrophoresis at low pH as a valid indication of homogeneity. BP isolated from guinea pig myelin with 0.1 N HCl also has a single band at that pH, but on electrophoresis at alkaline pH with or without urea several bands are visible. We have satisfied ourselves that these bands do not arise artifactually.

* L. C. Scheinberg, D. G. Kotsilimbas, R. Karpf, and N. Mayer, *Arch. Neurol.* **15,** 02 (1966).

† W. W. Tourtellote, J. A. Parker, R. M. Herndon, and C. V. Cuadros, *Neurology* **18,** 117 (1968).

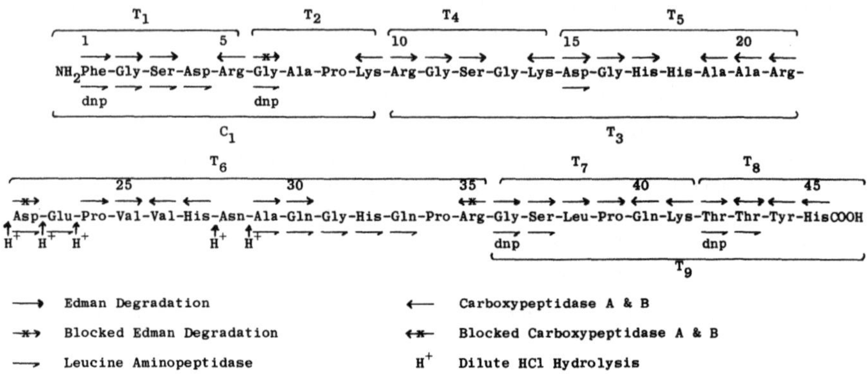

Fig. 3. Amino acid sequence of an encephalitogenic basic protein from bovine spinal cord.

ENG: Now, the molecular weight of this protein is about 9000 when one looks at it in ultracentrifuge by the Archibald principle, and if you calculate a minimum molecular weight from amino acid composition, it also comes out to be about 9000, and this accounts for about 95–100% of basic protein in myelin.

KIES: Your estimate of a minimum molecular weight of 9000 from ultracentrifugation data and amino acid analysis is puzzling. Our gel column data when interpreted with the proper standards (i.e., proteins consisting of a single polypeptide chain lacking disulfide cross-linkages and having a positive charge density at pH 2.0 similar to myelin basic protein) indicate a 17,000 molecular weight for human, bovine, rabbit, and guinea pig. Rat myelin BP is unique in that the heterogeneous population can be separated into two sizes, one of 17,000 and one of about 14,000. The amino acid data of other investigators—Eylar,[14] Carnegie,[9] and Nakao *et al.*[11]—all agree with a minimum molecular weight of 16,000–17,000.

SHAPIRA: The fragment of the basic protein Dr. Kibler and I are working with has a molecular weight of around 4500, depending upon the species of the material. Figure 3 shows the sequence of the bovine preparation. The N-terminal end is phenylalanine, the C-terminal end is histidine, and the protein is a continuous polypeptide chain of

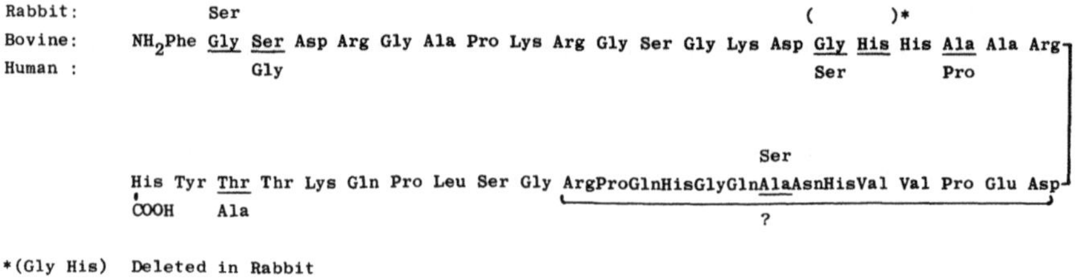

Fig. 4. Species variability.

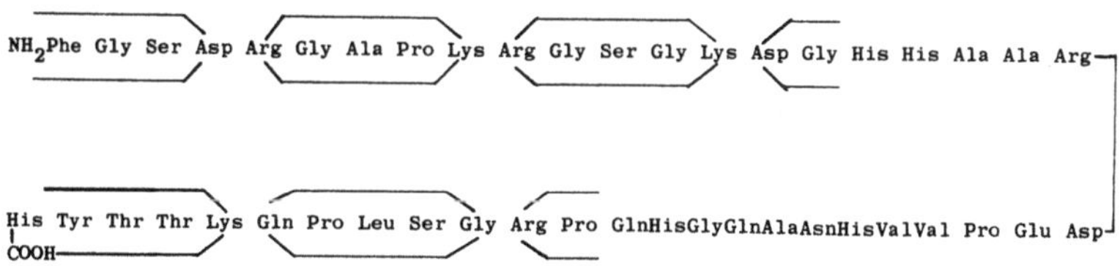

Fig. 5. Repeat unit in myelstone I.

45 amino acid residues. There are no sulfur-containing amino acids and there is no isoleucine in this preparation, although methionine and isoleucine are present in the larger encephalitogens. The fragment also contains one residue only of phenylalanine, tyrosine, and leucine. There is a lot of glycine, proline, and basic amino acids; in general, it looks like a histone, but that is as far as the similarity goes. Figure 4 shows the species differences. In the rabbit there are serines substituting for a glycine and an alanine, and a glycyl-histidine has been deleted. In the human substitutions so far determined include serine to glycine, glycine to serine, alanine to proline, and threonine to alanine. Figure 5 shows something that may bear on the binding of this membrane protein to lipid. There is a very strange situation. There seems to be three hydrophobic amino acids, and then two hydrophilic, in other words, three nonpolar and two polar, and this type of repeat unit seems to extend through quite a portion of the molecule. Now,

TABLE IV

Activity of the Low-Molecular-Weight Encephalitogen in the Rabbit

Source	Amount injected (μg)	$(N_{EAE}/N_{inj})^a$	Day of onset	
			Av.	Range
Bovine	200	27/27	13.5	(10–18)
		18/20	14.8	(11–18)
Rabbit	200	19/20	14.3	(11–18)
Bovine	100	4/4	14.2	(12–21)
Rabbit	100	4/4	17.0	(11–25)
		4/4	13.0	(11–17)
Bovine	50	4/4	13.2	(11–18)
Rabbit	50	4/4	12.8	(10–14)
Rabbit	25	3/4	17.0	(11–22)
Rabbit	15	3/4	20.3	(18–25)
		2/4	32.5	(19 & 46)

[a] (Number with EAE/number injected).

obviously, if you have a repeating unit of hydrophobic and hydrophilic amino acids, this would allow for very specific binding to lipid; in other words, the protein could easily be folded up in such a way that, say, all of the hydrophilic amino acids would be on one side, or you could figure out a way to make a meshwork where lipids could fit along the molecule. I also want to emphasize that this is an active preparation; 50 μg and larger doses produce EAE in over 90% of the rabbits, with around day 14 as an average day of onset. Below 50 μg, activity falls off, but the protein is still encephalitogenic at 15 μg. These findings are shown in Table IV. We are presently trying to determine where the encephalitogenic activity resides in the molecule.

Chapter 36

Transglutaminase Changes in the Brain and Other Tissues during Allergic Encephalomyelitis*

Isabel J. Wajda

New York State Research Institute for Neurochemistry and Drug Addiction Ward's Island, New York, N. Y.

INTRODUCTION

The specificity of an antigen such as protein in immunological reactions depends on its structure. Chemically induced alterations can change its antigenic specificity, as was shown by Landsteiner's studies on artificially conjugated proteins.[1] His experiments proved not only that large, heterologous molecules are able to produce specific immunological reactions, but that an introduction of a small foreign radical provides the protein with different specificity, typical of the introduced molecule. The effects of chemical treatment on immunochemical properties of proteins were summarized by Kabat and Mayer.[2] Of interest to us is their statement that the changes in the reactivity of native protein with antisera, which depend on the introduced radical, can sometimes result in a diminished reactivity, although the overall result is usually a profound change in the specificity of native protein. All these changes are recorded by standard immunological tests. Our interest in possible changes of proteins under the influence of transglutaminase were connected with autoimmune responses, reactions that are characterized by hypersensitive states and a lack of circulating antibodies.

A general view held at present is that immunological reactions are responsible for the development of autoimmune responses. During such response autologous proteins may be involved, as indicated by the ob-

* The study of transglutaminase during EAE was done jointly with Dr. Johanna Lee from the Department of Neurology, Albert Einstein College of Medicine. Our thanks are due to Dr. Amos Neidle for help in these studies, and to Mr. I. Manigault and Mrs. S. S. Keyes for their valuable assistance.

servations on animals suffering from experimental allergic encephalomyelitis (EAE); those observations confirm that the encephalitogen exists in the central nervous system. Administration of homologous or even autologous brain tissue results in damaging immunological reactions, which are confined to the central nervous system and finally kill the experimental animal by producing paralysis. It is generally accepted that in autoimmune response the animal produces antibodies to its own constituents. Whether some alterations of the constituent taking part in autoimmune manifestations occur during the process of the disease is still not known; however, the fact that the animal reacts to his own components strongly suggests such a possibility.

Also unknown is what happens to the injected antigen during development of the autoimmune response. Such an antigen is capable of influencing a specific target tissue, and it is puzzling why such an antigen is able to evoke EAE in some species more readily than in others. A wide margin of reactivity of different species used experimentally mainly parallels their susceptibility to hypersensitive reactions.

In an attempt to elucidate some of those questions, we studied the levels of transglutaminase in the brain, liver, and the reticuloendothelial cells in a time-course study during the development of EAE. The animals used in our experiments were two strains of mice: one a strain susceptible to bacterial and viral infections (bacteria-sensitive, virus-sensitive: BSVS), which also easily developes EAE; the other a resistant strain (BRVR), which, under identical experimental conditions, remains free of the disease.

EARLIER STUDIES OF TRANSGLUTAMINASE

Transglutaminase is capable of changing the primary structure of proteins without splitting the peptide bonds, by catalyzing the exchange

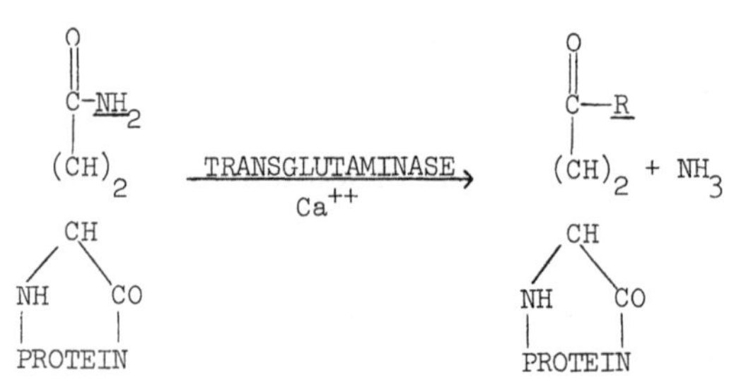

Fig. 1. Transglutaminase catalysis. R = OH (hydrolysis) or primary amine or ε-amino group of protein-bound lysine.

TABLE I
Substrate Activity of Proteins[a]

G.P. liver extract	+ + +	α-Globulin	+ +
Egg albumin	−	Casein	+ + + +
Bovine serum alb.	−	Pepsin	+ + + +
Human serum alb.	+	Ceruloplasmin	+ +
Fibrinogen	+ + +	Lysozyme	−
Bovine hemoglobin	(+)	β-Lactoglobulin	+ + + +
γ-Globulin	—	Edestin	+ +

[a] From Clarke *et al.*[5] Each tube contained 15 mg of the protein tested; 8 μM cadaverine-^{14}C (0.05 μCi/μM); 20 μM Tris buffer (pH 8.3); 10 μM $CaCl_2$; and 1.5 mg of the pH-6.5 extract protein as enzyme. Final volume: 2 ml. Incubations were carried out for 1 hr at 37°C.

of amide groups of protein-bound glutamine with primary amines.[3–5] In the absence of amine the ε-amino group of protein-bound lysine can enter this reaction to form an internal linkage. At slightly different pH hydrolysis occurs, and all such reactions are accompanied by evolution of ammonia (Fig. 1). The reaction is calcium dependent, and only protein- or peptide-bound glutamine serves as substrate.

Many proteins can be used as substrates for transglutaminase; the specificity of a series studied by Clarke *et al.*[5] is shown in Table I. It was of interest to us to note that among the serum proteins, fibrogen and α-globulin could serve as substrates, whereas γ-globulin proved to be rather inactive. Different affinities of amines were established by the same group, and a selection of amines and amino acids showing very varied reactivity is shown in Table II. Further studies from our laboratory provided data

TABLE II
Incorporation of Various Amines Into Heated Guinea Pig Liver pH-7.5 Extract[a]

Compound	μM /100 mg protein	Compound	μM /100 mg protein
Putrescine	8.9	Alaninamide	1.2
Phenylethylamine	6.3	Noradrenaline	0.89
Glycinamide	4.9	Serotonin	0.57
Histamine	4.2	*L*- or *D*-Lysine	0.56
Spermine	2.6	Aniline	0.14
Cadaverine	2.3	Tyrosinamide	—
Methylamine	2.3	Glycine	—
Ethanolamine	2.3	Glycyl-*L*-Leucine	—
Ammonia	1.5	γ-Aminobutyric acid	—

[a] Clarke *et al.*[5] Conditions of incubation: same as those in Table I, with heated guinea pig liver pH-7.5 extract as substrate.

on incorporation of different amines into insulin,[6] hemoglobin,[7] and the synthetically prepared glutamine peptides.[8] All these studies indicated that the reactivity of any particular glutamine residue depends upon the amino acid sequence in its immediate vicinity. They also disclosed the fact that glutamine, but not asparagine, enters the reactions catalyzed by transglutaminase.

Studies on the distribution of transglutaminase in different tissues revealed its presence in practically all organs; very high levels were present in liver, lung, kidney, and spleen. Negligible amounts were detected in serum. As far as the subcellular distribution is concerned, the enzyme is present in the soluble fractions of tissue homogenates. The distribution is not the same in different organs. Our latest experiments recorded in mice show that the content of the enzyme when assayed in total sucrose homogenates of the liver and compared with the soluble protein fraction (100,000 g supernatant) is about four times higher in the soluble fraction. A different relationship is found in lung tissue. Here, after centrifugation, the activity of the soluble fraction is nearly ten times lower than the one recorded in total homogenate. On the other hand, transglutaminase content in brain is of the same magnitude in both preparations, when total homogenates are compared with the soluble protein fractions. Species differences are considerable. Guinea pig livers contain large quantities of transglutaminase; only one-fifth as much was recorded in comparative studies of the livers of mice and rats. Several methods of purification are available, some giving 100-fold increase in specific activity.[9]

TABLE III

Transglutaminase Levels after Bacterial Endotoxin and Adrenocortical Hormones[a]

Exp.	Treatment	No. of animals	Transglutaminase specific activity[b]
1	Normal	8	18.0 ± 2.0
2	Adrenalectomy	18	25.3 ± 2.1
3	Sham-operated	12	18.7 ± 0.8
4	Cortisone	8	17.9 ± 2.0
5	*S. typhosa* LPS[c]	6	50.4 ± 6.2
6	*S. typhosa* LPS[c] and cortisone	10	24.9 ± 2.3

[a] From Wajda *et al.*[10] The number of animals in each experiment represents the number of enzyme assays. Cortisone, in doses of 2.5 mg/mouse, was injected subcutaneously for four consecutive days before administration of endotoxin, which was given simultaneously with the last dose of cortisone.

[b] Values ±SE of the mean: $0.05 > P > 0.025$ for experiments 1 and 2; $P = 0.001$ for experiments 5 and 6.

[c] *Salmonella typhosa* lipopolysaccharide.

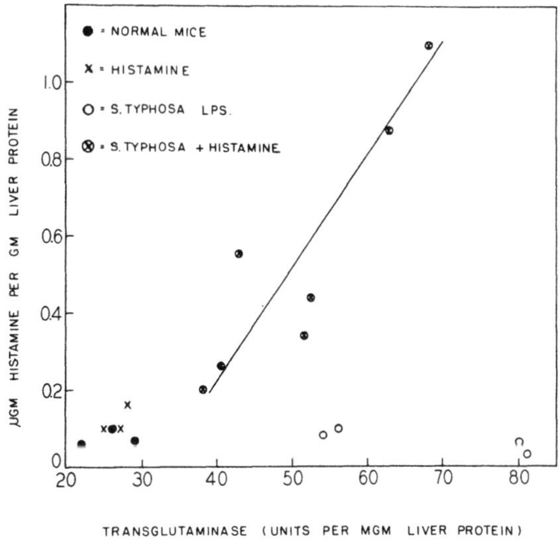

Fig. 2. Transglutaminase activity and protein-bound histamine in the livers of normal and endotoxin-treated mice. *S. typhosa* LPS: *Salmonella typhosa* lipopolysaccharide. (From Ginsburg *et al.*[11])

Before we started the studies on transglutaminase in autoimmune states, the levels of liver transglutaminase were studied in mice and rats subjected to different experimental procedures, and it was found that the normal values are very stable. However, treatment with bacterial endotoxins or pertussis vaccine resulted in a significant increase of liver enzyme. This increase could be prevented by a pretreatment of experimental animals with high doses of cortisone (Table III).[10] It is known that adrenal hormones suppress the formation of antibodies, but they also have a profound influence on the inflammatory processes. Therefore, it was difficult to draw any simple conclusion from our observation. At this time we also tested glutamine synthetase and protoinase in the livers of pertussis-treated mice and found their levels unchanged.

Although it is very easy to incorporate a suitable amine into a protein in a test tube by incubating at the required pH in the presence of the enzyme, calcium ions, and a reducing substance, it is rather difficult to show incorporation *in vivo*. Using mice treated with bacterial endotoxins, we were able to demonstrate the incorporation *in vivo* of injected radioactive histamine into liver protein.[11] The quantities of incorporated histamine correlated with the levels of the enzyme (Fig. 2). As the amounts of protein-fixed histamine were very small, we did not propose a causal

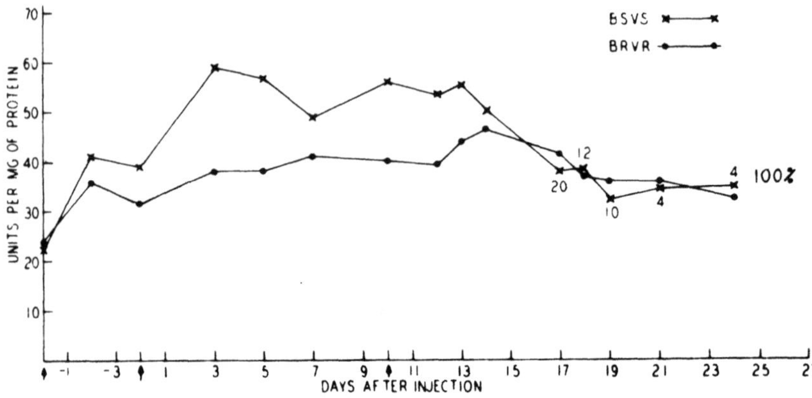

Fig. 3. Liver transglutaminase in mice during the development of EAE. The arrows on the abscissa indicate: (1) Intraperitoneal injection of 0.5 ml of pertussis vaccine, (2, 3) two intradermal administrations of 0.3 ml of homologous brain (14 mg), emulsified in complete Freund adjuvant. Each point represents averages of five separate estimations of the enzyme obtained from five animals. The numbers on the chart indicate mice that showed positive clinical signs of EAE. Protein was estimated by biuret method.

relationship between the increased enzyme activity and the levels of protein-bound histamine.

Also, the massive amounts of administered histamine necessary to obtain an increased fixation in liver protein suggested that an amine other than histamine might be a better physiological substrate of transglutaminase. Histamine was used in our experiments because it could be split unchanged from the protein during acid hydrolysis, and easily estimated.

TRANSGLUTAMINASE LEVELS IN EAE

The fact that transglutaminase can catalyze the incorporation of amines such as histamine, serotonin, or epinephrine into body proteins, or bind internally two chains of proteins by glutaminyl–lysine linkage, together with our observations of the changes occurring after use of bacterial endotoxins and cortisone, directed our attention to the role this enzyme system could have in immune, or even more, in autoimmune, responses. For the comparison of the enzyme levels during autoimmune disease, of particular interest were closely related animals of different susceptibility. Such an opportunity presented itself to us when Dr. Johanna Lee became interested in our studies, and cooperated with us in using her two genotypes of mice that differ in susceptibility to viral and bacterial infections and also to autoimmune diseases.[12]

Liver Enzyme Levels

Our first studies on transglutaminase levels recorded during the development of EAE included estimations of liver enzyme in BSVS and BRVR mice.[13,14] Figure 3 shows the variations in the enzyme levels assayed in the soluble fractions of liver homogenates during a time-course study of EAE evoked only in the sensitive (BSVS) strain. The brains of all the animals were examined histologically for the signs of the disease. Significantly higher levels of transglutaminase were found in the livers of the affected strain during the time preceding the clinical signs of the disease. During paralytic stages of EAE the liver enzyme returned to normal values. A 100% incidence of the disease was recorded in the sensitive brain, and the identically treated resistant strain was free of EAE. This work also included studies of liver enzyme in both strains of mice after administration of all encephalitogenic components given separately. EAE was not produced when pertussis vaccine alone, complete Freund adjuvant alone, or brain emulsion in incomplete Freund adjuvant were used. In these three experiments the enzyme levels in the liver were not significantly different between the two strains. We therefore came to the conclusion that a direct relationship exists between liver transglutaminase levels and the EAE response.

The method used for the estimations of liver enzyme was not sensitive enough to detect quantitatively the low levels of transglutaminase in brain. There is about twenty times less enzyme present in brain than in liver, and to be able to determine such low quantities a radiochemical assay was developed.[15] This procedure requires a synthetically prepared dipeptide and radioactive amine. Figure 4 represents the reaction occurring in the presence of transglutaminase.

Brain Enzyme Levels

Using this method we assayed transglutaminase levels in the brains of the two strains of mice subjected to complete encephalitogenic treat-

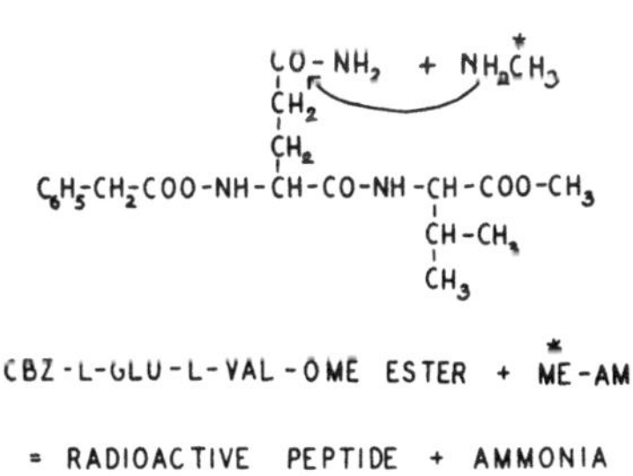

Fig. 4. Radiochemical assay reaction.

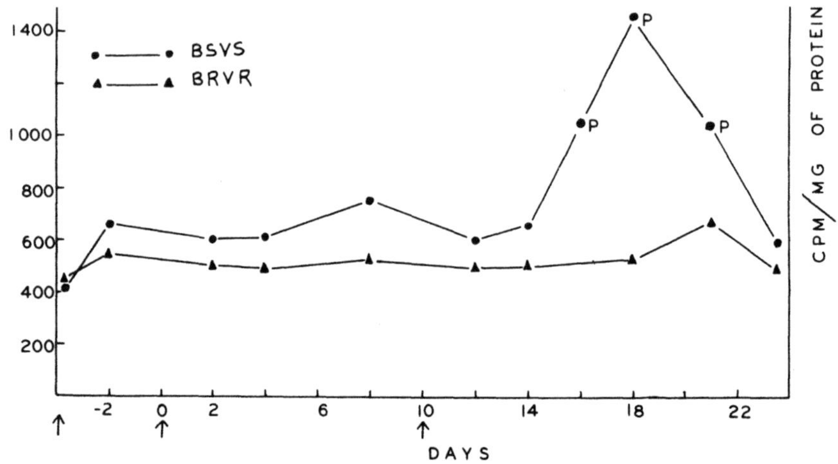

Fig. 5. Enzyme levels in brain. One hundred mice of each strain were subjected to the same treatment as described in Fig. 3. Each point represents an average of six individual estimations of transglutaminase. Half of each brain was homogenized in 0.25 *M* sucrose (1:6 w/v), and 0.3 ml of such homogenate was used for the estimation of enzyme. *P*: paralyzed animals. Protein was assayed by Lowry and micro-Kjeldahl methods.

ment.[12] The enzyme was estimated in whole homogenates; we used only half of the brain of each experimental animal, and the other half was fixed in formalin and used for histology. The activity of transglutaminase in the sensitive animals affected by the disease was very different from that of the resistant strain (Fig. 5). As soon as the clinical signs of EAE became evident the enzyme levels increased considerably, reaching levels three times higher than normal values. This happened only in the brains of the sensitive strain of mice at the time when they were showing all the signs of progressing disease. In the resistant strain, transglutaminase was at normal levels. The highest values in paralyzed mice were recorded on the 18th day of the experiment; to check on the possible leakage from the periphery through the circulation, we measured the enzyme levels in the serum pooled from six mice sacrificed on this day. The serum levels were very low, nearly at zero values.

White-Cell Enzyme Levels

Having established the fact that transglutaminase increase so closely follows pathological changes occurring in the target organ, we decided to look at the enzyme levels in reticuloendothelial cells. Again the same experimental procedure was followed: on predetermined days groups of ten mice were injected intraperitoneally with oil and sacrificed two days later. Peritoneal wash was obtained, and transglutaminase levels were assayed in the white cells of those exudates. The cells were washed according to procedure described by Bloom and Chase.[16] After an initial fall observed in both strains of mice, the enzyme levels started to increase in the non-reacting, resistant strain (Fig. 6). This rise was at its highest level at the time when the sensitive mice were showing all the signs of EAE. The enzyme in the sick BSVS mice was at the normal levels.

Summarizing the main points of this presentation:

1. Transglutaminase, an enzyme that is able to change the primary structure of proteins, shows characteristic patterns in different organs during the induction of autoimmune response in mice.

2. In the livers of BSVS mice that develop the disease the enzyme increases well before the clinical signs of EAE are in evidence. During the time of paralysis the enzyme levels return to normal or even below normal values. Similar, but significantly lower increase, is in evidence in BRVR (resistant) strain.

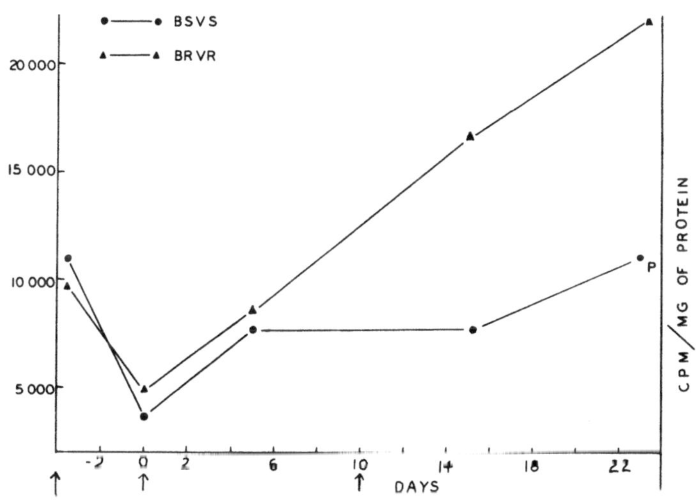

Fig. 6. Transglutaminase levels in white cells of peritoneal exudate. Ninety mice of each strain were used. Peritoneal cells were obtained from groups of ten animals. Two milliliters of oil were given intraperitoneally, and two days later the animals were killed, and the peritoneal cells were obtained using the procedure described by Bloom and Chase.[16] On the fourth and zero days ten mice represent each point. The three subsequent points on the chart are averages of two pooled groups of animals from each strain.

3. In brain, a striking rise in transglutaminase levels is found during the time of paralysis, reaching over 300% increase in the affected, sensitive mice. No change is recorded in the brains of the resistant strain.

4. The reverse is true for the changes in transglutaminase levels assayed in white cells obtained from the peritoneal wash. Here the enzyme doubles in the resistant, nonreacting strain, and normal values are recorded in the sick, sensitive animals.

DISCUSSION

Enzymatic reactions catalyzed by transglutaminase were also studied by others in important metabolic processes not related to immunology. The coupling of a diamine such as putrescine to tropomyosin, and linking of actin to tropomyosin through glutaminyl–lysine binding,[17] as well as similarities between fibrin-stabilizing factor and transglutaminase,[18,19] were reported. The last two observations deal with enzymatic activity that is easily detected in plasma, and also involves γ and β carbonyl groups of glutamine and asparagine;[18] those reactions distinguish the fibrin-stabilizing factor from transglutaminase. However, during local tissue damage, transglutaminase could accumulate locally in quantities large enough to contribute to the formation of a more stable blood clot.

Large quantities of transglutaminase present in white cells indicate its possible importance in immunological processes. According to present theories, lymphoid cells are involved in delayed hypersensitivity reaction, a condition that plays a major role in autoimmune responses. The mixture of white cells obtained from peritoneal exudates contains varying quantities of macrophages and lymphocytes. The macrophages have been shown to retain a small percentage of antigen in a form that was protected from the usual rapid breakdown and elimination.[20] The correlation of uptake, catabolism, and immunogenicity of antigen in macrophages, as studied by Unanue and Asconas,[20] might be different in such cells if they contain more transglutaminase than in those that have smaller quantities of this enzyme. Studies of such isolated systems and their importance in delayed hypersensitivity reactions was summarized in the Symposium on Delayed Hypersensitivity,[21] and it might be of interest to follow such experimental procedures with simultaneous estimations of transglutaminase.

The role played by mononuclear cells in allergic encephalomyelitis was stressed in studies demonstrating that the lesions in the central nervous system are initiated by perivenous infiltration of those cells;[22,23] such penetration is connected with an increased permeability of the brain-barriers,[24] a fact which also stresses the importance of a series of physiological conditions involved in the development of the disease. The increase in transglutaminase levels in the brains of the sensitive animals might represent an unsuccessful attempt to alter the penetrating material.

In an attempt to explain the possible role of transglutaminase changes during induction of EAE, one has to accept several situations in which this enzyme may be involved. It may act on endogenous protein, in which case such protein could acquire antigenic properties, but it may also act on the injected foreign protein, changing its properties, and possibly even decreasing its antigenicity. Both possibilities could be true if one could accept the idea that the enzyme is taking part in antigen-antibody reaction. In the sensitive, sick animals such a reaction is supposed to take place in the brain, and in the resistant strain it occurs in the peripheral reticulo-endothelial cells, providing the animals with resistance to EAE. Another possibility is that transglutaminase is not taking part causally in any of those processes, but accompanies generally changed metabolic activity of the tissues involved. At this time more experimental data are necessary to solve those problems. However, the fact remains that the characteristic pattern of behavior of transglutaminase during induction of EAE seems to locate an increased activity of this enzyme in the tissues involved in autoimmune reactions.

REFERENCES

1. K. Landsteiner, "The Specificity of Serological Reactions," Harvard Univ. Press, Cambridge, Mass. (1945).
2. E. A. Kabat and M. M. Mayer, *in*: "Experimental Immunochemistry," pp. 249–270, Charles C. Thomas, Springfield, Ill., (1958).
3. D. D. Clarke, A. Neidle, N. K. Sarkar, and H. Waelsch, *Arch. Biochem. Biophys.* **71**, 277–279 (1957).
4. A. Neidle, M. Y. Mycek, D. D. Clarke, and H. Waelsch, *Arch. Biochem. Biophys.* **77**, 227–229 (1958).
5. D. D. Clarke, M. Y. Mycek, A. Neidle, and H. Waelsch, *Arch. Biochem Biophys.* **79**, 338–354 (1959).
6. M. Y. Mycek, D. D. Clarke, A. Neidle, and H. Waelsch, *Arch. Biochem. Biophys.* **84**, 528–540 (1959).
7. J. H. Pincus and H. Waelsch, *Arch. Biochem. Biophys.* **126**, 34–43 (1968).
8. A. Neidle and G. Acs, *Fed. Proc.* **20**, 334 (1961).
9. J. E. Folk and P. W. Cole, *J. Biol. Chem.* **241**, 5518–5525 (1966).
10. I. J. Wajda, G. Acs, D. D Clarke, and H. Waelsch, *Biochem. Pharmacol.* **12**, 241–250 (1963).
11. M. Ginsburg, I. J. Wajda, and H. Waelsch, *Biochem. Pharmacol.* **12**, 251–264 (1963).
12. J. M. Lee and A. Schneider *J. Exp. Med.* **115**, 157–168 (1962).
13. I. J. Wajda, H. Waelsch, and J. M. Lee, *Life Sciences* **4**, 1053 (1965).
14. I. J. Wajda, J. M. Lee, and H. Waelsch, *J. Neurochem.* **14**, 389–399 (1967).
15. I. J. Wajda, J. M. Lee, and A. Neidle, *J. Neurochem.* **16**, 655–663 (1969).

16. B. R. Bloom and M. W. Chase, *Progr Allergy* **10**, 151–161 (1967).
17. N. Derrick and K. Laki, *Biochem. Biophys. Res. Commun.* **22**, 82–88 (1966).
18. J. Lorand (Bruner), T. Urayama, and L. Lorand, *Biochem. Biophys. Res. Commun.* **23**, 828–834 (1966).
19. L. Lorand and H. H. Ong, *Biochemistry* **5**, 1747–1752 (1966).
20. E. R. Unanue and B. A. Asconas, *J. Exp. Med.* **127**, 915–926 (1968).
21. Symposium on delayed hypersensitivity, *Fed. Proc.* **27**(1), 3–48 (1968).
22. T. U. Kosunen, B. H. Waksman, and I. K. Samuelsson, *J. Neuropathol. Exp. Neurol.* **22**, 367–402 (1963).
23. P. W. Lampert, *J. Neuropathol. Exp. Neurol.* **24**, 371–385 (1965).
24. R. W. P. Cutler, A. V. Lorenzo, and C. F. Barlow, *J. Neuropathol. Exp. Neurol.* **26**, 558–571 (1967).

DISCUSSION

VRBA: About 13 years ago I described deamidation in connection with physical exercise, and a short time ago in Strasbourg deamidation processes in brain protein when using barbiturates. It seems that any interference with the nervous system, be it by physical exercise or by an action of drugs, results in deamidation changes, and I wonder how far you feel it is right to consider this reaction as a special immunological process?

WAJDA: Transglutaminase is not only a deamidating enzyme. Of course, ammonia is produced as the side product, but the protein molecule accepts an amine, and therefore the process is not a simple deamidation. There are other deamidating enzymes, such as asparaginase, which was recently reported to increase in the brain during EAE. In a transglutaminase-catalyzed reaction we are interested in what happens to the remaining protein, because the acceptance of an amine, such as histamine or serotonin, which can be accepted by a covalent linkage, must obviously change the protein, and consequently the immunological properties of such protein.

KIES: Would it be of any use if you added some other autoimmune disease such as thyroiditis or adrenalitis?

WAJDA: We did try another autoimmune disease. Dr. Lee is still doing histological examinations of this experimental material.

LEE: To see if the increase in the enzyme was only connected with one autoimmune response, we produced allergic orchitis and we got exactly the same enzyme curve.

SELLINGER: Is the enzyme a soluble enzyme in terms of homogenate, and does the increase reflect in all fractions, or do you measure the increase only in the soluble fraction?

WAJDA: The liver enzyme was assayed in the soluble fraction of liver homogenates. The brain enzyme was assayed in the whole homogenate. In studying the distribution of transglutaminase, we find that its main bulk is in the cytoplasmic sap, but we are in the process of following the enzyme in different cellular subfractions, as it might partly adhere to some insoluble components of the cell, similarly to many other soluble enzymes.

FOLCH-PI: This is a very nice correlation and certainly worth studying, but in relation to the problems of EAE, I think you have fallen short of the necessary controls. I would not agree that the resistant strain is the proper control, because you do so many things You inject pertussis, then you inject brain emulsion. What do you use as control?

Furthermore, the actual amount of change in the tissue is only a very small fraction, and I doubt if any chemical change would reflect in the whole brain.

WAJDA: As far as the controls are concerned, histological examination reveals considerable changes in the brains of the sensitive strain which develops the disease, and this is accompanied by enzyme increase, whereas the resistant strain is practically devoid of histological pathology. Control experiments were done when we examined the liver enzyme, and this work is published. There were no changes in transglutaminase after pertussis alone, or Freund adjuvant alone, above the usual increase which occurs equally in both strains of mice, and which does not result in EAE. However, in the brain the normal values of transglutaminase do not increase at all in the resistant strain, even after the complete treatment, which results, in the sensitive strain, in 300% increase and EAE.

Chapter 37

Mescaline and Phenothiazines: Recent Studies on Subcellular Localization and Effects upon Membranes*

David N. Teller and Herman C. B. Denber

Research Division Biochemistry Laboratory
Manhattan State Hospital
Ward's Island, New York, N. Y.

Here we summarize some recent investigations of possible modes of action of two types of psychotropic drugs: psychotomimetics and phenothiazines, using studies with mescaline and chlorpromazine (CPZ), respectively, as examples. General reviews of their pharmacology have appeared, covering work done prior to 1962,[1] with reference to specific classes of psychotomimetic[2–4] or neuroleptic drugs,[5,6] but reviews of effects on biological membranes are less frequent.[7–13] [For reasons described below we have restricted our review of the effects of these drugs on subcellular organelles and the potential effects on central nervous system (CNS) protein metabolism to systems containing 50 μM or less psychotropic drug.]

Many different effects have been reported, usually at concentrations which greatly exceed those found in the CNS, when either psychotomimetic or tranquilized behavioral activity may be observed in clinical situations or in experimental animals. Furthermore, it has been known for some time that the maximum uptake of low doses of psychotropic compounds into the CNS occurs within 1 hr after injection.[14–17] The greatest uptake of mescaline into rat brain, occurring at 30–45 min after i.v. or i.p. injection, is less than 0.2% of the dose (0.5–40 mg/kg).[18] For phenothiazines maximum uptake is slightly slower and larger, reaching almost 1% of the dose (1–10 mg/kg CPZ) at 1 hr. The concentrations are below 10 μg/g of brain at these doses in all animals that we have studied. Domino *et al.*[5] review earlier

* Enthusiastic technical assistance was provided by S. Orenstein, and by B. Horn, and A. Sussman, Student Trainees supported by the N. Y. State Dept. of Mental Hygiene Summer Employment Program and by the Research Foundation for Mental Hygiene, Inc., Albany, respectively, during July and August, 1968.

work at higher doses; the upper limit of phenothiazine absorption was not reached in our studies.[19]

In our laboratory, with Wistar female rats, 200–250 g, 10 mg/kg of mescaline or 5–10 mg/kg of CPZ produces marked behavioral changes within 30 min that have been described elsewhere.[18,19] One half hour after injection the concentration of mescaline is 1.65 μg/g and of CPZ 2.6–5.3 μg/g brain tissue. Without regard for compartmentalization or localized binding, this represents 3 *μM* mescaline and 7–15 *μM* CPZ.

Here we disregard the 99% or more of the injected drugs which do not reach the CNS. The clinical and economic implications of these findings are discussed elsewhere,[20] and Forrest *et al.*[21] have recently reviewed CPZ metabolism in other body organs. We are concerned with concentrations of compounds in the brain just barely detectable by available chemical analyses, such as fluorometry and gas chromatography.[22] Most of the analytical data have been determined only recently by use of radioactive-tracer procedures.

Figure 1 shows the rate of mescaline and CPZ incorporation into rat brain 15–120 min after injection of 10 mg/kg of either drug alone. Figure 2 shows the effects of prior treatment with mescaline (10 mg/kg) upon the uptake of CPZ at 10 mg/kg, and CPZ (1 mg/kg, or thioperazine at 0.3 mg/kg) on brain mescaline. The two-phase uptake of CPZ is in agreement with data obtained by Steinecker *et al.*[23] at doses of 200 mg/kg. The amount of CPZ in the rat brain at 30 min after injection also agrees closely with data given by Symchowicz *et al.*[15] for perphenazine and Zehnder *et al.*[16] for thioridazine.

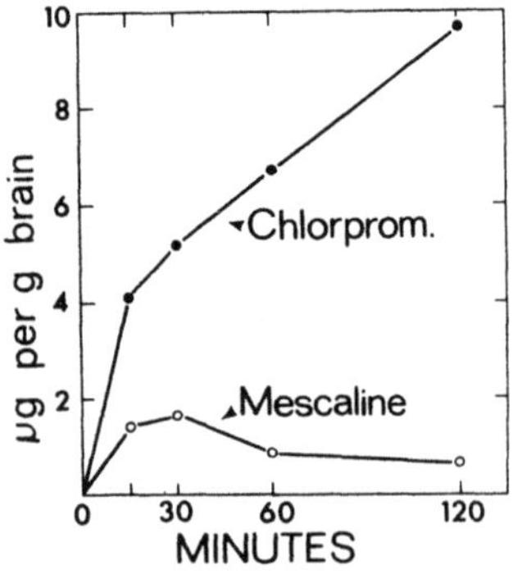

Fig. 1. Uptake of mescaline-8-^{14}C and chlorpromazine-9-^{3}H into rat brain. Mescaline HCl was injected intraperitoneally, CPZ HCl intramuscularly. The experimental procedure used was that of Denber and Teller,[18] with some modifications: 400 μl of 10% brain homogenate in 0.32 *M* sucrose was counted 40 min for tritium determinations. The CPZ-9-^{3}H originally contained 220 μCi/m*M* (from D. W. Blackburn, Smith–Kline and French laboratories) and was diluted with cold carrier CPZ HCl before use. The data points represent averages of results from four rats given either mescaline (10 mg/kg; open circles) or CPZ (10 mg/kg; closed circles). The ordinate is μg of drug per g brain (fresh); abscissa, minutes after injection.

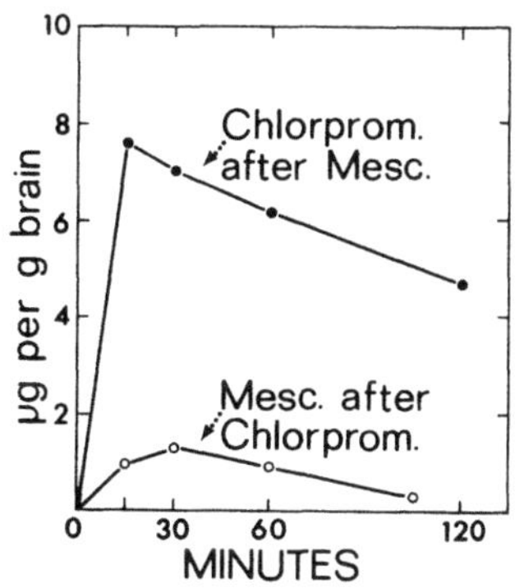

Fig. 2. Altered uptake of mescaline and of CPZ in the rat brain when rats are treated before the radioactive drug is injected with CPZ or mescaline, respectively. Experimental conditions were identical to those in Fig. 1, except that: either 30 min before 10 mg/kg mescaline-8-^{14}C HCl was injected i.p. (open circles) the rats received only 1 mg/kg CPZ HCl; or 30 min before 10 mg/kg of CPZ-9-^{3}H HCl (closed circles) was injected i.m. they received 10 mg/kg of mescaline HCl i.p. The abscissa is minutes after the second injection, the ordinate is µg of the second drug injected per g brain. The uptake of mescaline is slowed and reduced only slightly. However, the CPZ concentration is completely changed. It rises very rapidly above control levels, but without accumulating after 30 min. When CPZ is given 60 min after mescaline the phenothiazine concn. is reduced 50% compared to rats given CPZ alone at each interval after 15 min. These animals were housed in plastic cages with kiln-dried wood-chip bedding and were kept at 28–30° during the experiments.

All rats treated with CPZ were sedated[18] 20–30 min after the phenothiazine injection, without regard to the interval after mescaline injection. The animals were cataleptic by 60 min after 10 mg/kg CPZ, behaving similarly to controls that did not receive mescaline. A cataleptic rat's forepaws could be placed on a horizontal pencil 5–6 in. above the table surface and the animal would not move for 10–40 sec, or one paw would dangle while the other remained locked on the pencil. The reaction of rats treated with only 1 mg/kg CPZ before mescaline was not as severe.[18]

The mescaline concentration is reduced only 25% in rat brain by treatment with these phenothiazines and lysergic acid diethylamide (LSD) uptake is unaffected by CPZ.[24] However, behavioral effects of psilocybin,[25] LSD,[24] and mescaline[18] are completely blocked in rats and mice[26] by these phenothiazines. The blocking action of potent neuroleptics against mescaline in animals parallels previous clinical results.[27–32] In contrast, other rats with even lower mescaline brain concentrations (0.65 µg/g) untreated with phenothiazines show the typical mescaline behavioral changes. The symptoms due to 1.1–1.2 µg mescaline/g brain in rats are maximum at 30 min after injection. However, in rats treated with 1 mg CPZ/kg before 10 mg mescaline/kg, the uptake of only 1 2 µg CPZ/g of brain blocks the effects of the 0.9 μg mescaline/g of brain which is present.[18] There is the possibility that some of the mescaline (0.15–0.2 $\mu g/g$) which does not appear in the CPZ–mescaline-treated rats (the missing 25% of the usual mescaline uptake) is specifically responsible for activity. The remainder of the mescaline may be nonspecific.

This specificity is not yet apparent from gross anatomical distributions of either drug, alone or in combination. Both mescaline[13,18] and CPZ have 10:9:9:8 ratios of concentrations in (cortex):(basal ganglia, thalamus):(midbrain, pons):(cerebellum, medulla, and spinal cord). There are no significant differences in the gross anatomical localization of either CPZ or mescaline when the animals are treated $\frac{1}{2}$ hr earlier with mescaline or CPZ, respectively. Variations in local concentrations in the CNS reported in the literature can be due to differences in dose and route of administration,[18] as well as in analytical technique.

If we were to assume no specific molecular interactions leading to localized concentrations, then the relative concentrations in brain are similar: 1–15 μM for either drug. However, recent experiments with brain tissue have shown that most of the phenothiazines are tightly protein-bound (Teller *et al.*[33]), while little or none of the mescaline is bound.[13,18] Moreover, both drugs localize in nerve-ending particles, against concentration gradients. In nerve-ending particles from animals treated with neuroleptics *in vivo*, or after *in vitro* treatment, the mescaline content may actually increase threefold above that from control animal brains.[13] Therefore, if we assume that this compartment represents 10% of the brain volume, but contains 50% of the drugs, then the upper concentration limit will be approximately 30 μM. This upper limit can be exceeded by overdosing *in vivo*, but the effects of each drug are merely extended in time, or are superseded by toxic phenomena.[23,34–36]

Effects of phenothiazines on the structure of mitochondria have been studied for years,[37] and most recently were reviewed by Matsubara and Hagihara.[38] They showed a decrease in P:O ratios and increase in state 4 relative respiratory ratio due to changes in mitochondrial membrane structure. The respiration of submitochondrial particles is not stimulated by the phenothiazines. The lack of comparable effects on these units, derived from mitochondria whose membrane structure has been destroyed partially, is very significant. It indicates again the importance of the degree of membranal organization to the interpretation of results from *in vitro* studies with these drugs.[9,38,52] Moreover, above 50 μM the phenothiazines inhibit most enzyme systems,[7,9,11,39] with some exceptions.[40–42]

We have recently shown that this two-phase activity of the neuroleptics is concurrent with protein structure changes of myosin-B,[43,44] of nerve-ending particles, and other cell fractions.[33] Moreover, the phenothiazines often show protective effects *in vitro* at low concentrations when the system is later treated with another compound besides psychotomimetics. It requires at least millimolar CPZ alone to stop the active transport of *L*-DOPA into brain cortex slices (Kaniike *et al.*[45]) Protoveratrine, which is known as a membrane labilizer, at 1–5 μM reduces the accumulation of *L*-DOPA. However, 0.01–10 μM CPZ is capable of blocking the effects of protoveratrine.[46] Chlorpromazine between 10 and 50 μM is capable of blocking platelet aggregation produced by ADP.[46] At 3 μM CPZ or TPZ completely blocks the effects of up to 50 μM atropine on proteolipid "receptor protein" in chloroform–methanol solutions.[47] Weissmann[48] and Seeman[49] recently discussed some possible interpretations of this protective or lytic effect of phenothiazines and other drugs on lysosomal, mitochondrial, and erythrocyte membranes.

Mescaline has no effect upon ATPases or oxidative enzymes until 1–10 mM solutions are used.[50] It does not bind tightly to any subcellular fraction,[13] or to membrane subfractions from the synaptic endings (E. De Robertis, personal communication), or to plasma proteins.[51] Moreover, the duration of its behavioral effects are 3–5 times longer than the drug's presence in the rat CNS,[13,52] although humans vary more in retention.[53] For this reason we have assumed that mescaline has caused a structural change in the synapse whose repair is time-dependent.[52] The application of mescaline directly to cat cortex initially increases free NH_3, and decreases amide, lipid, and protein N.[54]

Although we have been able to demonstrate that relatively small doses of phenothiazines can block mescaline activity, it is not yet clear how this occurs. If the psychotomimetic effects are due to a change in membrane structure, phenothiazines may prevent this, perhaps by changing the same protein,[52] or by strengthening the synaptic membranes,[9] thereby decreasing the reactivity of the mescaline site of action. Now, there is evidence of only a fragmentary nature on the effects of CPZ upon CNS protein metabolism. In monkey brain 10 mg/kg CPZ had different effects depending upon the amino acid and on the anatomical area.[55] In otherwise normal rats low doses of CPZ have little effect on cerebral free amino acids[56] or upon amino acid synthesis from glucose,[57] nor is cerebral glucose metabolism altered, although in other species there may be greater sensitivity to CPZ.[58] Effects of CPZ (25 mg/kg) on glucose metabolism have been reviewed.[59] Lajtha and Toth[60] also found no effect of CPZ up to $10^{-4}M$ on amino acid uptake *in vitro*, while *in vivo* 50 mg/kg CPZ markedly inhibited Lys uptake, but mescaline and LSD had no effect on cyclo-Leu uptake. At present the picture is far from clear; however, by using the two antagonistic drugs as "titrating" agents, as acids and bases are used to establish the relative pH of complex solutions, we may be able to determine the specific CNS effects that are (1) due to these drugs at human-equivalent doses within the first hour after their administration and (2) causing the behavioral changes.

Further studies in progress will determine if mescaline does, indeed, cause structural changes in nerve-ending particle membranes leading to an uncontrolled release of neurotransmitter substances.[13,52] If this proposed structural change is prevented or reversed by the phenothiazines, we may be able to relate this work to endogenous disorders, as in schizophrenia. Certainly, the analysis of newly synthesized protein at the time *and after* the peak of mescaline-induced behavioral changes occur, and comparison with similar preparations from phenothiazine-"protected" animals, may offer direct evidence. We already have shown that disturbances in the circulating pool of amino acids by mescaline in humans[61] can be blocked by phenothiazine drugs.[30–32,62] However, in animals gross pharmacological changes due to sublethal dosage with either type of compound, and subtle ones, such as hypothermic effects of phenothiazines[63–65] upon amino acid incorporation, the loss of a fraction of these lipid-soluble drugs by partitioning from the aqueous phase,[66] etc., and inadvertant metabolic conditioning of rats by treatment of the animals and bedding with disinfectants, must be avoided. The elegant procedures for the analysis and characterization of protein-synthesis patterns described in other chapters of this volume will greatly help our understanding of the mode of action of psychotropic drugs, and, perhaps, of mental illness.

REFERENCES

1. S. Eiduson, E. Geller, A. Yuwiler, and B. T. Eiduson, "Biochemistry and Behavior," D. Van Nostrand Company, Princeton, New Jersey (1964).
2. L. G. Abood, *in*: "Drugs Affecting the Central Nervous System," (A. Burger, ed.), Vol. 2, pp. 127–167, Marcel Dekker, New York (1968).
3. S. Cohen, *Ann. Rev. Pharmacol.* **7**, 301–318 (1967).
4. A. Hofmann, *in*: "Drugs Affecting the Central Nervous System," (A. Burger, ed.), Vol. 2, pp. 169–235, Marcel Dekker, New York (1968).
5. E. F. Domino, R. D. Hudson, and G. Zografi, *in*: "Drugs Affecting the Central Nervous System," (A. Burger, ed.), Vol. 2, pp. 327–397, Marcel Dekker, New York (1968).
6. M. Gordon, *Med. Chem., Ser. Monogr.* **4**(II), 1–198 (1967).
7. P. S. Guth and M. A. Spirtes, *Int. Rev. Neurobiol.* **7**, 231–278 (1964).
8. H. Medina, A. Dmytraczenko, and M. Bacila, *Biochem. Pharmacol.* **13**, 461–467 (1964).
9. D. N. Teller, Kinetics of binding and inhibition of enzymatic activity by phenothiazine compounds, unpublished doctoral dissertation, New York University, 1964.
10. B. M. Bloom, *Ann. Rep. Med. Chem.* **1965**, 236–246.

11. P. M. Seeman, The erythrocyte as a model for studying membrane stabilization by tranquilizers, anesthetics, and steroids, unpublished doctoral dissertation, The Rockefeller University, 1966.
12. B. M. Bloom, *Ann. Rep. Med. Chem.* **1966,** 227–236.
13. H. C. B. Denber, Intracellular localization of psychotomimetic and psychotropic drugs, unpublished doctoral dissertation, New York University, 1967.
14. S. Symchowicz, W. D. Peckham, C. A. Korduba, and P. L. Perlman, *Biochem. Pharmacol.* **11,** 499–506 (1962).
15. S. Symchowicz, W. D. Peckham, M. Eisler, and P. L. Perlman, *Biochem. Pharmacol.* **11,** 417–422 (1962).
16. K. Zehnder, F. Kalberer, W. Kreis, and J. Rutschmann, *Biochem. Pharmacol.* **11,** 535–550 (1962).
17. F. Kalberer, W. Kreis, and J. Rutschmann, *Biochem. Pharmacol.* **11,** 261–269 (1962).
18. H. C. B. Denber and D. N. Teller, *Agressologie* **9,** 127–136 (1968).
19. R. P. Maickel, M. C. Gerald, D. M. Warburton, and M. A. Mahju, *Agressologie* **9,** 373–377 (1968).
20. D. N. Teller and H. C. B. Denber, *Dis. Nerv. Syst.* **30,** 15–18 (1969).
21. I. S. Forrest, A. G. Bolt, and M. T. Serra, *Biochem. Pharmacol.* **17,** 2061–2070 (1968).
22. S. H. Curry, *Agressologie* **9,** 115–121 (1968).
23. G. Steinecker, H. Schirardin, and P. Metais, *Ann. Pharm. Fr.* **25,** 315–323 (1967).
24. J. B. Appel and D. X. Freedman, *Biochem. Pharmacol.* **13,** 861–869 (1964).
25. H. K. Martin, *Int. J. Neuropsychiat.* **3,** 66–71 (1967).
26. C. Laville and J. Margarit, *Pathol. Biol. Semaine Hôp.* **13,** 762–766 (1965).
27. H. C. B. Denber and S. Merlis, *Psychiat. Quart.* **28,** 635–640 (1954).
28. H. C. B. Denber and S. Merlis, *J. Nerv. Ment. Dis.* **122,** 463–469 (1955).
29. S. Merlis and H. C. B. Denber, *J. Nerv. Ment. Dis.* **123,** 542–545 (1956).
30. H. C. B. Denber, *Psychiat. Quart.* **35,** 18–48 (1961).
31. P. Rajotte, H. C. B. Denber, and D. Kauffman, *in*: "Recent Advances in Biological Psychiatry" (J. Wortis, ed.), Vol. 4, pp. 278–286, Plenum Press, New York (1962).
32. H. C. B. Denber, *Riv. Neurobiol.* **10,** 1137–1168 (1964).
33. D. N. Teller, R. J. C. Levine, and H. C. B. Denber, *Agressologie* **9,** 167–189 (1968).
34. V. de P. Lynch, E. Clemente, and S. Carson, *J. Pharm. Sci.* **56,** 477–483 (1967).
35. V. Torjescu, *Ann. Pharm. Fr.* **26,** 311–322 (1968).
36. V. Torjescu and T. Severin, *Farmacia (Bucharest)* **15,** 741–749 (1967).
37. G. Aghajanian, *Biochem. Pharmacol.* **12,** 7–12 (1963).
38. T. Matsubara and B. Hagihara, *J. Biochem.* **63,** 156–154 (1968).
39. S. Okada, *Nara Igaku Zasshi* **19,** 139–146 (1968).
40. J. D. Robinson, J. Lowinger, and B. Bettinger, *Biochem. Pharmacol.* **17,** 1113–1116 (1968).
41. H. Balzer, M. Makinose, and W. Hasselbach, *Naunyn-Schmeidebergs Arch. Pharmakol. Exp. Pathol.* **260**(5), 444–455 (1068).
42. K. F. Gey and A. Pletscher, *Experientia* **24,** 335–336 (1968).
43. D. N. Teller, R. Levine, N. Wackman, and H. C. B. Denber, Proc. 7th. Int. Cong. Biochem., Abstracts, Vol. 5, No. 2, p. 1031 (1967).
44. R. J. C. Levine, D. N. Teller, and H. C. B. Denber, *Mol. Pharmacol.* **4,** 435–444 (1968).
45. K. Kaniike, S. Okada, and Y. Kurogochi, *Nara Igaku Zasshi* **19,** 155–158 (1968).
46. D. C. B. Mills and G. C. K. Roberts, *Nature* **213,** 35–38 (1967).

47. E. De Robertis, J. Gonzalez-Rodriguez, and D. N. Teller, *FEBS Letters* **4**, 4–8 (1969).
48. G. Weissmann, *in*: "The Interaction of Drugs and Subcellular Components in Animal Cells" (P. N. Campbell, ed.), pp. 203–212, Little, Brown and Company, Boston, Massachusetts (1968).
49. P. M. Seeman, *in*: "The Interaction of Drugs and Subcellular Components in Animal Cells" (P. N. Campbell, ed.), pp. 212–217, Little, Brown and Company, Boston, Massachusetts (1968).
50. T. Tsunoda, *Folia Psychiat. Neurol. Japon.* **14**, 156–166 (1960).
51. Y. H. Oh, W. E. Leitch, S. Axelrod, S. M. Small, R. J. Winzler, and B. E. Sanders, *Biochem. Pharmacol.* **16**, 849–861 (1967).
52. D. N. Teller and H. C. B. Denber, *Dis. Nerv. Syst.* **29**, 93–112 (1968).
53. K. D. Charalampous, K. E. Walker, and J. Kinross-Wright, *Psychopharmacologia* **9**, 48–63 (1966).
54. G. Badiu and N. Mison-Crighel, *J. Neurochem.* **13**, 1217–1222 (1966).
55. S. I. Singh and C. L. Malhotra, *J. Neurochem.* **14**, 135–140 (1967).
56. H. H. Tallan, *in*: "Amino Acid Pools" (J. T. Holden, ed.), pp. 465–470, Elsevier Publishing Co., Amsterdam (1962).
57. H. S. Bachelard, M. K. Gaitonde, and R. Vrba, *Biochem. Pharmacol.* **15**, 1039–1043 (1966).
58. S. Larsson, *Acta Physiol. Scand.* **53**, 68–74 (1961).
59. K. F. Gey, M. Rutishauser, and A. Pletscher, *Biochem. Pharmacol.* **14**, 507–514 (1965).
60. A. Lajtha and J. Toth, *Biochem. Pharmacol.* **14**, 729–738 (1965).
61. D. N. Teller and H. C. B. Denber, *in*: "Neuropsychopharmacology" (P. B. Bradley, F. Flugel, and P. Hoch, eds.), Vol. 3, pp. 423–426, Elsevier Publishing Co., Amsterdam (1964).
62. H. C. B. Denber, D. N. Teller, P. Rajotte, and D. Kauffman, *Ann. N.Y. Acad. Sci.* **96**, 14–36 (1962).
63. R. Molenda and A. Obrzut, *Przegl. Lek. Ser.* **2**(22) 544–547 (1966).
64. W. D. Ulrich, M. Maess, and J. Wirth, *Arch. Int. Pharmacodyn. Ther.* **167**, 80–89 (1967).
65. W. D. Ulrich, M. Maess, and J. Wirth, *Arch. Int. Pharmacodyn. Ther.* **167**, 90–100 (1967).
66. H. Balzer, M. Makinose, W. Fiehn, and W. Hasselbach, *Naunyn-Schmiedebergs Arch. Pharmakol. Exp. Pathol.*, **260**(5), 456–473 (1968).

DISCUSSION

SCHNEOUR: You suggested that the sulfur is a stabilizing part of the molecule. If this is true, then this whole molecule cannot lie flat. If you have fluorescence, the whole molecule will lie flat, that is to say, it has a double-bond intermediate that can exist, otherwise you cannot have the flat molecule.

TELLER: That is not true; dibenzoazepines with seven-membered central rings, which *are* forced to lie flat, have the same fluorescence and show the same fluorescence shift as phenothiazines upon partial protonation of the nitrogen. The "butterfly position" of the central ring has no relationship to the fluorescence. Pullman's group has shown how this ring system can fold.* From spectroscopy we know that some displacement

* B. Pullman, *Agressologie* **9**(1), 19–26 (1968).

of net charge must exist across the ring, and this is not irrelevant to the presence of a double-bond system here,[9,33,52] but I don't consider a bond actually shifting initially, rather that enough resonance energy is separated spatially so the ring N becomes slightly positive and the sulfur becomes negative. Indeed, the fluorescence energy changes. Chlorpromazine absorbs at 340–360 and emits at about 460 nm. When oxidized it absorbs at about 310 and then emits at 380 nm, indicating bond-length changes. Now, this emission shift into higher-energy wavelengths could be an indication of more energy in the nitrogen-orbital system in the protonized form. In contrast, the free radical absorbs at 260 and emits at 460 nm, which is a tremendous difference. An intramolecular bond shift is often characteristic of a stable free radical; however, the "butterfly" shape is probably a stabilized transient intermediate that normally would not accumulate in solution, and this work of others that you reported was done in solution.

SCHENOUR: But in this case, you are nevertheless dealing with a membrane structure, and you must assign some kind of charge to the membrane itself in order for this molecule to behave as you just said. The ESR studies of some of these compounds show that they do not exist in a free-radical fluorescent form. Do you have increased fluorescence upon binding or decreased fluorescence?

TELLER: These drug molecules are fluorescent and bind to proteins before any free radical forms. The free radical has entirely different optical properties and we try *not* to produce it, experimentally, fluorescent or not. There is no change in the fluorescence of the drug upon binding, unless it is oxidized, or, as in the case of phenazine methosulfate, reduced.

SCHNEOUR: Is Fig. 2 a fluorescent diagram?

TELLER: No, these are profiles of density gradients, scanned with the Aminco Bowman spectrophotofluorometer for light scattering and fluorescence emission of subcellular particles while the gradients were being pumped out of the ultracentrifuge tubes.

VRBA: What is the physiological significance of these experiments, and have you, on the basis of such experiments, a right to make any connection between the physiological action of phenothiazines and mescaline and those facts which you observed. For instance, Block, 15 years ago, after injection of C^{14} mescaline saw that a considerable amount of mescaline is bound to the liver and brain; only 2% of the count appeared in the brain. Perhaps by using a detergent instead of phenothiazine or mescaline we would find a very interesting way of binding this detergent to various protein structures, either in the brain or in any other tissue. It seems to be a general property of proteins, when such a foreign compound like phenothiazine or a detergent will enter, to bind it in some fashion before it can be detoxified.

TELLER: The phenothiazines are surface-active, that has been known for a number of years, and they bind tightly to membrane structures, but not as well to soluble proteins. The detergents have somewhat opposite characteristics in binding. Further, although the detergents may be carried into the central nervous system perhaps at a rate even higher than that of the phenothiazines (and I don't know if this is so), yet the phenothiazines are tranquilizers and the detergents are not. I prefer to work with the phenothiazines, since that is what our patients receive, not detergents. Finally, people have often said that it is unwise to speculate. I feel differently. As long as you label it as speculation, are ready to revise and readjust your opinion, without the bias of vested interest, there's nothing wrong with speculation. I feel that freedom to act upon such speculation has generated experiments which ordinarily would not have been done. For example, we never would have found out that an invertebrate muscle protein, myosin-B, bound the phenothiazines tightly, and that this caused a structural change in the protein which increased its enzyme activity, [43,44] if we hadn't speculated first upon what would happen if we brought together two substances and investigators from such widely separated research areas.

CLARKE: Dr. Schneour, regarding the ring flexure, I don't think the interaction of sulfur requires bending of the ring because of its *d* orbitals. The lack of π orbitals changes the picture completely. Now, Dr. Teller, why do you think that succinate dehydrogenase involves a hydride transfer?

TELLER: First, ring bending doesn't require the sulfur atom. Regarding π orbitals, there are completely different and opposite inferences from orbital calculations elsewhere (see reports by Pullman, and by Douzou.*) The succinate dehydrogenase reaction I learned about some years ago[9] seemed to involve a hydride transfer when coupled to phenazine methosulfate (PMS).

CLARKE: Recently Rittenberg claimed to have proven the opposite.

TELLER: I think that this would also depend on whether this was in solution or on the mitochondria, as to where the PMS-reducing electrons would come from.

BOGOCH: I wasn't so worried about the planar part of the tranquilizer, as that enormous tail. I wonder whether normal membranes can have holes large enough for a tail of that molecular size and clumsiness to fit in.

TELLER: Philip Seeman at the Rockefeller University used normal erythrocyte membranes as a model for the reaction of tranquilizers, steroids, alkaloids, etc. He found that enough of the phenothiazines could get into the surface of the erythrocyte membrane to increase the size by 15%. Whether the hole exists permanently, or is a discontinuity in the membrane, or whether something like mescaline has to make a hole first before the rest of the drug can get in, I don't know. This is just our speculation based upon the work of Cavanau and of Hechter that we reviewed recently.[52] I plan to see if there is a change in the protein *per se,* or only a change in the membrane configuration, after the mescaline is present, perhaps by using disc gel electrophoresis.

LAJTHA: I wonder how much we know about the binding constants, in two ways: pharmacological potency vs. binding, and antagonism. Do all those compounds that antagonize mescaline affect the binding constants?

TELLER: Some do. For example, lithium, calcium, and a number of other divalent metals will alter a behavioral response to psychototropic drugs, and these simple compounds do markedly affect the binding of the phenothiazines, but often by reacting with phenothiazines themselves, or they change the pH in the locale of the excited ion species. Further experiments on potency vs. binding were stopped because there is no animal brain where we can see the same reactions as occur from mescaline in patients, and, at present, we can't take normal animals treated with drugs and predict that this is the exact behaviorial change that occurs endogenously in human schizophrenics. Binding constants depend not only upon tissue source, but also on experimental protein concentrations. For instance, as much as 1 μg of phenothiazine drug is loosely absorbed per μg of brain protein. The bound drug will sediment with the protein, but it dissociates during gel column chromatography,[33] so that binding "constants" are almost fancies,

* J. Meunier, F. Leterrier, B. Viosat, J. Capette, and P. Douzou, *Agressologie* **9,** (1), 37–41 (1968).

particularly because of protein structure changes. When I found this out I started using very low concentrations of drugs, 10^{-9}–10^{-10} M, to try to approach the levels present in the central nervous system after a single injection. Now, I agree, this perhaps strikes most neurochemists as being restrictive, but this is the biological level of activity that we have to work by.

SELLINGER I agree with one of your premises, that one must look at the intracellular disposition of a drug if one wants to know how its going to be related to action. Recently, in collaboration with Dr. Domino, we injected radioactive hemicholinium into the ventricle of a dog, and looked at the intracellular disposition of the radioactive drug in the subcellular fractions of two regions: the caudate nucleus, and the hippocampus. To our amazement, the drug went into slightly different cell fractions in the two regions. In synaptososmal populations from both regions one could discriminate between various types of synaptosomes on the basis of their being able to bind hemicholinium. Overall, 11% of the dose in the caudate at the time of death (which is 4 hr after administration) will be very tightly bound to the membranes of the isolated synaptosomes. However, only 0.01% of morphine would be tied up with membranes of cerebral cortex in the rat. I think that the differences both in cell disposition and on the regional basis must be taken into consideration.

TELLER: Yes, part of the answer to reported discrepancies in drug uptake is often also in the method of administration. If I gave the phenothiazine, thioperazine, intracerebrally, the subcellular distribution was quite different than if it was injected intraveneously, and total uptake was different again after i.m. injection. However, with mescaline alone total brain uptake was similar after i.v. or i.p. injection, but if we gave phenothiazines or other drugs first, then there were marked differences in mescaline uptake i.v. vs. i.p.

BERL: It has been shown how reserpine disrupts membranes, the mitochondrial membranes, and endoplasmic reticulum, as well as the synaptic vesicles, even in very small doses; in fact, some animals are more sensitive than others, and in the cat the heart is particularly sensitive. We get mitochondrial disruption, but Bach has shown that this occurs in the brain too in the synaptic vesicles, and in the endoplasmic reticulum it seems to affect quite a number of membranes. I don't know how you could manipulate that structure with your theory. On the other hand, I think it's also confirmed that phenothiazine would also disrupt membranes of mitochondria of the heart. I think you're postulating that phenothiazines will stabilize membranes, whereas mescaline will disrupt them. Do you have any evidence of this?

TELLER: Others have also presented this concept of phenothiazine membrane stabilizing activity; Seeman[11,49] has worked on erythrocyte membrane stabilization, and Guth and Spirtes[7] presented work on whole erythrocytes; others who did work on the membrane stabilization concept have been reviewed in this paper and previously.[1–12,49,52] The differences may be due to doses. The concentrations at which a phenothiazine drug will disrupt the membrane is often more than 10–1000 times as high as the concentration at which it will stabilize. For all these compounds, including steroids, procaine, and a variety of alkaloids, one dosage will have a protective effect, and it depends on your personal interest and willingness to go to another, higher dosage at which the compounds will cause membrane disruption.

BERL: That might be true, because I think myocardial effects were found in patients who had probably died of an overdose of a phenothiazine, but on the other hand, do you have any evidence that mescaline disrupts?

TELLER: We are working on this. The only things that we have so far are a change in the sedimentation pattern of synaptosomes and their contents, and an increase of *in vitro* phenothiazine binding after *in vivo* mescaline treatment.*

SCHNEOUR: The action as detergent occurs at very high concentrations on the membrane of molecules per unit area. For these molecules you suggest essentially a single model, whose energy requirements seem to be extremely high compared to the energy available to that given molecule in a given area.

TELLER: Yes, here I agree. The energy requirements for these proposed changes *in vivo* do seem prohibitive at first glance. However, we have shown many times that only a few molecules may be involved. Now, as far as the concentration of phenothiazine in Fig. 15,[33] which showed the change in structure of synaptosomes, that is within the concentration range in the brain for chlorpromazine at the point where it blocks mescaline. This is about 40 molecules per cell. In comparison, the concentration for LSD when it is behaviorally active in man would be on the order of 1 molecule for every 100–500 synapses.

SCHNEOUR: And these protein structure changes developed at low concentration?

TELLER: Yes.

SCHNEOUR: Did you say that with mescaline there was a drop in the amino acids in the serum?

TELLER: That's right.

SCHNEOUR: Were they specific or general, and what were your proposed mechanisms for that?

TELLER: Well, at the time (1960–1961) I didn't know much about cell structure in terms of drug binding. This early work was a lab follow up of some extensive clinical experimentation by Dr. Denber.[30–32] We found that when the patients had a marked mescaline reaction the drop in amino acid level was severe.[62] At first we thought this was a stress reaction, and Dr. C. Cazullo obtained similar results with rabbits. We found out that the amino acids did not go from the serum into the urine, but into the tissues. However, we still don't know why. Possibly, they may be drawn upon by the tissues, or perhaps through the gastrointestinal membrane in response to other pharmacological effects of mescaline. Perhaps one way to find out would be to label amino acids in the serum, looking for specific incorporation into cellular and cerebral proteins, after mescaline activity declines. This would identify a "repair" protein synthesis.

SCHNEOUR: You didn't follow particular amino acids?

TELLER: We did distribution studies in human sera, and phenylalanine, for instance, totally disappeared from the serum.[61]

SCHNEOUR: I would have thought that if the membranes were disrupted, you would have the amino acids coming out from the tissues where the pool sizes were much larger.

TELLER: Yes. I don't know where this will lead to, unless the serum amino acids are necessary to help repair membranes affected by mescaline, all over the body.

* Figure 13 of Teller and Denber.[52] Also see D. N. Teller, H. C. B. Denber, and H. A. Charipper, *Fed. Proc.* **24**(2), 301 (1965).

COHEN: It might be possible at these low levels where you may have relatively few molecules in the area for a local change in just one small part of the membrane, which would modify synaptic transmission. In the amplification process (such as discussed in Chapter 15 by Dr. Nachmansohn), in this case by interrupting, sensitizing, or desensitizing such an amplifier, you would need just a small disturbance in a small region to get a marked effect.

TELLER: Yes, we have reviewed this concept previously,[52] and it is central to our thoughts.

Chapter 38

The Effects of Drugs on Protein Synthesis in the Nervous System*

Doris H. Clouet

New York State Research Institute for Neurochemistry and Drug Addiction Ward's Island, New York, N. Y.

INTRODUCTION

The effects which drugs have on protein synthesis in the nervous system of experimental animals fall into two general categories: (1) effects similar to those produced in other tissues by drugs which are well known as metabolic inhibitors or antibiotics, and (2) effects produced by drugs which have the nervous system as their probable site of action and which thus may be expected to have specific effects on metabolism in nervous tissue. In the present discussion the effects of metabolic inhibitors, antibiotics, and other, presumably nonspecific, drugs on protein synthesis or turnover in brain and nerve will be described briefly, with examples cited from the literature. Another class of drugs, those which affect a single enzyme system or classes of enzymes, which can be considered to be affecting a portion of the protein of the nervous system, will be discussed in terms of representative studies. Major attention will be given to studies from this laboratory on the responses in nervous tissue to the administration or addition of narcotic analgesic drugs.

The usual method of measuring protein synthesis in the brain of an experimental animal *in vivo* is to inject a labeled precursor amino acid either into the brain or peripherally, and after a period of time to sacrifice the animal, isolate the brain proteins, and measure the amount of labeled precursor incorporated into the various protein fractions. A variation of this method involves the use of radioautography to identify the site of protein

* The author wishes to acknowledge the assistance and cooperation of her colleagues and co-workers A. Neidle, M. Ratner, N. Williams, and J. Pecoraro in the studies reported from the author's laboratory.

synthesis in histological preparations. In protein synthesizing systems isolated from brain or nerve the effects of drugs may be studied by including the drug in the medium or by pretreating the animal with the drug before sacrifice and isolation and preparation of microsomal or mitochondrial protein-synthesizing system. Whenever the information is pertinent the method of measuring protein synthesis will be described.

METABOLIC INHIBITORS

The most active cell-free protein-synthesizing system which may be isolated from rat brain includes ribosomes, pH-5 enzymes from the soluble fraction, and an ATP-generating system.[1] This system is not altered by the addition of metabolic inhibitors unless the generation of ATP is inhibited.[2] However, it is inhibited by ribonuclease, presumably by the enzymatic hydrolysis of messenger RNA and ribosomal RNA.[2] A cell-free system prepared from brain mitochondria has quite different characteristics from microsomal or ribosomal preparations: ATP and pH-5 enzymes need not be added, and ribonuclease is without effect.[2,3] The mitochondrial system, however, is inhibited by the addition of dinitrophenol, rotenone or, antimycin A,[2] all of which inhibit or uncouple mitochondrial oxidative phosphorylation. This difference in inhibitor sensitivity between mitochondrial and microsomal protein-synthesizing systems has already been noted in similar preparations from other tissues.

ANTIBIOTICS

The nature of the interference in the protein-synthesizing apparatus of the cell by antibiotics has been elucidated using bacterial preparations, and by considering differences between bacterial and mammalian systems from comparable studies of liver ribosomal fractions. The results of the administration of antibiotics or the addition of antibiotics to brain preparations may be anticipated by analogy with liver. The administration of puromycin, which inhibits protein synthesis by causing a premature release of the nascent peptide chain from the ribosome, to a variety of animal species results in the inhibition of protein synthesis in the brains of the animals,[4–6] as does that of another well-known inhibitor of protein synthesis, acetoxycycloheximide,[7] which prevents peptide-chain elongation without release from the ribosome. Chloramphenicol has no effect on protein synthesis of mammalian ribosomes including brain ribosomes, but does inhibit protein synthesis in mitochondrial preparations, including those isolated from brain.[3] Chloramphenicol acts to prevent peptide chain elongation by inhibiting peptide transferase.

The administration of inhibitors of nucleic acid synthesis results in an eventual deficit in protein synthesis in brain and nerve: actinomycin D, an inhibitor of RNA polymerase, has produced an inhibition of protein synthesis in nerve,[9] and 8-azaguanine, a guanine competitor, inhibits in brain.[10]

AMINO ACIDS

The levels of amino acids have been shown to influence the rate of protein synthesis in brain.[11] Individual amino acids also have effects on the rate of synthesis. In brain slices the rate of incorporation of amino acids decreases when acidic amino acids are added to the incubation medium.[12] Phenylalanine acts as an inhibitor of protein synthesis in rabbit brain when the plasma levels of the amino acid are increased by the injection of large amounts of phenylalanine.[13] On the other hand, the addition of GABA to cell-free systems stimulates the rate of incorporation of amino acids into protein.[2] Whether GABA is acting as a neurotransmitter or as a substrate for citric acid-cycle metabolism is not clear. In general, the relative concentrations of groups of similar amino acids may be expected to have effects on both common carrier mechanisms and common metabolic pathways, which complicate the study of direct effects on protein synthesis.

ORGANOPHOSPHOROUS COMPOUNDS

A class of inhibitor drugs containing an organophosphorous moeity inhibit many esterases, including proteases and cholinesterases. The administration of many of these drugs peripherally has no effect on the nervous system, since they do not cross the blood-brain barrier. Several organophosphorous compunds are lyophilic, and so penetrate into the brain and cerebrospinal fluid. The administration of one such lyophilic compound, 217AO, inhibits the incorporation of ^{14}C-lysine into whole-brain proteins transiently *in vivo*,[14] as well as preventing acetylcholinesterase activity in nerve[15] and brain.[16] An eventual compensatory increase in protein synthesis to accommodate the replacement of irreversibly inhibited enzyme protein may be seen.[15]

SEDATIVES AND TRANQUILIZERS

The administration of pentobarbital to rats results in a decreased incorporation of ^{35}S-methionine into the brain proteins *in vivo*.[17] A similar inhibition of protein synthesis in brain is seen after the administration of chlorpromazine to rats.[18,19] However, in a detailed study of the effects of pentobarbital, reserpine, and chlorpromazine on the incorporation of tritiated histidine into brain protein *in vivo*, Shuster and Hannan[20] have correlated the inhibition of protein synthesis with the hypothermic response of rodents to the administration of these drugs. In addition to this general effect of sedatives on protein synthesis, chlorpromazine and other neurotropic drugs may be expected to alter the turnover of specific enzymes. Since catecholamines,[21] melatonin,[22] and Na^{+}-K^{+}-dependent ATPase[23] in brain are among the components altered by the administration of chlorpromazine, it is possible that changes in enzyme activity will be reflected by concurrent changes in the synthesis of these specific proteins in the nervous system.

NARCOTIC ANALGESIC DRUGS

The administration of morphine and other narcotic analgesic drugs to rats produces a transient inhibition in the rate of protein synthesis in whole

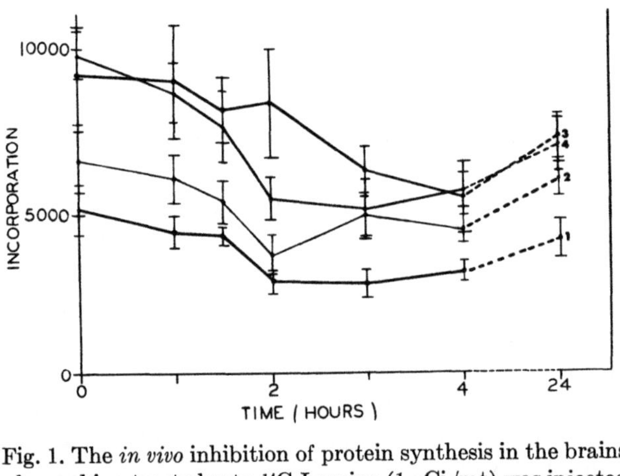

Fig. 1. The *in vivo* inhibition of protein synthesis in the brains of morphine-treated rats. ^{14}C-Leucine (1 μCi /rat) was injected intracisternally 30 min before sacrifice of the rats, which had been pretreated for varying periods of time (1–24 hr) with morphine (60 mg /kg). Brains were isolated and homogenized in order to fractionate the brain proteins into those derived from (1) nuclear, (2) mitochondrial, (3) microsomal, and (4) supernatant fractions. The incorporation of labeled leucine into protein is expressed as cpm /mg protein, with the standard deviation indicated by bars. The zero-time values are derived from the radioactivity of brain proteins of rats which had not received morphine.

TABLE I

The Effect of Hypothermia on Protein Synthesis in Brain *in Vivo*[a]

Conditions	Treatment				
	None	Morphine—3 hr		Morphine—24 hr	
Ambient temperature, °C	22	22	30	22	30
Average oral temp., °C	39.0	34.8	38.2	34.5	38.0
Number of rats	10	10	12	12	12
Protein synthesis[b]					
Nuclear	74	*51*	*50*	71	74
Mitochondrial	98	*71*	*75*	102	117
Microsomal	146	*101*	*109*	135	145
Supernatant	164	*88*	*92*	*119*	*130*
Total radioactivity[c]	605	*959*	*811*	669	692

[a] The dose of morphine was 60 mg/kg. A ^{14}C-leucine pulse was injected into the cerebrospinal fluid 30 min before sacrifice of each animal. Brain proteins were fractionated by centrifugation before measurement of radioactivity.
[b] CPM/mg protein $\times 10^{-1}$.
[c] CPM in whole-brain soluble fraction $\times 10^{-2}$.

brain.[24] The relation of the phenanthrene structure of narcotic drugs to that of steroid hormones suggests that the effect of these drugs in the nervous system may be analogous to that of steroid hormones in target tissues: namely, control of the rate of RNA and protein synthesis in the tissue. This hypothesis led to an examination of protein and RNA metabolism in the nervous system of drug-treated animals as compared to that in control animals.

For the *in vivo* experiments the adult male Wistar rat was injected with with a moderately effective dose of morphine (by pharmacological assay). When ^{14}C-leucine was used as the labeled precursor by intracisternal injection, a 30-min exposure to leucine resulted in less labeled brain protein when the rat was treated with morphine (Fig. 1).[24] Both particulate and soluble fractions of brain proteins had less label, with the time course of inhibition varying for the different fractions, with maximal inhibition at 2–4 hr after the injection of this dose of morphine.

Since hypothermia is one pharmacological response to the administration of morphine in rats, the relation of brain temperature to rate of protein synthesis was examined. Rats kept at an ambient temperature of 30° did not become hypothermic. When the rate of amino acid incorporation into protein in the brains of morphine-treated normothermic rats was com-

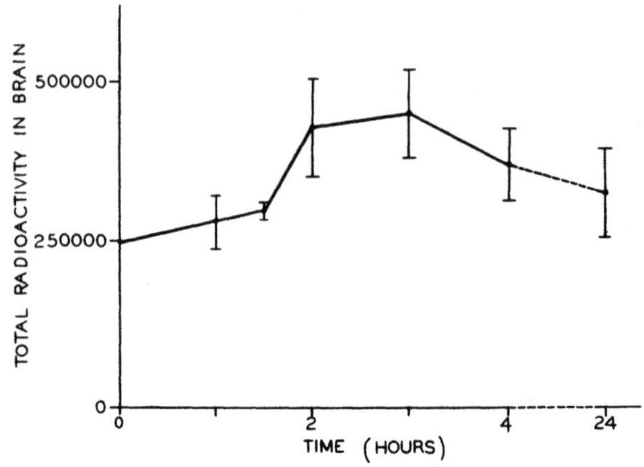

Fig. 2. The total soluble radioactivity in the brains of morphine-treated rats. The experimental details are as described in the legend to Fig. 1. The radioactivity is expressed as cpm/g brain.

pared with that of morphine-treated hypothermic animals the inhibition was present in each group of rats (Table I), in contrast to the lack of response to chlorpromazine and pentobarbital when hypothermia was prevented.[20]

A complication of the experiments in which labeled leucine was injected into the cerebrospinal fluid was that there was more nonprotein label in the brains of morphine-treated rats than in saline-injected control animals (Fig. 2). Since protein turnover is calculated in relation to the precursor specific radioactivity, differences in pool size of labeled precursor influence the calculated rate of protein synthesis. If the actual specific radioactivity of the precursor pool is used in the calculation, the inhibition of protein synthesis is greater than previously described. However, leucine is a precursor of citric acid-cycle intermediates through oxidation to acetyl CoA, as well as a protein precursor, so that it was possible that the oxidation of leucine was inhibited in morphine-treated animals, leading to more leucine remaining in the brain. Therefore, the nature of the ^{14}C-labeled, small-molecular-weight compounds (with carbon derived from uniformly labeled leucine) in the soluble portion of the brain was examined by means of an amino acid-analyzer separation with subsequent measurement of radioactivity of the separated compounds. The levels of glutamic acid, glutamine, aspartic acid, and leucine were similar in morphine- and saline-treated animals, but the specific radioactivity of the injected leucine and the glutamic acid and glutamine with carbon derived from leucine was higher in the brains of morphine-treated rats. Less than half the radioactivity remained in leucine 15 min after its injection into the cerebrospinal fluid in both treatment groups.[25] However, ^{14}C-leucine itself was responsible for the increased radioactivity in the brains of morphine-treated animals. The relative specific activity of glutamic acid and glutamine as compared to that of leucine was constant in the two groups of rats, so that no effect on leucine oxidation was demonstrated in the brains of morphine-treated rats (Fig. 3).

When protein synthesis was examined *in vitro* by isolating brain ribosomes by the desoxycholate method, a family of curves showed transient inhibition of protein synthesis in ribosomes prepared from the brains of

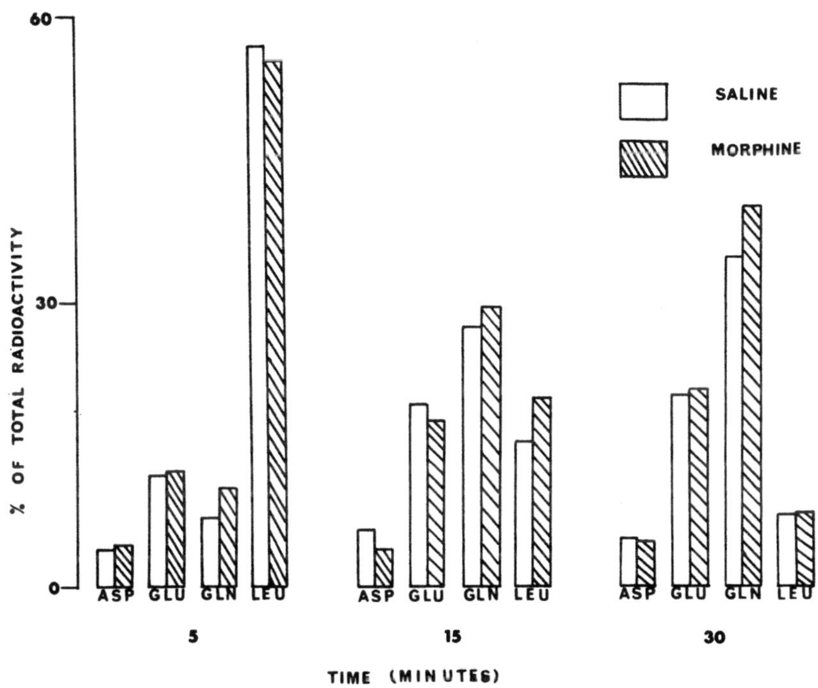

Fig. 3. The distribution of radioactivity of intracisternally administered ^{14}C-leucine in brain. ^{14}C-Leucine was injected into the *cisterna magna* of rats which were killed 5, 15, or 30 min later. The amino acid pool was examined by amino acid analyzer with subsequent determination of the radioactivity in each peak. The radioactivity in free aspartate (Asp), glutamate (Glu), glutamine (Gln), and leucine (Leu) as per cent of total radioactivity in the pool is shown for the three time periods of ^{14}C-leucine exposure, for rats injected with saline or morphine (hatched) 2 hr before sacrifice.

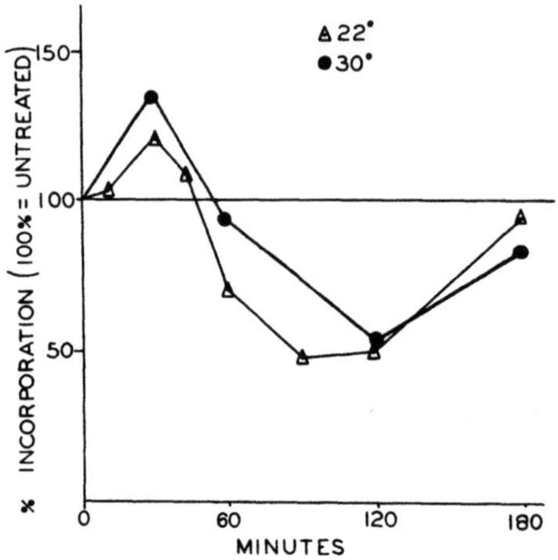

Fig. 4. Protein synthetic activity in ribosomes isolated from the brains of morphine-treated rats. Rats were killed at times from 10 min to 4 hr after an injection of 60 mg/kg morphine. Brain ribosomes were isolated using the desoxycholate method and assayed for the incorporation of ^{14}C-leucine into protein *in vitro.* Rats kept at 22° (triangles) were hypothermic, while the rats kept at 30° (circles) were normothermic after drug administration.

rats treated with varying doses of morphine[26] (see Fig. 4 for 60 mg/kg). Each component of the synthesizing system was examined to determine the nature of the biochemical lesion. Polysomal preparations from morphine-treated rats were incubated in a leucine-incorporating system before density-gradient centrifugation, and compared with similar preparations from saline-injected controls. In the usual 150-min centrifugation there were no differences in ribosomal patterns. However, when the polysomes were centrifuged for 16 hr in a sucrose density gradient two differences were apparent: the labeled material on the polysomes and the polysomes themselves were more stable in preparations from control animals (Fig. 5). The labeling patterns were similar for polysomes from the brains of saline- and levallorphan-treated animals, and diminished in the narcotic analgesic drug-treated animals, as was the distribution between gradient radioactivity and that sedimenting to the bottom of the tube.

In studies of the levels of messenger RNA, brain ribonuclease, and RNAase inhibitor, animal-to-animal variation was greater than differences

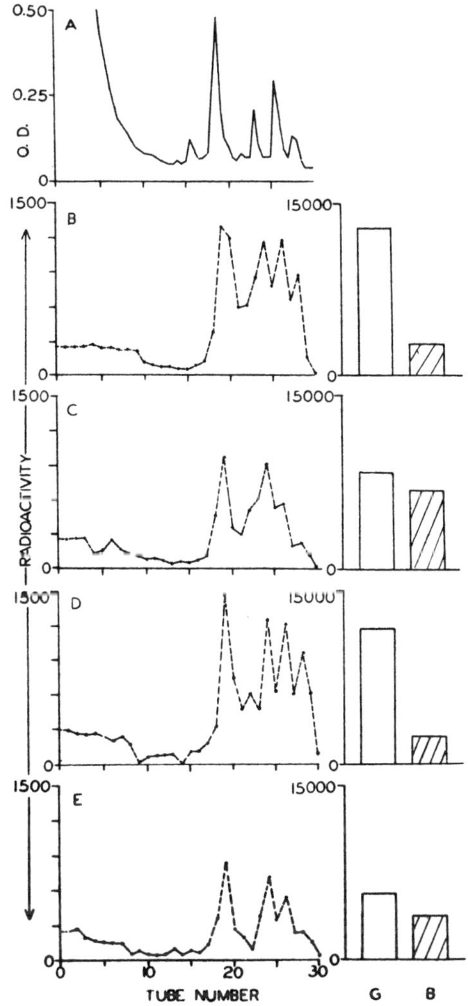

Fig. 5. Polysomal patterns after a long centrifugation. Polysomal preparations from rat brain were centrifuged through a 20–50% sucrose gradient for 16 hr. (A) typical density at 254 nm; (B)–(E) radioactivity patterns: (B) from rats receiving saline, (C) from rats receiving 60 mg/kg morphine, (D) from rats receiving 100 mg/kg of the antagonist levallorphan, and (E) from rats receiving 1 mg/kg morphine intracisternally. To the right *G* shows the amount of radioactivity still in the gradient, and *B* the amount sedimenting to the bottom of the tube.

in treatment groups, so that no specific effect due to narcotic drug administration was discernible.

The addition of morphine to the incubation medium had no effect on amino acid incorporation by brain ribosomes.[(26)] However, the addition of the antagonists levallorphan and nalorphine did show an inhibition of amino acid incorporation into protein *in vitro* (table II). A visible precipitate in the bottom of the reaction vessel led to the discovery that levallorphan plus sRNA formed a precipitate in the absence of any other incubation component, leading to an artifactual inhibition of protein synthesis by removal of a necessary factor from the incubation medium. The possibility arose that narcotic agonists and antagonists could be shown to interact with nucleic acids in the animal.

To this end, tritiated dihydromorphine was injected intracisternally in equi-analgesic doses with those used previously. The rats were killed and the homogenates of brain were divided into nuclear-mitochondrial and microsomal-supernatant fractions by a single centrifugation before the RNA was isolated by the phenol method. Both microsomal and nuclear RNA contained labeled drug (Fig. 6), with the nuclear RNA showing a time-dependent uptake. Dihydromorphine added to control brain homogenates before isolation and fractionation of RNA was found at the top of the gradient. The administered narcotic drug and brain RNA's were bound

TABLE II

The Inhibition of Amino Acid Incorporation by the Addition of Narcotic Drugs to the Incubation[a]

Drugs	Inhibition (%)			
	10^{-4} *M*	10^{-3} *M*	5×10^{-3} *M*	10^{-2} *M*
Morphine	3	3	12	12
Meperidine	—	5	—	13
Levorphanol	—	1	—	6
Nalorphine	—	+2	12	44
Levallorphan	5	21	55	96

[a] The drugs were added to an amino acid-incorporating system containing brain ribosomes in the final concentrations listed. The amount of a ^{14}C-amino acid incorporated during a 20-min incubation was measured in control assays to obtain 100% values. Nalorphine and levallorphan are antagonists, the other drugs agonists.

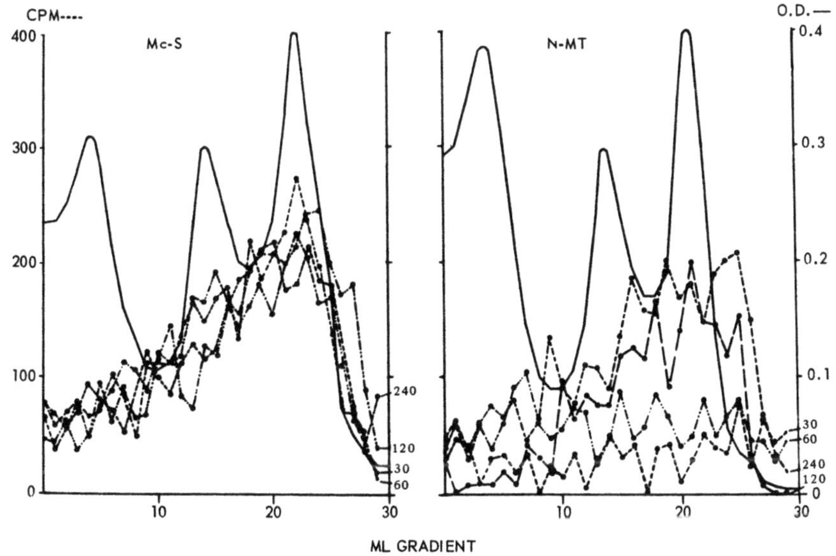

Fig. 6. ^{3}H-dihydromorphine binding to brain RNA. Tritiated dihydromorphine was injected intracisternally into rats which were sacrificed 30, 60, 120, or 240 min later. Brains were removed, homogenized in sucrose, and centrifuged at 9000 g/15 min in order to separate N-Mt (nuclear-mitochondrial) from Mc-S (microsomal-supernatant) fractions before preparing RNA. The RNA fractions were centrifuged in a density gradient and examined for radioactivity. The top of the gradient is to the left. The solid lines are recordings of optical density at 254 nm and the dotted lines represent radioactivity in the gradient. The length of time between drug administration and sacrifice of the animal is shown in small numbers indicating minutes of drug exposure *in vivo*.

tightly enough to survive alcohol precipitations and the density gradient centrifugation, but could be separated by exhaustive dialysis.

The inhibition of protein synthesis and the interaction of drug and RNA suggested the synthesis of these macromolecules may be involved in the pharmacological responses to narcotic administration, particularly, the phenomenon of tolerance. Other workers found that the administration of actinomycin D,[27] puromycin,[28] or 8-azaguanine[10,29] inhibited the development of tolerance to repeated doses of a narcotic drug. In the present experiments puromycin was injected into rats in combination with morphine or saline, and the incorporation of ^{14}C-lysine into brain protein was measured (Fig. 7). Morphine alone had its usual inhibitory effect. Puromycin at a dose of 0.8 mg/kg almost completely inhibited pro-

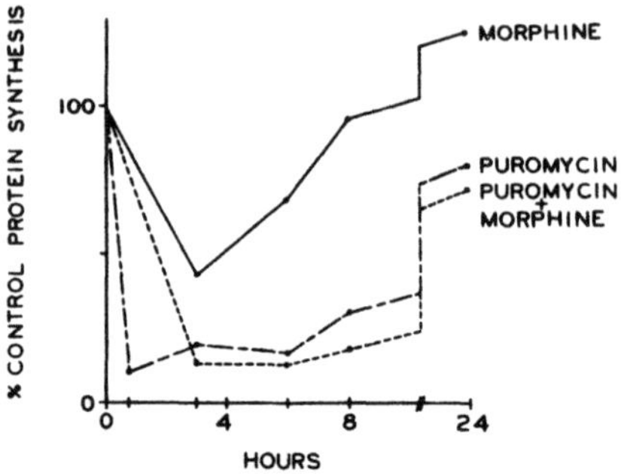

Fig. 7. The effect of puromycin and morphine on protein synthesis in rat brain. ^{14}C-Lysine (1 μCi/rat) was injected 30 min before sacrifice. In addition, the rats received 0.8 mg/kg puromycin or saline intraperitoneally and 60 mg/kg morphine at zero time. The incorporation of lysine into the brain proteins of saline-injected animals is shown as 100% incorporation.

tein synthesis in brain from 45 min after its injection to 6 hr later, so that the cumulative effect of morphine plus puromycin was not much greater than puromycin alone (Fig. 7). Two parameters of pharmacological response were used in these experiments in order to measure the effect of a second dose of morphine 5 hr (acute partial tolerance) or 24 hr (tolerance) after the initial injection: analgesia by hot-plate reaction time, and hypothermia by oral temperature recording. The per cent tolerance (second response/first) in each case was very similar in animals treated with saline or puromycin: 13% vs. 19% for hypothermia response, and 32% vs. 37% in the analgesic response to a second dose 5 hr after the first. For the 24 hr response the corresponding values are 12% vs. 16% and 47% vs. 42%. None of these values for puromycin-treated animals was significantly different from those from saline-treated animals. It should be noted that both the analgesic and the hypothermic responses were greater to the initial dose of morphine in puromycin-treated rats, so that the actual values for the second responses were greater in the puromycin-treated animals. The per cent tolerance values which were calculated from the actual initial response for each animal eliminated this difference in initial response to morphine.

A number of conclusions may be reached from the foregoing studies. There is an inhibition in protein synthesis in the brains of morphine-treated rats which is not dependent on hypothermia in the drug-treated animals. There is a concurrent inhibition of leucine exit from the cerebrospinal fluid in the drug-treated animals. The oxidation of leucine through the citric acid cycle in brain is not, however, impaired. Perhaps related to the inhibition of protein synthesis in the brains of morphine-treated rats is the binding of labeled dihydromorphine to brain RNA *in vivo*. The involvement of protein synthesis in the development of tolerance has not been demonstrated in the studies reported here, although other studies using different techniques seem to indicate that protein synthesis is necessary for the development of tolerance.

REFERENCES

1. D. H. Clouet, M. Ratner, and N. Williams, *Biochem. Biophys. Acta* **123**, 142–150 (1966).
2. M. K. Campbell, H. R. Mahler, W. J. Moore, and S. Tewari, *Biochemistry* **5**, 1174–1184 (1966).
3. H. S. Bachelard, *Biochem. J.* **100**, 131–137 (1966).
4. B. W. Agranoff, R. E. Davis, and J. J. Brink, *Proc. Natl. Acad. Sci. U.S.* **54**, 788–793 (1965).
5. S. H. Barondes and H. D. Cohen, *Brain Res.* **4**, 44–48 (1967).
6. S. H. Barondes and H. D. Cohen, *Proc. Natl. Acad. Sci. U.S.* **58**, 157–164 (1967).
7. L. B. Flexner and J. B. Flexner, *Proc. Natl. Acad. Sci. U.S.* **55**, 369–374 (1966).
8. A. J. Dunn and A. Korner, *Biochem. J.* **100**, 76P (1966).
9. A. Edström, *J. Neurochem.* **14**, 239–243 (1967).
10. M. T. Spoerlein and J. Scrafani, *Life Sci.* **6**, 1549–1564 (1967).
11. S. Roberts and B. S. Morelos, *J. Neurochem.* **12**, 373–387 (1965).
12. F. Orrego and F. Lipmann, *J. Biol. Chem.* **242**, 665–671 (1967).
13. K. F. Swaiman, W. B. Hosfield, and B. Lemieux, *J. Neurochem.* **15**, 687–690 (1968).
14. D. H. Clouet and H. Waelsch, *J. Neurochem.* **10**, 51–63 (1963).
15. D. H. Clouet and H. Waelsch, *J. Neurochem.* **8**, 201–215 (1961).
16. B. Holmstedt, M. Härkonen, G. Lundgren, and A. Sundwall, *Biochem. Pharmacol.* **16**, 404–406 (1967).
17. M. K. Gaitonde and D. Richter, *Proc. Roy. Soc. (London)* **145**, 83–92 (1965).
18. R. S. Piha, S. S. Oja, and A. J. Usitalo, *Ann. Med. Exp. Biol. Fenn.* **40**, Suppl. 5, 1–28 (1962).
19. A. J. Glasky, *Fed. Proc.* **22**, 272 (1963).
20. L. Shuster and R. V. Hannan, *J. Biol. Chem.* **239**, 3401–3406 (1964).
21. H. Nybäck and G. Sedvall, *J. Pharmacol. Exp. Ther.* **162**, 294–301 (1968).
22. R. J. Wurtman, J. Axelrod, and F. Anton-Tay, *J. Pharmacol. Exp. Ther.* **161**, 367–372 (1968).
23. P. Kraus and Z. Simane, *Experientia* **23**, 90–91 (1967).
24. D. H. Clouet and M. Ratner, *Brain Res.* **4**, 33–43 (1967).
25. D. H. Clouet and A. Neidle, unpublished experiments.
26. D. H. Clouet and M. Ratner, *J. Neurochem.* **15**, 17–23 (1968).
27. M. Cohen, A. S. Keats, W. Krivoy, and G. Ungar, *Proc. Soc. Exp. Biol. Med.* **119**, 381–384 (1965).
28. A. A. Smith, M. Karmin, and J. Gavitt, *Biochem. Pharmacol.* **15**, 1877–1879 (1966).
29. I. Yamamoto, R. Inoki, Y. Tamari, and K. Iwatsubo, *Japan. J. Pharmacol.* **17**, 140–146 (1967).

DISCUSSION

UNGAR: In regard to the correlation of binding with pharmacological action, I noticed that you chose a strain of rats that was relatively insensitive pharmacologically. If you had chosen a strain which gave a response with 5–6 mg/kg, would that then have the same amount of binding as the analgesic dose? The second question concerns the correlation with possible tolerance. All the experiments that you showed were very acute experiments; are there any data in rats that have received prolonged treatment, in which tolerance therefore developed?

CLOUET: As far as tolerance is concerned, if we inject the same amount every day for five days and do the same studies *in vivo*, protein synthesis is still inhibited at the same times. We haven't pursued the question further in this type of experiment; after 5 days the pharmacological responses to this dose are completely lacking, so there is tolerance in the two pharmacological parameters which we measure: analgesia and hypothermia. As far as the question of resistant animals is concerned, we've used narcotic drugs which were active over a fairly large range of dose and get the same effect, so that I don't think that it is dose-dependent.

RICHTER: You mentioned that morphine increased the amino acid pool size *in vivo*.

CLOUET: Morphine doesn't change the pool size, just the amount of the labeled material, which was a very small part of the amino acid pool, but the radioactivity of the injected material stayed in the brain.

RICHTER: Do you have any information as to why it should effect the pool size?

CLOUET: I think the effect is due to alterations in the rate of exit from the pool. According to our calculations it is not related to protein synthesis: the decrease in protein synthesis is not enough to account for the large lack of use of the ^{14}C-leucine that we injected.

BENUCK: You have reported a high rate of metabolism of leucine. Have you examined the labeled proteins to see which amino acids in the proteins had been labeled?

CLOUET: When we hydrolyzed the protein formed after a 30-min exposure to leucine in both the morphine-treated and saline-injected animals and examined the amino acids in the amino acid analyzer, only leucine was labeled. However, if glutamate had been labeled at 1/500 of the amount of leucine, we wouldn't see it in the protein hydrolyzate because you couldn't put that much amino acid on the gradient and still count it. I suspect that the limitations of the method mean that we just don't see some glutamate from protein that is also labeled.

PORCELLATI: Did you measure what the administration of morphine does to proteolysis?

CLOUET: No, we don't know anything about it.

GAULL: Leucine went into Krebs cycle intermediates easily.

CLOUET: We are trying to get the answer to why there is more radioactivity in the TCA-soluble pool, and we want to know what the specific radioactivity of the precursor leucine is. We found that leucine enters the tricarboxylic acid cycle and it enters very rapidly.

GAULL: Is this rapid entry into the cycle a peculiarity of the leucine or the branch chain amino acids?

CLOUET: We looked at ^{14}C-lysine, and I think that about 85% of ^{14}C was still in lysine after 30 min exposure in the same kind of experiment.

BERL: In answer to Dr. Gaull's question, this occurs very rapidly; within a few minutes you will find a rapid conversion of leucine to glutamic acid and glutamine. I think this also holds for a number of other amino acids, and, in addition, anything that will eventually be broken down to acetyl CoA or any of the Krebs cycle intermediates will of course end up in glutamic acid and glutamine, especially. I think it may be well to emphasize that in the study of protein turnover it may not be sufficient to count the protein, or radioactivity in the protein, but that one must also be careful to isolate the precursor. From the free precursor level, as well as from the protein, one may correlate the flux of counts into the protein. One can be fooled by just counting the isolated protein. This flow of radioactivity from leucine into glutamic acid and glutamine was also reported by Dr. Roberts, who has also found leucine oxidation in brain, and that there are some area variations, particularly in the cerebellum and caudate nucleus. You will find more of the radioactivity remaining in glutamic acid, but in most areas you will find that the specific activity of glutamine is greater than in glutamic acid. The ratio is greater in the cerebellum and in the caudate nucleus, where the ratio remains less than 1; and there is some variation and some difference in the metabolism of the acetyl CoA, which is one of the end-products of leucine metabolism.

MARKS: In view of the structure-activity relationship mentioned by Dr. Teller, it would be rather interesting to see the binding of some of the narcotics to RNA or nucleic acids. Do you think that it might be possible to examine further the relationship to other isomers?

CLOUET: I think that such a study is very important. So far the only structural relationships that we have is that antagonists which do not have pharmacological effects do not inhibit protein synthesis. At the moment we have no labeled antagonists, and I think that we certainly should obtain a few and do some of the same types of studies.

BOGOCH: You mentioned that with certain stimulating doses the inhibition was most marked in incorporation of leucine label. Did you mean behavioral stimulation?

CLOUET: Morphine in the rat is a depressant at low doses, and at higher doses is a stimulant. By stimulation I mean approaching the dose at which the animals go into convulsions; the rats die of overdose in convulsion.

BOGOCH: The reason I asked is that when our birds are very busy working the incorporation into carbohydrates is markedly decreased (we have not measured the incorporation of leucine), and I wondered whether there was something of interest here in terms of the hyperactivity of the animal and the decreased incorporation of leucine.

MANDEL: I am sure that you took care not to have any compounds other than leucine produced by radiolysis, because if you keep leucine for a couple of weeks, it can get many peaks, and these peaks might be exactly on some points where there are some amino acids, and you are measuring radioactivities and they might be artifacts.

CLOUET: Yes, the injected ^{14}C was 100% in leucine.

Index

MIX
Papier aus verantwortungsvollen Quellen
Paper from responsible sources
FSC® C105338

If you have any concerns about our products,
you can contact us on
ProductSafety@springernature.com

In case Publisher is established outside the EU,
the EU authorized representative is:
Springer Nature Customer Service Center GmbH
Europaplatz 3, 69115 Heidelberg, Germany

Printed by Libri Plureos GmbH
in Hamburg, Germany